Revaluation and Preservation of Indigenous Knowledge Systems in Modern Society

Tlou Maggie Masenya
Durban University of Technology, South Africa

Monicca Thulisile Bhuda
University of Mpumalanga, South Africa

IGI Global
Scientific Publishing
Publishing Tomorrow's Research Today

Vice President of Editorial Melissa Wagner
Director of Acquisitions Mikaela Felty
Director of Book Development Jocelynn Hessler
Production Manager Mike Brehm
Cover Design Phillip Shickler

Published in the United States of America by
IGI Global Scientific Publishing
701 East Chocolate Avenue
Hershey, PA, 17033, USA
Tel: 717-533-8845
Fax: 717-533-7115
Website: https://www.igi-global.com E-mail: cust@igi-global.com

Library of Congress Cataloging-in-Publication Data

Names: Masenya, Tlou, 1972- editor. | Bhuda, Monicca, 1992- editor.
Title: Revaluation and preservation of indigenous knowledge systems in modern society / edited by: Tlou Masenya, Monicca Bhuda.
Description: Hershey, PA : IGI Global Scientific Publishing, [2025] | Collection of essays by Sri Harshini Jupally and 14 others. | Includes bibliographical references and index. | Summary: "The overall objective of this book is to present scholarly perspectives, ideas and recommendations in addition to publishing excellent articles that explore the role and importance of indigenous knowledge in the modern society"-- Provided by publisher.
Identifiers: LCCN 2024055430 (print) | LCCN 2024055431 (ebook) | ISBN 9798369379646 (hc) | ISBN 9798369379653 (sc) | ISBN 9798369379660 (e)
Subjects: LCSH: Ethnoscience--Africa. | Ethnoscience--India. | Cultural property--Protection--Africa. | Cultural property--Protection--India. | Digital preservation.
Classification: LCC GN476 .R48 2025 (print) | LCC GN476 (ebook) | DDC 001.089--dc23/eng/20250118
LC record available at https://lccn.loc.gov/2024055430
LC ebook record available at https://lccn.loc.gov/2024055431
British Cataloguing in Publication Data
A Cataloguing in Publication record for this book is available from the British Library.

Table of Contents

Detailed Table of Contents

Despite being regarded as the most significant resource for indigenous indigenous people and their communities worldwide, the loss of Indigenous Knowledge Systems (IKS) has put these these most marginalised communities under a great deal of stress. Research has demonstrated that an unquantifiable amount of IKS is fading away since the beginning of the 20th century. The erosion and marginalization of IKS is the direct result of Westernization and colonization, all accompanies by the racist and colonial policies. The concern over the loss of this irreplaceable knowledge has thus raised a need to recover, reclaim and preserve it in the modern society. This chapter utilised literature review to critically analyse the resurgence and reclamation of IKS practices for the benefit of indigenous communities, the role of modern technologies in the recovery and preservation of IKS, the decolonial approaches for the revitalization and revaluation of IKS. Sociocultural learning theory and modernization theory were also used as underpinning theories to guide the study.

The paper seeks to investigate the benefits and challenges of digitally repatriating Zimbabwe's heritage resources housed in foreign museums. It also discusses the legal, ethical and cultural implications of such repatriations. Digital repatriation is the process of digitally returning heritage resources to their countries of origin. Worldwide, digital repatriations are emerging as a temporary solution and a means of providing access when physical repatriations are delayed or not possible. Zimbabwe lost a lot of cultural property under colonial rule and is still fighting for the return of its heritage resources. While appreciating that digital repatriation does not override the need for physical repatriation, the research evaluates the effectiveness of digital repatriations as a temporary solution to improving access to Zimbabwe's cultural heritage. This study used qualitative research method with a multiple case study research design.

This chapter examines narrative literature review concerning the indigenous data and knowledge sovereignty in the context of Southern African indigenous communities amidst technological dynamics. The study was also guided by the following research objectives: examine indigenous data and knowledge sovereignty among Southern African indigenous communities, examine national and international initiatives for data and knowledge sovereignty among indigenous communities, determine the role of intellectual property rights in the protection of indigenous data and knowledge, and suggest recommendations to enhance indigenous people's rights to data and knowledge sovereignty. Social Contract Theory (SCT) was also employed to underpin this study as it embraces the notion that a person's moral, political or legal responsibilities are reliant on their establishment of regulations or ideals that guide their society.

The chapter examines documentaries as repositories for the preservation, protection, promotion and dissemination of indigenous knowledge systems for future generations. This participatory and qualitative study, carried out among the Ndau of south-eastern Zimbabwe, utilised in-depth interviews and focus group discussions as interactive research methods where the indigenous knowledge holders, the old and the young, female and male and indigenous knowledge researchers were interviewed with the intention to investigate the extent to which documentaries can be reservoirs of indigenous knowledge. Documentaries were found to be effective repositories of indigenous knowledge but notwithstanding their myriad challenges, threats and risks as revealed by participants. The chapter argues that using a participatory approach and a high sense of respect towards knowledge holders, documentaries are a safety net for preserving indigenous knowledge facing extinction threats.

This chapter aims to answer this question: How can indigenous knowledge systems be re-appropriated and protected for the benefit of indigenous knowledge tribes? The chapter utilised literature review to explore the role of intellectual property rights laws, policy frameworks and protocols in the re-appropriation of indigenous knowledge systems in South Africa. The study was guided by the following objectives: the significance of indigenous knowledge among indigenous communities, the effect of cultural appropriation among indigenous communities, the anti-imperialism strategies for the re-appropriation and resignification of indigenous knowledge among indigenous tribes and the role of Intellectual Property Rights laws in the reclamation and protection of indigenous knowledge. Recommendations suggest the need for the protection of indigenous knowledge systems and promote access to indigenous knowledge through re-

Indigenous Knowledge (IK) has been faced with various challenges such as piracy and overall abuse of the rights of indigenous peoples towards their resources. This has led to communities being skeptical about their space and who to share their IK with This presented various problems leaving this IK at the periphery of the knowledge economy. It is pivotal to dwell into strategies that seek to preserve, protect IK as a resource which communities rely on for survival. This study applied participatory action research with an indigenous paradigmatic lens. The study focused on two Khoisan communities to unearth how this can benefit from developing their own Biocultural Community Protocols (BCPs). The data collected were analyzed using thematic analysis. The study finds that the development of BCPs will ensure that communities manage their IK thereby protecting and strengthening the intricate linkages between the cultural systems of knowledge and the management of land and natural resources. It is concluded that BCPs are a relevant resource to ensure that IK is protected, promoted and managed to drive sustainable development.

Due to the rapidly increasing population around the world, food access and availability has become a paramount concern to various developmental agencies and institutions. This is because food security and vital food systems are of great importance. One of the most significant current discussions in agriculture is the use and application of indigenous agricultural practices (IAPs) which have been used by local farmers to enhance household food security. Despite their significance, most of IAPs are at risk due to urbanization, cultural assimilation, and globalization. This study applies an explorative design, with an indigenous paradigmatic lens to illustrate the significance of these practices in ensuring food security. The data was collected through semi structure interviews with local farmers of Bakgatla-ba-Kgafela and was analyzed using thematic analysis. The study finds that the documentation and preservation of IAPs is crucial for ensuring sustainable food systems and food security. It is therefore concluded that new and innovative approaches to agriculture are needed.

It is becoming increasingly difficult to ignore cannabis as a resource that has the potential to drive socioeconomic development in indigenous communities. This resource has for a long time been an overlooked commodity within indigenous communities based on the negative perceptions and stigma that surround it. Cannabis has the potential to contribute towards the economy and various indigenous communities in South Africa such as AmaMpondo in the Eastern Cape Province are found to be the most producing area of cannabis for various purposes that contributes towards their local and household sustainability. This chapter applies a qualitative document analysis (QDA) to elucidate meaning on the trends and arguments within the agricultural cannabis thematic area.

Indigenous people of South Africa have for many years relied on Indigenous Knowledge to ensure food security. For example, they can identify different types of edible wild fruits, vegetables and roots, they know when to plant which types of food, how to plant them, when and how to harvest and process the food and how to preserve the food to ensure food security. Agriculture, based on Indigenous Knowledge, ensures the production of a variety of indigenous foods. However, the knowledge on these Indigenous foods is slowly being lost. This is mainly because indigenous knowledge is tacit and kept in the minds of knowledge holders whom most of them are no more. Without its protection and promotion, the knowledge is susceptible to change or completely being lost.

Indigenous knowledge (IK) of indigenous crops (IC) is deeply rooted in the history and environment of the Vatsonga people. It is crucial to preserve this knowledge, as it is part of their culture, language, and identity. IK, passed down through oral, observed, and written forms, includes unique farming practices specific to the community. However, with the rise of Western knowledge, conventional farming, and genetically modified foods, there is a decline in the use of indigenous crops, leading to the loss of Vatsonga IK. This study, using an indigenous paradigmatic lens and qualitative approach, explores how the current generation of Vatsonga applies their IK regarding indigenous crops. It finds that the Vatsonga of MTA have used these crops for centuries for survival, food access, income, and cultural identity. The study

Ndebele women have been creating murals and beading as a part of their cultural identity for decades and they have passed this expertise down to next generation. To create sustainable livelihoods, they have been traveling the world to showcase their skills and sell crafts to businesses and tourists. However, an existing challenge today is that most entrepreneurs are exploited of their knowledge due to lack of education on policies and poor government outreach initiatives on protection of Indigenous Knowledge (IK). Therefore, it is critical that IK entrepreneurs comprehend their rights and financial power over their expertise with government assistance. This study, which employed qualitative methods, was carried out in the KwaMhlanga area in Mpumalanga province, with Ndebele women who are artefacts entrepreneurs. It sought to examine existing IK policies that these entrepreneurs can use to protect their artefacts. The study also discussed government initiatives that guarantee entrepreneurs recognize their sovereignty over their expertise and receive a fair economic share.

The commercialization of indigenous Knowledge-based Businesses in South Africa is a crucial step towards economic empowerment and cultural heritage preservation. The government plays a crucial role in facilitating this process, providing financial support, capacity-building programs, and market access through government-led partnerships. However, challenges such as bureaucratic inefficiencies, limited awareness of government programs, and inadequate infrastructure hinder the full realisation of these opportunities. The tension between modern commercialisation demands and the preservation of traditional practices also complicates the process. The study underscores the need for more robust, culturally sensitive, and inclusive strategies to support indigenous businesses in South Africa.

This research emphasizes the importance of Indigenous data sovereignty (IDSov). It foregrounds the rights of Indigenous peoples to exercise control over their data while resisting data colonialism and its myriad harms. It argues that including Indigenous knowledge in the design, development, and implementation of data-based technologies is essential. The CARE Principles for Indigenous Data Governance are applied to underpin this. The four case studies from Canada, New Zealand, Australia, and the US illustrate the links between Indigenous peoples' rights and data and ecosystem protection and show how responsible data management can benefit people and the environment. Such an understanding of data sovereignty illuminates the exploration of data cooperatives. The focus is on data cooperatives as new institutional models enabling Indigenous communities to exercise digital sovereignty. This term refers to active control and ownership of digital assets and data. It then examines whether the data and sovereignty principles from the Indigenous context can be transferred to other contexts.

Foreword

This book, *Revaluation and Preservation of Indigenous Knowledge Systems in Modern Society*, is about what African scholars and activists in the struggle for the recognition, inclusion, preservation and promotion of Indigenous Knowledge Systems (IKS) have steadfastly said and written over the past decades. It arrives at a time when there are clarion calls for the decolonization and Africanization of knowledge and education in Africa. These calls come from and in various voices, whether young people protesting in the streets and in social media spaces, or from revolutionaries in academic terrains. At core of their call is a concern about the continued colonial traditions in education and knowledge production in Africa. The book follows in the tradition of numerous scholars who, over decades, have sought to remind us that the decolonial agenda remains incomplete. In fact, the editors of this book, Professor Tlou Maggie Masenya and Dr Monicca Thulisile Bhuda have in their own unique and yet unified ways persistently, unapologetically and unambiguously called for the decolonization of knowledge, vociferously foregrounding IKS as both a combative and remedial intervention against the continued cultural alienation and hollowness experienced by many Africans due to the lingering effects of western imperialism. This book should be best understood in the context of the uphill efforts toward the decolonization and Africanization of Africa. A documentation of indigenous crops utilized by Vatsonga households for food security, for example, as considered in this book, sheds light on one of the under-researched and therefore relatively unknown methods that Africans have always employed to ensure their livelihood in the land. By documenting these voices that have hitherto been in the margins and silenced, the pace towards decoloniality is doubtless accelerated. The discussion around the intellectual property protection of IKS for sustainable economic development is also an indicative of the fact that the book is best positioned as a midwife in the birth of a *New Africa*, which, of course, will remain incomplete until African ways of being, of knowing and of power are freed from the chains of coloniality through legal legitimization and protection.

IKS is desperately needed to center the discourse entirely on a decolonial base and perspectives on bridging heritage and enterprise to address socio-economic issues, with a precise emphasis on a policy analysis of IKS businesses is a heartily welcome pursuit. In making Africans realize that they have philosophical, spiritual, struggle and epistemological heritages, perhaps a resurgence, reclamation, and preservation of African IKS, which uses the past to look to the future is a necessary endeavor. Indeed, this book shows us that it is! There is enough room and lucid discourse in the book to accommodate indigenous data and knowledge sovereignty amidst technological dynamics to show that IKS can save, not only Africans, but the entire universe in this very modern society. Hence, significant insights have been proffered in this book to facilitate the transition of IKS through the Digital Transformation Framework. The book reminds us that we have entered a space, a place and a time in which we think and speak in lucid ways about the essence of intellectual tools and the generation of radical ideas in advancing Africa to the next dimension of decoloniality. Hence, unsurprisingly, the book imagines using biocultural community protocols to preserve and protect valuable IKS and environmental sustainability; the role of the Fourth Industrial Revolution (4IR) technologies in digital preservation and documentation of IKS, and using documentaries as reservoirs of IKS, all of which are anchored to apt technological and theoretical perspectives. Suffice it to say, we are at a critical juncture where, more than ever before, books must be written to (re)cognize, (re)cover, re(store) and (re)member often from obscurity the works of the indigenes, intellectuals and activists, all of whom thinking and speaking from indigenous paradigms have long made points and continue to do so in a manner that this timely book ever so proficiently does!

Moffat Sebola
University of Limpopo, South Africa

Preface

Knowledge has been affirmed as powerful resource which brings development in every human endeavour and it is critical to sustainable development of indigenous communities and society at large. Therefore, knowledge that is gained but is unavailable to others is wasted. Every society or community has local knowledge (also known as traditional or indigenous knowledge) which cuts across all aspects of human living on which livelihood and survival depend. Indigenous Knowledge is a way of life, skills, experiences, culture, insight and values embraced by people in local community. This knowledge has been the basis for agriculture, education, health care and the wide range of other activities that sustain a society and its environment in many parts of the world for many centuries. Although, indigenous knowledge often remains an irreplaceable resource, however, most of this knowledge is slowly disappearing in indigenous communities worldwide and get lost due to various factors such as death, sickness and memory loss, and also as a direct result of the pressures of modernization and globalization. Therefore, if memory institutions, cultural heritage institutions, governments, technology industry, individuals and all related stakeholders do not act now, we will lose our cultural heritage and our times will be a new dark age in history. Indigenous knowledge may be lost forever, if it is not properly documented and preserved. Greater efforts, initiatives or mechanisms must be established in different parts of the world to safeguard, document and preserve indigenous knowledge for the sustainable development of rural or indigenous communities and the use of the upcoming generations.

Preservation of knowledge has become a popular method for valorizing, safeguarding and enhancing access to indigenous knowledge in the digital era, characterized by globalization and knowledge-based economy. Researchers have therefore, advocated for the preservation and documentation of indigenous knowledge as important for recovery, sustainable development of rural communities and can help to build resilient communities. Enhancing long-term access to indigenous knowledge can also allow for identity building, reconciliation, creativity, innovation and many other activities that make societies stronger, richer and more peaceful. Word organizations such as UNESCO and the World Council of Indigenous People have recognized the need for preservation of indigenous knowledge and they have yet to devise formal, standardized and sustainable practices and proper policies for the purpose. Indigenist thinkers have also advocated for the recovery, promotion and safeguarding of indigenous knowledge systems as important in decolonizing indigenous nations. Indigenous knowledge thus needs to be safeguarded at all times, documented, preserved and shared in a way that benefit indigenous knowledge owners, indigenous communities and future generations. Documenting and preserving indigenous knowledge will make it easy to share and also help to ensure that indigenous communities are not disadvantaged because of the unique beliefs and folkways that pattern their lives. Despite the strategic role that indigenous knowledge plays in developing modern scientific knowledge and advancing science and technology, many rural communities in Africa and globally, research institutions, memory institutions and other relevant stakeholders do not have mechanisms or a means for safeguarding, documenting and preserving this vital knowledge. There is therefore a need to reimagine innovative strategies, techniques or mechanisms for documenting, safeguarding, preserving and sharing indigenous knowledge among indigenous communities to ensure a sustainable development and long-term knowledge access.

ORGANIZATION OF THE BOOK

Chapter 1: Resurgence, Reclamation, and Preservation of South African Indigenous Knowledge Systems: Using the Past to Look to the Future

Despite being regarded as the most significant resource for indigenous people worldwide, the loss of Indigenous Knowledge Systems (IKS) has put these communities under a great deal of stress. Research has demonstrated that an unquantifiable amount of IKS is fading away since the beginning of the 20th century. The erosion and marginalization of IKS is the direct result of Westernization and colonization, all accompanies by the racist and colonial policies. The concern over the loss of this irreplaceable knowledge has thus raised a need to recover, reclaim and preserve it in the modern society. This chapter utilised literature review to critically analyse the resurgence and reclamation of IKS practices for the benefit of indigenous communities, the role of modern technologies in the recovery and preservation of IKS, the decolonial approaches for the revitalization and revaluation of IKS. Sociocultural learning theory and modernization theory were also used as underpinning theories to guide the study.

Chapter 2: Digital Repatriations: A Temporary Solution to Improved Access to Zimbabwe's Heritage

The chapter seeks to investigate the benefits and challenges of digitally repatriating Zimbabwe's heritage resources housed in foreign museums. It also discusses the legal, ethical and cultural implications of such repatriations. Digital repatriation is the process of digitally returning heritage resources to their countries of origin. Worldwide, digital repatriations are emerging as a temporary solution and a means of providing access when physical repatriations are delayed or not possible. Zimbabwe lost a lot of invaluable cultural property to colonial violence and looting, including spirituality and cultural symbolism. This diverse moveable heritage carries significant values for source communities in Zimbabwe. The layers of their meaning now include the story of their expropriation and the continued fight for their return more than four decades after the attainment of independence in 1980. While appreciating that digital repatriation does not override the need for physical repatriation and not an end in itself, the research evaluates the effectiveness of digital repatriations as a temporary solution to facilitate access to Zimbabwe's cultural heritage. Repatriation is important because it restores cultural, historical, and spiritual heritage to its rightful owners, fostering healing, preservation, and respect for indigenous identities and histories. This study adopted a qualitative research method and a multiple case study research design.

Chapter 3: Indigenous Data and Knowledge Sovereignty Amidst Technological Dynamics: Reflections on Selected Southern African Indigenous Communities

This chapter examines narrative literature review concerning the indigenous data and knowledge sovereignty in the context of Southern African indigenous communities amidst technological dynamics. The study was also guided by the following research objectives: examine indigenous data and knowledge sovereignty among Southern African indigenous communities, examine national and international initiatives for data and knowledge sovereignty among indigenous communities, determine the role of intellectual property rights in the protection of indigenous data and knowledge, and suggest recommendations to enhance indigenous people's rights to data and knowledge sovereignty. Social Contract Theory (SCT) was also employed to underpin this study as it embraces the notion that a person's moral, political or legal responsibilities are reliant on their establishment of regulations or ideals that guide their society.

Chapter 4: Documentaries as Reservoirs of Indigenous Knowledge Systems for Ndau people of South-Eastern Zimbabwe

Indigenous knowledge systems have gained unprecedented traction in the past few decades. Discussions centre around their efficacy and agency in solving a variety of challenges ranging from environmental, agricultural, governance, educational, and/or health-related ones. While discourses on indigenous knowledge systems' usefulness are ongoing, they have tilted towards the preservation of them. This chapter focuses on documentaries as reservoirs of indigenous knowledge systems in a context of possible extinction due to a number of factors. Using in-depth interviews with indigenous knowledge practictioners and the postcolonial lens, the study found that documentaries are an effective way of preserving indigenous knowledge notwithstanding some challenges associated with documention and dissemination of indigenous knowledge in Africa.

Chapter 5: Intellectual Property Rights Laws in the Protection and Re-Appropriation of Indigenous Knowledge Systems Among South African Indigenous Tribes

Indigenous knowledge is mainly preserved in the minds of elderly people and shared through traditional cultural practices, manifestations or demonstrations and verbal transmission or communication. There is a need for epistemic justice and sovereignty as the vast amount of indigenous knowledge is presently unprotected by the current Intellectual Property Rights (IPR) laws and policy frameworks, that gives indigenous knowledge holders and their communities property rights to their original works and cultural heritage. This chapter aims to answer this question: How can indigenous knowledge systems be re-appropriated and protected by IPR laws for the benefit of indigenous knowledge owners and their communities? The chapter utilised literature review to critically analyse the protection of indigenous knowledge in South Africa, exploring the role of intellectual property rights laws, policy frameworks and protocols in the re-appropriation and resignification of indigenous knowledge systems. The study is guided by the following objectives: determine the significance of indigenous knowledge among indigenous communities, the effect of cultural appropriation among indigenous communities, the anti-imperialism strategies for the re-appropriation and resignification of indigenous knowledge among indigenous tribes and the role of Intellectual Property Rights (IPR) laws in the reclamation and protection of indigenous knowledge. Recommendations suggest the need for the protection, re-appropriation of indigenous knowledge systems and promote access to indigenous knowledge through reconstitutionalizing rights of intellectual property rights laws, policy, protocols, procedural practices, developing standards for ethical professional practice among indigenous tribes and ensuring that IPR laws meet international requirements

Chapter 6: Global Initiatives for Digital Preservation and Documentation of Indigenous Knowledge

This chapter aims to explore and analyze global initiatives in the preservation and documentation of Indigenous Knowledge Systems (IKS) through digital approaches. With increasing threats to the survival of IKS due to modernization and globalization, innovative and technology-based strategies are needed to support the preservation and dissemination of this vital knowledge. The methods used in this research include literature review, policy analysis, and evaluation of digital technologies such as artificial intelligence, blockchain, and online databases. Data was collected from various academic sources and global preservation projects to identify trends, challenges and solutions that have been implemented. The results show that the implementation of digital technologies plays a significant role in improving accessibility, sustainability and legal protection of IKS. In addition, the involvement of indigenous peoples in the digitization process is a crucial factor in the success of this initiative. Future research directions include the exploration of blockchain-based intellectual rights protection policies, the development of an AI platform for IKS classification, as well as a socio-economic impact study of IKS digitization. In conclusion, technology-based approaches provide great potential in preserving IKS effectively and sustainably, while respecting the principles of local wisdom and the rights of indigenous peoples.

Chapter 7: The Role of Kwa-Khaya Lendaba Cultural Village in the Preservation and Dissemination of Indigenous Knowledge

Cultural Villages (CVs) have existed in Africa for many years, primarily for tourist attractions and economic benefits. Many cultural villages deteriorate and are marginalized and vandalized by people who do not understand their educational value, as well as to appropriately conserve the Indigenous Knowledge (IK) contained within CVs. IK is recognised for its critical role as a strategic tool in developing people's lives worldwide. Therefore, the management and preservation of this knowledge is critical in academia. This research aims to underpin how the Kwa-Khaya Lendaba (KKL) cultural village preserves, manages and promotes the valuable IK that it embodies, including its dissemination using library databases. The data was generated by applying face-to-face, in-depth interviews and the Knowledge, Attitude, and Perceptions (KAP) Likert scale were used to collect data on local people's perceptions of IK, CV, and management strategies of IK at the cultural village. The data was analysed using thematic analysis, and SPSS was used to analyse the quantitative data. The study's findings revealed that cultural villages play a vital role in community development and that cultural villages depict diverse African cultures. However, the knowledge behind the cultural village is not documented, preserved, and managed for the community and learners to have access. The study concludes that the best way to preserve and disseminate IK is through ICTs and Libraries. Furthermore, there is a dire need to communicate the advantages of using knowledge management tools to Preserve and manage IK as a valuable resource that can drive development in many communities.

Chapter 8: Using Biocultural Community Protocols to Preserve and Protect Valuable IK: A Course to Develop an Extra Layer of Rights for Two Indigenous Communities

Indigenous knowledge (IK) and the systems thereof have been faced with various challenges such as piracy, misappropriation, misuse, and overall abuse of the rights of indigenous peoples towards their resources. This has led to various indigenous communities of South Africa being sceptical about who to let in into their space and who to share their valuable IK with, as an intangible resource. This presented various problems that range from a social to an economic magnitude thereby leaving this IK at the periphery of the knowledge economy. It is therefore pivotal to dwell into strategies that seek to preserve, protect and promote IK as a socioeconomic resource which indigenous communities rely on for survival. This study applied participatory action research with an indigenous paradigmatic lens to pinpoint how biocultural community protocols (BCPs) can be a viable tool for communities to reach protection and promotion of their IK. The study focused on two Khoisan communities based in Wupperthal and Bethelsdorp to unearth how this formerly marginalized indigenous communities can benefit from developing their own BCPs. The data collected through focus group discussions and in-depth interviews were analysed using thematic analysis. The study finds that biocultural community protocols are a vital tool that can be utilized to enforce the IK Act no 6 of 2019. The development of these BCPs will ensure that communities manage their IK thereby protecting and strengthening the intricate linkages between the cultural systems of knowledge and the management of land and natural resources. It is concluded that BCPs are a relevant resource to ensure that IK as a social capital of communities is protected, promoted and managed to drive sustainable development in communities.

Chapter 9: Documentation and Preservation of Indigenous Agricultural Practices Used to Enhance Food Security Among Bakgatla-Ba-Kgafela in Moruleng, North-west Province, South Africa

Due to the rapidly increasing population and the issue of global warming around the world, food access and availability has become a paramount concern to various developmental agencies and institutions. This is because food security and vital food systems are of great importance to everyone. The concern about the rapid growth of the populations further led governments and other institutions to look at means of ensuring the overall achievement of food security to the world's population. One of the most significant current discussions in agriculture is the use and application of indigenous agricultural practices (IAPs) which have been used by local farmers to enhance household food security. For decades, indigenous peoples in various countries have been utilizing IAPs to provide food for themselves and their families. Despite their significance, most of IAPs are at risk of being lost due to urbanization, cultural assimilation, and globalization. More recently, the emphasis is on the need to document and preserve indigenous agricultural practices to enhance food security and ensure sustainable agricultural practices. This study applies an explorative design, with an indigenous paradigmatic lens to illustrate the significance of these practices in ensuring food security. The data was collected through semi structure interviews with local farmers of Bakgatla-ba-Kgafela and was analyzed using thematic analysis. The study finds that the documentation and preservation of IAPs is crucial for ensuring sustainable food systems and food security since this approach utilizes valuable indigenous knowledge (IK) to foster meaningful and sustainable agriculture. It is therefore concluded that new and innovative approaches to agriculture are needed, and the readily available solution is indigenous innovations within the agriculture sector which are rooted in IK of communities

Chapter 10: A Pursuit to Mainstream Indigenous Cannabis Growers in the Cannabis Market to Foster Socioeconomic Development: A Theoretical Dialogue

It is becoming increasingly difficult to ignore cannabis as a resource that has the potential to drive socio-economic development in indigenous communities. Cannabis production has for a long time been an overlooked commodity within indigenous communities based on the negative perceptions and stigma that surround it. Cannabis has the potential to contribute towards the economy and various indigenous communities in South Africa such as AmaMpondo in the Eastern Cape Province are found to be the most producing area of cannabis for various purposes that contributes towards their local and household sustainability. However, legislation and policies hinder their production as there is vast legal obstacles that govern the cannabis industry and leaves the indigenous cannabis growers in difficult position. This paper applies a qualitative document analysis (QDA) to elucidate meaning on the trends and arguments within the agricultural cannabis thematic area. The findings of the study illustrate that although cannabis production is prohibited at a larger production scale, indigenous cannabis growers continue to make a living out of the production while policies and laws are not in favor of them. Furthermore, Cannabis production has the potential to grow the South African economy through industrial practices, health ministry etc. It is therefore concluded that mainstreaming indigenous cannabis growers in the cannabis production market will ensure that their rights are recognized, enhanced and valorized. IK plays a pivotal role in the knowledge economy and helps local communities to sustain their lives through different fields of production, policy and legislation.

Chapter 11: Documenting and Preserving the Indigenous Foods that the Moshana People in the Northwest Province, South Africa Utilize to Ensure their Food Security

Indigenous people of South Africa have for many years relied on Indigenous Knowledge to ensure food security. For example, they can identify different types of edible wild fruits, vegetables and roots, they know when to plant which types of food, how to plant them, when and how to harvest and process the food and how to preserve the food to ensure food security. Agriculture, based on Indigenous Knowledge, ensures the production of a variety of indigenous foods. However, the knowledge on these Indigenous foods is slowly being lost. This is mainly because indigenous knowledge is tacit and kept in the minds of knowledge holders whom most of them are no more. Without its protection and promotion, knowledge is susceptible to change or completely being lost. Documenting these indigenous foods is crucial as it provides a comprehensive record of indigenous knowledge, including methods of cultivation, preparation, and consumption. This is because Moshana community has a diverse indigenous plant species that are accessible and available. By doing this, the potential of these indigenous foods such as pumpkin, sorghum, sweet potato, amaranth and corn will be realized as well as their contribution to food security.

Chapter 12: Documentation of Indigenous Crops Utilized by Vatsonga Households for Food Security in Mhinga Traditional Authority (MTA), an Appraisal for Economic

Indigenous knowledge (IK) of indigenous crops (IC) is deeply rooted in the history and environment of the Vatsonga people. It is crucial to preserve this knowledge, as it is part of their culture, language, and identity. IK, passed down through oral, observed, and written forms, includes unique farming practices specific to the community. However, with the rise of Western knowledge, conventional farming, and genetically modified foods, there is a decline in the use of indigenous crops, leading to the loss of Vatsonga IK. This study, using an indigenous paradigmatic lens and qualitative approach, explores how the current generation of Vatsonga applies their IK regarding indigenous crops. It finds that the Vatsonga of MTA have used these crops for centuries for survival, food access, income, and cultural identity. The study concludes that indigenous crops are vital for food security, cultural preservation, and economic growth by creating jobs and promoting respect for their IK.

Chapter 13: The Role of the Fourth Industrial Revolution (4IR) Technologies in Digital Preservation and Documentation of Indigenous Knowledge Systems

The Fourth Industrial Revolution (4IR) heralds a new era of technological advancements that fundamentally alter how societies function, particularly in the realm of knowledge preservation and documentation. This study explores the application of 4IR technologies in the digital preservation and documentation of Indigenous knowledge. Indigenous knowledge systems, rich in cultural heritage and traditional practices, face the risk of erosion due to globalization and technological dominance. Leveraging 4IR technologies such as artificial intelligence (AI), blockchain, Internet of Things (IoT), and big data analytics presents a promising solution to this challenge. This study aims to examine how these technologies can be effectively utilized to safeguard and perpetuate indigenous knowledge for future generations. This chapter concludes by emphasizing the transformative potential of 4IR technologies in preserving and documenting IK. By bridging the gap between tradition and innovation, these tools can empower Indigenous communities, ensuring that their knowledge contributes to global solutions for health, sustainability, and cultural diversity. However, achieving these outcomes requires a balanced approach that addresses ethical considerations, fosters inclusivity, and prioritizes the rights and needs of Indigenous custodians. As the world navigates the complexities of 4IR, the digital preservation of IK stands as a testament to the value of blending ancient wisdom with cutting-edge innovation

Chapter 14: Using Social Media Networks in the Management of Indigenous Knowledge in Selected Public Libraries in Kwazulu-Natal, South Africa

Managing indigenous knowledge (IK) through social media is crucial for capturing, storing, and sharing IK effectively. Thus, the aim was to explore how social media networks are used in managing IK within public libraries in KwaZulu-Natal, South Africa. The study was guided by Nonaka's (1994) Knowledge Creation Theory (KC), using an interpretivism paradigm to understand participants' views and experiences regarding the management of Indigenous Knowledge (IK) via social media networks. Data was gathered through semi-structured interviews with librarians at the Pietermaritzburg Bessie Head Public Library in KwaZulu-Natal. Findings revealed a low level of awareness of using social media networks for IK management across KwaZulu-Natal public libraries, with only Bessie Head Public Library actively engaged in this area. The study contributes to existing IK literature, explores the relevance of knowledge creation theory for IK, and offers practical solutions for improving the use of social media in managing IK.

Chapter 15: Integrating Indigenous Knowledge Systems in the Fourth Industrial Revolution (4IR): Preserving Indigenous Knowledge through 4IR

The fourth industrial revolution (4IR) is transforming industries, economies, and societies through digital, biological, and physical system integration. While technologies such as artificial intelligence (AI), blockchain, and the Internet of Things (IoT) offer innovation opportunities, they also pose challenges for preserving indigenous knowledge systems (IKSs). As repositories of environmental stewardship and cultural heritage, IKSs face marginalisation due to rapid technological shifts and globalisation. This chapter examines how 4IR technologies can safeguard and enhance IKSs. It explores AI-driven language preservation, blockchain for verifying cultural artefacts, and IoT for monitoring holy sites. Ethical concerns surrounding data sovereignty, intellectual property rights, and cultural commodification are also addressed to uphold indigenous autonomy. Drawing on multidisciplinary perspectives and case studies, the chapter advocates for equitable collaboration between technology innovators and indigenous communities, ensuring emerging technologies empower rather than erode IKSs. This integration is crucial for cultural sustainability, social justice, and ethical technological advancement

Chapter 16: Harnessing 4IR Technologies: Innovating the Digital Preservation of Indigenous Knowledge

The revolutionary potential of Fourth Industrial Revolution (4IR) technologies in conserving and advancing indigenous knowledge systems is examined in this chapter. In the modern world indigenous knowledge which is frequently transmitted orally and through cultural practices is at serious risk of eroding especially as technological advancements and globalization change cultural landscapes. To preserve these knowledge systems for upcoming generations this chapter looks at the function of 4IR technologies such as blockchain artificial intelligence (AI) and digital archiving tools. The chapter stresses how crucial it is to respect cultural sensitivities integrate indigenous knowledge into the digital age and make sure that digital preservation techniques are consistent with traditional values. It talks about the difficulties and possibilities that digital technologies bring to the development of sustainable culturally relevant and easily accessible platforms for the documentation of indigenous knowledge.

Chapter 17: Indigenous Knowledge and Environmental Sustainability: A Technology Perspective

This qualitative case study explores the essence of indigenous knowledge (IK) and environmental sustainability from a Technology perspective. IK is linked to the natural environment because it is framed within cultural practices. However, IK is side-lined in education and seeking the environment's sustainable solutions, thus disregarding its critical contribution to preserving the environment. Eight Senior Phase Technology teachers and a Curriculum Subject Advisor (CSA) were purposively selected and interviewed individually. Document analysis was also conducted on the CAPS Technology, Sasol Inzalo textbooks, and Annual Teaching Plans (ATPs). Data was analysed following the inductive thematic analysis framework. It was learned in this study that teachers understood indigenous technology and articulated the importance of teaching about education for sustainable development (ESD). However, they did not integrate it into their teaching despite the opportunities in the analysed documents. The chapter contributes a more inclusive and culturally informed framework for teaching about SD and ESD.

Chapter 18: Bridging Heritage and Enterprise to address Socio-Economic Issues: A Policy Analysis of Indigenous Knowledge System Businesses In South Africa

This research examines the role of government policies in supporting the establishment and growth of Indigenous Knowledge Systems (IKS) in South Africa. It evaluates key legislative frameworks, such as the Indigenous Knowledge Systems Policy (2004) and the Protection, Promotion, Development, and Management of Indigenous Knowledge Act (2019), and provincial programs aimed at fostering IKS enterprises. The study also examines the effectiveness of these policies in promoting entrepreneurship, safeguarding Indigenous knowledge, and ensuring equitable benefit-sharing. It also examines the impact of these policies on marginalized communities, particularly in rural areas. The research also highlights the need for more inclusive, adaptive, and context-specific policies that integrate Indigenous knowledge into mainstream economic frameworks. The study advocates for policies that promote IKS businesses and contribute to a more sustainable, equitable, and culturally rich South Africa.

Chapter 19: The Commercialization and Intellectual Property Protection of Indigenous Knowledge: A look Into Ndebele Artefacts Entrepreneurs

Ndebele women have been creating murals and beading as a part of their cultural identity for decades and they have passed this expertise down to next generation. To create sustainable livelihoods, they have been traveling the world to showcase their skills and sell crafts to businesses and tourists. However, an existing challenge today is that most entrepreneurs are exploited of their knowledge due to lack of education on policies and poor government outreach initiatives on protection of Indigenous Knowledge (IK). Therefore, it is critical that IK entrepreneurs comprehend their rights and financial power over their expertise with government assistance. This study, which employed qualitative methods, was carried out in the KwaMhlanga area in Mpumalanga province, with Ndebele women who are artefacts entrepreneurs. It sought to examine existing IK policies that these entrepreneurs can use to protect their artefacts. The study also discussed government initiatives that guarantee entrepreneurs recognize their sovereignty over their expertise and receive a fair economic share.

Chapter 20: The Role of Government in the Commercialization of Indigenous Knowledge-based Businesses in South Africa Opportunities and Challenges: Opportunities and Challenges

The commercialization of indigenous Knowledge-based Businesses in South Africa is a crucial step towards economic empowerment and cultural heritage preservation. The government plays a crucial role in facilitating this process, providing financial support, capacity-building programs, and market access through government-led partnerships. However, challenges such as bureaucratic inefficiencies, limited awareness of government programs, and inadequate infrastructure hinder the full realisation of these opportunities. The tension between modern commercialisation demands and the preservation of traditional practices also complicates the process. The study underscores the need for more robust, culturally sensitive, and inclusive strategies to support indigenous businesses in South Africa.

Chapter 21: Indigenous Digital Sovereignty: Data Cooperatives as the New Standard for an Inclusive Digital Economy?

This research emphasizes the importance of Indigenous data sovereignty (IDSov). It foregrounds the rights of Indigenous peoples to exercise control over their data while resisting data colonialism and its myriad harms. It argues that including Indigenous knowledge in the design, development, and implementation of data-based technologies is essential. The CARE Principles for Indigenous Data Governance are applied to underpin this. The case studies illustrate the links between Indigenous peoples' rights and the protection of data and ecosystems and show how responsible data governance can benefit people and the environment. Such an understanding of data sovereignty illuminates the exploration of data cooperatives. The focus is on data cooperatives as new institutional models enabling Indigenous communities to exercise digital sovereignty. This term refers to active control and ownership of digital assets and data. It then examines whether the data and sovereignty principles from the Indigenous context can be transferred to other contexts

Chapter 22: Homogenizing Traditional Knowledge and Biodiversity Conservation: A Study Intersecting IP Rights and Sustainable Development

This chapter explores the nexus between biodiversity conservation and the preservation of traditional knowledge, focusing on the need for an integrated approach that is in line with sustainable development goals. This paper examines the mechanisms that exist for the protection of traditional knowledge and biodiversity by critically appraising the international frameworks such as the Convention on Biological Diversity and the Nagoya Protocol along with national legislations such as India's Biological Diversity Act, 2002. The emphasis here is on "biocultural heritage" and its integration into conservation frameworks that create traditional knowledge systems uniform. Fitted together with case studies and law contexts, this study highlights the important role that local communities are also assuming in custodianship over biodiversity and traditional knowledge. This research contributes to the broader discourse on environmental law and governance, including policy recommendations to harmonize biodiversity conservation with protection of traditional knowledge.

Chapter 23: Indigenous Knowledge and Sustainable Development: The Indian Story

This chapter explores the profound relationship between Indigenous knowledge and sustainable development in India. It highlights how traditional ecological knowledge, cultural practices, and community-driven initiatives of Indigenous populations contribute to environmental sustainability and socio-economic development. It delves into traditional ecological practices and community-driven initiatives that have sustained the environment and supported socio-economic development for centuries. By examining case studies from diverse regions of India, the chapter illustrates the successful integration of Indigenous knowledge with modern development practices. It also addresses the challenges and opportunities in leveraging Indigenous knowledge for sustainable development and offers policy recommendations to enhance its inclusion in development frameworks. This chapter aims to highlight the invaluable contributions of Indigenous communities to sustainable development and provide insights for policymakers, researchers, and development practitioners. (not well formatted, abstract, key term and definitions missing).

Chapter 24: In the Transit of Indigenous Knowledge Systems through Digital Transformation Framework

This chapter explores the transition of IKS through a digital transformation framework, emphasizing the need for culturally sensitive approaches that honour integrity and context of indigenous knowledge. This widens the potential of digital tools to enhance accessibility, foster innovation, and bridge generational knowledge gaps, while addressing critical issues such as intellectual property rights, data sovereignty, and ethical representation. Through a problem-based pedagogy, strategies for integrating IKS into modern digital ecosystems, promoting inclusive knowledge practices and global sustainability are outlined. The objectives propose a digital framework for transitioning IKS using digital platforms; to identify the role of digital transformation in preserving, transmitting, and revitalizing IKS; and to discuss ethical considerations, challenges and practical implementation in Higher Education (HE). This sparked debate about a redesigned role for digital framework education and authenticity. IKDTF is seen as having the ability to preserve, transmit, and revitalise IKS.

Chapter 25: Models and Frameworks for Digital Preservation of Indigenous Knowledge Systems

Preserving Indigenous Knowledge Systems (IKS) is essential for safeguarding cultural identity and heritage. This paper explores various models and frameworks for the digital preservation of IKS, emphasizing the importance of integrating modern technological advancements with traditional knowledge. The study highlights challenges such as data ownership, intellectual property rights, and the risk of knowledge misappropriation. It proposes a structured framework that incorporates ethical considerations, technological infrastructure, and community participation. The role of metadata standards, digital repositories, and access controls is analyzed to ensure the authenticity and security of indigenous knowledge. The findings suggest that collaborative approaches, involving indigenous communities and technological experts, are critical to the success of digital preservation initiatives. The study concludes that a balanced strategy combining technological innovation and cultural sensitivity can create a sustainable model for the preservation and transmission of Indigenous Knowledge Systems.

Tlou Maggie Masenya
Durban University of Technology, South Africa

Acknowledgment

First and foremost, I would like to thank the Almighty God, the Lion King of Judah who gave us wisdom and strength to make this book a reality. I would also like to acknowledge and thank all the authors, researchers, academics and indigenous knowledge experts for their significant contribution to this book "Revaluation and Preservation of Indigenous Knowledge Systems in Modern Society" through sharing their priceless wisdom, extensive experience, essential knowledge and considerable expertise. Without their contribution, this intellectual discourse would not have become a success. This legacy book went through a vigorous blind-peer review and editing by experts in accordance with the best research practices and global ethical standards and it would therefore not have been possible without the support and commitment of the reviewers and advisory board committee members drawn from different countries. I truly appreciate their thoughtful comments and efforts towards strengthening the quality of the book and for the priceless commitment and dedication that added value to the book. I am greatly indebted to IGI-Global development team for their persistent help and support throughout all the stages of editing and publishing this book. The editors of the book namely; Prof Tlou Maggie Masenya and Dr Thulisile, Monica Bhuda "who are regarded as true reflection of Mbokodo, meaning the Rock" in South Africa, have shown their bravery, courage and strength by grabbing the bull by its horns and accomplished this vital and impactful project. They worked tirelessly in ensuring that this book provide the best practices, current global innovative initiatives, strategies/techniques and approaches, preservation tools and technologies, theoretical and conceptual frameworks regarding the revaluation, reclamation, documentation and preservation of misappropriated indigenous knowledge /cultural heritage or cultural legacy among ethnocultural or indigenous Communities, globally. Women are underrepresented in academia, globally including South Africa. Book editors are commented for representing women by playing a pivotal leadership role in academic research and innovation as a way of bridging the gender equality gap and by also popularising and promoting research among women in academics in Higher Education Institutions. Special thanks to my family for their steadfast support, persistent love, encouragement, patience and I am forever grateful for immeasurable sacrifices they have made throughout my academic career and research journey.

Tlou Maggie Masenya
Durban University of Technology, South Africa

Chapter 1

Resurgence, Reclamation, and Preservation of South African Indigenous Knowledge Systems:
Using the Past to Look to the Future

Tlou Maggie Masenya
Durban University of Technology, South Africa

Monicca Thulisile Bhuda
https://orcid.org/0000-0002-8506-9562
University of Mpumalanga, South Africa

ABSTRACT

Despite being regarded as the most significant resource for indigenous indigenous people and their communities worldwide, the loss of Indigenous Knowledge Systems (IKS) has put these these most marginalised communities under a great deal of stress. Research has demonstrated that an unquantifiable amount of IKS is fading away since the beginning of the 20th century. The erosion and marginalization of IKS is the direct result of Westernization and colonization, all accompanies by the racist and colonial policies. The concern over the loss of this irreplaceable knowledge has thus raised a need to recover, reclaim and preserve it in the modern society. This chapter utilised literature review to critically analyse the resurgence and reclamation of IKS practices for the benefit of indigenous communities, the role of modern technologies in the recovery and preservation of IKS, the decolonial approaches for the revitalization and revaluation of IKS. Sociocultural learning theory and modernization theory were also used as underpinning theories to guide the study.

INTRODUCTION

Indigenous peoples hold an extensive wealth of knowledge known as indigenous or traditional knowledge, however, this vital knowledge has not always been recognized by post-industrial Euro-centric cultures, as noted by Battiste (2000). This unique knowledge which is restricted to a distinct group of people, culture or society (Senanayake, 2006), has been previously ignored and denigrated by colonisers.

DOI: 10.4018/979-8-3693-7964-6.ch001

In this modern world whereby colonization, westernization and globalisation dominate, we have become increasingly dismissive of indigenization of knowledge or indigenous ways of being and knowing. Indigenization is the process by which Indigenous ways of being, knowing and doing are incorporated into the cultural, educational, organizational and social structures of the institution. However, the indigenization of knowledge is often regarded by Western scholarship as primitive and therefore redundant in the face of modernity (Murove, 2018). Indigenous knowledge (IK) has been viewed as irrelevant to the sustainable development, and the colonialists regarded this knowledge as primitive, uncivilized, backward, superstitious and savage (Briggs & Sharp, 2004). This mindset has led to the devaluation and marginalisation of Indigenous Knowledge Systems that have been passed down through generations. Colonial powers used knowledge systems to alienate Africans indigenous communities from their own knowledge and traditions. The colonial operationalizing of these knowledge systems aimed at both the alienation of Africans from their own knowledges, existential realities, traditions and the integration of the few into the colonial apparatuses (Murove, 2018). This has also prevented Africans indigenous communities from indigenising, intellectualising and Africanising the knowledge they encountered on their own terms and for their own purposes.

According to 2017 estimates, in South Africa, the indigenous groups or people known as Khoisan, comprising the San and the Khoikhoi, represent approximately 1% of the total population of around 59 million. Khoisan people are not generally considered black due to racism, while they are designated as "coloured", the term, originally coined by the British and was used during apartheid to label citizens who did not fit the binary race model including most Afrikaans-speaking non-whites and mixed-race children. South African government does not acknowledge them as the country's first peoples, and their land was mostly taken long before the apartheid era. The South African ruling party, which is African National Congress (ANC) has embarked on a mission to redistribute land, since the end of apartheid in 1994. However, the land distribution process has largely excluded the Khoisan who were the first inhabitants of Southern Africa and one of the earliest distinct groups of Homo sapiens, enduring centuries of gradual dispossession at the hands of every new wave of settlers, including the Bantu people. This amorphous categorization condemned much of the Khoisan's history to oblivion and facilitated the theft of their land. and their land restitution was conceived to benefit black South Africans. As noted by Murove (2018) indigenous people and their communities in African have experienced the current dominant knowledge systems that have impacted and dominated the African epistemic terrain, in the post-colonial era, as continuing to remain Eurocentric and not African.

African Indigenous, for example, the Khoisan are not formally recognized in terms of national legislation as a customary community and as a result, South African government has voted in favour of the adoption of the UN Declaration on the Rights of Indigenous Peoples and the protection, promotion, development and management of the communities' indigenous knowledge systems. The South African Parliamentary Committee on Cooperative Governance and Traditional Affairs presented the Traditional and Khoi-San Leadership Bill (B23-2015), in 2016, aimed at providing for the establishment and operation of a National Office of Indigenous Knowledge Systems and the management of the rights of indigenous knowledge holders or owners and their communities. The bill also provides formal recognition and open opportunities for access to justice for the historic communities of Khoisan, for the first time in the last 300 years. Additionally, the bill further allows Khoisan to be included in the governmental administrative processes within the various ministries while also allowing these ministries to make specific provisions for the social, economic and cultural priorities of the Khoisan communities.

Deterritorialization, misappropriation and insignificance of indigenous wisdom is not only unjust but also deprive indigenous communities and the society at large, an opportunity to address some of the most pressing socio-economic challenges they are faced with. Therefore, best practices, initiatives and strategies must be implemented by indigenous knowledge owners, indigenous researchers, academics, cultural heritage institutions, policy makers, government and relevant stakeholders to recover, reclaim and safeguard Indigenous Knowledge Systems. A growing movement of indigenous activists also believes the time has come to take back what's theirs and they are now exercising their rights to self-identification and recognition. The research objectives formulated for this chapter were to:

- Determine the resurgence and reclamation of indigenous knowledge practices for the benefit of indigenous communities.
- Examine the role of modern technologies in the recovery and preservation of Indigenous Knowledge Systems.
- Determine the decolonial approaches for the revitalization and revaluation of Indigenous Knowledge Systems.

THEORETICAL FRAMEWORK

Sociocultural Learning Theory and Modernization Theory were used as a guide or roadmap for applying best rediscovery, reclamation and digital preservation strategies for Indigenous Knowledge Systems.

Sociocultural Learning Theory

Although, culture is everything that symbolic communication does, however, as stated by Valter and Kull (2014) the development of any culture cannot be understood without knowing what the people knew about their culture. Self-awareness has been regarded as an essential step in becoming culturally aware as people should not only be aware of what is around them but also be aware of their awareness (Bender, Negi & Fowler, 2010). Theoretical knowledge including Indigenous Knowledge Systems and cultural heritage studies can play a remarkable role in cultural self-awareness. Knowledge is influenced by culture and environmental constructs and is shaped by people's interactions. Sociocultural learning theory, developed by Vygotsky (1978) regarded the culture as the primary determining factor for knowledge construction and individuals learn through cultural lens by interacting with societal members, guided by the rules, skills, and abilities shaped by the culture of that society. This theory stresses on the importance of interaction between people and the culture in which they live, and it embraces how the world and people in it shape learners' knowledge and skills development through meaningful interactions and the use of language and signals. Knowledge experts often guide the exploration of ideas while learning leads to development through the use of language, which strongly influences thought (Follari, 2007; Davidson, 2010). As stated by Vygotsky (1978) the younger generation is completely dependent on elder people, especially on parents, who initiate his decisions while instructing him what to do, how to do it and what not to do, in early stages.

Vygotsky (1978) further noted that the child receives the knowledge through the contacts and interactions with people or elders, and then assimilates this knowledge by also adding the personal values in it. Youth or young generation learn from their families and the communities and the culture they imbibe influences the way they grow, reason and develop, while the norms and values they acquire significantly influence their path of development in all spheres of life. This is thus a transfer of what has been learned from the interaction to the personal values and it is realized through language which plays a big role in the way youth adapts to the social inheritance. Ellis (2000) concur that the social-cultural theory of learning starts during interaction. As also stated by Pionke and Browdy (2008) people learn through what begins as an individual's collective ability to experience and interpret the world in a community context and this can be referred to as learning by doing.

Knowledge acquisition, sharing and learning process starts through interactions and communications with other individuals and it is also influenced by the social environment in which people live. All these processes thus took place as a result of social interaction whereby community members are interactively involved in these processes through discussion, collaboration and feedback. However, as also observed by Bandura (1971) people can learn through watching the behaviour of those around them, which comes in the form of individual performance. The socio-cultural learning theory thus views the indigenous knowledge owners and their communities as a social fabric for learning leading to collective knowledge (Pionke & Browdy, 2008). Therefore, indigenous knowledge holders or owners, caregivers, experts, teachers, peers, parents or elders are all responsible for the development of an individual's higher order functions. However, the socio-cultural learning theory works well in the scenario whereby the one possessing indigenous knowledge acts as a mentor to the one interested in acquiring this knowledge.

Modernisation Theory

This chapter looks at internal dynamics while referring to social and cultural structures and the variation of digital media technologies for social change and development (Janos,2000). Modernisation theory was one of the central models in the disciplines of the social sciences between the 1950s and 1960s, and re-emerged after 1991 through the fourth industrial revolution period (Janos, 2000). This theory is grounded on the impression that people must accept the use of new technological innovations and digital media platforms in the digital transformation era, for a society to increasingly transit from a 'pre-modern' to a 'modern'. This could assist to build on a modern society that is innovatory, more advanced and highly developed. This theory further proposes that the marginalised societies would be more knowledgeable, develop and progress well as they adopt the new moral values and standards, modernised and contemporary practices (Janos, 2000) and new ways of being. The indigenous people and their communities can thus develop, transform or grow through the acceptance of the modernised practices and processes of knowledge gain and sharing in the digital revolution. Modernisation theory also confirms that various technological innovations such as an advanced medical technique, a new production system or a new agricultural technique as well as the need to update traditional techniques such as socialization, communication and protection processes make transformation, revitalization and modernisation more necessary than ever in this digital era. Mdhluli et al. (2021) indicated that the technique that indigenous peoples use to manage, preserve and share knowledge is reasoned to have contributed to their knowledge being threatened with extinction. Digitisation and modernisation initiatives should thus be fully introduced to preserve and safeguard the indigenous knowledge against the intellectual theft and allow knowledge holders or owners from marginalized communities to benefit from such initiatives.

METHODOLOGY

This chapter utilized literature review to critically analyse the resurgence and reclamation of indigenous knowledge practices for the benefit of indigenous communities, the role of modern technologies in the recovery and preservation of Indigenous Knowledge Systems, the decolonial approaches for the revitalization and revaluation of Indigenous Knowledge Systems, using qualitative content analysis, following the guidelines advanced by Kitchenham (2004). The review process begun with the inclusion criteria to ensure inclusion of all the research objectives of the study in the content analysis, which aimed to identify studies that provide direct evidence about the research objectives. The documents were identified and selected on the basis of their usefulness and relevance to the study. The following search terms were used to find published articles reporting on resurgence of Indigenous Knowledge Systems practices, the role of modern technologies in digitizing and preserving Indigenous Knowledge Systems and strategies for the revival and revaluation of Indigenous Knowledge Systems. The following databases that provide access to publications in a variety of fields were used: ScienceDirect, Web of Science, EBSCOhost, Emerald, Springer, Wiley, Google Scholar, and Scopus. These databases are considered a suitable choice for systematic literature reviews as it allows using complex search strings and filters, which makes it easy to apply complex selection criteria as noted by Wang and Noe (2010). The selection criteria included studies published in peer-reviewed journals focusing on the resurgence of Indigenous Knowledge Systems practices in different parts of the world. The thematic analysis technique or process developed by Braun and Clarke (2006) was also used to systematically analyze the qualitative data or text extracted directly from the literature review and outlined as follows.

Familiarization With the Data

Data familiarization was developed by reading the articles selected for the review using the "repeated reading" approach to search for meanings and patterns.

Generation of Themes

Generation of potential themes was research-objectives driven and were developed through capturing the aspects of resurgence of Indigenous Knowledge Systems practices, the role of modern technologies in digitizing and preserving Indigenous Knowledge Systems and the decolonial strategies for the revival and revaluation of Indigenous Knowledge Systems.

Reviewing Themes

All of the themes were defined and common characteristics of the themes were outlined as per the research objectives of this chapter. The resurgence of Indigenous Knowledge Systems practices was a common thread connecting all the themes, which led to the development of main themes, such as the resurgence and reclamation of indigenous knowledge practices, the role of modern technologies in the recovery and preservation of Indigenous Knowledge Systems and the decolonial approaches for the revitalization and revaluation of Indigenous Knowledge Systems.

Producing the Written Analysis

The analysis process resulted in the identification of resurgence and reclamation of indigenous knowledge practices, and decolonial approaches for the revitalization and revaluation of Indigenous Knowledge Systems explored in previous studies and the potential research gaps needing further investigation.

FINDINGS OF THE STUDY

Given the analysis that unfolded, the study reports on the following themes arranged according to the objectives that informed the study.

Resurgence and Reclamation of Indigenous Knowledge Systems Practices Among Indigenous Communities

Some African governments have been reluctant to acknowledge the existence of indigenous communities within their territories although indigenous practices play a very important role in the lives of the African people. As noted by Kanyandago (2007) indigenous peoples in African countries cannot ignore the fact that they live in the 21st century and that Africa is part of the global scene and at the same time their cultural heritage cannot be ignored in the creation of African leadership. As stated by Slater (2019) at least 24 out of the 54 African countries suffer from the brutality of dictators while they claim that they abide by the rule of democracy. Some of the Westerners still fail to recognise non-Westerners as fully human beings, particularly African indigenous communities due to paternalistic colonialism and condescending views of Africa which are still overtly racist (Slater, 2019). Masango (2002) maintains that indigenous people's inability to cope with the vestiges of colonialism resulted in African leaders to cover up their leadership styles and inabilities by reacting and blaming the Western world and previous colonial powers. O'Donohue (2007) further argued that African indigenous peoples have no confidence in themselves, and they are lazy, fatalistic, take the past as the model, believe in instant solutions, do not experiment, and do not use their creative faculties. However, whatever the indigenous communities in Africa have lost or been deprived of can be reclaimed, revived, preserved and perpetuated. Indigenous peoples and their communities may not know where they are going unless they know where they are coming from and they thus need to take steps backwards to reclaim the past so as to understand the present and to realise why and how they have come to be where they are and who they are.

The indigenous African philosophy of Sankofa ('go back and take it') can be adopted in reclaiming their cultural heritage or learn from the past. Sankofaism is referred to as a revaluation of our indigenous way of life and a new search for our true identity as people. Slater (2019) described the Sankofa symbol as a mythical bird flying forward with its head turned backward and the thought behind this is that there is wisdom in learning from the past, both to understand the present and shape the future. It means "Go back and get it" or "Go back and take it." The Sankofaism has also been used as a symbol whereby knowledge of the past can be utilised to map out the future; or where a mistake has been made, the wrong can be rectified and lessons can be learnt from the experience (Slater, 2019). The idea is of looking to the past with the understanding that both the good and the bad have formed the present situation. Galloway (2004:3) described the egg in the mouth of the bird as representing the 'gem' or knowledge of the past upon which wisdom is based while also signifies the generation to come that would benefit from the wisdom. The forward and backward gaze of the Sankofa bird is based on the Akan proverb, 'Se wo wen? *ti na wosankofa a yenki* which translates to 'It is not wrong or shameful to go back for something you have previously forgotten (Quan-Bafour, 2012). The documentation and preservation of indigenous knowledge need to be performed in line with Sankofaism as it provides a renaissance launch pad for the preservation of treasured past (Eshun 2011). However, indigenous knowledge is mainly shrouded in secrecy and the hope of it being disseminated to everyone and utilised for the public good is almost impossible. It is because of this that indigenous peoples' way of preserving, managing and sharing knowledge is argued to have contributed to their knowledge being threatened with extinction. Traditional health practitioners noticed that some of indigenous knowledge were regarded as sacred as it was merely passed to certain groups of people and not to everyone, and were also involved the interference of the ancestors in governing its management. The young generation also regard some of these indigenous practices as being outdated and need to be modernised as they no longer live in the past (Mdhluli et al., 2021). Some of African churches have preached against indigenous practices, destroyed them and calling its followers 'heathens'. Blacking (1967) further noted that songs, dances, games and other means that were once a bedrock for indigenous ways of transmitting indigenous knowledge were also condemned and deserted. Indigenous people who repented and denounced the so-called 'backward and uncivilised' practices were accepted into Western schools and secured formal jobs. Some Africans have been ashamed of the indigenous African way of life and adopted the Western way of life to the extent that they rejected their cultures and have become agents for the propagation of the white nations' culture and their way of life. The corollary to this attitude is that the white man's way of life is a model to be copied and appropriated by Africans (Dzobo 1976). The disregard and rejection of Indigenous Knowledge Systems as being primitive by colonialists have thus led to the extinction of this unique and vital knowledge.

The Role of Modern Technologies in the Recovery and Preservation of Indigenous Knowledge

A concern over the loss of Indigenous Knowledge Systems, control and rights over this knowledge due to westernization, colonization and globalization has raised a pressing need to share and preserve this knowledge in digital formats (Masenya, 2024). The chapter thus addresses the urgent need to reclaim and preserve Indigenous Knowledge Systems and traditions using digital media technologies and digital libraries. In this digital age, leveraging digital media technologies offers a promising avenue for the revitalization and preservation of Indigenous Knowledge Systems and languages. Modern technologies are bringing about a revolution in society by enabling the preservation and promotion of indigenous knowledge through tangible objects created with innovative materials. As digital media technologies represent a major era characterized by profound technological advancements reshaping global societies and industries, it is vital to acknowledge their influence (Oosthuizen, Ungerer & Volschenk, 2023). Mdhluli et al. (2021) asserts that the speedy use of digital media in the modern, dynamic world gives valuable opportunities to facilitate the process of preserving, managing, and sharing knowledge that is unique to South African communities. Hinton et al. (2016) further notes that digital media technology enables meticulous documentation, comprehensive archiving, and seamless digitization of indigenous languages, extending their accessibility to a broader and more diverse audience. Advocates of indigenous knowledge can foster collaborations, share valuable resources, and revitalize their languages in unprecedented ways, by harnessing digital tools and platforms (Gwerevende & Mthombeni, 2023).

Decolonial Approaches for the Revitalization and Revaluation of Indigenous Knowledge in the Modern Society

The current approaches to indigenous knowledge development are dominated by the kind of exploitation that maximises gains for the corporate world while limiting benefits for the indigenous knowledge holders or owners and their communities (Masenya, 2021). Indigenization and intellectualization of knowledge, indigenization of technology, collaborative partnerships, and afrocentrism approach have been identified as some of the decolonial strategies for revitalization and revaluation of indigenous knowledge.

- **Indigenization and intellectualization of Indigenous Knowledge Systems**

Indigenization promotes the culture of science and technology in the country and it leads to a sense of accomplishment and independence among indigenous communities. The indigenisation of knowledge must be pursued under the presumption of a recognition that all knowledge is cultural or context specific to some degree. As such, the multicultural nature and plurality of knowledge systems formations should be acknowledged, as well as the fact that all knowledge production includes an aspect of the indigenisation of knowledge (Murove, 2018). Murove (2018) further notes that the indigenisation of knowledge systems does not only aim at correcting misperceptions that were created by Western scholars about African societies, especially during the time of colonisation. It also aims at displacing erroneous conceptions of notions of objectivity and neutrality in knowledge production, and the grounding of knowledge within African existential realities of self-actualisation and self-understanding (Masolo, 2010). Indigenous knowledge systems discourse recognises that knowledge intellectualisation and knowledge production should not be divorced from contextual nor cultural inclinations or experiences.

- **Indigenization of technology**

Irani et al. (2010) highlighted how technology is designed in the West and with Western values, and when it migrates elsewhere and encounters different cultures, it can perpetuate coloniality. Many technology-enabled and self-determined initiatives in the digital era are not centered on community or societal needs. Technology should meet the specific needs and demands of the indigenous communities so that they will be able to operate more proficiently for sustainable development. Indigenous communities should have full ownership of technology to improve scope of their research and development and this will also enable these communities to take up more innovative projects. This ultimately leads to technological advancement of indigenous communities which further contributes to the enhancement and improvement of the economy. Indigenous communities have long been engaged in the process of ensuring that technology platforms reflect and respond to their traditional ways, cultures and languages (Carpenter et al., 2017). Taylor et al. (2019) concur that technology use that is responsive to the local Indigenous community can foster more ethical relationships and relational language technologies. For example, the indigenous language learners need to find technology culturally engaging and integrated in an accessible, current and user-friendly way that enables language learning beyond isolated words or decontextualized phrases, as noted by Pitawanakwat (2018).

Human Computer Interaction (HCI) scholars have adopted both decolonial and postcolonial perspectives to empirically study the relationship between technology and local and indigenous practices (Dosono & Semaan, 2020). Indigenous content developers are becoming increasingly influential in decolonizing the digital landscape to better serve indigenous peoples and their communities. As noted by Meighan (2021) indigenous creators and developers are being firm in applying indigenous thought and practice into the design and construction, even if some underlying software is in English. Lewis et al. (2020) further stated that there are indigenous peoples working to ensure that the rapidly evolving future of artificial intelligence technologies has an ethical foundation rooted in and reflective of indigenous worldviews and languages. The sociotechnical systems play an important role of helping people manage life change, especially through the opportunities afforded in seeking social support and in developing communities (Das & Semaan, 2022). Indigenous people can also use socio-technical systems to decolonize their identities and thus reclaim narrative agency over how they see themselves and others. In examining the uses of sociotechnical systems for identity work, Human computer Interaction (HCI) scholarship has focused on people's uses of online communities and social media during life changes, such as recovering from experiences with domestic abuse, gender transition and job loss (Burke & Kraut, 2013; Haimson et al., 2015). Researchers demonstrate how unique local practices emerge around technologies that promotes sustainability and how technologies work as a medium of imposition of Western standards on the Global South, building upon Irani and colleagues' work on postcolonial computing (Ravishankar, Pan& Myers, 2013; Rifat, Toriq & Ahmed, 2020; de Castro Leal, 2021). Decolonial computing scholars further studied the innovation practices in the Global South that reflects local needs and values (Bidwell, 2021). However, only the digital natives may fully understand the technological specifications and optimum calibrations required to be able to operate in the regional conditions. HCI studies from postcolonial perspectives have also continued to scrutinize the utopian promise of development through technology and often found those to be unsuccessful to solve complex social problems with overly simplistic solutions (Philip, Irani & Dourish, 2012).

- **Collaborative partnerships approach**

Collaborative partnership initiatives are essential for developing and maintaining relationships between memory institutions and various indigenous communities, and provide an opportunity for indigenous communities to increase their exposure (Masenya,2023). Diverse skills and knowledge is achieved through this collaborative approach, resulting in the success of indigenous knowledge preservation initiatives. As stated by Sithole (2007) collaboration between libraries and local communities is also essential for ensuring that indigenous knowledge is digitally preserved and protected. Collaboration with technology partners enables the digitization and knowledge sharing through digital platforms, mobile applications, social media platforms and online repositories. As stated by Liu et al. (2019) collaborative partnerships encompass a wide spectrum of activities, ranging from the creation of user-friendly interfaces to the development of mobile apps and online platforms dedicated to language learning and documentation. Indigenous communities should actively participate in defining project goals and shaping the direction of initiatives with such community involvement fostering ownership and empowerment while strengthening cultural identity and revitalization efforts (McCarty, 2018).

As noted by Reijerkerk and Nyitray (2023) Libraries, Archives and Memory (LAM) institutions can collaborate with indigenous communities to preserve indigenous knowledge and traditions in many ways, including: organizing cultural programming and events in collaboration with local indigenous communities to demonstrate and celebrate their cultural heritage; engaging in community-based research projects in collaboration with local indigenous communities, exploring topics related to traditional ecological knowledge, cultural practices and local histories; working with indigenous communities to document and archive indigenous knowledge, including traditional stories, oral histories, cultural practices, and artefacts; supporting language revitalization initiatives led by local indigenous communities by offering resources and programs aimed at preserving and promoting indigenous languages.

Afrocentrism Approach

Indigenous decolonizing methods suitable to the African contexts including Ubuntu and Afrocentrism approach can be used to challenge the Eurocentric research methods that undermine the local knowledge and experiences of the marginalized population groups, as noted by Smith (1999). The Ubuntu concept expresses the African philosophy of humanness, that a person is a person through other people (Murithi, 2006). Molefi Kete Asante founded the Afrocentric school of thought in communication studies and developed Afrocentricity paradigm that emphasizes the importance of viewing African people and culture as central to their own history and narrative. The Afrocentric is a philosophical and theoretical perspective that suggests cultural and social immersion as the best approach to understand African phenomena as opposed to scientific distance (Mkabela, 2005). Asante (2016) defined Afrocentrism as a mode of thought and action in which the centrality of African interests, values and perspectives predominate. It is a philosophical and theoretical perspective that suggests cultural and social immersion as the best approach to understand African phenomena as opposed to scientific distance (Mkabela, 2005). Afrocentricity means "African centeredness", according to which Africans should be given their intellectual pride as the originators of civilisation (Chukwuokolo, 2010). This approach places African people at the centre of any analysis of African phenomena in terms of action and behaviour and it is intended as an answer to the intellectual colonialism that undergirds and serves to validate political and economic colonialism (Chawane, 2016). Afrocentrism has also been described as a devotion to the idea that what is in the best interest of African consciousness is at the heart of ethical behaviour and seeks to cherish the idea that "Africanness" itself is an ensemble of ethics (Stikkers, 2008). Afrocentrism, as an ideology it represents the continued longing among Africans for some set of ideas that would bind them together as a community and offer some alternative to an assimilation that is either excluded by Europeans or seen by Africans as an admission of inferiority and defeat. In essence, Afrocentrism asserts and celebrates the right of people of African descent to strive for self-determination, which in turn implies a need for Afrocentric ontologies, epistemologies, methodologies and metrics.

CONCLUSION AND RECOMMENDATIONS

The national and international best practices and decolonial approaches were reviewed in this chapter in order to understand the resurgence, reclamation and digital preservation of indigenous knowledge for the benefit of indigenous knowledge owners and their communities. So far, little progress has been made in the recovery and reclamation of indigenous knowledge in African countries, particularly in South Africa. The inheritance of colonial Eurocentric's perspectives and values is believed to be hindering this progress. Although national initiatives, programs, policy and procedural practices in safeguarding indigenous knowledge were developed by the government and other stakeholder, however, some substantial challenges remain. It has been accorded in literature review that there is also a need to follow proper protocols when dealing with the indigenous people concerned. Indigenous knowledge researchers, academics and archivists should appropriately align the ways in which they engage with communities so that they are respectful of and responsive to sociocultural contexts. South African government and cultural heritage institutions should play a role in the recovery and preservation of indigenous knowledge and build collaborative partnerships networks with various indigenous communities and relevant stakeholders in the recovery, reclamation and preservation of indigenous knowledge. Knowledge should be indigenized and be shared in ways that benefit indigenous communities through consultations with research participants and culturally relevant protocols. This chapter recommends the:

- Development of new models, frameworks and policies to guide and address the challenges associated with resurgence, reclamation and preservation Indigenous Knowledge Systems.
- Development of standards for ethical professional practice among indigenous communities, in line with the international standards.
- Review of current Indigenous Knowledge Systems policy, protocols, intellectual property rights laws, copyrights laws, and ensuring that they all meet the international requirements.

FUTURE RESEARCH DIRECTIONS

Indigenous Knowledge Systems and cultural manifestations need to be appropriated and appreciated so that they can be re-expressed and preserved digitally in our Modern society. This chapter viewed decolonial strategies as best ways to recover, reclaim and preserve Indigenous Knowledge Systems for the benefit and acknowledgement of Indigenous Knowledge owners or holders and their communities. Future research should thus focus more on developing new models, framework and policies are needed to cope with the challenges associated with resurgence, reclamation and preservation indigenous knowledge systems.

REFERENCES

Asante, M. K. (2016). *Afrocentricity and its critics.* Available at: https://science.jrank.org/pages/8216/Afrocentricity-Afrocentricity-Its-Critics.html. Accessed on 2 April 2016.

Balogun, T., & Kalusopa, T. (2021). Developing a framework for web archiving of Indigenous Knowledge Systems (IKS) in selected repositories in South Africa. *In C. Mnubi-Mchombu.,* Nengomasha, C., & Uutoni, W. *(Ed.),Proceedings of the 24th Standing Conference of Eastern, Central and Southern African Library and Information Associations*, pp. 293- 300.

Bastos, F. M. C. (2009). Art education in the spirit of Sankofa. *Art Education, 62*(2), 1–5. DOI: 10.1080/00043125.2009.11519005

Battiste, M., & Henderson, J. H. (2000). *Protecting indigenous knowledge and heritage: A global challenge*. Purich. DOI: 10.59962/9781895830439

Bender, K., Negi, N., & Fowler, D. N. (2010). Exploring the Relationship between Self-Awareness and Student Commitment and Understanding of Culturally Responsive Social Work Practice. *Journal of Ethnic & Cultural Diversity in Social Work, 19*(1), 34–53. DOI: 10.1080/15313200903531990 PMID: 23255873

Bidwell, N. J. (2021). Decolonising in the Gaps: Community Networks and the Identity of African Innovation. In *Re-imagining Communication in Africa and the Caribbean* (pp. 97–115). Springer. DOI: 10.1007/978-3-030-54169-9_6

Blacking, J. (1967). *Venda children's songs: A study in ethnomusicological analysis*. Witwatersrand University Press.

Braun, V., & Clarke, V. (2006). Using thematic analysis in psychology. *Qualitative Research in Psychology, 3*(2), 77–101. DOI: 10.1191/1478088706qp063oa

Briggs, J. (2005). The use of indigenous knowledge in development: Problems and challenges. *Progress in Development Studies, 5*(2), 99–114. DOI: 10.1191/1464993405ps105oa

Burke, M., & Kraut, R. (2013). Using Facebook after losing a job: Differential benefits of strong and weak ties. *InProceedings of the 2013 conference on Computer supported cooperative work.* 1419–1430. DOI: 10.1145/2441776.2441936

Carpenter, J.. (2017). *Digital access for language and culture in first Nations communities*. Heiltsuk Arts UBC.

Chawane, M. (2016). The Development of Afrocentricity: A Historical Survey. *Yesterday and Today, 16*(16), 78–99. DOI: 10.17159/2223-0386/2016/n16a5

Chukwuokolo, J.C. (2009). Afrocentrism or Eurocentrism. *New Journal of African Studies,* 32.

Chukwuokolo, J. C. (2010). Afrocentrism or Eurocentrism: The Dilemma Of African Development. *OGIRISI a New Journal of African Studies, 6*(1). Advance online publication. DOI: 10.4314/og.v6i1.52333

Das, D., & Semaan, B. (2022). Collaborative Identity Decolonization as Reclaiming Narrative Agency: Identity Work of Bengali Communities on Quora. In *CHI Conference on Human Factors in Computing Systems (CHI '22)*. New Orleans, LA, USA. ACM, New York, NY, USA. DOI: 10.1145/3491102.3517600

de Castro Leal, D., Duarte, A. M. B., Krüger, M., & Strohmayer, A. (2021). Into the Mine: Wicked Reflections on Decolonial Thinking and Technologies. *In C&T'21:Proceedings of the 10th International Conference on Communities & Technologies-Wicked Problems in the Age of Tech.* 269–280. DOI: 10.1145/3461564.3461578

Dei, G. S. (2012). Indigenous anti-colonial knowledge as "heritage knowledge" for promoting Black/African education in diasporic contexts. *Decolonization, 1*(1), 102–119.

Dosono, B., & Semaan, B. (2020). Decolonizing Tactics as Collective Resilience: Identity Work of AAPI Communities on Reddit. *Proceedings of the ACM on Human-Computer Interaction 4,* CSCW1, 1–20. DOI: 10.1145/3392881

Dzobo, N. K. (1976). The courage to be an African in education. *The Oguaa Educator, 7*(1), 132–138.

Eshun, G. (2011). *Ecotourism development in Ghana: A postcolonial study withfocus on Boabeng-Fiema monkey sanctuary and Kakum National Park.* University of Leicester.

Galloway, D. (2004). *Sankofa, Sasa and Zamani: The Ongoing Influence of Spirituals on African American Poetry.* http://ctl.du

Gwerevende, S., & Mthombeni, Z. M. (2023). Safeguarding Intangible Cultural Heritage: Exploring the Synergies in the Transmission of Indigenous Languages, Dance and Music Practices in Southern Africa. *International Journal of Heritage Studies, 29*(5), 398–412. DOI: 10.1080/13527258.2023.2193902

Haimson, O. L., Brubaker, J. R., Dombrowski, L., & Hayes, G. R. (2015). Disclosure, stress, and support during gender transition on Facebook. *InProceedings of the 18th ACM Conference on Computer Supported Cooperative Work & Social Computing.* 1176–1190. DOI: 10.1145/2675133.2675152

Hinton, L., Huss, L., Roche, G., & Tsosie, E. (2016). Language Revitalization in the Digital Age: Promises and Challenges. In *The Routledge Handbook of Language Revitalization, edited by L. Hinton, L. Huss, and G. Roche, 371-84.* Routledge.

Irani, L., Vertesi, J., Dourish, P., Philip, K., & Grinter, R. E. (2010). Postcolonial computing: a lens on design and development. *InProceedings of the SIGCHI conference on human factors in computing systems.* 1311–1320. DOI: 10.1145/1753326.1753522

Janos, A. C. (2000). *East Central Europe in the Modern World.* University of California Press.

Kanyandago, P. (2007). From Negative to Positive Universality: Anthropological and Ethical Implication for Africa. *In Universality; From Theory to Practice: An Intercultural and Interdisciplinary Debate about Facts, Possibilities, Lies and Myths, 25th Colloquium of the Swiss Academy of Humanities and Social Sciences.* Fribourg: Academic Press, 316.

Kitchenham, B. (2004). *Procedures for Performing Systematic Reviews.* Keele University.

Leonard, W. Y. (2017). Producing language reclamation by decolonising 'language'. *In Leonard, W.Y & De Korne, H. (eds) Language Documentation and Description,* 14. London: EL Publishing. pp. 15-36.

Lewis, J. E.. (2020). *Indigenous protocol and artificial intelligence position paper*. Indigenous Protocol and Artificial Intelligence Working Group.

Masango, M. J. S. (2002). Leadership in the African Context: Words on Leadership. *Verbum et Ecclesia, 23*(3), 707–718. DOI: 10.4102/ve.v23i3.1234

Masenya, T. M. (2021). Digital Preservation of Indigenous Knowledge in South African Rural Communities. In Tshifhumulo, R., & Makhanikhe, T. (Eds.), *J: Handbook of Research on Protecting and Managing Global Indigenous Knowledge*. IGI Global Scientific Publishing.

Masenya, T. M. (2022). Decolonization of Indigenous Knowledge Systems in South Africa: Impact of Policy and Protocols. *International Journal of Knowledge Management, 18*(1), 1–22. DOI: 10.4018/IJKM.310005

Masenya, T. M. (2023). Valorization and Digital Preservation of Indigenous Knowledge Systems in South African Indigenous Communities: Best Practices in the Digital Transformation Era. In Masenya, T. M. (Ed.), *Digital Preservation and documentation of global indigenous knowledge systems*. IGI Global Publishers. DOI: 10.4018/978-1-6684-7024-4.ch002

Masenya, T. M. (2024). Revitalization and Digital Preservation of Indigenous Knowledge Systems for Sustainable Development of Indigenous Communities in South Africa. *The Serials Librarian, 84*, 5–8, 86–102.

Masolo, D. (2010). *Self and Community in a Changing World.* Indiana University Press. DOI: 10.2979/5989.0

McCarty, T. L. (2018). Community-Based Language Planning Perspectives from Indigenous Language Revitalization. In Hinton, L., Huss, L., & Roche, G. (Eds.), *The Routledge handbook of language revitalization* (pp. 22–35). Routledge. DOI: 10.4324/9781315561271-4

Mdhluli, T. D., Mokgoatšana, S., Kugara, S. L., & Vuma, L. (2021). Knowledge management: Preserving, managing and sharing indigenous knowledge through digital library. *Hervormde Teologiese Studies, 77*(2), a6795. DOI: 10.4102/hts.v77i2.6795

Meighan, P. J. (2021). Decolonizing the digital landscape: The role of technology in Indigenous language revitalization. *Alternative, 17*(3), 397–405. DOI: 10.1177/11771801211037672

Mkabela, Q. (2005). Using the Afrocentric Method in Researching Indigenous African Culture. *Qualitative Report, 10*(1), 178–189.

Msuya, J. (2007). Challenges and Opportunities in the Protection and Preservation of Indigenous Knowledge in Africa. *International Journal of Information Ethics, 7*, 346–353. DOI: 10.29173/irie38

Murove, M. F. (2018). Indigenous Knowledge Systems Discourse and Inclusionality: An Afro-centric Quest for Recognition in a Globalised World. *Journal for the Study of Religion, 31*(1), 159–176. DOI: 10.17159/2413-3027/2018/v31n1a9

O'Donohue, J. (2007). *The Problem of Africa. From Negative to Positive Universality: Anthropological and Ethical Implication for Africa*. Academic Press.

Oosthuizen, J. H., Ungerer, M., & Volschenk, J. (2023). A Fourth Industrial Revolution Integrated Intelligence Taxonomy for Top Management. *J. Contemp Management*, *20*(1), 404–443. DOI: 10.35683/jcm1018.202

Philip, K., Irani, L., & Dourish, P. (2012). Postcolonial computing: A tactical survey. *Science, Technology & Human Values*, *37*(1), 1, 3–29. DOI: 10.1177/0162243910389594

Pitawanakwat, B. (2018). Strategies and methods for Anishina abemowin revitalization. *Canadian Modern Language Review*, *74*(3), 460–482. DOI: 10.3138/cmlr.4058

Quan-Baffour, K. P. (2012). Sankofa: 'Gazing Back' to Indigenous Knowledge and Skills for Socio-economic Development of Ghana. *Studies of Tribes and Tribals*, *10*(1), 1–5. Advance online publication. DOI: 10.1080/0972639X.2012.11886637

Ravishankar, M. N., Pan, S. L., & Myers, M. D. (2013). Information technology offshoring in India: A postcolonial perspective. *European Journal of Information Systems*, *22*(4), 387–402. DOI: 10.1057/ejis.2012.32

Rifat, M. R., Toriq, T., & Ahmed, S. I. (2020). Religion and Sustainability: Lessons of Sustainable Computing from Islamic Religious Communities. Proceedings of the ACM on Human-Computer Interaction 4, CSCW2 (2020), 1–32.

Sarkhel, J. K. (2017). Strategies of indigenous knowledge management in libraries. *Qualitative and Quantitative Methods in Libraries*, *5*(2), 427–439.

Senanayake, S. G. J. N. (2006). Indigenous Knowledge as a Key to Sustainable Development. *Journal of Agricultural Sciences*, *2*(1), 87–94. DOI: 10.4038/jas.v2i1.8117

Sithole, J. (2007). The Challenges Faced by African Libraries and Information Centres in Documenting and Preserving Indigenous Knowledge. *IFLA Journal*, *33*(2), 117–123. DOI: 10.1177/0340035207080304

Slater, J. (2019). Sankofa-the need to turn back to move forward: Addressing reconstruction challenges that face Africa and South Africa today. *Studia Historiae Ecclesiasticae*, *45*(1), 1–24. DOI: 10.25159/2412-4265/4167

Smith, L. (1999). *Decolonizing methodologies: Research and indigenous peoples*. Zed Books.

Stikkers, K. W. (2008). An outline of methodological Afrocentrism, with particular application to the thought of W.E.B. DuBois. *The Journal of Speculative Philosophy*, *1*(22), 40–49. DOI: 10.2307/25670688

Taylor, J. L.. (2019). A relational approach to designing social technologies that foster use of the Kuku Yalanji language. *Proceedings of the 31st Australian conference on human computer-interaction* (pp. 161–72). DOI: 10.1145/3369457.3369471

Wang, S., & Noe, R. A. (2010). Knowledge sharing: A review and directions for future research. *Human Resource Management Review*, *20*(2), 115–131. DOI: 10.1016/j.hrmr.2009.10.001

KEY TERMS AND DEFINITIONS

Afrocentrism: It is a perspective that prioritizes African culture, values and history, often as a counterpoint to Eurocentric views that have historically marginalized or devalued African contribution.

Colonization: It occurs when one nation subjugates another, conquering its population and exploiting it, often while forcing its own language and cultural values upon its people.

Decolonization: It is the recovery from colonial impact and restoration of indigenous people's identities, languages and experience. It involves actively deconstructing and challenging colonialism and its systems that do not account for the lives of Indigenous peoples and their communities.

Digital Preservation: It is the formal process to ensure that digital knowledge or information of continuing value remains accessible and usable over a long period of time.

Indigenization: It is the process of making practices of a particular culture more indigenous or native, often involving adapting foreign ideas or practices to local contexts. It can also be described as the process whereby Indigenous ways of being, knowing, doing are incorporated into educational, organizational, cultural and social structures of the institution.

Indigenous Knowledge: It is regarded as undocumented knowledge that exists in the minds of community people which is passed from one generation to the other through word of mouth.

Indigenous Communities: Are social and cultural groups that share collective ancestral ties to the lands and natural resources that are distinct from those of the dominant societies in which they live.

Sankofaism: It is a philosophy advocating for the reclamation and valuation of African cultural heritage and indigenous knowledge to guide the present and future. It derived from the Akan word "Sankofa" meaning "go back to reclaim the past."

Chapter 2
Digital Repatriations: A Temporary Solution to Improve Access to Zimbabwe's Heritage

Pauline Chiripanhura
https://orcid.org/0009-0005-9643-3488
Great Zimbabwe University, Zimbabwe

Munyika Sibanda
https://orcid.org/0000-0002-5253-7948
Great Zimbabwe University, Zimbabwe

Farai Chabata
National Museums and Monuments of Zimbabwe, Zimbabwe

ABSTRACT

The paper seeks to investigate the benefits and challenges of digitally repatriating Zimbabwe's heritage resources housed in foreign museums. It also discusses the legal, ethical and cultural implications of such repatriations. Digital repatriation is the process of digitally returning heritage resources to their countries of origin. Worldwide, digital repatriations are emerging as a temporary solution and a means of providing access when physical repatriations are delayed or not possible. Zimbabwe lost a lot of cultural property under colonial rule and is still fighting for the return of its heritage resources. While appreciating that digital repatriation does not override the need for physical repatriation, the research evaluates the effectiveness of digital repatriations as a temporary solution to improving access to Zimbabwe's cultural heritage. This study used qualitative research method with a multiple case study research design.

INTRODUCTION AND BACKGROUND

The colonial era witnessed a substantial plundering of cultural objects from Africa and their appropriation or retention in western museums and private collections. With the end of colonization, the former colonies began to seek, mainly through diplomatic engagements, the return of their looted material culture. As awareness continued to increase regarding the historical and cultural significance of

DOI: 10.4018/979-8-3693-7964-6.ch002

these cultural objects, there has been an increasing global recognition of the importance of repatriating them to their countries of origin. One of the highlights of such calls was the International Conference on African Cultures hosted by Zimbabwe in 2021, whose main focus was repatriation and restitution of cultural property unjustly or illicitly trafficked from the African continent. Repatriation is regarded by Shylon (2014) as the return of cultural heritage properties to the source or descendant communities. Unconditional repatriations are considered as one of the processes of decolonising museum spaces. Decolonisation is defined as the process of removing or reducing colonial structure and influence on the greatest extent possible (Betts, 2012). However, a close evaluation of the scale of colonial plunder and looting in Africa against what has been returned so far reveals a glaring imbalance and testament that the repatriations effected thus far are just a tip of the iceberg. There is an untenable disjuncture between the growing number of lodged repatriations claims and the painstakingly slow pace at which the negotiations have progressed of extended periods that, in some instances, span over several decades and in a worst-case scenario over a century. For instance, the Aksum obelisk, a funeral style approximately 1700 years old and a symbol of the Ethiopian people's identity was returned to Ethiopia after 68 years of being kept in Rome, Italy (Mariam, 2009). The bottom half of the Zimbabwe bird was returned after it had been kept in Museum für Völkerkunde in Berlin, Germany for close to 100 years. The museum records showed that the lower half of the Zimbabwe bird was legally purchased from the missionary Karl Axenfeld of a Missionary Museum in Berlin in 1906 (Dewey, 2006). The Stone Birds of Zimbabwe are monumental soapstone carvings of birds found within Great Zimbabwe, and these divine soapstone figurines are testimony to the use of the site as a place of worship spanning from the ancient past to the present day (Dewey, 2006; Matenga, 2011).

Commenting on the pace of repatriations, Shylon (2014) opines that African countries that are seeking the return of their cultural objects should have 'the endurance of the long-distance runner'. The implications of such statement can be taken to mean that African nations should never be tire in their quest for the return of their cultural property. The repatriation examples given above have undoubtedly shown that the process of repatriation is as complex issue as it is time consuming. Yet, in some cases, the repatriation claims might not even be honoured. Kaplin (2016) is of the view that the former colonial powers and other rich European countries in particular museum professionals and their boards or trustees remain unwilling to part with any material culture even if they remain locked up in stores and warehouses. Zimbabwe, like any other former colony, has a lot of cultural materials that were looted by the former colonial masters and is struggling with their return. Taking into consideration Batt (2021) and Felch and Frammolino (2011)'s sentiments that the rise of western museums and the building of their collections in the 19th Century was made possible partly by the confiscation and appropriation of cultural artefacts from territories that western countries conquered and colonised across the world, it is highly possible that most of Africa's restitution claims might never be fulfilled. Returning these material objects will to some extent appear as an attempt to partly dismantle or close some sections of these western museums, as they were built purposely for the housing of African material objects.

As it is, the Universal Museums Declaration and the travelling exhibitions concept can be viewed as strategic attempts by western museums to secure a firm grip over African material culture. The Declaration of the Importance and Value of Universal Museums was issued in December 2002 by the directors of 18 museums located throughout Europe and the United States. Among these museums are the Louvre Museum of France; the Metropolitan Museum of Art, USA; the British Museum (United Kingdom) and the American Museum of Natural History (USA). Kaplin (2016) and Adams (2022) hold the view that the declaration which gave birth the notion of Universal Museum (a museum holding a broad all-encompassing exhibition of cultures from every continent) was an attempt to defend against demands for restitution of museum objects as this declaration coincided with a rise in restitution demands. Reacting to excessive backlash and criticism against universal museums, the same western museums proposed increased access to collections in their holding through travelling exhibitions. However, these exhibitions are yet to reach the African continent. The end result has been continued deprivation of source communities and/or countries. Against such a background of enduring imperial approaches to museum practice, this paper explores the possibility of digital repatriations becoming a temporary solution for Zimbabwe to access heritage in foreign countries.

PROBLEM STATEMENT

Zimbabwe, like many former western colonies, lost substantial material culture to former colonial masters. The collectors registered these cultural heritage resources into private estates or in museums run by autonomous trustees with the capacity to use western private property laws to justify their continued possession of artefacts looted from former colonies like Zimbabwe. Some of the artefacts lost had been used by source communities for their socio-economic and political sustenance because of the spiritual, symbolic and utilitarian values they carry. The former colonisers continue to drag their feet in responding to Zimbabwe's call for the unconditional return of its original material culture. The nation's heritage authenticity and integrity has been tucked in hidden cultural treasure in foreign countries whose governments continue to perpetuate colonial injustices under the guise of respecting private property rights enshrined in their constitutions. The potential of digital repatriation to improve access to Zimbabwe's heritage kept in museums outside the country has not yet been fully explored, given the complexities of legal and diplomatic processes that precede any repatriations. There have been many explanations proffered on the cultural implications of the continued delays in securing such repatriation. Furthermore, in as much as Zimbabwe needs its original cultural material returned, not much effort has been invested in appreciating the possible *stolen* heritage that can be digitally repatriated. The research aims of this study sought to address the following three key questions:

a) Which aspects of Zimbabwe's heritage can or cannot be digitally repatriated?
b) How can digital repatriations enhance access to Zimbabwe's stolen heritage?
c) What are the legal, ethical, and cultural implications of digitally repatriating Zimbabwe's stolen heritage?

LITERATURE OVERVIEW ON DIGITAL REPATRIATIONS

Before focusing on digital repatriation, it is noble to define repatriation and briefly discuss its importance. Bouquet (2012) is of the view that repatriation is an umbrella term which, when applied to museums, implies the act or process of restoring or returning cultural objects to the country of origin. Onibere and Edewor (2024) are of the view that all the material culture that were acquired through various unethical means, among them colonization, looting or illegal trading, should be returned to their rightful owners, museums or communities. The key issue here is the return of the original cultural objects to their countries of origin. The physical return of material culture is most preferred because returning cultural objects is regarded as an acknowledgement of the historical wrongs of colonization and the recognition of the importance of cultural diplomacy in shaping individual and collective identities and promoting a more equitable future (Appiah, 2016; Ngoepe & Netshakhuma, 2018). Sullivan (2005) further argues that repatriation create a platform for communities to reconnect with their cultural past, strengthen cultural practices, and preserve their heritage for future generations. Regarding human ancestral remains, Onibere and Edewor (2024) considers their repatriation as an important act of cultural revitalization and reclamation of heritage for many indigenous communities around the world, as it provides them an opportunity to reconnect with their ancestors and heal from historical trauma.

Digital repatriation can be described as a process in which cultural heritage objects, physically located outside of countries or communities of their origin, are returned to the originating cultures in a digital form (Christen, 2011; Salmond, 2012). This implies that the cultural heritage materials are made available digitally to the originating culture. The National Museum of Australia (2020) stresses that with digital repatriation, it is the copies of images and documentation held by the foreign museums/ heritage institutions that are provided to cultural custodians. This means that the original material objects are duplicated through the creation of digital photographs, 3D scans, audio recordings, and virtual exhibits. It is these duplicates that are then given to countries requesting their return, while the original objects remain in the current museum. For example, had a digital repatriation been opted for with respect to the Zimbabwe bird that was returned from Germany, the original bird would still be in Germany with Zimbabwe in possession of maybe 3D images of the bird. Bell *et al.* (2013) consider digital objects as a surrogate that is not meant to replace the physical object but rather an alternative form of and dynamic life for certain physical objects. While digital repatriation appears at first glance as appropriate for certain types of material culture such as records, it does not give exclusive ownership rights to the originating culture. It should be remembered that repatriation is not only about the return of material culture, but also the return of authority and responsibility for what happens to them in the future. Bell, Christen and Turin (2013) argue that restitution is the best option as it can encompass both the return of the original material and/ the provision of copies of materials, such as photographs, documents, and sound and film media. Furthermore, digital repatriation is not fully compatible with the repatriation of human ancestral remains and sacred objects because these items hold deep spiritual, cultural, and ancestral significance. For many Indigenous and traditional communities, the physical presence of these objects and remains is essential for fulfilling important rituals, ceremonies, and spiritual practices that connect individuals to their ancestors and heritage. While digital representations or archives of these items may be valuable for educational or preservation purposes, they cannot replace the profound, irreplaceable role that the actual physical items play in the community's spiritual life. The return of these items is often seen as a necessary step in healing, restoring cultural identity, and maintaining ongoing relationships with ancestors, which cannot be fully realized through digital means.

Although acknowledging that digital technologies allow materials to be repatriated quickly, circulated widely and annotated endlessly, Bell *et al.* (2013) argue that these technologies pose challenges to indigenous communities who wish to maintain traditional cultural protocols for viewing, circulation and reproduction of these new cultural materials. However, it should be noted that even though the physical return of cultural remains appropriate and necessary for some objects, there are certain circumstances that might render digital repatriations the only possible alternative. Rossi (2017) argues that digital repatriations are ideal in situations where the receiving community either lack proper storage facilities for returning objects or internal politics are not conducive for the safe return of the physical objects. While digital repatriations are disputed, the creation of digital copies of cultural objects is being welcomed by researchers as they increase access to images, objects and information, thus overcoming cultural barriers to conducting research (Phillips, 2011; Rowley, 2013). A good example is the Electronic Reciprocal Research Network project that was launched by the University of British Columbia, which enabled collaborative research between museums and Columbia native communities and to link collections of north-western objects scattered around the world (Rossi, 2017).

From this section, it is clear that while digital repatriations may not be compatible with certain forms of material objects, such as human ancestral remains and sacred objects that are integral to spiritual practices, there are other types of cultural and historical artifacts that can be digitally repatriated without significant challenges. For instance, archival materials, photographs, manuscripts, and other non-sacred objects could be digitized and shared with communities, providing access to important cultural heritage without diminishing the value of the original items. However, it is important to emphasize that digital objects, while valuable in preserving knowledge and promoting accessibility, do not necessarily replace the physical items themselves. The physical presence of an object often carries symbolic, cultural, and emotional significance that a digital version cannot replicate. With this understanding, we now turn to discussing digital repatriations within the Zimbabwean context, where the implications of digital technologies in the return of cultural heritage are particularly pertinent. The discussion will explore the opportunities, challenges, and cultural considerations involved in digitally repatriating Zimbabwe's cultural artefacts, and how such efforts could impact both the preservation of heritage and the revitalization of cultural practices within the country.

- **The politics of digital repatriation**

Digital repatriation has emerged as a form of restitution aimed at rectifying historical wrongs, including colonial plunder, wartime looting, and illegal antiquities trade. Unlike physical repatriation, digital repatriation does not involve the return of physical objects but rather provides access to digital records and images, which can hold cultural and historical significance. Yet, the reluctance of the holders of these records seem to understand and fear the litigation backlash that is likely to follow the release of any record or image that is likely going to solve a heritage provenance or biographical gap that has existed for several years. One of the central political themes surrounding digital repatriation is the tension between accessibility and ownership, as well responsibility and accountability. Scholars such as Bell, Christen and Turin (2013) have highlighted that digital repatriation serves as a middle ground between the impossibility of returning certain items and the need to honour the cultural rights of indigenous and local communities. The digital repatriation process is seen as a way to "democratise" access to heritage while also acknowledging the significance of cultural ownership. While some scholars emphasise the benefits of digital access for marginalised communities (Banton, 2017), others caution digital repatriation risks further exploitation of cultural resources by external actors without proper recognition of the communities' rights to control their heritage (Snowball, Collins & Nwauche, 2021).

- **Colonialism and power dynamics**

The digital repatriation of heritage must be understood in the context of colonial histories and the ongoing power imbalances between Global North and Global South. Many of the objects and materials that are the subject of digital repatriation were historically taken without consent or under unclear circumstances during the colonial era. The legacies of this brazen colonial looting activities continue to shape contemporary discussions on restitution and the control of cultural knowledge. Guo (2024) argues that digital repatriation should be viewed as a continuation of post-colonial struggles for cultural sovereignty, where the return of digital data is a form of symbolic reparations. The idea of "knowledge sovereignty" emerges in this context, wherein indigenous peoples and nations assert control over how their heritage is represented and shared globally. According to Krupa and Grimm (2021) there is concern that digital repatriation can perpetuate colonial structures if it is managed by Western institutions that hold significant technological and archival power. This situation may reinforce existing inequalities, where marginalised communities are offered digital access without corresponding control over the means of distribution and preservation (Krupa & Grimm, 2021). In this context, digital repatriation has been critiqued for potentially offering a superficial remedy that fails to address the deep structural inequalities underlying heritage theft. Breckenridge (2014) explores how digital repatriation, while beneficial in terms of providing access to global archives, may also deepen the divide between those who control digital technologies and the community's seeking restitution. The political economy of digital heritage determines how digital records are stored, managed, and accessed and also reflects the continued dominance of Western institutions, despite the claims to universality of digital platforms.

- **Ethical considerations in digital repatriation**

There are ethical concerns which are central to debates on digital repatriation. The primary ethical question revolves around who decides what constitutes "repatriation" and how it is implemented. The potential for the exploitation of digital heritage is significant, as digital reproductions can be easily disseminated and manipulated. Mahachi (2011) argues that digital copies do not fully replace the physical objects in their cultural significance, and their replication can create issues of misrepresentation or unauthorised use. Thus, while digital repatriation offers some form of restitution, it cannot fully replicate the complex, lived connections that local communities have with their physical cultural heritage. Further ethical dilemmas include concerns about consent, privacy, and intellectual property. For instance, many indigenous groups and communities have raised concerns about digital copies being distributed without proper consultation or informed consent. Mazarire (2004) points out that some objects might hold sacred or spiritual significance, and their digital reproduction could be seen as a form of cultural exploitation, even when it is done with good intentions. This leads to a broader discussion about the need for ethical frameworks in digital heritage management, especially when the digital repatriation of materials intersects with traditional knowledge and community values. The control of digital reproductions by Western academic and museum institutions raises significant concerns for Indigenous people and communities, particularly around issues of cultural ownership, autonomy, and the ethical handling of their heritage (Snowball, Collins & Nwauche, 2021). Historically, many Indigenous cultures have experienced systemic marginalization, colonization, and exploitation of their cultural and spiritual assets by external institutions. As digital records and reproductions are often managed and controlled by these Western institutions, Indigenous communities may face challenges in asserting their rights over their own cultural heritage in the digital realm. First and foremost, the question of who owns the digital reproductions is a critical issue. When Western institutions hold the rights to digital copies of Indigenous cultural materials, they have the power to determine how these materials are distributed, licensed, and shared, often without consultation or consent from the communities to which they belong. This can lead to a situation where Indigenous peoples are excluded from decision-making processes that directly affect the accessibility and representation of their cultural heritage. As such, these communities may feel that their cultural assets are being commodified, used for profit, or presented in ways that do not align with their values or needs.

Moreover, the Western institutions' control over digital reproductions may limit Indigenous peoples' ability to reclaim and manage their cultural knowledge (Akena, 2012). In some cases, the very act of digitizing cultural materials can be seen as a form of appropriation or distortion of cultural meaning. Objects, stories, or practices may be extracted from their cultural context and repackaged for academic or commercial purposes, potentially stripping them of their original significance and misrepresenting them in ways that are harmful or offensive to Indigenous communities. Ethical considerations in the negotiation of digital rights and cultural ownership must prioritize respect for the autonomy and self-determination of Indigenous communities. This includes the need for consultation, consent, and collaboration with Indigenous groups before digital reproductions are made, ensuring that the decision-making power remains with the communities from which these materials originate. Additionally, there must be safeguards to prevent the misuse or exploitation of digital cultural materials, ensuring that the digital records are used in ways that align with Indigenous cultural values and that communities can retain control over how their heritage is shared and represented.

- **Technological challenges**

Digital repatriation is deeply connected to advances in digital technologies, including digitization, cloud storage, and the use of blockchain for ensuring authenticity and provenance. Chiwara (2019) suggests that technology holds the potential to revolutionise how cultural heritage is repatriated by making it easier to preserve and share digital copies of artefacts and documents. In this way, technology can bypass logistical and financial barriers to physical repatriation, providing marginalised communities with access to resources that were previously inaccessible. However, there are also significant challenges related to the technological infrastructure required for digital repatriation. Chiwara (2019) and Magama (2018) discuss the digital divide between developed and developing nations, which may hinder the ability of certain communities to fully benefit from digital repatriation. Furthermore, the preservation of digital records over time presents issues related to data longevity, security, and the sustainability of digital archives. Another challenge is the risk that the digitization process might inadvertently lose important contextual information, thus diminishing the object's significance. Scholars like De la Porte & Higgs. (2019) caution that digitization cannot fully capture the multisensory and relational aspects of heritage, which are critical to its meaning and importance within local communities. The politics of digital repatriation is a complex and evolving field that intersects with issues of cultural sovereignty, ethical stewardship, and technological access. While digital repatriation presents opportunities for providing global access to cultural heritage, it also raises critical questions about power dynamics, control, and the ethics of cultural restitution. The promise of digital repatriation lies not just in the accessibility of digital records, but in the ability to empower communities to assert ownership over their cultural narratives.

METHODOLOGY

This study adopted qualitative research method with a multiple case study research design. The qualitative methodology was chosen as the study sought to collect the views, perceptions and feelings of people towards digital repatriation. Interviews and desktop research were used to gather data. The study purposively targeted participants working in museums, archives and art galleries as they understand issues of digital repatriation. All the targeted participants were interviewed. Data was presented in the form of narratives or descriptive text. In addition, the data is presented and analysed thematically based on the study's objectives.

FINDINGS AND DISCUSSIONS

Possibilities and Constrains in Digitally Repatriating Zimbabwe's Heritage

The concept of digital repatriation has been increasingly significant in discussions about cultural heritage, particularly when considering the diverse forms of heritage in Zimbabwe. Digital technologies provide new opportunities to preserve and share cultural resources, but questions arise about which aspects of Zimbabwe's heritage can be successfully digitised and shared, and which ones may be unsuitable for digital reproduction. Zimbabwe's heritage is multifaceted, encompassing tangible artefacts, intangible cultural practices, oral traditions, and landscapes that require different approaches to repatriation in the digital age.

- **Heritage forms that can be digitally repatriated**

Participants reported that paper records and artefacts, such as pottery, sculptures, and stone tools, have been digitised in the past using high-resolution photography and 3D scanning. These digital records make it possible for people, especially those in Zimbabwe, to access detailed representations of these items, despite the physical artefacts being housed in museums and archives abroad. Digital repatriations do not only give access and preserve these objects but also offer educational and research opportunities that would have been limited by the geographical distance between the artefacts and the communities from which they originated. Photographic archives and documents have also been successfully digitised. The National Archives of Zimbabwe holds a wealth of historical photographs, manuscripts, and colonial-era documents. Through interviews, it was reported that many of these materials have already been digitised and can be made accessible through online platforms, ensuring that Zimbabweans and scholars worldwide can access these crucial historical records. The digitisation of these documents, including rare photographs and written accounts of Zimbabwe's colonial and post-colonial history, the delineation reports of chiefs allow for a broader and more inclusive engagement with the nation's history, especially for those who may have been historically excluded from accessing such archives (Mazarire, 2004).

Zimbabwe's oral traditions and traditional knowledge provide another area where digital repatriation has shown promise. The rich oral storytelling traditions in Zimbabwe, which have preserved history, customs, and indigenous knowledge, are increasingly being recorded and digitised (Kiwa, Chimhundu, Tsvere & Nleya, 2023). For instance, many projects have focused on recording the oral histories of communities, traditional games, including stories, proverbs, and local narratives. These are then transcribed, archived, and made available for future generations, both locally and globally. This type of digital repatriation ensures that these voices are preserved in a way that transcends generations and geographical boundaries. Similarly, Zimbabwean music and performative arts can also be digitally preserved. The mbira, a traditional Zimbabwean instrument, along with various forms of music, song, and dance, has been recorded and digitised for broader access (Matiure, 2023). These recordings, often made available through platforms such as YouTube or music archives, allow Zimbabwe's rich cultural heritage in the realm of music to reach a global audience. These digital records not only preserve the sounds and rhythms integral to Zimbabwean identity but also facilitate cultural exchange and education.

- **Heritage forms that cannot be digitally repatriated**

Participants also reported that while many aspects of Zimbabwe's heritage can be digitised, there are certain elements that cannot be effectively represented through digital means. One of the most reported significant heritages forms that cannot be repatriated is the sacred rituals and ceremonies integral to Zimbabwean spiritual practices. For example, traditional healing ceremonies, spirit possession rituals, and other rites, such as those associated with spirit possession, cannot be replicated through digital technology. These practices are deeply experiential, communal, and spiritual, requiring direct participation and engagement that cannot be captured in digital formats (Shylon, 2014). The essence of these ceremonies lies in their lived experiences, which are not translatable to digital representations (Aça, 2023). Digitising these rituals risks reducing their significance and possibly misrepresenting them, making it difficult to maintain their cultural integrity.

Similarly, the physical heritage sites and monuments of Zimbabwe, such as the Great Zimbabwe ruins, cannot be fully experienced through digital reproductions. While 3D scans, high-resolution images, and virtual tours can provide an impression of these sites, they fail to capture the full sensory experience of visiting these heritage sites. The impact of standing in the ruins, feeling the weight of history, and engaging with the surrounding landscape cannot be communicated through a screen (De la Porte & Higgs, 2019). Moreover, elements such as the environmental context, seasonal changes, and the social dynamics of these places are lost in digital representations, limiting the depth of engagement with the heritage they embody. Another reported area where digital repatriation falls short is with cultural practices that are dependent on context, such as traditional dance and performative arts. While some dances and performances can be recorded and digitised, the full meaning and significance of these cultural expressions lie in their context that is the specific rituals, the participants, and the community setting in which they occur. Traditional dances performed during ceremonies or celebrations are not just artistic performances, they are social practices imbued with deep cultural meaning (Kiwa, Chimhundu, Tsvere & Nleya, 2023). Attempting to digitise these performances risks distorting the cultural significance of these practices, as it reduces them to mere spectacle, divorced from their living context.

Living cultural practices, such as traditional agricultural methods, were also reported to be presenting challenges for digital repatriation. Zimbabwe's agricultural practices are deeply intertwined with the land and local knowledge systems. These practices, including the cultivation of crops and the ceremonies associated with farming, are dynamic and deeply connected to the environment (Mapara, 2009). While some aspects of these practices can be documented through film or photography, the lived experience of participating in these activities, the relationship between people and the land, and the intergenerational transfer of knowledge cannot be fully captured digitally. These practices evolve over time, and their richness is rooted in physical engagement with the land and community, making them difficult to convey through digital formats.

Digital Repatriation and the Dynamics *of* Stolen Heritage Access

Digital repatriation can be a powerful tool for improving access to Zimbabwe's cultural heritage that has been taken or displaced, particularly those items housed in institutions outside the country. Through digitisation, cultural heritage that was once inaccessible could be shared with wider audiences, enabling greater engagement, education, and preservation. In this context, digital repatriation offers Zimbabweans opportunities to reclaim and reconnect with their heritage, even when the physical objects remain abroad (Batt, 2021). Below, is a presentation of the benefits of digital repatriation in improving access to our heritage.

- **Breaking Geographical Barriers**

One of the most significant ways digital repatriations improve access to Zimbabwe's stolen heritage is by breaking down geographical barriers. Many important cultural artefacts, archival materials, and historical documents have been removed from Zimbabwe and are now housed in foreign museums and archives (Mazarire, 2004). Physical access to these items can be difficult, particularly for people in Zimbabwe or other parts of Africa. However, through digitisation, these objects are made accessible online, allowing anyone with internet access to view high-resolution images, 3D scans, and detailed descriptions (Krupa & Grimm, 2021). This means that Zimbabweans, as well as people worldwide, can engage with these artefacts, regardless of where they are located. Digital repatriation also plays a vital role in enabling remote access for research and education. Many of Zimbabwe's stolen heritage items, such as photographs, manuscripts, and colonial-era documents, have been digitised and made available online. This opens up new opportunities for scholars, students, and educators to study Zimbabwe's history, regardless of their physical location. Digital archives and repositories allow researchers to access rare materials that would otherwise be difficult to obtain, especially for those who cannot afford to travel abroad or who face logistical barriers (Mazarire, 2004).

- **Preserving cultural heritage for future generations**

Another crucial benefit of digital repatriation is its role in preserving Zimbabwe's cultural heritage for future generations. Many of Zimbabwe's cultural artefacts, particularly those that were taken during colonial times, are at risk of deterioration due to age, environmental factors, or misuse. Digitising these artefacts ensures they are preserved in a digital format, safeguarding them from further damage. High-resolution images and 3D scans of these objects can be stored indefinitely, providing a lasting record of Zimbabwe's cultural heritage. For example, while the physical artefacts from the Great Zimbabwe site may be exposed to wear and tear over time, their digital counterparts can serve as a permanent record, ensuring that future generations will be able to access them even if the originals are damaged or lost (Chiwara, 2019).

- **Empowering Zimbabwean communities**

Digital repatriation can also empower Zimbabwean communities by enabling them to reconnect with their cultural heritage. Many of the artefacts and historical materials that were removed from Zimbabwe are deeply tied to the identity and pride of local communities. By making these materials accessible through digital platforms, communities can regain a sense of ownership and control over their cultural heritage (De la Porte & Higgs, 2019). This process of reclaiming heritage through digital means is particularly meaningful for Zimbabweans who may have felt disconnected from their history due to the removal of their cultural assets. Digitising Zimbabwe's oral traditions and indigenous knowledge is another form of empowerment. Stories, proverbs, and local histories, often passed down through generations, can be recorded, archived, and made available to communities in Zimbabwe (Kiwa, Chimhundu, Tsvere & Nleya, 2023). This ensures that traditional knowledge is preserved and accessible, especially for younger generations who may not have access to these oral histories in their everyday lives.

- **Raising public awareness and engagement**

Digital repatriation also raises public awareness and fosters engagement with Zimbabwe's stolen heritage. By making cultural heritage accessible through online platforms, Zimbabweans and people around the world can learn about the richness of the country's history and culture. Virtual exhibitions, digital storytelling, and social media campaigns can spark interest and curiosity about Zimbabwe's cultural legacy, encouraging people to engage with heritage in a new way (Kiwa, Chimhundu, Tsvere & Nleya, 2023). Public access to these digital collections can lead to a greater appreciation for Zimbabwe's contributions to global history and help to highlight the importance of protecting and preserving cultural heritage. For example, virtual exhibitions of Zimbabwe's heritage can be showcased online, allowing visitors to interact with digital versions of artefacts and learn more about their cultural significance (Chiwara, 2019). These online experiences can reach a wide audience, spreading awareness about Zimbabwe's stolen heritage and the need for its preservation.

- **Facilitating collaboration and partnerships**

Digital repatriation also opens the door for collaboration and partnerships between Zimbabwean institutions and global museums, universities, and cultural organisations. Through digitisation, Zimbabwe's cultural heritage can be shared in a way that encourages joint efforts in preserving and promoting these materials. Partnerships with international institutions can help Zimbabwean museums and cultural organisations gain the expertise, resources, and funding needed to digitise and share their heritage more effectively. Such collaborations can also ensure that digitised materials are culturally sensitive and accurately represent the heritage of Zimbabwe.

- **Creating virtual heritage spaces**

One of the more innovative uses of digital repatriation is the creation of virtual heritage spaces. Virtual reality (VR) and augmented reality (AR) technologies allow for the recreation of heritage sites and environments that may no longer exist in their original form or are difficult to access. Through 3D modelling and virtual tours, digital repatriation can offer immersive experiences that give users a sense of place and help them engage with Zimbabwe's heritage in meaningful ways (Guo 2024). For example, a digital reconstruction of the Great Zimbabwe Ruins using VR technology can provide users with the experience of "walking" through the site as it appeared during its peak. This allows for a more interactive and engaging exploration of the site, even for those who cannot physically visit.

- **Protecting against further loss or destruction**

Digital repatriation offers a form of protection against further loss or destruction of Zimbabwe's cultural heritage. Physical objects and artefacts are vulnerable to theft, environmental damage, or the effects of time. However, digital records can serve as a safeguard, ensuring that, even if the original objects are lost or damaged, a digital version of the artifact remains. This backup ensures that Zimbabwe's cultural heritage is not completely erased, even if the physical objects themselves are no longer accessible. In the event of further loss or destruction, such as the damage to a cultural artifact housed in a foreign museum, its digital representation remains intact, providing both a historical record and a means of preserving the heritage for future generations.

Ethics and Legality in Digital Repatriations

The digital repatriation of Zimbabwe's stolen heritage presents significant legal, ethical, and cultural challenges, each of which must be carefully considered, to ensure the process respects Zimbabwe's history and its people's cultural rights. As Zimbabwe seeks to recover and preserve its cultural heritage through digital means, it is crucial to address the implications these efforts have in terms of legal frameworks, ethical considerations, and cultural sensitivities.

- **Legal implications: International laws and conventions**

A central legal concern in digital repatriation is the question of intellectual property (IP) rights over digital versions of cultural heritage. The digitization of heritage objects creates new questions about ownership: who owns the digital reproductions? When cultural artefacts are digitised by institutions abroad, the intellectual property rights to those digital versions may be retained by the institutions holding the originals, complicating Zimbabwe's access. Perullo, (2018) argues that these issues can make it difficult for the country to control the use and distribution of these digital assets. Clear agreements are needed to ensure that Zimbabwe has the rights to use, share, and preserve these digital versions. In addition, once objects are digitised, they are susceptible to unauthorized reproduction, distribution, or manipulation. As outlined by Corbett & Boddington (2011), it is essential to establish legal protections to safeguard digital heritage, preventing the exploitation of these materials without proper consent from the originating nation or community. International laws, such as UNESCO's (1970) Convention on the Means of Prohibiting and Preventing the Illicit Import, Export, and Transfer of Ownership of Cultural Property, primarily focus on the physical repatriation of cultural heritage. However, these conventions also provide some context for the repatriation of digital assets, as cultural institutions globally are bound by ethical standards that influence the digitization of cultural property (UNESCO, 1970). The challenge lies in translating these legal frameworks into the digital domain, where the concepts of ownership and custody are less straightforward. Moreover, the transnational nature of digital content introduces further complications. If a Zimbabwean artefact is digitised and stored in a foreign server, questions arise regarding jurisdiction and legal responsibility for its protection, use, and distribution (Dewey, 2006). These concerns necessitate international cooperation to create clear, cohesive legal standards for digital cultural property.

- **Data protection and privacy**

Data protection is another significant legal concern when digitizing cultural materials. Sensitive information, particularly oral histories or sacred traditions, may be contained within digitised materials. Without proper safeguards, this information could be misused or disseminated beyond the control of Zimbabwean communities. Corbett & Boddington (2011) noted that cultural materials must be treated with the same level of confidentiality as personal data, and protective measures must be in place to ensure privacy and prevent the exploitation of such materials.

- **Ethical implications**

Ethical issues can make or break the success of any digital repatriation. It is vital that digital repatriation respects the cultural significance of heritage materials. Some items may have sacred, spiritual, or deeply personal meanings for local communities and may not be suitable for digitization or wide dissemination. Perullo, (2018) alludes to the risk that digitising such materials without consulting the relevant communities may violate cultural norms or lead to the exploitation of sacred knowledge. In Zimbabwe's case, consulting with local custodians, cultural leaders, and indigenous communities before digitising sacred items or practices is crucial to avoid ethical violations. Ethical repatriation involves seeking the informed consent of these communities and ensuring that they have a say in what is digitised and shared. In seeking consent, attention ought to be paid to the contested and multivocality that may be associated with the heritage to be repatriated. For after repatriation of cultural materials, there may be need to explore their restitution to source communities.

- **Cultural ownership and control**

The question of ownership and control over digital representations of cultural artefacts raises significant ethical concerns. While digital repatriation offers the possibility of reclaiming cultural heritage, there is also the risk that digital artefacts may be misused or misrepresented in ways that do not align with the values of the community. For instance, cultural symbols or practices could be commodified or distorted when shared online. Ethical repatriation requires that Zimbabwean cultural authorities oversee the digitisation process to ensure that cultural materials are represented accurately and respectfully. Ethical considerations also include ensuring that digital repatriation efforts are equitable. Not all Zimbabweans have equal access to digital technologies, especially in rural areas or among marginalised populations. Without addressing this digital divide, digital repatriation risks privileging urban or more connected populations over others (Mazarire, 2004). It is important to make sure that digital heritage is accessible to all Zimbabweans, regardless of their geographic or socio-economic status. Digital heritage projects must be supported by adequate resources and infrastructure to ensure they remain accessible and relevant for future generations.

- **Cultural implications**

Digital repatriation provides an opportunity for Zimbabweans to reclaim their cultural identity and heritage, particularly in the wake of the colonial appropriation of cultural materials. As Mahachi (2011) argues, the ability to access and control cultural heritage is a powerful tool for communities to define and assert their identity. For Zimbabweans, digitised heritage materials offer a chance to reconnect with their cultural roots, especially for those in the diaspora who may have been disconnected from their heritage due to displacement. However, digital repatriation may not fully capture the lived experiences, spiritual meanings, and deeper significance of certain cultural practices. Some aspects of Zimbabwe's heritage, particularly sacred rituals, are grounded in physical and communal experiences that cannot be fully conveyed through digital media. While digitisation can preserve and share Zimbabwe's history, it should not be seen as a replacement for the cultural practices and meanings embedded in the physical objects themselves.

- **Authenticity and cultural context**

When dislocated from their cultural context and digitisation may do just that, Cultural could lose some of their authenticity and significance when removed from their original context. For example, artefacts and rituals tied to specific cultural settings can be misinterpreted if presented outside their cultural context. It is essential that Zimbabwean communities have a role in curating the digital representation of their heritage to ensure it remains authentic and culturally accurate. Moreover, Zimbabwe's cultural custodians must ensure that digital representations are contextualized in ways that respect their cultural importance. Without such safeguards, there is a risk that these materials could be misunderstood or misused by external audiences. The digitisation of cultural practices, especially those related to living traditions, carries risks. Some practices might be commercialised, appropriated, or diluted once they are digitised and shared with a global audience. This raises concerns about the ethical implications of making such practices available without sufficient control over their use. Snowball, Collins & Nwauche (2021) notes that cultural practices that are sacred or personal may be irreversibly altered or exploited once digitised. Zimbabwean communities must retain the ability to control how their cultural practices are shared, ensuring that they are not commodified or taken out of context

CONCLUSION AND RECOMMENDATIONS

In conclusion, the digital repatriation of Zimbabwe's stolen heritage represents a significant step to restore cultural assets, deepen the connection to the country's history, and expand access to heritage for both local and global audiences. While this process is complicated by a myriad of challenges such as determining which forms of heritage can be digitized and addressing the legal, ethical, and cultural concerns, it also offers considerable potential for preserving and making heritage more widely accessible. Establishing a comprehensive national framework, engaging local communities, and ensuring that digital content is carefully curated with respect to cultural sensitivity and scholarly rigour, Zimbabwe can overcome these obstacles. Moreover, combining digital repatriation with efforts to physically return cultural objects will reinforce the country's cultural sovereignty and allow Zimbabweans to engage more profoundly with their heritage. Ultimately, digital repatriation is a vital step in the ongoing efforts to reclaim and safeguard Zimbabwe's cultural legacy while ensuring it is shared and appreciated globally. Zimbabwe should create a comprehensive framework that outlines clear guidelines for the digital repatriation of cultural heritage. This framework should be developed in collaboration with cultural leaders, heritage professionals, and legal experts to ensure that both tangible and intangible cultural heritage is handled properly. It should include criteria for determining which forms of heritage are suitable for digitisation, while considering cultural sensitivities, particularly for sacred or sensitive materials. The following recommendations are made:

Involve Local Communities in the Digitization Process

Ethical considerations in the digital repatriation process require active, inclusive, multivocal and respectful participation from local communities Chipangura and Seabela (2024). Involving cultural custodians, indigenous groups, and heritage experts will ensure that digitized materials are accurately represented and culturally respectful. Informed consent must be sought for the inclusion of sacred or sensitive heritage elements, ensuring that digital representations align with community values and traditions. The desire for repatriation should be used as a smoke screen for pervasion of cultures.

Create a Collaborative Digital Archive

Establishing a centralized digital archive for Zimbabwe's heritage could serve as a global repository for cultural materials that have been repatriated digitally. This archive should be publicly accessible and managed by local institutions or Zimbabwean authorities to guarantee that control over the content remains in Zimbabwean hands. The digital archive could include high-quality images, virtual exhibitions, and even digitized performances to engage audiences, with proper context provided to avoid misinterpretations. This would deviate from the tendency of creating a digital sepulchre for the heritage that has tended to define museum storage areas.

Enhance Access to Technology to Overcome the Digital Divide

The success of digital repatriation largely depends on the infrastructure available to source communities. Zimbabwe should prioritize improving internet access, especially in rural areas, and create public spaces, such as libraries or community centres, equipped with digital tools to ensure that all citizens can benefit from the digital heritage resources. Additionally, digital literacy programs could be launched to help people effectively navigate and engage with the digital collections.

Establish Legal Protections for Digital Heritage

Zimbabwe needs to develop legal measures that safeguard its digital cultural heritage. This includes securing intellectual property rights over digitized materials and ensuring these digital resources are not exploited without proper authorization. Engaging in international legal frameworks that address the unique challenges of digital repatriation will help ensure that Zimbabwe's rights are respected and upheld globally.

Curate Digital Content Ethically and Responsibly

Ethical curation of digital materials is essential. Digital representations must be accompanied by thorough contextual information to avoid cultural misinterpretations. This includes ensuring that culturally significant or sacred items are digitized with care, and that any digital dissemination is done in a way that reflects the material's original cultural context. Collaboration with cultural and academic experts will be key in presenting materials in an accurate, responsible, and culturally sensitive manner.

Complement Digital Repatriation With Physical Repatriation

While digital repatriation is valuable, it should not replace efforts to physically return cultural artifacts. Both forms of repatriation should work together to restore Zimbabwe's heritage. Physical repatriation remains crucial for deepening the cultural connection to these objects and ensuring their protection in their original context. Together, digital and physical repatriation can strengthen Zimbabwe's cultural legacy.

Monitor the Impact of Digital Repatriation

It is important to regularly evaluate the impact of digital repatriation on local communities. This includes tracking how the digitized materials are accessed, whether they are being used appropriately, and if they contribute to the preservation or revitalization of cultural practices. Implementing feedback mechanisms would help improve the process, ensuring it continues to meet the needs of Zimbabweans.

Advocate for Global Support and Awareness

Zimbabwe should raise both local and international awareness on the importance of digital repatriation and advocate for global cooperation in returning its stolen heritage. Building partnerships with museums, universities, and cultural organizations could facilitate the digital repatriation process. Promoting Zimbabwe's rich cultural history on the global stage would foster a broader understanding and respect for its heritage. Repatriation offers an opportunity for cultural diplomacy among co-operating nations.

Secure Sustainable Funding for Digital Repatriation

To ensure the long-term success of digital repatriation efforts, Zimbabwe must identify sustainable funding sources. This could include government funding, international partnerships, or collaborations with private sectors. Investing in cost-effective methods for digitization will ensure that Zimbabwe's cultural heritage can be preserved and made accessible for generations to come.

The Limits of Access and Representation in Digital Repatriation

This chapter would be incomplete without highlighting some of the issues or debates around limits of digital repatriation. One could argue that in the African context, there are limitations that inevitably make digital repatriation inadequate to become an end in itself. Source communities may not have access to requisite digital tools to be able to access and reconnect with their lost treasure. In some instances, what repatriation means to source communities may not necessarily lie in the pomp and fanfare of digital repatriation. Some communities may wish to have a touch of the material culture and, perhaps, conduct a cultural ceremony. That may be all they want. There is, therefore, need to understand the expectations of source communities so that they are not sacrificed at the altar of national pride and expediency.

As highlighted by Lucarelli (2023), digital repatriations seldom answer some very difficult questions, such as questions of intellectual property rights and accessibility of the virtual platforms where the replicas are shared. The need for the physical object has not become obsolete simply because digital surrogates can be created (Carlton, 2010). This raises the potential exploitation of both cultural resources and the source communities by the emerging metaverse which is evolving into a complex digital economy and the commoditisation of virtual experience where virtuality is a commercial commodity, with good and services that can be traded and bought using digital currencies and cryptocurrencies. It is plausible to argue that digital repatriations can easily been viewed as part of the tell signs of the emergence of exploitative digital economies, which are not favourable for fair trade. In the worst-case scenarios, such digital vices can outweigh the educational and publicity benefits that make the façade of the converse realities of digital repatriation. Thus, the conversation of ethical issues in digital repatriations should remain ongoing, perhaps, at a faster rate than the repatriations themselves. For Zimbabwe to benefit meaningfully from digital repatriations, these limitations and ethical issues must be addressed. Without such reflections and interventions, repatriation risk becoming overly superfluous and a new form of colonial exploitation of cultural resources. In any case, the meaning of repatriation, for some communities in Zimbabwe, is vested in physical contact accompanied by a cultural ceremony. For some of these looted or expropriated cultural resources are of spiritual value and in them source communities visualize their past, their ancestors and derive identities and meanings to life. Digital repatriations may be offensive to source, if access to such cultural resources is virtually and unethically granted without their express, informed consent.

FUTURE RESEARCH DIRECTIONS

Future research on the digital repatriation of stolen property in Africa must focus on leveraging technologies like blockchain and AI to track and recover cultural assets, as well as developing secure digital archiving and storage solutions. It must also explore creating legal frameworks that address digital ownership and intellectual property, with an emphasis on ethical considerations, especially in relation to indigenous knowledge. Research must further investigate capacity building within African nations to support digital preservation and repatriation efforts, while promoting international collaboration and public awareness. Additionally, it needs to examine the economic impact of digital repatriation, ensuring that communities benefit ethically from the return of cultural assets.

REFERENCES

Aça, M. (2023). The Role of Rituals and Ceremonies in the Identity Struggle of a Society that Has Lost Its Language: the Example of Lithuanian Tatars. Lietuvos istorijos studijos, (Spec/2023), 93-108.

Adams, G. K. (2022). Does the argument that museums hold collections on behalf of the world still stand? Museums Association. https://www.museumsassociation.org/museums-journal/news/2020/01/17012020-does-universality-argument-still-stand/# Accessed 28/11/24.

Akena, F. A. (2012). Critical analysis of the production of Western knowledge and its implications for Indigenous knowledge and decolonization. *Journal of Black Studies, 43*(6), 599–619. DOI: 10.1177/0021934712440448

Banton, M. (2017). Displaced Archives in the National Archives of the United Kingdom. In Lowry, J. (Ed.), *Displaced Archives*. Routledge. DOI: 10.4324/9781315577609-4

Batt, F. (2021). *The repatriation of African heritage: shutting the door on the imperialist narrative.* African Human Rights Yearbook.

Bell, J., Kimberly, C., & Turin, M. (2013). After the Return: Digital Repatriation and the Circulation of Indigenous Knowledge Workshop Report. 1. 195–203. DOI: 10.3167/armw.2013.010112

Betts, R. F. (2012). Decolonization A brief history of the word. In Bogaerts, E., & Raben, R. (Eds.), *Beyond Empire and Nation: The Decolonization of African and Asian societies, 1930s-1970s*. Brills. DOI: 10.1163/9789004260443

Breckenridge, K. (2014). The Politics of the Parallel Archive: Digital Imperialism and the Future of Record-Keeping in the Age of Digital Reproduction on JSTOR. *Journal of Southern African Studies, 40*(3), 499–519. https://www.jstor.org/stable/24566488. DOI: 10.1080/03057070.2014.913427

Chipangura, N., & Seabela, M. G. (2024). Community Collaborations and Social Biographies of Museum Collections from Colonial Contexts: Meanings of Zulu Beadwork. *Museum Worlds, 12*(1), 16–30. DOI: 10.3167/armw.2024.120103

Chiwara, D. (2019). Documentation: A Security Tool for the Identification and Repatriation of Illicitly Trafficked Objects from Museums with Particular Reference to the National Gallery of Zimbabwe. *Heritage, 2*(1), 390–399. DOI: 10.3390/heritage2010027

Corbett & Boddington. (2011). Copyright Law and the Digitisation of Cultural Heritage. SSRN Electronic Journal. DOI: 10.2139/ssrn.1806809

de la Porte, B., & Higgs, R.De la Porte & Higgs. (2019). Challenges in digitisation of cultural heritage material in the Western Cape, South Africa. *South African Journal of Information Management, 21*(1). Advance online publication. DOI: 10.4102/sajim.v21i1.1104

Dewey, W. J. (2006). Repatriation of a Great Zimbabwe Bird. *Presentation to the Society of Africanist Archaeologists (SAFA)*.safa.rice.edu/WorkArea/DownloadAsset.aspx. (Accessed 28/11/24).

Felch, J., & Frammoline, R. (2011). *Chasing Aphrodite: The Hunt for Looted Antiquities at the World's Richest Museum*. Mifflin Press.

Gunderson, F., Lancefield, R. C., & Woods, B. (Eds.). (2019). *The Oxford handbook of musical repatriation*. Oxford University Press.

Guo, J. (2025). Digital cultural heritage challenges, solutions, and future directions: edited by Vinay Kukreja, Amitoj Singh, Deepinder Kaur, and Jagpuneet Kaur Bajwa, London and New York, CRC Press, 2025, 320 pp., $154.80 (hardback), ISBN 978-1-032-63054-0.

Kaplan, I. (2022). The Case against the Universal Museum. https://www.artsy.net/article/artsy-editorial-the-case-against-the-universal-museum (Accessed 28/11/24).

Kiwa, F. J., Chimhundu, H., Tsvere, M., & Nleya, S. (2023). Transmission of Traditional Culture through Digital Technology in Zimbabwe. *East African Journal of Education and Social Sciences*, *4*(3), 1–6. DOI: 10.46606/eajess2023v04i03.0271

Krupa KL & Grimm KT. (2021). Digital Repatriation as a Decolonizing Practice in the Archaeological Archive. ATD, VOL18(ISSUE1/2)

Lucarelli, R. (2023). From Virtual Reality to virtual restitution: How 3D-Egyptology can contribute to decolonizing the field and the question of digital copies vs the original. In Palladino, C., & Bodard, G. (Eds.), *Can't Touch This*. Ubiquity Press., DOI: 10.5334/bcv.i

Mahachi, G. (2011). *Zimbabwe Report on the UNESCO Workshop on Prevention and Fight Against Illicit Traffic of Cultural Goods in Southern Africa: Current Situation and Way Forward*; UNESCO: Windhoek, Namibia, 14–15 September 2011.

Mapara, J. (2009). Indigenous knowledge systems in Zimbabwe: Juxtaposingpost-colonialal theory. *The Journal of Pan African Studies*, *3*(1), 139–155.

Mariam, H. (2009). The Return of the Axum Obelisk: The cultural benefits of the return of the axum obelisk. *Museum International*, *61*(1), 48–51. DOI: 10.1111/j.1468-0033.2009.01684.x

Matenga, E. (2011). *The Soapstone Birds of Great Zimbabwe: Archaeological Heritage, Religion and Politics in Postcolonial Zimbabwe and the return of Cultural Property*. Uppsala University.

Matiure, P. (2023). Preserving the Tangible Material Culture of the Shona Traditional Music Legacy: An Applied Ethnomusicological Report. *Collections*, *19*(4), 553–570. Advance online publication. DOI: 10.1177/15501906231189216

Mazarire, G. (2004). Ethical considerations for Zimbabwean archives and the digital challenge. *Innovation.*, *24*(1). Advance online publication. DOI: 10.4314/innovation.v24i1.26445

Ngoepe, M., & Netshakhuma, S. (2018). Archives in the trenches: Repatriation of African National Congress liberation archives in diaspora to South Africa. *Archival Science*, *18*(1), 51–71. DOI: 10.1007/s10502-018-9284-9

Perullo, A. (2018). *Digital Repatriation: Copyright Policies*. Fair Use, and Ethics.

Shylon, F. (2014). Repatriation of Antiquities to sub-Saharan Africa: The Agony and the Ecstasy. *Antiquity*, *19*, 121–132.

Snowball, J., Collins, A., & Nwauche, E. (2021). Ethics, values and legality in the restoration of cultural artefacts: The case of South Africa. *International Journal of Cultural Policy*, *28*(5), 513–530. DOI: 10.1080/10286632.2021.1995377

ADDITIONAL READINGS

Batt, F. (2021). The repatriation of African heritage: Shutting the door on the imperialist narrative. *African Human Rights Yearbook*, *5*, 328–350.

Breckenridge, K. (2014). The Politics of the Parallel Archive: Digital Imperialism and the Future of Record-Keeping in the Age of Digital Reproduction on JSTOR. *Journal of Southern African Studies*, *40*(3), 499–519. https://www.jstor.org/stable/24566488. DOI: 10.1080/03057070.2014.913427

Ngoepe, M., & Netshakhuma, S. (2018). Archives in the trenches: Repatriation of African National Congress liberation archives in diaspora to South Africa. *Archival Science*, *18*(1), 51–71. DOI: 10.1007/s10502-018-9284-9

Shylon, F. (2014). Repatriation of Antiquities to sub-Saharan Africa: The Agony and the Ecstasy. *Antiquity*, *19*, 121–132.

Snowball, J., Collins, A., & Nwauche, E. (2022). Examining restitution and repatriation options for cultural artefacts: An empirical enquiry in South Africa. *International Journal of Cultural Policy*, *28*(5), 531–545. DOI: 10.1080/10286632.2021.1995378

KEY TERMS AND DEFINITIONS

Access: It refers to the ability of individuals or communities to engage with, experience, and benefit from cultural, historical, or natural resources that are considered part of their collective heritage. This includes both physical access to heritage sites (such as museums, landmarks, or archives) and intellectual access, which involves the availability of information, knowledge, and interpretations related to heritage.

Digital repatriation: It refers to the process of returning digital copies or reproductions of cultural heritage materials (such as artefacts, documents, images, audio recordings, or videos) to the communities or nations from which they originated. This is often applied in the context of objects or cultural items that are held in museums, libraries, or archives, sometimes in foreign institutions far from their original place of origin.

Expropriation: It refers to the wrongful or unjust removal, appropriation, or seizure of cultural heritage objects, artefacts, or human remains from their communities of origin, often without consent or fair compensation.

Cultural Heritage: It refers to the cultural, historical, and natural assets that are passed down from previous generations and are valued by individuals, communities, or nations. It encompasses a wide range of elements that contribute to a community's identity, continuity, and collective memory.

Restitution: It refers to the act of returning or restoring something to its rightful owner or original condition. In the context of heritage, restitution typically involves the return of cultural property, artefacts, or human remains that were wrongfully taken, often through expropriation, colonialism, looting, or illicit trade, to the communities, nations, or indigenous groups from which they originated.

Chapter 3
Indigenous Data and Knowledge Sovereignty Amidst Technological Dynamics:
Reflections on Selected Southern African Indigenous Communities

Collence Takaingenhamo Chisita
https://orcid.org/0000-0002-7375-8627
University of South Africa, South Africa & Durban University of Technology, South Africa

Tlou Maggie Masenya
https://orcid.org/0000-0001-5400-0549
Durban University of Technology, South Africa

ABSTRACT

This chapter examines narrative literature review concerning the indigenous data and knowledge sovereignty in the context of Southern African indigenous communities amidst technological dynamics. The study was also guided by the following research objectives: examine indigenous data and knowledge sovereignty among Southern African indigenous communities, examine national and international initiatives for data and knowledge sovereignty among indigenous communities, determine the role of intellectual property rights in the protection of indigenous data and knowledge, and suggest recommendations to enhance indigenous people's rights to data and knowledge sovereignty. Social Contract Theory (SCT) was also employed to underpin this study as it embraces the notion that a person's moral, political or legal responsibilities are reliant on their establishment of regulations or ideals that guide their society.

INTRODUCTION

Dealing responsibly with data in an era of digitisation poses a dilemma. On the one hand, the processing of personal data by public and private institutions adds a tangible and easily comprehensible value to the individual (Hummel,2021). Firstly, it is possible to personalize products and services, refine healthcare, provide favourable insurance rates to individuals with low-risk profiles, and offer benefits

DOI: 10.4018/979-8-3693-7964-6.ch003

based on the accessibility of sensitive data (Hummel,2021). Secondly, it is imperative to ensure that individuals, social groups or communities should hold power over how their data is handled. It is increasingly recognized that indigenous knowledge and materials should be handled ethically and responsibly, and that this vital knowledge must be properly recognized in library repositories, databases and other information systems (Allison-Cassin, Bateman, Bhardwaj & Chisita, 2023). There are numerous standard techniques and practices associated with information management that do not appropriately handle indigenous materials, for example, colonial cataloguing and classification syastems. Indigenous knowledge and data sovereignty raise numerous issues including legal and ethical considerations for data storage, ownership, access and permission, as well as intellectual property rights and practical data use in policy, research and everyday life (Kukutai & Taylor, 2016).

Librarians need to provide more case studies, examples and recommendations in the research literature as change is occurring within the profession of librarianship as a whole (Allison-Cassin et.al., 2023). Comparably, the indigenous data ecology is expansive, comprising data generated or stored by indigenous people and organisations, governments, the public sector, worldwide government agencies, non-governmental organisations (NGOs), research institutes and business enterprises. Indigenous communities have difficulty in accessing materials and data because non-indigenous researchers, archives, libraries, governments and other organizations collect and maintain indigenous data for their own interests. Specifically, indigenous data is any information or knowledge about indigenous peoples or their lives, regardless of format or medium (Maiam Nayri Wingara and Australian Indigenous Governance Institute, 2018).

Oguamanam (2018) described data sovereignty as the right of states to govern the collection, ownership, access and use of data residing within their jurisdictions about other states. Public and private organisations (including governments as well as important private-sector actors) can leverage a wide range of high-tech breakthroughs to make use of big data. Meanwhile, open data is getting closer to big data as part of the openness or open access movement and its continuing metamorphosis as a malleable approach towards enhancing the flow of information, reducing the costs of access, and maximising its public harm in contrast to a closed proprietary approach (Oguamanam, 2018). Allison-Cassin et.al (2023) argued that various library practices, such as metadata, descriptive standards, databases, repositories and other library practices, embedded colonial knowledge structures and concepts of intellectual property thus making it cumbersome for indigenous people to access their data and knowledge.

CONTEXTUAL BACKGROUND

As noted by Ouzman (2004) the San, Khoi and Bantu indigenous peoples are regarded as the first nations of Southern Africa, dating back to (178,000 BCE). In addition to being the source (birthplace) of the modern human species (Oliver, 2014), the original San hunter-gatherer groups live in Southern Africa 'South of the Zambezi Valley' (Adhikari, 2010). Giliomee and Mbenga (2007) asserted that the San only entered this part of Africa 14 000 years ago and also referred to as Khoisan (Meyer, 2012). The Khoisan group is composed of the San, whom the Dutch derogatorily referred to as Bushmen (Giliomee & Mbenga, 2007) and the Khoi-Khoi also known as Khoekhoe or Khoekhoen, who the Dutch referred to as Hottentots (Giliomee & Mbenga, 2007). Khoe and San people can be found in Angola, Botswana, Namibia, South Africa, Zambia, and Zimbabwe, though other groups like the Himba and Vadema claim indigenous status (Daniels, 2004; Harring, 2004; Akpan et al., 2004). Approximately 25 to 20,000 years ago, Southern African hunter-gatherers still followed cultural practices such as making ostrich eggshell beads, shell ornaments, rock art, bows and arrows which were all still practised to this day (Choudhury, 2021). The San hunter-gatherer population once covered the entire African subcontinent, extending from the Zambezi River in Angola and Zambia to Cape Town in the south (Hitchcock, 1999).

There are various theories about Bantu-expansion routes based on linguistics and archaeology (Schlebusch & Jakobsson, 2018) nevertheless, archaeological and linguistic conclusions disagree on several points. Schlebusch and Jakobsson (2018) dated the Bantu spread to the 30th and 40th centuries in Western Africa. The expansion of the Bantu in the Southern African Region is commonly referred to as the Bantu migration and it serves as the foundation of the greatest civilisations in the region. The Bantu people have always been moving due to climatic changes and depletion of resources or conflicts. The word "Bantu" denotes a linguistic family consisting of a set of languages related to the Benue Congo branch of the Niger-Kordofanin supra-group along with the people who originated in what is now the Nigeria-Cameroon border about 5000 years ago (Cavalli-Sforza et al., 1994). Central Africa, parts of East Africa, West Africa and Southern Africa are dominated by Bantu and their distribution patterns are closely related to Africa's thousands of years of migration (Liu, 2019). Several hypotheses exist about Bantu origins, the first of which holds that they originated near the Great Lakes in East Africa; the second claims, the Bantu originated in the Congo River Basin and the third claims the Bantu originated on the Cameroon Plateau, a region in West Africa that is now part of the Nigeria-Cameroon border (Liu, 2019; Greenberg,1972). Arguably the main significant distinction between precolonial Africa's history and that of Europe and most of Asia is population movement, particularly among farmers. African farmers with tropical and subtropical soils were more mobile than those in Europe or Asia who practised intensive agriculture, replenishing soil fertility on the same piece of ground annually (Vansina, 1992). This was not feasible across most of Africa, thus agriculture had to be broad, with farmers migrating from field to field each year, resulting in a population that was less attached to specific parcels of land and more mobility than elsewhere (Vansina, 1992).

In other instances, autochthones implying the indigenous people were slowly ousted, as may have happened when south -eastern Bantu speakers confronted San hunters and Khoi herders, and there is also proof of absorption (Vansina, 1992). The aforementioned indigenous groups including the First Nations (San and Khoisan) and the Bantu have been generating indigenous data and knowledge for centuries as evidenced by their rich oral-based culture. However, the advent of Open Science while providing opportunities, also presents challenges regarding data and knowledge sovereignty. Indigenous Peoples are uncomfortable in open data contexts because persistent encounters with settler colonisation and past power inequalities impede prospects for sustainable growth and ethical engagement in the knowledge economy (Kukutai, Russo Carroll & Walter, 2020). Open Data is expected to help leverage indigenous data and knowledge globally, but traditional data ecosystems do not consider indigenous worldviews or Indigenous data requirements when planning data services (Kukutai, Russo Carroll & Walter, 2020). These issues will be further explored in subsequent sections.

THE CONCEPT OF INDIGENEITY WITHIN AN AFRICAN CONTEXT

Although there is no single, agreed-upon definition of indigenous peoples, Saugestad (2004) mentions in the book "The indigenous peoples of Southern Africa: An Overview" that the four most frequently invoked elements are: (a) a priority in time (b) the voluntary perpetuation of cultural distinctiveness (c) a sense of subjugation, marginalisation and dispossession and (d) a sense of self-determination. Hitchcock and Vinding (2004) claimed that the concept of indigeneity in Africa refers to people and groups who are offspring of the initial populations (that is the "first nations") residing in a country and that causes special challenges in Africa. As mentioned by Hitchcock and Vinding (2004) Africa is the continent with the longest record of human occupation as well as the broadest variety of human genetic and cultural diversity. In many cases, determining antecedence is difficult due to the movement of various populations into and out of local areas over time (Hitchcock & Vinding,2004). Numerous indigenous individuals in Southern Africa encounter prejudice, loss of ancestral lands and deprivation of resources. While some countries such as South Africa, have taken steps to acknowledge their rights, others have violated them (Hitchcock & Vinding, 2004).

Werner (2023) stated that across the age of colonialism, the expression indigenous was commonly employed to refer to anyone other than white immigrants who resided in nations that were annexed and colonised by European powers of imperialism. The idea of indigeneity can be traced back to the Pan-American Union's statement in 1938 that indigenous peoples were the offspring of the original inhabitants of the land, and thus it was necessary to provide them exclusive protection by public authorities. In 1949, the United Nations (UN) General Assembly proposed the investigation of the 'Social Problems of the Aboriginal Populations and Other Under-Developed Social Groups of the American Continent', which also concentrated on promoting their integration (Daes, 1995; Werner, 2023). Throughout the 1952 General Assembly debate on decolonisation and indigeneity, Belgium advocated that Indigenous people be broadened to consist of 'backward indigenous peoples' living in autonomous countries around the world, rather than just the American continent or colonies. Despite these changes at the UN, Lightfoot (2016) argued that in 1957, the International Labour Organisation (ILO) decided to enforce Convention No. 107 on the protection and integration of indigenous and tribal peoples in independent states, which did not focus solely on colonies abroad but differentiates between indigenous and tribal communities in independent nations while providing them with unique shared rights and privileges.

The second part of the 20th century noticed an upsurge in indigenous campaigning around the world; for example, the World Council of Indigenous Peoples was founded in 1974 and the United Nations Sub Commission on the Prevention of Discrimination and the Protection of Minorities commissioned a study on discrimination against indigenous peoples in 1971. José Martínez Cobo launched his report in 1983, nearly ten years after receiving his very first commission. The study demonstrated the challenges in determining indigenous peoples. Cobo (1986) examined clarifications from many nations and discovered commonalities, resulting in a novel and broad definition based on genealogy or descent from pre-colonial or before the invasion societies, distinct culture and most importantly self-identification and acknowledgement by the members of the indigenous group (Cobo, 1986). The United Nations (UN0 established the Permanent Forum for Indigenous Issues in 2000, with participation recruited from various indigenous groups (Werner, 2023).

The Forum's membership lasted from 2020 to 2022. Despite being late to the debate, African indigenous activists were participating in the global indigenous movement in the 1990s (Kipuri, 1999). As a result, certain African countries became increasingly interested in concerns involving the Declaration on the Rights of Indigenous Peoples. South Africa showed support for the draft declaration at Working Group meetings and referenced indigenous peoples in Africa at African Commission on Human and Peoples' Rights (ACHPR) sessions that began in 2000 (Kipuri, 1999). African states have asserted that self-government is only accessible to peoples under colonial and/or foreign rule, and that providing autonomy to groups within states would be destructive to the state's political unity and sovereignty (Werner, 2023). During its 28th regular session in 2000, the ACHPR adopted a resolution on the Rights of Indigenous Peoples' Communities in Africa (ACHPR/Res.51(XXVIII00), which established a Working Group of Experts on the Rights of Indigenous Peoples. Daes (2008) proposed the following criteria for identifying indigenous peoples:

- Occupation and use of an explicit territory
- Voluntary preservation of cultural diversity, which may include traits of language, social organisation, religion and spiritual values, means of production, laws and institutions
- Self-recognition as a distinct group and recognition by others
- Exclusion, marginalisation, dispossession, subjugation or exclusion

Data sovereignty has emerged as an essential component of discussion on the creation, deployment, and adjusting of new data-driven technologies and infrastructures (Hummel, 2021). The term 'data sovereignty' is a new term emanating from the sudden growth of information on the internet and mobile devices (Hummel, 2021). In its simplest form, data sovereignty involves handling information in line with the laws, customs and conventions of the country in which it is located (Snipp, 2016). Data sovereignty is an acknowledgement of responsibility for ensuring that information is managed in ways that are consistent with the laws, practices and traditions of the nation-state upon which it is located (Snipp, 2016). The control of data flows through national jurisdiction is an essential component of data sovereignty; nevertheless, studies by Baezner and Robin (2018) and Floridi (2020) emphasised the complications associated with converting state sovereignty to cyber sovereignty.

Idealistically cyberspace is a borderless zone of shared freedoms but in reality, it has its challenges regarding competing interests. It is imperative to address strains between definitions and other calls for data sovereignty that include a changeability of topics, including the authority of national governments over data stored in domestic or foreign clouds (Irion, 2012). Research and surveillance should consider indigenous data sovereignty (Kukutai & Taylor, 2016), as well as patient-data sovereignty (German Ethics Council, 2017). Allison-Cassin et.al (2023) argued that the indigenous peoples are severed from their data and cultural materials due to the impact of colonisation, making it essential to not only provide them with access to these materials but also to support and enable indigenous communities to govern those materials, as well as to support data stewardship (Allison-Cassin et.al, 2023). Given that the majority of indigenous data is controlled by non-indigenous institutions, there is a responsibility to engage respectfully with those communities to ensure the use of indigenous data supports capacity development, increasing community data capabilities and the strengthening of indigenous languages and cultures (Carroll et.al., 2021:2). It is critical for indigenous communities to be able to collect their own data or simply deal with the limits enforced by the colonial state, but there are obstacles such as a lack of expertise and internet access, that impede this process. Indigenous people may have more control over their data as a result of the digital transformation, such as increasing access to mobile devices and the internet (Snipp, 2016).

RESEARCH PROBLEM

The current interconnected world presents numerous problems and opportunities for the safeguarding of indigenous people's data and knowledge. With the introduction of Open Scholarship and its corollaries, including Open Data, there are growing concerns about the way these will apply to indigenous people in terms of their data and knowledge that is sui generis or collectively owned. Indigenous data and knowledge have been overlooked for decades due to colonialism, and African governments continue to grapple with the issue of protecting indigenous peoples' rights as enshrined in the Article 31 of the United Nations Declaration on the Rights of Indigenous Peoples (UNDRIP) (2007). Collaboration among Indigenous cultural centres, libraries and archives enables indigenous knowledge systems to shape library and archive practice for a future void of marginalisation and epistemological inequalities (Callison et al., 2023).

Indigenous populations may experience both benefits and drawbacks from the global data revolution and related technology, and as noted by Kukutai and Taylor (2016), positive effects can be achieved if indigenous peoples have control over what and how data and knowledge are collected, examined, and documented, as well as their transmission and use. The lack of strong rules and policies for those who use indigenous people's data and knowledge is a serious threat that should be addressed through the collaboration of all stakeholders, including libraries and related institutions, governments, indigenous people, internet service providers and the indigenous people. The governments sluggish response to serious concerns involving indigenous people's data and knowledge is likely to increase biopiracy, intellectual theft and the extinction of such valuable knowledge systems.

RESEARCH OBJECTIVES

The chapter was guided by the following research objectives, which are to:

- examine indigenous data and knowledge sovereignty among Southern African indigenous communities.
- identify the national and international initiatives for data and knowledge sovereignty among indigenous communities.
- determine the role of intellectual property rights in the protection of indigenous data and knowledge.
- recommend indigenous people’s rights to data and knowledge sovereignty.

THEORETICAL FRAMEWORK

The study employed a Social Contract Theory (SCT), that argues that a person's moral, political or legal responsibilities are dependent on the establishment of principles or regulations that govern their community. The social contract unites societal groups and the government to achieve mutually advantageous results in the form of stability, as stated below:

> “the entirety of explicit or implicit agreements between all relevant societal groups and the sovereign (the government or any other actor in power), defining their rights and obligations towards each other” (Loewe et al., 2020:3)

The SCT stems from the Hobbesian notion premised on the idea that to be free from the state of nature, it is normal for human beings to surrender their freedom and liberty and to reciprocally concede them to others for security and self-preservation under an absolute authority (Hobbes, 1651/1985). Indigenous people are part of the social contract with their responsible governments and the latter is obliged to ensure that indigenous data and knowledge are protected from those with an intent to exploit and misappropriate their cultural heritage. Friend (2004) proposes that humans surrender some freedoms in exchange for societal benefits, which is the root for the Social Contract Theory (SCT). Indigenous people in Southern Africa are recognised by their national constitutions as citizens (South African constitution, 1996; Zimbabwean constitution, 2023). Human beings were described by Hobbes (1651) as misanthropic, deprived, ruthless, tiny and undeserving of regulations in his pessimistic view of life. In Hobbes' (1651) view, citizens must give up some liberties to the sovereign state, typically a dominion, in order to avoid civil anarchy. As a reward for sacrificing civil liberties, the sovereign state would guarantee everyone's peace and security.

Unlike Hobbes (1651), Locke (1689), a proponent of the SCT, gave a more optimistic view of human nature, perceiving individuals as naturally anthropical, lucid and mindful of fellow citizens' rights. Locke (1689) stated that the absence of a dispassionate authority for conflict settlement in a pre-governed state necessitates the establishment of a government. South Africa and Zimbabwe have democratic constitutions built on a social contract between government and citizens, regardless of how these ideals are upheld. Locke (1689)'s vision of the social contract required citizens to give up some of their freedoms, while the government is legally obligated to look after the individuals' greatest interests. Rousseau (1762), SCT's other proponent, espoused beliefs similar to Locke's (1689), but opposed in his thesis that humans as a whole were unsocial, tranquil and free of social corruption. Rousseau (1762) believed that the emergence of private property forced mankind to enter into a social contract to safeguard their property rights, which resulted in the establishment of a government. Indigenous people's right to data and knowledge sovereignty is exacerbated by the fact that such knowledge is owned communally and privatising it would be an anathema of the indigenous people's rights.

As stated by Loewe et al. (2021), the critical aspects of the SCT are spatial, temporal and substantive. The spatial facet of a societal contract specifies who is going to participate, which can include legislatures, social groupings and people, while the aspect relates to the participants' authority. The absence of strong laws and structures to safeguard the privacy of indigenous data and knowledge becomes weaker indigenous data sovereignty. The extent of the agreement can vary from subnational to national to transnational to supranational (Loewe et al., 2021). Considering the temporal element of social contracts, the shifting dynamics of contractual relationships over time are investigated because the duration and precise moment of renegotiation might vary significantly (Abbas et al., 2024). This sort of contract strengthens state-society ties, but renegotiations and modifications are required because of power of imbalances or the perceived incapacity of states to meet obligations (Loewe et al., 2021). Loewe et al. (2021) stated that the substantive feature of SCT is the 3Ps, namely: protection, provision and participation, which define social contracts. Hickey (2011) emphasised the necessity of protecting and maintaining inherent rights, whereas provision refers to how a government gives services and resources to society. Citizens who participate actively engage in public issues and government procedures (Loewe et al., 2021). Furness and Trautner (2020) stated that the delivery of the 3Ps, as well as people's expectations and allegiance to the state can be seen in the three aspects of SCT, which are stressed as follows:

a) First, the spatial dimension illustrates the territorial limitations of the social compact and its building parts, which include political settlements, horizontal linkages between and within groups, and citizen identity. The spatial dimension highlights place-based governance and acceptance of indigenous people's right to data and knowledge sovereignty;

b) Second, the substantive aspect includes all of the 3Ps: protection, provision and participation, in exchange for acknowledgement and engagement. This aspect argues that governments should go beyond mere acknowledgement of indigenous people but instead enact laws to ensure data and knowledge sovereignty is realised; and
c) Third, the temporal dimension explores how conflict experiences and peacebuilding processes affect societal expectations and trust within government institution (Furness & Trautner, 2020). This dimension obliges government to continuously rethinking and realignment of laws to protect indigenous data and knowledge.

The CARE (collective benefit, authority to control, responsibility, ethics) and FAIR (findable, accessible, interoperable, reusable) principles for Indigenous Data Governance (GIDA, 2019; Boeckhout, Zielhuis; Bredenoord, 2018) serve as practical ethical guidelines for data sharing with a potential to protect indigenous data and knowledge from the caprices of business data sharing or data capitalism (Chisita, Durodolu & Rusero, 2025). Understanding and appreciating the context where data sharing happens is vital. For example, business data sharing operates in three primary ways: structure, system and marketplace (van den Broek & van Veenstra, 2015). In a hierarchical mode (i.e. supply chains), focal collaborators establish data sharing through regulated, consolidated regulation. The network mode can be distinguished by lateral connections among data ecosystem members, demonstrating social contracting and collaborative methods (Jarke, Otto & Ram, 2019). The network mode disadvantages indigenous knowledge because of the digital divide, thus leading to digital exclusion. The key actors in indigenous data ecosystem include, indigenous communities and nations, knowledge custodians, government and policy makers, researchers and academics, technology providers, data collectors and independent data collectors. It is imperative to consider way forward in deconflating the commoditisation data shared as a product for profit through official agreements via data marketplaces and the interests of indigenous people regarding data sovereignty (Spiekermann, 2019).

METHODOLOGY

Given the volume and speed with which articles are published, a periodic synthesis of knowledge is essential (Bolderston, 2008; Green, Johnson & Adams, 2001). Such a synthesis can be fulfilled through conducting a narrative review. The demand for a literature review may come from the volume of information, different viewpoints, or a lack of common ground on an issue. Unlike original publications, literature reviews aim at analysing previously published material and provide the most recent evidence available. A review is characterised as a 'secondary research' study since it builds on 'primary research' findings, and also summarising prior studies, reducing repetitions, and discovering current subject matter that have not yet been covered (Pautasso, 2013). Narrative reviews enable researchers to explain what is known about a topic while executing a subjective examination and critique of an entire body of literature (Sukhera, 2022). For example, researchers may explain the topic's current status while offering insights on advancing the field, new theories or current evidence from unique or unusual. A narrative review method allowed for a thematic study of the many use cases and the review included peer reviewed journal articles, book chapters and conference proceedings (Green, Johnson & Adams, 2006). The majority of narrative reviews follow subjectivist and interpretivist paradigms, which emphasise that reality is subjective and changeable (Sukhera, 2022).

Narrative reviews thus contrast with the positivist and post-positivist worldviews that underpin systematic reviews, which are founded on the idea that empirical investigation can reveal a single reality. Green, Johnson and Adams (2001:103) suggest that narrative literature reviews can be excellent venues to present philosophical perspectives in an equilibrated way. In contrasting systematic reviews, Greenhalgh, Thorne and Malterud (2018) claimed that narrative reviews are a distinct kind of intellectual interest that may integrate with systematic reviews. Narrative review procedures align with subjectivist and interpretivist research frameworks, that foreground the subjective nature of reality and its ongoing change (Sukhera, 2022). Search terms relating to data sovereignty, indigenous knowledge, indigenous data and intellectual property rights were used to collect data from relevant electronic resources such as Research Gate, Web of Science, and Google Scholar while titles of possibly relevant articles were also reviewed.

LITERATURE REVIEW

The following section critically analyzed the literature review with specific reference to sources that explored indigenous data and knowledge sovereignty amidst technological dynamics.

Indigenous Data and Knowledge Sovereignty in Southern African Indigenous Communities

Indigenous data has been obtained and controlled by international information merchants including Google, Facebook, Twitter, Instagram, e-Bay, Amazon, and Wikipedia (Oguamanam 2018). Indigenous data relates to all information, in any format, on indigenous peoples, lands, resources, communities, lifeways, and cultures (Rainieetal, 2019). Carroll et.al (2021) argued that indigenous data encompasses information and knowledge that affects indigenous peoples, nations, and communities at both individual and group levels. It encompasses data on resources and environments, as well as data regarding individuals and collectives. This data encompasses an extensive amount of data, typically in digital form, that is obtained by governmental organisations, such as departments of the government, government regulatory bodies, and data agencies (Walter, et.al., 2021). Indigenous data sovereignty is centred on the UN Declaration on the Rights of Indigenous Peoples (UNDRIP), which affirms their right to control data related to their people, territory and resources (UNDRIP, 2007). The Global Indigenous Data Alliance (GIDA), an officially recognised worldwide network founded in 2015, is a pioneering institution in indigenous data sovereignty (Bowman & Bremner, 2024). As stated by Rainie et al. (2019) indigenous data sovereignty provides a model for maximising the positive effects of open data for indigenous peoples and other indigenous data consumers, as well as encouraging data stewardship in general

Indigenous knowledge, that is passed down from generation to generation has the potential to boost scientific research and improve science's public image. However, this form of knowledge is frequently dismissed as "unscientific," notwithstanding being the case that it serves as an underlying basis for our present-day scientific understanding (Willoughby, 2019). The increasing popularity of big data technology, and also the associated encourage for access to publicly owned data (open data), for example, generate new places of indigenous data tension. Open Data is part of Open Scholarship which aims to make scholarly outputs to be democratically accessed by anyone, anytime, anywhere to support teaching, learning and research. Rainie, et al (2019) highlighted that Open data offers possibilities for sustainable development for indigenous peoples, but also encounters difficulties because of colonisation, bias and lack of education about indigenous rights. Big data is defined as massive-volume, high-velocity and high-variety information assets (Lefèvre, 2018) that surpass the capacities provided by conventional or isolated data processing applications and can be used for diversified and far-reaching uses by extremely endowed organisations. These groups, which include governments and significant private-sector actors may utilise a wide range of high-tech developments to exploit big data. Dunbar and Scrimgeour (2017) characterise risk in terms of Western ethical and privacy worries, along with the need to safeguard indigenous knowledge. In the past ten years, the concept of open data, a belief that data should be accessible to be freely accessed, re-used, and disseminated (ODC, 2015), has gained approval globally, and the International Open Data charter, to which many countries are signatory in 2017 (Walter et.al., 2021).

The Open Data Charter (ODC)(2020) describes itself as a collaboration between over 100 governments and organisations working to open up data based on a shared set of principles. ODC (2020) push for policies and practices that enable government to collect, share and use well-governed data, to respond effectively and accountably to our most pressing, social, economic and environmental challenges. The ODC 's six pillars demand that data should be open by default, timely and thorough, readily available and useful, comparative and interconnected, for better oversight and citizen involvement, and for inclusive development and innovation (ODC, 2020). The cultural, social and economic worth of indigenous knowledge systems in South Africa makes their safeguarding as well as oversight critical (Zwane & Masipa, 2021). Similarly, Zimbabwe experiences similar threat to its indigenous cultural heritage including the data and knowledge. The two countries boast of wealth of indigenous data and knowledge, for example, oral traditions or classical African Art of conservation, traditional agricultural practices and indigenous medicine, environmental management, conflict resolution, and industry and craftmanship. Oguamanam (2020) bemoans that indigenous communities have and continue to be "researched to death" and this undermines the concept of indigenous peoples right to data and knowledge sovereignty. Indigenous people in Zimbabwe and South Africa are not the only victims of data capitalism. Blair (2015) raised similar concerns regarding the experiences of the indigenous people of Australia subjected to unethical research practices. The new oil is currently credited with data as a vital asset that propels the knowledge economy, similarly to how oil fuelled the industrial economy (Hirsch, 2015). Data is the next phase of digital transformation, especially in the fourth industrial revolution, which needs combining digitisation with physical production equipment, as stated by Naudé (2018). In Africa, participation with the global cause has failed achieved critical mass. One of the likely clarifications for the delayed pace of the movement in Africa compared to the Canada, Australia New Zealand and United States (CANZUS) states is the highly different and confusing history of the instrumentalization of the idea of indigeneity on the continent (Hirsch,2015; Oguamanam & Jain, 2017). With regards to Africa, engagement with the international movements championing indigenous data sovereignty has not yet gained deeper roots because of the confusing history of indigeneity and use of the term for socio-economic and political expedience by the ruling elites in post-colonial Africa (Oguamanam, 2020).

National and International Initiatives for Data and Knowledge Sovereignty Among Indigenous Communities

There is significant progress in enacting laws that govern data use as will be highlighted in this section. Furthermore, international initiatives have also given impetus and guidelines regarding how data should be managed. Intense efforts have been made globally, over the past few years to promote the rights of indigenous data and knowledge for the benefit of indigenous communities (Masenya, 2022). The United Nations (1992) formed a Convention on Biodiversity (CBD) and Nagoya Protocol focusing on Access and Benefit Sharing to genetic resources associated to indigenous knowledge, within a framework of fair and equitable sharing of benefits. The CBD's definition of biotechnology refers to any technological application that uses biological applications, living organisms and derivatives to make or modify products or processes for specific use (UN, 1992). World Intellectual Property Organization (WIPO) and World Trade Organization (WTO) also considered the protection of indigenous data and knowledge for the benefit of local communities that have contributed to an invention or intellectual property development (Gorjestani, 2000).WIPO (2002) further established an Intergovernmental Committee (IGC) to protect the indigenous knowledge, genetic and biological resources and folklore, and developed a sui generis system for the protection of indigenous knowledge.

United Nations Educational, Scientific and Cultural Organization (UNESCO) (2003) formulated the Convention on the Protection of the Diversity of Cultural Contents and Artistic Expressions aimed at empowering member states to consider protection of traditional knowledge using intellectual property systems. International Indigenous Data Sovereignty Interest Group (IDSIG) further issued the International Indigenous Data Sovereignty Charter Statement, which states that the indigenous nations need data about their citizens and communities to make informed decisions. However, the state, local governments and other private companies have primarily collected indigenous data and knowledge for their own use. Research Data Alliance (RDA) and International Data Week convened a global workshop in Gaborone, Botswana, in 2018 focusing on "Indigenous Data Sovereignty Principles for the Governance of Indigenous Data", convened in 2018, developed two guiding principles of the global indigenous data sovereignty movement, namely: FAIR (findable, accessible, interoperable, reusable) and CARE (collective benefit, authority to control, responsibility, ethics) principles for Indigenous Data Governance (GIDA, 2019). FAIR primarily focus on characteristics of data that will facilitate increased data sharing among entities and its framework is oriented towards the collective benefit of indigenous peoples achieved through the use of data for inclusive innovation and self-determined development (Boeckhout, Zielhuis & Bredenoord, 2018). This principle also supports the authority of indigenous peoples and their communities to control their territories, resources, land, data. knowledge, and language, and to ensure culturally sensitive data governance models. CARE is people and purpose-oriented, reflecting the crucial role of data in advancing innovation, governance of data following data governance protocols, and self-determination among indigenous peoples (Carroll, et al., 2021). This principle complements the existing data-centric approach represented in the FAIR (Boeckhout, Zielhuis & Bredenoord, 2018), and its goal is that stewards and other users of indigenous data will 'be FAIR and CARE.' It strongly supports ethical considerations across the life cycle of indigenous data, including reducing real and perceived harms, optimising benefits, and promoting human rights, including the rights espoused in United Nations Permanent Forum on Indigenous Issues (UNDRIP) (Boeckhout, Zielhuis & Bredenoord, 2018).

The South African constitution (1996) recognises indigenous languages that are carriers of indigenous data, information and knowledge. According to the aforementioned constitution, indigenous languages have historically been marginalised and used less; hence, the state should take proactive initiatives to enhance their status and promote their usage. The South African Constitution (1996) recognises the need to maintain lawful administration of indigenous forests and to recognise indigenous law. Furthermore, Act No. 6 of 2019: Protection, Promotion, Development and Management of Indigenous Knowledge Act, 2019 which aims to:

a) To provide for the protection, promotion, development, and management of indigenous knowledge
b) To provide for the establishment and functions of the National Indigenous Knowledge Systems Office
c) To provide for the management of rights of indigenous knowledge communities
d) To provide for the establishment and functions of the Advisory Panel on indigenous knowledge
e) To provide for access and conditions of access to knowledge of indigenous communities
f) To facilitate and coordinate indigenous knowledge-based innovation, as well as address related issues

It is the responsibility of the National Indigenous Knowledge Systems Office (NIKSO) to ensure the compliance with the Protection, Promotion, Development and Management of Indigenous Knowledge Act, to maintain and recognize indigenous knowledge as belonging to people of indigenous descent, and to provide the rights and benefits that were previously denied to indigenous people. In accordance with the aforementioned act, NIKSO encourages and coordinates the development of ancestral knowledge, as well as building and managing the credentialing of assessors and practitioners of indigenous knowledge. NIKSO is responsible for ensuring that indigenous knowledge is properly recognized and used for cultural and economic benefit by the indigenous population through education and promotion of awareness; establishing criteria for granting licences to use indigenous competences and establishing certification license agreements for indigenous knowledge usage.

South African government has also recognised a need to protect and preserve indigenous knowledge and enacted a policy to protect indigenous knowledge systems including the intellectual property system, databases, sui generis laws and registers (Department of Science and Technology, 2004). The Patents Amendment Act, 2005 was also initiated by the Department of Trade and Industry (2005) in South Africa and it is being used by the WTO and WIPO as legislation model under the law of patents. Southern African Indigenous communities may therefore use the laws of patents to protect and commercialise their indigenous knowledge. In South Africa, the Department of Science and Technology (DST) (2011) created the Bio-Innovation programme with the goal of integrating indigenous knowledge-based concepts within the national system of innovation and facilitating community-based technology transfer. Indigenous Navigator was also established in 2014 as a framework and set of tools to systematically monitor the level of recognition and progress towards indigenous empowerment and the enforcement regarding their rights in six countries: Peru, Kenya, Cameroon, Suriname, Thailand, and Nepal (Gilbert & Lennox, 2019). The second phase of Indigenous Navigator was launched in 2017, involved 11 countries including other countries such as Colombia, Tanzania, Bangladesh, Philippines, Bolivia and Cambodia. The Constitution of Zimbabwe (2023) Chapter 2, Section 33 concerning the preservation of traditional knowledge, is expected to take steps to conserve, safeguard and develop indigenous knowledge systems, encompassing knowledge of the medicinal and other properties of animal and plant life that local groups and people held. Zimbabwe remains one of Africa's few countries with legal and policy structures functioning that advocate local community rights and the preservation and protection of indigenous knowledge (Chibememe et al., 2014). Chapter 2 of the Constitution recognises local communities' rights to access and benefit from natural resources, laying down the foundation for their involvement in biodiversity conservation and environmentally friendly growth. The same constitution Chapter 16, Section 295 recognises indigenous peoples right to land and compensation in the case of dispossession. According to Section 116 of the Environmental Management Act (Chapter 20:27), Zimbabwe's significant Minister must execute initiatives that preserve ecosystem diversity and achieve its CBD obligations. One of the key approaches is to safeguard indigenous groups' property rights in relation to bio-diversity (Chibememe et al., 2014).

The National Environment Policy (NEP) and Strategies of 2009 of Zimbabwe is designed to unify all environmental sectoral strategies and ensure biodiversity conservation and sustainable resource use are made possible by local communities including indigenous people. According to Guiding Principle 28, communities and individuals have the right to keep and share indigenous technical knowledge about natural resources, and should receive equitable benefits from its utilisation. The document advocates for equitable sharing of benefits from indigenous technical knowledge and practices, as well as the creation of monitoring and enforcement mechanism to ensure progress (Chibememe et al., 2014). Zimbabwe further approved the Data Protection Act in December 2021, which includes provisions addressing cybersecurity and cybercrime. However, the law is not explicit on indigenous people's data and knowledge. For a number of benefits, it is crucial to improve the implementation and supervision of the Statutory Instrument 61 of 2009 (Access to Indigenous Genetic Resources and Genetic Resource-based Knowledge), regulations, as well as to align it with the Nagoya Protocol and constitutional requirements. The State, its agencies and interested parties must take steps to ensure that local communities receive equitable benefits from their resources (Chibememe et al., 2014).

The Role of Intellectual Property Rights in the Protection of Indigenous Data and Knowledge

Indigenous peoples are faced with some challenges on how their data is generated, accessed, shared, applied and owned due to colonial, neo-colonial and unitary national frameworks. As part of their quest for greater self-determination, the indigenous peoples and their political entities have recently begun staking their claims in the data space, as part of the broader decolonisation project (Open North and British Columbia First Nations Data Governance Initiative, BCFNDGI, 2017). The demand for indigenous data and knowledge is increasing as indigenous people and communities engage in social, economic and cultural development. However, the indigenous data and knowledge have been typically owned by non-indigenous actors (particularly researchers and corporations) or the state in which indigenous peoples are embedded. Among the most abuses of indigenous data and knowledge by non-indigenous actors are those perpetrated via the application of intellectual property rights, for example, via private-sector patenting, pharmaceuticals, nutraceuticals and agricultural or food products developed with the use of indigenous knowledge or genetic resources sourced from indigenous lands (Robinson, 2014). Indigenous nations and communities are thus positioned to claim a seat at the table to ensure indigenous peoples are directly involved in efforts to promote data equity in indigenous communities (IDSIG, 2017).

The importance of indigenous data and knowledge has thus led to demands for epistemic justice and its sovereignty in developing countries, particularly in Southern African indigenous communities. Simoene (2004) also stated some of the reasons why indigenous data and knowledge should be protected, namely: promoting respect for indigenous knowledge, deterring the misappropriation of indigenous knowledge, empowering indigenous knowledge owners and communities who are typically marginalised and protecting tradition-based innovation. This has underscored the need for its protection to be included in intellectual property rights (IPR) laws which aimed at decolonizing human rights and providing indigenous knowledge holders and their communities property rights to their indigenous data and knowledge. As noted by Hossain and Ballardini (2021) indigenous knowledge owners or people and their communities deserve the right to control how their data and knowledge is used. IPR include involving indigenous peoples and their communities in a proactive decision-making role, requiring their consultation and consent for decisions affecting them. However, the existing intellectual property law in South Africa only addresses the laws from a western perspective, while indigenous communities, who comprise 79.6% of the population are not catered for, as reported by the Statistics South Africa (2007). Vast amount of indigenous knowledge is presently unprotected by the current intellectual property law and policy frameworks in Southern African countries (McManis, 2007). There is, therefore, an urgent need to examine the intellectual property law and policy frameworks in Southern Africa and to ascertain whether the question of ownership of indigenous data and knowledge is addressed adequately. However, this needs to be done through applying values such as fairness, social justice and the right to culture associated with indigenous peoples' right to self-determination, as noted by Hossain and Ballardini (2021). Fairness and social justice are some of the values rooted in good governance, which refers to assuring intellectual property rights laws as both a substantive right and a procedural right.

It is also imperative that the indigenous peoples have control over the data representations of their genetic resources, indigenous data and knowledge by effectively participating in access and benefit sharing (ABS) (Oguamanam, 2018). Indigenous data can therefore be harnessed in a manner that mitigates the potential for ABS abuses through compilation of indigenous knowledge databases that provide the provenance of the knowledge. For example, India's Traditional Knowledge Digital Library (TKDL) was established in 2001 by the Indian Government's Council of Scientific and Industrial Research (CSIR) and the project has structured and classified the Indian Traditional Medicine System, enhancing the quality of search and examination of prior art with respect to patent applications in the area of traditional knowledge (Oguamanam, 2018). The indigenous knowledge documented in TKDL is made available to patent offices around the world in order to protect Indian traditional medicinal knowledge while also preventing its misappropriation.

The social contract theory as espoused by (Hobbes,1651; Locke,1689; and Rousseau, 1762) is manifested in the constitutions of South Africa and Zimbabwe as clearly articulated in their preambles, rights and freedoms, government structure and the rule of law. However, what is lacking in the aforementioned constitutions is clarity regarding indigenous people's data and knowledge sovereignty. Through the right to the franchise, citizens of South Africa and Zimbabwe are entitled to elect a government that will guarantee that their interest are taken care of, however, it will be a long to go before the indigenous people especially the first nations realise their genuine aspirations regarding sovereignty over their data and knowledge. While citizens have capitulated their liberty exclusively to the sovereign (Hobbes, 1651), usually the dominion to prevent self-interest from encouraging societal anarchy, they still have to realise total sovereignty over their data and knowledge. In principle governments have enacted laws and policies to promote indigenous knowledge but more needs to be done to deal with issues regarding data and knowledge sovereignty. Even though governments pledge to serve citizens honestly there is lack of precise policies regarding indigenous people including first nations.

In terms of the social contract, it is crucial to stress that geographical aspects, as well as the offering, protection, and participation of the social contract, must be considered in exchange for recognition and engagement (Furness &Trautner, 2020). Despite the clearly expressed aspirations within the individual countries, the practicalities of execution demand greater clarity, most notably in terms of local knowledge and data sovereignty. There is need to heighten advocacy and awareness amongst all indigenous people to ensure effective dialogue, as highlighted by (Loewe et al., 2021). The South African NIKSO should be seen to be actively involved in engaging indigenous people regarding sovereignty over the data and knowledge in line with the social contract. The commodification and monetisation of data and knowledge to private independent information professionals raises concerns over data security (Spiekermann, 2019). Rousseau (1762) viewed the introduction of private property as the raison d'etre why people yield to a social contract to defend their property rights, subsequently establishing a government.

CONCLUSION AND RECOMMENDATIONS

The chapter critically reviewed narrative literature to analyze indigenous data and knowledge sovereignty in the context of Southern African indigenous communities amid technological dynamics. The study identified the need for indigenous people's rights to control and govern their data, lands and resources, that all relevant stakeholders must prioritize. Intellectual property laws, sovereign rights, policy frameworks, protocol and ethical procedural practices should be put in place to assist in realizing inherent rights attached to indigenous people, ensuring respect for diversity, non-discrimination and access to social protection. Library, Archives and memory institutions have a critical role in ensuring that the indigenous data and knowledge is recovered, digitally preserved and protected. The protection of indigenous data and knowledge requires the systematic and holistic integration of regimes linking key concepts of private law, such as property, ownership, and exclusivity, with human rights and biocultural rights. Ethical principles such as informed consent should also be implemented in the procedural practices to promote best practices for access to and protection of indigenous data and knowledge. This chapter thus recommends that:

a) Governments in Southern Africa should prioritise research on indigenous data and knowledge sovereignty, as well as undertake further scientific probes based on Afrocentric perspectives
b) Indigenous data and knowledge sovereignty initiatives at the national and international levels should be evaluated and updated to ensure that indigenous people have sovereign rights to their data and information
c) International Property Rights regimes should empower indigenous people to control and own their data and knowledge and have a say regarding how such resources are used
d) Collaborative partnerships between government, organisations and indigenous communities to lobby for policies that support indigenous people's data and knowledge sovereignty
e) National laws should be revisited to ensure that they are explicit with regards to the indigenous people's rights to data and knowledge sovereignty

FURTHER RESEARCH DIRECTIONS

It is imperative to explore the notion of indigenous people data and knowledge sovereignty with regards to Southern Africa. Furthermore, academics and indigenous researchers/scholars should explore future studies or research on indigenous people's data and knowledge sovereignty underpinned by Afrocentric paradigms that resonate with the lived experiences of indigenous people. It is also imperative to consider how the social contract can be leveraged to ensure equity and inclusivity. Research on community-centric or community-led research also serves as a fertile or fecund ground for further studies on indigenous data and knowledge sovereignty.

REFERENCES

Abbas, A. E., van Velzen, T., Ofe, H., van de Kaa, G., Zuiderwijk, A., & de Reuver, M. (2024). Beyond control over data: Conceptualizing data sovereignty from a social contract perspective. *Electronic Markets, 34*(1), 20. DOI: 10.1007/s12525-024-00695-2

Adhikari, M. (2010). A total extinction confidently hoped for: The destruction of Cape San society under Dutch colonial rule, 1700–1795. *Journal of Genocide Research, 12*(1-2), 19–44. DOI: 10.1080/14623528.2010.508274 PMID: 20941880

AIGI. (2018). Indigenous Data Sovereignty Communique. Australian Indigenous Governance Institute. https://aigi.org.au/news/indigenous-data-sovereignty-communique

Akpan, J., et al. (2004). Human rights and participation among southern African indigenous peoples. *IWGIA document,* (110), 184-201.

Allison-Cassin, S., Bateman, R., Bhardwaj, R. K., & Chisita, C. T. (2023). Introduction. *The Serials Librarian, 84*(5–8), 62–65. DOI: 10.1080/0361526X.2024.2313400

Baezner, M., & Robin, P. (2018). *Cyber sovereignty and data sovereignty*. CSS Cyber Défense Project.

Blair, N. (2015). Researched to death: Indigenous peoples talkin'up our experiences of research. *International Journal of Qualitative Research, 8*(4), 463–478. DOI: 10.1525/irqr.2015.8.4.463

Boeckhout, M., Zielhuis, G. A., & Bredenoord, A. L. (2018). The FAIR guiding principles for data stewardship: Fair enough? *European Journal of Human Genetics : EJHG, 26*(7), 931–936. DOI: 10.1038/s41431-018-0160-0 PMID: 29777206

Bolderston, A. (2008). Writing an effective literature review. *Journal of Medical Imaging and Radiation Sciences, 39*(2), 86–92. DOI: 10.1016/j.jmir.2008.04.009 PMID: 31051808

Bowman, D. N., & Bremner, L. (2024). Indigenous Data Sovereignty: Applying It By, With, For and Through Indigenous Evaluators and Evaluations. *The Canadian Journal of Program Evaluation, 39*(2), 265–287. DOI: 10.3138/cjpe-2024-0039

Capgemini Research Institute. (2020). *Digital Mastery: How organizations have progressed in their digital transformations over the past two years*. Available at: https://www.capgemini.com/insights/research-library/digital-mastery/

Carroll, S. R., Herczog, E., Hudson, M., Russell, K., & Stall, S. (2021). Operationalizing the CARE and FAIR Principles for Indigenous data futures. *Scientific Data, 8*(1), 108. DOI: 10.1038/s41597-021-00892-0 PMID: 33863927

Cavalli-Sforza, L. L. (1994). *The history and geography of human genes*. Princeton UP.

Chibememe, G., (2014). *Review of National Laws and Policies that support or undermine Indigenous Peoples and Local communities-Zimbabwe.* Available at: https://naturaljustice.org/wp-content/uploads/2015/09/Zimbabwe-Legal-Review.pdf (Accessed 23 December, 2024).

Chisita, C. T., Durodolu, O. O., & Rusero, A. M. (2025). Data capitalism in the milieu of the surveillance economy: What can libraries do? *IFLA Journal*, •••, 03400352241286170. DOI: 10.1177/03400352241286170

Choudhury, A., Sengupta, D., Ramsay, M., & Schlebusch, C. (2021). Bantu-speaker migration and admixture in Southern Africa. *Human Molecular Genetics*, *30*(R1), R56–R63. DOI: 10.1093/hmg/ddaa274 PMID: 33367711

Cobo, J. R. M. (1986). *Study of the problem of discrimination against indigenous populations.*

Couture, S., & Toupin, S. (2019). What does the notion of "sovereignty" mean when referring to the digital? *New Media & Society*, *21*(10), 2305–2322. DOI: 10.1177/1461444819865984

Cyber and Data Protection Act [*chapter 12:07*] Available at: https://zimlii.org/akn/zw/act/2021/5/eng@2022-03-11 (Accessed 23 December, 2024).

Daes, E. I. A. (1995). *Standard-setting activities: Evolution of standards concerning the rights of indigenous people. United Nations. Sub-Commission on Prevention of Discrimination and Protection of Minorities.* Working Group on Indigenous Populations.

Daes, E. I. A. (2008). An overview of the history of indigenous peoples: Self-determination and the United Nations. *Cambridge Review of International Affairs*, *21*(1), 7–26. DOI: 10.1080/09557570701828386

Deacon, H. J., & Deacon, J. (1999). *Human Beginnings in South Africa: Uncovering the Secrets of the Stone Age. In David Philip.* Cape Town, South Africa.

Dunbar, T. & Scrimgeour, M. (2017). LSIC: Procedural ethics through an Indigenous ethical lens. *Indigenous Children Growing Up Strong: A Longitudinal Study of Aboriginal and Torres Strait Islander Families,* 61-78.

Floridi, L. (2020). The fight for digital sovereignty: What it is, and why it matters, especially for the EU. *Philosophy & Technology*, *33*(3), 369–378. DOI: 10.1007/s13347-020-00423-6 PMID: 35194548

Friend, C. (2004). *Social contract theory*. Internet Encyclopaedia of Philosophy.

Furness, M., & Trautner, B. (2020). Reconstituting social contracts in conflict-affected MENA countries: Whither Iraq and Libya? *World Development*, *135*, 105085. DOI: 10.1016/j.worlddev.2020.105085

German Ethics Council. (2017). *Big Data and Health: Data Sovereignty as the Shaping of Informational Freedom (Executive Summary & Recommendations)*. German Ethics Council.

Giliomee, H. B., & Mbenga, B. (2007). *New History of South Africa*. Tafelberg.

Government of the Republic of Zimbabwe. (2023). As amended up to 20th June, 2023 https://veritaszim.net/sites/veritas_d/files/Constitution%20Consolidated%20%282023%29.pdf

Green, B. N., Johnson, C. D. & Adams, A. (2001). Writing narrative literature reviews for peer-reviewed journals: secrets of the trade. *Journal of sports chiropractic and rehabilitation,* 15(1), 5-19.

Green, B. N., Johnson, C. D., & Adams, A. (2001). Writing narrative literature reviews for peer-reviewed journals: Secrets of the trade. *Journal of Chiropractic Medicine*, *5*(3), 101–117. DOI: 10.1016/S0899-3467(07)60142-6 PMID: 19674681

Green, B. N., Johnson, C. D., & Adams, A. (2006). Writing narrative literature reviews for peer-reviewed journals: Secrets of the trade. *Journal of Chiropractic Medicine*, *5*(3), 101–117. DOI: 10.1016/S0899-3467(07)60142-6 PMID: 19674681

Greenberg, J. H. (1972). Linguistic Evidence Regarding Bantu Origins. *Journal of African History*, *13*(2), 189–216. DOI: 10.1017/S0021853700011427

Greenberg, J. H. (1972). Linguistic evidence regarding Bantu origins. *Journal of African History*, *13*(2), 189–216. DOI: 10.1017/S0021853700011427

Hickey, S. (2011). The politics of social protection: What do we get from a 'social contract' approach? *Canadian Journal of Development Studies*, *32*(4), 426–438. DOI: 10.1080/02255189.2011.647447

Hirsch, D. D. (2013). The glass house effect: Big Data, the new oil, and the power of analogy. *Me. L. Rev.*, *66*, 373.

Hobbes, T. (1651). *Leviathan*. London: Penguin Books (reprint 1985).

Hossain, K., & Ballardini, R. (2021). Protecting Indigenous Traditional Knowledge Through a Holistic Principle-Based Approach. *Nordic Journal of Human Rights*, *39*(1), 51–72. DOI: 10.1080/18918131.2021.1947449

Hummel, P., Braun, M., Tretter, M., & Dabrock, P. (2021). Data sovereignty: A review. *Big Data & Society*, *8*(1), 2053951720982012. DOI: 10.1177/2053951720982012

International Open Data Charter. (2015). Available at: https://opendatacharter.org/wp-content/uploads/2023/12/opendatacharter-charter_F.pdf (Accessed 8 December 2024).

Irion, K. (2012). Government cloud computing and national data sovereignty. *Policy and Internet*, *4*(3-4), 40–71. DOI: 10.1002/poi3.10

Jarke, M., Otto, B., & Ram, S. (2019). Data sovereignty and data space ecosystems. *Business & Information Systems Engineering*, *61*(5), 549–550. DOI: 10.1007/s12599-019-00614-2

Kipuri, N. (1999). The human rights situation of indigenous peoples in Africa. *Indigenous Affairs*, (1), 18–25.

Kukutai, T., & Cormack, D. (2018). Census 2018 and Implications for Māori. *New Zealand Population Review*, *44*, 131–151.

Kukutai, T., Russo Carroll, S., & Walter, M. (2020*). Indigenous Data Sovereignty. Indigenous World.* https://researchcommons.waikato.ac.nz/server/api/core/bitstreams/6796content (Accessed 10 March 2025)

Kukutai, T., & Taylor, J. (2016). *Indigenous data sovereignty: Toward an agenda*. ANU press. DOI: 10.22459/CAEPR38.11.2016

Lefèvre, T. (2018). Big data in forensic science and medicine. *Journal of Forensic and Legal Medicine, 57*, 1–6. DOI: 10.1016/j.jflm.2017.08.001 PMID: 29801942

Lightfoot, S. (2016). *Global Indigenous Politics: A Subtle Revolution*. Routledge. DOI: 10.4324/9781315670669

Liu, J. (2019). The impact of Bantu migration on agriculture in sub-Saharan Africa. *International Journal of New Developments in Engineering and Society*, *3*(2), 28–34.

Locke, J. (1689). *Two Treatises of Government*. New Haven: Yale University Press (reprint 2003).

Malaccorto, R. B. (2023). *ODC's contribution to the Global Digital Compact- Report Open Data and the Right to Privacy in a global digital strategy*. Available at: https://www.un.org/techenvoy/sites/www.un.org.techenvoy/files/GDC-submission_Open-Data-Charter.pdf

Masenya, T. M. (2022). Decolonization of Indigenous Knowledge Systems in South Africa: Impact of Policy and Protocols. *International Journal of Knowledge Management, 18*(1), 1–17. DOI: 10.4018/IJKM.310005

Meyer, A. (2012). Suid-Afrika se oerverlede. *In F. Pretorius F (ed.), Geskiedenis van Suid-Afrika: van voortye tot vandag,* pp. 17–36, Tafelberg, Cape Town.

Naudé, W. (2018). Brilliant technologies and brave entrepreneurs. *Journal of International Affairs*, *72*(1), 143–158.

Oguamanam, C. (2018). ABS: Big data, data sovereignty and digitization: A new Indigenous research landscape. Genetic Resources, Justice and Reconciliation: Canada and Global Access and Benefit Sharing, ed (Cambridge: Cambridge University Press, 2018). *Ottawa Faculty of Law Working Paper,* (2019-14).

Oguamanam, C. (2020). Indigenous Peoples, Data Sovereignty, and Self-Determination: Current Realities and Imperatives. *The African Journal of Information and Communication, 16*(26). Advance online publication. DOI: 10.23962/10539/30360

Oguamanam, C., & Jain, V. (2017). Access and benefit sharing, Canadian and Aboriginal research ethics policy after the Nagoya Protocol: Digital DNA and transformations in biotechnology. *Forthcoming, JELP*, *3*(1), 79–112.

Oliver, W. H. (2014). Reconsidering the skin colour (race) of Jesus and his ancestors. *Journal of Early Christian History*, *4*(2), 95–115. DOI: 10.1080/2222582X.2014.11877306

Open Data Cube (ODC). (2020). *An Open Source Geospatial Data Management & Analysis Platform.*

Ouzman, S. (2004). Silencing and sharing southern African indigenous and embedded knowledge. In *Indigenous Archaeologies* (pp. 195–212). Routledge.

Pautasso, M. (2013). Ten simple rules for writing a literature review. *PLoS Computational Biology*, *9*(7), e1003149. DOI: 10.1371/journal.pcbi.1003149 PMID: 23874189

Protection, P.Development and Management of Indigenous Knowledge Act. (2019). https://www.gov.za/sites/default/files/gcis_document/201908/42647198act6of2019protectpromodevelopmanagementindigenousknowledgeact.pdf

Rainie, S., (2019). Indigenous data sovereignty. Issues in open data: Indigenous data sovereignty. In Davies, T., Walker, S., Rubinstein, M., & Perini, F. (Eds.), *The state of open data: Histories and horizons* (pp. 300–319). African Minds and International Development Research Centre., http://stateofopendata .od4d.net

Robinson, D. F. (2014). *Biodiversity, access and benefit-sharing: Global case studies*. Routledge., DOI: 10.4324/9781315882819

Rousseau, J.-J. (1762). *Du Contrat Social: Ou, Principes du Droit Politique*. Chez Marc Michel Rey.

Schlebusch, C. M., & Jakobsson, M. (2018). Tales of human migration, admixture, and selection in Africa. *Annual Review of Genomics and Human Genetics*, *19*(1), 405–428. DOI: 10.1146/annurev-genom-083117-021759 PMID: 29727585

Simeone, T. (2004). *Indigenous traditional knowledge and intellectual property rights*. Available: https://www.parl.gc.ca/information/library/PRBpubs/prb0338-e.htm. (Accessed 10 September 2009).

Snipp, C. M. (2016). What does data sovereignty imply: what does it look like? *Indigenous Data Sovereignty: Toward an agenda*. Available at: https://www.scielo.org.za/scielo.php?pid=S2077-72132020000200003 &script=sci_arttext

Sobhy, H. (2021). The lived social contract in schools: From protection to the production of hegemony. *World Development*, *137*(1), 1–15. DOI: 10.1016/j.worlddev.2020.104986

Spiekermann, M. (2019). Data marketplaces: Trends and monetisation of data goods. *Inter Economics*, *54*(4), 208–216. DOI: 10.1007/s10272-019-0826-z

Sukhera, J. (2022). Narrative reviews: Flexible, rigorous, and practical. *Journal of Graduate Medical Education*, *14*(4), 414–417. DOI: 10.4300/JGME-D-22-00480.1 PMID: 35991099

Tamuno, P. (2017). New human rights concept for old African problems: An analysis of the challenges of introducing and implementing indigenous rights in Africa. *Journal of African Law*, *61*(3), 305–332. DOI: 10.1017/S0021855317000249

The Constitution of the Republic of South Africa. (1996) Available at: https://www.saflii.org/za/legis/num_act/cotrosa1996423/. (Accessed 23 December 2024).

United Nations. (2007). *United Nations Declaration on the Rights of Indigenous Peoples: Resolution/ Adopted by the General Assembly*, 2 October 2007, A/RES/61/295.

United Nations General Assembly. (2007). United Nations Declaration on the Rights of Indigenous Peoples. United Nations., Retrieved October 2, 2007, from.

van den Broek, T., & van Veenstra, A. F. (2018). Governance of big data collaborations: How to balance regulatory compliance and disruptive innovation. *Technological Forecasting and Social Change*, *129*, 330–338. DOI: 10.1016/j.techfore.2017.09.040

Vansina, J. (1992). Population movements and emergence of new socio-political forms in Africa. *General History of Africa*, *5*, 46–73.

Wadley, L., Hodgskiss, T., & Grant, M. (2009). Implications for complex cognition from the hafting of tools with compound adhesives in the middle stone age, South Africa. *Proceedings of the National Academy of Sciences of the United States of America, 106*(24), 9590–9594. DOI: 10.1073/pnas.0900957106 PMID: 19433786

Walter, M., Lovett, R., Maher, B., Williamson, B., Prehn, J., Bodkin-Andrews, G., & Lee, V. (2021). Indigenous data sovereignty in the era of big data and open data. *The Australian Journal of Social Issues, 56*(2), 143–156. DOI: 10.1002/ajs4.141

Werner, K. (2023). Who is indigenous in Africa? The concept of indigeneity, its impacts, and progression. *Millennium, 51*(2), 379–404. DOI: 10.1177/03058298221135666

Willoughby, S. (2019). In Davies, T., Walker, S., Rubinstein, M., & Perini, F. (Eds.), *Open data and the environment. The State of Open Data: Histories and Horizons* (pp. 103–118).

Zwane, E. M., & Masipa, M. P. (2021). The importance of indigenous knowledge system for food security and conservation of natural resources. *Indilinga, 20*(1), 124–140.

KEY TERMS AND DEFINITIONS

Data Capitalism: An economic system that commodifies data form the benefit independent information enterprises at the disadvantage of the rights of those who have generated and own the data

Data Sovereignty: The handling information in accordance with the laws, norms, and conventions of the nation-state in which it is located.

Decolonisation: A process includes reversing colonial rule, the way imperial governments created and predominated foreign territories, and the changing of colonies into independent states, as well as the reclamation of cultural, economic, and political autonomy by previously colonised populations.

Indigenous Knowledge: Indigenous knowledge refers to the ancient knowledge, customs, and values handed along throughout centuries within a society. It covers an extensive number of subjects, encompassing farming, health care, handling the environment, and relationships.

Social Contract: An agreement between all relevant societal groups and the sovereign (the government or anyone else in power) defining the rights and obligations of each group.

Chapter 4
Documentaries as Reservoirs of Indigenous Knowledge Systems for Ndau People of South-Eastern Zimbabwe

Tenson Muyambo
https://orcid.org/0000-0001-6765-5034
Great Zimbabwe University, Zimbabwe

ABSTRACT

The chapter examines documentaries as repositories for the preservation, protection, promotion and dissemination of indigenous knowledge systems for future generations. This participatory and qualitative study, carried out among the Ndau of south-eastern Zimbabwe, utilised in-depth interviews and focus group discussions as interactive research methods where the indigenous knowledge holders, the old and the young, female and male and indigenous knowledge researchers were interviewed with the intention to investigate the extent to which documentaries can be reservoirs of indigenous knowledge. Documentaries were found to be effective repositories of indigenous knowledge but notwithstanding their myriad challenges, threats and risks as revealed by participants. The chapter argues that using a participatory approach and a high sense of respect towards knowledge holders, documentaries are a safety net for preserving indigenous knowledge facing extinction threats.

INTRODUCTION

Discourses on indigenous knowledge systems have, of late, gained traction (Bhuda & Gumbo, 2024, Reprint; Muyambo & Marashe, 2020; Shizha, 2016; Dei, 2014; Odora Hoppers, 2002). The reason for this sudden upsurge is because indigenous knowledge has proved over time its agency in all spheres of humankind. Indigenous knowledge systems have an overarching role in education, food security, health, sustainable development, just to mention a few, to such an extent that the World Health Organisation (WHO) has accepted the utilisation of indigenous knowledge systems in all sectors including the health sector. In the implementation of the WHO Traditional Medicine Strategy 2014- 2023, WHO acknowledges the efficacy of indigenous knowledge on health matters and advocates for its promotion (WHO

DOI: 10.4018/979-8-3693-7964-6.ch004

) Similarly, there are calls and efforts towards indigenising higher education in Africa (Bhuda, 2021; Shizha, 2016, Dei 2000;2014). South Africa, for example, has taken indigenous knowledge seriously, evident in curriculum transformation in its institutions of higher learning. Higher education in South Africa has mainstreamed indigenous knowledge in their teaching and learning processes.

It is against such 'winds of change' that we have witnessed movements in mainstreaming indigenous knowledge from all sectors of human life. The epistemic hegemony of the West is being challenged as people of Africa and other indigenous people, whose knowledge had been pushed to the margins, are now recentring their knowledge systems. This is evidently clear in Namibia's use of the hoodia plant (Chinsembu, 2015). This points to the efficacy of indigenous knowledge systems as humanity searches for solutions to a plethora of challenges faced in 21st century Africa. The recent COVID-19 pandemic (from 2019) was not only an awakening in terms of the geopolitics of knowledge but a reckoning of the agency of indigenous knowledge as communities try to come to terms with epidemics and pandemics. Surely, such knowledge cannot be left to communities but must become global knowledge, hence the need for documentaries that share and disseminate indigenous knowledge as 'experiential and experimental knowledges' (Ndlovu-Gatsheni, 2020:372). Admittedly, indigenous knowledge has been and continues to be the subject of research from many researchers of different orientations. One thing that features most in these researches is that indigenous knowledge systems are domiciled in rural communities with the elderly members of the communities as the knowledge holders. The knowledge holders pass on this knowledge orally and there are various efforts to document the indigenous knowledge systems. Given that the elderly members of these communities are passing on, indigenous knowledge systems are at risk of being lost for good. Relying on memory may eventually prove very difficult and the fear is that we may lose a lot through these deaths. The death of an elder in a community is equated to the burning of a library. Knowledge hegemony from the West is yet another subtle subjugation of knowledges of the Global South.

Given the foregoing, this chapter asks: How can we ensure that we do not lose indigenous knowledge systems through deaths? How do we disseminate indigenous knowledge of one community to another, especially in this Fourth Industrial Revolution, where the whole world has become a village? Ndlovu-Gatsheni's (2020:382-383) concept of decanonisation from his model of the 10Ds- *deimperialisation, deWesternisation, departiachisation, deracialisation, debourgeoisement, decorporalisation, democratisation, decanonisation, deborderisation* and *desecularisation* is quite insightful in that it advocates for the de-universalising and decanonising Western knowledge whose efficacy has been heavily challenged by the COVID-19 pandemic, and was found wanting, hence the justification of the use of African knowledge (Ndlovu-Gatsheni, 2020). By using documentaries as a way of promoting and preserving this knowledge, we are not only repushing the knowledge to the centre but also making it accessible for whoever cares to use it as it shares the global epistemic space. This creates what Escobar (2018) calls 'pluriversity', a world where many worlds coexist- where different epistemologies coexist for human flourishing.

There is thus, a need to continue on this trajectory of the 10Ds 'until a pluriversal decolonial world is constructed' (Gatsheni-Ndlovu 2020:383), a "multipolar world order" (Mignolo, 2018:92) is realised. One way of realising Ndlovu-Gatsheni's model of the 10Ds is through the wide use of documentaries as repositories of indigenous knowledge of the Global South. By accepting this knowledge pluriversity we usher ourselves into what Ndlovu-Gatsheni (2020:384) refers to as "decolonial love" where humanity flourishes with no epistemology that 'inferiorises' others as it 'superiorises' itself. It is against this backdrop that I undertook this study with a view to assess people's views and perceptions of documentaries as repositories of indigenous knowledge systems of the Global South at a time when the echoing voices for the use of IKS are loud and clear. It goes without saying that what the chapter proposes is equally challenging as shall be illustrated in sections that follow.

OBJECTIVES OF THE STUDY

Taking cognisant of the questions raised above, this chapter aims to explore the dynamics of documentaries as possible avenues of documenting IKS as a way of preserving, protecting, promoting and disseminating the knowledge. In order to achieve the above aim, this chapter's objectives are to:

- Conceptualise documentaries as repositories of communities' indigenous knowledge systems.
- Identify the merits of documentaries in the preservation and dissemination of indigenous knowledge systems.
- Gather views and perceptions of selected elderly, young community members and academics on the feasibility of documenting indigenous knowledge for a wider usage and for future generations.

THE CONTEXT OF THE NDAU PEOPLE OF SOUTH-EASTERN ZIMBABWE

The Ndau people are (mis)construed as wizards with regard to issues of healing, use of magic (not in its negative sense) and many such mystical events. The name, Chipinge by which the Ndau people of South-Eastern Zimbabwe are known, has erroneously become synonymous with having people who excel in *uroyi* (witchcraft). Other tribes in Zimbabwe frequent the place for various reasons and some genuinely looking for assistance from the local traditional healers while others look for *uroyi* of which to my best knowledge, the Ndau traditional healers do not practice medicine that harm others but practice medicine that promotes human flourishing. This does not mean that there are no bogus traditional healers bent on making money by lying to unsuspecting clients. The fact is that the Ndau people, as a traditional ethnic group, have mastered the art of traditional medicine and its associated functions such that whoever seeks help in that regard gets not only professional but efficacious remedies to his/her problem. It is for such indigenous knowledge and the existence of decorated traditional healers that the Ndau people of Chipinge have earned for themselves an unpalatable nickname, *ndau yeuroyi nevaroyi* (place and people of witchcraft) which are unfounded. Whoever, has been wronged and wants to retaliate tends to think of Chipinge for magic to harm his/her opponent. This has made the Ndau people wherever they are, be it at workplaces throughout Zimbabwe be treated with awe, suspicion and fear as experts in black magic. While this may appear true, it is a misconception of proportional magnitude. While some Ndau people have taken this misconception as a blessing in disguise to ward off would be opponents by bragging of coming from Chipinge, others are irked by this as it becomes a source of not only discrimination at workplaces but a form of blackmailing.

Be that as it may, I have taken pains to explain this in this chapter to illustrate that the Ndau people of Chipinge are a proud ethnic group whose knowledge on a number of health-related challenges is second to none. This is no mean achievement that the people have been associated with for centuries. The indigenous knowledge held by these people on several existential challenges is passed on from generation to the other orally. The knowledge is experiential and experimental knowledge (Ndlovu-Gatsheni, 2020). It is out of this that I found the Ndau people suitable co-researchers for this study as they harbour lots of indigenous knowledge that we must not lose but also share with others through documentaries for the achievement of sustainable goals 2, 3 and 13. We must do away with the half truths peddled by missionaries who failed to comprehend, holistically the Ndau (Muyambo, 2019) but let us embrace, as advised by Ndhlovu-Gatsheni (2020), the 'decolonial love' where indigenous knowledge unchains itself from colonial chains-which have been unfortunately embraced by some among us. Ndau IKS promotes human prosperity and flourishing. In this chapter, therefore the case of the Ndau people of south-eastern Zimbabwe is used as a microcosm of African communities whose IKS are quintessential.

THEORETICAL THEORY: POSTCOLONIAL THEORY

In order to put the study into proper context the lens for this study is the postcolonial one. According to the well-known postcolonial scholar, Edward Said (1978), the focus of postcolonial studies is to examine the long-standing impact of cultures in contact with each other and with a particular emphasis on Western colonial interactions with non-western countries. The theory aims to uncover local knowledge, indigenous knowledge in this respect, which has been pushed to the margins by the knowledge hegemony of the West. Quayson (2002:2) conceptualises postcolonial theory as a theory that "involves a studied engagement with the experience of colonialism and its past and present effects, both at the local level of ex-colonial societies as well as at the level of more general global developments thought to be the after effects of empire." In other words, postcolonial theory is a means of defiance by which any exploitative and discriminative practices, regardless of time and space, can be challenged (Rukundwa & van Aarde, 2007). It is a theory that offers a critical perspective to the study of social phenomena particularly issues surrounding power relations between various groups in society (Velautham, 2015). For Quayson (2002) the effects of colonial power did not end with independence. Instead, as Velautham (2015) argues, current social practices which have been established as a result of the colonial encounter such as inherited governments, law, media, journalism, business and a range of other interactions are fraught with issues that stem from unequal power relations. The postcolonial theory is an area of cultural and critical theory that has been used in the study of literary texts as well (Mapara 2009). In literal terms, postcolonial theory has been widely used by writers in previously colonised countries. It is used to correct the distortions of the colonised people's experiences and realities where the colonised are portrayed as inferior while the coloniser is superior (Mapara, 2009). In other words, the postcolonial theory is about the colonised and formerly colonised announcing their presence and identity as well as reclaiming their past that was lost or distorted because of being 'Othered' by colonialism (Mapara, 2009). It is a realisation by the colonised or formerly colonised that what had been said about them is incorrect and hence the need for the colonised or former colonised to correct the misinformation, untruths and half-truths about them. It is a theory that does not only focus on the dilemmas of the colonised in attempting to develop national identities after the demise of colonial rule, but also deals with the way in which writers articulate and celebrate that identity (Mapara, 2009). Postcolonial is a theory anchored on binary oppositions, where white is presented as superior and black as inferior. The choice of this lens is because we are discussing a knowledge system which was disenfranchised during the colonial period and rendered useless in the face of the epistemic hegemony of the West. At colonisation knowledge was viewed from a binary point of view where normative and genuine knowledge was from the West and other knowledges from the *rest* (emphasis mine) were treated as *quasi* knowledge or alternative knowledge. Using this lens, I intend to demystify and decolonise IKS. In other words, the idea is to reclaim a 'lost' knowledge system and reaffirm it so that it takes its rightful position in this knowledge-based economy. This can only happen unless and until we safeguard indigenous knowledge in documentaires from extinction for usage by current and future generations .

METHODOLOGY

This study adopted a participatory and qualitative approach in which interactive research methods such as in-depth interviews and focus group discussions were used to interact with the knowledge holders and practitioners in the study community (Dhlamini & Kaya, 2016) and academics on IKS. This was done to ensure a partnership between the researcher and those being researched (Kaya & Koitsiwe, 2016) as opposed to never involve the community knowledge holders (Macaulay 2007). The total number of participants was 29 comprisng six academics (one female and five males), five key infromants consisiting of four traditional healers (two females, two males) and one male culturalist. In addition 18 participants took part in three focus group discussions made up of three males and three females per each focus group. In order to contextualise this chapter, the researcher used a case study focusing on the Ndau people of south-eastern Zimbabwe for two reasons. First, the Ndau people attract a lot of interest from IKS researchers because of the special indigenous knowledge they hold, hence were found more appropriate for this study. Second, the researcher is a member of this group and most of the participants in the key informant interviews (KIIs) and the focus group discussions (FGDs) conducted call the researcher, 'our boy' to mimic Duma Gideon Boko's speech at his inauguration as the new president of the Republic of Botswana. Being a Ndau people's 'boy' made me have easy access to the participants but, of course, not blinded by my positionality. To counter the biases associated with positionality, the phenomenology of religion was quite instructive in this regard.

Pinned under the phenomenology of religion's *epoche*, the researcher made use of online interviews through the WhatsApp platform on the efficacy of documentaries as reservoirs of indigenous knowledge systems. The WhatsApp interviews were conducted with known researchers and scholars of indigenous knowledge systems whom are called academics in the context of this study. These academics were holders of Doctor of Philosophy (PhD) and some were professors with an impeccable research record in IKS. The target was six of them from universities in Zimbabwe, South Africa and as far as Sweden, and United Arab Emirates.The academics from abroad were African IKS scholars working in universities there. A list of these academics and their contact details were sought. They were then asked to respond to the questions posed on their individual WhatsApp platforms. These questions were either asked in a written form or through audios and the responses adopted the same format. Countries covered through interviewing the academics included Zimbabwe, where the bulk of the academics come from, and South Africa. The delimitation to IKS academics was premised on the idea that they are experts in the area and need no introduction to issues at stake on IKS. The researcher also carried out interviews with the Ndau IKS holders. These were done in two forms. Firstly, 5 key informants were identified as given above aged 60+ and the director of Ndau Festival of the Arts, a haven of Ndau IKS. The participants in the key informant interviews (KIIs) were purposively sampled where decisions to select the individuals were based on the specialist knowledge of the research issue (Oliver 2006). The in-depth face to face interviews were carried-out with selected participants. The questions for the interviews were around their understanding of documentaries, their views on having their knowledge documented through videos, cameras (filming) as well as on how amenable they were to have their knowledge shared widely with the rest of the world. To make follow ups on what came out from these key informant interviews, the researcher used method triangulation by using three focus group discussions with selected elderly and young members of community. Each group had up to six members of both genders, old and young. This was done to get group dynamics on the subject under discussion. Topical in the discussions facilitated by the researcher were issues around documentaries as repositories of IKS, their efficacy, challenges, threats and risks, just to mention a few. Written consent was obtained from academics while for the rest of the participants it was obtained verbally after the purpose of the study was fully explained to them. Participants' privacy and confidentiality of the data they provided was assured. The participants were informed that they could withdraw from participating at any point of the interviews without hard feelings from the researcher. The researcher used codes and pseudonyms to conceal the identity of the participants. The data gathered were transcribed and thematically analysed.

FINDINGS AND DISCUSSION

This section presents and discusses the findings of the study based on the following themes that emanated from the data collected:

- Conceptualisations of documentaries
- Documentaries as reservoirs in the preservation and dissemination of indigenous knowledge,
- Challenges of documentaries of indigenous knowledge
- Protection of Intellectual Property Rights

THE CONCEPTUALISATION OF DOCUMENTARIES

According to Eitzen (2018) documentaries in the forms of films have surged in popularity in the last three decades, reaching new heights with the rise of digital media. Documentaries' ability to candildly portray people's experiences captures attention, drawing audiences due to their educational and enlightening appeal (ibid). All the six academics that were involved in this study revealed that they were very conversant with documentaries and their benefits. The general academics' understanding of documentaries was revealed by E1 who had this to say:

> Documentaries are non-fictional audio-visual productions that aim, essentially to educate, inform or raise awareness about particular topics, issues or cultural practices. They use various story-telling techniques. They use archival footage; they use interviews and enactments to convey factual information (E1,2024).

This resonates with what J1 says:

> Generally, a documentary is a story, or film based on official evidence or information, usually provided by officials. There is here the use of images (still and motion) as well as interviews with real people to give factual insights (J1, 2024).

D1 defined documentaries as "non-fictional TV programmes or films. These educate people on certain real issues affecting people". All the submissions indicate to the idea that documentaries are a form of capturing people's lived realties as they express themselves in their natural settings. The key informant, director of the NdaFA (P1), categorically conceptualised documentaries as ways of capturing information that pertains to people's livelihoods, what surrounds us. These are films which can be visual, in audio, pictorial forms and written documents. The same sentiments were echoed by the young in the focus group discussions who generally viewed documentaries as the capturing of information that depicts a people's life in real time. They understood documentaries as portrayals of a people's life through digital gadgets such as cameras where video recordings are done with people showcasing their day-to-day existential activities such as farming, celebrating their culture, and tending for their environment and its nature. These resonate with Eyo (2024) who argues that documentaries portray real life scenarios that serve to educate, inform and entertain audiences. According to Dowling and Miller (2019) documentaries essentially serve as a powerful means to communicate and engage viewers. Adam, Othman and Halim (2022:164) concur when they state that documentaries of indigenous knowledge are " a process of identifying, collecting, organizing, registering, or recording indigenous knowledge as a means to dynamically maintain, manage, use, disseminate, and/or protect indigenous knowledge according to specific goals of communities". This points to the meticulous process of ensuring that documentaries of indigenous knowledge take note of the interests of the indigenous knowledge holders, the communities.

The above understanding of documentaries was not equally shared by the four traditional healers in the KIIs and the elderly in the focus group discussions. Some of them had no clue as to what these were but when they listened to what the young were saying, they too began to share some important insights. For example, two elders in the focus group discussions shared that they remembered having been pictured when they were helping their clients. One of them shared an incident where a cameraman was taking photos when they were holding their annual dances within their communities. The NdaFA director revealed that while some elders are still conservative when it comes to documenting their IKS, others are beginning to be receptive, especially if this is genuinely and honestly done. Understanding of documentaries was a bit ambivalent from the traditional healers and the elderly who revealed their lack of exposure to digitalisation while the young in the focus group discussions had a fair understanding of documentaries and their advantages in preserving community held knowledge. They were more amenable to the digitalisation of IKS.

The conspicuous variance in understanding of documentaries exhibited above can be explained in terms of exposure. While the academics and the young have a clear understanding of documentaries, the same cannot be said for the traditional healers and the elderly. The former have been more exposed to the widespread use of digital media as a tool for documentation, creation, preservation and sharing of audio-visual content (Borish et.al, 2021) than the latter. The academics pointed to videos, films, pictures or images (still or motion) as the various forms of documentaries where "interviews with real people to give factual insights" (J1) are conducted.

DOCUMENTARIES AS RESERVOIRS IN THE PRESERVATION AND DISSEMINATION OF INDIGENOUS KNOWLEDGE

On the prospects of documentaries as reservoirs of IKS, the academics' views were that documentaries can be "useful in as far as they document what was and what is from the knowledge holders and practitioners" (J1, 2024). J1 indicated, that:

> They (documentaries) can be useful in as far as they document what was and what is from the knowledge holders and practitioners. The best contributors to IK documentaries and documentation are the indigenous people themselves.

E1 revealed that the Ndau, upon which this study was conducted, have missed some opportunities that other communities enjoyed and continue to enjoy. Being a marginalised community, the Ndauland and the Ndau culture have remained very tangential not only to research but also development. Hence, documentaries remain an essential tool for capturing oral traditions, rituals and practices that would be otherwise be lost. He further pointed out that documentaries are a tool for transmitting knowledge to future generations. It was also revealed that documentaries make IKS accessible, that is making IKS widely available across geographical and cultural boundaries. E1 said "Documentaries promote the recognition, the respect and integration of IKS into broader knowledge systems". The same views were shared by P1 when he intoned:

> We have a future to save and this can only happen when we document that which makes us a people. That which makes us a people are our traditions, rituals and practices. These are endangered as we continue to lose them through deaths of the knowledge holders. As authors of our future, we should preserve the best knowledge that we feel can save our future to be observed by future generations. IKS are foundational and we rely on them, Hence, preserving them through documentaries is long overdue. (P1, 2024).

Closely related to the foregoing was the revelation by one academic, M1 who linked documentaries to preserving a people's identity. For him, IKS are firmly rooted to a particular place and a particular people. Filming this in a documentary means preserving the particular place and the particular people's IKS for future generations. On the effectiveness of documentaries as IKS repositories, participants revealed that indeed, they are. E1 succinctly puts that they are highly effective in disseminating IKS globally provided certain measures are taken. There is need for community involvement in the production of documentaries to maintain authenticity. This calls for inclusivity, where the old, the young, the less privileged, the female and the male participate in producing documentaries. The exclusion of any member of the community upholding the knowledge is an infringement on the rights of the knowledge holders. J1 sums it up saying that documentaries can be an important way of reaching out to the world in the search for epistemic justice. The participants were clear and their views resonated with Bali and Kofinas' (2014) argument that videography preserves a people's legacy and serves as a repository of traditional knowledge in changing times.

Literature points to the fact that in recent years, documentaries have gained traction as increasing powerful strategies for research (Munro & Bilbrough, 2018; Brandt et al, 2016; Petrarca & Hughes, 2014). Eyo (2024:151) confirms this by observing that documentaries "serve as platforms for information dissemination, behaviour shaping and reinforcement of viewers' beliefs". Eyo (2024) further posists that documentaries are essential tools for raising awareness and fostering cultural consciousness and countering the erosion of local traditions and values. In the same vein Adam et al (2022) argue that documentaries are deemed as a plausible means to preserve indigenous knowledge given that the dissemination and transfer of indigenous knowledge relies heavily on oral traditions which are susceptible to being lost. While scholarship is agreeable as to the agency of documentaries as repository of indigenous knowledge, it can be a miscarriage of justice if this chapter does not also discuss the challenges that confront documenting in documentaries of indigenous knowledge. The section below just does that.

CHALLENGES OF DOCUMENTARIES OF INDIGENOUS KNOWLEDGE

Inasmuch as most participants' understanding of documentaries and their agency was evident, they also pointed out the challenges, threats and risks of documentaries in the preservation of indigenous knowledge systems. One academic interviewee stated that the challenge was who would fund the budgets of developing these documentaries. He argued that African governments who should fund efforts to preserve, protect, promote and disseminate IKS prioritise the comfort of politicians and not the truth zones created by academia. For him the West would not likely support projects that they may perceive as undermining their epistemological hegemony. In a study by Adam et al. (2022) in Malaysia findings were that most documentaries of indigenous knowledge lacked financial and expert support from authorities . The Dusun and the Bajau people under study revealed that documentaries that document indigenous knowledge is a large-scale project that requires a series of fielwork that must be supported by the labour of expertise from different backgrounds. It is not done in isolation from others and therefore requires financing which most authorities are not prepared to give a hand. Accroding to Adam et al. (2022: 168) documentation of indigenous knowledge is " a multidimensional effort that demands engagement and participation from different levels of individuals, including indigenous communities, researchers, experts and government bodies". Bringing together all these stakeholders calls for huge investments in terms of finances and time which are not easy to find. Another challenge that most of the participants were quick to indicate was that there are certain qualitative attributes of a place and people that cannot be captured in a documentary. Findings revealed that documentaries were only useful when we look at a people's tangible heritage but very difficult to use in capturing a people's intangible heritage. Participants revealed that a people's identity, for example, keeps on evolving due to a number of factors and is never static. Documentaries are static while a people's IKS are constantly in motion as people adapt to changing circumstances. Examples cited by participants were that there were certain knowledge systems that communities were taught to abandon during the COVID-19 pandemic such as observing funeral rites, communitarian gatherings and shaking hands as ways of expressing solidarity. The COVID-19 pandemic was a gamechanger. While communities are adaptive to circumstances, documentaries are not and as a result may pass the wrong signals if not correctly revised to suit prevailing situations. Similarly, J1 revealed that IKS are constantly evolving and makes it difficult to capture images of a lived experience.

PROTECTION OF INTELLECTUAL PROPERTY RIGHTS

Ngulube ((2002) posited that the major contestation issue in the management and preservation of indigenous knowlwdge is the protection of intellectual property rights. Intellectual Property Rights (IPRs) have been defined as "legal rights that are attached to information emanating from the mind of a person if it can be applied to making a product that is made distinctive and useful by that information" (Pusey & Dutfield, 1996:230). The issues of management, control and ownership of documentaries were raised. Echoing some of the same sentiments Mudiwa (n.d) argues that while it may be easy to have documentaries of tangible heritage, it becomes extremely difficult to do so with intangible people's heritage. No one holds absolute ownership of this knowledge.The participants indicated that in most cases it is not the knowledge holders who own, manage and store the documentaries. The documentaries are owned by foreign individuals like media houses. The geopolitics of power comes into play here. One of the academics, J1, cautioned that "some documentaries lose credibility due to official interference". He revealed that the problem with officials is that they are not paid to tell the truth, but to push the agenda of their respective institutions or the powers that be at any particular time. Similarly, E1 had this to say:

> There are risks of misrepresentation or oversimplification of IKS for entertainment and commercial appeal. This is where some of the materials captured are selected with in mind the commercial use, the commercial value of the documented materials. Sometimes some of the original aspects of these IKS are being threatened by the desire to commercialise the practices involved, ending up with performances that do not reflect well on the roots of that practice (E1, 2024)

This misrepresentation and/or oversimplification takes away the value associated with the practice by its holders, resulting in the loss of power, control and ownership of both IKS and the documentaries. The United Nations Draft Declaration on the Rights of Indigenous Peoples underscores the fact that the indigenous peoples have the right to own and control their cultural and intellectual property pertaining to their sciences, technologies, seeds, medicines, knowledge of flora and fauna, oral traditions, designs, art and performances (United Nations, n.d). This calls for protection of IK owned by individuals or communities from biopiracy, where some ill-intentioned individuals or companies exploit indigenous knowledge and resources for personal gain (Adam eta al., 2022). According to Efferth et al (2016:166) biopiracy is " the use of biological resources and/or knowledge of indigenous tribes or communities without allowing them to share the revenues generated out of the economic exploitation or other non-monetary incentives associated with the resource/knowledge". There was also a concern by the traditional healers interviewees who indicated that most researchers and documentary officials short change them. They indicated that they were very reluctant to have their knowledge be collected in documentaries by people they do not know. They expressed fear that their knowledge would be used to benefit others at their own expense. This resonates with research that has indicated that a people's IKS have been stolen by big pharmaceutical companies benefitting at the expense of the people. Namibia's case as given by Chinsembu (2015) is evident of the fraudulent nature of IKS research where the knowledge holders lose. The knowledge is patented by the companies as theirs. This, in the views of the participants, was not only unethical but deprives the knowledge holders of the right to own, control, manage and benefit from their knowledge. Apart from benefitting from their knowledge, the knowledge holders are even dispossessed of their knowledge. This results in communities shunning sharing their knowledge. The traditional healers were very sceptical about having their knowledge paraded in documentaries. To confirm this concern on ownership and IPR, E1 said:

> Ownership can sense issues around intellectual property and control over how indigenous knowledge is presented, especially when we talk about communities where knowledge or knowledges are co-owned by the community. When this material is brought into a documentary, sometimes there is no acknowledgement of indigenous communities who own the kind of knowledge depicted in a documentary (E1,2024).

The foregoing points to the threats and risks involved in documenting IKS. Piracy is a big threat that may jeopardise the wide use of documentaries as repositories of IKS in Africa. Researchers are accused of cheating the knowledge holders. Instead of doing research with the knowledge holders (Macaulay 2007), researchers 'mine' from communities and are never back to share what they would have done with the knowledge they collected. This creates mistrust between knowledge holders and researchers. This estranged relationship hinders the use of documentaries as repositories of IKS, despite their efficacy in the preservation, protection, promotion and dissemination of IKS. It was also revealed that while academic and young participants to the study were quite amenable to the use of documentaries, the old were a bit hesitant to fully commit themselves. The old participants indicated limited knowledge about documentaries and were suspicious of having their knowledge in the public domain. The traditional healers talked of their specialised knowledge that they gain through either dreams, visions or apprenticeship. They hold this knowledge as highly secretive and making it public does not only flout the taboos associated with it but angers the spiritual world. M1 confirmed this when he said "Some of the knowledge holders do not want their knowledge video recorded or pictured. For example, spirit mediums detest the use of digital objects on them" (S1, 2024). This resonates with Muyambo's (2019) findings that traditional healers only apprenticeship those close to them and their knowledge is kept a top secret. Herbalists, for example do not share uses of herbs to everyone for that jeopardises their trade, which has become a source of livelihood.

THE WAY FORWARD

Coming to what needs to be done in terms of ensuring that documentaries are accepted by the indigenous knowledge holders, the academics proposed a number of initiatives. They all agree that there is a need to conscientize communities and knowledge holders on documentaries as a form of promoting and preserving IKS. S1 indicated that takers of documentaries should guarantee knowledge holders that their knowledge is copyright protected and this has to be respected. M1 talked of access whereby documentaries are inaccessible to knowledge holders. The indigenous knowledge holders do not have access to journals, Youtube and many other platforms where these documentaries are placed. This resonates with Hamwaalwa (n.d) who revealed that IKS in Namibia are not documented making them inaccessible to wider audiences. This is the reason why the Directorate of Namibia Library and Archives Service embarked on a project, 'Preserve Namibia's indigenous knowledge', with the aim to contribute and enrich the documentation and preservation of indigenous knowledge in Namibia. There is need to ensure accessibility in terms of where we store the documentaries and the language used in the documentaries must be mutually intelligible to the knowledge holders if successful documentation of IKS is to be achieved. Another initiative was having participatory approach where researchers and indigenous knowledge holders coalesce and share. This allays fears and the suspicions between the researchers of IKS and IKS holders. As they mingle and mix a sense of community is created, good rapport developed and finally a heightened sense of trust is established. IKS researchers must be co-consumers of the IKS they are researching on and must go to communities and share the beauty in the community's knowledge. Suggestions were that through the participatory approach knowledge holders may warm up to the technology based IKS preservation methods. The above submissions point to cultural sensitivity, where the issue is about potential appropriation or exploitation. Some materials are quite sensitive and may not be meant for public consumption. For example, documenting how traditional leaders like chiefs are buried may prove a difficult task. Chiefs' mortuary practices are highly cultural and done in secrecy from the public eye. Documentaries have no limit as the use of such highly secretive practices hence documenting them may result in cultural exploitation and abuse. Another aspect that came out during the focus group discussions was that while the young members of the focus group discussions were aware of documentaries and their efficacy, they were ambivalent as whether IKS should be preserved, protected, promoted and disseminated at all since they are localised epistemologies. One of the participants stated thus:

While I do not dispute the wide use of documentaries, I am a bit unsure if at all there is need to make localised knowledges global. There is no point in having IKS documentaries for they are not global by nature so why widely disseminating them. For what? The above sentiment was challenged by another young member of the focus group discussion who said: While IKS are localised forms of knowledges there is evidence that they can be useful outside their community. We have seen people from other nations such as South Africa, Botswana and many others coming here to seek assistance from our traditional healers, as an example. I also feel that we need documentaries of our IKS for purposes of maintaining our identity as a people. These documentaries are essential for future generations so that they won't get lost as to who they are.

The above submissions by the two young participants indicate that while some young people have lost sense of what it means to preserve, protect, promote and disseminate IKS, there are others who still find pride in their IKS. The clarion call by the South African higher education students for the inclusion of IKS in curricula acts as a reminder that a majority of young people feel that what they get as part of their education is not relevant to their existential challenges. The Rhodes must fall campaigns that gripped most of South Africa's universities are indicative of the need to decolonise higher education (Cabe, 2023). The students called for the need to indigenize higher education by incorporating local knowledges that students can associate with from their various backgrounds. To further illustrate the positives of documentaries of IK, Nhambura (2023) opines that despite the fear that once the knowledge is available for wider dissemination and access where there are greater chances of abuse and misappropriation of the knowledge by foreigners, indigenous knowledge should be documented for easier transmission and patenting. Success stories where local communities' consent was sought in the documenting of their indigenous include the Himbas of Namibia and the Masaai of East Africa, especially the Masaai of the Rift Valley of Kenya (Semali & Asino, 2013). The films are in the public domain as cinema.

FUTURE RESEARCH DIRECTIONS

Documentaries of indigenous knowledge systems has the potential to promote, protect, preserve and disseminate this local knowledge. While this study largely focused on the possibility of using documentaries as repositories of indigenous knowledge, it did not examine how the documentaries should be done and who to involve when making the documentaries including how the issues of IPRs and patenting of IK can be enhanced. A future study focusing on the above issues as well as the ethical procedures that need to be observed when doing IK documentaries can be carried out as they can plug some of the grey areas that remain less studied on discourse of indigenous knowledge systems.

CONCLUSION

The chapter interrogated the efficacy of documentaries as repositories on a people's indigenous knowledge systems focusing on the Ndau people of south-eastern Zimbabwe. It argued that documentaries are an effective way of preserving, protecting, promoting and disseminating indigenous knowledge systems. Documentaries do not only preserve IKS for future generation but are a platform through which IKS can be shared globally, This chapter also brought out the challenges, threats and risks that confront documentaries as reservoirs of IKS, chief among them being the question of ownership, that is the issue of intellectual property rights. The knowledge holders have been cheated previously by researchers resulting in their reluctance to freely share their indigenous knowledge and their argument is that once beaten twice shy. Documentaries are also a power game where the knowledge holders are treated as subjects in the production of the documentaries not co-producers. The chapter also captured that documentaries are reservoirs of IKS under threats and risks such as deaths. Although IKS are a localised knowledge base, they can be shared with other communities as coping strategies for a number of existential challenges provided there are ethical considerations made in documenting a people's indigenous knowledge. There were also a raft of measures that were suggested in order to upscale the uptake of documentaries as repositories of IKS, but of course with several cautions proffered.

REFERENCES

D1, interviewed on 19 November, 2024

M1, interviewed on 20 November, 2024

S1, interviewed on 21 November, 2024 J1, interviewed on 23 November, 2024

P1, interviewed on 23 November, 2024 E1, interviewed on 24 November, 2024

Adam, A. A., Othman, N., & Halim, A. A. (2022). Indigenous Knowledge Documentation: Perspectives of Dusun and Bajau communities in Kota Belud, Sabah, Malaysia. *International Journal of Academic Research in Business & Social Sciences, 12*(10), 163–175. DOI: 10.6007/IJARBSS/v12-i10/15205

Bali, A., & Kofinas, G. P. (2014). Voices of Caribou people: A participatory videography method to document and share local knowledge from the North-American human *Rangifer* systems. *Ecology and Society*, *19*(2), 1–13. DOI: 10.5751/ES-06327-190216

Bhuda, M. T. (202). "Making a Case for Indigenous Education Systems in South Africa." *African Journal of Development Studies* (2), 67–82.

Bhuda, M. T., & Gumbo, M. T. (2024). Developments Towards Integrating African Indigenous Knowledge in South African Universities. Reprint. Pretoria. University of South Africa Press. DOI: 10.25159/UnisaRxiv/000082.v1

Borish, D., Cunsolo, A., Mauro, I., Dewey, C., & Harper, S. L. (2021). Moving Images, Moving Methods: Advancing documentary film for qualitative research. *International Journal of Qualitative Methods, 20*, 1–14. DOI: 10.1177/16094069211013646

Brandt, H. M., Freedman, D. A., Friedman, D. B., Choi, S. K., Seel, J. S., Guest, M. A., & Khang, L. (2016). Planting healthy roots: Using documentary film to evaluate and disseminate community-based participatory research. *Family & Community Health, 39*(4), 242–250. DOI: 10.1097/FCH.0000000000000120 PMID: 27536929

Cabe, M. (2023). 'How Rhodes must fall amplified calls to decolonise'. https://newint.org/features/2023/08/21/how-rhodes-must-fall-amplified-calls-decolonize. Accessed on 29 Nov. 2024.

Chinsembu, K. C. (2015). 'Bioprospecting for 'green diamonds': Medicinal plants used in the management of HIV/AIDS related conditions'. In Chinsembu, K.C., Cheikhyoussef, A., Mumbengegwi, D., Kandawa-Schulz, M., Kasanda, Choshi, D. and Kazembe, L (eds) *Indigenous Knowledge of Namibia*, Windhoek: UNAM Press, 9-40.

Dei, G. J. S. (2000). Rethinking the role of Indigenous knowledges in the academy. *International Journal of Inclusive Education, 4*(2), 111–132. DOI: 10.1080/136031100284849

Dei, G. J. S. (2014). Indigenizing the curriculum: The case of the African University. In Emeagwali, G., & Dei, G. J. S. (Eds.), *African Indigenous Knowledge and the Disciplines* (pp. 165–180). Sense Publishers.

Dhlamini, N., & Kaya, H. O. (2016). Environmental Security, Indigenous Knowledge Systems and implications for food security in South Africa. *Journal of Human Ecology (Delhi, India)*, *53*(2), 135–140. DOI: 10.1080/09709274.2016.11906965

Dowling, D. O., & Miller, K. J. (2019). Immersive Audio Storytelling: Podcasting and Serial Documentary in the Digital Publishing Industry. *Journal of Radio & Audio Media*, *26*(1), 167–184. DOI: 10.1080/19376529.2018.1509218

Eitzen, D. (2018). The Duties of Documentary in a Post-Truth Society. In *Cognitive Theory and Documentary Film*. Palgrave Macmillan. DOI: 10.1007/978-3-319-90332-3_6

Escobar, A. (2018). *Designs for the pluriverse: Radical interdependence, autonomy and the making of worlds*. Duke University Press. DOI: 10.1215/9780822371816

Eyo, N. A. (2024). Influence of documentary films on learning cultural awareness among university students in Uyo, Nigeria. *Quantum Journal of Social Sciences and Humanities*, *5*(2), 151–171. DOI: 10.55197/qjssh.v5i2.359

Hamwaalwa, N. (n.d). Indigenous knowledge management: Roles of public library and archives in preserving and providing access to indigenous knowledge (IK) in Namibia: Case study of 'Preserve Namibia Indigenous Knowledge' project. https://www.scecsal.org/publications/papers2021/scecsal_papers2021_hamwaalwa.pdf

Kaya, H. O., & Koitsiwe, M. (2016). African Indigenous Knowledge Systems and Natural Disaster Management in North West Province, South Africa. *Journal of Human Ecology (Delhi, India)*, *53*(2), 101–105. DOI: 10.1080/09709274.2016.11906961

Macaulay, A. C. (2007). Promoting participatory research by family physicians. *Annals of Family Medicine*, *5*(6), 557–560. DOI: 10.1370/afm.755 PMID: 18025494

Mignolo, W. D. (2018). On pluriversality and multipolar world order: Decoloniality after decolonization; dewesternization after the Cold War. In Reiter, B. (Ed.), *Constructing the pluriverse: The geopolitics of knowledge* (pp. 90–116). Duke University Press. DOI: 10.1215/9781478002017-005

Munro, K., & Bilbrough, P. (2018). An ecology of relationships: Tensions and negotiations in documentary filmmaking practice as research. *Media Practice and Education*, *19*(3), 256–269. DOI: 10.1080/25741136.2018.1511361

Muyambo, T. (2019). Indigenous Knowledge Systems of the Ndau people of Manicaland province in Zimbabwe: A case study of *Bota reshupa*. PhD Thesis, University of KwaZulu Natal, Pietermaritzburg, South Africa.

Muyambo, T. M., & Marashe, J. (2020). Indigenous knowledge systems and sustainable development: a case of Zunde raMambo (*Isiphala seNkosi*) as food security in Chipinge, Zimbabwe. *Indilinga: Journal of Indigenous Knowledge Systems*19(2), 232-244. https://journals.co.za/toc/linga/19/2

Ngulube, P. (2002). Managing and Preserving Indigenous Knowledge in the Knowledge Management Era: Challenges and opportunities for information professionals. *Information Development*, *18*(2), 95–101. DOI: 10.1177/026666602400842486

Nhambura, C. (2023). Indigenous People and the Protection of Intellectual Property Rights (IPRs) in Zimbabwe. In Barnabas, S. G. (Ed.), *Indigenous and Minority Populations - Perspectives From Scholars and Writers across the World*. Nigerian Turkish Nile University. DOI: 10.5772/intechopen.108167

Odora Hoppers, C. A. (Ed.). (2002). *Indigenous Knowledge and the integration of knowledge systems: Towards a philosophy of articulation*. New African Books.

Oliver, P. (2006). *Purposive Sampling*. Sage Publishers.

Petrarca, D. M., & Hughes, J. M. (2014). Mobilizing knowledge via documentary filmmaking—Is the academy ready? *McGIill Journal of Education* 49(3). https://mje.mcgill.ca/article/viewFile/9074/7009

Posey, D. A., & Dutfield, G. (1996). *Beyond intellectual property*. International Development Research Centre.

Semali, L. M., & Asino, T. I. (2013). Decolonizing cultural heritage of indigenous people's knowledge from images in global films. *Deconization: Indigeneity & Society*, *2*(2), 25–56.

Shizha, E. (Ed.). (n.d.). *(2016). African Indigenous knowledge and the sciences: Journeys into the past and the present. Boston: Sense Publishers United Nations (UN)*. Draft Declaration on the Rights of Indigenous Peoples. Prepared by the Aboriginal and TorresStrait Islander Commission. [Online], http://www.gcc.ca/Political_Issues/international/

WHO. https://www.who.int/activities/implementation-of-the-WHO-traditional-medicine-strategy-2014-2023. Accessed 16 November 2024

KEY TERMS AND DEFINITIONS

Indigenous Knowledge: It refers to as knowledge that is locally developed and locally accumulated over time by a community, and passed on from one generation to the next. It is grounded in most facets of human activities, that involve agriculture, traditional and alternative medicine, human and animal health, forestry and botany.

Indigenous Knowledge Systems: It is the cumulative skills and collective philosophies that are developed by societies which have protracted histories of interacting with their immediate natural environs.

Documentaries: It is a creative treatment or interpretation of actuality, and this is where indigenous knowledge of indigenous people is filmed for purposes of capturing it *in situ* with a view to either share with others or preserve for future use.

Intellectual Property Rights: Are legal rights that are attached to information emanating from the mind of person if it can be appled to making a product that is made distinctive and useful by that information.

Ndau People: Are said to be a transnational ethnic group predominantly found in South-eastern of Zimbabwe, who are socially, economically, politically and developmental marginalised, and are very rich in terms of their indigenous knowledge on traditional healing, agriculture and environment.

Chapter 5 Intellectual Property Rights Laws in the Protection and Re-Appropriation of Indigenous Knowledge Systems Among South African Indigenous Tribes

Tlou Maggie Masenya
https://orcid.org/0000-0001-5400-0549
Durban University of Technology, South Africa

ABSTRACT

This chapter aims to answer this question: How can indigenous knowledge systems be re-appropriated and protected for the benefit of indigenous knowledge tribes? The chapter utilised literature review to explore the role of intellectual property rights laws, policy frameworks and protocols in the re-appropriation of indigenous knowledge systems in South Africa. The study was guided by the following objectives: the significance of indigenous knowledge among indigenous communities, the effect of cultural appropriation among indigenous communities, the anti-imperialism strategies for the re-appropriation and resignification of indigenous knowledge among indigenous tribes and the role of Intellectual Property Rights laws in the reclamation and protection of indigenous knowledge. Recommendations suggest the need for the protection of indigenous knowledge systems and promote access to indigenous knowledge through re-constitutionalizing rights of intellectual property rights laws, policy and developing standards for ethical professional practice among indigenous tribes.

INTRODUCTION AND BACKGROUND

The growing awareness of indigenous knowledge has led to the demands for epistemic justice and sovereignty for knowledge in developing countries. This also underscore the needs for its protection to be included in exclusive property regime types such as intellectual property rights (IPR). IPR laws encourage and protect innovation, creativity, economic freedom and decolonization of human rights. It thus gives indigenous knowledge holders and their communities property rights to their original works

DOI: 10.4018/979-8-3693-7964-6.ch005

and cultural heritage. IPR include involving indigenous peoples and their representative institutions in a proactive decision-making role, requiring their consultation and consent for decisions affecting them (Hossain & Ballardini, 2021). Intellectual property rights are thus designed to protect commercial innovations and have been deemed inappropriate for protecting traditional knowledge (Sillitoe 2006). However, as reported by the Statistics South Africa (2007), the existing intellectual property laws in South Africa only addresses the laws from a western perspective, while indigenous communities, who comprise 79.6% of the population are not catered for. In developing countries such as South Africa, vast amount of indigenous knowledge which is presently unprotected by the current intellectual property laws and policy frameworks (McManis, 2007). Cultural heritage institutions, policy makers and other relevant stakeholders also neglect this vital knowledge that can be the basis for developing practical, action-oriented and innovative approaches to sustainable development of indigenous communities. There is, therefore, an urgent need to examine the current state of intellectual property laws and policy frameworks in South Africa and to ascertain whether the question of ownership of indigenous intellectual property is addressed adequately. This undertaking should be able to influence both indigenous policy and law-makers to reconsider the role and position of the intellectual property laws or legislation in the re-appropriation and protection of indigenous knowledge among South African indigenous tribes.

Simoene (2004) stated some of the reasons why traditional or indigenous knowledge should be protected, namely: promoting respect for traditional knowledge; deterring the misappropriation of traditional knowledge; empowering traditional knowledge holders who are typically marginalized indigenous people and protecting tradition-based innovation. Hossain and Ballardini (2021) further noted that the protection of indigenous knowledge or traditional knowledge requires the systematic and holistic integration of regimes linking key concepts of private law, such as property, ownership, and exclusivity, with human rights and biocultural rights. However, this needs to be done through applying values such as fairness, social justice and the right to culture associated with indigenous peoples' right to self-determination, as noted by Hossain and Ballardini (2021). Fairness and social justice are some of the values rooted in good governance, which refers to assuring intellectual property rights laws as both a substantive right and a procedural rights. Ethical principles such as informed consent should also be implemented in the procedural practices to give effect to IRP and to promote best practices for access to and protection of indigenous knowledge.

PROBLEM STATEMENT

Indigenous knowledge is essential to the overall well-being of communities and members, and it has wide-ranging scientific and environmental value (Heritage Chairs of Australia and New Zealand) (HCOANZ) (2020). In the 1970s, there was a shift among academics and environmentalists in the perception of how indigenous peoples interacted with the environment, towards a belief that these peoples often lived "in harmony with nature" and a realisation that their knowledge had been undervalued for too long (Huntington, 2005: 29). Koptie (2010) further noted that despite its importance to sustainable and equitable development, indigenous knowledge or heritage have largely been neglected, marginalized and suppressed due to arrogance and ignorance of politics and the dominant ideology of a particular historical period. Disparaging and misappropriating indigenous knowledge from its holders, and social and political context thus have serious implications for its accessibility and usefulness. As noted by Masenya and Chisita (2025), being misappropriated from the culture can contribute to a breakdown of health and well-being, especially if being mostly involved in rituals connected to religious activities, traditions and values, which are the essence of identity, health and well-being. However, indigenous researchers, scholars and academics have been engaging with the knowledge without having to bother about the indigenous knowledge holders or owners who live that knowledge, and who may have rights for its usage. There have also been debates among scholars concerning indigenous peoples' rights and whether they are entitled to privileged and sometimes exclusive rights at all. The disconnection of knowledge from its holders also frees indigenous researchers and academics from their responsibilities to analyse the discourse on indigenous knowledge in its colonial and post-colonial context (Simpson, 2004: 378).

The Southern African region was devastated by the arrival of the Portuguese, Dutch and British colonisers from 1400 to 1800, in terms of control of their indigenous heritage (Masenya & Chisita, 2025). Indigenous peoples globally, are increasingly becoming aware of the dangers of disconnecting them from their knowledge and insist on reclaiming misappropriated indigenous knowledge and remaining in control of their indigenous heritage. The herders in Laponia argue that they themselves must be given the formal responsibility for the protection and development of Sami culture (Mijá ednam, 2000: 71) and that "information and education about the Sami and Sami conditions should be transmitted by Sami people or in cooperation with Sami people" (Mijá ednam, 2000: 66). Although, the Sami herders were concerned about the need for sufficient control, however, much more attention has been given to difficulties in getting access to their indigenous knowledge or to the problem of translating it into forms that are useful for resource managers (Dahlström, 2009). Indigenous communities in Southern African countries such as Botswana and Namibia are also among the most marginalized communities, misappropriated and deteritorialised cultures in many societies and they are often excluded from participation in socio-economic life and rarely have access to political power while also encountering obstacles to manifesting their identity (United Nations High Commissioner for Refugees) (UNHCR) (2018).

Re-appropriation and reclamation of indigenous knowledge should thus be centred on indigenous value, intellectual property rights laws, indigenous policy and protocols. The academic community has been debating about indigenous knowledge rights since the mid-1960s, but to date, the indigenous knowledge policies and intellectual property rights laws are still based on very negative assumptions about the protection and access to indigenous communities. The development of Intellectual property right laws, policy frameworks and protocols provided by government and other relevant stakeholders are thus crucial in the protection and re-appropriation of indigenous knowledge systems. However, these IPR laws, indigenous policy and protocols needs to be well understood by indigenous communities. This chapter thus looked into Intellectual property right laws, policy and protocols aimed at protecting indigenous knowledge and its impact in re-appropriating indigenous knowledge systems in South Africa. Research objectives formulated for this chapter were to:

- Determine the significance of indigenous knowledge among indigenous tribes.
- Examine the effect of cultural appropriation among South African indigenous tribes.
- Ascertain the anti-emperialism strategies for the re-appropriation and resignification of indigenous knowledge among South African indigenous tribes.
- Determine the role of Intellectual Property Rights laws in the re-appropriation and protection of indigenous knowledge among South African indigenous tribes.

METHODOLOGY

The chapter critically reviewed literature in order to investigate the protection of indigenous knowledge among indigenous tribes in South Africa by exploring the role of intellectual property rights laws, policy frameworks and protocols in the re-appropriation and resignification of indigenous knowledge systems. This chapter thus employed the qualitative research method grounded on document analysis approach, which has also been used as a stand-alone research approach while also served mostly as a complement to other research methods. This approach was used in analyzing various qualitative data such as those collected during web-based documentary research, unstructured or semi-structured interviews. This implies that the study relied on the secondary sources in drawing the conclusions and providing recommendations. For this current study, document analysis was thus applied for reviewing literature following the guidelines of 5-stage scoping review framework by Arksey and O'Malley (2005). The stage 1, includes the identification of the key variables that are the focus of the topic or research objectives under investigation, using the following search terms: "intellectual property rights AND indigenous knowledge systems", "protection of indigenous knowledge systems", "antiemperialism strategies AND indigenous knowledge systems", "significance of indigenous knowledge OR indigenous knowledge systems", "reappropriation of indigenous knowledge," "constitutionalizing the rights of indigenous intellectual property". In stage 2, the retrieved articles were screened, whereby first round was focusing on titles and abstracts and the full text of the selected articles were screened to determine if they met the outlined criteria and to delete duplicates, in the second round. Literature review on intellectual property rights laws in the protection and reappropriation of indigenous knowledge were conducted in major databases such as SCOPUS, EBSCOhost, Web of Science, ScienceDirect, Springer and Google Scholar to ensure inclusion of all relevant studies in the document analysis. Stage 3 involves the identification and selection of documents on the basis of their usefulness and relevance to the study. In stage 4, the researcher further reviewed all relevant articles that met the inclusion criteria using the thematic analysis technique developed by Braun and Clarke (2006), and removed all the duplicates which considerably reduced the sample size. In Stage 5, collected and extracted data from included articles was summarized to provide and develop contextual meaningful understanding of issues under investigation. The thematic analysis process was research objectives driven and all potential themes were generated in line with the research objectives of the chapter. The findings were thus organized into four broad themes, namely: the significance of indigenous knowledge among indigenous communities, the effect of cultural appropriation among South African indigenous tribes, the anti-imperialism strategies for the protection, re-appropriation and resignification of indigenous knowledge among South African indigenous tribes and the role of Intellectual Property Rights laws in the protection, reappropriation and resignification of indigenous knowledge among South African indigenous tribes.

Inclusion/Exclusion Criteria

The aim of inclusion criteria as described by Kitchenham (2004) is to identify studies that provide direct evidence about the research question. The types of studies considered for inclusion included all qualitative, quantitative and mixed methods studies relevant to the study. However, the book reviews were excluded from the current study as they do not include original research and the study considered only peer reviewed journal articles, as they represent a major mode of communication among researchers and they were therefore taken as unit of analysis. The selection criteria thus included: empirical studies published in a peer-reviewed journals, focusing on intellectual property rights in the protection and reappropriation of indigenous knowledge systems. Although many relevant articles were retrieved, however some of the articles were removed after thorough review of all the articles, mainly due to its irrelevance and lack of quality.

FINDINGS FROM LITERATURE REVIEW

The findings are presented under the following themes, based on research objectives of the study: the significance of indigenous knowledge among indigenous tribes, the effect of cultural appropriation among indigenous tribes; ant-imperialism strategies for the reappropriation and resignificance of indigenous knowledge for the benefit of indigenous tribes and the role of Intellectual Property Rights laws in the reclamation and protection of indigenous knowledge.

The Significance of Indigenous Knowledge Among Indigenous Tribes

Indigenous knowledge encompasses a vast array of skills, practices, beliefs, and insights related to the environment, agriculture, health, spirituality and social organization (Ajani et al., 2024). This vital knowledge has become increasingly important to the global economy as it increases the economic value of biodiversity, Hosain and Ballardini (2021), and has also been the basis for education, health care, agriculture and the wide range of other activities that sustain a society in many parts of the world for many centuries (Senanayake, 2006). This vital knowledge defines a worldview of people and provides direction for their survival economically, socially, spiritually and politically. It also links the survival of every human being to the wholeness of nature that support life, provides the concrete situations of indigenous communities in relation with the environment and provides practical solutions to the problems of the indigenous people (Worldbank, 1998). As also noted by Worldbank (1998) the key significance of indigenous knowledge is the development of peoples' capability to understand the world in a very simple manner as understanding the human self. Indigenous Peoples have repeatedly shown how valuable their knowledge and how successful their management of socio-ecological systems can be (Ogar, Pecl & Mustonen, 2020). For example, the Skolt Sámi have been able to choose a regional functional approach to restoring habitats that were degraded by human actions in the past, for the resilience, food security, and survival of both humans and non-humans in the present, through community-led self-reflection (Ogar, Pecl & Mustonen, 2020). Therefore, this vital resource cannot be simply ignored or undermined due to the fact that indigenous people and their generations are using it survived for survival. This knowledge also has a profound social value as it fosters community cohesion and identity, and from a historical perspective, indigenous languages and traditions that are threatened by globalization and modernization, provide vital insights into the histories and experiences of indigenous communities (Ajani, et al., 2024). The United Nations Declaration on the Rights of Indigenous Peoples also emphasized the rights of Indigenous peoples to maintain, revitalize, and transmit their languages (Hohmann & Weller 2018). Indigenous communities have been living peacefully with their knowledge of local ecosystems, sustainable agriculture, natural environments and resource management invaluable in addressing contemporary environmental challenges, for centuries.

Indigenous knowledge also provides insight into how to mitigate climate change, protect biodiversity, and preserve ecosystems (UNDP, 2024), and thus preserving this knowledge is crucial for promoting responsible environmental stewardship. For instance, the traditional ecological knowledge of indigenous peoples can inform sustainable agricultural practices and the conservation of endangered species. Indigenous knowledge systems often include a rich repository of traditional medicines and healing practices (Otang-Mbeng et al., 2023), with these systems based on extensive knowledge of local flora and fauna and their medicinal properties. As stated by Bird et al. (2019) the importance of this knowledge is increasingly recognized as it offers alternative or complementary approaches to modern medicine, moreover, the development of new pharmaceuticals and treatments can benefit from indigenous knowledge. For instance, drugs like quinine, derived from the cinchona tree, were originally used by indigenous people in South America to treat malaria (Bird et al. 2019). Given that this knowledge is of utmost importance to indigenous people and their communities, it thus needs to be recovered, revive, preserved, protected and be accessible for future generations. Digital preservation of indigenous knowledge has become a vital tool in safeguarding the cultures and identities of indigenous people and tribes. Digital preservation of indigenous knowledge is not only essential for long- term access and cultural continuity of indigenous communities but also holds significant relevance in our modern world (Masenya, 2024).

The Effect of Cultural Appropriation Among Indigenous Tribes

Cultural appropriation refers to the use of elements of one culture by members of another culture, in ways that are perceived as unacknowledged or inappropriate (Young & Brunk, 2009). It is when a dominant group uses an element of a marginalized group's heritage without their permission whereby the appropriators are using cultural artefacts of minority cultures for their own benefit, whether it be monetary or social, without compensating or acknowledging them. Cultural appropriation is thus the act of appropriating symbolic features or cultural elements (ie. themes, symbols, artefacts, customs, styles) that belongs to minority culture by the dominant culture. The dominant culture or the appropriators adopt language, behaviour, clothing, or tradition belonging to a minority culture or social group in a way that is exploitative, disrespectful, or stereotypical. As noted by Ziff and Rao (1997) cultural appropriation discourse frames the "taking" of elements from another culture in problematic terms as a harmful act of distortion, decontextualization, and domination. Siems (2019) described cultural appropriation as the taking of intellectual property, cultural expressions or artefacts, history, and ways of knowledge. Cultural appropriation thus deprives the original cultural proprietors of their commercial property rights and cultural materials by utilizing imitation and distortion, and as noted by Young (2011) it causes offences, sacredness desecration and the authentic culture's wrong reproduction. The denial of cultural appropriation thus refers to denied cultural property ownership claims, and as stated by Lalonde (2021) non-recognition within cultural appropriation as a phenomenon encapsulated in voice lessness and invisibility. From the Marxist philosophy, cultural appropriation concept is creating cultural conflicts while dissolving them. There are four types of appropriation by Rogers (2006), namely: cultural exchange, dominance, exploitation and transculturation.

- ***Cultural exchange***: it is the reciprocal exchange of symbols, artefacts, rituals, genres, and/or technologies between cultures with roughly equal levels of power.

- ***Cultural dominance:*** it is the use of elements of a dominant culture by members of a subordinated culture in a context in which the dominant culture has been imposed onto the subordinated culture, including appropriations that enact resistance.
- ***Cultural exploitation:*** it is the appropriation of elements of a subordinated culture by a dominant culture without substantive reciprocity, permission and compensation. It is when a more powerful group or individual uses the cultural elements of a less powerful group (Luedicke, 2015).
- ***Transculturation:*** refers to cultural elements created by multiple cultures such that identification of a single originating culture is problematic e.g. multiple cultural appropriations structured in the dynamics of globalization and transnational capitalism creating hybrid forms.

Cruz, Seo and Scaraboto (2024) also identified assimilation as another element of cultural appropriation, and that is when members of a less powerful group (i.e immigrants) use the elements of a more powerful group (i.e. indigenes) For example, immigrants encounter and make use of dominant cultural elements that are not theirs but that nonetheless exert a powerful force in their daily lives (Weinberger, 2015). The cultural appropriation discourse foregrounds the harm that can be perpetuated through appropriative acts and outlines when these harms are more likely to occur (Cruz, Seo & Scaraboto, 2024). However, as stated by Cruz, Seo and Scaraboto (2024) appropriative acts are harmful when they involve inhabiting a cultural voice or performing a cultural practice in a way that erodes a marginalized culture's distinctive identity, misrepresents or profoundly offends another's culture, infringes cultural property rights without compensation and fetishizes another's culture as a commodity. The consumption of cultural difference is rendered in more sinister shades, as an act of misrecognition, infringement, and exploitation (Lalonde, 2021). Cultural heritage has thus been misrecognised, misappropriated and exploited by individuals outside the communities that own the knowledge, simply because they have codified it. Therefore, if Africans do not look back to reclaim, reappropriate and protect their past, they will perpetuate poverty, disease and instability. Antiempirialism strategies or techniques can thus be used in the reappropriation and resignification of indigenous knowledge systems among South African indigenous tribes. Reappropriation and digital preservation processes should be conducted in an equitable manner and guarantee protection and effective knowledge sharing. Indigenous knowledge owners and their communities should thus be seen as equal partners in all matters concerning the protection, reappropriation, reclamation, access and use of this vital knowledge.

Anti-Imperialism Strategies for the Re-Appropriation and Resignification of Indigenous Knowledge Among Indigenous Tribes

Cultural Exchange and Appreciation

Cultural exchange involves learning and sharing of customs, experiences, values, ideas, traditions, and other aspects of culture among people from different ethnic groups. This can occur both within one's own community or across borders, and this interaction helps in breaking down cultural barriers and stereotypes. Cultural elements offer open affordances for ongoing processes of creative and transformative reconfiguration (Cayla & Eckhardt, 2008). As noted by Schneider (2003) culture exchange is not only endemic but beneficial because the appropriative acts can promote the recognition of another culture's practices. Schneider (2003) further stated that the use of elements from other cultures is a transformative and creative practice, expanding the horizons of those who appropriate, while extending the social life of an appropriated practice, artefact, or text by investing it with renewed significance. The exchange and use of elements from other cultures is thus viewed as a depoliticized, even desirable, practice that promotes cultural adaptation, diffusion, learning, and blending (Cruz, Seo, & Scaraboto, 2024). The following are the importance of cultural exchange from different perspectives by FasterCapital (2024) and these include: promotes mutual, understanding and respect; promotes diversity and inclusivity, enhances cultural awareness and promoting tolerance towards different cultures; build bridges between nations by promoting cultural diplomacy and fostering good relations between nations; boost economic ties through exchange of ideas, knowledge, and skills that can lead to the development of new products and services, which can also create new business opportunities. On the other hand, cultural appreciation discourse frames the "making one's own" of another culture's elements as an unproblematic process of cultural diffusion and blending (Young & Brunk, 2009). The outcome of the cultural appreciation discourse is that it sustains the consumption of difference by constituting individuals as ethnic consumer subjects (Veresiu & Giesler, 2018). The loss of self or cultural identity and self-determination can thus have negative impacts on indigenous people and their communities, including feelings of disconnection and loss of cultural pride.

Creating Knowledge Sharing Platforms

Knowledge sharing is a process of communication and interaction between the owner of knowledge and the receiver, with knowledge as the core, to improve the utilization value of knowledge (Usoro, Sharratt &Tsui, 2007). Indigenous knowledge can be shared through virtual community of practices and social media platforms, although is communicated orally. Lave and Wenger (1991) has coined the term "community of practice" and define it as an activity system about which participants share understandings concerning what they are doing and what that means in their lives and for their community. Community of practice is created by people who engage in a process of collective learning in a shared domain of human endeavour such as a tribe learning to survive, a band of musicians trying to find new forms of expression, a group of defining their identity in school etc (Wenger, McDermott & Snyder, 2002). In communities of practice, less experienced members of the community learn from social interactions with more experienced members and experts of a specific knowledge domain (Lave & Wenger, 1991). The most widely recognized benefit of community of practices is their ability to allow for the generation and dissemination of tacit knowledge which is hard to communicate as it is mostly intuitive and embedded in a specific context (Nonaka, 1994). Engaging community of practices, and immersive experiences allow indigenous communities to showcase their diversity and richness of their cultural heritage to a wider audience, and fostering cross-cultural understanding and appreciation (Chaturvedi, 2023).

Virtual Community of practices, which make use of online collaborative technologies are becoming an increasingly popular way for knowledge sharing among geographically dispersed employees (Hildreth et al., 2000). Indigenous people and tribes can share their stories, cultural traditions and practices that have been overlooked or marginalized through digital media technologies, while challenging stereotypes and asserting their cultural identity. Marginalised indigenous communities can protect their cultural legacy and share it with the world by harnessing the power of digital media platforms while fostering creativity and be empowered. Proper knowledge sharing is likely to be enabled by social media platforms and digital technologies which open up new possibilities, helps overcome geographical remoteness by ensuring faster communication, provide more access to knowledge and enabling users to collaborate and interact with each other in virtual communities and networks (Masenya, 2023).

Social media platforms have played a pivotal role in uniting communities around endangered languages. Social media platforms such as Facebook offers features for building virtual communities among dispersed indigenous peoples (Masenya, 2023). The study conducted by San Nicolas-Rocca and Parrish (2013) on indigenous knowledge sharing among the Chamorro peoples through social media platforms revealed that Facebook belongs to the top social networking websites that are used by Chamorros people to learn and share indigenous knowledge. Facebook has also become a major domain of communication and expression among Brazilian Amazonian indigenous groups, as noted by Virtanen (2015). Therefore, dedicated pages or channels on platforms such as Facebook, Instagram and YouTube serve as virtual hubs where speakers of endangered traditions can share content, stories and engage with others who are passionate about learning their language and cultural practice. These virtual platforms foster traditional practices, cultural exchange, and networking opportunities, contributing to the revitalization of cultural heritage or indigenous knowledge.

Promoting Cultural Continuity Among Indigenous Tribes

Brown et al. (2007) described cultural continuity as the ability to preserve the historical traditions of a culture and carry them forward with that culture into the future, and it is closely linked to the concept of cultural identity that has a major influence on our confidence and self-esteem. Cultural disconnection and lack of cultural continuity creates a loss of confidence at the individual level in understanding how to live life and make decisions (MacNeil, 2008). MacNeil (2008) further stated that the colonial policies which have led to an erosion of language and cultural traditions have created a loss of cohesion and identity in indigenous people and their communities. Morris, et al (2007) also identifies the devastating effects of cultural loss on youth and there is consensus among these young people that they are suffering from a loss of culture and a loss of family. Morris, et al (2007) further noted that there is a disconnect from the cultural values and traditions espoused in communities due to colonization and cultural assimilation. Indigenous tribes have lost the power and sense of cultural continuity and it became practically impossible to recover lost or misappropriated heritage. Intergenerational transmission of cultural knowledge, cultural exchange, creating a sense of belonging and cultural identity among indigenous people, interaction with other civilizations and cultural ceremonies may be regarded as one of the strategies to promote cultural continuity. The cultural continuity can be sustained through the adaptation of the ways of communication, for example, folk media that encompasses oral traditions including songs of happiness and sadness, dances, drama etc.

The Role of Intellectual Property Rights Laws in the Reclamation and Protection of Indigenous Knowledge

Indigenous tribes are rapidly losing their indigenous knowledge, natural resources and traditional environmental knowledge, worldwide and these valuable resources are in danger of becoming extinct due to cultural erosion, cultural appropriation, deterritorialization and cultural acculturation. As observed by Msuya (2007), most indigenous knowledge in developing countries is not legally protected and this leaves much of this vital knowledge open to bio-piracy and other forms of misappropriation. The indigenous people and tribes prefer a moral recognition of their contribution to the development of the knowledge and be given credit for their work, however, their indigenous knowledge has been used without their consent and acknowledgement. Eason, Brady and Fryberg (2018) called on indigenous owners and their communities to resist non-recognition, misrecognition, misappropriation, suppression and exploitation of their indigenous knowledge systems by asserting and reaffirming their right to self-determination. Masango (2010) added that indigenous knowledge needs to be protected because Western science makes a lot of money from medicinal plants without the consent of the possessors of the resources and knowledge.

McConkey (2004) also argued that epistemic injustice results from voice lessness in non-recognition since cultural members disregard epistemic contributions about their culture and cultural property. In many instances, indigenous tribes are more concerned with protecting their cultural integrity which is threatened when their indigenous knowledge is not protected, and are about monetary gain. For example, health and pharmaceutical industries benefit financially when they exploit the medicinal plants used by indigenous tribes in treating chronic illnesses such as cancer, high blood, sugar diabetes etc. without acknowledging where the knowledge from such medicinal plants comes from (Masango, 2010). This is a clear example of bio piracy which should be prevented by prohibiting unauthorized parties from using this knowledge through the implementation of intellectual property and copy rights laws, and by ensuring that indigenous communities are remunerated or granted some benefits for the use of such knowledge. The Nagoya Protocol, adopted in 2010 further highlights the need for a repository of Indigenous biocultural community protocols conceived of as norms connected to societal structures and processes, and it eventually contributed to precedent-setting legislation i.e. prohibiting biopiracy (Hosain & Ballardini, 2021). The African Regional Intellectual Property Organization (ARIPO) (2010) has also urged member states to protect and stop the bio-piracy of traditional or indigenous knowledge. There is thus a need for re-appropriation and protection of African indigenous knowledge that has been misappropriated due to the effect of biopiracy, cultural appropriation and deterritorialization. Protecting indigenous knowledge involves prohibiting unauthorized parties from unfairly acquiring Intellectual property rights to indigenous peoples' knowledge, customs and innovations. Andanda (2012) concur that protecting indigenous or traditional knowledge through the Intellectual Property regime carries out the essential function of preventing third parties from using the knowledge inappropriately. The intellectual property protection of indigenous knowledge among African indigenous communities is thus of the utmost importance. One of the main objectives of this chapter is thus to determine the role of intellectual property rights laws in the re-appropriation, resignificance and protection of indigenous knowledge. The Department of Trade and Industry (DTI) (2010) in South Africa has developed national intellectual property policy which explains how indigenous knowledge can be protected through the use of patents, trademarks, designs and copyrights. Department of Science and Technology (DST) (2013) also established the National Indigenous Knowledge Systems Office (NIKSO) aimed at protecting intellectual property rights of indigenous communities and ensure equitable sharing of resources.

Although protocols for handling indigenous knowledge help to uphold its interests by developing standards for ethical professional practice, however, they do not provide legal protection for indigenous communities nor do they provide any legal framework for those who want their cultural and intellectual property rights protected (Masenya, 2022). It has also been accorded in literature review that the current South African intellectual property laws do not adequately address the specific needs of indigenous intellectual property (Masenya, 2022) and do not provide effective protection to indigenous knowledge. The modern intellectual property regimes failed to protect Indigenous knowledge rights and the solutions are less forthcoming in African countries, particularly in South Africa. South African government needs to reposition itself to solve this legal gap by affirming, recognizing and protecting the rights to indigenous knowledge under indigenous knowledge Act No 6 of 2019. Sui generis approach to intellectual property rights should also be implemented and adopted as a way to benefit South African indigenous tribes as this implies that adding indigenous knowledge to the national and international databases automatically constitutes establishing a legal claim over it while also preventing poaching of this knowledge (Masenya, 2023).

CONCLUSION AND RECOMMENDATIONS

The chapter critically reviewed literature to analyses the intellectual property rights laws, policy, protocols, and anti-empirialism strategies relating to the re-appropriation and resignificance of indigenous knowledge systems. The study has confirmed that the misappropriation of cultural heritage among South African indigenous tribes is a reality that all relevant stakeholders must deal with. Intellectual Property Rights laws should play a role in the recognition of indigenous peoples' institutions at the national level and accommodating their customary rules within the national legal framework. Policy frameworks, protocol and procedural practices should also be put in place at that will help in realizing inherent rights attached to indigenous tribes, guaranteeing respect for diversity, non-discrimination, and access to social protection, and ensuring that Indigenous agencies have the possibility of realizing economic rights in which fairness and social justice are rooted. Memory and Cultural heritage institutions have a critical role in ensuring that the collective indigenous knowledge and South African indigenous people's memories are recovered, re-appropriated, protected and preserved. This chapter recommends that:

- There is a need for more scientific research and studies focusing on the reclamation of cultural heritage among indigenous tribes, premised on research methodologies that are deeply rooted in Sankofaism and Afrocentric paradigms.
- More studies should be undertaken concerning cultural reappropriation and cultural continuity in all aspects of socio, economic and political activity.
- There is a need for re-constitutionalizing intellectual property rights laws, policy, protocols, procedural practices
- The development of standards for ethical professional practice among indigenous tribes and ensuring that intellectual property rights laws meet international requirements.

FURTHER RESEARCH DIRECTIONS

Cultural heritage and traditional cultures need to be re-appropriated and protected for the benefit of indigenous tribes. Indigenous researchers, indigenous knowledge owners, indigenous tribes, academics, government and cultural heritage institutions need to work together in dismantling the imperialism and colonial manifestations by engaging in anti-imperialism and decolonial strategies for the protection of the destroyed and misappropriated indigenous knowledge. This chapter focused on determining the role of intellectual property rights laws in the protection and reappropriation of indigenous peoples' cultural heritage in South Africa. Future studies or research should thus focus more on reclamation and re-constitutionalization of indigenous knowledge rights through adopting the principles of Sankofaism and Afrocentric paradigms.

REFERENCES

Ajani, Y.A., Aladokun, B.D., Agboola, O.S. & Amaechi, M.N. (2024). Revitalizing Indigenous Knowledge Systems via Digital Media Technologies for Sustainability of Indigenous Languages. *Preservation, Digital technology & culture,* DOI:.DOI: 10.1515/pdtc-2023-0051

Andanda, P. (2012). Striking a Balance between Intellectual Property Protection of Traditional Knowledge, Cultural Preservation and Access to Knowledge. *SECO/WTI Academic Cooperation Project Working Paper Series 2012/02.* Available at SSRN: http://dx.doi.org/DOI: 10.2139/ssrn.2718887

Arksey, H., & O'Malley, L. (2005). Scoping Studies: Towards a Methodological Framework. *International Journal of Social Research Methodology*, *8*(1), 19–32. DOI: 10.1080/1364557032000119616

Bird, S., Shobbrook, K., Flick, D., Carroll, S., Austin, A., & Lim, A. (2019). Technology for Indigenous Language Revitalization: The State of the Field. *Language*, *95*(S1), e1–e29.

Braun, V., & Clarke, V. (2006). Using thematic analysis in psychology. *Qualitative Research in Psychology*, *3*(2), 77–101. DOI: 10.1191/1478088706qp063oa

Brown, J., Knol, D., Prevost-Derbecker, S., & Andrushko, K. (2007). Housing for Indigenous Youth in the Inner City of Winnipeg. *First Peoples Child & Family Review*, *3*(2), 56–64. DOI: 10.7202/1069464ar

Cayla, J., & Eckhardt, G. M. (2008). Asian Brands and the Shaping of a Transnational Imagined Community. *The Journal of Consumer Research*, *35*(2), 216–230. DOI: 10.1086/587629

Chaturvedi, A. (2023). *Importance of media in preserving and promoting cultural heritage*. Medium.

Cherid, M. I. (2021). Ain't Got Enough Money to Pay Me Respect: Black fishing, Cultural Appropriation, and the Commodification of Blackness. *Cultural Studies, Critical Methodologies*, *21*(5), 359–364. DOI: 10.1177/15327086211029357

Cruz, A. G. B., Seo, Y., & Scaraboto, D. (2024). Between Cultural Appreciation and Cultural Appropriation: Self-Authorizing the Consumption of Cultural Difference. *The Journal of Consumer Research*, *50*(5), 962–984. DOI: 10.1093/jcr/ucad022

Dahlström, A. N. (2009). The Two-Way Appropriation of Indigenous Knowledge: Environmental Management Policies and the Laponia Process. *Journal of Northern Studies*, *2*, 39–57.

Department of Science and Technology. (2013). *The Bio-economy Strategy*.

Eason, A. E., Braddy, L. M., & Fryberg, S. A. (2018). Reclaiming Representations & Interrupting the Cycle of Bias Against Native Americans. *Daedalus*, *147*(2), 70–81. DOI: 10.1162/DAED_a_00491

FasterCapital. (2024). *Cultural Exchange: Bridging Nations: BRICS and Cultural Exchange*. Nad Al Sheba.

Figueiredo, B., Larsen, H. P., & Bean, J. (2021). The Cosmopolitan Servicescape. *Journal of Retailing*, *97*(2), 267–287. DOI: 10.1016/j.jretai.2020.09.001

Heritage Chairs of Australia and New Zealand. (HCOANZ). (2020). Dhawura Ngilan: a vision for Aboriginal and Torres Strait Islander heritage in Australia and the best practice standards in cultural heritage management and legislation, *Australian Government Department of Agriculture, Water and the Environment,* Canberra.

Hildreth, P., Kimble, C., & Wright, P. (2000). Communities of practice in the distributed international environment. *Journal of Knowledge Management*, *4*(1), 27–38. DOI: 10.1108/13673270010315920

Hohmann, J., & Weller, M. (2018). The UN Declaration on the Rights of Indigenous Peoples: a commentary, Oxford commentaries on international law. Oxford: Oxford University Press, pp. 63-86. DOI: 10.1093/law/9780199673223.001.0001

Hossain, K., & Ballardini, R. M. (2021). Protecting Indigenous Traditional Knowledge Through a Holistic Principle-Based Approach. *Nordic Journal of Human Rights*, *39*(1), 51–72. DOI: 10.1080/18918131.2021.1947449

Huntington, H. P. (2005). 'We dance around in a ring and suppose.' Academic engagement with traditional knowledge. *Arctic Anthropology*, *42*(1), 29–32. DOI: 10.1353/arc.2011.0101

Kitchenham, B. (2004). *Procedures for Performing Systematic Reviews*. Keele University.

Koptie, S. W. (2010). Indigenous Self-Discovery: "Being Called to Witness". *First Peoples Child & Family Review*, *5*(1), 114–125. DOI: 10.7202/1069068ar

Lalonde, D. (2021). Does Cultural Appropriation Cause Harm? *Politics, Groups & Identities*, *9*(2), 329–346. DOI: 10.1080/21565503.2019.1674160

Lave, J., & Wenger, E. (1991). *Situated Learning: Legitimate Peripheral Participation*. Cambridge University Press. DOI: 10.1017/CBO9780511815355

Luedicke, M. K., Craig, J. T., & Markus, G. (2010). Consumer Identity Work as Moral Protagonism: How Myth and Ideology Animate a Brand-Mediated Moral Conflict. *The Journal of Consumer Research*, *36*(6), 1016–1032. DOI: 10.1086/644761

MacNeil, M. (2008). An Epidemiologic Study of Aboriginal Adolescent Risk in Canada: The Meaning of Suicide. *Journal of Child and Adolescent Psychiatric Nursing*, *21*(1), 3–12. DOI: 10.1111/j.1744-6171.2008.00117.x PMID: 18269407

Masango, C. A. (2010). Indigenous Traditional Knowledge Protection: Prospects in South Africa's Intellectual Property Framework? *South African Journal of Library and Information Science*, *76*(1). Advance online publication. DOI: 10.7553/76-1-88

Masenya, T. M. (2022). Decolonization of Indigenous Knowledge Systems in South Africa: Impact of Policy and Protocols. *International Journal of Knowledge Management*, *18*(1), 1–22. DOI: 10.4018/IJKM.310005

Masenya, T. M. (2023). Valorization and Digital Preservation of Indigenous Knowledge Systems in South African Indigenous Communities: Best Practices in the Digital Transformation Era. In Masenya, T. M. (Ed.), *Digital Preservation and Documentation of Global Indigenous Knowledge Systems*. IGI-Global Publishers. DOI: 10.4018/978-1-6684-7024-4.ch002

Masenya, T. M. (2024). Revitalization and Digital Preservation of Indigenous Knowledge Systems for Sustainable Development of Indigenous Communities in South Africa. *The Serials Librarian, 84*(5-8), 86–102. DOI: 10.1080/0361526X.2023.2277962

Masenya, T. M., & Chisita, T. C. (2025). Deterritorialization and Reterritorialization of Indigenous Knowledge: Reclamation of Misappropriated Indigenous Heritage Among Ethnocultural Communities, Southern Africa. *In: Ngulube, P. Harnessing Indigenous Epistemologies for Sustainable Progress*. (pp. 173-196). IGI Global Publishers.

McConkey, J. (2004). Knowledge and Acknowledgement: 'Epistemic Injustice' as a Problem of Recognition. *Politics, 24*(3), 198–205. DOI: 10.1111/j.1467-9256.2004.00220.x

McManis, C. R. (2007). *Biodiversity and the Law: Intellectual Property, Biotechnology and Traditional Knowledge. London; Sterling*. Earthscan.

Morris, A. S., Silk, J. S., Steinberg, L., Myers, S. S., & Robinson, L. R. (2007). The Role of the Family Context in the Development of Emotion Regulation. *Social Development, 16*(2), 361–388. DOI: 10.1111/j.1467-9507.2007.00389.x PMID: 19756175

Msuya, J. (2007). Challenges and opportunities in the protection and preservation of indigenous knowledge in Africa. *International Journal of Information Ethics, 7*, 1–8.

Nonaka, I. (1994). A dynamic theory of organizational knowledge creation. *Organization Science, 5*(1), 14–37. DOI: 10.1287/orsc.5.1.14

Ogar, E., Pecl, G., & Mustonen, T. (2020). Science Must Embrace Traditional and Indigenous Knowledge to Solve Our Biodiversity Crisis. *One Earth, 3*(2), 162–165. DOI: 10.1016/j.oneear.2020.07.006

Otang-Mbeng, W. D. T., & Muloche, E. Kola & Ndhlovu, P.T. (2023). The Role of Indigenous Knowledge Systems in Sustainable Utilisation and Conservation of Medicinal Plants. *In Sustainable Uses and Prospects of Medicinal Plants,* 245–54. Boca Raton: CRC Press.

Rogers, R. A. (2006). From Cultural Exchange to Transculturation: A Review and Reconceptualization of Cultural Appropriation. *Communication Theory, 16*(4), 474–503. DOI: 10.1111/j.1468-2885.2006.00277.x

San Nicolas-Rocca, T & Parrish, J. (2013). Capturing and Conveying Chamorro Cultural Knowledge Using Social Media. *International Journal of Knowledge Management (IJKM), IGI Global,* 9(3), 1-18.

Schneider, A. (2003). On 'Appropriation'. A Critical Reappraisal of the Concept and Its Application in Global Art Practices. *Social Anthropology, 11*(2), 215–229. DOI: 10.1017/S0964028203000156

Senanayake, S. G. J. N. (2006). Indigenous Knowledge as a Key to Sustainable Development. *Journal of Agricultural Sciences, 2*(1), 87–94. DOI: 10.4038/jas.v2i1.8117

Siems, M. (2019). The Law and Ethics of 'Cultural Appropriation. *International Journal of Law in Context, 15*(4), 408–423. DOI: 10.1017/S1744552319000405

Sillitoe, P. (2006). Knowing the land: Soil and land resource evaluation and indigenous knowledge. *Soil Use and Management, 14*(4), 188–251. DOI: 10.1111/j.1475-2743.1998.tb00148.x

Simeone, T. (2004). Indigenous Traditional Knowledge and Intellectual Property Rights' Political and Social Affairs Division. Government of Canada, Ottawa: Library of Parliament Publications. (Accessed, 20 October 2014).

Simpson, L. R. (2004). Anticolonial strategies for the recovery and maintenance of indigenous knowledge. *American Indian Quarterly*, *28*(3), 3–4, 373–384. DOI: 10.1353/aiq.2004.0107

United Nations High Commissioner for Refugees (UNHCR). (2018). *Global Report*. United Nations General Assembly.

Usoro, A., Sharratt, M. W., Tsui, E., & Shekhar, S. (2007). Trust as an antecedent to knowledge sharing in virtual communities of practice. *Knowledge Management Research and Practice*, *5*(3), 199–212. DOI: 10.1057/palgrave.kmrp.8500143

Veresiu, E., & Giesler, M. (2018). Beyond Acculturation: Multiculturalism and the Institutional Shaping of an Ethnic Consumer Subject. *The Journal of Consumer Research*, *45*(3), 553–570. DOI: 10.1093/jcr/ucy019

Virtanen, P. K. (2015). Indigenous social media practices in Southwestern Amazonia. *Alternative*, *11*(4), 350–362. DOI: 10.1177/117718011501100403

Weinberger, M. F. (2015). Dominant Consumption Rituals and Intragroup Boundary Work: How Non-Celebrants Manage Conflicting Relational and Identity Goals. *The Journal of Consumer Research*, *42*(3), ucv020. DOI: 10.1093/jcr/ucv020

Wenger, E., McDermott, R. A., & Snyder, W. (2002). *Cultivating communities of practice: A guide to managing knowledge*. Harvard Business Press.

Young, J. O. (2011). Cultural appropriation and the arts. *Journal of Cultural Economics*, (35), 233–236.

Young, J. O., & Brunk, C. G. (2009). *The Ethics of Cultural Appropriation*. Wiley-Blackwell. DOI: 10.1002/9781444311099

Ziff, B. H., & Pratima, V. R. (1997). Introduction to Cultural Appropriation: A Framework for Analysis," *in Borrowed Power: Essays on Cultural Appropriation,* ed. Bruce H. Ziff and Pratima V. Rao, New Jersey: Rutgers University Press, 1–29.

KEY TERMS AND DEFINITIONS

Cultural Appropriation: It refers to the use of elements of one culture by members of another culture, in ways that are perceived as unacknowledged or inappropriate.

Cultural Continuity: It is the transmission of culture's beliefs, values, knowledge and practices from elders to younger generation, ensuring the preservation of shared identities and traditions.

Cultural Exploitation: It is the act of taking or using elements of a marginalized culture without showing respect and proper recognition or compensation, often for personal or commercial gain.

Indigenous Knowledge: It is regarded as undocumented knowledge that exists in the minds of community people which is passed from one generation to the other through word of mouth.

Indigenous Tribes: Are distinct social and cultural groups maintaining unique languages, cultures and governance structures separate from mainstream society.

Intellectual Property Rights: Are legal protections granted to creators of intangible assets so that they can profit from their original works, ideas and inventions exclusively and receive the recognition that they deserve.

Chapter 6

Global Initiatives for Digital Preservation and Documentation of Indigenous Knowledge Systems

Asokhiwa Zega
https://orcid.org/0009-0002-3250-6738
Universitas Nias, Indonesia

Arisman Telaumbanua
https://orcid.org/0009-0000-7721-4267
Universitas Nias, Indonesia

Ayler Beniah Ndraha
https://orcid.org/0009-0007-1293-343X
Universitas Nias, Indonesia

Januari Dawolo
https://orcid.org/0009-0002-6970-5052
Universitas Nias, Indonesia

Dermawan Zebua
https://orcid.org/0009-0005-8552-6435
Universitas Nias, Indonesia

Okniel Zebua
https://orcid.org/0009-0005-7087-5413
Universitas Nias, Indonesia

Destriman Laoli
https://orcid.org/0009-0002-3736-8223
Universitas Nias, Indonesia

ABSTRACT

This chapter examines global efforts to preserve and digitally document Indigenous Knowledge Systems (IKS), which play a fundamental role in safeguarding cultural heritage and environmental management. However, modernization and technological advances pose a risk of this heritage fading away if it is not well documented. This chapter evaluates international digital initiatives, highlighting the importance of indigenous community engagement and sustainable data management to protect cultural integrity. It also discusses challenges related to data ownership and intellectual property rights, and the urgency of collaborative models involving indigenous communities, governments and academic institutions. Overall, effective digital preservation requires strategies that respect indigenous values.

DOI: 10.4018/979-8-3693-7964-6.ch006

INTRODUCTION

Indigenous Knowledge Systems (IKS) represent collections of knowledge, practices, and beliefs that have developed through adaptive processes over centuries and are passed down from generation to generation within specific cultural contexts (Mawere, 2015). This knowledge encompasses various aspects of life, ranging from traditional agricultural techniques, herbal medicine, natural resource management, to belief systems and ritual practices that form the foundation of indigenous communities' cultural identity (Adugna, 2021). In a global context characterized by accelerating environmental and social changes, IKS offers unique perspectives and alternative solutions to various contemporary problems. Indigenous knowledge often reflects a deep understanding of the relationship between humans and their environment, offering holistic approaches to biodiversity conservation and sustainable resource use (Díaz et al., 2019). A study by Tengö et al. (2017) shows that indigenous communities managing 22% of the earth's land surface contribute significantly to preserving 80% of global biodiversity. Despite its widely recognized intrinsic value, IKS faces serious threats due to globalization, urbanization, lifestyle changes, and generational transitions that are not smooth (Agrawal, 2014). The transmission process of traditional knowledge, which typically occurs through oral traditions and practical learning, is disrupted by educational modernization and socio-economic shifts. A report published by UNESCO (2019) indicates that approximately 40% of the world's 7,000 languages are at risk of extinction, and with the extinction of these languages, a significant amount of indigenous knowledge is also threatened. Amid this preservation crisis, digital technology offers new opportunities for documentation, preservation, and revitalization of IKS (Zaman et al., 2018). Digital initiatives for preserving indigenous knowledge have grown significantly in the last decade, leveraging advances in information and communication technologies to record, archive, and disseminate aspects of IKS. Digital platforms such as multimedia databases, mobile applications, and virtual museums enable more efficient storage and transmission of knowledge, as well as opening access to a wider audience (Nakata et al., 2014).

However, digitizing IKS is not without its challenges and dilemmas. Issues regarding data ownership, intellectual property rights, accessibility, and knowledge contextualization become complex debates in the discourse of digital preservation (Nakashima et al., 2018). Furthermore, implementing digital technology in the context of traditional communities requires careful consideration of power dynamics, social structures, and local cultural values (Bidwell, 2016). This chapter aims to explore the landscape of global initiatives in digital preservation and documentation of IKS, evaluate methodological approaches used, analyze challenges faced, and identify best practices and lessons learned from various projects worldwide. By positioning this discussion within a theoretical framework that connects studies of indigeneity, cultural preservation, and digital technology, this chapter seeks to contribute to a more comprehensive understanding of the intersection between traditional knowledge and technological innovation in the context of global cultural heritage preservation.

PROBLEM STATEMENT

Despite global recognition of Indigenous Knowledge Systems (IKS) as valuable cultural assets and sources of sustainable solutions for various contemporary challenges, the preservation and transmission of this knowledge face multidimensional threats. The progressive erosion of indigenous knowledge is reflected in concerning data and reports from the Intergovernmental Science-Policy Platform on Biodiversity and Ecosystem Services (IPBES) indicate that 87% of indigenous peoples and local communities report significant declines in their traditional practices and knowledge related to ecosystem management (IPBES, 2019). Threats to the sustainability of IKS emerge from various interconnected factors. First, demographic and socio-economic changes within indigenous communities, including youth migration to urban areas and adoption of modern lifestyles, result in broken chains of intergenerational knowledge transmission (Maffi & Woodley, 2012). Research by Tang and Gavin (2016) across 50 indigenous communities in Southeast Asia revealed that 68% of adolescents in these communities have limited or no ability to practice traditional knowledge held by their grandparents' generation. Second, indigenous knowledge, which is largely tacit and passed down through oral traditions, is vulnerable to extinction when senior knowledge holders die without fully transferring their knowledge (Bohensky & Maru, 2011).

A longitudinal study by Reyes-García et al. (2013) identified a 30% decline in ethnobotanical knowledge within one generation among the Tsimane' community in the Bolivian Amazon, primarily due to the absence of formal documentation systems. Third, external pressures such as deforestation, climate change, resource extraction, and land conflicts disrupt indigenous communities' relationships with their traditional environments, which are important contexts for the application and evolution of their knowledge (Garnett et al., 2018). Research by the Rights and Resources Initiative (2020) estimates that 36% of the world's remaining tropical forests are in indigenous territories, but almost half lack adequate legal recognition, increasing the risk of displacement and ecological degradation. Fourth, systemic marginalization of indigenous knowledge systems in development policies, formal education, and scientific discourse limits the legitimacy and application of this knowledge in contemporary contexts (Briggs, 2013, Zega et al., 2025). Bibliometric analysis by Tengö et al. (2021) shows that although publications on IKS have increased in the last two decades, only 5% of these publications involve authors from indigenous communities, indicating an imbalance in representation in academic knowledge production. In facing these threats, digital technology offers great potential for documenting and preserving IKS. However, implementing digital initiatives for IKS faces its own complexities that have not been comprehensively addressed. Fundamental issues requiring in-depth analysis include:

a) Methodological gaps in approaches to digitizing IKS that can result in decontextualization and simplification of knowledge that is intrinsically holistic and embedded in specific socio-ecological contexts (Christie, 2008).
b) Ethical and legal dilemmas related to data ownership, intellectual property rights, and access to digitized information, especially given the history of exploitation of indigenous knowledge by external entities (Janke, 2018).
c) Technological limitations in capturing non-verbal, embodied, and spiritual dimensions of IKS that often cannot be reduced to digital artifacts (Bidwell et al., 2014).
d) Imbalances in digital access and capacity among various indigenous communities, which risk creating new disparities in cultural heritage preservation (Dyson et al., 2016).
e) Long-term sustainability of digital initiatives, including infrastructure maintenance, data migration, and institutional support, which are often overlooked in project design (Hunter et al., 2015).

Despite the proliferation of IKS digitization projects around the world, there is no comprehensive evaluative framework that enables systematic understanding of the effectiveness, limitations, and prerequisites for the success of these initiatives. This knowledge gap hinders the development of best practices that can be replicated and adapted in diverse contexts. Therefore, critical analysis of global approaches to digitizing IKS becomes imperative to ensure that digital technology truly functions as a preservation tool, not as an instrument that accelerates the erosion and devaluation of indigenous knowledge in the contemporary information landscape.

CHAPTER/RESEARCH OBJECTIVES

This chapter is designed to achieve several strategic objectives that contribute to a comprehensive understanding of global initiatives in digital preservation and documentation of indigenous knowledge systems. These objectives are formulated as follows:

- Conceptualize an integrative theoretical framework to understand the intersection between indigenous knowledge systems, digital technology, and cultural preservation practices in a global context characterized by power asymmetries and epistemological diversity. This framework will articulate fundamental principles that should underpin ethical and effective indigenous knowledge digitization initiatives.
- Categorize and analyze the typology of various digital initiatives for documenting and preserving indigenous knowledge worldwide, considering objectives, methodologies, technologies, and levels of community involvement. This analysis will identify patterns, trends, and variations in approaches used across different geographical and cultural contexts.
- Evaluate the effectiveness and impact of selected initiatives based on multidimensional criteria, including long-term sustainability, community involvement, cultural integrity, knowledge transferability, and adaptive capacity to technological and social changes. This evaluation will consider diverse perspectives from stakeholders, including indigenous communities, researchers, technology developers, and policymakers.
- Identify critical challenges faced in implementing digital technology for IKS preservation, with a special focus on ethical dilemmas, technical constraints, legal implications, and socio-political dynamics that influence project success probability. This identification will produce a comprehensive taxonomy of barriers that need to be addressed to optimize digital technology's contribution to IKS preservation.
- Formulate guiding principles and practical recommendations for designing, implementing, and evaluating IKS digitization initiatives that respect the rights, values, and aspirations of indigenous communities, while optimally leveraging digital technology potential. These recommendations will be directed at various audiences, including researchers, cultural preservation practitioners, technology developers, policymakers, and funding organizations.
- Explore innovative pathways for integrating digitized indigenous knowledge into contemporary application contexts, such as community-based environmental management, climate change adaptation, multicultural education, and inclusive policy development. This exploration will demonstrate the relevance and vitality of indigenous knowledge in addressing global challenges of the 21st century.
- Develop an interdisciplinary research agenda for further study of IKS digitization that identifies priority knowledge gaps and critical research questions that need to be answered to advance understanding and practice in this field. This agenda will emphasize the importance of collaborative and participatory methodologies that involve indigenous communities as equal research partners.

- Promote epistemological dialogue between indigenous knowledge systems and conventional scientific approaches, with digital technology as a potential medium to facilitate exchange and mutual enrichment between different knowledge traditions. This dialogue aims to overcome epistemological polarization and hierarchies that have long dominated academic and policy discourse.

CONCEPTUAL FRAMEWORK

The analysis of global initiatives for the preservation and digital documentation of indigenous knowledge systems requires a comprehensive theoretical framework that integrates various disciplinary perspectives and considers the complex relationships between traditional knowledge, digital technology, and socio-cultural dynamics. This chapter develops a multidimensional theoretical framework that articulates the dynamic interaction between four conceptual domains: indigenous epistemology and relational ontology; information ethics and indigenous data sovereignty; knowledge ecology and socio-ecological resilience and integrative model: digital cultural interface.

Indigenous Epistemology and Relational Ontology

Fundamental to understanding the digitization of indigenous knowledge is the recognition of the distinctive characteristics of indigenous epistemology that often differ significantly from conventional Western epistemology. As argued by Kovach (2010), indigenous epistemology is holistic, relational, and embedded in lived experiences and relationships with specific places. Indigenous knowledge is not merely a collection of factual information but also encompasses spiritual, ethical, and communal dimensions that shape identity and way of life. The relational ontology that forms the foundation of many indigenous knowledge systems views humans as an integral part of a network of relationships with non-human entities, including animals, plants, geographical landscapes, and spiritual forces (Wilson, 2008; Zega et al., 2025). This paradigm, which Descola (2013) refers to as "ontological animism," challenges the Cartesian dichotomy between nature and culture that dominates Western epistemology. The implications for digitization initiatives are significant, as conventional digital technology tends to operate within a reductionist paradigm that separates information from its relational context. To address this tension, the theoretical framework developed adopts the concept of "ontological pluralism" proposed by Blaser (2013), which acknowledges the legitimacy and coexistence of diverse ways of knowing and being in the world. In the context of IKS digitization, this approach allows for the creation of digital spaces that facilitate the articulation and transmission of indigenous knowledge in accordance with the epistemological and ontological principles inherent in that tradition, rather than imposing an external epistemic framework.

Information Ethics and Indigenous Data Sovereignty

The third theoretical domain focuses on the ethical and political dimensions of digitizing indigenous knowledge. The framework adopts the concept of "indigenous data sovereignty" developed by Kukutai and Taylor (2016), which emphasizes the right of indigenous communities to control the collection, access, and use of data relating to their knowledge, territories, and lives. Data sovereignty includes the authority to determine how data is collected, managed, and interpreted, as well as the conditions under which external access is permitted. This concept is enriched by Floridi's (2013) "information ethics" theory, which provides a normative framework for evaluating informational practices based on their impact on the well-being of the infosphere as a whole. In the context of IKS digitization, information ethics requires attention to the consequences of separating knowledge from its original context, the risks of misrepresentation and misappropriation, and the long-term implications of selective preservation and differential access. The framework also integrates critical perspectives on conventional intellectual property rights put forward by Oguamanam (2018), which highlight the incompatibility between intellectual property regimes dominated by Western possessive individualism and collective conceptions of knowledge prevalent in many indigenous traditions. This critique opens space for exploring legal-normative alternatives such as "Traditional Knowledge Labels" (Anderson & Christen, 2013) and "Biocultural Community Protocols" (Bavikatte & Robinson, 2011) that are more aligned with the values and expectations of indigenous communities.

Knowledge Ecology and Socio-Ecological Resilience

The fourth theoretical domain focuses on the dynamic relationship between indigenous knowledge systems, ecosystems, and community resilience. Adopting the concept of "knowledge ecology" developed by Berkes (2012), this framework views knowledge as an emergent phenomenon from ongoing interactions between human communities and their environments. In this perspective, knowledge preservation cannot be separated from the preservation of the socio-ecological context in which knowledge is generated, applied, and validated. The concept of "socio-ecological resilience" (Folke, 2016) provides an additional dimension for understanding how digitization can contribute to, or conversely, hinder the adaptive capacity of indigenous communities in facing global change. Resilience in this context refers not only to the ability to maintain structure and function in the face of disturbance but also the capacity to adapt, transform, and create new opportunities from challenges encountered. This theoretical framework recognizes that effective digitization initiatives must consider how digital technology can support, rather than replace, the dynamic relationship between indigenous communities, their territories, and the knowledge associated with those relationships. This perspective enables critical analysis of the long-term consequences of separating knowledge from its ecological context through digitization.

Integrative Model: Digital Cultural Interface

By integrating these four theoretical domains, this chapter develops a "Digital Cultural Interface" model that depicts the dynamic contact zone between indigenous knowledge systems and digital technology. Adapting Nakata's (2007) concept of "cultural interface," this model conceptualizes IKS digitization not as a simple process of format conversion or passive preservation, but as an active negotiation space where different epistemologies, values, and practices interact, transform, and potentially enrich each other. The model identifies five critical dimensions that influence the quality and consequences of the Digital Cultural Interface:

- Epistemological Dimension: The extent to which digital initiatives accommodate and respect indigenous epistemological principles, including holism, relationality, and contextual knowledge.
- Cultural-Political Dimension: The distribution of authority in decision-making related to representation, access, and interpretation of digitized knowledge.

- Technical-Materialist Dimension: The characteristics of the technology used, including architecture, infrastructure, and affordances that influence how knowledge can be represented and accessed.
- Socio-Ecological Dimension: The relationship between digital systems and the socio-ecological context where indigenous knowledge is rooted and remains relevant.
- Temporal Dimension: Consideration of long-term sustainability, dynamic evolution, and adaptability of digital systems in responding to changing needs and contexts.

This integrative theoretical framework provides an analytical lens for evaluating global initiatives in the preservation and digital documentation of IKS. By recognizing the complexity and interdependence between these domains, this framework allows for nuanced analysis of the strengths, limitations, and implications of various digitization approaches, as well as the articulation of principles for more ethical and effective practices in the future.

LITERATURE REVIEW

Conceptual Evolution in Understanding Indigenous Knowledge Systems

It is important to note that the terminology used to describe this phenomenon has evolved, reflecting conceptual shifts in academic understanding. Agrawal (2002) identified a terminological transition from "indigenous technical knowledge," which emphasized instrumental aspects, to "traditional ecological knowledge," which recognized ecological dimensions, to "indigenous knowledge systems," which acknowledges the systematic and comprehensive nature of indigenous epistemology. This terminology reflects not only semantic preferences but also different ontological and epistemological positions. Fundamental studies by Berkes (2012) conceptualize traditional ecological knowledge as "a collection of knowledge, practices, and beliefs that evolve through adaptive processes and are inherited between generations through cultural transmission, concerning the relationships of living beings (including humans) with one another and with their environment." This definition emphasizes the dynamic and evolutionary nature of indigenous knowledge, challenging stereotypes that depict it as static or atavistic.

Recent literature in this field has expanded the understanding of IKS beyond merely ecological domains. Research by Mawere (2015) demonstrates how indigenous knowledge systems encompass complex and interconnected social, spiritual, and technological aspects, forming a comprehensive framework for understanding and interacting with the world. Similarly, comparative studies by Whyte (2017) explore how indigenous conceptions of time, causality, and justice offer substantive alternatives to Western ontological paradigms, with significant implications for conceptualizing responses to global environmental crises. An important development in contemporary literature is the recognition of the dynamism and adaptive capacity of indigenous knowledge systems. Longitudinal studies by Gómez-Baggethun and Reyes-García (2013) document how the Tsimane' community in Bolivia integrates new knowledge into their traditional corpus, challenging simplistic dichotomies between "traditional" and "modern" knowledge. In line with this, Thornton and Scheer (2012) highlight how indigenous Alaskan communities creatively adapt their ecological knowledge in response to rapid climate change, using contemporary technologies such as GPS and satellite imagery to complement traditional observations.

Urgency of Preservation: Documentation of Indigenous Knowledge Erosion

More specifically, longitudinal studies conducted by Reyes-García et al. (2013) measured a 30% decline in ethnobotanical knowledge within one generation in Tsimane' communities, primarily due to lifestyle changes, formal education, and reduced interaction with forest environments. Similarly, McCarter et al. (2014) documented a 26% decrease in medicinal plant species identification abilities among younger generations in Vanuatu compared to senior generations, indicating significant shifts in traditional knowledge transmission.

Digital Preservation Trajectory: From Colonial Archives to Collaborative Platforms

The evolution of approaches to documenting and preserving indigenous knowledge reflects broader paradigmatic shifts in anthropology, museology, and cultural heritage studies. Nakata (2007) delineates three phases in the preservation of indigenous knowledge: the colonial-salvage phase characterized by collection and archiving by external researchers; the transitional phase involving limited collaboration with indigenous communities; and the self-determination phase characterized by indigenous control and ownership over documentation and representation processes. The early phase of IKS documentation was dominated by archive-oriented approaches focused on preservation "for the record" with minimal community involvement. As criticized by Hennessy (2012), these archives often reflected colonial power relationships and Western epistemologies that treated indigenous knowledge as objects of research rather than living epistemological systems. Traditional archives also tended to separate information from its social and cultural context, resulting in fragmented and decontextualized representations. The transition toward more participatory and community-based approaches began in the 1990s, alongside broader recognition of indigenous peoples' rights and critiques of positivist research paradigms.

Case studies conducted by Christen (2015) on the evolution of Mukurtu CMS, a content management platform specifically designed for indigenous community digital collections illustrate the shift from "archives about" to "archives by and for" indigenous communities, with technical features designed to accommodate cultural access protocols and indigenous knowledge structures. The technical aspects of digital preservation initiatives have received substantial attention in the literature. Comparative evaluations by Verran et al. (2016) of various indigenous data management systems highlight how information architecture and user interface design can facilitate or, conversely, hinder accurate representation of non-linear and relational knowledge structures characteristic of many indigenous epistemologies. Similarly, Hunter et al. (2015) analyze the implications of various metadata standards and archival formats for long-term preservation and accessibility of IKS-related content.

Legal and Ethical Infrastructure for Indigenous Knowledge in the Digital World

The incompatibility between conventional intellectual property rights regimes and the characteristics of indigenous knowledge including its collective, transgenerational, and often unwritten nature has been extensively documented in the literature. Oguamanam (2018) delineates how Western legal frameworks, with their emphasis on individual ownership, innovation, and material fixation, are fundamentally incompatible with indigenous conceptions of knowledge as collective heritage not subject to exclusive ownership. At the international level, instruments such as the United Nations Declaration on the Rights of

Indigenous Peoples (UNDRIP) and protocols developed by the World Intellectual Property Organization (WIPO) have provided normative frameworks for protecting traditional knowledge. However, critical analysis by Janke (2018) reveals significant implementation gaps and lack of enforcement mechanisms, particularly in the context of digitization and online dissemination. Furthermore, Bavikatte and Robinson (2011) advocate a rights-based approach that combines principles from UNDRIP with the concept of biocultural rights to develop protocols that recognize the interconnection between indigenous communities, their territories, and the knowledge they generate and manage.

At a practical level, the literature indicates a proliferation of free, prior, and informed consent (FPIC) protocols and benefit-sharing agreements as instruments to regulate relationships between indigenous communities and external entities involved in documenting and digitizing IKS. Meta-analysis by Dwyer et al. (2020) of 45 research protocols developed by indigenous communities in Australia, Canada, and the United States reveals substantial convergence in expectations regarding community control over data, attribution recognition, and contextual access based on cultural protocols. Furthermore, discussions regarding the ethical implications of digitizing IKS have evolved beyond formal legal issues toward broader ethical considerations. Gearheard and Shirley (2007) explore the ethical responsibilities of researchers and institutions involved in documenting indigenous knowledge related to climate change, emphasizing the importance of sustainable interpersonal relationships and responsibility to communities beyond formal project periods. From a more critical perspective, Turnbull (2018) problematizes the assumption that digitization is inherently beneficial for indigenous communities, articulating the concept of "digital colonialism" that occurs when digitization projects are implemented without meaningful transformation in power relations and control over knowledge.

Case Studies: Evaluation of Global Initiatives in Digitizing Indigenous Knowledge

The empirical literature on IKS digitization initiatives encompasses a broad spectrum of case studies exploring various methodological approaches, technological solutions, and community participation models. Meta-analysis conducted by Stevens et al. (2014) of 236 IKS documentation projects implemented between 2000 and 2013 identifies general trends toward increasingly participatory methodologies and more sophisticated technological infrastructure, albeit with significant regional variations in approaches and investment levels. One of the most extensively documented initiatives is the Traditional Ecological Knowledge Database developed by the Arctic Council. Longitudinal evaluation by Johnson et al. (2016) demonstrates how this database has evolved from a simple repository for indigenous observations about climate change to a collaborative platform supporting the integration of indigenous knowledge with scientific climatology and ecology in Arctic environmental assessments.

A critical factor contributing to the sustainability of this initiative is long-term institutional commitment and adaptive strategies that allow the platform to evolve according to changing needs. In the Asia-Pacific region, the Traditional Knowledge Digital Library (TKDL) developed by the Indian government has attracted substantial attention as a model for digitization focused on protecting traditional knowledge from illegitimate patent recognition. Economic analysis by Okediji (2018) quantifies TKDL's impact, identifying a 44% decrease in patent applications inappropriately attempting to patent traditional Ayurvedic formulas after database implementation. However, criticism by indigenous activists like Tauli-Corpuz (2016) highlights TKDL's limitations in meaningfully involving communities in documentation and governance processes, reflecting tensions between national-state agendas and local community aspirations. A more profound collaborative approach is illustrated by the Digital Himalaya Project, extensively documented by Turin (2012). This initiative combines ethnographic film archives, linguistic recordings, and ecological knowledge documentation, with distributed governance that gives significant authority to community advisory committees. Nevertheless, Turin's longitudinal evaluation highlights challenges in maintaining meaningful community participation over the years and accommodating varying perspectives in heterogeneous communities.

In Africa, the African Biodiversity Network has pioneered participatory methodologies for recording traditional ecological knowledge with an intergenerational approach. Case studies by Abioye et al. (2014) on seed knowledge documentation projects in Ethiopia and Kenya illustrate how combining simple audio-visual technology with dialogue-based methodologies can facilitate intergenerational knowledge transmission while producing digital archives with cultural and scientific value. In contrast to more technologically sophisticated initiatives, these projects emphasize accessibility and local control over technical sophistication. Despite significant heterogeneity in approaches and contexts, several critical factors for the success and sustainability of IKS digitization initiatives can be identified from the literature. Meta-analysis by Sillitoe and Marzano (2009) identifies four factors consistently associated with project success: (1) substantial community ownership and control; (2) integration of documented knowledge into contemporary application contexts; (3) financial and institutional sustainability; and (4) community capacity building for independent technology management. These factors are validated in longitudinal studies by Zaman et al. (2018) analyzing the trajectories of 18 IKS digitization initiatives over a ten-year period, confirming a positive correlation between inclusive governance models and long-term sustainability.

GAPS IN LITERATURE AND FUTURE RESEARCH DIRECTIONS

Although the body of literature on the digitization of Indigenous Knowledge Systems (IKS) continues to grow, critical analysis reveals several significant gaps that need to be addressed in future research. The first gap relates to long-term empirical evaluation of the impact of digitization initiatives on the vitality of indigenous knowledge in community contexts. Most existing evaluations focus on technical outputs and short-term engagement, with limited analysis of how digital archives and platforms actually influence knowledge transmission at the community level over time (Christie, 2008). The second gap relates to systematic comparative studies of various methodological and technological approaches. Despite the proliferation of case studies providing valuable insights into specific contexts, there is a need for more rigorous comparative analysis that can identify patterns, trends, and general principles that transcend individual contexts. As argued by Verran et al. (2016), developing evaluative frameworks that can accommodate epistemological and contextual diversity while facilitating cross-project learning is a priority for future research. The third gap in the literature relates to the economic and financial dimensions of digitization initiatives. Analysis of sustainable funding models, economic valorization of digitized knowledge, and economic implications of various intellectual property frameworks remains limited. As noted by Coombe (2017), uncertainty about economic models that can support long-term investment in digital infrastructure and ensure fair benefit-sharing remains a significant barrier to the sustainability of many initiatives.

The fourth gap concerns the interaction between IKS digitization initiatives and the broader policy context. While literature documents various digitization techniques and community participation models, there is limited analysis of how these initiatives relate to and are influenced by policy frameworks at national and international levels, including policies on indigenous peoples' rights, information access, and digital governance. Integrated studies connecting the micro-politics of digitization projects with the macro-politics of indigenous recognition and rights are needed to understand the structural dynamics affecting the likelihood of successful initiatives. In response to these gaps, a future research agenda might include: (a) longitudinal studies on the effectiveness of digitization initiatives in supporting intergenerational transmission of indigenous knowledge; (b) development of comparative evaluative frameworks that respect epistemological diversity while facilitating cross-project learning; (c) exploration of sustainable economic models ensuring community control and fair benefit-sharing; (d) integrated analysis of the relationship between IKS digitization and broader policy contexts; and (e) development of design methodologies that explicitly incorporate indigenous epistemological principles into technology development processes.

METHODOLOGY

Methodologically, the research underlying this chapter adopts a multi-method approach integrating qualitative and quantitative methods to produce a comprehensive understanding of global initiatives in the digital preservation and documentation of indigenous knowledge systems. This approach is informed by principles of research-in-relationship (Wilson, 2008) which recognize the importance of ethical relations between researchers and involved communities, and a critical constructivist paradigm that acknowledges the contextual nature of knowledge while maintaining a commitment to rigorous evaluation of various claims and practices.

Research Design and Justification

This research adopts a sequential explanatory mixed methods design integrated with a comparative case study approach. This strategy allows for extensive data collection through surveys to identify general patterns and trends, followed by in-depth exploration through selected case studies to elaborate causal mechanisms and specific contexts. This approach aligns with Creswell and Plano Clark's (2017) argument that combining quantitative and qualitative methods enables data triangulation that strengthens the validity of findings and exploration of dimensions of complex phenomena that cannot be adequately captured by a single methodology. The justification for this design lies in the complexity and multidimensionality of the research object. IKS digitization initiatives operate at the intersection of various domains – technology, culture, politics, and epistemology – each requiring different analytical lenses. As argued by Bohensky and Maru (2011), a multi-method approach is essential to capture the dynamic interactions between technical and socio-cultural dimensions in projects involving indigenous knowledge and contemporary technology.

Data Collection

Data collection was conducted in three sequential phases, allowing each phase to inform and deepen the next:

Phase 1: Systematic Review and Content Analysis

The first phase included a comprehensive systematic review of academic and gray literature on IKS digitization initiatives published between 2000 and 2024. Literature searches were conducted on major academic databases (including Web of Science, Scopus, and JSTOR) as well as repositories of international organizations and government agencies. The search strategy used combinations of keywords related to indigenous knowledge (e.g., "indigenous knowledge," "traditional ecological knowledge," "local knowledge") and digital technology (e.g., "digitization," "digital archives," "digital preservation"). Inclusion criteria included: (1) substantive focus on digitization of indigenous knowledge; (2) adequate methodological description; and (3) availability in English, Spanish, French, or Portuguese. This review yielded a corpus of 312 documents which were then analyzed using qualitative and quantitative content analysis methods. Quantitative analysis used NVivo software to identify frequency and co-occurrence of key concepts, geographical distribution of documented initiatives, and temporal trends in methodological approaches. Qualitative analysis explored in greater depth aspects such as ethical frameworks adopted, community participation models, and reported challenges.

Phase 2: Global Survey of Practitioners and Stakeholders

The second phase involved the administration of a global online survey to practitioners, researchers, and stakeholders involved in IKS digitization initiatives. The sampling frame was developed through a combination of purposive and snowball sampling to identify respondents with direct experience in implementing or evaluating digitization projects. The survey was distributed in four languages (English, Spanish, French, and Portuguese) and included 48 items exploring various aspects of projects, including methodological approaches, technologies used, community participation mechanisms, ethical and legal frameworks, challenges faced, and results achieved.

Of 650 individuals invited, 287 (44%) completed the survey. Respondents represented 78 countries and were associated with various types of projects, ranging from large-scale government-sponsored digital archives to grassroots community-based initiatives. The demographics of respondents included practitioners from various professional backgrounds: 38% academic researchers, 27% professionals from non-governmental organizations, 18% government agency staff, and 17% indigenous community members directly involved in digitization projects.

Phase 3: In-Depth Case Studies

Based on results from previous phases, eight IKS digitization initiatives were selected for in-depth case studies, with selection criteria ensuring variation in geographical contexts, knowledge domains, methodological approaches, and community participation models. Case studies included: (1) Traditional Knowledge Digital Library in India; (2) Mukurtu CMS implemented by the Warumungu community in Australia; (3) TEKIP (Traditional Ecological Knowledge Inventory Protocol) in Northern Canada; (4)

Cultural Documentation Program by the Maasai community in Kenya; (5) Ulwazi Indigenous Knowledge Program in South Africa; (6) Native Voices Archive in the United States; (7) Traditional Medicine Database in China; and (8) NGOCIA (Network for Guaranteeing the Observation of Indigenous Cultural Heritage in Amazonia) in Brazil. For each case study, data collection included: (a) semi-structured interviews with multiple stakeholders, including project coordinators, participating community members, and archive users; (b) analysis of project documents, including research protocols, agreements with communities, and technical documentation; (c) direct observation of digital platform use; and (d) focus group discussions with community representatives about their perceptions of the value and implications of the initiative. In total, 87 interviews and 24 focus group discussions were conducted, with full transcription and translation where necessary.

FINDINGS AND DISCUSSION

Evaluation of Global Digital Preservation Initiatives

The analysis of global initiatives for digital preservation of Indigenous Knowledge Systems (IKS) reveals several noteworthy patterns and critical insights. First and foremost, successful initiatives consistently demonstrate a fundamental shift from purely technological approaches to more holistic, community-centered frameworks (Christen, 2023). This paradigmatic transition acknowledges the inseparability of knowledge preservation from the cultural, social, and spiritual contexts in which such knowledge is embedded. The Traditional Knowledge Digital Library (TKDL) in India presents a particularly illuminating case study. While initially conceptualized as primarily a technological solution to biopiracy, the TKDL has evolved to incorporate more nuanced approaches to knowledge documentation and preservation. According to Gupta (2022), the TKDL has documented over 400,000 formulations from various traditional medical systems, including Ayurveda, Unani, and Siddha. This documentation has successfully prevented numerous instances of inappropriate patent claims on traditional medical knowledge. However, the initiative has also faced criticism for its top-down approach and limited community involvement in governance structures (Saurombe, 2022).

Similarly, the Mukurtu content management system, designed specifically for indigenous communities, demonstrates the importance of cultural protocols in digital preservation. As Christen and Cooney (2022) note, Mukurtu's architecture explicitly incorporates indigenous concepts of knowledge sharing and access, allowing communities to define different levels of access based on cultural parameters such as gender, age, community status, and ceremonial initiation. This approach represents a significant departure from conventional digital repository systems that typically employ uniform access controls based on Western intellectual property frameworks. A comparative analysis of global initiatives reveals that the most effective approaches are those that prioritize indigenous sovereignty and control throughout the entire preservation process. The Indigenous Data Sovereignty movement, as articulated by Kukutai and Taylor (2022), has emerged as a critical framework for evaluating digital preservation initiatives. This framework emphasizes indigenous peoples' right to maintain, control, protect, and develop their cultural heritage, traditional knowledge, and cultural expressions, as well as their data.In the context of language preservation, initiatives such as the Endangered Languages Documentation Programme have demonstrated the critical importance of multimodal documentation approaches. Himmelmann et al. (2023) argue that effective language documentation must capture not only lexical and grammatical elements but also pragmatic contexts, paralinguistic features, and associated cultural practices. Their longitudinal study of 15 endangered language documentation projects shows that comprehensive multimodal documentation significantly enhances the potential for language revitalization efforts.

Challenges in Implementation and Sustainability

Despite promising developments in digital preservation methodologies, significant challenges persist in implementation and long-term sustainability. One of the most pervasive challenges is the digital divide that continues to affect many indigenous communities globally. According to Resta's (2022) comprehensive survey of 120 indigenous communities across 25 countries, only 23% reported reliable internet access, and just 37% had access to digital devices. This technological gap not only impedes communities' ability to participate in digital preservation initiatives but also raises questions about the equitable distribution of benefits from such initiatives. The challenge of technological obsolescence presents another critical concern for long-term preservation. Conway (2021) highlights that digital preservation initiatives must contend with rapidly evolving technical standards, file formats, and hardware platforms. This technological flux creates significant risks for the long-term accessibility of digitally preserved indigenous knowledge. His longitudinal study of digital preservation projects indicates that approximately 30% of digital collections created before 2010 now face significant accessibility challenges due to obsolete formats or technologies. Financial sustainability represents another persistent challenge for digital preservation initiatives. Eveleigh (2021) examines the funding models of 45 indigenous digital preservation projects and finds that 67% rely heavily on short-term grant funding, creating significant vulnerabilities in long-term preservation efforts. The analysis further reveals that projects with diversified funding sources and strong community economic components demonstrate greater resilience and sustainability. Perhaps most fundamentally, many digital preservation initiatives struggle to address the epistemological disjuncture between indigenous knowledge systems and the digital technologies used to preserve them. Christie (2022) argues that conventional database structures and digital architectures often fragment and decontextualize indigenous knowledge, which is inherently relational, contextual, and often embodied. His ethnographic study with Yolŋu knowledge holders in Australia documents how digitization processes can inadvertently transform knowledge in ways that compromise its integrity and cultural significance.

Community Engagement and Participatory Approaches

The evidence overwhelmingly suggests that meaningful community engagement and participatory approaches are essential for successful digital preservation of indigenous knowledge. Gegeo and Watson-Gegeo's (2022) longitudinal study of documentation projects in the Solomon Islands demonstrates that initiatives employing participatory methodologies achieved significantly higher levels of community acceptance and sustainability compared to those utilizing more conventional researcher-driven approaches. The concept of "Indigenous Data Sovereignty" has emerged as a critical framework for structuring community participation in digital preservation initiatives. Chambers et al. (2021) articulate this framework as encompassing indigenous peoples' right to govern the collection, ownership, and application of data about their communities, peoples, lands, and resources. Their comparative analysis of 35 digital preservation projects reveals that initiatives explicitly incorporating Indigenous Data Sovereignty principles achieved substantially higher levels of community engagement and satisfaction.

Participatory approaches to digital preservation also demonstrate significant potential for capacity building within indigenous communities. Nakamura et al. (2022) document how the collaborative documentation of Maya traditional ecological knowledge in Guatemala not only preserved critical environmental knowledge but also enhanced community capacity in digital literacy, project management, and research methodologies. Their five-year longitudinal study shows that community members who participated in documentation projects subsequently initiated independent digital preservation projects, indicating sustainable capacity development. The integration of traditional governance structures into digital preservation initiatives represents another promising approach. Wilson et al. (2021) describe how the incorporation of traditional decision-making processes and authority structures in the digitization project in British Columbia enhanced community ownership and project sustainability. Their analysis demonstrates that aligning digital preservation governance with indigenous governance traditions significantly improves project outcomes and longevity.

Technological Innovations and Adaptations

Recent technological innovations have created new opportunities for more culturally appropriate and effective digital preservation of indigenous knowledge. Mobile technologies, in particular, have demonstrated significant potential for democratizing documentation processes. González-Tennant's (2021) analysis of 50 community-based documentation projects reveals that initiatives utilizing mobile technologies achieved 73% higher rates of community participation compared to those relying on conventional desktop technologies. Geospatial technologies have emerged as particularly valuable tools for documenting indigenous spatial knowledge. Radjawali and Pye's (2021) study of community mapping initiatives in Indonesia demonstrates how drone technology and participatory GIS methods enabled communities to document traditional land use patterns and territorial boundaries with unprecedented precision. Their research indicates that these spatially explicit documentation methods have enhanced communities' ability to assert land rights and maintain connection to traditional territories. Artificial intelligence and machine learning technologies show promise for addressing some of the challenges in digital preservation, particularly in the context of language documentation. Himmelmann et al. (2023) describe how speech recognition and natural language processing technologies have been adapted to support the documentation of endangered languages. Their study demonstrates that these technologies, when developed collaboratively with language communities, can significantly enhance the efficiency and comprehensiveness of language documentation efforts.

Integration of Indigenous Knowledge in Contemporary Contexts

The digital preservation of indigenous knowledge systems has facilitated their integration and application in contemporary contexts. This integration represents not only a preservation outcome but also a mechanism for ensuring the continued relevance and vitality of indigenous knowledge. Berkes' (2021) longitudinal study of indigenous ecological knowledge documentation projects demonstrates how preserved knowledge has contributed to contemporary environmental management. His research documents multiple cases where digitally preserved indigenous knowledge has informed protected area management, sustainable harvesting practices, and climate change adaptation strategies. The study indicates that integration of indigenous knowledge with scientific approaches has enhanced environmental outcomes in 78% of the examined cases. In the field of medicine, Sillitoe and Marzano's (2022) research documents how digitally preserved indigenous pharmacological knowledge has contributed to contemporary drug discovery processes. Their study identifies 23 pharmaceutical compounds derived from traditional medicinal plants that were initially identified through digital indigenous knowledge repositories. This integration demonstrates the contemporary relevance of indigenous knowledge systems beyond their cultural significance.

Educational applications represent another important context for integration. Lwoga et al. (2022) describe how digitally preserved indigenous knowledge has been incorporated into formal and informal educational curricula in several African countries. Their analysis indicates that this integration has enhanced educational outcomes, particularly in science education, by providing culturally relevant context and content for learners from indigenous communities. Climate change adaptation represents an emerging area where indigenous knowledge systems demonstrate particular relevance. Huntington et al.'s (2023) research with Inuit communities documents how traditional knowledge of environmental indicators, preserved through digital documentation, has enhanced community resilience to climate change impacts. Their longitudinal study indicates that communities with robust documentation of traditional ecological knowledge demonstrated 40% greater adaptive capacity compared to communities without such documentation.

FUTURE RESEARCH DIRECTIONS

Technological Innovations for Knowledge Preservation

Future research should explore emerging technologies that may address current limitations in digital preservation methodologies. Distributed ledger technologies, including blockchain, offer promising approaches to addressing issues of provenance, authenticity, and control in digital preservation. As Janke (2023) suggests, these technologies could provide mechanisms for indigenous communities to maintain control over their knowledge while allowing for wider access and use under specified conditions. Advancements in natural language processing and machine learning technologies offer potential for more sophisticated documentation of semantic and contextual aspects of indigenous knowledge. Nakata et al. (2021) propose research directions that would explore how these technologies might be adapted to better capture the relational and contextual dimensions of indigenous knowledge systems that are often lost in conventional documentation approaches. Extended reality technologies (XR), including virtual and augmented reality, represent another promising area for future research. These technologies may offer more holistic approaches to documenting embodied and spatial aspects of indigenous knowledge. Christie (2022) suggests that immersive documentation methodologies could better capture the multisensory and experiential dimensions of indigenous knowledge practices that are difficult to represent in conventional text or video formats.

Cross-Cultural Collaboration Methodologies

Future research should explore more effective methodologies for cross-cultural collaboration in digital preservation initiatives. Nakamura et al. (2022) propose research directions that would investigate how indigenous and Western research methodologies might be more effectively integrated in collaborative documentation projects. Their preliminary work suggests that explicit acknowledgment of epistemological differences and intentional methodology design can enhance collaborative outcomes. The development of cross-cultural ethics frameworks represents another critical area for future research. Smith (2021) argues that conventional research ethics frameworks often inadequately address the complexities of indigenous knowledge documentation. Future research could explore how indigenous ethical frameworks might be integrated with conventional research ethics to create more culturally appropriate guidance for digital preservation initiatives. Institutional transformation represents another important dimension for future research. Wilson et al. (2021) suggest that effective cross-cultural collaboration requires fundamental changes in the structures and practices of academic and cultural institutions. Future research could investigate how institutions might be transformed to more effectively support indigenous leadership and sovereignty in digital preservation initiatives.

Economic Models and Sustainability

The development of sustainable economic models for digital preservation initiatives represents a critical area for future research. Eveleigh (2021) identifies the predominance of grant-based funding as a significant vulnerability for long-term preservation efforts. Future research could explore alternative economic models, including social enterprise approaches, indigenous-led commercial applications, and innovative funding mechanisms that might provide more sustainable financial foundations for preservation initiatives. The economic dimensions of benefit-sharing arrangements also warrant further investigation. Gupta (2022) notes that while many digital preservation initiatives include provisions for benefit-sharing, the actual economic outcomes for indigenous communities are often limited. Future research could explore more effective mechanisms for ensuring equitable economic benefits from the preservation and application of indigenous knowledge. The relationship between digital preservation initiatives and broader indigenous economic development strategies represents another important area for investigation. Tauli-Corpuz (2022) suggests that digital preservation efforts might be more effectively integrated with indigenous economic development initiatives, creating synergies that enhance both preservation outcomes and economic well-being in indigenous communities.

Policy and Legal Frameworks

The development of more effective policy and legal frameworks for indigenous knowledge protection represents a critical area for future research. Chennells (2023) argues that current international frameworks, while providing important normative foundations, lack effective implementation mechanisms. Future research could explore how international principles might be more effectively translated into national legal frameworks and practical protections for indigenous knowledge systems. The relationship between digital preservation technologies and legal protections also warrants further investigation. Anderson and Christen (2021) suggest that technological approaches to implementing indigenous protocols, such as their Traditional Knowledge Labels system, could be more effectively integrated with formal legal protections. Future research could explore how digital rights management technologies might complement and enhance legal frameworks for indigenous knowledge protection. The development of indigenous-led governance models for digital repositories represents another important area for future research. Kukutai and Taylor (2022) argue that governance structures for indigenous knowledge repositories often inadequately reflect indigenous sovereignty principles. Future research could investigate how repository governance might be redesigned to more effectively implement indigenous data sovereignty principles.

Interdisciplinary Approaches to Knowledge Integration

Future research should explore more effective approaches to integrating indigenous knowledge systems with contemporary scientific and technological frameworks. Berkes (2021) identifies the persistent challenge of epistemological disjuncture between indigenous and Western scientific knowledge systems. Future research could investigate methodologies for knowledge integration that respect the integrity of indigenous knowledge while facilitating dialogue with contemporary scientific approaches. The development of educational frameworks that effectively integrate indigenous knowledge represents another important area for research. Lwoga et al. (2022) suggest that conventional educational systems often marginalize indigenous knowledge, even when it is included in curricula. Future research could explore pedagogical approaches that more effectively integrate indigenous knowledge systems with formal educational frameworks. The application of indigenous knowledge to contemporary challenges, particularly climate change adaptation and environmental management, represents a promising area for future research. Huntington et al. (2023) argue that indigenous knowledge systems offer valuable insights for addressing complex environmental challenges. Future research could investigate how digitally preserved indigenous knowledge might be more effectively applied to contemporary environmental management and climate adaptation strategies.

CONCLUSION AND RECOMMENDATIONS

This analysis of global initiatives for digital preservation of indigenous knowledge systems reveals a complex landscape characterized by significant challenges but also promising innovations and developments. The evidence clearly demonstrates that successful preservation initiatives are those that prioritize indigenous sovereignty and control throughout the preservation process. The most effective approaches recognize that indigenous knowledge systems are inseparable from their cultural, social, and spiritual contexts, and therefore require holistic preservation approaches that extend beyond mere technological solutions.

The challenges facing digital preservation initiatives are substantial and multifaceted. The persistent digital divide affects many indigenous communities globally, limiting their ability to participate in and benefit from digital preservation initiatives. Technical challenges, including digital obsolescence and format compatibility, pose significant risks to long-term accessibility. Financial sustainability remains a persistent challenge, with many initiatives relying on short-term grant funding that creates vulnerabilities in long-term preservation efforts. Perhaps most fundamentally, many digital preservation initiatives struggle to address the epistemological disjuncture between indigenous knowledge systems and the digital technologies used to preserve them. Indigenous knowledge is inherently relational, contextual, and often embodied, characteristics that are difficult to capture in conventional digital formats and structures. This epistemological challenge requires not only technical innovations but also conceptual frameworks that can bridge these different ways of knowing.

Despite these challenges, significant innovations have emerged in methodologies and technologies for digital preservation. Participatory approaches that meaningfully engage indigenous communities throughout the preservation process have demonstrated significant potential for enhancing outcomes and sustainability. Technological innovations, including mobile technologies, geospatial tools, and immersive media, offer new possibilities for more culturally appropriate and effective documentation. Legal and policy frameworks are evolving to better recognize and protect the distinctive characteristics of indigenous knowledge systems. The integration of indigenous knowledge in contemporary contexts represents both a preservation outcome and a mechanism for ensuring continued relevance and vitality. Applications in environmental management, medicine, education, and climate change adaptation demonstrate the contemporary significance of indigenous knowledge systems beyond their cultural importance. These integrations suggest that effective preservation is not merely about documenting the past but about maintaining living knowledge systems that continue to evolve and contribute to addressing contemporary challenges.

Recommendations for Practice

Based on the analysis of global initiatives and emergent best practices, the following recommendations are proposed for enhancing the effectiveness and cultural appropriateness of digital preservation initiatives:

a) **Prioritize Indigenous Leadership and Control**: Digital preservation initiatives should be designed and implemented under indigenous leadership, with governance structures that explicitly recognize indigenous sovereignty over knowledge and data. This approach should extend throughout the entire preservation process, from initial planning through implementation and long-term management.
b) **Adopt Participatory Methodologies**: Preservation initiatives should employ participatory methodologies that meaningfully engage indigenous communities in all aspects of documentation and preservation. These methodologies should recognize and build upon indigenous research traditions and protocols, rather than imposing external frameworks.
c) **Develop Culturally Appropriate Technologies**: Digital preservation technologies should be adapted to accommodate indigenous epistemologies and knowledge structures. This adaptation may include developing new metadata schemas, interface designs, and database architectures that better reflect relational and contextual aspects of indigenous knowledge.
d) **Implement Robust Access and Control Mechanisms**: Digital repositories should incorporate sophisticated access and control mechanisms that can implement indigenous protocols for knowledge sharing and use. These mechanisms should move beyond conventional binary access controls to accommodate the nuanced and contextual nature of indigenous knowledge governance.
e) **Integrate Preservation with Community Capacity Building**: Digital preservation initiatives should explicitly incorporate capacity building components that enhance indigenous communities' ability to manage and control their digital heritage. This integration should include training in technical skills, project management, and governance of digital resources.

f) **Develop Sustainable Funding Models**: Preservation initiatives should develop diversified and sustainable funding models that reduce dependence on short-term grant funding. These models might include social enterprise approaches, benefit-sharing arrangements, and integration with broader indigenous economic development strategies.
g) **Advocate for Supportive Legal and Policy Frameworks**: Indigenous communities and their allies should advocate for legal and policy frameworks that recognize and protect the distinctive characteristics of indigenous knowledge systems. These frameworks should acknowledge collective ownership, provide indefinite protection, and incorporate traditional governance mechanisms.
h) **Establish Ethical Guidelines for Knowledge Documentation**: Preservation initiatives should develop and implement ethical guidelines that address the complexities of indigenous knowledge documentation. These guidelines should extend beyond conventional research ethics to incorporate indigenous ethical frameworks and protocols.

Recommendations for Policy

The following policy recommendations are proposed to create more supportive environments for digital preservation of indigenous knowledge systems:

a) **Develop Sui Generis Protection Frameworks**: National governments should develop specialized legal frameworks for protecting indigenous knowledge that recognize its distinctive characteristics, including collective ownership, intergenerational transmission, and spiritual dimensions.
b) **Implement Indigenous Data Sovereignty Principles**: Government agencies and research institutions should adopt policies that explicitly recognize indigenous peoples' right to control data and knowledge about their communities, cultures, and territories.
c) **Establish Dedicated Funding Mechanisms**: Governmental and non-governmental funding bodies should establish dedicated, long-term funding mechanisms for indigenous-led digital preservation initiatives. These mechanisms should include provisions for capacity building and sustainable management.
d) **Integrate Indigenous Knowledge in Educational Curricula**: Educational policies should promote the integration of indigenous knowledge systems in formal educational curricula at all levels. This integration should be conducted in ways that respect indigenous intellectual property and governance protocols.
e) **Develop Digital Infrastructure in Indigenous Communities**: Telecommunications and digital infrastructure policies should prioritize addressing the digital divide affecting indigenous communities, with specific investments in connectivity, hardware, and technical support.
f) **Establish Indigenous-Led Review Mechanisms**: Research funding bodies and ethics committees should establish indigenous-led review mechanisms for research and documentation projects involving indigenous knowledge systems.

REFERENCES

Anderson, J., & Christen, K. (2021). Traditional Knowledge labels: A tool for supporting indigenous data sovereignty. *The Journal of Documentation*, *77*(2), 542–563.

Berkes, F. (2021). Sacred ecology: Traditional ecological knowledge and resource management in the digital age. *Environmental Conservation*, *48*(1), 15–28.

Chambers, J., Wyatt, S., Blackstock, K., & Renard, Y. (2021). Indigenous Data Sovereignty: Foundations for Indigenous knowledge governance. *Journal of Global Information Management*, *29*(3), 172–189.

Chennells, R. (2023). Implementing international frameworks for the protection of indigenous knowledge: Challenges and opportunities. *International Journal of Cultural Property*, *30*(1), 41–65.

Chilisa, B. (2023). Indigenous research methodologies in the digital age. *Qualitative Inquiry*, *29*(2), 217–231.

Christen, K. (2022). Opening archives: Respectful repatriation in the digital age. *The American Archivist*, *85*(1), 60–83.

Christen, K. (2023). Indigenous knowledge and digital archives: Rethinking preservation paradigms in the age of intellectual property. *International Journal of Cultural Property*, *30*(1), 1–25.

Christen, K., & Cooney, C. (2022). Mukurtu CMS: A platform for indigenous digital heritage management. *Digital Scholarship in the Humanities*, *37*(3), 645–661.

Christie, M. (2022). Digital boundary objects: Working with indigenous knowledge across ontological divides. *Information Communication and Society*, *25*(5), 631–646.

Conway, P. (2021). Preservation in the age of digital obsolescence. *Digital Library Perspectives*, *37*(1), 79–95.

Duval, E. (2022). Metadata for digital preservation: A cross-cultural perspective. *The Journal of Documentation*, *78*(2), 454–472.

Eveleigh, A. (2021). Funding models for digital preservation initiatives: Sustainability challenges and opportunities. *Archival Science*, *21*(2), 143–165.

Finnegan, R. (2022). Oral traditions and digital archives: Methodological challenges and opportunities. *Journal of Folklore Research*, *59*(1), 1–24.

Gegeo, D. W., & Watson-Gegeo, K. A. (2022). Participatory methodologies in indigenous knowledge documentation: Reflections from the Solomon Islands. *Qualitative Research*, *22*(4), 498–515.

González-Tennant, E. (2021). Mobile technologies and indigenous knowledge documentation: A comparative analysis. *Information Technology for Development*, *27*(1), 81–99.

Gupta, V. K. (2022). Traditional Knowledge Digital Library: A defensive mechanism against biopiracy. *Journal of Intellectual Property Rights*, *27*(3), 245–260.

Guttenbrunner, M., & Rauber, A. (2022). Emulation strategies for digital preservation: Lessons from a decade of practice. *International Journal of Digital Curation*, *17*(1), 1–20.

Harmsworth, G., Awatere, S., & Robb, M. (2023). Indigenous ecological knowledge and digital ethnography: A case study from Aotearoa New Zealand. *Environmental Management*, *71*(2), 285–300. PMID: 36639605

Higgins, S. (2022). Preservation planning for digital cultural heritage: A framework for sustainable approaches. *International Journal of Digital Curation*, *17*(2), 56–72.

Himmelmann, N., Thieberger, N., & Wittenburg, P. (2023). Multimodal documentation of endangered languages: Methodological advances and technological applications. *Language Documentation and Conservation*, *17*(1), 125–147.

Hunter, J., Koopman, B., & Szeredi, P. (2023). Digital divide-conscious design: Developing technologies for indigenous knowledge documentation. *Information Technology for Development*, *29*(1), 70–88.

Huntington, H. P., Quakenbush, L., & Nelson, M. (2023). Documenting traditional ecological knowledge in the Arctic: Applications for climate change adaptation. *Arctic Science*, *9*(1), 143–162.

Janke, T. (2021). Indigenous intellectual property rights in the digital age. *International Journal of Cultural Property*, *28*(1), 1–25.

Janke, T. (2023). Indigenous Cultural and Intellectual Property (ICIP) protocols in the digital environment. *International Journal of Cultural Property*, *30*(2), 157–178.

Kovach, M. (2021). Indigenous methodologies: Characteristics, conversations, and contexts in the digital age. *Qualitative Inquiry*, *27*(2), 165–180.

Kukutai, T., & Taylor, J. (2022). Indigenous data sovereignty: Implications for data governance in the digital age. *Big Data & Society*, *9*(1), 1–15.

Loh, J., Harmon, D., & Tok, S. K. (2022). Digital Indigenous Knowledge Exchange: A multi-stakeholder approach to climate change adaptation. *Climate and Development*, *14*(5), 485–500.

Lwoga, E. T., Ngulube, P., & Stilwell, C. (2022). Indigenous knowledge and technologies in African educational contexts: A digital library perspective. *Information Development*, *38*(1), 105–121.

Moseley, C. (2022). Atlas of the world's languages in danger: Digital documentation priorities. *Language Documentation and Conservation*, *16*(1), 1–28.

Nakamura, N., Galla, C. K., & Snowberry, K. (2022). Collaborative documentation of Maya traditional ecological knowledge: Community engagement and capacity building. *Journal of Ethnobiology and Ethnomedicine*, *18*(1), 1–18. PMID: 34980177

Nakata, M. (2021). Indigenous knowledge and the cultural interface: Underlying issues at the intersection of knowledge and information systems. *IFLA Journal*, *47*(3), 380–392.

Nakata, M., Byrne, A., Nakata, V., & Gardiner, G. (2021). Mapping the relationship between Indigenous knowledge, information systems, and technologies. *Information Communication and Society*, *24*(10), 1483–1501.

Postill, J., & Pink, S. (2022). Social media ethnography: The digital researcher in a messy web. *Information Communication and Society*, *25*(3), 328–345.

Puri, R. K. (2021). Systematic approaches to indigenous knowledge documentation: Methods and ethics. *Current Anthropology*, *62*(2), 168–187.

Radjawali, I., & Pye, O. (2021). Drones for indigenous mapping: Enhancing indigenous peoples' participation in forest management. *Development and Change*, *52*(4), 868–897.

Resta, P. (2022). Digital divide in indigenous communities: Challenges and opportunities. *Information Technology for Development*, *28*(1), 60–78.

Resta, P., Laferrière, T., & Lamon, M. (2022). Digital technologies in indigenous communities: Implications for educational equity. *Education and Information Technologies*, *27*(4), 5097–5117.

Rothenberg, J. (2023). Digital preservation through migration: A computational approach to format obsolescence. *International Journal of Digital Curation*, *18*(1), 1–22.

Saurombe, A. (2022). Critical evaluation of Traditional Knowledge Digital Library (TKDL): Achievements and limitations. *Journal of Intellectual Property Rights*, *27*(4), 301–315.

Sillitoe, P., & Marzano, M. (2022). Ethnobotanical knowledge documentation in Papua New Guinea: Digital approaches and indigenous participation. *Journal of Ethnobiology and Ethnomedicine*, *18*(1), 1–20. PMID: 34980177

Smith, L. T. (2021). Decolonizing methodologies: Research and indigenous peoples in the digital age. *Qualitative Inquiry*, *27*(1), 141–156.

Tauli-Corpuz, V. (2022). Indigenous networks for knowledge exchange: Global experiences and lessons learned. *International Indigenous Policy Journal*, *13*(2), 1–25.

Terchilova, E. (2022). Intellectual property rights and indigenous knowledge: Identifying incompatibilities and potential solutions. *Journal of Intellectual Property Law*, *29*(1), 67–94.

Verran, H., Christie, M., & Anbins, B. (2021). Ontological transparency and indigenous knowledge representation in digital environments. *Information Communication and Society*, *24*(5), 627–643.

Weibel, S. (2022). Dublin Core and indigenous knowledge: Towards culturally sensitive metadata. *International Journal of Metadata, Semantics and Ontologies*, *17*(1), 1–15.

Wendland, W. (2022). Protecting traditional knowledge: The WIPO Intergovernmental Committee on Intellectual Property and Genetic Resources, Traditional Knowledge and Folklore. *International Journal of Cultural Property*, *29*(2), 117–136.

Wilson, K., Bowles, N., & Bell, L. (2021). Indigenous-led digital heritage initiatives: Lessons from the Indigitization project. *Information & Culture*, *56*(3), 217–238.

Zega, A., Ndraha, A. B., Laoli, D., Zebua, R. D., & Telaumbanua, B. V. (2025). Designing and managing deep sea biodiversity. In *Technological advancements for deep sea ecosystem conservation and exploration* (pp. 99–128). IGI Global Scientific Publishing.

KEY TERMS AND DEFINITIONS

Indigenous Knowledge Systems (ICS): Knowledge passed down from generation to generation within an indigenous community that reflects their understanding of the environment, culture and social life.

Digital Transformation: The process of applying digital technologies to improve or change the way a system works, in this context related to the documentation and preservation of IKS.

Blockchain: A decentralized technology that enables secure and immutable data storage, used to protect intellectual property rights over IKS.

Artificial Intelligence (AI): Technology that enables machines to learn, understand and make decisions based on given data, potentially used in the management and classification of traditional knowledge.

Intellectual Property Rights: Legal rights that protect the ownership and use of intellectual works, including documented traditional knowledge.

Community Engagement: The active participation of local communities in documenting, managing and reusing their traditional knowledge through digital technologies.

Digital Technologies: Various computer-based tools and systems used to document, store and manage traditional knowledge data.

Safeguard Policies: Strategies and regulations implemented to ensure the sustainability and legal protection of IKS.

Sustainability: Efforts to maintain the continued use and conservation of IKS in the long term through digital approaches.

Socio-economic Impact: The influence of IKS digitization on the social and economic life of indigenous communities.

Technological Innovation: The development and application of new technologies to improve the efficiency of IKS documentation and preservation.

Cultural Conservation: Efforts to maintain, preserve and transmit culture and traditions from generation to generation through digital methods.

Chapter 7
The Role of Kwa-Khaya Lendaba Cultural Village in the Preservation and Dissemination of Indigenous Knowledge

Kgakgamatso Suprice Boikanyo
https://orcid.org/0000-0003-4673-4926
North-West University, South Africa

Mothusiotsile Edwin Maditsi
https://orcid.org/0000-0001-5796-0174
North-West University, South Africa

Motheo Thomas Koitsiwe
North-West University, South Africa

ABSTRACT

Cultural villages have existed in Africa for many years, mainly for tourist attractions and for economic benefits. Many cultural villages deteriorate, are marginalized and vandalized by people who fail to understand their educational value, as well as to appropriately conserve the Indigenous Knowledge (IK) Contained within Cultural Villages (CVs). Therefore, the management and preservation of this knowledge is critical in academia. This research aims to underpin how Kwa-Khaya Lendaba (KKL) cultural village preserves, manages and promotes the valuable IK that it embodies including its dissemination using databases at libraries. The data was generated using in-depths interviews and the Knowledge, Attitude, and Perceptions (KAP) Likert scale was used to collect data on local people's perceptions of IK, CV, and management strategies at the cultural village. The data was analyzed using thematic analysis and Statistical Package for the Social Sciences (SPSS) for quantitative data. The study's findings revealed that cultural villages play a vital role in the community development and that CVs depict diverse African cultures. However, the knowledge behind the cultural village is not documented, preserved and managed.

DOI: 10.4018/979-8-3693-7964-6.ch007

INTRODUCTION

South Africa boasts a rich cultural diversity; it is a vibrant multicultural society made up of various ethnic groups with different traditions and customs. Tisdell (2005) argues that human perceptions of the living world are shaped by their cultural understanding of what this world entails, and the human impact on the environment is also informed by this cultural perspective. Activities within local communities contribute to the deterioration of numerous cultural villages, which are not always sufficiently safeguarded by the local population. The Indigenous knowledge produced and upheld in these cultural villages is regarded as a reflection of the societal cultural dynamics. Zimu-Biyela (2016) argues that Indigenous Knowledge (IK) has gained momentum due to its significance in numerous aspects of life, such as poverty alleviation, healthcare, natural resource conservation, decision-making, and problem-solving at local, national, and global levels. Indigenous Knowledge (IK) serves as a crucial strategic resource for development in the lives of a large portion of the world's population, enhancing its value. Indigenous knowledge is anchored in specific locales and the experiences cultivated by the people inhabiting those areas. It is transmitted orally or through imitation and demonstration. Nevertheless, despite the oral transmission of this knowledge, much Indigenous knowledge is lost owing to the encroachment of foreign technologies and developmental concepts that promise immediate benefits and solutions to challenges but lack the capability for sustainability.

Cultural villages play a crucial role in local communities, and they are tourist destinations in rural areas that display traditional homesteads, cultural performances, and local crafts, such as traditional dance and music, as well as restaurants and other live cultural displays. These cultural villages typically cater for international tourists, offering an experience of local culture. In contrast, some domestic tourists visit cultural villages as part of educational and cultural tours and learn more about their own cultures; the primary demand comes from foreign visitors seeking an authentic cultural experience as part of all-inclusive tours (Hughes, 2003). However, indigenous knowledge found in cultural villages is lost due to a lack of documentation. As a consequence of this loss, cultures are diluted, and knowledge cannot be preserved. UNEP (2008) confirms that this is due to loss of knowledge over time because the knowledge is shared orally and partially documented or published in the form of books or journals. The stark absence of research completed on Credo Mutwa's cultural villages suggests that such knowledge is not documented and could get lost. When such knowledge is lost because of a lack of preservation and documentation, then the younger generation cannot access the history, traditions, heritage, customs, and practices. Many cultural villages are also dilapidated, and the IK behind these cultural villages is not documented, archived, or managed.

Zimu-Biyela (2016) states that Indigenous Knowledge (IK) is tacit and embedded in the holder's mind. The preservation and continuity of IK have been affected by several factors like migration, urbanisation, limited or lack of social cohesion, death of the knowledge holder and limited or lack of documentation. Nzewi (2010) argues that certain aspects of the IKS might have been lost through oral transmission, notwithstanding the very crucial aspects of Africa's cultural heritage have been preserved. Preserving Indigenous Knowledge at cultural villages is crucial as these institutions are not only for tourist attractions but also for educational value. Cultural villages portray cultural life; in many cases, these villages are spaces where certain people would never encounter a specific culture or part of history. The culture itself and the employees of the cultural village possess various levels of IK (Mearns et al., 2006).

The chapter's objectives are as follows:

- To document Indigenous Knowledge and practices found at Kwa-Khaya Lendaba cultural village (KKL).
- Identify the Knowledge Management tools and methods used to preserve, manage and disseminate Indigenous Knowledge at the KKL cultural village.
- Assess the challenges and opportunities of Indigenous Knowledge Management at Kwa-Khaya Lendaba cultural village.
- Explore the Knowledge, Attitudes, and Perceptions (KAP) of the local people and visitors regarding the use of the KKL cultural village and how this cultural village contributes to the people's lives.

The research questions are as follows:

- Which Indigenous Knowledge practices are practiced at the KKL cultural village?
- Which knowledge management tools and methods are used to manage, preserve, and disseminate Indigenous Knowledge at the KKL cultural village?
- What are the challenges and prospects of managing and preserving the Indigenous knowledge at KKL cultural village?
- What are the locals' and visitors' attitudes and perceptions towards the KKL cultural village and the preservation of Indigenous Knowledge?

THEORETICAL FRAMEWORK

The theory underpinning the study is the Indigenous Wholistic theory. Indigenous people recognise a spiritual connection with the environment or land (Chilisa, 2012), wherein the spiritual is "one's internal sense of connection to the universe," which may include one's personal connection to a higher being or humanity or the environment. The research is mixed-method research, integrating qualitative and quantitative, so that the researcher generates empirical data based on the study.

Indigenous Wholistic Theory

Kathy Absolon developed the Indigenous Wholistic theory in 2010, encompassing the spiritual, emotional, mental, and body aspects of being. The past, present, and future are fully recognised. It is holistic and multi-layered. The theory is relevant to this study because it looks at the whole person, not just their mental health but their emotions, physical, mental, and spiritual aspects. The study examines the Indigenous Knowledge found at the KKL cultural village. Before understanding the cultural village, we need to grapple with an understanding of the progenitor of the cultural village. The study reflects on Credo Mutwa as a whole, looking at his teachings, beliefs, and spiritual, psychological, and emotional orientations that he held and furthered in establishing the KKL. According to Absolon (2010), Indigenous wholistic theory is rooted in the earth and derived from the teachings of the land, sun, water, sky, and all of creation. It is a philosophical and theoretical framework that originates from indigenous knowledge systems and worldviews of the indigenous people. The theory emphasises the interconnectedness and relatedness of all living things and seeks to promote a holistic understanding of the world. Credo Mutwa believed in fully preserving African culture, religion, healing and IK practices.

The KKL cultural village depicts the history, diversity of cultures in South Africa, reality, and Credo Mutwa's relationship with the environment. The village is filled with mythical creatures and various sculptures, including Zulu chiefs, African art, culture, folklore, and architecture. It is spirit-filled and spooky but a remarkably interesting place for relaxing and reviving one's connectedness to the holistic world. The KKL cultural village was created based on the beliefs, traditions, culture, and prophecies that Credo Mutwa had. The village encompasses aspects of the late Credo Mutwa, the folklores, tales, ceremonies and rituals, artwork, paintings, and sculptures in which he found solace. He is recognised for his African indigenous teachings, religion, and extra-terrestrial encounters, as he strongly connected with the natural environment and the spirit world.

THE ROLE OF CULTURAL VILLAGE IN THE PRESERVATION, MANAGEMENT, AND DISSEMINATION OF INDIGENOUS KNOWLEDGE

Indigenous knowledge is a collection of profound, detailed, and shared beliefs and rules about physical resources, social norms, health, ecosystems, culture, and the livelihood of people interacting with the environment in rural and urban settings. It has served as the foundation for community decision-making in agriculture, health care, and other fields (Warren, 1991). According to Mosimege (2004), cultural villages serve educational purposes rather than just being tourist destinations. A cultural village is regarded as a cultural tourism enterprise, and it is often located on or near established tourist routes in rural areas. The typical exhibit includes a homestead, an arena for live cultural performances, a restaurant, and a craft centre. Each cultural village employs some people who are knowledgeable about cultural activities and can be able to demonstrate them to visitors. However, the knowledge in the cultural village must be preserved and managed through academic libraries and information services. Sarkhel (2017) stated that libraries should implement programs to collect, preserve, manage, and disseminate IK resources and invite IK holders to tell their stories.

Libraries can plan, gather, codify, organise, protect, use, control, and disseminate IK efficiently and effectively. A library with locally relevant content will encourage communities to use library services, particularly if they are allowed to contribute to developing the content. According to Krubu and Osawaru (2011), information and communication technology (ICT) has introduced unprecedented changes and transformations to academic libraries and information services. Local communities should be included in the library content-making, and IK should be stored in university libraries so that students can easily access Indigenous community knowledge. Libraries can be helpful in using ICTs to capture, preserve, and manage vital information on natural resource management, rituals, proverbs, myths, and ceremonies (Adam, 2007). However, capturing knowledge of proverbs, rituals, and ceremonies is challenging because some ceremonies cannot be coded, and others are deemed sacred.

Digital libraries can facilitate the transmission of culturally unique knowledge between generations. The use of digital media offers new opportunities for preserving, managing, and disseminating knowledge to South African communities. The digitalisation of IK ensures that knowledge is protected and preserved from becoming extinct and will allow communities near the cultural village or worldwide to have access to the information (Mdhluli et al., 2021). According to Podolecka (2018), Credo Mutwa strove to promote his culture by producing books titled *Africa Is My Witness and My People and* constructing 'cultural villages' for tourists. His ideas were destroyed, and he was accused of backing apartheid and altering African history, yet one of his cultural villages outside Johannesburg is now a tourist destination. Therefore, there is a need to explore and document the IK at KKL cultural village, one of Mutwa's crusades and achievements.

CONTEXTUAL SETTING

The study was carried out in the Gauteng Province, specifically in Soweto, on Bochabela Street at the Kwa-Khaya Lendaba cultural village, which is also known as the Credo Mutwa cultural village (Figure 1).

*Figure 1. Study site in Gauteng province, South Africa (**A**), Gauteng province Soweto(**B**), Soweto, central western Jabavu, Kwa-khaya Lendaba cultural village (**C**).*

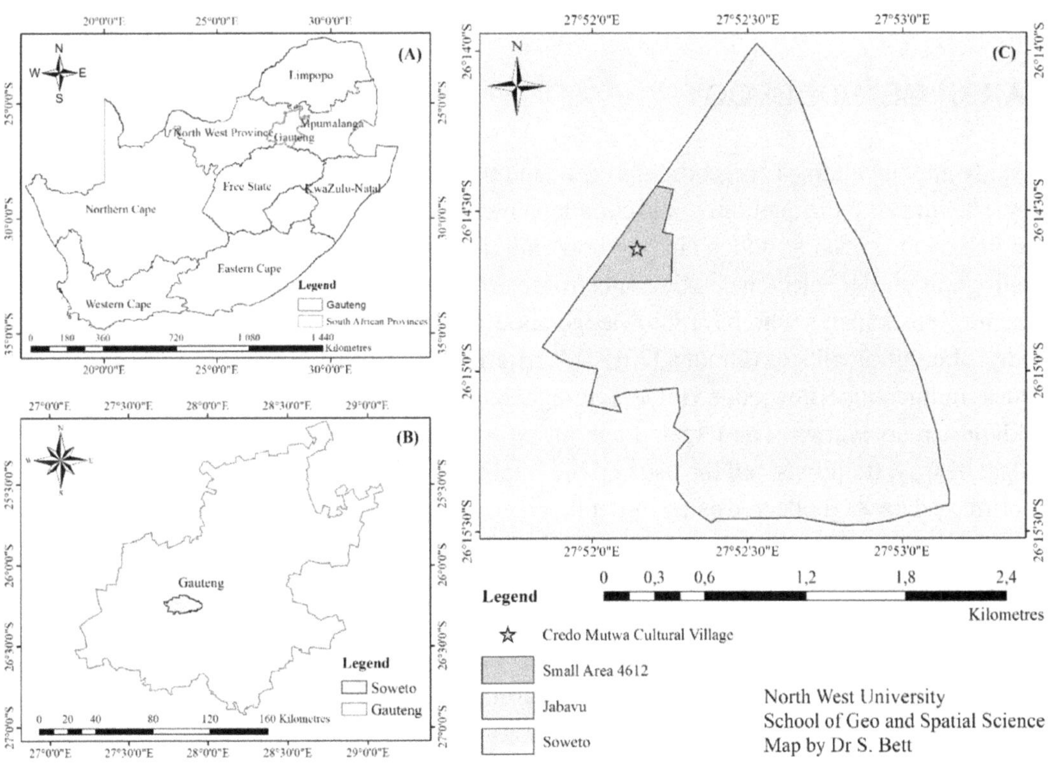

Kwa-Khaya Lendaba cultural village is one of the cultural villages created by Mkhulu Credo Mutwa. The cultural village is known as the "Home of the stories." Soweto houses surround the CV. The population of Jabavu is 45 198, 51% females and 49% males. The district has different ethnic groups, which are dominated by Zulu-speaking people (43), Sesotho (25), Setswana (8), isiXhosa (8), Sepedi (4) and Tsonga (4) people (STATS SA, 2011).

Historical Background of Kwa-Khaya Lendaba Cultural Village

Kwa-Khaya Lendaba cultural village, meaning "Home of the story", is pivotal in preserving and disseminating Indigenous Knowledge in South Africa. The cultural village was established by the well-known Credo Mutwa in 1974; the outdoor museum is located in Soweto and is a repository for African cultural practices, folklore, artefacts, and traditions. The village displays various gigantic sculptures and traditional homesteads representing multiple Southern African ethnic groups: the Nguni clan, Basotho clan and the Arabs, thus fostering an appreciation for the continent's rich cultural tapestry. The significance of Kwa-Khaya Lendaba extends beyond mere preservation as it promotes the transmission of traditional knowledge to future generations. The Kwa-Khaya Lendaba was established to instil pride in African heritage among the younger population and the community at large through folklore, storytelling sessions, and guided tours led by knowledgeable community members. However, the Kwa-Khaya Lendaba cultural village is essential in preserving Indigenous knowledge through artefacts, storytelling, and traditional practices.

RESEARCH METHODOLOGY

The study applied a mixed-methods approach, and the research design used was descriptive and exploratory. The study is more qualitative and complemented by quantitative. Ethnography is the description of social behaviour aimed at understanding how individuals perceive and explain their surroundings. Snowballing was used to select the participants from the targeted population. The participants were asked to choose new participants who have knowledge about the history, culture, traditions, and Indigenous Knowledge of cultural villages (Omona, 2013). A criterion that was used to select participants included tour guides, Indigenous Knowledge Holders, Indigenous historians, and participants who have worked with Mkhulu Credo Mutwa. The targeted age group was based on participants' understanding of the knowledge, history, traditions and the use and the importance of cultural villages.

The primary data was collected using in-depth interviews to collect data on the Indigenous knowledge practices found at Kwa-Khaya Lendaba cultural village and focus group discussions (FGDs) with cultural village members validated the data collected from in-depth interviews. The data was collected using both local languages: IsiZulu and English. The participants responded in Setswana, Sesotho, IsiZulu, English and Tsotsitaal. The in-depth interview guide was divided into two sections: Section A focused on the demographic details of the participants, such as name, gender, age, and level of education; Section B focused on the Historical importance of Indigenous knowledge and practices found at Kwa-Khaya Lendaba cultural village. The focus group discussion schedule was divided into two sections: Section A the names of the participants, and Section B focused on the history, traditions, knowledge management tools and methods used in the management and preservation of indigenous knowledge and the challenges they faced to preserve and manage IK found at Kwa-khaya Lendaba. Quantitative data (Likert scale), The Knowledge, Attitude and Perception (KAP) questionnaire were utilised to collect data about the participants' knowledge, perception, and attitude towards the use of cultural villages and preservation. The participants who had been to cultural villages or heritage sites were purposively selected. The questionnaire contained closed-ended questions. The participants were offered a choice of four pre-coded responses: strongly agree, agree, uncertain and disagree. The in-depth data was analysed using thematic analysis, transcribing, organising, coding, and identifying themes, and SPSS was used to analyse the

KAP Likert scale data, graph, and chart to present the data. The study aimed to identify different preservation methods that can be used to preserve, manage, and disseminate Indigenous knowledge found at Kwa-Khaya Lendaba cultural village and how students and the community can access the knowledge via ICTs and Libraries. Participants included the members of the cultural village, knowledge holders, caretakers, and tour guides.

RESULTS AND DISCUSSION

Demographic Details of Participants

Table 1. Demographic characteristics of participants at Kwa-khaya Lendaba cultural village

Name of Participants	Age Group	Code
Curator/Knowledge holder/ Tour guider	40-45	Participant 1
Knowledge holder/curator/ tour guider	45-50	Participant 2
Knowledge holder Caretaker	45-55 30-35	Participant 3 Participant 4
Tour Guide	35-40	Participant 5

The IK and Practices Found at Kwa-Khaya Lendaba Cultural Village

One of the primary functions of Kwa-Khaya Lendaba is to function as a repository for traditional practices or Indigenous knowledge practices. The cultural village hosts activities and workshops that educate learners and visitors about the skills we found at the site, such as painting, pottery, music, dance, stories, and teachings on traditional healing (traditional medicine) and midwifery. According to participants, traditional medicine was the primary health care before going for Western medicine. Most individuals in local communities prefer traditional medicine rather than conventional medicine. Below are the IK and practices found at Kwa-Khaya Lendaba cultural village:

Conflict Resolution and Peacebuilding

The cultural village features a historic house known as the Mother Nature House, which served as a conflict resolution centre for women in the community. It was specifically used to settle disputes between two women who were not getting along. The underlying principle is rooted in the African cultural belief that women are nurturers tasked with building peaceful homes and maintaining respectful relationships.

This idea is encapsulated in a Sotho idiom, "*Mmangwana o tshwara thipa ka bohaleng*", which means "the mother of a child holds the dagger on the sharper edge. "According to Mhandara (2020), reconciliation is a crucial mechanism used for ending conflict or violence and fostering a new relationship that benefits both the perpetrators and victims. This process promotes healing, rebuilds relationships and maintains mutual respect among community members, including both the perpetrators and victims. Ultimately, this approach aims to promote equality, not only among women but also among men, to create a harmonious society.

Traditional Healing and Midwifery

Figure 2. The KKL cultural village surrounded by attractive and vast trees, used for traditional medicine

Traditional healing practice is holistic and focuses on the individual's general well-being. It places the body, ego, and society within a dynamic equilibrium framework. The holistic method considers a person's ideals, passions, beliefs, social interactions, and spiritual orientation (Dalal, 2011). The Kwa-Khaya Lendaba cultural village is known for its traditional healing. Credo Mutwa was a Sanusi healer and prophet. He is well-known for his spiritual gifts and belief in traditional teaching, folklore, dance, and healing. The cultural village features a fertility/labour room where women receive traditional fertility treatment to conceive. This same room served as a prenatal care facility, providing traditional antenatal treatment and addressing other gynaecological services. According to the participant, the fertility room is built in a round shape, which symbolises women. The design serves as a tribute to the womb and womanhood. The other participants alluded that according to what they were taught, the first person to walk on earth was Ma, known as Nomkhubulwane. She was ordered to come to earth and cultivate the earth as it was rocky. It is believed that as the earth produces, so is a woman. Both mother and child were protected in the fertility room. Men were not permitted to see their infants. This practice is still used in some cultures, where a father should wait for a week to see his infant, as the father's parent only visits the child to confirm the child's identity if the child truly belongs to the clan and delivers some gifts.

Furthermore, Kwa-Khaya Lendaba is instrumental in fostering cultural pride among local communities while simultaneously promoting awareness among visitors through guided tours, storytelling sessions and ceremonies led by knowledgeable practitioners such as Sangoma, known as traditional healers, tour guides and knowledge holders. Attractive enormous trees surround the village. All the trees in the cultural village are medicinal plants used to extract herbal medicine. Most traditional healers use traditional medicine to cure diseases and illnesses by collecting tree bark and herbs, which are mixed to create traditional medicine. For years, across the countries, people have resorted to traditional healers, home remedies, and historical medicinal knowledge to meet their well-being and health issues (WHO, 2023).

Figure 3. The inside and outside of fertility room found at Kwa-Khaya Lendaba cultural village

THE USE OF CULTURAL VILLAGES AS A TOOL TO PRESERVE AND MANAGE INDIGENOUS KNOWLEDGE

Kwa-Khaya Lendaba cultural village is used as an entity to preserve IK through Ceremonies, oral tradition, and apprenticeships. Kwa-Khaya Lendaba uses ceremonies to preserve Indigenous knowledge, ceremonies that include training, initiation ceremony and childbirth. Credo Mutwa was known as a Sanusi, and he trained numerous traditional healers known as "*amathwasana.*" Oral tradition was and is still used for cultural preservation. It provides insight into how people entertained themselves through centuries and understanding their physical world (Foley, 2012). Oral tradition at Kwa-Khaya Lendaba is transmitted orally from one generation to another through speech, ancestral utterance, traditional medicine, dance, songs, proverbs, poetry, storytelling and folklore. Oral traditions are used to teach morals, cultural values, and the importance of identity, knowledge and belonging to a specific ethnic group. According to Lilomaiava-Doktor (2020), oral traditions are believed to be the heart of African Indigenous cultures; they are a part of Africans way of sharing knowledge, cultural heritage, history, and beliefs; she further states that oral tradition is mainly shared through storytelling and proverbs. Kwa-Khaya Lendaba transmits their oral tradition through ceremonies, storytelling, folklore, training, and cultural heritage events. They share knowledge about their history, traditions, and beliefs, and as Africans, we believe in the supreme God, the creator. Parangu and Salim (2018) attest that oral tradition is commonly inherited from generation to generation through oral literature that contains cultural values and local wisdom that characterise the area itself. The participant explained that the cultural village is not just a mere museum or a tourist attraction; it is a holy place where African culture, religions and indigenous science are preserved and recorded in pictures and sculpture forms. Below is one of the sculptures that is found at Kwa-Khaya Lendaba cultural village

Figure 4. Mother Nature seating behind Lerato, holding a spear and two boys

Preservation refers to the process of preventing or slowing deterioration (Ogunmodede, & Ebijuwa, 2013). Preservation professionals in libraries, archives, and museums share a commitment to protect information on culture in all formats, including the built environment that stores raw materials of history and evidence that feeds our memories (Parangu and Salim, 2018). Kwa-Khaya Lendaba serves as an educational platform for tourists seeking authentic experiences related to the history of the cultural village and the Indigenous culture. The guided tours give insight into historical contexts and cultural significance behind different traditions practised within the cultural village. This experience promotes awareness and encourages respectful engagement with Indigenous cultures, fostering an appreciation rather than appropriation. The knowledge found at Kwa-Khaya Lendaba is tacit; according to Adebayo and Adeyemo (2018), Indigenous knowledge and practices are usually unwritten and rely on oral transmission and human memory. Most of the cultural villages are tacit knowledge; it is informal and subjective. Sometimes, it is difficult for participants to share or express their opinions because it is affected by personal beliefs and values. Indigenous knowledge found at Kwa-Khaya Lendaba is embedded in people's minds, photos, videos, and newspapers exhibited through artefacts, and much is not documented in books. The cultural village will continue serving as a vital hub for preserving and disseminating Indigenous knowledge among local communities. However, intensive research needs to be conducted by scholars who will focus on the preservation and management of Kwa-Khaya Lendaba.

THE IMPORTANCE OF INFORMATION AND COMMUNICATION TECHNOLOGIES (ICTS) AND LIBRARIES IN KNOWLEDGE MANAGEMENT

Sarkhel (2017) states that libraries should implement programmes to collect, preserve, manage, and disseminate IK resources and invite IK holders to tell their stories. Libraries could plan, gather, codify, organise, protect, use, control, and disseminate IK efficiently and effectively. A library with locally relevant content will encourage communities to use library services, particularly if they are allowed to contribute to developing the content. According to Kiarie (2024), digital strategies such as virtual reality and digital archiving can be crucial in documenting, preserving, and promoting cultural heritage. These technologies can assist in capturing and storing detailed information about cultural artefacts and practices, making it available to a worldwide audience. Furthermore, digital platforms can enable virtual tours and educational courses involving younger generations and a wider audience who may not have direct access to heritage sites. Preservation efforts can be extended beyond physical bounds by harnessing digital technologies, ensuring that cultural heritage is valued and preserved worldwide, especially in the face of globalisation challenges.

Knowledge management in universities refers to the systematic strategy of capturing, structuring, managing, and disseminating knowledge throughout the university (Jumba et al., 2020). However, the impoverished rural people who lack access to Information and Communication Technologies (ICT) and own IK may be marginalised once more (Sarkhel, 2017). ICTs have significance in all areas of national life, including politics, economics, and social and cultural development. It is rapidly changing how people conduct business, obtain information and services, communicate with one another, and share ideas (Jumba et al., 2020). The participant suggested that the best way to preserve and disseminate IK found at Kwa-Khaya Lendaba is through libraries with ICTs, where the students can easily access the information.

CHALLENGES AND OPPORTUNITIES IN THE PRESERVATION AND MANAGEMENT OF IK AT KWA-KHAYA LENDABA Cultural Village

The challenge of preserving and managing IK at Kwa-Khaya Lendaba cultural village is that the knowledge is orally shared and is embedded in people's memories. It is not documented, and it can be easily forgotten. The challenge is that most of the knowledge is tacit knowledge, experiences, personal beliefs, personal encounters, extra-terrestrial encounters, teachings, and values of Credo Mutwa, which was presented as sculptures and not writing. However, some tacit knowledge can be recorded, documented, and easily accessed by researchers and libraries for preservation. There are, therefore, extensive opportunities for documenting, recording, and preserving IK at Kwa-Khaya Lendaba cultural village. Libraries could be extended in the versions of recordings and transcriptions here so that students can easily access IK found at the KKL cultural village. Okonkwo (2017) states that shortage of funding, lack of public awareness, and the illegal trafficking of cultural artefacts are some of the challenges of preserving and managing Indigenous Knowledge. However, while Kwa-Khaya Lendaba cultural village is essential in preserving Indigenous knowledge, its impact is intricately tied to broader socio-economic factors. Heritage tourism's growth presents opportunities and challenges for cultural sites like Kwa-Khaya Lendaba cultural village. However, leveraging international collaborations and community-based initiatives can help to strengthen preservation efforts. Increasing public awareness campaigns, enhancing legal frameworks, and fostering international partnerships to combat the illicit trade of cultural artefacts (Okonkwo, 2017).

Community participation and empowerment are essential for practical cultural heritage preservation in the face of globalization. Local communities are the custodians of their tradition and cultural heritage, possessing invaluable knowledge and practices that are essential for authentic and sustainable preservation. Engaging communities in heritage preservation initiatives ensure that preservation efforts are culturally essential and respectful and strengthens community identity and cohesion. Empowered communities can lead grassroots preservation efforts, document management, and revive traditional practices, as well as promote the protection of their heritage. Policies and programs that support community leadership, provide resources for local initiatives, and foster partnerships between communities and external stakeholders are essential for successful cultural heritage preservation (Kiarie, 2024). The preservation and management of Indigenous knowledge and cultural heritage are crucial for Indigenous communities' survival and development. However, according to Kamwi (2017), there are significant challenges to overcome, including language barriers, cultural appropriation, limited resources, climate change and aging population. Despite these challenges, there are opportunities for the effective prevention and management of cultural heritage and Indigenous knowledge. UNESCO (2028) emphasizes the importance of community-led initiatives, education, and digital preservation. By leveraging these opportunities, Indigenous communities can take control of their cultural heritage ensuring its preservation for future generations and promote cross-cultural understanding, morals, and respect.

Table 2. The demographic details of participants for Likert scale

Feature	Frequency (n)	Percentage (%)
Age group		
18-35	19	48
36-49	15	36
50-65	6	16
Formal Education		
Knowledge holders		20
Indigenous historians	8	15
High school	6	30
Secondary	12	15
Tertiary	6	20
	8	
Gender	22	55
Male		45
Female	18	

The KAP Likert Scale had 40 male and female participants, with 22 males (55%) and 18 females (45%). The 40 participants were chosen for the study because they had previously visited cultural villages, museums, and heritage sites and understood Indigenous knowledge. The age group ranges from 18-35 (48%) youth, youth adults 36-49 (36%) and adults 50-65 (16%). Snowballing was used to select who would participate in the study. 20% of the participants had a tertiary level of education. They included Indigenous historians 15%, secondary 15%, high school 30% and IK holders 20%. The participants were asked several questions regarding the use of cultural villages, the Indigenous knowledge found in cultural villages, the importance of preserving and managing Indigenous Knowledge through libraries and ICTs, and the role of communities in documenting and preserving IK. Figure 5 shows participants' responses regarding the significance of preserving Indigenous Knowledge in cultural villages. 80% of the participants strongly agreed that IK at the cultural villages is vital and should be preserved for future use. They state that CVs play a significant role in education as people learn about Indigenous knowledge, culture, and traditions. 20% of the participants agreed, though stating that, as much as it is essential, not much is done about preserving the IK. Most of the IK found in cultural villages is preserved through media, photos, and video and is not documented in books.

Figure 5. The response of the participants on the importance of the preservation of IK in cultural villages

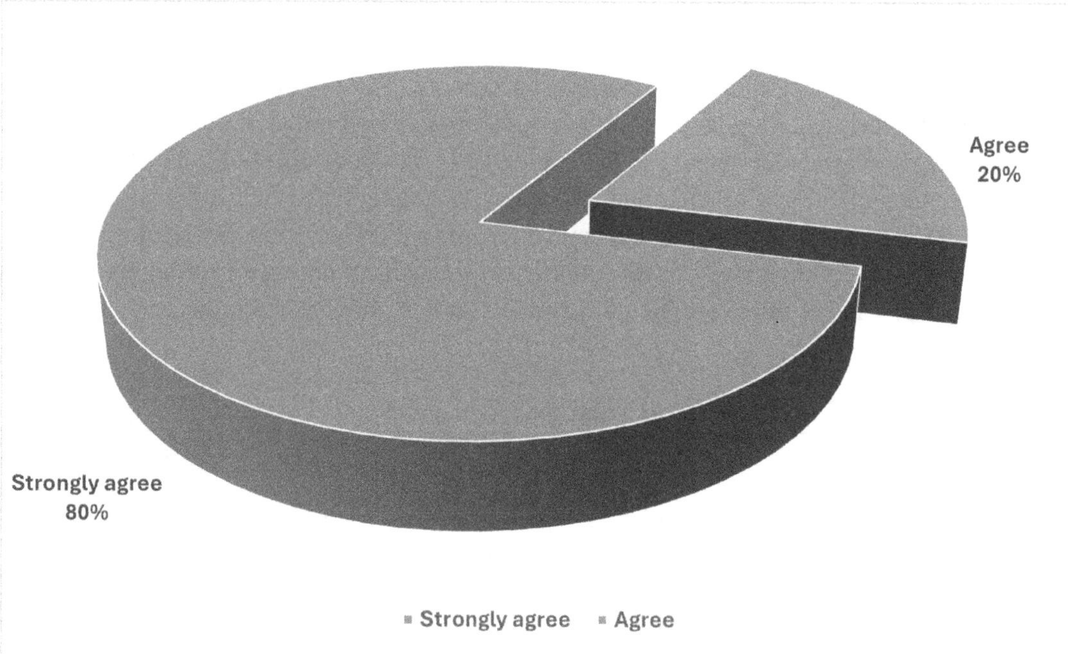

Figure 6 indicates that 45% of participants opposed the notion that communities contribute to preserving IK in cultural villages, indicating that most cultural villages are privately owned and protected by the owners. The community can only help in terms of research and documentation of knowledge. 40% agreed, stating that they help by writing academic papers, media, and social platforms. 15% are not sure about the contribution of the community in preserving the IK in cultural villages because caretakers undertake this. 10% strongly agreed that the communities do help a little, and the community needs to be educated about the importance of CVs in society. 80% of the participants expressed their satisfaction with the existence of cultural villages because they discovered and learned new information, particularly regarding the culture of many ethnic groups. 20% agreed that they are happy that CVs exist; they are important for leisure, to get closer to nature and the aura, and to be revived.

Figure 6. The response of the participants on the contribution of communities to preservation of IK at cultural villages

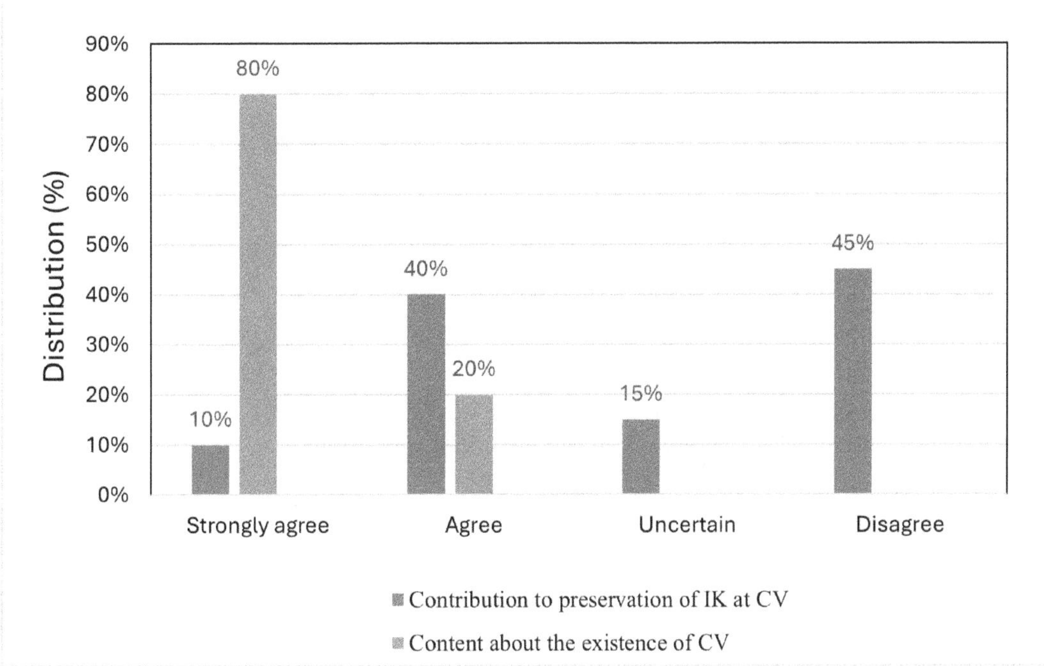

Figure 7 shows that 90% of participants contended that Indigenous knowledge in cultural villages would not be accessible in the forthcoming years without preservation, asserting that the absence of documentation, preservation and management would result in the loss of knowledge, thereby denying future generations access to it. So, it is imperative to document IK found in cultural villages to make it readily available in future. 10% of the participants indicated that they are unsure if the information would be available without proper documentation, especially when it is stored in the people's minds.

Figure 7. The participants response on the availability of IK at cultural villages without preservation

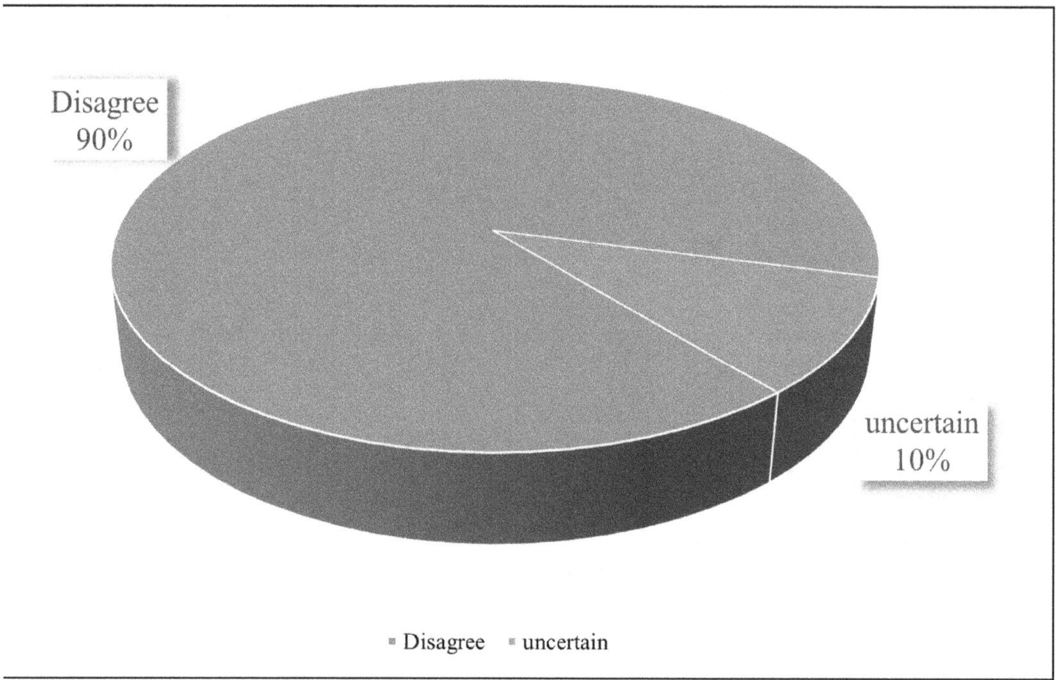

Figure 8 shows that 50% of the participants strongly agreed that cultural villages preserve IK through sculptures, art, paintings, buildings, and music. 40% of the participants consented that cultural villages do not fully preserve IK ultimately, and the information needs to be recorded. 10% of the participants disagreed, claiming that using cultural villages does not guarantee the preservation of IK entirely due to their dilapidation. Unless the IK is documented, there is no guarantee that the IK will be preserved for future use.

Figure 8. The participant's response on the use of cultural villages as preservation of IK

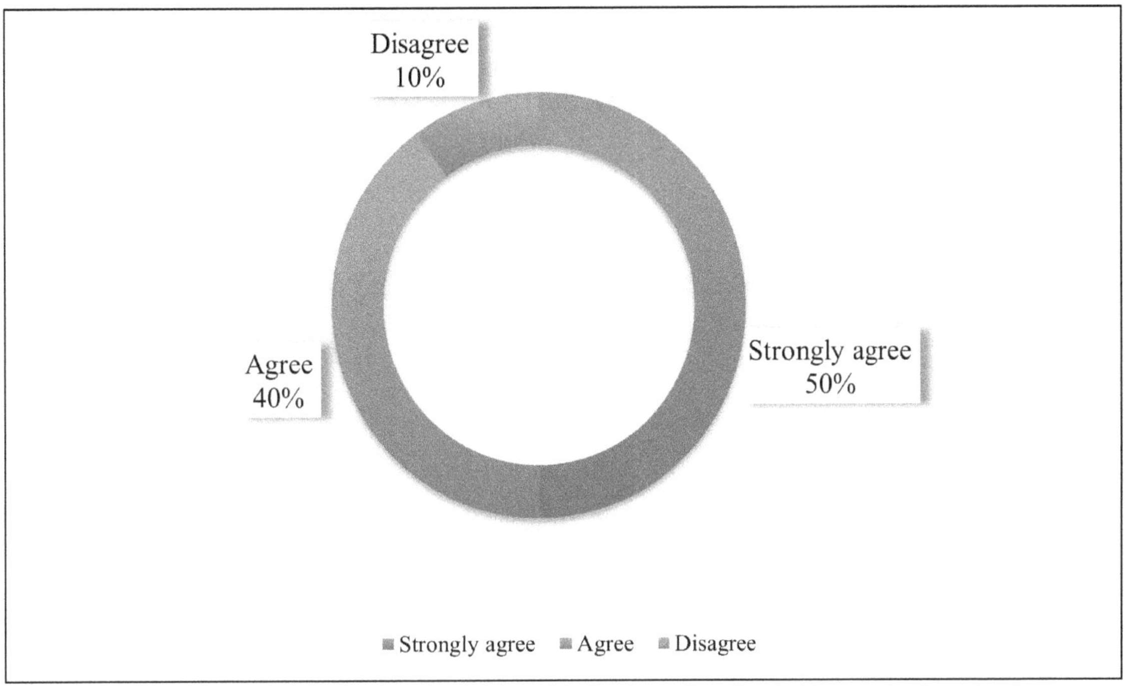

The results of the KAP Likert scale questionnaire demonstrated the importance of documenting, preserving, and managing IK found in cultural villages. Engaging communities in the preservation of IK in cultural villages is critical. Sustainable preservation programs will necessitate a collaborative approach emphasising education, respect for tradition and community involvement. Cultural heritage preservation through workshops, education and community engagement ensures that future generations inherit a rich tradition and identity.

ETHICAL CONSIDERATIONS

This research adhered to the ethical guidelines of the Faculty of Natural and Agricultural Sciences (FNAS) Research Committee. The researcher presented her proposal to the research committee and obtained a research ethics certificate. The researcher obtained ethical clearance from the Faculty of Natural and Agricultural Sciences Ethics Committee (FNASREC) at North-West University, N W U-0 1 2 7 5 - 2 2 - A 9.

- ***Access and Permission***

The village leaders granted permission to conduct research at Kwa-Khaya Lendaba (KKL) cultural village. A letter from the Indigenous Knowledge Systems (IKS) centre detailing the researcher's information and the study's purpose was provided for data collection.

- *Informed Consent*

Consent to collect data at the KKL cultural village was obtained by telephone. Participants were contacted, requested to participate, and informed about the research purpose and location. Interviews were scheduled based on participants' availability, and Informed consent was granted. Participants were assured of their right to participate willingly and to withdraw without consequences.

- **Confidentiality and Anonymity**

A non-disclosure clause was presented to participants to ensure confidentiality and anonymity. Although participants did not sign the consent form, they verbally agreed to record, photograph, and disclose their names. Key participants consented to be mentioned in the study.

- *Cultural Protocol*

The researcher adhered to the cultural protocol established at KKL cultural village, respecting values, customs, norms, and cultural practices. Recognizing the village as sacred, the researcher honoured its protocols by not littering and respecting the land's sanctity.

FUTURE RESEARCH DIRECTIONS

The study will contribute to the work of Credo Mutwa, and it will add to the body of Indigenous knowledge. The study is timely as it strives to add value to Credo Mutwa's work, demonstrating the relevance of preserving IK for teaching and learning. The study assists in keeping Credo Mutwa's legacy alive. The purpose of the study is to document, manage and disseminate Indigenous knowledge discovered at Kwa-Khaya Lendaba cultural village, as well as to avoid the extinction of Credo Mutwa's knowledge that faces demise because there is no documentation or record information on his cultural villages. As a future study, the study will also contribute to the community through researched and well-documented information that will be available through the digitalised libraries and serve as a resource for learning. Documenting IK will be beneficial for future generations and the community at large.

CONCLUSION

Kwa-Khaya Lendaba is crucial in safeguarding Indigenous knowledge systems while facilitating their dissemination through educational initiatives. As a living testament to Africa's rich heritage, it preserves artefacts and cultivates a deeper awareness of cultural diversity among its visitors. However, external difficulties associated with tourism and globalisation may compromise preserving Indigenous African knowledge efficiency. Continued support from both local communities and national policies will be essential for maintaining the integrity and authenticity of such cultural institutions. Kwa-Khaya Lendaba Cultural Village stands as a house for preserving Indigenous knowledge in South Africa. Its commitment to maintaining cultural heritage through education and community engagement ensures that future generations can access rich knowledge. Thus, it highlights the importance of localised efforts in safeguarding diverse cultures against threats posed by global homogenisation. It plays an indispensable role in preserving Indigenous knowledge through its commitment to maintaining traditional practices and history, promoting language education, and providing educational opportunities for visitors. As globalisation still threatens cultural identities worldwide, initiatives like those at Kwa-Khaya Lendaba become increasingly crucial in safeguarding heritage for future generations. African teachings, religion, art, and culture are among the most critical aspects of a cultural village because they help Indigenous people re-awaken from their colonised minds and break free from mental bondage and slavery while also preserving Africa's past. Additionally, the purpose is to assist Indigenous People in understanding their identity, values, customs, and culture and recognise and comprehend their sense of belonging.

SOLUTIONS AND RECOMMENDATIONS

African religion and teachings need to be included in school curricula so that students can receive both formal and informal education that could assist them in understanding their culture's origins, values, and traditions. The data collected and analysed illustrated that the knowledge and teachings of Credo Mutwa live inside the people who have lived close to them, but much is not written in the form of books or documents. Most of the information regarding his cultural villages is found in the form of videos, pictures, and the internet. Henceforth, it is vital that the IK of the KKL cultural village should be recorded, preserved, and managed through Indigenous knowledge management tools to benefit society. The participants in the study indicated that CVs are essential in local communities. They display the diversity of cultures and art. They are institutions for communities to learn about African knowledge, which is not taught in the classroom. The majority of the participants revealed that it is vital that the IK of cultural villages be preserved through IK management systems and ICTs to avoid loss of knowledge.

The study also confirmed that awareness is needed to educate people about the importance of cultural villages, especially those around the cultural villages. Sustainable preservation programs will necessitate a collaborative approach emphasising education, respect for tradition and community involvement. This would help maintain and understand them as institutions of learning and not only as tourist attractions. The study verified the challenges and prospects of preserving and managing knowledge at the KKL cultural village. The knowledge is shared orally, preserved through the collective memory of the people, and not in a written form, which exposes the knowledge to distortion and disruption. However, the opportunities are that the knowledge is still available and shared amongst local people. Information is available through videos, pictures, and audio recordings of the late Credo Mutwa and his cultural villages. The study further established that the extent to which IKS are preserved in cultural villages is poor. The best place to preserve and manage IK is through libraries at the institution, through ICTs and databases so that the information can be accessible to the students at universities worldwide, and sustainable preservation programs will necessitate a collaborative approach that emphasises education, respect for tradition and community involvement.

REFERENCES

Absolon, K. (2010). Indigenous wholistic theory: A knowledge set for practice. *First Peoples Child & Family Review*, *5*(2), 74–87. DOI: 10.7202/1068933ar

Adam, L. (2007, April). Information and communication technologies, knowledge management and Indigenous knowledge: Implications to livelihood of communities in Ethiopia. In *a Workshop on" The Role of ICT in Preserving and Disseminating Indigenous Knowledge", Addis Ababa.*

Adeyemo, A. A., & Adebayo, J. O. (2018). Documentation and Dissemination of Indigenous Knowledge by Library Personnel in Selected Research Institutes in Nigeria. Library Philosophy & Practice.

Chilisa, B. (2012). *Indigenous research methodologies*. University of Botswana: SAGE.

Dalal, A. K. (2011). Folk wisdom and traditional healing practices: Some lessons for modern psychotherapies. *Foundations of Indian psychology, 2*, 21-35.

Foley, J. M. (2012). *Oral tradition and the Internet: Pathways of the mind*. University of Illinois Press. DOI: 10.5406/illinois/9780252037184.001.0001

Hughes, H. (2003, December). What is remembered and what forgotten: a decade of redefining culture and heritage for tourism in South Africa. In *Conference on Developing Cultural Tourism,* University of Nottingham, Nottingham, UK (pp. 470-479).

Jumba, H., Bundi, D., & John, J. (2020). Use of ICTs in Knowledge Management for Enhanced Institutional Sustainability. *Regional Journal of Information and Knowledge Management*, *5*(1), 1–17. DOI: 10.70759/kpbgc061

Kamwi, J. (2017). Indigenous knowledge and cultural heritage: Challenges and opportunities for preservation and management. *Journal of Cultural Heritage*, *23*, 1–9.

Kiarie, E. (2024). *Cultural Heritage Preservation in the Face of Globalization. International Journal of Humanity and Social Sciences*. Catholic University of Eastern Africa.

Krubu, D. E., & Osawaru, K. E. (2011). The impact of information and communication technology (ICT) in Nigerian university libraries. *Library philosophy and Practice, 2011*(583), 1-19.

Lilomaiava-Doktor, S. I. (2020). Oral traditions, cultural significance of storytelling, and Samoan understandings of place or fanua. *Native American and Indigenous Studies*, *7*(1), 121–151.

Mdhluli, T. D., Mokgoatšana, S., Kugara, S. L., & Vuma, L. (2021). Knowledge management: Preserving, managing, and sharing Indigenous knowledge through digital library. *Hervormde Teologiese Studies*, *77*(2). Advance online publication. DOI: 10.4102/hts.v77i2.6795

Mearns, M. A., Du Toit, A. S. A., & Mukuka, G. (2006). Conservation of Indigenous knowledge at cultural villages: An exploratory study. *Mousaion*, *24*(1), 23–50.

Mhandara, L. (2020). Community-based reconciliation in practice and lessons for the National Peace and Reconciliation Commission of Zimbabwe. *African Journal on Conflict Resolution*, *20*(1), 35–58.

Mosimege, M. (2004). Mathematical knowledge and its use in daily activities of workers at South African cultural villages. In *10th International Congress on Mathematical Education,* Copenhagen, Denmark.

Nzewi, O. E. S. (2010). *The use of performance composition on African music instruments for effective classroom music education in Africa*. University of Pretoria.

Ogunmodede, T. A., & Ebijuwa, A. S. (2013). Problems of conservation and preservation of library resources in African academic libraries: A review of literature. Greener Journal of Social Sciences, 3(1), 050-057.

Okonkwo, E. (2017). Challenges and opportunities in cultural heritage preservation in Nigeria. *Journal of Cultural Heritage*, *24*(3), 27–34.

Omona, J. (2013). Sampling in qualitative research: Improving the quality of research outcomes in higher education. *Makerere Journal of Higher Education*, *4*(2), 169–185. DOI: 10.4314/majohe.v4i2.4

Parangu, R. N. A., & Salim, T. A. (2018). Indigenous knowledge preservation of oral Literature "Hahiwang" in West Lampung. *PEOPLE: International Journal of Social Sciences*, *4*(2), 1221–1232. DOI: 10.20319/pijss.2018.42.12211232

Podolecka, A. (2018). Vusamazulu Credo Mutwa–Sangoma, African sage, fraud or a new age shaman? *Studies in African Languages and Cultures*, *52*, 131–153. DOI: 10.32690/SALC52.6

Sarkhel, J. K. (2017). Strategies of Indigenous knowledge management in libraries. *Qualitative and Quantitative Methods in Libraries, 5*(2), 427-439.Stats, S. A. (2011). Statistics south Africa. *Formal census, 1*, 79-106.

Tisdell, C. A. (Ed.). (2005). *Economics of environmental conservation* (Vol. 1). Edward Elgar Publishing. United Nations Environment Programme (UNEP). (2008) Indigenous Knowledge in Disaster Management in Africa. Nairobi, Kenya.

UNESCO. (2018). *Indigenous knowledge and cultural heritage*. UNESCO Publishing.

Warren, D. M. (1991). *Using Indigenous knowledge in agricultural development* (No. 127, p. 46pp).

Wildwood, C. (2003). *Mood Enhancing PLANTS*. Bloomsbury publishing plc.

World Health Organization. (2023). Traditional medicine has a long history of contributing to conventional medicine and continues to hold promise. *World Health Organization [Internet]. [cited 25 Feb 2024]. Available:*https://www. who.int/newsroom/feature-stories/detail/traditional-medicine-has-a-long -history-o*contributing-to-conventional-medicine-and-continues-to-hold-promise*.

Zimu-Biyela, A. N. (2016). *The management and preservation of indigenous knowledge in Dlangubo Village in KwaZulu-Natal, South Africa* (Doctoral dissertation, University of South Africa).

KEY TERMS AND DEFINITIONS

Indigenous Knowledge: It is a unique knowledge developed by local people in a specific geographical area.

Preservation: It is documenting, storing, and safeguarding traditional knowledge and practices for future generations.

Knowledge Management: It is the process of collecting, organizing and sharing information.

Dissemination: It involves sharing knowledge and practices with communities, organizations or individuals.

Library: It is a collection of resources, such as book, documents, digital materials that are organized and made available for people to access and use.

Kwa-Khaya Lendaba: It is a Zulu phrase that means "the home of stories" or "the place of story-telling" where stories, traditions, and cultural knowledge are shared and preserved.

Cultural Village: It is a physical space that showcases the traditions, customs and practices of a particular culture or community. Its often a tourist destination or educational centre where visitors can learn about and experience the local culture.

Chapter 8
Using Biocultural Community Protocols to Preserve and Protect Valuable IK:
A Course to Develop an Extra Layer of Rights for Two Indigenous Communities

Mothusiotsile Edwin Maditsi
https://orcid.org/0000-0001-5796-0174
North-West University, South Africa

Modjadji Mandy Rasehlomi
https://orcid.org/0000-0003-4736-9238
North-West University, South Africa

Thato Hazel Seemise
https://orcid.org/0009-0003-1212-3884
North-West University, South Africa

Kamogelo Macdonald Noge
https://orcid.org/0000-0003-4203-3153
North-West University, South Africa

ABSTRACT

Indigenous Knowledge (IK) has been faced with various challenges such as piracy and overall abuse of the rights of indigenous peoples towards their resources. This has led to communities being skeptical about their space and who to share their IK with This presented various problems leaving this IK at the periphery of the knowledge economy. It is pivotal to dwell into strategies that seek to preserve, protect IK as a resource which communities rely on for survival. This study applied participatory action research with an indigenous paradigmatic lens. The study focused on two Khoisan communities to unearth how this can benefit from developing their own Biocultural Community Protocols (BCPs). The data collected were analyzed using thematic analysis. The study finds that the development of BCPs will ensure that communities manage their IK thereby protecting and strengthening the intricate linkages between the

DOI: 10.4018/979-8-3693-7964-6.ch008

cultural systems of knowledge and the management of land and natural resources. It is concluded that BCPs are a relevant resource to ensure that IK is protected, promoted and managed to drive sustainable development.

INTRODUCTION

One of the most significant current discussions in the field of indigenous knowledge (IK) is its protection, preservation and overall management of this resource for sustainable development, especially in rural communities. Budeli et al (2022) provides that IK is a systematic body of knowledge that encompasses skills, philosophies, techniques and experiences which are developed by a close group of people, mostly those who share a common ethnicity or culture. It has therefore become increasingly difficult to ignore how this knowledge system has been marginalized and in turn suffering from possible erosion and dilution due to the transcendence and domination of Eurocentric epistemologies, pedagogies and knowledge economy at large. Biyela et al (2016) posits that the marginalization of IK has been publicized by various scholars across the globe as well as Africa and this is evidence that that there is a huge concern regarding the preservation and protection of valuable IK to drive various developmental sectors.

Mdhluli et al (2021) contends that IK is the primary property of indigenous communities, and this knowledge is packaged and shared during appropriate circumstances and under specific standards. These indigenous communities are mostly rural and have less development because of their geographical location. Despite the safety utilization and use of this IK by these ethnic communities, this knowledge is in danger of suffering from several major drawbacks, especially in this current knowledge economy. Indigenous knowledge has received much attention from government and private sectors world over (Correa, 2001; Boven & Morohashi, 2002). It is a known fact that IK endured and continues to endure massive losses thereby leaving indigenous communities and peoples vulnerable to exploitation and other forms of misuse of their knowledge. Neglecting of IK, marginalization of indigenous practices and beliefs have also perpetuated this loss of IK.

Furthermore, South Africa has become signatory to various international treaties and legal prescripts and has further developed its own legally binding documents that are set to protect this IK. This is because the country has realized the great beauty, power and utility of this underutilized knowledge which has the potential to drive sustainable development, not only in South Africa, but the world in its entirety. Western scientific knowledge which is globalized has led to IK being put in the margins of the knowledge economy even though this western knowledge has failed to sustain resource use and led to the erosion of biological diversity (IKS Policy, 2004).

The following study's objectives were formulated to:

- Review the legal instruments that seek to address the preservation and protection of IK
- Examine the success stories of Biocultural Community Protocols developed in other countries and organizations
- Explore the potential of biocultural community protocols in providing the preservation and protection of IK
- Examine the role and voice of indigenous communities in the development of BCPs

Given the above objectives, the following questions are asked:

- What are the legal instruments that seek to address the preservation and protection of IK?

- What are the success stories of BCPs developed in other countries and organization?
- How effective are BCPs in preserving and protecting IK in two South African communities?
- What is the role and voice of indigenous communities in the development of BCPs?

THEORETICAL FRAMEWORK

Self-determination theory (SDT) is applied in this study. This is a theoretical approach to human motivation and personality that uses traditional empirical methods while employing an organismic metatheory that highlights the importance of humans evolved inner resources for personal development and behavioural self-regulation (Maditsi, 2018). The theory provides that human, because of their self-motivation and personality integration, have growth tendencies that drive their innate psychological needs as well as conditions to foster positive processes (Ryan & Deci, 2009). SDT is essential for facilitating optimal functions for growth and integration, as well as for constructive social development and personal well-being. This theory is relevant to the current study because it underpins how ethnic communities were able to determine what they needed and how this was going to be of use to them. Ethnic communities were self-motivated and driven to lead their lives using local knowledge systems. Most of these communities are still self-driven under the leadership of traditional leaders, community elders and councillors.

TOWARDS THE PRESERVATION AND PROTECTION OF IK IN SOUTH AFRICA

Debate continues about the best strategies to employ in the endeavor to preserve and protect valuable IK of indigenous communities. Verhoef (2023) provides that since the late 20th century, there have been various instruments that have been developed at an international and regional level which seek to focus and dwell on the protection and preservation of valuable IK of various ethnic communities. Furthermore, Verhoef (2023) presents that the development of these various legal instruments has shined the light on the importance of IKS and of indigenous peoples and local communities alike. This recognition has grown to be received and incorporated into the development agendas of various national governments and international organizations who are now part of the collective advocating for the protection and preservation of IK. The momentum to pinpoint the relevance of IK as a strategic resource for development in the lives of the majority of the world's population (Zimu-Biyela, 2016) has strengthened the call for this knowledge to be preserved for the future generation and protected from being pirated and misappropriated by third parties who do so to make financial gain without the indigenous communities benefiting from them. Despite various instruments developed to focus and dwell on IK and IKS globally, regionally and in South Africa particularly, there have been vast challenges that have hindered the full realization of this preservation and protection of IK.

BIOCULTURAL COMMUNITY PROTOCOLS AS VIABLE TOOLS TO PRESERVE AND PROTECT VALUABLE IK

A biocultural community protocol is a community developed instrument that outlines clear terms and conditions for engaging with third parties or external actors (Norms, 2012). This includes but not limited to government departments, companies, research institutions and etc. The engagement that the BCP is mainly focused on is access to local resources, knowledge, customary laws, values and decision-making processes of communities. BCP are normally developed utilizing a participatory approach by all community members of the community to ensure that the legal document reflects the unique biocultural context and values of the community. The protocol is meant to steer the activity of an 'outsider' in a manner that is beneficial to the community as a whole and provides greater protection to the intellectual heritage within the product. It's a recognition of the fact that 'researcher interest' or 'private enterprise' are not bad for communities in themselves, but historically their interaction with indigenous and local societies has often led to the dissipation of IK with little benefit to its bearers, not to mention adverse effects on the environment (Bavikatte et al, 2015).

A major component of the BCP is giving prior informed consent (PIC) for any utilization of IK thereby determining the terms of access and expected outcomes. PIC embodies a 'no surprises' policy that ensures the community will not be confronted with unexpected discoveries of what an outside person is doing with their knowledge and resources and why. It also ensures that everything a person is doing with the knowledge is the result of a negotiation and not just free assumption of a right to use it. This is important because every use of IK can affect the environment and/or lead to some kind of product and the original knowledge bearers may not know what the person using it has in mind. This process sets a solid foundation for negotiating access and benefit-sharing by allowing both parties to think about what they are getting into and what they want to achieve (Jonas, Shrumm & Bavikatte, 2010). The scope of BCPs is wide-ranging and involves various elements. Some of which are as follows: to document and legally protect the collective rights of indigenous and other traditional communities to their cultural knowledge. BCPs seek to maintain and restore customary laws and sustainable resource use systems to conduct systematic inventories and impact assessments of external (national and international) development projects and conservation programmes which may affect the resources, knowledge or rights of the communities concerned.

METHODOLOGY

The study was explorative in nature as there are only two BCPs developed thus far in South Africa. The one developed by Traditional Health Practitioners (THPs) in Bushbuckridge, Mpumalanga and the one developed for the rooibos. The exploratory design further enables the researchers to unearth new information about the development of BCPs with an intent to safeguard and preserve valuable IK. An indigenous research approach is applied to harness rich data and information which strengthened the study. Furthermore, the study is qualitative in nature. This is because when engaging with participants it grants them to convey and exchange knowledge at free flow and without disturbance from the researchers. This study applied an indigenous paradigm (Chilisa, 2012) to elucidate meaning and practices as per the provision of indigenous communities. The indigenous paradigm affords the researchers a lens which is devoid of western influence thereby approaching this study with an intent to receive decolonizing data and knowledge, Furthermore, it provided insignts to develop and implement BCPs in indigenous communities successfully. The lekgotla method together with in-depth interviews were utilized to collect the data for the study. The lekgotla method was applied to involve communities, key knowledge holders and stakeholders of the community while the in-depth interviews were conducted with traditional leaders, community leaders and youth involved in IK development. The recruitment of participants was done through snowballing. Key informants provided insights as to which other people are knowledgeable and active in IK issues within the communities thereby providing us with a total sample of 15 participants per community. Which means the total number of participants was 30 for both communities combined. The data was analysed using Braun and Clarke (2006) steps of thematic analysis and themes identified presented herein as finding of the study.

LEGAL FRAMEWORKS GIVING EFFECT TO THE DEVELOPMENT OF BIOCULTURAL COMMUNITY PROTOCOLS IN SOUTH AFRICA

There are several legal frameworks that are considered when a BCP is developed. This is because a BCP is meant to be a legal instrument that underpins how communities want to be addressed by any third party and their local laws that need to be followed by any person who needs or wants to establish a relationship with communities thereby accessing and utilizing their indigenous knowledge (IK). Therefore, a BCP is not developed devoid of all other legal documents, from international to national level.

THE CONVENTION ON BIOLOGICAL DIVERSITY

The aim of the Convention on Biological Diversity (CBD) is the conservation of biological diversity, the sustainable use of its components and the fair and equitable sharing of the benefits arising from the use of genetic resources (CBD, 1992). This legally binding document of which South Africa is a signatory, is the first document which has focused on IK with the intent of making sure that benefits accruing from genetic resources that are associated with IK are shared with the communities with which are custodians of such knowledge. This is demonstrated by this Convention, in which it recognizes the close and traditional dependence of various indigenous and local communities embodying traditional lifestyles on biological resources and recognizes the need for a fair sharing of the resulting benefits. The CBD is viewed as the first steppingstone in the protection of IK for sustaining local communities. This is seen as a first legislative document which has recognized and affirmed IK as a resource of indigenous or local communities which must be protected and developed with the overall aim of equitably sharing benefits with the holders and propagators of this IK.

Article 8(j) of the CBD stipulates that the traditional knowledge of indigenous and local communities is to respect, preserve and maintain knowledge, innovations and practices of indigenous and local communities that embody traditional lifestyles necessary for conservation and sustainable use of biodiversity and to promote its wider application with the consent and involvement of the holders of such knowledge, innovations and practices and promote the fair sharing of the benefits arising from the use of such knowledge, innovations and practices (CBD, 1992). Davis (1998) points out that through bioprospecting, it may be possible to achieve the twin goal of conservation and protection of the rights of IK holders and innovators. This is because bioprospecting is consistent with the implementation of Article 8 (j) of the CDB. Therefore, the CBD recognizes the significance of indigenous and local communities to the conservation and sustainable utilization of biological diversity. UNEP (1997) states that the Convention also requires in Article 10(c) that the customary use of biological resources should be protected and promoted in accordance with traditional cultural practices. Information on traditional knowledge (TK) and technologies should be part of the information to be exchanged and, where possible, repatriated (Article 17(2)), while technological cooperation between Parties should also include cooperation on indigenous and traditional technologies (Article 18(4)).

INTERNATIONAL LABOUR ORGANISATION (ILO) CONVENTION 169

Although not yet ratified, South Africa has considered ILO Convention 169, which deals with the rights of indigenous and tribal peoples. Article 1 of the convention applies to the following:

(a) tribal peoples in independent countries whose social, cultural and economic conditions distinguish them from other sections of the national community, and whose status is regulated wholly or partially by their customs or traditions or by special laws or regulations;
(b) peoples in independent countries who are regarded as indigenous on account of their descent from the populations which inhabited the country, or a geographical region to which the country belongs, at the time of conquest or colonisation or the establishment of present state boundaries and who, irrespective of their legal status, retain some or all of their own social, economic, cultural and political institutions.

Given the above article, it is deemed vital that a BCP be developed in various indigenous communities to pinpoint and underpin how these communities wish to engage or be engaged by third parties, especially those who seek to access their IK, IKS and genetic resources. A Biocultural Community Protocol enables an indigenous community to provide clear terms and conditions that any third party must abide by.

AFRICAN CHARTER ON HUMAN AND PEOPLES' RIGHTS

South Africa is a party to the African Charter, which recognises the rights of individuals and communities, including the right to culture. The African Charter of 1981 (in force since 1986) captures the essence of a BCP and the stance of the indigenous communities well in articles 20 and 22. Article 20 provides the following:

(1) All peoples shall have the right to existence. They shall have their unquestionable and inalienable right to self-determination. They shall freely determine their political status and shall pursue their economic and social development according to the policy they have freely chosen.

While article 22 stipulates the following.

(1) All people shall have the right to their economic, social and cultural development with due regard to their freedom and identity and in the equal enjoyment of the common heritage of mankind.

(2) *States shall have a duty, individually and collectively, to ensure the exercise of the right to development.*

The African Charter therefore shows that it values the people and views that all people have the right to participate in the economy. This may be by utilizing their cultural, social and IK mandates to do so. This is a legal framework that seeks to show inclusivity and ensures that everyone's right is taken into consideration.

THE NAGOYA PROTOCOL

The Nagoya protocol on access to genetic resources and fair and equitable sharing of benefits arising from the utilization of biological resources was successfully adopted by the plenary of the tenth session of the conference of parties to the convention on biological diversity (CBD COP10) on the 30th of October 2010 (Buck & Hamilton, 2011). Various countries are signatory to these protocols as they viewed it as one element of the Nagoya package. The Nagoya Protocol's main aim is fair and equitable sharing of benefits arising from the utilization of genetic resources, contributing to the conservation of biological diversity and the sustainable use of its components (Article 1).

Access to Genetic Resources: Access to genetic resources is subject to the prior informed consent of the Party providing such resources, ensuring legal certainty, clarity, and transparency (Article 6). Benefits arising from the utilization of genetic resources and traditional knowledge must be shared fairly and equitably with the providing Party and indigenous and local communities, based on mutually agreed terms (Article 5). Parties must take measures to ensure compliance with domestic legislation on access and benefit-sharing and monitor the utilization of genetic resources through designated checkpoints (Articles 15, 16, and 17). The Protocol emphasizes the need for capacity-building, technology transfer, and cooperation to support developing countries, particularly the least developed countries and small island developing States, in implementing the Protocol (Articles 22 and 23).

HOW DOES THE PROTOCOL ADDRESS INDIGENOUS KNOWLEDGE ASSOCIATED WITH GENETIC RESOURCES?

The Nagoya Protocol addresses traditional knowledge associated with genetic resources through several key provisions:

Prior Informed Consent: Access to traditional knowledge held by indigenous and local communities must be obtained with their prior informed consent or approval and involvement (Article 7).

Benefit-Sharing: Benefits arising from the utilization of traditional knowledge associated with genetic resources must be shared in a fair and equitable way with the indigenous and local communities holding such knowledge, based on mutually agreed terms (Article 5).

Compliance: Parties are required to take measures to ensure that traditional knowledge associated with genetic resources is accessed in accordance with prior informed consent and that mutually agreed terms have been established (Article 16).

Customary Laws and Protocols: In implementing their obligations, Parties must take into consideration the customary laws, community protocols, and procedures of indigenous and local communities (Article 12).

Awareness and Capacity-Building: The Protocol encourages the development of community protocols, minimum requirements for mutually agreed terms, and model contractual clauses to support the fair and equitable sharing of benefits. It also emphasizes the need for awareness-raising and capacity-building initiatives involving indigenous and local communities (Articles 12 and 21).

THE INDIGENOUS KNOWLEDGE SYSTEMS POLICY OF 2004

The Indigenous Knowledge Systems (IKS) Policy was approved by Cabinet in November 2004, marking the first significant breakthrough in South Africa's attempts to recognize, affirm, develop, promote, and protect indigenous knowledge systems. The IKS policy aims to enable an improved understanding of the historic, cultural context and value of indigenous and local communities. It is deemed a cornerstone policy that produces key drivers that are catalysts for the advancement and economic feasibility of IK holders and practitioners. It also aims to respond positively to a quickly changing environment and to enable indigenous and local communities and individuals to fairly participate in South Africa's social and economic opportunities. The preamble to the policy states that the Government of the Republic of South Africa expresses its devotion to the recognition, promotion, development, protection and affirmation of IKS. The National IKS Policy (2004) considers that indigenous biological resources such as traditional medicine and indigenous agricultural practices represent a vital economic activity in South Africa. This is because IK continues to play a critical role in the sustainable livelihoods of a large percentage of South Africa's population.

The policy additionally articulates that indigenous agriculture which is dominantly practiced in indigenous local communities is an environmentally tolerant and resilient crop production technique that has illustrated sustainability over long periods. The policy posits that this agriculture practice drives sustainable development in many indigenous communities as this practice optimises production security through adaptation to the local environment. Crop security is assured through the growth of an intricate system involving such determinants as diversity of crops, well-dispersed plantings, heterogeneous genetic resources, minimum tillage, and varying fallow, as well as sharing of food and labour. Such practices tend to be rational responses to local conditions and are logical variations to risks (National IKS Policy, 2004).

In addition, IKS is owned by and provides services to people who are prone to unemployment. Therefore, consideration needs to be given to the role of IKS as a job creation mechanism. The National IKS Policy (2004) provides that IKS based projects and programmes be developed as these will drive the sustainability of such communities. The Policy outlines that to deploy the economic potential of IKS, three points should be considered. The points are outlined below as follows:

- The creation of incentive mechanisms to promote IKS innovation
- The promotion of IKS in the context of sustainable development
- The promotion of IKS as an employment generator

THE INDIGENOUS KNOWLEDGE ACT NO 6 OF 2019

All the above-mentioned laws gave rise to the IK Act no 6 of 2019. This law is a sui-generis law, meaning that it is one of a kind to be developed. The IK Act (2019) was designed to ensure the protection, promotion, development, and management of indigenous knowledge. This is due to the value that IK has in developing and advancing the holders and disseminators of this knowledge. The knowledge is collective property, meaning that the law provides for a collective benefit to the community(s). The law also provides for the establishment and functions of the National Indigenous Knowledge Systems Office (NIKSO), which is responsible for managing the rights of indigenous knowledge communities (IK Act, 2019). In addition, the Act confers powers and privileges on communities by providing for the establishment and functions of the Indigenous Knowledge Advisory Board. This body consists of IK holders, practitioners and other relevant stakeholders who view IK as a catalyst for sustainable development. The law also provides for access and conditions for access to knowledge of indigenous communities. This will be implemented through a Biocultural Community Protocols (BCPs) being developed by various local/indigenous communities of South Africa. The law recognizes that IK is a national asset and that it is in the country's interest to protect and promote IK through law, policy and both public and private sector programs. Therefore, the law envisages the facilitation and coordination of indigenous knowledge-based innovations that will serve as pillars of the sustainable development of various indigenous communities.

SUCCESS STORIES OF OTHER DEVELOPED BIOCULTURAL COMMUNITY PROTOCOLS ACROSS THE GLOBE

South Africa is not the first country to be busy assisting with the development of BCP's for various indigenous communities. There are other countries that have already embarked in the development of these legal instruments and that has led to these communities protecting their valuable IK and making sure that it works for them to meet their socioeconomic needs. South Africa therefore saw it befitting to develop this BCP for communities to assist them safeguard their IK. Indigenous knowledge for many of these communities is a source of wealth, a social capital that makes them drive sustenance in their local communities and survive.

INTER-COMMUNITY AGREEMENT FOR EQUITABLE BENEFIT-SHARING IN THE POTATO PARK, PERU

Argumedo (2008) posits that through the assistance received from ANDES and IIED, six Quechua communities founded the Potato Park near Cusco, Peru in 2002. Presently, the park encompasses 8240 hectares and more than 6000 inhabitants, comprising the common ground of the Amaru, Chawaytire, Pampallaqta, Paru Paru, and Sacaca communities. This Andean region is recognized as a micro-center of potato origin and diversification. With 700 native cultivars, 410 varieties repatriated from the International Potato Centre (IPC), and an additional 151 types conserved for the Ayacucho and Apurimac people, the park functions as a genetic reserve. There are also a lot of indigenous Andean foods and medicinal plants growing there. Context Six Quechua communities founded The Potato Park in Cusco, Peru, in The Association of Communities The Inter-community Agreement, which was created over the course of two to three years through a comprehensive participatory process led by Quechua community researchers, offers a general framework for the fair distribution of all benefits that the Potato Park receives, whether they are directly or indirectly attributable to its biocultural resources. Through an intercommunity fund, the advantages of various economic collectives are divided and put back into bolstering the biocultural system.

Reciprocity, duality, and equilibrium are the three fundamental customary law principles that sustain biocultural systems. Based on prevailing local norms and practices, these principles' derivatives were found and utilized to expand the benefit-sharing framework (Taylor, 2008). Through participatory action research, the Potato Park Inter-Community Agreement was developed, which has led to learning about how to create suitable procedures for equitable benefit sharing. The researchers and community members involved in this study firmly believe that preconceived notions about access and benefit sharing agreements and the procedures for obtaining prior informed consent must be abandoned to design appropriate mechanisms to implement sui generis systems that are both practical and efficient, while also being consistent with the aspirations, values, and beliefs of indigenous and local communities. Analysing access agreements and consent processes from the viewpoint of the communities themselves is a crucial first step in creating sui generis systems. Customary norms that have so far guided the preservation and upkeep of local traditional knowledge (TK) should be used as the primary lens for this analysis (Norms, 2012;Frison et al, 2022).

BUSHBUCKRIDGE TRADITIONAL HEALTH PRACTITIONERS (BTHP)

According to Sibuye et al (2012), the Kukula traditional medicine practitioners of Bushbuckridge, South Africa, have a history of unpaid bioprospecting and are currently dealing with ecological crises brought on by overharvesting medicinal plants in addition to marginalization. However, they have made a direct claim to rights by creating a biocultural community protocol (BCP). Their BCP is a charter or community document that affirms their customary roles that have been practiced for a long time within their communities and their responsibilities to preserve the natural resources and knowledge that they depend on. The BCP calls on external agents, including corporations and the government, to protect health practitioners' rights over their land, resources, and expertise by identifying and outlining the issues they face. Members of the community were involved in the development of the BCP. Sibuye et al (2012) concludes that the Kukula Traditional Healers of Bushbuckridge have found that the BCP has been a valuable tool in helping them define themselves as a community with shared values. Their BCP lays out in unambiguous terms their shared perspectives on preservation and the long-term usage of medicinal plants, including the use of their traditional knowledge. What was once a tiny group of people has grown to about 300 members. Regarding land allocations for their livelihood and conservation projects, the association is now in communication with traditional authorities.

FINDINGS OF THE STUDY

The primary data collected within the two communities provided the following findings that are reported as themes. Bavikatte and Jonas (2009) posit that biocultural community protocols, also called BCPs, are tools that set clear terms for governments and the private, research and non-profit sectors to engage with indigenous communities and access their local resources and knowledge. They are developed through culturally anchored, participatory decision-making processes within communities and are based on the norms, values, and laws common in the community. Although each is adapted to its local context, Shrumm and Jonas (2012) outline that a biocultural community protocol is generally:

- Determined by a self-defined community with a close connection to a specific territory or area, which forms the basis of their identity, culture, language and way of life
- Documented, developed and used in a participatory manner by the community and wherever appropriate, with the support of trusted and long-standing organizations
- The aim is to promote appropriate recognition and support for community-specific customary ways of life and the management of their territory or area
- Based on established values, standards, procedures, rights, and responsibilities This is anchored in customary, national and international laws and guidelines (Shrumm & Jonas, 2012)

RECIPROCITY AND MUTUAL UNDERSTANDING AS A GATEWAY TO BCPS

The *Lekgotla* method and in-depth interviews (*puisano*) revealed that accessing indigenous knowledge from communities has never been a problem if it was on a mutual understanding of the terms of the agreement upon the access of the knowledge. Participant X from one of the communities elucidated as follows:

"Indigenous peoples have always had a reciprocal relationship with their environment, believing that taking care of the environment and its resources meant that the environment would take care of them".

This symbiotic relationship that communities have with their communities enable them to further develop their IK which in many cases gets violated and pirated by third parties. Hence, this is why it is vital for third parties such as development agencies wanting to access IK and indigenous resources to speak and engage with the members of the communities. The protection of their indigenous knowledge depended on that, as it meant that members of communities would be part of decisions made in their communities.

Whilst being indigenous, communities that have lived in harmony for centuries until colonization stripped them of their rights over their knowledge and resources. The experiences have pushed these communities millions of steps backward. From their knowledge being deemed invalid whilst being stolen, to them not benefiting and having a say in the decisions regarding their resources. Participant P exclaims as follows:

"As a first nation, we have been stripped of our ancestral land rights, and everyone knows that for us to engage, develop and manage our IK we require land access"

The above statement proves that the fight to protect indigenous knowledge has been long and frustrating to such an extent that most communities feel as though their voices are not heard and that their cries have been falling on deaf ears.

ADVOCACY OF COLLECTIVE AND INDIGENOUS RIGHTS AND ETHICS

Protecting indigenous knowledge means advocating for the rights of indigenous communities as well as paying attention and respecting their values, principles, cultures and traditions. Participants Z from one community explains as follows:

> "Protecting our IK means involving every member of the community and informing them of the decisions that involve their knowledge and resources, from the youth, elders as well as the traditional councils".

The above on its own is a thorn to so many leaders of the communities and third parties that want to gain access to the knowledge and resources of the community. Whilst there is a gospel that indigenous knowledge is knowledge that is communally owned, there are still many community members who have no idea as to how decisions are made on their behalf regarding their resources and are not benefiting anything. Biological resources are often accessed and used up by third parties, yet the communities remain poor. Participant S from one of the communities echoes as follows:

> "The knowledge and resources of communities have the potential to stimulate economic and sustainable development, and yet for many communities this is a dream".

Biocultural Protocols are enforced by the CBD, the Nagoya Protocol and the IK Act, and they are tools that provide governments and the private, research and non-profit sectors with clear conditions for engaging with and accessing indigenous and local communities including their resources and their knowledge. They are developed through participatory decision-making processes within communities and are based on the norms, values, and laws common in the community.

BCPs are for communities because they ensure that every member of the community is involved, it leaves no one behind. They are the steppingstone towards the protection of indigenous knowledge and promote community development. With BCPs, community members are guaranteed and involvement in the decisions that will be taken regarding their knowledge and resources. The protection of indigenous knowledge cannot be established without involving the holders of the knowledge and cannot be established without being informed by the laws, values and regulations that played a role in the development of the knowledge.

RESPECT FOR INDIGENOUS WAYS OF LIFE AND VALORISATION OF IK

For centuries communities wanted to be acknowledged as the holders of their knowledge and that is what BCPs are doing for them. The protection of indigenous knowledge means carrying the pride of the indigenous knowledge holders and communities high and showing them how valid their knowledge is. It means ensuring that there will be no unauthorised use of the knowledge either commercially or personally. It means development for communities. BCPs will instruct third parties of the way of living of the communities, who they are, where they come and how they want their knowledge to be used and what they expect from the use. For years' indigenous knowledge has been used by third parties without fully involving the people who hold the knowledge. And BCPs are putting a stop to that. There are ways in which communities want to be approached, and every stakeholder should respect those ways. There are several ways that communities want to benefit from the use of their knowledge and those ways should be respected. BCPs are not only going to benefit communities but they are going to make sure that the knowledge is documented, and a knowledge documented means a knowledge protected and not forgotten.

The major challenge experienced in the protection of indigenous knowledge was that it was not well documented and what is not documented could not be protected. For communities to have BCPs, it means documentation, not just of the laws and regulations that enforce their knowledge but of the knowledge as well. Community laws and regulation have always protected indigenous knowledge in their own way, hence, it was not lost but shared from generations to generations in ways that they understood. BCPs are going to make sure that all those ways and relationships that communities have with their resources are well documented so they can be protected.

DEVELOPING BCPS THROUGH DIALOGUE AND ACTIVE PARTICIPATION TO DRIVE ETHICAL AND SUSTAINABLE DEVELOPMENT AGENDAS

The Khoi Communities identified in the study echo that they are commitment to locally defined development plans and priorities. This stems from a deep understanding of the intrinsic link between their biocultural heritage and sustainable development. As stewards of a rich tapestry of traditions, values, and aspirations, the community recognises the need for initiatives that align with their unique identity. This intricate relationship that they have with their land and resources including their local knowledge therefore needs to be harnessed and preserved properly to advance their developmental mandates. Participant A elucidates as follows:

> "The development of BCP's for our community will enable us to safeguard our knowledge and utilize it to move forward the community development endeavours that will see us benefit from this knowledge and knowledge associated with our resources".

Though they are not yet affiliated with any structural affiliations, these communities are in the pursuit of locally driven development, and the community engages in a participatory and inclusive process. This involves extensive dialogue among community members to identify and prioritise development goals. By ensuring that every voice is heard, the community creates a comprehensive understanding of their collective needs and desires. This inclusive approach not only strengthens the social fabric but also empowers community members to actively contribute to their development. Participant E explains as follows:

> "The development plans of our community are intricately woven into the fabric of our community's values, traditions, and environmental ethos".

Traditional practices and ecological wisdom are integral components, ensuring that any developmental interventions harmonise with the natural environment. The community sees sustainable development not as a separate endeavour but as an organic extension of their longstanding connection with the land. Moreover, economic goals are carefully aligned with cultural preservation. Economic initiatives are designed to uplift the community without compromising their cultural integrity. This approach fosters a balance between progress and tradition, allowing the community to navigate the challenges of the modern world while preserving the essence of their identity. Participant Z explain as follows

> "The impact of our locally-defined development plans goes beyond economic prosperity".

What this implies is that their development plans serve as a mechanism for safeguarding the intricate Indigenous Khoi heritage. This is achieved by grounding development in their cultural roots, the community ensures that each step forward is a conscious choice to preserve the legacy for future generations. This commitment to sustainable and culturally rooted development positions the Indigenous Khoi Communities as custodians not only of their destiny but also a living, breathing cultural tapestry.

FREE PRIOR INFORMED CONSENT (PIC) AND BENEFIT SHARING

The principle of Prior Informed Consent (PIC) is foundational to the Indigenous Khoi Communities' engagement with external entities. This principle reflects a deep-seated respect for their biocultural resources and traditional knowledge (TK). The communities insist on a comprehensive and transparent consultation process before any external project proceeds, ensuring that they are fully informed about the nature, scope, and potential impacts of proposed initiatives. This commitment to PIC is a testament to the community's agency in decision-making. It empowers them to actively shape their interaction with the external world, preserving their autonomy and protecting their heritage. The thorough consultation process facilitates a mutual understanding between the community and external entities, fostering a collaborative spirit rooted in respect.

In addition to PIC, the community places significant emphasis on a fair and equitable Benefit-sharing mechanism. The recognition that their traditional knowledge and resources contribute to the success of external projects underscores the importance of reciprocity. Participants P and Z exclaims as follows:

> "The Indigenous Khoi Communities expect tangible benefits that accrue to the community as a whole and directly enhance the well-being and autonomy of community members".

What this implies is that these benefits will cut across all borders to equally benefit the entire community. These take various forms, including financial compensation, capacity-building opportunities, or other support structures. These Communities leadership in line with its leadership structure perform a needs analysis to establish each branch's requirements to sufficiently allocate and distribute benefits and resources fairly.

The community envisions a partnership where external entities not only acknowledge the value of their contributions but actively contribute to the community's well-being and sustainable development. This reciprocal arrangement strengthens the bonds between the Khoi Community and external collaborators, fostering relationships built on fairness, trust, and shared prosperity.

DISCUSSION OF FINDINGS

It is vital for any developmental agency and even governments to recognize the value of IK of communities (Priyadarshini and Abhilash, 2019) in providing core solutions to developmental agendas. Advocating for this IK to be mainstreamed and considered as a knowledge that can complement the mainstream knowledge economy can provide fruitful results that can assist in driving the sustainability agenda. This is because the use of this knowledge can have a positive impact on important sectors that stimulate development such as: agriculture, forest and biodiversity management, infrastructure, and fisheries (Berkes, 2009). The importance of indigenous communities and IK has gradually gained prominence in the global community. This is evidenced by various IK repositories in different parts of the world which have positively impacted major economic sectors such as agriculture (Priyadarshini and

Abhilash, 2019). IK has helped various developmental agencies to cope with the sustainability demand especially in Africa generally, and South Africa specifically.

The importance of traditional practices and IKS in the context of sustainable development and for the timely implementation of the SDGs is obvious. Traditional practices and indigenous knowledge of different ethnic communities are quite closely linked to the principles of sustainable development (Priyadarshini and Abhilash, 2019). Partnerships between government, private sector and civil society are crucial to any sustainable development agenda. However, experience in Africa has shown that the involvement of local communities in development projects is rarely interactive or empowering (Maditsi & Materechera, 2021). Indigenous knowledge systems (IKS) refer to the complex set of knowledge, skills and technologies that exist and develop under specific conditions among indigenous populations and communities in a given geographical area (Noyoo 2007). This is a form of knowledge that is deeply rooted in community practices, institutions, relationships, and rituals (Materer et al. 2001). IKS is informed about and relates to all areas of life and the environment. It can be summarized as a body of knowledge and practices, whether explicit or implicit, used in managing socio-economic, environmental, and spiritual aspects of life (Osman, 2010). Recently, great interest in IKS has emerged from the international community. This interest has led to IKS being seen as a solution to many problems faced by development agencies (Magni 2017). Awareness of the value of IKS has increased among many stakeholders and development agencies, making IKS an important catalyst for sustainable development, particularly in Africa. This is primarily because Africa remains a goldmine of indigenous knowledge (IK) for sustainable development (Akinwale 2013). An evolution to a networked, global knowledge economy will be one of the most vital economic and social changes of the next upcoming years and hence this study is aiming to articulate how this IK can be harnessed to foster and drive sustainable development.

FUTURE RESEARCH DIRECTION

This study has investigated the potential of BCP in protecting and promoting IK within indigenous communities. The paper has show that BCP are vital tools that each community needs to have in order to guide their engagements with third parties. Given that, the upcoming researchers can focus on how the legal instruments can be enhanced and how on the same can they have the voices of communities themselves. These documents are meant to serve the community in a legal manner; however, those who develop these instruments do so devoid of community leaders, elders, knowledge holders and IK propagators. Therefore, it is imperative that these legal instruments be analyzed using an indigenous lens to pinpoint if they have silenced the voices of the communities and if they do enshrine the ethos, values and standards of the indigenous populations.

CONCLUSION AND RECOMMENDATIONS

Policies and laws that are developed for the protection of indigenous knowledge are the steppingstones towards the sustainable development of local communities. These legal documents ensure that there is better management of indigenous knowledge. Repeatedly, when indigenous knowledge is put on a trial of tests, it continues to prove just how crucial it is in developing communities. Its potential to grow the economy of communities cannot be overlooked and avoided. For years, indigenous knowledge communities have been oppressed and their worlds have been taken over by the Western knowledge, and despite that, indigenous knowledge continues to rise above the challenges thrown at it. As a science, indigenous knowledge can contribute towards the global challenges and whilst managing it creates relevant sustainable systems for communities, it is crucial to promote the protection and development of the knowledge to ensure that it is not misused and abused by outside stakeholders. Poor communities are the evidence of what leaving out local knowledge in discussions of policies can do. Different communities have their own visions of how development should be and involving them in the discussions means understanding their needs and how to better meet those needs.

The protection and preservation of indigenous peoples' knowledge must be recognized to achieve sustainable development. Development and policy programming must enable sustainable and genuine participatory participation processes of knowledge multiplicity systems. It is important to promote indigenous knowledge to promote the formation of sustainable development and prevent knowledge from being abandoned. Development and government agencies in Africa need to plan and revitalize development programs, institutional and legal frameworks at local and central government levels, and local institutions to improve the development and promotion of sustainable development and indigenous knowledge systems. It is critical for Africa to transfer, adapt and preserve this knowledge. The following recommendations are made by this paper:

- Stakeholders should promote public awareness and show support to the government commitments and sustainable development through capacity building, campaigns, information sharing, and policy dialogue
- Stakeholders and development agencies should assess and analyze the impact of finance and policy to increase positive sustainable impacts, informing decision making, and avoiding negative potential impacts towards IK
- Government should involve all stakeholders within the IK sector when developing policies and legislation aimed at managing, preserving and protecting IK
- Sustainable development does not exist outside IK, and working with local communities will provide a clear understanding on how to best bring solutions to indigenous communities' needs

REFERENCES

African Charter on Human and Peoples' Rights. https://au.int/sites/default/files/treaties/36390-treaty-0011_-_african_charter_on_human_and_peoples_rights_e.pdf

Akinwale, A.A. (2013). Digitisation of indigenous knowledge for natural resources management in Africa.

Argumedo, A. (2008). The Potato Park, Peru: Conserving agrobiodiversity in an Andean indigenous biocultural heritage area. *Protected landscapes and agrobiodiversity values, 1*, 45-58.

Bavikatte, K., & Jonas, H. (2009). Bio-cultural community protocols: a community approach to ensuring the integrity of environmental law and policy. *Square Brackets*, (2).

Bavikatte, K. S., Robinson, D., & Oliva, M. J. (2015). Biocultural community protocols: dialogues on the space within. *IK: Other Ways of Knowing*, pp.1-31.

Berkes, F. (2009). Indigenous ways of knowing and the study of environmental change. *Journal of the Royal Society of New Zealand, 39*(4), 151–156. DOI: 10.1080/03014220909510568

Biyela, N., Oyelude, A., & Haumba, E. (2016). Digital preservation of indigenous knowledge (IK) by cultural heritage institutions: A comparative study of Nigeria, South Africa and Uganda.

Boven, K., & Morohashi, J. (2002). *Best practices using indigenous knowledge.* Nuffic.

Chilisa, B. (2019). *Indigenous research methodologies.* Sage publications.

Clarke, V., & Braun, V. (2017). Thematic analysis. *The Journal of Positive Psychology*, *12*(3), 297–298. DOI: 10.1080/17439760.2016.1262613

Clarke, V., & Braun, V. (2017). Thematic analysis. *The Journal of Positive Psychology*, *12*(3), 297–298. DOI: 10.1080/17439760.2016.1262613

Correa, C. M. (2001). Traditional knowledge and intellectual property. *Geneva: The Quaker United Nations Office (QUNO), 17.*

Edgar, B. A., Eva, M. P., & Lee, K. S. (2022). Disappearance of African Indigenous Knowledge of Water Conservation and Management in Limpopo Province of South Africa: An IKS Perspective. *African Journal of Development Studies*, *12*(2).

Frison, C., Parks, L., & Tsioumani, E. (2022). Biocultural Community Protocols: Making Space for Indigenous and Local Cultures in Access and Benefit Sharing? In *Access and Benefit Sharing of Genetic Resources, Information and Traditional Knowledge* (pp. 177-190). Routledge.

Indigenous Knowledge Act no 6 of 2019, Government Gazette, Cape Town, 19 August 2019

Indigenous Knowledge Systems Policy. (2004), Department of Science and Technology, Pretoria International Labour Organization (ILO) Convention 169: https://www.ilo.org/dyn/normlex/en/f?p=NORMLEXPUB:12100:0:NO:P12100_ILO_CO DE:C169

Jonas, H., Shrumm, H., & Bavikatte, K. (2010). Biocultural community protocols and conservation pluralism. *Policy Matters*, *17*, 102–112.

Jonas, H., Shrumm, H., & Bavikatte, K. (2010). Biocultural community protocols and conservation pluralism. *Policy Matters*, *17*, 102–112.

Maditsi, M. E. (2018). *The role of traditional leaders in postapartheid local governance in South Africa: a case study of the Bapong 2 community, North West Province, South Africa* (Doctoral dissertation, North-West University (South Africa)).

Maditsi, M. E., & Materechera, S. A. (2021). The Role of Indigenous Communities of Practice in Fostering Community Engagement and Partnerships for Sustainable Development in Africa. In *Sustainable Development in Africa* (pp. 3–18). Springer. DOI: 10.1007/978-3-030-74693-3_1

Magni, G. (2017). Indigenous knowledge and implications for the sustainable development agenda. *European Journal of Education*, *52*(4), 437–447. DOI: 10.1111/ejed.12238

Materer, S. M., Valdivia, C., & Gilles, J. L. (2002). *Indigenous knowledge systems: Characteristics and importance to climatic uncertainty* (No. 1717-2016-140130).

Mdhluli, T. D., Mokgoatšana, S., Kugara, S. L., & Vuma, L. (2021). Knowledge management: Preserving, managing and sharing indigenous knowledge through digital library. *Hervormde Teologiese Studies*, *77*(2). Advance online publication. DOI: 10.4102/hts.v77i2.6795

Norms, Q.C., 2012. *Community Biocultural Protocols.*

Noyoo, N. (2007). Indigenous knowledge systems and their relevance for sustainable development: A case of Southern Africa. *Tribes and Tribals*, *1*, 167–172.

Osman, A. (2010). *Indigenous knowledge in Africa: Challenges and opportunities. Inaugural Lecture.* University of the Free State.

Priyadarshini, P., & Abhilash, P. C. (2019). Promoting tribal communities and indigenous knowledge as potential solutions for the sustainable development of India. *Environmental Development*, *32*, 100459. DOI: 10.1016/j.envdev.2019.100459

Ryan, R. M., & Deci, E. L. (2024). Self-determination theory. In *Encyclopedia of quality of life and well- being research* (pp. 6229–6235). Springer International Publishing.

Shrumm, H., & Jonas, H. (2012). *Biocultural Community Protocols: A Toolkit for Community Facilitators*. Natural Justice.

Sibuye, R., Uys, M.-T., Cocchiaro, G., & Lorenzen, J. (2012). The Bushbuckridge BCP: Traditional Healers Organise for ABS in South Africa in Participatory Learning and Action, Biodiversity and Culture: Exploring Community Protocols, Rights and Consent.http://pubs.iied.org/pdfs/G03403.pdf

Taylor, E.C. (2008). Sharing of benefits or enclosure of the commons? investigating the compatibility of national and local access and benefit sharing mechanisms in Peru.

UN, I. (1992). Convention on biological diversity. *Treaty Collection.*

Verhoef, G. (2023). *The public domain and the legal protection of Indigenous Knowledge Systems* (Doctoral dissertation, North-West University (South Africa).

Verhoef, G. (2023). *The public domain and the legal protection of Indigenous Knowledge Systems* (Doctoral dissertation, North-West University (South Africa).).

Zimu-Biyela, A. N. (2016). *The management and preservation of indigenous knowledge in Dlangubo Village in KwaZulu-Natal, South Africa* (Doctoral dissertation, University of South Africa).

KEY TERMS AND DEFINITIONS

Indigenous Knowledge: It is the unique body of knowledge, practices, and beliefs developed by indigenous communities through generations of interaction with their environment and culture, encompassing everything from traditional ecological knowledge to cultural traditions and spiritual beliefs.

Ethics: Are moral principles that govern a person's behaviour or the conducting of an activity.

Policy: It is a course or principle of action adopted or proposed by an organization or individual.

Legal Protection: It is the safeguarding of an individual's or entity's rights and interests through legal means, encompassing laws, regulations, and legal processes designed to prevent harm and provide recourse in case of disputes.

Prior Informed Consent: It is obtaining explicit consent from stakeholders or rights holders for access to knowledge or genetic resources, based on truthful information about the use of those resources, before the access takes place.

Biocultural Community Protocols: Are instruments developed by communities to outline their core ecological, cultural, and spiritual values, customary laws, and terms for engaging with external actors

Chapter 9
Documentation and Preservation of Indigenous Agricultural Practices Used to Enhance Food Security Among Bakgatla-Ba-Kgafela in Moruleng, Northwest Province

Thato Hazel Seemise
https://orcid.org/0009-0003-1212-3884
North-West University, South Africa

Mothusiotsile Edwin Maditsi
https://orcid.org/0000-0001-5796-0174
North-West University, South Africa

Hangwani Tshisevhe Muedi
Department of Agriculture, Land Reform, and Rural Development, South Africa

ABSTRACT

Due to the rapidly increasing population around the world, food access and availability has become a paramount concern to various developmental agencies and institutions. This is because food security and vital food systems are of great importance. One of the most significant current discussions in agriculture is the use and application of indigenous agricultural practices (IAPs) which have been used by local farmers to enhance household food security. Despite their significance, most of IAPs are at risk due to urbanization, cultural assimilation, and globalization. This study applies an explorative design, with an indigenous paradigmatic lens to illustrate the significance of these practices in ensuring food security. The data was collected through semi structure interviews with local farmers of Bakgatla-ba-Kgafela and was analyzed using thematic analysis. The study finds that the documentation and preservation of IAPs is crucial for ensuring sustainable food systems and food security. It is therefore concluded that new and innovative approaches to agriculture are needed.

DOI: 10.4018/979-8-3693-7964-6.ch009

INTRODUCTION

Recent developments in agriculture have heightened the need for indigenous agricultural practices (IAPs) to be adopted by local communities. The utilization of IAPs is also hindered by the change in weather patterns and increased temperatures, unstable weather conditions as well as altered precipitation cycles. Indigenous agricultural practices are regarded as crucial part of cultural heritage and identity. However, the issue of climate change has resulted in extreme temperatures towards the crops and the change in precipitation rate which can lead to drought or floods. This concept has recently received some attention mainly because IAPs rely on indigenous knowledge (IK), mostly in agricultural system to preserve biodiversity, ecosystem and it is also useful in maintaining sustainable human health and food (Sharma et al., 2020). in addition to these practices, they address nature conservation and rural development which contribute to maintaining ecosystem of a specific community which result in sustainable food security. in various local communities of south africa, farmers use ik in management of livestock which decrease the dependency of external inputs by using different renewable farm resources as agricultural practices (Kumar & Chander, 2017). For centuries, indigenous communities used their IK to attain food security. This knowledge has been passed down from one generation to the other and it is deeply rooted in natural heritage and culture of local communities.

Indigenous farmers in local communities often use their IK to assist them in tracking early warning signs of unstable weather conditions by observing the moon and the behavior of specific animals. For instance, IK provides weather variable and reliable forecast of the probability of rainfall which indigenous farmers may then prepare their farming activities to maximize the productivity of crops and minimize the risks. The value and importance of indigenous agricultural practices are limited on national level in South Africa, whereas scientific and western modes of crop production dominate. This has affected the value of indigenous agriculture and food security negatively in South Africa, and specifically in the North-West Province (Cloete, 2010). In most developing and developed countries, the importance of IAPs is disregarded, and its documentation and transmission continue to provide significant difficulties. Indigenous agricultural practices are not only promoting food security, but they also play a crucial role in biodiversity conservation, enhancing rural livelihoods, and preserving cultural heritage for future generations. Therefore, the documentation and transmission of this knowledge and practices demand special attention. This might serve as a platform for researchers and development intermediaries to affirm, support and valorize IK with cutting-edge technology to withstand the harshness of nature, and promote their role in achieving food security (Wesley & Faminow, 2014).

Farmers in developing countries can manage their agricultural practices by carefully managing the nutrients, soil, water, and local resources to ensure sustainable farming. However, the utilization of indigenous agricultural practices has been threatened heavily by technological interventions. For instance, the use of technological inputs like chemicals, genetic modification and machinery use has been utilized by many people as they think it saves time and they prefer fast growing and high-yielding crops, neglecting the use of indigenous agricultural practices. Understanding IAPs through IK reveals crucial ecological clues for the development of sustainable agriculture to achieve the food security of the resource-poor farmers. Farmers play a significant role in managing crops and maximizing food security by producing subsistence crops and local crop varieties that can be preserved over an extended period during storage (Kazmi et al., 2014).

The Batswana of Moruleng (*Bakgatla- ba- Kgafela*) comprise a community of 350 000 and were split between Kgatleng in Botswana and Moruleng in the North-West Province due to historical reasons. Moruleng is a rural village found along the slopes of Mmammitlwa Mountains in the North-West Province. It is an ancestral home of the *Bakgatla- Ba- Kgafela* people. Furthermore, the Batswana of Moruleng (BBKs) were the main participants of this study as they are still rooted in their culture and identity.

PROBLEM STATEMENT

Indigenous agricultural practices are regarded as intangible heritage of local communities that are orally transmitted and through practice. Despite their importance, there is a lack of documentation and recognition of the knowledge associated with the IAPs. This result in knowledge loss, neglect, and have further led to undervalued IAPs. There is also an absence of preservation, dissemination and documentation of IAPs, which can hinder the continuation of intergenerational knowledge transfer. Due to modernization and globalization, younger generations who are supposed to be taught these practices are moving to cities and they further regard IAPs as old-fashioned Indigenous communities often experience the issue of climate change which result in escalated temperatures, change in precipitation which hinder their ability to produce food.

Chapter Objectives

- To document indigenous agricultural practices utilized by Bakgatla-ba-Kgafela to enhance food security
- To investigate how Bakgatla-ba-Kgafela preserve indigenous agricultural practices to foster food security
- To explore how indigenous agricultural practices drive and ensure sustainable food security among the Bakgatla-ba-Kgafela of Moruleng
- To identify new strategies that can be used to preserve and promote indigenous agricultural practices to enhance food security.

Research Questions

- Which indigenous agricultural practices are utilized by Bakgatla to enhance food security?
- How does Bakgatla-ba-Kgafela preserve indigenous agricultural practices to foster food security?
- How does the indigenous agricultural practices drive and ensure sustainable food security among the Bakgatla-ba-Kgafela of Moruleng?
- What are the new strategies that can be used to preserve and promote indigenous agricultural practices to enhance food security?

LITERATURE REVIEW

The Significance of Indigenous Agricultural Practices in Enhancing Food Security

Indigenous agriculture has been a prominent source of food in indigenous communities, where the community members would use subsistence farming to sustain their families and communities (Pimentel & Burgess, 2014). Indigenous agricultural practices are not just a method of survival, but it has also contributed towards sustaining the natural environment as this method is organic. There is not much documented on IAPs as compared to conventional practices, thus resulting in a reduction of the use of indigenous farming methods over the years (Korat & Mathukia, 2022). Agriculture is the main source of food production in many countries including South Africa. It also plays a major role in keeping people's food secure. For decades, indigenous peoples in various countries have been practicing indigenous agriculture mainly to provide food for their families. Their methods and practices work with nature; however, they are often excluded from mainstream agriculture. Malapane et al. (2024) explained that the Vhavenda community in the Musina Local Municipality have seen potential with the utilization of indigenous agricultural practices such as crop rotation, livestock farming, seasonal agriculture, polyculture and intercropping. Furthermore, the above-mentioned practices were not only used to promote food security, but to reduce poverty and Enhance their livelihoods. Agriculture is one of the earliest job inventions by the indigenous peoples. The life skill education of these indigenous peoples allows them to immediately utilize their environment to work and produce food (Arcand et al., 2020). Despite IAPs being sustainable, the methods are not well documented, and therefore they are not widely disseminated. In South Africa, IAPs that are adopted by local communities depend more on IK to preserve biodiversity and ecosystem, as they are significant in maintaining human health and sustainable food. There is a need for enhancing smallholder indigenous agriculture in local communities of South Africa to improve rural livelihood sustainability, attain food security, and reduce the change in food prices as well as to address extensive rural unemployment in the country (Notsi, 2012).

Indigenous agricultural methods offer a few opportunities and benefits for enhancing household food security. These traditional farming methods can be adapted to the environment. Indigenous farming methods are frequently created and improved over many generations to fit the unique soil, climatic, and biological circumstances of a place (Xia, 2020). These methods provide greater resilience to environmental changes and variations since they are tailored to local resources including land, water, and biodiversity. As a result, even in difficult circumstances, indigenous farmers can continuously produce food. After utilizing synthetic fertilizers for a while, the soil starts to lose its fertility, which leads to several other issues such as soil erosion and nutrient mining. A wide range of techniques is offered by organic farming to boost agriculture, as a result, even in difficult circumstances, indigenous farmers can have food access, adequacy, availability and sustainability (Chappel & La Velle, 2011).

Indigenous Agricultural Practices and Sustainable Development Goals (SDG)

Sustainable Development Goals is the centerpiece of sustainable development 2030 agenda which was approved by UN. SDGs consist of 17 interconnected goals with corresponding indicators and targets that must be accomplished by 2030. The 2030 agenda was made as an action plan to enhance the human development in priority field such as prosperity, people, peace, partnerships and planet (Shahmohamadiloo et al., 2021). Sustainable Development Goals provide execution and planning framework of worldwide strategies that are aiming in ensuring prosperous and healthy future for upcoming generation. These goals serve as a breeding foundation for new products and ideas. Borthakur and Singh (2021) articulates that the goals were also virtualized from an economic perspective which protects the nation's resources and the environment, while prioritizing the raising of 'standard of living' for all citizens as well as for future generations.

The planned objectives of SDGs are aiming at pointing out issues and resolve problems that have been specified and identified within a specific target. Indigenous agricultural practices have the potential to gear the agricultural sector towards sustainable agricultural production which promotes SDG no. 2 (zero hunger) and 12(responsible consumption and production) (Dasgupta, et al., 2021). Furthermore, IAPs deals with crop production and indigenous methods of farming which improves the chances individuals to attain food nutrition and security. Indigenous agricultural practices also assist in achieving the United Nations Sustainable Development Goals number 1 and 2, which requires safe and nutritionally adequate food supply at the household level (Kumar, et al., 2016).

Food Security in South Africa

The concept food security is referred to the capacity to meet the energy and nutrient needs of the population by the national food supply (Leroy et al., 2015). Food safety and security are two complementing elements of world sustainable future. Additionally, it was discovered by Misselhorn et al. (2012) that availability of food was not the only determinant of food security but also such factors as economic situation and behaviour of people affected. Food security may be at different levels which involves the individual, household, national, regional and global as well. Food security has four major elements such as food availability: The availability of adequate supplies of food of the right quality, whether they are imported or produced domestically (including food aid). Food access: Individuals' ability to access sufficient financial resources (entitlements) to buy the right foods for a balanced diet. Given the legal, political, economic, and social structures of the community in which a person resides, entitlements are defined as the collection of all commodity bundles over which a person can exercise control (including traditional rights such as access to common resources). Utilization: Utilizing food to fulfil all physiological demands through a healthy diet, access to clean water, sanitary conditions, and medical care. This highlights the value of non-food inputs in ensuring food security (Havas & Salman, 2011).

Household food security systems are important and has positive impact on livelihoods, and it has different determinants that are not the same as national level. These include the population density of the household, employment and income status of the household, health status of the household occupants and its own food production. South Africa is currently facing the burden of food and nutritional security as well as diet-related diseases and bilateral malnutrition due to single-family malnutrition and micronutrient deficiencies (Ingram, 2011). According to the General Survey of 2017, 1, 7 million people were found to be food insecure in the North-West Province. In 2019, North-West was one of the most affected by food insecurity. Addressing this issue requires strategies that affect both a healthy lifestyle and a healthy diet (Berry et al., 2015). There is hidden hunger as most of the South Africans does not consume sufficient nutrient content.

Indigenous Knowledge as a Cornerstone of Agricultural Growth in Society

Indigenous knowledge is a significant and critical aspect of technology and culture of any indigenous community. Indigenous knowledge is defined by Sohail, (2022) as a systematic body of knowledge acquired by the indigenous peoples through the accumulation of informal experiment, experiences, and the development of an intimate understanding of the environment of the given culture. Indigenous knowledge is also holistic, transmitted orally and passed down from generation to generation. Indigenous knowledge contributes towards enhancing livelihoods and further provide insights into the preservation of natural resources. This knowledge promotes sustainable technology innovation and acceptance. Furthermore, indigenous farmers in various communities have proven the importance of indigenous knowledge in the agricultural sector, through the incorporation of local practices and strategies for both indigenous and commercial agriculture (Mutengwa et al., 2023).

Indigenous knowledge plays a crucial role in various farming seasons and periods, as it provides local communities with problem-solving strategies, more especially the small-scale indigenous farmers. Through indigenous knowledge, indigenous farmers are exposed to crucial methods of controlling weeds, pests, and aphids, and the methods are confirmed as a significant tool for enhancing crop yields and health (Petzold, et al, 2020). Through indigenous knowledge, small-scale indigenous farmers know how to read and interpret weather patterns. For instance, among the Batswana communities in the North-West Province, South Africa, they can interpret weather patterns through the observation of lunar cycles, seasonal patterns, and stars, which create and provide a space for agricultural growth.

Indigenous farmers in Disaneng community of Ratlou Municipality in the North-West Province, can also read the colour of the clouds in order to determine clouds that might carry hailstones, destroying the crops (Ekobi et al, 2023). Indigenous knowledge also enables indigenous famers to observe the phase of the moon and easily indicate whether it is going to rain. In case when they experience heavy hailstorm, they use hail net to protect their crops. Indigenous knowledge has created a conducive condition for the improvement of agricultural potential, through the interpretation and reading of weather. Onyaniha (2022) asserts that this knowledge has also guided indigenous farmers in Disaneng community to detect if there is going to be drought or rainfall in a specific farming season, assisting them to prepare for catastrophes like food shortages and draught. Furthermore, Leal Filho et al (2022) posits that indigenous knowledge associated with weather forecast assist indigenous farmers in local communities to determine specific crops needed to be sowed. For instance, drought resistance crops or crops that need a lot of rainfall.

Prospects Associated With Indigenous Agricultural Practices

Indigenous agricultural practices play a huge role in the livelihood of indigenous communities. Indigenous agricultural practices have an ability to maximize the income made from surplus crops. For instance, indigenous farmers would grow crops to feed themselves and their families while the remaining outcomes from the harvest are sold and utilized as a way to generate income (Babalola, 2012). Indigenous communities can make a living through the use of indigenous agricultural practices, allowing them to benefit more from their own indigenous knowledge and strength. These practices increase the yield and food production in local communities. Furthermore, indigenous agricultural practices also increase food security at household level. According to Modi, (2003), indigenous farmers in Msinga and Embo communities (KwaZulu-Natal) were utilizing indigenous agricultural practices and they experienced that the practices are environmentally friendly and some natural resources and techniques are used to deal with diseases and pests

Indigenous farmers produce crops that are nutritious and ready for immediate consumption, and crops that are considered the best in having a long-term storage. Indigenous agricultural practices has a potential to enhance the stabilization of food production through the reduction and diversification of reliance towards market prices. Through the use of indigenous agricultural practices, indigenous farmers are able to manage or control various crops simultaneously. These practices are also valuable and capable of managing pests and insects, while fertilizing the soil in preparation for planting. One of the importance of indigenous agricultural practices is that they are cost-effective, they often do not require money to be implemented (Apraku, 2021)

Challenges Associated With the Use of Indigenous Agricultural Practices at Household Level

Despite the good prospects that IAPs offer in the livelihood of indigenous communities, there are also some challenges associated with them. One of the challenges associated with indigenous agricultural practices is that they are characterized by high competition of water (Hamadani et al., 2021). Utilization of these practices require indigenous farmers to have water. Some of the local communities in South Africa experience scarcity of water. As a result, it is quiet challenging for indigenous farmers to implement IAPs in their farms and backyard gardens. Mixed cropping is considered as one of the important agricultural practice. However, it may result in a risk of minimized crop production because of large number of crops grown, leading to completion of space and water (Huynh et al., 2020)

It was stated by Fox et al, (2018) that indigenous communities in Cambodia such as Kuy and Bunong have been depending on rainfed agriculture and shifting cultivation because of the issue of climate change, resulting in unpredictable precipitation and prolonged droughts which hinder the process of indigenous agricultural practices. While in Kenya, the Maasai pastoralists often experience extreme temperatures and recurring droughts which pose a negative impact on the effectiveness of the agricultural practices (Galvin et al., 2020). The Maasai pastoralists have been dependent on their indigenous knowledge. As a result, they were able to overcome challenges and maneuver through the use of indigenous knowledge to manage their water resources. One of the challenges associated with indigenous agricultural practices is that they require a space or land. Soil erosion is also one of the challenges associated with the use of IAPs, as it makes it difficult for indigenous farmers to use these practices (Borbale, 2021). Additionally, indigenous agricultural practices cannot be utilized on eroded soil and even crops cannot grow on eroded soil. One of the major challenges of IAPs is the lack of documentation, which may lead to younger generation not appreciating and valuing the oral transmitted knowledge from generation to generation.

METHODOLOGY

This study applied the indigenous paradigm lenses. Indigenous research paradigm entails concerning and acknowledging the expertise and knowledge that Bakgatla-ba-Kgafela in Moruleng possess regarding their challenges associated with indigenous agricultural practices. Instead of imposing western methodologies, Kovach, (2021) articulates that indigenous research paradigm encourages the researcher to approach the study with cultural and humility sensitivity. Moreover, qualitative research approach was used in this study, as it allowed researcher to gain the understanding of what the participants were saying about the indigenous agricultural practices as well as the participant's attitude, beliefs and motivation towards the utilization of indigenous agricultural practices (Wilson, 2020). Additionally, the researcher discovered that there are only few studies that were done on the use of indigenous agricultural practices to enhance food security among the Bakgatla-ba-Kgafela of Moruleng, thus the study is exploratory in its nature. A semi-structured interview guide was used as a tool to collect data from the participants. Furthermore, the targeted population was indigenous farmers and the indigenous knowledge holders. This is because the targeted population were suitable for the study, to provide the researcher with their knowledge and understanding of the utilization of IAPs. The sampling was purposive, with a snowballing technique.

The sample size were 13 participants, and the researcher reached saturation on the 12th participant. This implies that the researcher got the same data from the 12th participant to the 13th. The age of the participants involved elders who ranged from 40-70 years (both men and women), and the range of youth were 28-30 years. The collected data was analyzed through thematic analysis (Pham, 2018). The researcher read all the transcribed data from in-depth interviews. After the researcher was familiarized with data, the researcher then identified concepts and phrases that were considered to be important and were then grouped into themes (Clarke et al., 2015). The themes emerged from this study are reported as the study results.

FINDINGS AND DISCUSSION

Indigenous Agricultural Practices Used to Enhance Food Security Among the Bakgatla-Ba- Kgafela in Moruleng

a) Crop rotation

The study found out that crop rotation is one of the indigenous agricultural practices used by the Bakgatla Ba Kgafela in Moruleng. They plant various types of crops in the same space across seasons. The participants engaged in this study emphasized that they cannot plant one crop for the whole season in the same area, and further repeat the same crop in the upcoming seasons. They also stated that if they plant one crop season after season, soil degradation will occur and all the nutrients in the soil will be depleted; thereby hindering crop sustainability and production. Within the Bakgatla ba Kgafela, crop rotation is mainly utilized to assist indigenous farmers in producing high yields of different crops to have an easy access of food to feed themselves and the community at large.

This study has found that crop rotation allows the participants to experience various nutrients needs which lead to a well-maintained soil fertility and increased productive fields. The participants have explained that it is very crucial for one to pay attention when applying crop rotation in the backyard garden. Furthermore, crops are planted according to their food group. The following statement was said by the participants on how crop rotation is done in their surroundings: *"Gore re kgone mofuta wa go fetola dijalo, re jala dijalo go ya ka dikarolo tse di maleba, jaaka go jala dijalo tsa medi, morago ga ditlha tse dingwe re bo re jala merogo, morago ga merogo ke maungo, go be go latela dinawa. Fa re dira jalo, re fitlhelela palo e e kwa godimo ya dijo le go nonofatsa mmu"*. English Translation:" For us to succeed in crop rotation, we sow according to the crop families. We start with root crops and replace them with leafy vegetables in the next season. Another season we sow oil seed crops. This assist us in producing high yield and this process also improve the soil health".

b) Mixed cropping

The Bakgatla ba Kgafela in Moruleng prefer using mixed cropping as their indigenous agricultural practice to attain food security. They have stated that it is called mixed cropping because it involves planting two or more crops in the same field or space during a growing season. The study has found out that growing two or more crops in the same field can minimize the risks of crop failure due to unstable weather conditions. This indigenous agricultural practice has been a significant tool for the participants to have food access and food availability. Additionally, they have alluded that mixed cropping enables them to have access to a constant food supply throughout the season. The participants have also mentioned that the body needs nutrients and mixed cropping can assist them by providing a varied diet, as their mixed cropping include legumes, staples and vegetables.

Among the Bakgatla ba Kgafela, indigenous farmers have mentioned that Moruleng village is a community that is more likely to have pests and termites. This means that they often lose some of their crops. Therefore, mixed cropping assists them to avoid losing all their yield. Data collected from the participants have shown that soil fertility is important in farming. If the soil is not fertile, the yield will be low, and sometimes crop failure occurs. Due to crop differences, some crops might need to be left for a while for the soil to regain its fertility after being harvested, while others help with fertilizing the soil. They also explained that if they cultivate crops that grow out of the soil this time, they would plant crops that grow into the soil in the upcoming season, and that is another way of sustainably utilizing the fertility of the soil.

c) Agroforestry

Only few participants from the collected data implemented agroforestry in their farms. They have both crops and trees situated in the same farm. For example, there is a plot for spinach, onions, carrots, maize, tomatoes, nuts, beans. Alongside the mentioned crops, there were lemon trees, orange trees, avocado trees, morula and *mokgalo* trees. The participants have explained that they did not plant morula and mokgalo trees, they just found them in their farms and infuse crops in the same space, thereby providing a shade to the crops. They have found agroforestry crucial because they experience extreme temperatures and these trees serves as a shell for the crops not to be affected by too much sunlight and heavy rainfall.

Intergenerational Knowledge Transfer

- **Knowledge from generation to generation**

The knowledge associated with indigenous agricultural practices has been passed on from one generation to the next. Additionally, the knowledge is transmitted orally by elders (indigenous knowledge holders) to the young generation in the community. This is one of the main reasons that has ensured that there is food security in the households of Moruleng, since the community is working together to fight food insecurity. For centuries, indigenous agricultural practices have helped in maintaining food security within the Bakgatla Ba Kgafela, making it possible for them to have accessibility and availability of nutritious food, as well as culturally appropriate food. The participants are still valuing and appreciating the utilization of indigenous farming practices in their community, thereby generating income and for the sustenance of their families.

The participants in the study have exhibited character traits in their indigenous agricultural practices that is embedded in the culture and traditions of their community. From the interviews between the researcher and the participants, some of the indigenous farmers stated that their relationship with the environment, knowledge and practice of indigenous farming started from a young age and the ways of utilizing those indigenous farming methods was shared by their grandparents and parents. Some of them also alluded that they have lost their grandparents but they are still grateful for the knowledge that was shared prior their passing. Moreover, it is clear from the observation that their grandparents taught them how to fertilize the soil, plant crops, manage and maintain crops, as well as the harvesting process. Participants also told the researcher that they are not just utilizing these practices to only put food on the table, but they are also using it so as to teach the younger generation about the importance of producing their own food.

- ***Passion and Hard work for indigenous farming***

The Bakgatla Ba Kgafela in Moruleng made it clear to the researcher that what they do as indigenous farmers, is driven by passion. They said that they do not just grow crops using indigenous agricultural practices to grow food but this is also followed by their love to these practices as well as their benefits. However, other participants explained that they only started to grow crops to produce food and they never liked growing food from scratch because they see it as a tiring process. One of the participants articulated to the researcher that he used to do agriculture in high school and he feel in love with it to a point where he had to make his own garden to produce vegetables, He also stated that his passion for farming has turned to be something big because he sees himself as a future large-scale farmer. In addition to this, the participant said that he is paving his route in selling vegetables to retail stores in their nearest mall. He has turned his passion into progress.

Indigenous agricultural practices are considered to be implemented manually using hands and minimal gardening tools. Thus, requiring hard work and dedication from indigenous farmers. The hard work consists of preparing the soil for the planting season such as digging the soil and dividing sections of the garden so as to plant various crops. Participants explained that they utilize ashes from fire to fertilizer the soil prior planting as well as dry grass from the fields to fertilize the soil. The types of fertilizers they use depends on seasons and crops. Furthermore, this require them to work hard and ensure that they produce good crops for a good yield. One of the participants stated that hard work contributes towards ensuring that the crops are well taken care of and protected from pest infestations.

Indigenous Ways of Soil Fertilization

- ***Animal manure***

Majority of indigenous farmers in Moruleng use animal manure, especially cow dung to fertilize the soil. The cow dung is collected from the kraal, placed in the sun to dry then sprinkled into the soil while turning it over and adding water. According to the participants, this is a way of preparing land prior planting. Moreover, soon after preparation process, planting process begins. Some of the indigenous farmers prefer to use the animal manure when it is dry, while others prefer to combine it with water. However, almost 90% of the participants explained to the researcher that animal manure works best when it is dry. It was said by the participants that using cow dung does not only fertilize the soil but it also helps to destroy pests and diseases because it attracts insects that can deal with pests. The gathered data from the participants reveal the use of animal manure activate and maintain sustainable indigenous agricultural practices.

- ***Mulching***

Dry grass might be a waste to be thrown away. However, indigenous farmers in Moruleng explained that dried grass is one of the significant elements in making their indigenous farming successful. They also explained that combining the soil with dried grass enhanced its fertility and the dried grass has a lot of nutrients which is needed for the soil. The participants also said that they have been using this method since they started backyard farming, and they attested that the method is reliable and effective. In this process, the grass is mixed with the soil a few days prior ploughing. Sometimes they combine dead leaves with dried grass to make it more effective. According to the participants, this is an old method of fertilizing the soil and it has never disappointed them.

Prospects of Indigenous Agricultural Practices

- ***Access to nutritious food***

Access to food at all times is one of the elements of food security whereby can access healthy and nutritious food at all times. All participants interviewed stated that the main reason for using indigenous agricultural practices in their home gardens was to have easier access to food. The participants further stated that they have these gardens in their backyards to avoid purchasing from retail stores. It was explained by the participants that as they are now aged, the body needs nutritious food since they often get sick. They produce food that has the potential to boost their immune system when consumed. The participants also explained to the researcher that they eat organic fruits and vegetables from their home gardens, which are rich in vitamins and iron. They further outlined that they rarely seek medical attention.

The participants further explained how the use of indigenous agricultural practices has prompted them to use organic strategies of farming and thus a great overflow of organic food that they feel will not be harmful not only to their health but their environment as well. According to them, organic food is the healthiest food one can access and it is of great advantage to access such foods in their backyards. Moreover, retail store vegetables are genetically modified and some of the commercial farmers that sell vegetables to the retail stores use fertilizers infused with certain chemicals to increase production faster. With that being said, they prefer to grow their own organic foods.

- ***Indigenous agricultural practices are cost-effective***

It was mentioned by the participants that when growing their own food in their gardens, they do not have to buy from retail stores. One of them additionally asked that why they have to buy from retail stores when they have the knowledge and hands to grow their own food. They further said that sometimes they do not have money to buy vegetables at retail stores as the prices are now escalating hence, they grow their own food to always have access to them even when they cannot afford to purchase. Indigenous agricultural practices have not only helped them to feed themselves, but they are also able to generate income to survive.

Challenges Faced by Indigenous Farmers in Moruleng

There are numerous challenges experienced by indigenous farmers in Moruleng community. One of them is the issue of water around the area they live in. The participants stated that sometimes they can spend the whole month without having access to water, and they have to make a plan. As a result, there is no way they can grow crops without having water availability. One of them explained that not all farmers in a community have boreholes in their yards. Therefore, the situation hinders the process of the use of indigenous agricultural practices. They are now depending on rainwater to irrigate their crops and if there is no rain, nothing will be produced. They also stated that sometimes they do ask for water from their neighbors, but they cannot ask for water every day. One of the indigenous farmers mentioned that she is passionate about indigenous farming but because of the lack of water, she turns to lose interest in farming. She further said that they are trying by all means to feed themselves and their families through farming as they do not have money to buy in retail stores. They further ask themselves on how they are going to pursue that while they have scarcity of water.

Unstable weather conditions are hindering the process of growing crops and the use of indigenous agricultural practices in Moruleng community. Participants elaborated that they experienced extreme temperatures for the past months and the situation is recurring, which resulted in crop extinction. During winter, they lost large number of leafy vegetables because of extreme cold temperatures. The local farmers also explained that even heat affects the whole farming process. In this case, they use nets to protect crops from being extinct. Despite the use of nets, sometimes the winds, rainfall and heat can be severe and erode all the surface crops. Indigenous farmers in Moruleng explained that they experience pests, termites and weeds, which often lead to crop disease and failure. They also stated that they have ran out of options to kill them, as they tried with organic insect repellents but failed. They have tried avoiding pests by creating a statue in their gardens to serve as a threat to birds, but birds are already used to them to a point they no longer become scared of the statues. With regard to the existing challenge, they alluded that they use ash to kill termites, but the strategy is no longer working for them. Furthermore, they use various types of cans with different colours and open small holes. A wire is connected to holes and the whole setup is used to surround the crops. This is a method they use to chase pests, as they would pull the wire; thereby ensuring that irritating sound and movement of cans scare pests. According to the gathered data, weeds serve as a disadvantage to their crops. Additionally, weeds disturb crops from growing as they compete with crops on water. The water that should be absorbed by crops, is absorbed by weeds. In this statement, weeds absorb more water, affecting crops to grow freely. They often try to remove weeds by hand, but they still grow alongside crops. They are unable to produce more food because of weeds as it reduces their crop yields. Consequently, weeds consume more nutrients which should be consumed by crops, thus reducing the nutrients and quality of crops.

Communities have created complex farming systems that are regionally suited to assist them manage a range of settings to meet their sustenance demands after generations of cultural and biological evolution (Netting 1993). Rural communities receive food supplies from indigenous crop cultivation adds (Gao, Rao et al. 2001). Among Bakgatla-ba-Kagfela in Moruleng, certain households have their own type of indigenous agricultural practices and knowledge embedded in them because it has been information that was passed down from generation to generation, experts to novices via word of mouth, participation, observation which is what has kept the participants to promote food security, because having home gardens/farms have been a way to sustain themselves regardless of how indigenous knowledge seems to disappear, not being fully recognized and acknowledge but still grow indigenous crops with the knowledge that has been preserved in the community for years. Indigenous agricultural practices are practices adopted and utilized by various indigenous communities, and the knowledge associated with these practices are commonly passed down from one generation to the other. The Bakgatla-Ba-Kgafela in Moruleng use indigenous agricultural practices such as mixed cropping, crop rotation, agroforestry and inter cropping. The knowledge on the application of these practices to their farming systems was passed on to them by their parents or elders. All these practices combined have all the needed elements for a farm to give a good agricultural output (Baye &Teshome, 2020).

These indigenous agricultural practices, especially mixed cropping is now viewed as a promising technique to develop sustainable farming systems. Mixed cropping and crop rotation result in better yields, better pest, weeds, and disease control, control of soil erosion, and improvement and maintain soil erosion (Gururani, Sood et al., 2021). These were the reasons why the participants in Moruleng continued utilizing these practices. These practices allow them to continue practicing indigenous farming and maintain their food system and promote food security even when they cannot afford western farming tools and equipment. The indigenous farmers of Moruleng have explained that with crop rotation, crops are selected by looking at whether they complement each other so that they grow properly. This is asserted by (Sharma, Kanta et al. 2020) that crops are carefully chosen based on their relationship, and they complement or supplement each other rather than competing. Intercropping deals away with most of the challenges that might be faced with crops, pests, and the soil (Mousavi & Eskandari, 2011).

In order for people to be able to live an active and healthy life at all times, they must have access to sufficient, safe, and nutritious food at all times, regardless of their dietary needs and preferences. It is said to be food security at all levels, at home, national, regional, and global levels (Boussard, Trouvé et al., 2010) It is evident that the community of Moruleng priorities with farming and gardening is to ensure that there is access to enough food at all times to have and lead an active healthy lifestyle. Food security was acknowledged to be a serious issue by the middle of the 1990s, 46 ranging from a personal to a global scale. However, access now required enough food, showing that protein-energy malnutrition is still a concern. The concept was expanded to include both food safety and nutritional balance in response to worries about food composition and the minimal dietary requirements for an active and healthy life. Socially or culturally influenced food choices are increasingly considered. The concept had lost its simplicity and was no longer a goal in and of itself, but rather an intermediary collection of acts that contribute to an active and healthy life, according to the potentially high degree of context specificity, Boussard, Trouvé et al. (2010), this is seen how the participants responded, that access to food is not just about any food but food that is appropriate for them holistically. In choosing the types of vegetables to plant, the participants looked at the nutritional benefits first, as some stated that they battled sicknesses that required them to eat healthy food.

FUTURE RESEARCH DIRECTIONS

From the above sections, the existing literature presented out some areas where further research is needed. In-depth exploration and discussion of the integration of indigenous agricultural practices with conventional farming practices would be helpful. For instance, further research can compare and contrast indigenous agricultural practices with the conventional practices, as well as looking at their nature and how they are applied in various communities to ensure sustainable agriculture. It would also be helpful for future research to focus also on various policies that support, manage and protect the knowledge associates with indigenous agricultural practices. Further research should not only focus more on food security, but also on nutritional security.

CONCLUSION AND RECOMMENDATIONS

Indigenous agricultural practices play a major role in the livelihoods of small and large-scale farmers among the Bakgatla ba Kgafela in Moruleng. These practices have allowed them to remain food secure and generate income from the surplus crops. Although these were discarded on national level as civilization emerged and conventional agricultural practices took over. However, the practices remained their shield in preventing food insecurity and poverty to the Bakgatla Ba Kgafela in Moruleng. The food produced by the indigenous farmers also contribute to their healthy living as they only use organic fertilizers made from wood ash, animal manure and decomposed grass. The local farmers of Moruleng community presented the evidence of the relevance and significance of indigenous farming practices, hence there is a need for recognition, documentation and preservation of these practices.

The aim of the indigenous farmers and knowledge holders was not only to grow crops and sell afterwards, but to feed themselves and the community. As a result, their creativity towards the use of indigenous agricultural practices such as opening Pholo Modi wa Sechaba program, have assisted the community and ensure nutritious food supply throughout the year. Accessibility of food continues to be an urgent issue that needs to be addressed from both the individual and community perspectives. Population continues to grow rapidly, which may result in more food needed to accommodate people. This can be achieved when a community is united and tackle food insecurity, ensuring improved food accessibility and availability on daily basis.

The study revealed that indigenous agricultural practices are still valuable in the community and deserves to be recognized. The knowledge associated with indigenous agricultural practices is occurring as it has been passed down from one generation to the other within the community. There is a need to document the knowledge associated with these practices. Indigenous agricultural practices play a crucial role in promoting food security among the Bakgatla Ba Kgafela in Moruleng village. Although there are some challenges faced by famers, having accessibility and availability of food is a top priority among farmers and this implies that there is a future behind the utilization of indigenous farming practices. Indigenous agricultural practices can make a huge contribution in sustainable development. Therefore, it is recommended that policy makers should promote the utilization of indigenous agricultural practices as well as to acknowledge their potential contribution to the agricultural sector. It is recommended that:

- Agricultural Development Projects should engage with farmers in Moruleng and provide them with financial assistance and conduct workshops on farming

- Water is a basic need, therefore households in Moruleng should be supplied with water regularly and built well operating reservoirs for the community at large.
- There should be bigger land allocated for farming in the community where indigenous farmers are going to be given enough space to grow crops
- The knowledge associated with indigenous agricultural practices should be documented, disseminated, recognized and valued
- There is a need for Policy makers to promote the use of indigenous farming practices as well as to acknowledge their potential contribution to the agricultural sector.

REFERENCES

Apraku, A., Morton, J. F., & Gyampoh, B. A. (2021). Climate change and small-scale agriculture in Africa: Does indigenous knowledge matter? Insights from Kenya and South Africa. *Scientific African, 12*, e00821. DOI: 10.1016/j.sciaf.2021.e00821

Arcand, M. M., Bradford, L., Worme, D. F., Strickert, G. E., Bear, K., Johnston, A. B. D., Wuttunee, S. M., Gamble, A., & Shewfelt, D. (2020). Sowing a way towards revitalizing Indigenous agriculture: creating meaning from a forum discussion in Saskatchewan, Canada, *Canadian Science Publishing* 1840 Woodward Drive, Suite 1, Ottawa, ON K2C 0P7. 5: 619-641.

Babalola, O. O., & Glick, B. R. (2012). Indigenous African agriculture and plant associated microbes: Current practice and future transgenic prospects. *Scientific Research and Essays, 7*(28), 2431–2439.

Baye, B., & Teshome, W. (2020). The role indigenous knowledge in agricultural farming practices: The case of Gonder and Gojam area, Amhara regional state, Ethiopia. *International Journal of Advanced Research in Biological Sciences, 7*(12), 106–112.

Berhane, G., Berhane, G., Ragasa, C., Abate, G. T., & Assefa, T. W. (2018). *The state of agricultural extension services in Ethiopia and their contribution to agricultural productivity*. Intl Food Policy Res Institution.

Berry, E. M., Dernini, S., Burlingame, B., Meybeck, A., & Conforti, P. (2015). Food security and sustainability: Can one exist without the other? *Public Health Nutrition, 18*(13), 2293–2302. DOI: 10.1017/S136898001500021X PMID: 25684016

Borbale, M. K., Pohekar, A. U., Nistane, J. R., Nichit, N. S., Yesankar, T. I., & Kaware, A. G. (2021). Study of traditional method of weeding and fertilizing and their drawbacks. *International Research Journal of Engineering and Technology, 8*(3), 236–242.

Borthakur, A. & Singh, P. (2021). Indigenous agricultural knowledge towards achieving sustainable agriculture. *Sustainable Agriculture Reviews 50: Emerging Contaminants in Agriculture*, 401-413.

Boussard, J., Choplin, G., Strickner, A., & Trouvé, A. (2010). *World Food Summit Plan of Action, World Food Summit,* 13-17 Novembre, Roma http://www. fao. org/wfs/index_en. Htm.

Chappell, M. J., & LaValle, L. A. (2011). Food security and biodiversity: Can we have both? An agroecological analysis. *Agriculture and Human Values, 28*(1), 3–26. DOI: 10.1007/s10460-009-9251-4

Clarke, V., Braun, V.& Hayfield, N. (2015). Thematic analysis. *Qualitative psychology: A practical guide to research methods,* 222(2015), 248.

Cloete, P. C. (2010). *Economic growth and development through agriculture: the case of the North West Province of South Africa* (Doctoral dissertation, University of the Free State).

Dasgupta, R., Dhyani, S., Basu, M., Kadaverugu, R., Hashimoto, S., Kumar, P., Johnson, B. A., Takahashi, Y., Mitra, B. K., Avtar, R., & Mitra, P. (2021). Exploring indigenous and local knowledge and practices (ILKPs) in traditional jhum cultivation for localizing sustainable development goals (SDGs): A case study from Zunheboto district of Nagaland, India. *Environmental Management*, •••, 1–13. PMID: 34342685

Ekobi, G.A., Mboh, L. & Tanga, P. (2023). Contribution of Indigenous Knowledge to Agricultural Growth in South Africa: A Case of Disaneng Community in the Ratlou Local Municipality. *African Journal of Inter/Multidisciplinary Studies, 5*(1), 1-12.

FAO. (2009). The State of Food Insecurity in the World. Economic crises-impacts and lessons learned, FAO, Roma. *Agriregionieuropa (Ancona)*, •••, 30.

Fox, J.. (2018). Shifting cultivation and climate change in Southeast Asia. *Nature Climate Change*, *8*(1), 1–3.

Galvin, K. A.. (2020). Climate change and pastoralism: Traditional coping mechanisms and sustainable adaptation in Maasai communities. *Global Environmental Change*, *64*, 102–110.

Gao, W., Rao, V. R., & Zhou, M. (2001). *Plant genetic resources conservation and use in China*. Bioversity International.

Gonçalves, C. D. B. Q., Schlindwein, M. M., & Martinelli, G. D. C. (2021). Agroforestry systems: A systematic review focusing on traditional indigenous practices, food and nutrition security, economic viability, and the role of women. *Sustainability (Basel)*, *13*(20), 11397. DOI: 10.3390/su132011397

Gururani, K., Sood, S., Kumar, A., Joshi, D. C., Pandey, D., & Sharma, A. R. (2021). Mainstreaming Barahnaja cultivation for food and nutritional security in the Himalayan region. *Biodiversity and Conservation*, *30*(3), 551–574. DOI: 10.1007/s10531-021-02123-9 PMID: 33526962

Hamadani, H., Rashid, S. M., Parrah, J. D., Khan, A. A., Dar, K. A., Ganie, A. A., Gazal, A., Dar, R. A., & Ali, A. (2021). Traditional farming practices and its consequences. *Microbiota and Biofertilizers. Ecofriendly Tools for Reclamation of Degraded Soil Environs*, *2*, 119–128.

Havas, K., & Salman, M. (2011). Food security: Its components and challenges. *International Journal of Food Safety, Nutrition and Public Health*, *4*(1), 4–11. DOI: 10.1504/IJFSNPH.2011.042571

Huynh, C. V., Pham, T. G., Nguyen, T. Q., Nguyen, L. H. K., Tran, P. T., Le, Q. N. P., & Nguyen, M. T. H. (2020). Understanding indigenous farming systems in response to climate change: An investigation into soil erosion in the mountainous regions of Central Vietnam. *Applied Sciences (Basel, Switzerland)*, *10*(15), 5091. DOI: 10.3390/app10155091

Ingram, J. (2011). A food systems approach to researching food security and its interactions with global environmental change. *Food Security*, *3*(4), 417–431. DOI: 10.1007/s12571-011-0149-9

Jones, A. D., Ngure, F. M., Pelto, G., & Young, S. L. (2013). What are we assessing when we measure food security? A compendium and review of current metrics. *Advances in Nutrition*, *4*(5), 481–505. DOI: 10.3945/an.113.004119 PMID: 24038241

Kazmi, T., Ghafoor, A., Chaudhry, A. A., & Khan, S. E. (2014). Farmers beliefs about indigenous farming practices and sustainable agricultural development. *Pakistan Journal of Agricultural Research*, *27*(1).

Kom, Z., Nethengwe, N. S., Mpandeli, S., & Chikoore, H. (2023). Indigenous knowledge indicators employed by farmers for adaptation to climate change in rural South Africa. *Journal of Environmental Planning and Management*, *66*(13), 2778–2793. DOI: 10.1080/09640568.2022.2086854

Korat, H. & R. Mathukia. (2022). *Effect of natural farming, organic farming and conventional farming on soil physical, chemical and biological properties.*

Kovach, M. (2021). *Indigenous methodologies: Characteristics, conversations, and contexts*. University of Toronto press.

Kumar, G., & Chander, H. (2017). Documentation of indigenous agricultural implements, practices and other conservation techniques in subtropical climatic zone of Shivalik hills, North Western Himalaya. *J. Biol. Chem. Chron, 3*(2), 15–23.

Kumar, S., Kumar, N., & Vivekadhish, S. (2016). Millennium development goals (MDGS) to sustainable development goals (SDGS): Addressing unfinished agenda and strengthening sustainable development and partnership. *Indian Journal of Community Medicine : Official Publication of Indian Association of Preventive & Social Medicine, 41*(1), 1. DOI: 10.4103/0970-0218.170955 PMID: 26917865

Leal Filho, W., Barbir, J., Gwenzi, J., Ayal, D., Simpson, N. P., Adeleke, L., Tilahun, B., Chirisa, I., Gbedemah, S. F., Nzengya, D. M., Sharifi, A., Theodory, T., & Yaffa, S. (2022). The role of indigenous knowledge in climate change adaptation in Africa. *Environmental Science & Policy, 136*, 250–260. DOI: 10.1016/j.envsci.2022.06.004

Leroy, J. L., Ruel, M., Frongillo, E. A., Harris, J., & Ballard, T. J. (2015). Measuring the food access dimension of food security: A critical review and mapping of indicators. *Food and Nutrition Bulletin, 36*(2), 167–195. DOI: 10.1177/0379572115587274 PMID: 26121701

Malapane, O. L., Musakwa, W., & Chanza, N. (2024). Indigenous agricultural practices employed by the Vhavenda community in the Musina local municipality to promote sustainable environmental management. *Heliyon, 10*(13), e33713. DOI: 10.1016/j.heliyon.2024.e33713 PMID: 39040358

Misselhorn, A., Aggarwal, P., Ericksen, P., Gregory, P., Horn-Phathanothai, L., Ingram, J., & Wiebe, K. (2012). A vision for attaining food security. *Current Opinion in Environmental Sustainability, 4*(1), 7–17. DOI: 10.1016/j.cosust.2012.01.008

Modi, A. T. (2003). What do subsistence farmers know about indigenous crops and organic farming? Preliminary experience in KwaZulu-Natal. *Development Southern Africa, 20*(5), 675–684. DOI: 10.1080/0376835032000149306

Mousavi, S. R., & Eskandari, H. (2011). A general overview on intercropping and its advantages in sustainable agriculture. *Journal of Applied Environmental and Biological Sciences, 1*(11), 482–486.

Mutengwa, C. S., Mnkeni, P., & Kondwakwenda, A. (2023). Climate-smart agriculture and food security in southern Africa: A review of the vulnerability of smallholder agriculture and food security to climate change. *Sustainability (Basel), 15*(4), 2882. DOI: 10.3390/su15042882

Netting, R. M. (1993). *Smallholders, householders: farm families and the ecology of intensive, sustainable agriculture*. Stanford University Press. DOI: 10.1515/9781503622067

Notsi, L. (2012). African indigenous farming methods used in the cultivation of African indigenous vegetables: A comparative study of Tsitas Nek (Lesotho) and Mabeskraal village (South Africa). *Conference on Strategies to Overcome Poverty and Inequality: Towards Carnegie III at University of Cape Town,* South Africa.

Onyancha, O. B. (2022). Indigenous knowledge, traditional knowledge and local knowledge: What is the difference? An informetrics perspective. *Global Knowledge, Memory and Communication*, *73*(3), 237–257. DOI: 10.1108/GKMC-01-2022-0011

Petzold, J., Andrews, N., Ford, J. D., Hedemann, C., & Postigo, J. C. (2020). Indigenous knowledge on climate change adaptation: A global evidence map of academic literature. *Environmental Research Letters*, *15*(11), 113007. DOI: 10.1088/1748-9326/abb330

Pham, L. T. M. (2018). *Qualitative approach to research a review of advantages and disadvantages of three paradigms: Positivism, interpretivism and critical inquiry*. University of Adelaide.

Pimentel, D., & Burgess, M. (2014). An environmental, energetic and economic comparison of organic and conventional farming systems. Integrated Pest Management, Springer: 141-166. DOI: 10.1007/978-94-007-7796-5_6

Ponge, A. (2011). Integrating indigenous knowledge for food security: perspectives from the millennium village project at Bar-Sauri in Nyanza Province in Kenya. *International Conference on Enhancing Food Security in the Eastern and Horn of Africa regions, held at the Imperial Royale Hotel in Kampala, Uganda*

Richards, P. (2023). *Indigenous agricultural revolution: ecology and food production in West Africa*. Routledge. DOI: 10.4324/9781003383734

Rosenberg, M. (2021). Thoughts about food security, food loss and waste and what has to be done. *AIMS Agriculture and Food*, *6*(3), 797–798. DOI: 10.3934/agrfood.2021048

Shahmohamadloo, R. S., Febria, C. M., Fraser, E. D., & Sibley, P. K. (2021). The sustainable agriculture imperative: A perspective on the need for an agrosystem approach to meet the United Nations Sustainable Development Goals by 2030. *Integrated Environmental Assessment and Management*, *18*(5), 1199–1205. DOI: 10.1002/ieam.4558 PMID: 34821459

Sharma, I. P., Kanta, C., Dwivedi, T., & Rani, R. (2020). *Indigenous agricultural practices: A supreme key to maintaining biodiversity*. Microbiological Advancements for Higher Altitude Agro-Ecosystems & Sustainability.

Sohail, M. T., & Chen, S. (2022). A systematic PLS-SEM approach on assessment of indigenous knowledge in adapting to floods; A way forward to sustainable agriculture. *Frontiers in Plant Science*, *13*, 990785. DOI: 10.3389/fpls.2022.990785 PMID: 36092446

Trigo, A., Marta-Costa, A., & Fragoso, R. (2021). Principles of sustainable agriculture: Defining standardized reference points. *Sustainability (Basel)*, *13*(8), 4086. DOI: 10.3390/su13084086

Wesley, A., & Faminow, M. (2014). Background paper: research and development and extension services in agriculture and food security. *Asian Development Bank Economics Working Paper Series*, (425).

Wilson, S. (2020). *Research is ceremony: Indigenous research methods*. Fernwood publishing.

Xia, Y., Zhang, M., Tsang, D. C., Geng, N., Lu, D., Zhu, L., Igalavithana, A. D., Dissanayake, P. D., Rinklebe, J., Yang, X., & Ok, Y. S. (2020). Recent advances in control technologies for non-point source pollution with nitrogen and phosphorous from agricultural runoff: Current practices and future prospects. *Applied Biological Chemistry*, *63*(1), 1–13. DOI: 10.1186/s13765-020-0493-6

ADDITIONAL READINGS

Kazmi, T., Ghafoor, A., Chaudhry, A. A., & Khan, S. E. (2014). Farmers beliefs about indigenous farming practices and sustainable agricultural development. *Pakistan Journal of Agricultural Research, 27*(1).

Lwoga, E. T., Ngulube, P., & Stilwell, C. (2010). Managing indigenous knowledge for sustainable agricultural development in developing countries: Knowledge management approaches in the social context. *The International Information & Library Review*, *42*(3), 174–185. DOI: 10.1080/10572317.2010.10762862

Masekoameng, M.R. & Molotja, M.C. (2019). The role of indigenous foods and indigenous knowledge systems for rural households' Food Security in Sekhukhune District, Limpopo Province, South Africa. *Journal of Consumer Sciences*.

Ngema, P. Z., Sibanda, M., & Musemwa, L. (2018). Household food security status and its determinants in Maphumulo local municipality, South Africa. *Sustainability (Basel)*, *10*(9), 3307. DOI: 10.3390/su10093307

Wilson, S. (2020). *Research is ceremony: Indigenous research methods.* Fernwood publishing.

KEY TERMS AND DEFINITIONS

Food Security: It is defined as the ability of an individual to always have economic and physical access to nutritious food to meet their dietary needs at all times.

Indigenous Agriculture: It is referred to as an agriculture that follows indigenous methods and techniques of farming, that are developed in a specific society or culture through generations.

Indigenous Knowledge Systems: It is a term used to describe multiple and holistic knowledge systems created and transmitted through the generations within indigenous societies.

Indigenous Agricultural Practices: Are farming methods and practices that are developed by indigenous communities to enhance their livelihoods

Sustainable Agriculture: It is defined as farming in sustainable ways to meet textile needs and society's available food without the ability to compromise current and future generations.

Chapter 10
A Pursuit to Mainstream Indigenous Cannabis Growers in the Cannabis Market to Foster Socioeconomic Development:
A Theoretical Dialogue

Kamogelo Macdonald Noge
https://orcid.org/0000-0003-4203-3153
North-West University, South Africa

Mothusiotsile Edwin Maditsi
North-West University, South Africa

ABSTRACT

It is becoming increasingly difficult to ignore cannabis as a resource that has the potential to drive socio-economic development in indigenous communities. This resource has for a long time been an overlooked commodity within indigenous communities based on the negative perceptions and stigma that surround it. Cannabis has the potential to contribute towards the economy and various indigenous communities in South Africa such as AmaMpondo in the Eastern Cape Province are found to be the most producing area of cannabis for various purposes that contributes towards their local and household sustainability. This chapter applies a qualitative document analysis (QDA) to elucidate meaning on the trends and arguments within the agricultural cannabis thematic area.

INTRODUCTION

For many centuries, indigenous communities have sought to utilize their knowledge to survive drastic conditions that did or do not favor them. Given that, there have been recent developments in the mainstream field of agriculture as it is one of the biggest means of production. Indigenous knowledge (IK) plays a vital role in the knowledge economy as well as to foster social and economic development, yet it is marginalized. An example of such marginalization can be drawn from indigenous cannabis

DOI: 10.4018/979-8-3693-7964-6.ch010

growers (Moahi, 2007). These indigenous cannabis growers are people within various communities who produce, process, market and sell cannabis strains, products and ointments which people purchase. From an indigenous perspective, there is a relationship between indigenous knowledge systems (IKS), indigenous knowledge (IK), and indigenous agricultural practices (IAP). Indigenous knowledge system (IKS) is an umbrella field which comprises of systems with which IK is embedded on through skills and productions. Indigenous cannabis growers depend on IK and the skills transferred from generations to generations (Abah *et al*, 2015). Moreover, indigenous cannabis growers fall within the indigenous community due to how they produce or practice their cannabis distribution into their community economy. Although they are regarded as unregulated traders, the cannabis market is rapidly expanding which now puts the policy makers in light to reconstruct the policies that are not in favor of the cannabis growers.

STATEMENT OF THE PROBLEM

African indigenous agricultural practices are viewed as backward and out of context due to the shifting of times however they still are relevant in communities that have no total access to basic services (Bebbington, 2017). Little has been documented towards indigenous agricultural practices on cannabis farming because of the politics surrounding this matter. In addition, it has caused indigenous cannabis growers on cannabis to be side-lined in the academic spaces and economic dispensation (Ncube, 2018). In some instances, Indigenous knowledge is not viewed as science on its own yet carries multi facets of sustainable development that can bring about change in local indigenous communities, only the Western science is deemed relevant. There are prospects that cannabis farming can enhance such as economic growth, lower unemployment rates, health improvement, preserve indigenous agricultural practices and skills transfer. These are the factors that the institutions of higher learning and government are neglecting, not prospecting from the indigenous point of view. Moreover, challenges such as legal frameworks are hindering the growth of cannabis market. This is prohibited through the then Medical, dental and pharmacy Act (No. 13 of 1928) which was imposed by the neo-liberal organizations such WHO and its accomplices (Parle, 2019). Illegal nature of cannabis possession or large production appears to be a key determinant towards contribution to households and poverty alleviation in most provinces where they produce cannabis.

The Chapter's Objectives Are as Follows:

- To explore the role of indigenous cannabis growers in ensuring socio-economic development in local indigenous communities.
- To investigate the role and use of indigenous agricultural practices on cannabis production.
- To explore challenges and prospects faced by indigenous cannabis growers in producing cannabis.
- To examine the legal frameworks giving effect to the preservation and protection of IK held by indigenous cannabis growers.
- To describe indigenous measures that can be applied to promote, preserve and mainstream indigenous agricultural practices (IAP) of cannabis production.

The Research Questions Are as Follows:

- What role do indigenous cannabis growers play in ensuring socioeconomic development in local communities?

- What is the role and use of indigenous agricultural practices on cannabis production?
- What are the challenges and prospects faced by indigenous cannabis growers in producing cannabis?
- Which legal frameworks give effect to the preservation and protection of IK of indigenous cannabis growers?
- Which indigenous measures can be applied to promote, preserve and mainstream indigenous agricultural practices of cannabis production?

THEORETICAL FRAMEWORK UNDERPINNING THE RESEARCH

The theoretical framework that underpins this chapter is critical theoretical framework, indigenous standpoint and African renaissance. According to Odora (2021) critical theory aims to decolonize and provide an analysis for making dimensional powers visible. Critical theory has created a space within the Western worldviews because its basis is to voice, represent and reveal the truth about certain facts that have not been challenged within Western science (Singh & Major, 2017). It is relevant for this chapter as it created critical gaps from Western thoughts or research that continues to enjoy the high ranking of academia, sidelining other useful perspectives from different settings. Critical theory is relevant for this chapter as it underpins issues of multiple realities influenced by human beings towards their economy, politics, rights, and values (Ryan, 2018). The reality of the matter is that indigenous farmer/ indigenous cannabis growers have their economic space, political willingness, and the rights to operate their businesses within the legal frameworks of the South African constitution and policies. Critical theory tackles the issues of legalizing cannabis because it is not challenged thus far, growing the cannabis industry to sustain the economy by doing away with international policies imposed by neo-liberal organizations. The need to reclaim and revive the indigenous knowledge practices from different streams of interest is gradually growing; therefore, indigenous standpoint theory is also be employed in this paper (Chilisa, 2017).

According to Nakata (2007) indigenous standpoint theory demonstrates the position of a researcher towards their experiences in terms of political, social, and cultural stance. This theory is relevant for this paper as it is one of the theories that gives voice to the minority groups of indigenous peoples and supports the notion of reclaiming the ways of knowing in marginalized communities. Kovach (2015) posits that indigenous ways of knowing have always been overlooked due to the imposed Western methodologies that continue to suppress the indigeneity of African research. Furthermore, there is a need to build African frameworks and methodologies that will fit within the line of academic space to restore the indigeneity of African people with their practices. These two theories are relevant for this chapter because they both challenge the Western methodologies imposed on African studies. Indigenous cannabis growers have sought to be recognized as people who can contribute towards economic freedom, sustainable development, and contribution towards the agricultural sector.

Lastly, African renaissance theory complements the mentioned theories as it alludes that Africans must be guided or live within their African values and norms because they are the building blocks of an African child (Hoppers, 2002b). Indigenous knowledge systems as a complex set of skills with innovations is complicit with the African renaissance theory in its ability of flexible change that initiates the social and value structures of an African. Moreover, indigenous cannabis growers distribute cannabis for poverty reduction therefore African values such as ubuntu must be seen as imperative within the social boarders of poverty reduction policies by freely allowing the cannabis market to expand legally and to enhance sustainable development goals.

HISTORICAL BACKGROUND OF CANNABIS PRODUCTION AROUND THE WORLD

South Asia is where cannabis originated and developed, and there is a long history there. Approximately a millennium ago, the psychotropic strain made its way to both East and North Africa. Through trade networks, it gradually and unevenly expanded over Africa. Cannabis users throughout Asia consumed the drug through various pipes, but Africans who smoked the plant invented the modern way of using it, which produced a much stronger impact much faster (Ambler, 2020). Many cannabis users were regular workers, especially the couriers who travelled great distances to deliver things. There were several functions that smoking served. It lightened the monotony and weight of the labor, and it further intensified the pleasure while also calming smokers down.

Cannabis is not only for smokers but there is a good potential on the entrepreneurial part where a lot of people have seen an opportunity to start developing skin care products. This is one of the factors that need government consultation to find out the key issues that can be addressed to help cannabis cultivators (farmers) have a free market growth without any legal disputes. Historically, Rastafarians have been perceived as a group of people with psychological problems due to fact that they smoke cannabis as part of their healing and spiritual source (Paruk, 2022). However, the Rastafarians falls under the umbrella of indigenous peoples as a religion. Furthermore, the United Nations Declaration on the rights of Indigenous Peoples (UNDRIP) conforms to the freedom of indigenous peoples to enact reconciliation with the resources they have rights on (Gover, 2015). This is the framework that seeks to win the rights to the legalization of cannabis, traditional knowledge, cultural practices, and political freedom. In respect of the legal framework mentioned above, it includes among others traditional healers, religious leaders, and traditional healers. Part of conducting such study, it is to demonstrate that not only Rastafarians are the users of different types of cannabis but also farmers who saw a potential in healing the society through their practices (Hanson, 2020). From growing to harvesting, curing to usage, even though very little investment of financial capital is involved at least for very localized and personal consumption cannabis farmers persist. Combining this with the ancient knowledge that cannabis is a plant of mystery, a guide to the beyond and a tool for exploring alternative dimensions in a society (Malabadi & Chalannavar, 2023).

CONTEMPORARY UNDERSTANDING OF CANNABIS IN SOUTH AFRICA

In the Eastern Cape province (South Africa), cannabis is grown for recreational uses and on the other hand indigenous practitioners utilize it for different purposes. Literature shows that households in the province have principles of adherence or social differences in terms of how they distribute and use (Mbhele & Eric, 2023). Some use it to generate income, animal stamina, medicinal purposes and to appease their ancestors as a tool of communication which they call it in Xhosa (ukuphahla). One of the most common examples is the Mlanjeni village, there are three categories of indigenous cannabis growers situated in the village. The first category is the ones that grow and smoke for their personal satisfactory, the second is those that only smoke and distribute and the last one are those who only distribute or use for spiritual purposes, medicinal purposes such as traditional healers (Semenya & Potgieter, 2017).

As mentioned above, there are principles that are followed towards the trading of cannabis and growing. In an indigenous perspective this can be named as indigenous principled measures that guide the production of cannabis within the villages. According to Lum (2013), when a boy within the household wants to produce cannabis, he must cultivate it next to the river or at the mountain because the culture that is built in within the household permits only elders to grow cannabis in the yard. However, the planting of cannabis next to the rivers and mountain, it is a security measure to avoid being arrested by the police due to not having permits. This is why there is a need to decriminalize cannabis production because it has the potential for economic growth. As cannabis cultivation moves from the black markets to the multi-billion-dollar industries, indigenous farmers of cannabis remain excluded and isolated (Black, 2020). The contribution of indigenous farmers goes unnoticed due to the approval of policies that continues to discredit their systems of production on cannabis sativa. The role that indigenous cultivation practices have played in the protection of cannabis genetic resources and related traditional knowledge has not been adequately addressed (Hasan & Abdullah, 2015). In the Eastern Cape, cannabis farmers cultivate mostly in the hidden bushes and indoors due to a serious prohibition, quasi-legal status, and the need to hide growing sites.

While species of cannabis grows in varied and harsh environmental conditions, cannabis growers in the Eastern Cape have sought to plant it in the nearest bushes to avoid imposed legal enforcements (Maroyi, 2017). The soil there tends to be fertile and rich in nutrients although remaining dry throughout the farming period with no rainfall. Some farmers grow cannabis sativa next to the riverbanks where alluvial soil dominates mostly. Literature shows that in China (Shandong province), it was observed that a suitable soil for cannabis growth should at least be PH 6.0-7.5, it must be deep, rich in nutrients and good water holding capacity to allow the smooth growth of the type of cannabis to be planted (Amaducci, 2022). With other agricultural crops, practices differ due to the mentioned positive and negative implications of growing cannabis sativa in the local rural areas of the Eastern Cape. Indigenous knowledge is applied through local decision making of the people or farmers deciding to begin with cultivation of cannabis. According to Ebhuoma and Simatele (2019), there is a wealth of local knowledge on predicting the weather/climate and farmers have developed intricate systems for gathering, predicting, interpreting and decision making in relation to weather. This helps them in controlling the production cycle of cannabis sativa. However, these indigenous practices are also applied to any other crops within the indigenous farming production.

INDIGENOUS KNOWLEDGE SYSTEMS AND SOCIO-ECONOMIC BENEFITS OF CANNABIS FARMING

Historically, cannabis was grown for various reasons which was of importance to the indigenous peoples or general cannabis growers. This industrial or agricultural crop was not only used for smoking in Africa but for economic development such as product making, fiber manufacture and trading (Rehman et al, 2021). Firstly, the people who grew up growing this plant had to find a special place whether indoors or outdoors due to the kind of type they wanted to grow. The land was supposed to be fertile, with availability of water or access to sufficient water, controlled conditions in terms of light and darkness. These are the factors that has the effects on the growth and quality of cannabis post-harvest. One of the most important aspects of strategic means of income generation is the commercial cultivation of cannabis and its trading that is overlooked due to the frontiers. Cannabis farming has the potential to contribute towards the economy, agricultural heritage, and health ministry (Santoro, 2022). It can further contribute towards the decrease of unemployment in South Africa as the farmers will actively be on duty to produce cannabis for various purposes. Notably, South Africa's land is arable and can grow cannabis on a massive scale hence there is need for a policy shifting to allow free market flow (Hall, 2004).

Cannabis economic value has the potential to mitigate the gap towards the economic dispensation, poverty reduction and social injustices in rural indigenous communities. It is estimated that the production of cannabis for sale in rural communities has the potential of contributing 27 billion in the year 2023 if the production is at a large scale (Raihan & Bijoy, 2023). The argument from different scholars reaffirms that cannabis is not only important towards religion, cultural practices, and health but also important towards socio-economic benefits (Siddiqui et al, 2022). The negative impacts of criminalizing cannabis in the economic space have caused a huge damage towards the participation of indigenous communities. These socio-economic benefits of cannabis will only be achievable through the legalization of cannabis to expand the economy. Legalization of cannabis to foster socio-economic benefits rely heavily on political measures, social, technology, legislation, and environmental aspects (Junior et al., 2023). These factors are the ones that can pave a way for legalization of cannabis at a commercial production level which will enhance sustainable development in various indigenous communities that produce cannabis. This will expand the agricultural heritage as the land will have to be cultivated and prepared to produce cannabis at a larger scale.

THE ROLE OF INDIGENOUS CANNABIS GROWERS IN ENSURING SOCIOECONOMIC DEVELOPMENT IN LOCAL COMMUNITIES

A clear understanding is needed when dealing with issues that focus on IK because it is the basis for local level on decision making particularly on agricultural practice (Nyong et.al., 2007). However, it is not limiting the practices of knowledge on agriculture only, but also on health, science, and arts. Indigenous knowledge systems enhance the generic understanding of IK as a complex set of skills, innovations and practices applied for sustainable development among indigenous communities (Hoppers, 2002). A topic on cannabis such as this one is sensitive given the centred factors around it such as politics, legality, social, psychological, and economical. The mentioned above factors may affect the gradual growth of cannabis farming in a negative or positive way due to the general views of the society and current jurisdictions.

Indigenous knowledge systems enhance human practices in its different forms of fields for sustainable development (Ngulube, 2015). Cannabis as an industrial plant, falls between the two spectrums of knowledge within the IK which is indigenous health and agriculture. Indigenous cannabis growers cultivate the plant to grow to the stage where it can be used for various purposes by applying indigenous farming practices (Schaffrick, 2023). Through crop selection, crop rotation and crop protection they can achieve a good production. From the indigenous health perspective, cannabis is used to treat certain ailments such as skin problems, influenza, and digestive issues. Although IK is undervalued in other parts of the world, its practices are used to generate income and make a better living in rural areas (Arnold, 2021). Without indigenous agricultural heritage, the production of cannabis in indigenous communities would not be active as it depends on indigenous knowledge practices (Buthelezi, 2014). This is because the mainstream agricultural institutions ignore the link between indigenous knowledge practices within the IKS. Indigenous knowledge systems do not limit the free marketing of cannabis within communities because it is complicit with knowledge economy that to assist indigenous communities to generate household income.

INDIGENOUS AGRICULTURAL PRACTICES OF CANNABIS SATIVA PRODUCTION

Indigenous knowledge plays a pivotal role in the rural areas such as Nyandeni in the Eastern cape and other neighbouring villages although it is seen as backward knowledge. This is one of the most common challenges that continues to discredit the knowledge of indigenous peoples in the villages and other knowledge systems enjoy superior ranking (Paterson, 2009). One of the objectives of this study is to enquire about the challenges that indigenous cannabis growers face in terms of the current laws and policies of cannabis production from the neoliberal organizations globally to the government of South Africa. There is a huge gap between the governing bodies of leadership that needs to coexist to boost the South African and local economy through cannabis production. Literature shows that the potential economic boost of cannabis production is worth billions of rands if taken into consideration but there is no legislation and enforcement that supports the discourse (Quarshie, 2022).

Cannabis sativa is one of man's oldest cultivated plants. Botanically it is distinct from all other plants and readily recognized. Yet among individual plants it is extremely variable. Across the world, cannabis sativa is mainly common in South Africa and dominant in the Eastern Cape province (Dumani *et al*, 2024). Pre cultivation of cannabis sativa depends on the seed distribution among the cannabis farmers. Without the pre practices such as land tenure, skills, materials and availability of other resources, production phase cannot begin. Preparation of land, irrigation, security, and other measures are the most common to the people who grow cannabis on unregulated markets. During the process of the cannabis sativa growth, the indigenous farmers continue to observe how the plant grows with also the factors that affect its growth such as sunlight, veldfires, heavy rainfall and windy conditions. Some farmers grow cannabis sativa with crops such as pumpkins, spinach, tomatoes, and watermelons. This is to help them hide the plant from the law enforcements or trespassers, they also continue to trim the plant as and when it grows taller. Post cultivation practices are the most important as these practices assist in producing good and quality cannabis sativa.

Indigenous cannabis growers on post practices, harvest the leaf's for drying using sunlight, mulching, packaging, and storing them for trading. This study aims at exploring the knowledge that these cannabis cultivators use to complete the whole cycle of growing cannabis sativa until its distribution. According to Jones (2024) the long history of cannabis cultivation, was valued for its psychoactive and non-psychoactive properties alike widely used for its fibres (usually referred to as hemp), as a food, and for a wide range of ceremonial, recreational, therapeutic, and medicinal applications across cultures.

THE CHALLENGES AND PROSPECTS FACED BY INDIGENOUS CANNABIS GROWERS IN PRODUCING CANNABIS

Legal Frameworks Underpinning Cannabis Market and Production

It is important to note that acts, laws, and policies take time to be amended and to be implemented due to heavy consultations that takes time. There is a need to legalize cannabis in South Africa to uplift the health ministry and the indigenous health practitioners (Manu et.al, 2021). Cannabis production is only allowed for medicinal purposes by the health ministry through pharmaceutical companies that continues to manipulate the cannabis cultivators which are the indigenous cannabis farmers using indigenous farming practices in South Africa. The Inkatha freedom party is one of the movements that has tried it best in making the government realize the need to legalize medical cannabis as well as its extractions (Parry & Myers, 2014). Due to the satisfaction of western powers whose grip is on the economy, the movement pleaded with at least 10 African countries to enforce policy shifts to implemented strategic ideas for this discourse. This is one of the examples that showed a clear power control over African countries by the neo-liberal organisations such as World health organisation (WHO), World trade organisation (WTO) and other technical agencies in the United Nations (UN) (Brock-Utne, 2008). Given the legal status of cannabis in the Eastern Cape province of South Africa, indigenous cannabis farmers have the rights to produce on their land. This is underpinned by the international labour organization (ILO) on convention no.169 which was adopted in the early 1950's and still active to date (Yupsanis, 2012). It states in articles 6, 27 and 28 that the indigenous peoples have the right to be consulted and participate in the extractive activities within their communities. Through the means of production in the communities the people shall be consulted with relevant channels and provisions. The current suppression of the indigenous peoples' rights shows that active laws that recognizes indigenous peoples are sidelined. The UNDRIP of 2007 advocates for the indigenous cannabis farmers to exercise their knowledge on production of cannabis through Article 5 which states that the indigenous people have the rights to maintain their political, agricultural, cultural, and economic expansion (Champagne, 2013). This allows them to have the rights to their land to freely participate in cannabis free-flowing markets through these legal prescripts. Moreover, the UNDRIP of 2007 is complicit with the South African freedom charter that states that the people shall govern particularly on their land (Suttner, 2015). Indigenous communities strive through the Ubuntu concept which encourages productions that allows access to benefit sharing (Letseka, 2014). Legalization of cannabis at a larger scale will foster access to benefit sharing among the indigenous communities. However, this can be achieved through self-determination by not allowing anyone to take a decision on behalf of them with their extractive activities. The current legal status of cannabis production on commercial level is still unclear due to the infiltration of manipulative legislative systems that supress its production hence indigenous cannabis farmers produce in private spaces.

The relationship between cannabis production and biodiversity goes a long way which carries multifacets of indigenous agricultural practices that cuts across the protection of the biodiversity through the convention on biological diversity (CBD) (Hens, 2006). The Rio declaration principle 22 states that the government and all other institutions should support the indigenous communities and be in full partnership with equal power that will benefit all the involved participators (Shelton, 2008). Whatever extract from the community should be on contractual basis to avoid manipulation and rights degradation. Lastly, the South African constitution along with the IK act of 2019 uphold the protection of indigenous knowledge as a property. Section 25 of the South African constitution states that no one should be denied of property in terms of economic values that are movable or immovable (Muller, 2011). Through the IK Act of 2019, on the protection of indigenous knowledge clause 13 provides the basis of the rights of indigenous communities that any benefit or proceeds arising from the commercial production shall be shared among the individuals who worked the land.

Cannabis Legalization

Cannabis has always been one of the medicinal plants that were neglected and still neglected by different communities based on their views and beliefs. During pre-democratic era, there was a different narrative about the usage of cannabis among communities as compared to today's communities (Ahrens, 2020). The reason behind that was the kind of traditional leadership that governed the setting that was found within communities (Mohashoa, 2018). This was favouring both the supporting people on cannabis and the unsupportive on cannabis as they were dictated by the authority governing around the communities. Most of the groups that were in support of the uses of cannabis were traditional healers as they were the ones who used them as part of their healing source (Mothibe & Sibanda, 2019). There are barriers that contributed or still contribute towards the official legalization of cannabis in South Africa which is the religious sector, current traditional leaderships, and existing neo-liberal organizations (Gwala, 2023). These are ones of the few groups that continue to create barriers for the official legalization of cannabis. Although there are still exceptions on cannabis farming within other communities, it does not guarantee a full free market of cannabis distribution as it is a private cultivation. These systems that created a segregation of African medicinal practices were brought by the colonial imposters which left a heavy segregation among black communities. The colonial powers were in control and still in control of the traditional leaderships within different communities which dictates perceptions and beliefs towards traditional practices. The black administration Act of 1927 was imposed on African communities which led to strong political pressures that caused severe conflict among black people practicing their indigenous medicinal practices (Delius, 2008). This Act brought about the political pressures among people who practiced African medicine. Furthermore, other Acts such witchcraft act was imposed towards traditional healers and traditional leadership as part of ending the practices of medicinal healing. There are organizations that still have control over the states within Africa. These organizations are permeating policies over the nations that perpetuate privatization of public sphere (Alvarez, 2012). This caused the manipulation of African indigenous healers by pharmaceutical companies sidelining the participation of indigenous medicinal practices (Reihling, 2008). The western model begins on the premise of the textbook idea of perfect competition while its actual manifestation is characterized by monopoly power, and deep-seated exploitation to feed the industrial and commercial complex. This continues to weaken the mandate to legalize the cannabis within the business industry and allow a free market for them.

Attitudes and Perceptions Towards Cannabis in Society

The society has put a negative narrative towards the usage of cannabis within the society due to the lack of morality among the youth of South African communities. According to Mphahlele (2020) cannabis use contributes towards criminal activities within the society and subject to drug offence. Mphahlele (2020) further argues that cannabis usage affects academic performance and non-completion of important activities hence there are a lot of dropouts. This clarifies that there is a need to reshape the educational background among the youth in terms of the usage of cannabis. The perception of religious sector particularly Christianity, views cannabis usage as part of the contributing source of ungodly practice among communities (Mothibe & Sibanda, 2019). This further shows that collective worldviews can dictate the attitude of the whole community towards a certain phenomenon. However, there are other groups of African people who are subscribing to Christian practices yet accept African indigenous medicinal practices. These African people fall within the brackets of African indigenous churches that also accepts traditional healing practices (Meyer, 2004). Christianity on its own carries a wide range of bias Western frameworks that contains certain narratives hence the ridicule of African indigenous medicinal practices among the communities (Cohen, 2010).

The Ghanian state has followed the linear western paradigm of development over the past 60 years (Konadu-Agyemang & Takyi, 2018). This archetype consigns the economic and social fortunes of the country to the periphery as producers of raw materials while actively exploiting its resources for the satisfaction of western powers whose grip on the economy has never loosened for the past 400 years or more (Quarshie & Alagidede, 2020). In turn, the economy is a dumping ground for all kinds of products from so-called developed markets. This is one of the examples of manipulation of the economic system among African countries which continues to hoover behind closed doors.

INDIGENOUS PHILOSOPHICAL UNDERPINNINGS OF THE STUDY

a) Ontology

According to Kovach (2012) and Wilson (2008), ontology outlines the essential features and questions relating to what it means to exist in a certain setting. This further confirm that human beings have the relationship with their land, ecosystems and other beings. Bryman (2001) alludes that there are two types of ontologies emerging from the multiple sets of worldviews within different societies, on how they view a certain phenomenon according to their values and norms. Objectivism and constructivism are the ones that Bryman (2001) highlights, in contrast to emphasize the social actors that dictates whether a certain phenomenon have an existence that is independent (Objective ontology) or annually constructed through constant state of revision (constructive ontology).

The reality is that cannabis production existed independently for centuries, until indigenous agricultural practices were applied to it. Ontology as a form of questioning the existence of cannabis production, opened fields of research that ended up opening avenues that are potentially contributing to the economic sector. Relatively, indigenous cannabis growers, cannabis, indigenous agricultural practices and indigenous knowledge became a compact that seeks to being about sustainable development among indigenous communities. This complements the relationship that the indigenous cannabis growers have with their land and ecosystems that thrives to provide a meaningful livelihood.

b) Epistemological Grounding

Epistemology refers to the expression of African indigenous ways of knowing or the relationship between the knower and the known. The following three questions address the basic meaning of epistemological grounding to a certain phenomenon, what is knowledge, what is knowing and what is known. Epistemology is the crucial component of philosophy within research disciplines as it cuts across sets of multiple realities that entail the ways and practices of doing and how one view the universe. From an indigenous point of view, indigenous cannabis growers carry multiple facets of tacit knowledge of agriculture which they use into producing cannabis explicitly. The knowledge then becomes explicit, practiced with skills and innovations then transferred to the next generations. Indigenous research methodologies underpin one's epistemological grounding to enquire what knowledges they favour and practice within their productions such as indigenous cannabis growers in rural areas like Nyandeni, Eastern Cape.

The important part to be noted from an indigenous perspective is that philosophical contribution of epistemologies has no formal authority on knowledge creation and transfer however broadens the scope of knowledge to be acquired by individuals. By this process, new paradigms emerge which are the building blocks of social constructs that are constructed by the individuals or communities. On the other hand, indigenous cannabis growers are subjected to two epistemological positions such as interpretivism and positivism. This is due to their different views, application of the methods to the natural sciences (interpretivism) and strategies that are employed which respects people and their surroundings (positivism).

c) Cosmology

Cosmology in simple terms refers to the connection between aspects of unity, peace and solidarity. In its deep etymology, refers to the universe and science. In this instance, the researcher subscribed the term cosmology to African cosmology to link the beliefs of the researched indigenous cannabis to the phenomenon that is also being investigated. African cosmology refers to how Africans are cognizant, conceive and contemplation of how they view their universe. These are the beliefs, values and norms that are embedded within their cultures as a way of life.

Indigenous cannabis growers have their own agricultural practices that are interlinked to their beliefs. There are certain times that they do not harvest, touch or temper with the process of growth. They harvest cannabis, sell to the traditional healers within the community of Nyandeni. These traditional healers use cannabis to appease their ancestors in a way they called "Ukuphahla". Certain ailments are also cured with the use of cannabis such as influenza, skin conditions and other related concoctions. African cosmology is connected to indigenous knowledge, the ideas, decision making strategies, rituals, social conducts and norms that are applied on daily basis by the indigenous knowledge holders.

d) Axiology

The researcher selected methodology and axiology based on the indigeneity of the study conducted, that seeks to decolonize the unfavourable frameworks manipulating indigenous communities and their practices. Axiology refers to the values, ethics and protocols to be observed when conducting a study in an indigenous setting. In this study, the researcher followed a proper protocol to gather participants and conduct interviews about the researched phenomenon. Part of the protocol was the applied research methodology that was used to frame the study based on systems, standards and responsibility. Due to the nature of the axiological position, morals needed to be an integral part of the methodology to conduct this study. Traditional authorities were consulted before the research was conducted in Nyandeni village in the Eastern Cape. With the methods that were built between the study and the researcher, they were constructed in such a way that it's respectful towards the traditional authorities and the participants. The researcher began by introducing or explaining who they are, where they are coming from, the purpose of conducting the research and their interest in conducting such research on the gaps identified. Relational ethical framework was applied to respect the rights of the participants on cannabis production, anonymity, confidentiality with their knowledge and accountable responsibility of the data collected.

African Indigenous Worldviews and Paradigms

African worldviews play a significant role towards understanding factors that affect lives on daily basis whether positive or negative (Bhuda & Koitsiwe, 2022). Cannabis farming practices can be viewed from an African perspective as it was one of the most agricultural crops that was used for various activities. Indigenous people were cropping through the practices that were unique and produced a quality flower. In all that, the people were working together in a state of relatedness through decision making, skills transfers and spiritual knowledge. The kind of views they had, were of spiritual importance as they were subjecting cannabis practices as their health resource not anything else. The views of African indigenous communities or people were infiltrated by the Western frameworks which later instilled its methods and practices into African communities with the intentions of disrupting the practices of indigenous communities (Bakwesegha, 2007). Moreover, this affected their views to date; the narrative about cannabis practices from the indigenous point of view has suddenly changed. Indigenous knowledge research perpetuates a paradigm that seeks to infuse connectedness into the society by mainly decolonizing the curriculums that still oppose good discourses around research and communities. In this study, transformative paradigm was adopted to transform the knowledge about cannabis.

African Indigenous Languages and Culture in Research

Language plays an important role in research generally, communication flows by words that create a language. Indigenous knowledge research is enriched by African indigenous languages because that is where knowledge holders freely give knowledge using language (Manyike & Shava, 2018). In simple terms, knowledge is embedded in language and research is dependent on language. It is important for researchers particularly the ones from Indigenous knowledge system background to note that embracing African research, language is of paramount and significant. Knowledge transfer is dependent on language, cannabis growers depend on language to communicate, indigenous famers and knowledge holders depend on language to transfer skills. Without the aid of language, knowledge will not move from place to the other. Indigenous cannabis growers in Nyandeni, eastern cape make use of their language to communicate the skills and innovations to be implemented in their production. They speak in IsiXhosa language, one of the Nguni languages spoken heavily in the eastern part of South Africa. The place was called Transkei during the colonial era, with Xhosa people who are heavily culturally rooted and follow their axiological positions.

Indigenous Research Methodologies

According to Chilisa (2012), it is important for researchers to proudly begin to recover, value, and internationalizing Indigenous research methodologies. African indigenous cannabis growers were colonized and never had a fair of going back at their practices due to colonial oppression. In this instance, indigenous postcolonial research paradigms critique the Euro-Western methods of practices through its Eurocentric research that promotes individualism. The need to decolonize and indigenize is high as researchers keeps on revealing methodologies that will criticize how the Western world views indigenous farming methods on cannabis. Perceptions of cannabis practices were intentionally imposed by the Euro western researchers and farmers to manipulate the growth of sustainable developments in African indigenous communities. The sole purpose of indigenous research methodologies is to voice out the voices of the marginalized indigenous communities such as Nyandeni in the Eastern Cape province. Moreover, to revitalize the African value systems, identities, restore and preserve indigenous knowledge in a multiculturalism context. In an indigenous research, indigenous research methodologies condemn an idea of a single universal reality in a community however it promotes a multi-unity of individuals that seeks to find solutions to their current challenges such indigenous cannabis growers. The production of cannabis has been sidelined for years from an indigenous point of view or academic landscape due to politics that favours other foreign frameworks that undermines indigenous knowledge holders.

METHODOLOGY

This research, which is based on the qualitative approach seeks to underpin the role, use and value of indigenous agricultural practices used by the indigenous cannabis growers. Qualitative research approach enables the researcher to actively involve participants in information rich cases (Merriam, 2019). Furthermore, qualitative research approach helps in acquiring deeper understanding of political, economic, and social issues surrounding the production of cannabis Sativa from an indigenous perspective. This approach is more focused on concrete description, interpretation, social and natural surroundings therefore it is relevant for this study. Given the qualitative research approach, the researcher used an exploratory design to explore indigenous agricultural practices embedded from the targeted population in their body of knowledge. The research employed an indigenous paradigm to interpret the current occurrences towards cannabis production and the those who produce it which is the indigenous cannabis growers. This paradigm is suitable for this research as it focuses on aspects of transformation towards the critiques of undermined indigenous peoples and their knowledge. The Indigenous paradigm is further used to reconstruct the body of IK that continues to be manipulated thereby emphasizing the multiple sets of realities that are connected and that human beings should have a relationship with their environment.

FUTURE RESEARCH DIRECTIONS

This study contributes to the indigenous communities that seeks to be recognized as people who contributes to the knowledge economy and sustainable development within their setting. It also serves as a resource towards new avenues that can be mainstreamed within the cannabis market. By providing discussions that supports the notion that cannabis is becoming a commodity on its own, production of it at a larger scale can reduce unemployment and address sustainable development goal's agenda for indigenous communities.

CONCLUSION AND RECOMMENDATIONS

In conclusion, the relationship between indigenous knowledge practices in farming and cannabis production goes a long way in the Eastern Cape province of South Africa. Institutions of higher learning have a great mandate to absorb and mainstream studies were historically marginalized in rural areas. They carry rich knowledge that continues to enhance the educational curriculum. Cannabis production as overrated as it is, produces billions of rands in the black market and ought to be regulated to increase the South African revenue in various sectors. The following are the recommendations;

- There is a need for a policy shift, specifically towards indigenous cannabis growers to be regulated as legal farmers who contributes hugely on the economy.
- Mainstream indigenous cannabis growers in the strategic means of production to fast-track sustainable development discourse
- Preserve the knowledge that the indigenous cannabis growers have in the academic dispensation
- Foster engagements that would lead the participation of indigenous cannabis growers in the agricultural sector
- Craft the legal frameworks that will perpetuate the production of cannabis farming at larger scales
- Encourage co-participation between the health ministry and indigenous cannabis growers for health knowledge contribution

REFERENCES

Abah, J., Mashebe, P., & Denuga, D. D. (2015). Prospect of integrating African indigenous knowledge systems into the teaching of sciences in Africa. *American Journal of Educational Research*, *3*(6), 668–673.

Amaducci, S. (2022). Adapting the cultivation of industrial hemp (Cannabis sativa L.) to marginal lands: *A review.Global Change Biology. Bioenergy*, *14*(9), 1004–1022. DOI: 10.1111/gcbb.12979

Ambler, C. (2020). The African Roots of Marijuana by Chris S. Duvall. *Bulletin of the History of Medicine*, *94*(2), 303–304. DOI: 10.1353/bhm.2020.0045

Arnold, S. E., Elisante, F., Mkenda, P. A., Tembo, Y. L. B., Ndakidemi, P. A., Gurr, G. M., Darbyshire, I. A., Belmain, S. R., & Stevenson, P. C. (2021). Beneficial insects are associated with botanically rich margins with trees on small farms. *Scientific Reports*, *11*(1), 15190. DOI: 10.1038/s41598-021-94536-3 PMID: 34312457

Black, B. E. (2020). *Creating a qualified cannabis workforce: How higher education can support cannabis career pathways*. Kansas State University.

Buthelezi, N. N. (2014). Indigenous knowledge systems and agricultural rural development in South Africa: Past and present perspectives. *Indilinga*, *13*(2), 231–250.

Gover, K. (2015). Settler–state political theory, 'CANZUS'and the UN declaration on the rights of indigenous peoples. *European Journal of International Law*, *26*(2), 345–373. DOI: 10.1093/ejil/chv019

Hall, R. (2004). A political economy of land reform in South Africa. *Review of African Political Economy*, *31*(100), 213–227. DOI: 10.1080/0305624042000262257

Hanson, V. J. (2020). Cannabis policy reform: Jamaica's experience. In *Legalizing Cannabis* (pp. 375–389). Routledge. DOI: 10.4324/9780429427794-21

Hasan, M., & Abdullah, H. M. (2015). Plant genetic resources and traditional knowledge: emerging needs for conservation. *Plant genetic resources and traditional knowledge for food security*, pp.105-120.

Hoppers, O. (Ed.). (2002a). *Indigenous Knowledge and the Integration of Knowledge Systems: Towards a philosophy of articulation*. New Africa Books.

Jones, M. (2024). *'O Cannabis': A History of Cannabis Legalization in Canada* (Doctoral dissertation, University of Victoria).

Junior, A. A., Gana, D., Adukeye, E., & Ahuchaogu, C. (2023). Unlocking the Socioeconomic Advantages of Cannabis Sativa L. Legalization in African Countries: An In-Depth Review. *Global Journal of Agricultural Research*, *11*(1), 32–52. DOI: 10.37745/gjar.2013/vol11n13252

Kepe, T. (2003). Cannabis sativa and rural livelihoods in South Africa: Politics of cultivation, trade and value in Pondoland. *Development Southern Africa*, *20*(5), 605–615. DOI: 10.1080/0376835032000149252

Lum, N. R. (2013). Community perceptions of Cannabis sativa (dagga) in the Mzintlava location, Ingquza Hill Local Municipality, Eastern Cape province [master's dissertation]. *Walter Sisulu University, Umtata*.

Malabadi, R. B., Kolker, K. P., & Chalannavar, R. K. (2023). Medical Cannabis sativa (Marijuana or Drug type); The story of discovery of Δ9-Tetrahydrocannabinol (THC). *International Journal of Innovation Scientific Research and Review*, *5*(3), 4134–4143.

Maroyi, A. (2017). Diversity of use and local knowledge of wild and cultivated plants in the Eastern Cape province, South Africa. *Journal of Ethnobiology and Ethnomedicine*, *13*(1), 1–16. DOI: 10.1186/s13002-017-0173-8 PMID: 28789692

Mbhele, T. P., & Eric, N. (2023). The role of supply chain technology in the growth of the cannabis industry. In *Rapid Innovation and Development in the Global Cannabis Market* (pp. 194–214). IGI Global. DOI: 10.4018/978-1-6684-6346-8.ch011

Merriam, S. B. (Ed.). (2019). *Qualitative research in practice: Examples for discussion and analysis*. John Wiley & Sons.

Moahi, K. H. (2007). Globalization, knowledge economy and the implication for indigenous knowledge. *International Journal of Information Ethics*, *7*, 55–62. DOI: 10.29173/irie13

Ngulube, P. (2015). Towards a cartography of indigenous knowledge systems in library and information science training and education in Anglophone eastern and southern Africa. *Indilinga*, *14*(2), 145–168.

Nthutang, S. T. (2015). *Issues affecting e-learning practices at Northwest University, Mafikeng Campus* (Doctoral dissertation).

Nyong, A., Adesina, F., & Osman Elasha, B. (2007). The value of indigenous knowledge in climate change mitigation and adaptation strategies in the African Sahel. *Mitigation and Adaptation Strategies for Global Change*, *12*(5), 787–797. DOI: 10.1007/s11027-007-9099-0

Paruk, Z. (2022). *Exploring indigenous beliefs of black Africans in the South African workplace* (Doctoral dissertation, University of Pretoria (South Africa).

Paterson, C. (2009). *Prohibition & resistance: a socio-political exploration of the changing dynamics of the Southern African cannabis trade, c. 1850-the present* (Doctoral dissertation, Rhodes University).

Raihan, A., & Bijoy, T. R. (2023). A review of the industrial use and global sustainability of Cannabis sativa. *Global Sustainability Research*, *2*(4), 1–29. DOI: 10.56556/gssr.v2i4.597

Rehman, M., Fahad, S., Du, G., Cheng, X., Yang, Y., Tang, K., Liu, L., Liu, F. H., & Deng, G. (2021). Evaluation of hemp (Cannabis sativa L.) as an industrial crop: A review. *Environmental Science and Pollution Research International*, *28*(38), 52832–52843. DOI: 10.1007/s11356-021-16264-5 PMID: 34476693

Santoro, A. (2022). Agricultural heritage systems and agrobiodiversity. *Biodiversity and Conservation*, *31*(10), 2231–2241. DOI: 10.1007/s10531-022-02460-3

Schaffrick, M., Perreault, M. L., Jones, A. M. P., & Illes, J. (2023). Understanding and Rebalancing: A Rapid Scoping Review of Cannabis Research Among Indigenous People. *Cannabis and Cannabinoid Research*, *8*(3), 426–433. DOI: 10.1089/can.2022.0251 PMID: 36454240

Semenya, S. S., & Potgieter, M. J. (2014). Bapedi traditional healers in the Limpopo Province, South Africa: Their socio-cultural profile and traditional healing practice. *Journal of Ethnobiology and Ethnomedicine*, *10*(1), 1–12. DOI: 10.1186/1746-4269-10-4 PMID: 24410790

Siddiqui, S. A., Singh, P., Khan, S., Fernando, I., Baklanov, I. S., Ambartsumov, T. G., & Ibrahim, S. A. (2022). Cultural, social and psychological factors of the conservative consumer towards legal cannabis use—A review since 2013. *Sustainability (Basel)*, *14*(17), 10993. DOI: 10.3390/su141710993

ADDITIONAL READINGS

Douglas, M., Manu, E., & Ntsaba, M. J. (2019). Adolescent illicit marijuana practices in selected communities in the Eastern Cape province of South Africa: Implications for public health. [AJPHES]. *African Journal for Physical Activity and Health Sciences*, *25*(4), 490–507.

Gerwel, H. (2018). *Institutional innovation and pro-poor agricultural growth:* cannabis cultivation in the Eastern Cape province of South Africa as fertile opportunity.

Lum, N. R. (2013). Community perceptions of Cannabis sativa (dagga) in the Mzintlava location, Ingquza Hill Local Municipality, Eastern Cape province [master's dissertation]. *Walter Sisulu University, Umtata.*

Ngobeni, N. D., Mokoena, M. L., & Funnah, S. M. (2016). Growth and yield response of fibre hemp cultivars (Cannabis sativa L.) under different N-levels in Eastern Cape Province of South Africa. *African Journal of Agricultural Research*, *11*(2), 57–64. DOI: 10.5897/AJAR12.0675

KEY TERMS AND DEFINITIONS

Cannabis: A tall plant that is used to produce hemp fibre, used to cure certain ailments used by knowledge holders or health practitioners.

Indigenous Agricultural Practices: The practices that are applied by indigenous farmers to produce food, often used to protect the biodiversity and soil sustenance

Indigenous Cannabis Growers: A group of people that applies indigenous farming practices to produce cannabis strands to make a living, these are seen as farmers

Indigenous Knowledge Systems: It is an all-inclusive collection of skills, practices and knowledge that is traced back to a certain or specific community

Sustainable Development: The developments that are created to meet the needs of the people without jeopardizing the future, maximizing sustainability and minimizing profit.

Chapter 11
Documenting and Preserving the Indigenous Foods to Ensure Food Security by the Moshana People in the Northwest Province, South Africa

Nthabiseng Mmathari Maepa
https://orcid.org/0000-0002-8755-1328
North-West University, South Africa

Aviwe Noxolo Matsoso
https://orcid.org/0009-0005-2624-8316
North-West University, South Africa

Hangwani Muedi
Department of Agriculture, Land Reform, and Rural Development, South Africa

ABSTRACT

Indigenous people of South Africa have for many years relied on Indigenous Knowledge to ensure food security. For example, they can identify different types of edible wild fruits, vegetables and roots, they know when to plant which types of food, how to plant them, when and how to harvest and process the food and how to preserve the food to ensure food security. Agriculture, based on Indigenous Knowledge, ensures the production of a variety of indigenous foods. However, the knowledge on these Indigenous foods is slowly being lost. This is mainly because indigenous knowledge is tacit and kept in the minds of knowledge holders whom most of them are no more. Without its protection and promotion, the knowledge is susceptible to change or completely being lost.

DOI: 10.4018/979-8-3693-7964-6.ch011

INTRODUCTION

Indigenous plants have long been used by humans as food and medicine in practically all communities. They grow naturally in ecosystems and are known to be potent sources of nutrients, therefore better for health. Forager and hunter-gatherer societies historically relied only on local flora for their diets. Currently, to combat food insecurity, many people rely on a varied diet that includes edible plant material harvested from the wild as well as edible plant material grown in backyard gardens. For instance, on the African continent, more than 95% of households use grass, roots, and leaves from wild plants to supplement their diets (Rankoana, 2021). To maintain food self-sufficiency and the continued existence of millions of people, these communities cultivate and consume native fruits and vegetables (Takaidza, 2023). In South Africa, there is a wide variety of locally grown food as well as collected in the wild that is not only nutritious but also effective in preventing chronic illness. Indigenous plant meals have long played a significant role in folklore. Many people in rural areas still harvest and eat both cultivated foods and wild vegetation. There is evidence from studies conducted in South Africa and around the world, that indigenous food is consumed by many households, especially in areas where they are available, and thus can play an important role in alleviating food insecurity and contribute to dietary diversity. These include indigenous grain crops, vegetables, and fruits (Mbenyane, 2017).

The ability of indigenous food systems to contribute to food security and the eradication of hunger and poverty is one of the most fundamental reasons for their importance. Many indigenous communities have discovered strategies to cultivate and harvest climate-resistant crops, and as such, they always had access to nutritious food. Indigenous food crops play a vital role in the livelihood of over 60% of rural African communities in tropical Africa (Pichop *et al.*2014). Socio-economic surveys on traditional vegetables conducted in Central, Western and Eastern Africa indicated that indigenous vegetables are important commodities in household food security (Weinberger & Pichop, 2009). Some indigenous fruits are harvested and processed into products such as jams and wines, and these are of great importance in economic empowerment of rural communities, and improvement of livelihoods. Domestication of indigenous fruit trees would ensure their contribution to income generation in rural communities. Indigenous foods provide nutritional and therapeutic benefits. Most of these crops are undervalued because they are considered weeds or poor man's food, lowering their economic position. Indigenous foods are gaining popularity due to the need to promote food security. These foods nutritional components and bioactive substances are determined using advanced techniques. Some of these foods, like cowpea, have been shown to have significant nutrition content but are limited by anti-nutritional factors (Takaidza, 2023).

THE CONCEPT OF FOOD SECURITY

According to Du Toit (2011), Food Security is the ability of individuals to obtain enough food on a daily basis. These definitions place special emphasis on availability, accessibility, use, and stability which are the four main variables used to assess food security. Food availability suggests that a nation has to consistently have enough food accessible for both households and the country as a whole. The ability of a country and its families to obtain enough food in a sustainable manner is implied by the term "food access." The term "food use" describes how food should be used in accordance with fundamental understanding of nutrition and care, as well as access to clean water and sanitation. However, food stability is the consistency of food supply and access (Du Toit, 2011). Since the methods used to evaluate

Food Security at the national and household levels are not the same, there are differences in food security at those levels (Du Toit, 2011). At the national level, Food Security refers to the ability of a country to produce, import, maintain, and sustain the food necessary to meet the basic nutritional criteria for each member of the population. At the local level, Food Security suggests that rural areas are capable of producing, preserving, and storing an adequate amount of food for later consumption. In addition to considering the years that lie ahead should a dry period occur (Masekoameng, 2007). According to the findings of a 2013 study by Ndwandwe on the role of Indigenous Knowledge practices in home food production and Food Security, indigenous Household Food Security is attained when consistent and sufficient food is available, obtained, and appropriately used at the household level. Making food at home is crucial for improving the financial status of rural families since their main sources of employment are farming and resources in their immediate vicinity (Ndwandwe, 2013).

RESEARCH METHODOLOGY

The research approach that was used for this paper is a Qualitative Research Approach as it helped the researcher to discover new thoughts and views by letting the researcher engage herself with the study through the ways in which the research will be done by having open questions and interviews. The Indigenous research Paradigm was applied in this paper. This paradigm is appropriate for the research since it relies on the basic idea that knowledge is relational, shared by all creations, and hence cannot be owned or discovered (McGuire-Adams, 2019). To gather knowledge on these indigenous foods, a focus group interview guide was used use. Focus group interview data was analyzed using a thematic analysis. The data collected from the voice recorders was transcribed. Once the researcher has become familiar with the data, the researcher began to identify key concepts and phrases, which will then be categorized into themes (Braun & Clarke, 2006). The Batswana people from Moshana in the North-West province of South Africa, Bahurutshe, are a part of the broader Batswana ethnic group, defined by their language, culture, history, and way of life. They speak Setswana, with possible dialectal variations, and belong to clans with unique totems that symbolize their ancestry. This community has a strong sense of identity and tradition. The Traditional leadership of the Moshana community plays a significant role in governance and cultural preservation. Many community members of Moshana engage in subsistence farming, growing crops such as maize and sorghum, while cattle herding remains an important economic and social practice. The study focuses on this particular community because of this rich knowledge that they have of indigenous foods that helps with food security, therefore the researcher aimed to document their indigenous knowledge in order to preserve and protect it.

IMPORTANCE OF INDIGENOUS FOODS IN MAINTAINING FOOD SECURITY

Indigenous food plants can be an extremely important part of the local diet, for a number of reasons. Indigenous food plants tend to be very nutritious, often more nutritious than some introduced (and perhaps more popular) plants. Indigenous crops are well-adapted to the regions where they originate. They often are free of pests and diseases and are able to grow in extremely difficult conditions (in very dry areas or in very acidic or alkaline soil, for example). The diversity provided by eating indigenous vegetables contributes to food security. Rather than relying on one food crop, a wide range of food plants are eaten and enjoyed. If one fails to grow and produce, others can be eaten instead (Mijatović et al., 2013). On an environmental level, indigenous foods have the potential to improve environmental sustainability by contributing to agrobiodiversity, strengthening climate resilience and because of often lower natural resource requirements compared to more commercial and intensively farmed crops. Indigenous vegetables can be harvested more times per year compared to many other crops, they support the resilience of farmers who, due to climate change impacts, are more often faced with failed harvests. Several research projects gained relevant insights into the use of organic fertilizers and pest management tools, water saving techniques, or the conservation of local forest resources (Mijatović et al., 2013). The consumption of Indigenous foods acts as a cultural diet for the indigenous people as they are freely accessible, and the knowledge have been used by knowledge holders. Indigenous foods have the highest value in nutritional needs, so they have an important role in lifting those vulnerable to low diets and illness. Moreover, people who consume food plants can obtain a good supply of proteins, carbohydrates and vitamins (Jena et al., 2018).

Table 1. Most common indigenous foods of South Africa

Main categories	**Scientific name**	**Common name**	**Nutritional/Health benefits**	**Main uses**	**Main Production areas**
Grain crops	*Pennisetum Glaucum*	Pearl millet	High in fibre, protein, minerals, and fatty acids. Gluten-free	Ground flour, grainlike rice. Feed for poultry	Limpopo, KwaZulu Natal (KZN), Free State
	Sorghum Bicolour	Sorghum	High in carbohydrates. Small amounts of micronutrients. Anti-inflammatory and anticancer properties.	Ground flour	Mpumalanga, Free State
	Vigna Unguiculata	Cowpea	High in energy, fibre, protein, vitamin A, C, B vitamins. bioactive compounds protective against chronic diseases	Leaves and pods are eaten as vegetables.	Mpumalanga, Limpopo, KwaZulu-Natal, Gauteng, North-west
	Vigna Subterranea	Bambara groundnut	High in starch, protein, iron, zinc	Ground flour	Mpumalanga, KwaZulu-Natal, Limpopo
	Vigna Radiate	Mung bean	High in starch, protein, zinc, high in bioavailable iron	The base for soup, flour, bean sprouts	Limpopo, Mpumalanga
Vegetable crops	*Cleome Gynandra*	Spider plant	High in calcium, iron, magnesium, and zinc	Leaves are eaten as a vegetable	Limpopo, Free State, KwaZulu Natal, Northern Province, and North-West
	Amaranthus	Aramanths	Rich iron, beta-carotene, vitamin A, vitamin C, and	Leaves are eaten as vegetables or soup	Limpopo, North West,
Main categories	**Scientific name**	**Common name**	**Nutritional/Health benefits**	**Main uses**	**Main Production areas**

continued on following page

Table 1. Continued

Main categories	Scientific name	Common name	Nutritional/Health benefits	Main uses	Main Production areas
	Bidens Pilosa	Black jack	Rich in essential amino acids, iron, and zinc more than 40 bioactivities protective against chronic disease	Leaves and shoots are eaten as a vegetable. Medicine	KwaZulu-Natal, Limpopo
	Corchorus Olitorius	Jew's mellow	Rich in vitamin A and C, folic acid, beta-carotene, iron, calcium	Leaves are eaten as a vegetable	Limpopo, Gauteng, Eastern Cape, and Mpumalanga
	Manihot Esculenta	Cassava	High in starch, calcium, vitamin C	Roots are eaten as a vegetable	Mpumalanga, KZN
	Colocasia Esculenta	Amadumbe	Significant amounts of B vitamins iron, phosphorus, zinc, vitamin C, niacin, potassium, copper, and manganese	The root is eaten as a vegetable	Mpumalanga, KZN
Fruit crops	*Sclerocarya Caffra*	Marula fruit	Rich vitamin C, calcium, potassium, and magnesium.	Eaten raw Products made include jams and wines	Limpopo, KZN, Eastern Cape, Mpumalanga
	Mimusops Zeyher	Red milkwood	Vitamin C	Eaten raw	Limpopo, Gauteng, North West, Mpumalanga, KZN, Free State
Main categories	**Scientific name**	**Common name**	**Nutritional/Health benefits**	**Main uses**	**Main Production areas**
	Parinari Curatellifolia	Mabolo plum	Rich vitamin C, calcium	Eaten raw, cooked as porridge, fermented into beer	Limpopo, Mpumalanga
	Carissa Macrocarpa	Amatungulu, numnum	High in vitamin C, iron, calcium, and essential fatty acids	Eaten raw Used for making jam	KZN
	Dovyalis Caffra	Kei apple	High in vitamin C	Eaten as fruit Fruit juice	Limpopo, Mpumalanga, Eastern Cape, KZN
	Strychnos Spinos	Monkey orange	Rich fibre, vitamin C, and B vitamins	Eaten raw, dried to make fruit rolls, jams, and win	Eastern Cape, KwaZulu Natal, and Limpopo

(Bobo, 2021)

THE USE OF INDIGENOUS FOODS IN SOUTH AFRICA

Even in South Africa, which satisfies all requirements for national food security and provides everyone with access to sufficient nutrition, there are still significant differences in the availability of food both within and between households. Inadequate food access affects about 20.2% of families and 23.8% of people (Larson *et al.* 2020). Furthermore, according to Statistics South Africa (2018), about

70% of the poorest families are found in rural areas. While there are additional reasons that have led to food insecurity and poverty in South Africa, such as high unemployment rates, inflated food costs, and ethnic inequities, the primary issue these families face is availability to food (Van der Merwe et al., 2016). Although the average household size in South Africa is 3.49 people, research indicates that the majority of informal and rural areas have typical household sizes that double to six or seven people because of extended family members (Statistics South Africa, 2018). Given that food makes up 31% of total consumption, these families are the most susceptible to increases in food prices (Statistics South Africa, 2018). In South Africa, there is a serious problem with food affordability. For these less fortunate individuals, the availability of appropriate resources has a significant impact on consumer decisions.

Food security in underprivileged communities is greatly aided by the informal food sector. But the cost of food has increased drastically, particularly when it comes to fresh vegetables, meat, and animal goods. Consequently, individuals turn to unhealthy coping strategies including consuming highly processed and refined foods that are deficient in nutrients (Schönfeldt et al., 2017). However, these foods are still extremely refined and processed. Though lack of knowledge, limited access to a variety of food options, and restricted financial resources sometimes limit consumer consuming choices. The purpose of public health communications in South Africa is to increase the availability, accessibility, and usage of safe and nutritious foods through dietary variety (Ruel, 2003). However, the focus has mostly been on promoting the consumption of conventional and commercial staple foods and vegetables, with minimal attention paid to wild vegetables, also referred to as indigenous foods (Schönfeldt et al., 2017).

Agriculture has been the foundation of food systems for millennia, and for many rural people, indigenous foods are the primary source of nutrition. According to Mabhaudhi et al. (2018), indigenous crops are those that either originated in the area or are exotic crops that have been cultivated there for many years. They have been shown to supply vital nutrients needed by human physiology, and their nutritional composition is usually better than that of conventional vegetables. According to Mbhenyane (2017), foods that are indigenous are high in iron, zinc, calcium, protein, and vitamins A, C, and A. Back then, these crops were the cornerstone of local food systems, adding nutrition to meals and offering much-needed support in times of need, like droughts. According to van der Hoeven et al. (2013), the consumption of wild vegetables that sustain nutritional status has dropped, while the consumption of a select few commercial plant staples has increased.

THE CONNECTION BETWEEN FOOD SECURITY AND INDIGENOUS KNOWLEDGE SYSTEMS

Indigenous knowledge (IK) is that of the indigenous people that has been passed down through the ages and has proven to be sufficiently adaptive to deal with change (Eyong, 2007). IK is the body of knowledge and skills that rural women have directly gathered from their immediate surroundings, and which are widely recognized and applied in food processing and preservation operations. Communities develop and disseminate IK over time in order to adjust to their distinct socioeconomic and agroecological contexts (Fernandez, 1994). Over the past 20 years, there has been a significant increase in the focus on the role that IK plays in genuinely participative ways to address food security for sustainable development in African communities (Tikai & Kama, 2010). Recent research has provided important insights into how indigenous people adjust and improve, for example, natural resource management using their own locally produced knowledge. The section on agriculture offers a useful example. Farmers employ a wide variety of traditional farming methods that are derived from years of experience, unplanned experimentation, and an in depth understanding of their environment (Tikai & Kama, 2010). African women possess IK skills that help them deal with the difficult situations they face on a daily basis. Women, for instance, employ IK methods and technology to prepare, process, and store food for their household's food security during seasonal food shortages as the only supply until the start of harvest later in the year (Virginia & Pallavi, 2016; Ibnouf, 2012). In Odisha, India, a 2015 study by Panda et al. (2015) found that customs regarding food and IK continue to play a big role in rural life and culture. As a result, the IK of food preparation and preservation was handed down from one generation to the next through imitation and observation of the elders (Ibnouf, 2012). Without any written documentation, oral tradition is the primary method used to pass along the traditional knowledge of indigenous wild vegetables from generation to generation. Many rural and indigenous tribes throughout the world continue to engage in such practices (Kumar, 2013). The link between indigenous foods and food security is that indigenous foods can influence food security in a number of ways, for example through changes in and access to income, food prices and their variability, labour allocations, particularly of mothers and caregivers; energy and nutrient expenditure and nutrient composition of foods (Masekoameng, 2007). When indigenous agricultural products such as millet, sorghum, cowpeas and Bambara ground nuts are sold in the marketplace, the income earned may be used to purchase goods and services that contribute to nutritional well-being. Similarly, when those products are consumed directly by members of the producing household, their dietary intake is improved. It would seem, therefore, that indigenous agricultural food production, processing, preservation and utilization can make a direct contribution to improving nutrition (Masekoameng, 2007).

DIFFERENT TYPES OF INDIGENOUS FOODS USED IN MOSHANA COMMUNITY FOR FOOD SECURITY

In the Moshana community, indigenous foods play a pivotal role in ensuring food security and maintaining it. The diverse array of indigenous crops and wild foods, adapted to local environmental conditions, provides essential nutrients and supports traditional farming practices. There is a variety and diverse indigenous foods that are used and found in the Moshana community. These foods are either planted or are found and grow naturally, an example of those that are grow naturally are wild roots and fruits. Those that are produced and planted are *mabele, dinawa and mmidi.* According to the participants from the focus group, indigenous foods are healthy than todays foods, mainly because the food people eat currently are not healthy and durable. Indigenous foods can be kept for a long time because there is no use of fertilizers in planting these indigenous foods. Participant 1 expressed that there is a wide range of indigenous foods that are used for food security.

> Dijo tse di amanang le setso jaaka mabele a re a lemang di fa dikotla tse di kgethegileng tse di amanang le boitekanelo jwa motho. Mabele a setswana a le proteini le vitamini tse di amanang le go tokafatsa boitekanelo. Morula o na le vitamini e thusang masole a mmele gore a lwantshe malwetsi.. Ka go fa mefuteng e amanang le boitekanelo, dijo tse di thusa go oketsa boitekanelo jwa motho ka go oketsa dikotla tse di tlhokegang mo mmeleng. Go dirisa dijo tse tsa selegae go re thusitse gore re kgone go nna le dijo tse di nang le dikotla ka dinako tsotlhe.

Meaning, Indigenous foods such as millet and morula fruit provide essential nutrients that are critical for health and well-being. Millet is rich in proteins, vitamins, and minerals, making it a valuable staple that supports overall nutritional needs. Morula fruit is high in vitamins, which help boost the immune system and prevent deficiencies. By offering a diverse range of nutritious options, these foods contribute to a balanced diet and enhance the community's ability to maintain food security.In the case of the people that use indigenous foods in the Moshana community, indigenous foods are primarily sourced from their local environment through foraging. This practice involves gathering wild plants, fruits, nuts, and vegetables that naturally grow in the community. For instance, fruits like *morula, motlhatswa, seretologa* are harvested from trees and shrubs that thrive in the local climate. The process of foraging involves a deep understanding of the local ecosystem, including seasonal patterns and the specific habitats where different foods can be found.

STORYTELLING AND ORAL TRADITION AS MECHANISMS FOR KNOWLEDGE TRANSMISSION IN THE MOSHANA COMMUNITY TO IMPROVE FOOD SECURITY

Oral tradition and storytelling have long been vital means of passing along knowledge, especially critical knowledge about food security, among the Moshana community. This custom, which is ingrained in the community's culture, is a valuable source of knowledge and direction that may be used to improve food security. The participants of the study indicated that the knowledge they possess of Indigenous foods was passed on from them, from their parents or relatives to them. Most of the participants said that if it was not for their parents and efforts, they would not know anything about Indigenous foods and its farming. Participant said:

Ke ithutile ko gae ke bonela mogo mme motsadi, ke gola ke di ja dijo tse, ke godile molelapeng la temo, ka jalo ka itsi le go ithuta go lema ka yone Nakong yeo.

The participant means that she was taught by her mother, she would see from her mother and that is how she acquired her knowledge. She grew up in a household where they practiced farming and farmed indigenous foods, she grew up eating them.

Storytelling is essential to preserving the agricultural knowledge in the Moshana community. Elders tell stories that highlight the importance of different planting seasons and the kinds of crops that work well at each time. These stories not only helping the people of Moshana to feel more connected to the land, but they also offer useful knowledge about indigenous foods. A story about an abundant harvest, for example, can emphasize traditional crops that are high in nutrition and resistant to climate change, highlighting their importance for food security. Younger generations are taught to value and use these indigenous foods through these engaging stories, preserving important knowledge on indigenous foods and their contribution to food security. Participant said:

Nna kena le tsela e farologaneng e ke rutang le go tlotlela bana ka temothuo kgotsa ka dijo tsa Setswana tse re di lemang le botlhokwa ga mmogo le dikotla tsa tsone. Bana fa e le ba gago o a baitsi. nna ke rutile le go tlotlela bana ka go tlhopha, o ke bonang okare o kgatlhegela go lema.

This literally means, I have unique way of teaching and telling stories to my children about farming of indigenous foods we grow or harvest in the wild, and their importance and nutritional value. You know your children, I taught mine through selection, when you seem to be interested in farming and learning about indigenous foods''.

Indigenous food stories help pass on knowledge about indigenous foods and their nutritional significance, which encourages the next generation to value and include them in their diets. This is why telling stories about indigenous foods is essential to maintaining food security. Moreover, oral histories that describe previous food scarcity difficulties help people become resilient and adaptive, preparing them to address present and future food security challenges. Telling stories promotes the preservation of these foods. Ultimately, storytelling serves as a potent tool to preserve and improve communal food security.

ENVIRONMENTAL AND CLIMATIC VULNERABILITIES AFFECTING INDIGENOUS PEOPLE OF MOSHANA'S INDIGENOUS FOODS AND FOOD SYSTEMS

Climate change poses significant challenges to the indigenous foods and food systems of the Moshana community. One of the most pressing issues is temperature variability, which can alter growing seasons and affect crop yields. As temperatures rise, certain indigenous crops may struggle to thrive, resulting in reduced harvests and diminished food availability. This shift not only impacts the community's diet but also threatens their cultural practices tied to these foods. Additionally, the increasing frequency of extreme weather events such as droughts, floods, and storms devastate agricultural production. Prolonged droughts lead to water scarcity, making it difficult for farmers to irrigate their crops. Conversely, heavy rains and flooding can wash away seeds and damage established plants. These unpredictable weather patterns create an environment of uncertainty, making it harder for the community to rely on their traditional food systems for sustenance and economic stability. Participant said:

> Pula e kgona go na mme nako engwe fa e nele e senya dijwalo tsa rona ka gore di tseiwa ke metsi. nako engwe re a lema mme ebe pula e sa ne, lefatshe le omelela fela e be dijwalo tsa rona di swa ka gore a gona pula.

Meaning, Rain can sometimes lead to flooding that washes away our seeds and crops, severely damaging our harvest. Conversely, there are times when we plant but receive no rain, leaving the ground dry and causing our crops to wither and die from lack of water.

Rainfall can create significant challenges for our agricultural practices. In some instances, heavy rains lead to flooding that washes away the seeds and crops, resulting in substantial losses. On the other hand, there are times when the farmers plant crops but do not receive any rain, leaving the ground parched and causing plants to wither and die. This unpredictability in weather patterns poses serious threats to the community's food security and overall livelihoods, making it difficult to rely on traditional farming methods.

The Moshana community, like many indigenous communities worldwide, has long relied on indigenous foods as a fundamental part of their food security systems. These foods, which range from cultivated crops to wild-harvested fruits and roots, provide vital nutrition and contribute to the preservation of cultural practices. The role of indigenous foods in promoting food security in the Moshana community cannot be understated, especially as modern agricultural practices and global food systems face increasing pressures from climate change. This discussion will explore the significance of indigenous foods for food security in the Moshana community, the role of knowledge transmission through oral traditions and storytelling, and the challenges posed by environmental and climatic vulnerabilities. Ultimately, the discussion will focus on how these factors are interlinked and essential for ensuring long-term food security and resilience in the Moshana community.

Indigenous foods, such as millet (mabele), morula, motlhatswa, and seretologa, are deeply embedded in the Moshana community's food systems. These foods, which are either cultivated or foraged from the local environment, provide essential nutrients and serve as a foundation for community nutrition. According to participants in the study, indigenous foods are considered healthier than modern foods due to their natural cultivation without reliance on fertilizers or pesticides, which often compromise the nutritional quality of contemporary crops. Mabele, for example, is a rich source of protein, vitamins, and minerals, which makes it an important staple in the diet of the Moshana community. Similarly, the morula fruit is high in vitamins, which strengthens the immune system and contributes to overall health. These foods not only fulfill nutritional needs but also contribute to the community's cultural heritage and self-sufficiency (Khoddami et al., 2023). The nutritional value of indigenous foods is one of the key reasons they are prioritized within the Moshana community. These foods, rich in proteins, vitamins, and minerals, support the health and well-being of individuals and the community as a whole. By providing essential nutrients, they help in the prevention of nutrient deficiencies and contribute to the maintenance of good health. Importantly, indigenous foods are typically well adapted to the local environmental conditions, which makes them more resilient to climate stressors compared to many modern crops. This resilience is especially critical in light of the changing climate and the increasing unpredictability of weather patterns. In times of drought or floods, the Moshana community is better positioned to rely on these hardy, locally adapted crops, which require fewer external resources and inputs to thrive.

The transmission of knowledge about indigenous foods and farming practices in the Moshana community is primarily done through oral traditions and storytelling. This form of knowledge transfer is essential for maintaining the community's connection to its agricultural heritage and ensuring the continued cultivation and use of indigenous foods. As noted by the participants in the study, many of the elders learned about indigenous foods from their parents or grandparents, who passed down wisdom about planting, harvesting, and preparing these foods. Through stories, knowledge about the nutritional value, cultural significance, and best practices for growing indigenous crops is transmitted to younger generations (Zimu-Biyela, 2016). Storytelling plays a central role in the Moshana community's agricultural education. Elders use stories to convey not only practical knowledge, but also cultural values associated with food production. For example, stories may emphasize the importance of planting certain crops during specific seasons or the techniques for ensuring healthy soil. These stories are often rich with lessons drawn from past experiences, including challenges related to food security. The story of a bountiful harvest, for instance, might highlight the resilience of particular crops to climate fluctuations, while a story about food scarcity may stress the importance of adapting to environmental changes and diversifying the food supply. The value of oral traditions in preserving agricultural knowledge cannot be overstated. In the Moshana community, storytelling fosters a sense of continuity and community cohesion. By learning through stories, children and younger generations not only acquire practical knowledge about food security but also develop a deeper connection to their cultural identity. These narratives help reinforce the importance of indigenous foods and the practices that have sustained the community for generations. The participants in the study highlighted that it was through this process of learning from parents and elders that they gained a profound understanding of the role of indigenous foods in promoting health and food security. Additionally, oral traditions are particularly valuable in the context of climate change. As environmental conditions become more unpredictable, the ability to adapt and respond to changing climates becomes increasingly important. Through storytelling, the Moshana community can pass on adaptive strategies and lessons from the past, helping younger generations build resilience in the face of climate challenges. Stories about past hardships, such as droughts or poor harvests, offer valuable insight into how the community has responded to these challenges and what strategies were effective in ensuring food security.

The Moshana community's traditional food systems face significant challenges due to environmental and climatic vulnerabilities. Climate change is causing temperature variability, altered rainfall patterns, and extreme weather events, all of which directly affect agricultural productivity (Thornton *et al.,* 2014) These climatic changes disrupt the growing seasons of indigenous crops, making it difficult for farmers to predict when to plant and harvest. Increased temperatures, for instance, can lead to shorter growing seasons or affect the availability of water for irrigation, while erratic rainfall patterns result in both droughts and floods, each of which can severely damage crops. The impact of these climatic vulnerabilities on the Moshana community's food security is significant. As mentioned by participants in the study, rainfall is crucial for agriculture, and its unpredictability is a major concern. On the one hand, excessive rainfall can lead to flooding, washing away crops and damaging seeds. On the other hand, the absence of sufficient rainfall can lead to drought conditions, leaving the ground dry and making it impossible for crops to grow. These unpredictable weather patterns create an atmosphere of uncertainty, making it difficult for the community to rely solely on traditional food systems.

The ability of indigenous foods to withstand some of these climatic challenges, however, offers hope. Many indigenous crops, such as millet, are known for their resilience to drought and poor soil conditions. These crops are adapted to the local environment and can survive under conditions that may be too harsh for modern, industrial crops. For instance, millet is drought-tolerant, and its ability to thrive in arid conditions makes it an ideal crop for regions facing water scarcity. By prioritizing these hardy crops, the Moshana community can increase its resilience to climate change and ensure a stable food supply in the face of unpredictable weather patterns.

SOLUTIONS AND RECOMMENDATIONS

The challenges posed by climate change underscore the need for adaptive strategies that can help the Moshana community maintain food security in the long term. This includes not only preserving indigenous knowledge but also integrating it into contemporary agricultural practices. By combining traditional ecological knowledge with modern agricultural innovations, the Moshana community can develop more resilient food systems that are capable of adapting to climate variability and ensuring food security for future generations. The preservation and promotion of indigenous foods, along with sustainable farming practices, are critical to achieving food security in the Moshana community. By investing in local food systems and supporting the cultivation of drought-resistant crops, the community can mitigate the risks associated with climate change. Additionally, fostering the transmission of knowledge through oral traditions and storytelling will help ensure that younger generations continue to value and use indigenous foods. This knowledge transfer, coupled with climate-smart agricultural practices, will help the Moshana community build resilience and adaptability, ensuring that they can continue to rely on indigenous foods to meet their nutritional needs and maintain food security.

FUTURE RESEARCH DIRECTIONS

Future research on the traditional foods of the Moshana people in the Northwest Province of South Africa should focus on several key areas to help preserve and protect their food systems. First, researchers should study and document the different plants used for food, including how they are prepared and their nutritional value. This can help understand how these foods support the community's food security. It's also important to study how climate change might affect these traditional food systems and how they can adapt to new challenges. Another area of research should look into the cultural and social importance of these foods, as they play a key role in the community's identity and unity. Understanding how traditional food preservation methods, such as drying and fermentation, work can help keep these practices alive. Research should also look at how modern farming and food systems impact traditional food practices, and how indigenous knowledge can be used in modern agriculture. Finally, collaborating with local communities, governments, and organizations is essential to ensure that the Moshana people's food practices are recognized and supported in food policies. By focusing on these areas, future research can help protect and strengthen the Moshana people's food systems.

CONCLUSION

The Moshana community, like many other indigenous groups, has a rich history and deep-rooted knowledge of food production systems that have allowed them to thrive for generations. Indigenous foods, such as millet *(mabele), morula, motlhatswa, and seretologa*, are vital components of the community's food security. These foods provide essential nutrients that contribute to a healthy and well-balanced diet, which is key to the community's overall health and resilience. The Moshana community's dependence on indigenous foods highlights their crucial role not only in addressing the nutritional needs of the population but also in supporting traditional agricultural practices that have stood the test of time. Indigenous foods in Moshana are grown using traditional methods that do not rely on modern fertilizers or intensive chemical inputs, making them environmentally sustainable and more resistant to the challenges posed by contemporary agricultural practices. The community's reliance on locally sourced, indigenous foods that grow naturally, such as wild roots and fruits, further highlights the resilience of the traditional food system in the face of environmental challenges. By foraging for wild plants, fruits, and nuts, the Moshana people ensure that their diets remain diverse and nutritionally rich throughout the year, even in times of hardship.

However, despite the importance of these foods, climate change poses significant challenges to the sustainability of the community's food systems. The increased unpredictability in weather patterns, including prolonged droughts, flooding, and temperature fluctuations, affects crop yields and threatens the stability of the local food supply. Such climatic vulnerabilities jeopardize the community's ability to maintain food security in the long term. Extreme weather events can wash away crops, deplete water resources, and disrupt the growing seasons, creating a volatile and uncertain environment for food production. This makes it crucial for the Moshana community to adapt to these changing conditions while preserving their cultural practices and indigenous knowledge. The transmission of knowledge through oral traditions and storytelling plays a critical role in ensuring that the younger generations are aware of the nutritional value of indigenous foods, the best agricultural practices for cultivating them, and how to adapt to the challenges brought about by climate change. Through storytelling, elders impart knowledge about the significance of indigenous foods, their health benefits, and the traditional methods that have helped the community maintain food security for centuries. The process of learning through storytelling ensures that this knowledge is passed down through generations, maintaining continuity in agricultural practices and food security strategies.

Moreover, the community's commitment to preserving indigenous knowledge and agricultural practices is vital in enhancing resilience to climate change. By continuing to farm and forage indigenous crops, such as millet and morula, the Moshana people preserve biodiversity and protect local food systems from the potential threats of monoculture farming and industrial agriculture. The importance of these foods lies not only in their nutritional value but also in their resilience to changing climatic conditions. For example, millet is a hardy crop that can withstand drought and poor soil conditions, making it an ideal staple in regions prone to erratic weather patterns. The knowledge transmission process within the Moshana community, particularly through the medium of storytelling, also helps the community adapt to the evolving challenges they face. These stories often contain lessons from past experiences, such as the hardships of food scarcity, drought, and poor harvests. By recalling these past struggles, the community is reminded of the importance of food resilience, which strengthens their ability to adapt and face new challenges. Storytelling also fosters a sense of community cohesion, encouraging collective action and collaboration in times of need. In this way, oral traditions not only preserve the history and culture of the Moshana people but also provide practical guidance for addressing contemporary challenges related to food security. The inclusion of indigenous knowledge and agricultural practices in contemporary food security strategies can help strengthen the resilience of rural communities like Moshana in the face of global challenges such as climate change and environmental degradation. The preservation of indigenous foods, along with the continuation of sustainable farming practices and the transmission of traditional knowledge, will ensure that communities can maintain food security and health in an ever-changing world. Furthermore, the integration of indigenous knowledge into national and global food security policies can provide valuable insights into sustainable agricultural practices that prioritize both human health and environmental conservation.

The Moshana community's reliance on indigenous foods, supported by oral traditions and storytelling, is central to their food security and resilience. The community's sustainable farming practices, which prioritize biodiversity and the use of indigenous crops, provide an invaluable model for other communities facing similar challenges. However, to safeguard this heritage and ensure the continued availability of these foods, it is imperative that the Moshana people address the emerging threats posed by climate change. By strengthening local food systems, preserving traditional knowledge, and adapting to the changing climate, the Moshana community can continue to thrive while maintaining a deep connection to their land and culture. The preservation of indigenous foods and agricultural knowledge is not only a matter of sustaining the community's food security but also an essential component of protecting the cultural identity and resilience of indigenous peoples globally.

REFERENCES

Africa, S. S. (2018). Men, Women and Children: Findings of the Living Conditions. *Survey (London, England)*, *2014*(15).

Barnhardt, R., & Oscar Kawagley, A. (2005). Indigenous knowledge systems and Alaska Native waysof knowing. *Anthropology & Education Quarterly*, *36*(1), 8–23. DOI: 10.1525/aeq.2005.36.1.008

Béteille, A. (2013). What should we mean by "indigenous people"? In *Indigeneity in India* (pp. 19–31). Routledge.

Bleibleh, S., & Awad, J. (2020). Preserving cultural heritage: Shifting paradigms in the face of war, occupation, and identity. *Journal of Cultural Heritage*, *44*, 196–203. DOI: 10.1016/j.culher.2020.02.013

Bobo, Z. N. (2021). *Promoting traditional and indigenous foods in South Africa: a desktop review.* Doctoral dissertation, Stellenbosch: Stellenbosch University.

Braun, V., & Clarke, V. (2006). Using thematic analysis in psychology. *Qualitative Research in Psychology*, *3*(2), 77–101. DOI: 10.1191/1478088706qp063oa

Deaton, B. J., & Scholz, A. (2022). Food security, food insecurity, and Canada's national food policy: Meaning, measures, and assessment. *Outlook on Agriculture*, *51*(3), 303–312. DOI: 10.1177/00307270221113601

Du Toit, D. C. (2011). Department of Agriculture, Forestry and Fisheries. *Food security*. Pretoria.

Eyong, C. T. (2007). Indigenous knowledge and sustainable development in Africa: Case study on Central Africa. *Indigenous knowledge systems and development: Relevance for Africa.Tribes and Tribals*, *1*, 121–139.

Ezeanya-Esiobu, C. (2019). *Indigenous knowledge and education in Africa.* Springer Nature. DOI: 10.1007/978-981-13-6635-2

Fernandez, M. E. (1994). Gender and indigenous knowledge. *Indigenous knowledge and development monitor* 2(3), 6-7.

Ibnouf, F. O. (2012). The Value of women's' indigenous knowledge in food processing and preservation for achieving household food security in rural Sudan. *Journal of Food Research*, *1*(1), 238–253. DOI: 10.5539/jfr.v1n1p238

Jena, A. K., Deuri, R., Sharma, P., & Singh, S. P. (2018). Underutilized vegetable crops and their importance. *Journal of Pharmacognosy and Phytochemistry*, *7*(5), 402–407.

Khoddami, A., Messina, V., Vadabalija Venkata, K., Farahnaky, A., Blanchard, C. L., & Roberts, T. H. (2023). Sorghum in foods: Functionality and potential in innovative products. *Critical Reviews in Food Science and Nutrition*, *63*(9), 1170–1186. DOI: 10.1080/10408398.2021.1960793 PMID: 34357823

Kumar, A. (2013). Ethno-botanical study of wild vegetables used by rural communities of Kanaujidistrict, Uttar Pradesh, India. *Emirates Journal of Food and Agriculture*, *25*(10), 760–766. DOI: 10.9755/ejfa.v25i10.16403

Langer, J. (2023). From acting to understanding: The comparative development of meaning. In *The nature and ontogenesis of meaning* (pp. 191–213). Routledge. DOI: 10.4324/9781003428152-9

Larson, N., Laska, M. N., & Neumark-Sztainer, D. (2020). Food insecurity, diet quality, home food availability, and health risk behaviors among emerging adults: Findings from the EAT 2010–2018 study. *American Journal of Public Health, 110*(9), 1422–1428. DOI: 10.2105/AJPH.2020.305783 PMID: 32673120

Lugo-Morin, D. R. (2020). Indigenous communities and their food systems: A contribution to the current debate. *Journal of Ethnic Foods*, *7*(1), 1–10. DOI: 10.1186/s42779-019-0043-1

Mabhaudhi, T., Chibarabada, T. P., Chimonyo, V. G. P., Murugani, V. G., Pereira, L. M., Sobratee, N., Govender, L., Slotow, R., & Modi, A. T. (2018). Mainstreaming underutilized indigenous and traditional crops into food systems: A South African perspective. *Sustainability*, *11*(1), 172. DOI: 10.3390/su11010172 PMID: 37681213

Masekoameng, M. (2007). *Indigenous knowledge systems in food gathering and production in selected rural communities in Sekhukhune* District of the Limpopo Province. Polokwane: University of Limpopo.

Mbenyane, X. G. (2017). Indigenous foods and their contribution to nutrient requirements. *The South African Journal of Clinical Nutrition : SAJCN : the Official Journal of the South African Society of Parenteral and Enteral Nutrition and the Association for Dietetics in South Africa*, *30*(4), 5–7.

McGuire-Adams, T. D. (2019). Paradigm shifting: Centering Indigenous research methodologies, an Anishaabe perspective. *Qualitative Research in Sport, Exercise and Health*, *12*(10), 34–47.

Mijatović, D., Van Oudenhoven, F., Eyzaguirre, P., & Hodgkin, T. (2013). The role of agricultural biodiversity in strengthening resilience to climate change: Towards an analytical framework. *International Journal of Agricultural Sustainability*, *11*(2), 95–107. DOI: 10.1080/14735903.2012.691221

Ndwandwe, S. (2013). *The Contribution of indigenous knowledge practices to household food production and food security: A case of UKhahlamba local municipality, South Africa*. University of KwaZulu-Natal.

Panda, T., Mishra, N., Pradhan, B. K., & Mohanty, R. B. (2015). Diversity of leafy vegetables and its significance to rural households of Bhadrak District, Odisha, India. *Scientia Agrícola*, *11*(3), 114–124.

Pichop, G. N., Abukutsa-Onyango, M., Noorani, A., & Nono-Womdim, R. (2014). Importance of indigenous food crops in tropical Africa: Case study. In *XXIX International Horticultural Congress on Horticulture: Sustaining Lives, Livelihoods and Landscapes (IHC2014): 1128*, 315-322.

Rankoana, S. A. (2021). Indigenous plant foods of Dikgale community in South Africa. *Journal of Ethnic Foods*, *8*(1), 1–8. DOI: 10.1186/s42779-021-00080-w

Schönfeldt, H. C., Hall, N. & Pretorius, B. (2017). Nutrition-sensitive agricultural development for food security in Africa: a case study of South Africa. *International Development*, 1.

Takaidza, S. (2023). *Indigenous South African Food: Nutrition and Health Benefits*. Vaal University of Technology.

Thornton, P. K., Ericksen, P. J., Herrero, M., & Challinor, A. J. (2014). Climate variability and vulnerability to climate change: A review. *Global Change Biology*, *20*(11), 3313–3328. DOI: 10.1111/gcb.12581 PMID: 24668802

Tikai, P., & Kama, A. (2010). A study of indigenous knowledge and its role to sustainable agriculture in Samoa. *Open Journal of Social Sciences*, *3*(1), 65–79.

Van der Hoeven, M., Osei, J., Greeff, M., Kruger, A., Faber, M., & Smuts, C. M. (2013). Indigenous and traditional plants: South African parents' knowledge, perceptions and uses and their children's sensory acceptance. *Journal of Ethnobiology and Ethnomedicine*, *9*(1), 78. DOI: 10.1186/1746-4269-9-78 PMID: 24274394

Van der Merwe, J. D., Cloete, P. C., & Van der Hoeven, M. (2016). Promoting food security through indigenous and traditional food crops. *Agroecology and Sustainable Food Systems*, *40*(8), 830–847. DOI: 10.1080/21683565.2016.1159642

Virginia, P., & Pallavi, S. (2016). Role of women's indigenous knowledge in food processing and preservation for enhancing household food security in Allahabad District. *World Journal of Pharmacy and Pharmaceutical Sciences*, *5*(11), 729–743.

Weinberger, K., & Pichop, G. N. (2009). Marketing of African Indigenous Vegetables along Urban and Peri-Urban Supply Chains in Sub-Saharan Africa1. In *African Indigenous Vegetables in Urban Agriculture* (pp. 225–244). Routledge.

World Food Summit. (1996). Rome Declaration on World Food Security.

Zimu-Biyela, A. N. (2016). The management and preservation of indigenous knowledge in Dlangubo Village in KwaZulu-Natal, South Africa. Doctoral dissertation, University of South Africa.

KEY TERMS AND DEFINITIONS

Indigenous Foods: Are plant and animal-based foods that are naturally existing and produced in specific locations and consumed as part of traditional diets.

Food Insecurity: It is the condition of not having access to sufficient food, or food of an adequate quality, to meet one's basic needs.

Indigenous Peoples: Are distinct social and cultural groups that share collective ancestral ties to the lands and natural resources where they live, occupy or from which they have been displaced.

Indigenous Knowledge: It is seen to be collective, holistic, and adaptable knowledge that is unique to a specific geographical or cultural context.

Indigenous Knowledge Systems: It is the corpus or bodies of knowledge, technologies and beliefs produced and handed down from generation to generation within a specific location such as rural or urban.

Food Security: It exists when all people always have physical and economic access to sufficient, safe and nutritious food that meets their dietary needs and food preferences for an active and healthy life.

Preservation: It refers to the act or process of maintaining or protecting something in its existing state. It often applies to cultural, natural, or historical resources, aiming to protect them from damage or decay.

Traditional Food Systems: Are composed of all the biophysical, economic, technological, and sociopolitical components that uphold the traditional ways of producing, gathering, preparing, distributing, consuming, and using traditional foods for survival and health.

Chapter 12

Documentation of Indigenous Crops Utilized by Vatsonga Households for Food Security in Mhinga Traditional Authority (MTA): An Appraisal for Economic Development

Vanessa Risimati
https://orcid.org/0000-0002-4639-5293
North-West University, South Africa

Mothusiotsile Edwin Maditsi
https://orcid.org/0000-0001-5796-0174
North-West University, South Africa

ABSTRACT

Indigenous knowledge (IK) of indigenous crops (IC) is deeply rooted in the history and environment of the Vatsonga people. It is crucial to preserve this knowledge, as it is part of their culture, language, and identity. IK, passed down through oral, observed, and written forms, includes unique farming practices specific to the community. However, with the rise of Western knowledge, conventional farming, and genetically modified foods, there is a decline in the use of indigenous crops, leading to the loss of Vatsonga IK. This study, using an indigenous paradigmatic lens and qualitative approach, explores how the current generation of Vatsonga applies their IK regarding indigenous crops. It finds that the Vatsonga of MTA have used these crops for centuries for survival, food access, income, and cultural identity. The study concludes that indigenous crops are vital for food security, cultural preservation, and economic growth by creating jobs and promoting respect for their IK.

DOI: 10.4018/979-8-3693-7964-6.ch012

INTRODUCTION

Indigenous crops have always been a part of indigenous peoples since the beginning of time. Where they would use indigenous crops for community consumption and their animal consumption. With indigenous people having connections to the environment, their food preferences not only represent a symbol of self- sustenance but also a representation of how indigenous people interact with their environment as well as the significance and potentials embedded in the use of indigenous crops. Therefore, indigenous peoples have safe guided their indigenous crop knowledge for centuries despite the influences of the world's constantly evolving food trends. This use of indigenous crops in Africa can be traced as far back as 1000 B.C.E to the Bantu group in central Africa (Lui, 2019). Who were known for manufacturing agricultural tools and practicing indigenous crop cultivation methods that overtime have spread throughout the continent of Africa as they migrated (Phillipson, 1977).While some scholars may argue that indigenous crop cultivation methods did not originate from the Bantu ethnic group, but were a combination of skills they were taught by their neighbouring ethnic communities and the Bantu only contributed in the spread and practice of the knowledge within the African diaspora as they migrated, settled, shared their knowledge of indigenous crops cultivation and various consumption methods of the crops with various communities (Isern and Fort, 2019; Bostoen, 2018). Despite the argument of where the indigenous knowledge of indigenous crops cultivation originated. It continues to remain a fundamental component of indigenous crops use and perception in this current day and age. To some indigenous communities, the use of indigenous crops within their communities is a representation of who they are, their lineage, their sense of belonging and to those who are descendants of the Bantu. The similarities of the indigenous crop cultivation to the ancient Bantu from 3000 years ago serves an ancestral connection that is emanant in their planting methods, their interaction with the environment, harvesting methods, harvest thanksgiving ceremonies, weather and climate patterns, community development as well as indigenous crops preservation methods (Smith, 2019; Murray, 1970; Holden, 2017). Additionally, regardless of the various spatial population dynamics of the groups of descendants of the Bantu, where Vatsonga located in the Southern hemisphere are descendant of, the knowledge of indigenous crops has been passed on for generations through oral orientations and practices. Therefore, Indigenous crops play a fundamental role in maintaining food security and sustaining livelihoods of the indigenous people within indigenous communities. As they are abundant in indigenous communities. Indigenous crops consist mainly of crops that are accessible and indigenous to an area (Ayanwale et al. 2016; Akinola et al 2020). They constitute of natural resources that most Africans utilize at a household level for food and profit. These crops are considered indigenous because of their availability in the area, how well they grow, the community's acceptance, them being a staple and the connections that they have to the indigenous people of that region (Balasubramanian, 2014).

Indigenous crops can also be defined as plants that are native to a place or as plants that were introduced to a place and recognised as crops that are indigenous to a place because of the cultural connections they embed. These crops have different personal, cultural appropriations and preferences aligned with the way indigenous people consume them (Olum, et al.,2017). These crops form part of culture and heritage of indigenous communities and plays a role in the cultural identities of different ethnic groups. It is estimated that over 10,000 years ago, indigenous communities across the globe started forest gardening as way of sustaining its people and maintaining their food security (Kuhnlein et al,2019). This was informed by the need for food security which in turn influenced the norms of indigenous communities' survival and the way they value indigenous crops. Subsequently as the human population grew, farming expanded into agricultural activities that contributed towards the economy's development of indigenous communities.

It is against this backdrop that shows that indigenous crop production greatly contributed to the food system of third world countries in the last 50 years (Turner and Turner, 2007). Thus, (Bvenura and Afolayan, 2015; Akinola, et al, 2020) allude that agriculture in the Green Revolution during the 1960's was primarily focused on the development of standard horticultural crops and cereal, emanating in conventional food becoming more popular and replacing most indigenous crops produced locally resulting in indigenous crops being undervalued. Indigenous crops within various indigenous communities are facing indigenous crop knowledge extinction (Leyew, 2024), also, as a result of the influence of the Eurocentric food markets, conventional food consumption trends and commercial access by the public. This book chapter aims to contribute to documenting indigenous crops from a Vatsonga perspective. While correcting some literature discrepancies such as the indigenous crop names to their English names and adding knowledge of indigenous crops that Vatsonga use that has not yet been documented using community-based knowledge from MTA knowledge holders. The chapter will serve as a literature resource from a Vatsonga perspective in indigenous crop agriculture and the indigenous knowledge of crops and in community maintenance. While putting into perspective the importance of indigenous crops and their economic potentials from a view of MTA's day to day life.

PROBLEM STATEMENT

Indigenous crop production has been used for years by communities across the globe. Where even presently in some communities it is still practiced. With Vatsonga ethnic group from MTA in Limpopo have been consuming indigenous crops in their different communities for centuries (Kock, 2020). They have been using different indigenous crops to maintain food security and for their local economic development. As a result of the world having shifted and is now living in a western dominate system of life and knowledge, indigenous ways of life and indigenous food production tends to be given less recognition (Vorster, Stevens and Steyn, 2008). Resulting in a decline of indigenous knowledge on indigenous crops. This decline is because of the under-documentation of indigenous Vatsonga cultivation and consumption methods. Moreover, with this indigenous knowledge being passed on through oral orientations and practices. It has resulted in this knowledge extinction as time progresses. Additionally with food security being a challenge because of climate change, conventional farming has taken-over the indigenous food markets as conventional crops can easily meet the consumer demand and supply as compared to indigenous crops. Thus, making them a more preferred method by commercial food producers for consumers (Van de Merwe, Cloete and Van der Hoeven, 2016). While these cash crops can meet the demand and supply for consumers, they are also influencing a reduction of indigenous dietary diversity. Moreover, with indigenous crops taking time to cultivate and harvest, they are now slowly nicely being packaged, rebranded and sold at higher price as luxury healthy food products that can only be afforded by consumers with more money than a basic income consumer (Socha, at al, 2011). As a result, everyday indigenous consumers who are facing climate change have limited access to indigenous crop products such as Sorghum and millet to sustain them and their families for a month as compared to conventional food products with GMOs such as Maize meal and rice (Kristen, et al, 2017).

Consequently, indigenous knowledge has become sidelined and young food consumers have limited continuation of Vatsonga indigenous farming methods as these methods are slowly being less practiced. With the marginalization contributing towards less consumption of indigenous crops and loss of the indigenous links between the people and the indigenous crops, limited documentation of the knowledge of indigenous crops from a Vatsonga perspective also limits the potentials that indigenous crops may have in the conventional food markets such as in conventional supermarkets. Thus, limiting the benefits and potentials in the economy for indigenous crops. With Sections 26 and 27 of the South African Constitutional law of 1996 stating that citizens have the right to access sufficient food (Constitution of the Republic of South Africa, 1996) and (Mapiye, Chikwanha, et al., 2020) stating that about 70% of Southern Africans in the year 2020 depended directly on agricultural activities for food, income, nutrition as well as health maintenance. Documenting indigenous knowledge on indigenous crop cultivation practices, nutritional benefits and economic benefits can assist in promoting indigenous food policies that will ensure that there is responsible consumption, communities benefit from their knowledge. While promoting right to access to food that is adequate. Additionally, the benefits of indigenous crops and economic potentials of indigenous crops across South Africa not being fully recognised and explored by the western body of knowledge (Mbhenyane, 2017). Consequently, the indigenous knowledge on crop farming of the Vatsonga ethnic community is least documented. With most of the knowledge being lost as the world changes. Further, contributing towards an underutilization of the indigenous methods that contribute towards helping local economies grow. As most of the world has shifted from the indigenous methods of farming to using conventional farming practices.

RESEARCH OBJECTIVES AND QUESTIONS

The Objectives of the Study

- To document the types of indigenous crops used within Vatsonga households in Mhinga Traditional Authority
- To explore the production of indigenous crops used within Vatsonga households in Mhinga Traditional Authority
- To investigate the utilization of indigenous crops within Vatsonga households in Mhinga Traditional Authority
- To examine the effects of indigenous crops of Vatsonga in enhancing the local economy of Mhinga Traditional Authority

The Research Questions of the Study

- What are the types of indigenous crops used within Vatsonga households of Mhinga Traditional Authority?
- How are the indigenous crops used within Vatsonga Households of Mhinga Traditional Authority produced?

- How are the indigenous crops utilized within Vatsonga Households of Mhinga Traditional Authority?
- How does indigenous crops of Vatsonga enhance the local economy of Mhinga Traditional Authority?

THEORETICAL FRAMEWORK

Self Determination

Self-determination theory focuses on human being's personalities as well as the motivation that influences how human beings interact. This theory is influenced by the social environment and socioeconomic effects (Vallerand, et al,2008; Deci, 2017). It is a sense of several mini theories that are put together to result in comprehensive understanding of the human function as well the movement (Vasconcellos et al, 2020). Thus, the self-determination theory is based on the core humanistic notion of Vatsonga from MTA that people naturally and actively strive for self-organization and growth. Furthermore, by integrating new experiences, cultivating their needs such as Theka (*Amaranthus spp*), desires, and interests, and connecting with others and the outside world, people's attempt to extend and comprehend themselves. Within the community of Mhinga T.A the need to have enough food for nutrition with this theory is exhibited more on how they use indigenous crops for food. Further how they evolved into the need for food security, nutrition and generating household income from their indigenous crops and contributing towards the growth of the economy of Mhinga T.A. Moreover, with some of the components being choice, decision making, problem-solving and self-advocacy (Reeve,2002). The study looks at the relatedness and competence of Mhinga T.A with this theory, as the study seeks to document indigenous crops utilized by Vatsonga households for food security in Mhinga Traditional Authority (MTA) as well as the economic connections that indigenous crops with the communities of MTA. Additionally, the reasons behind the need to plant, harvest, preserve and sell.

New Growth Theory

Paul Romer created this theory about the economic growth and the connection that it has to the desires of human beings (Nelson, 1997). New growth theory focuses on improving and increasing capital. Through emphasizing the development and prosperity of the economy. By demonstrating the effective production and the general growth of the economy that is directly linked to people. This theory is deeply rooted in the purchasing, investing, decision-making and investments (Rima, 2004). In addition, the theory's primary pursuit is for personal profit generation by promoting and emphasizing innovation, knowledge, as well as entrepreneurship. Additionally, this theory further highlights the knowledge that individuals possess and how they use that knowledge to generate income through buying, selling, and investing (Solow, 1994; Banerjee and Duflo, 2005). In this study the new growth theory exhibits the economical connections that the indigenous communities of MTA have to indigenous crops as well as their ability to contribute to the local economy of MTA using indigenous crops. Furthermore, this theory ensures that it clearly exhibits the effectiveness and growth of the economy linked to the people and indigenous crops the indigenous examined. As well as what has influenced the community members that sell their surplus crops to start selling as well as how this practice has emanated into the communities of MTA developing. With a percentage of the income generated in the development being from selling indigenous crops such as guxe (*Vigna subterranean*)

LITERATURE REVIEW

The Importance of Documenting Indigenous Crops in Africa

Indigenous crops have been the primary source of food in indigenous communities for centuries at end. This knowledge of indigenous crops has always been passed on from one generation to the next. Indigenous crops are planted, and some grow naturally (Tshidzumbza, 2020). They are used for both nutrition and medicinal purposes. Concurrently, the knowledge of indigenous crops is embedded in the oral traditions and practices of the indigenous peoples of a given location where the crops originate (Agrawal, 1995). As a result, they are considered as wealth for indigenous communities as they are useful in maintaining food security, health maintenance, soil management, water management and income generation (Haumba and Kaddu, 2017). They are linked to indigenous communities and the relationship that people have with food and their environment. Further emanating into the complex relationships that people have with animals, cosmology and the earth (Nyota and Mapara, 2008). Thus, shaping the way people value their indigenous crops, the naming of the crops, how to plant, harvest and consume. Moreover, shaping economic developments of indigenous communities with the knowledge of indigenous crops being embedded in the elderly.

Puoane et al (2006) asserted that in indigenous households and the elderly are ones who cook and consume indigenous crops. This is because of the knowledge that elders amassed from their parents when growing up. Furthermore, they have knowledge and expertise of how to prepare traditional meals using indigenous crops, know the cooking time and know the value of indigenous crops in their bodies when consumed. Scientifically, indigenous crops are rich in nutrients and comprise of protein, dietary fibre, carbohydrates, minerals (including calcium, iron, zinc, magnesium and potassium) and vitamins (including A, C, D, K and B) (Luthria and Pastor-Corrales, 2006). Thus, these known as traditional crops, have been cultivated and consumed by local communities for generations and contribute to agricultural biodiversity, helping to maintain a wide variety of nutrients in the diet. Its dietary diversity provides unique flavours and textures that can make meals more interesting and nutritious (Mavengahama, 2013). This diversity can enhance the overall nutritional quality of diets. Since they are rich in essential nutrients such as vitamins, minerals, and antioxidants. For an example, African leafy vegetables like Amaranth are high in vitamins A and C, calcium, and iron (Mbhenyane, 2017). Making them good for consumption, subsequently, providing a stable source of food more than what conventional crops fail to provide.

Additionally, indigenous crops are often well-suited to local environmental conditions and are resilient to drought, pests, and diseases, making them an important food source, especially in regions with challenging agricultural conditions and climatic conditions (Nesamvuni, Steyn & Potgieter, 2001). Furthermore, adapted to the local environment and often require fewer inputs, such as water, fertilizer, and pesticides, which can reduce the environmental impact of agriculture (Njeme, Goduka & George, 2014). Concurrently, enhancing food security, particularly in regions prone to climate change and extreme weather. Further, indigenous crops are often grown using traditional, sustainable farming practices, which can have a lower environmental impact compared to modern or conventional farming and GMO food cultivation. These indigenous crops are versatile as one crop species can be prepared in multiple methods resulting in a nutritional balanced meal such as sweet potatoes being consumed from the roots to the leaves. As a result, the elderly value their indigenous crops as they are cost effective and are aware of the benefits that these crops contain. However, as the elderly consume indigenous crops in numbers, the younger generation tends to consume less and have limited knowledge on the indigenous crops and their benefits.

Consequently, the indigenous knowledge (IK) on indigenous crops is disappearing amongst the youth. For this reason, it is vital to document and preserve this knowledge from an indigenous perspective before it extinct in the community and becomes unavailable for the next generation as well as for the maintenance of the community's indigenous knowledge (Uusiku et al, 2010). Moreover, in literature there is limited knowledge on the indigenous crops used by across various communities. With some of the limited knowledge that is documented having some discrepancies such as the indigenous name of the crops not matching with the scientific names of the crops (Rivera, et al, 2014). Therefore, resulting in the loss of the correct names of the indigenous crops and IK embedded in the crop species. Furthermore, it has resulted in indigenous crop name adoptions from other cultures that do not represent the authentic language, traditions and indigenous crop knowledge embed in indigenous crops (Lara, et al, 2023; Ardhede, 2007). The adoption of names of indigenous crops in place of the indigenous names already given by the ancestors because of less information by some of the current generation has resulted in the loss of the rhetorical power connected to the traditions and the culture of the indigenous people and their food (Kalland,2000). Subsequently, there are some indigenous crops that are used by the with communities that have not been yet documented. Documenting the knowledge of indigenous crops will help in providing a detailed background of the crops and their agricultural developmental importance while assisting in preserving the knowledge of these undervalued crops as most of the population in the community is influenced by the western ways of orientation because of less use and less knowledge of the crops (Bush, 2005). Documenting the indigenous crops used by the people will help in preserving and exploring the importance of using indigenous crops and how they are used from the community's unique knowledge.

Impact of Indigenous Crops Used for Food Security and Household Income Generation

Food security is a state that exists when people have access to food that is nutritionally balanced (FAO, 2002; Ferranti, Berry, and Jock, 2018). This ensures that all people always have food that has sufficient nutritious and is safe, while meeting their dietary requirements and food preferences that are healthy (Pinstrup-Andersen, 2009). Indigenous crops are the backbone of food security and are at the core of humanity's survival for the rest of continent. As African's have been using indigenous crops for centuries, for consumption, clothing, and food for animals. According to Mabhaudhi, Chibarabada, et al (2019) the agricultural food system for more than 10 000 years has shifted as human activities also changed, from hunting and gathering to cultivating and growing crops for food. Indigenous crops in Africa were used for the purpose of sustaining the growing population. As a way of ensuring that individuals and the collective within the community are food secure. Further, playing a role in household food security throughout the continent.

The diversity of food crops such as traditional food and indigenous food has influenced how people consume their food, maintaining nutrition and making profit out of the surplus from their yield. With these crops being at the core of nutritional security, poverty reduction and the financial stability of households (Aliber and Hall, 2012). Indigenous crops are defined as consumable plants that are native to a place, these plants may be in a form of tubers, stems, leaves and legumes. With these either being cultivatable or growing naturally (Elizer, Ravichandran, and Antony, 2019). The indigenous crop usage in Africa is important as it forms part of the continent's economy through the sale of food from indigenous crops, raw materials as well as fuel. With indigenous crops being at the core of food security and reduction of poverty in Africa, it is important for the indigenous crops to be preserved and used in a sustainable way (Ruelle, 2015).

Additionally, with indigenous crops being sustainable and can be managed using natural resources they ensure that communities are able to grow in revenue. From small-scale farming that community members have in their backyards to larger farm plots. These subsistence farms assist in combating hunger, illnesses caused by lack of nutritional balance from the food they consume as well as help them generate household income. Moreover, with them being resilient they can withstand harsh climate conditions caused by climate change. It helps in promoting food security for the locals and poverty reductions (Webber, et al,2014). Moreover, with Africa's human population growing at a rapid rate the use of indigenous crops is significant because of its easy access to the indigenous people, it is cheap to purchase from the local vendors and in some cases, it can be accessed free of charge (Poulton, Dorward and Kydd, 1988). Furthermore, one indigenous crop species can be prepared in different ways from the leaves to the tubers resulting in a nutritionally balanced meal, food preservation and a months' worth of food supply. Moreover, maximizing the profits of those who sell their surplus crops. The income generated ensures that households can maintain themselves and, in some cases, also be used for wealth generation and a sustainable community development that is mainly reliant on the profits from yearly and seasonal indigenous crops. Promoting proper use of the crops without wasting the crops and the nutrients they possess. Therefore, maximising the profit that indigenous crop sellers can gain during sales and catering for community members with adequate nutritionally balanced meals that do not harm consumers and their natural surroundings.

South Africa's Indigenous People's Interactions of Culture, Landscapes and Their Dietary Preferences

South Africa is a country with diverse agricultural landscapes. Landscapes that have allowed its people and biodiversity to thrive. The landscapes have allowed cultivation and an abundance of both indigenous and conventional crops. With the of perspectives on indigenous crops being influenced by a variety of factors, including cultural, economic, and environmental considerations (Rindos, 2013). Where culture and traditions play a significant role. As indigenous crops have deeper connection to cultural, community practices and traditional significance in South Africa. These crops have been cultivated and consumed by various ethnic groups for generations. They play a crucial role in preserving cultural heritage and traditional food practices (Pinhasi and Stock, 2011). In addition, they contribute to biodiversity and sustainability in South Africa's communities. These indigenous crops are often better adapted to local environmental conditions and require fewer inputs, such as water and pesticides, making them more resilient during climate change problems and resource constraints. With many indigenous crops being highly nutritious and offer valuable dietary options. For an example, Amaranth (Bangala na Theke) and various traditional grains to use to add nutritional value to each meal. These crops provide essential vitamins, minerals, and fibre and a well-rounded diet (Emmanuel and Babalola, 2022).

Indigenous crops have economic potential for South Africa. They serve as a source of income for small-scale farmers, the unemployed and contribute to food security. With of South Africa’s indigenous crops, like rooibos and honeybush, having gained international recognition and contribute to South Africa's exports and economic growth (Zadik,1986; Marnewick, et al, 2019). Concurrently, there are efforts to conserve and promote indigenous crop varieties, both for cultural preservation and the unique food genetic diversity. With some even aiming to document and protect indigenous knowledge associated with these indigenous crops and ensure their long-term sustainability (Elmore, 1996). Despite their many benefits, indigenous crops face challenges. Some indigenous crop varieties are underutilized or at risk of disappearing due to changing dietary preferences, urbanization, climate change, knowledge holders dying with the knowledge and conventional agricultural practices favouring modern, high-yielding farming benefits and its possibilities for the economy.

Thus, government and various food organizations need to be more involved in promoting awareness about the importance of indigenous crops and the need to preserve traditional knowledge (Muchineripi, 2008: Eliazer, et al, 2019). Events, such as Indigenous Food gatherings such as festivals need organized to showcase the cultural, culinary and nutritional value of indigenous crops. Lastly, South Africa should recognize the significance of indigenous crops in terms of culture, sustainability, and economic potential guided by the IKS Act of 2019 and the adopted or implemented food security policies and legislations. As the efforts made can conserve and promote these crops without facing a lot of challenges as the country's agricultural landscape evolves (Cleveland and Murray, 1997). As indigenous crops remain an essential and significant part of South Africa's agricultural and cultural heritage.

Indigenous Farming Practices Embedded in Indigenous Crop Production in South Africa

Indigenous farming practices in South Africa have a long history, a diverse tradition and connected the cultivation of indigenous crops. These practices are passed down through generations and play a crucial role in the sustainable production of indigenous crops (Denison and Wotsehela, 2012). With South Africa being home to a variety of indigenous crops, including grains like millet, sorghum as well as legumes like cowpeas and Bambara groundnuts, tuber, and leaves (Zondi, et al,2022) . The cultivation of these crops has been essential to the nutrition, food security, and cultural identity of many communities (Mgwenya, et al, 2025). Indigenous knowledge on indigenous farming practices is transferred through oral traditions and practical experiences. These practices are aligned to an understanding of the local ecosystems, climate, and soil conditions, which is crucial for successful indigenous crop cultivation (Seko, Bain and Maponya, 2021). Moreover, indigenous communities have been using different methods of farming such as indigenous communities having long practiced crop rotation to maintain soil fertility. This practice involves rotating crops in a specific order to improve soil health and reduce the risk of pest and disease (Dury, et al, 2012). Additionally, mixed cropping or intercropping is a practice where multiple crop species are grown in proximity to maximize land use and reduce vulnerability to environmental stressors (Hiddink and Termorshuizen, 2010).

Indigenous farmers have developed sustainable water management systems, including saving rainwater in tanks and dams, indigenous irrigation to capture and store rainwater for crops (Ncube, 2018). Subsequently, with indigenous farmers being skilled in selecting and saving seeds from their best harvest of plants it has led to indigenous crop varieties being adapted to local conditions. As indigenous farming practices are inherently organic and do not use synthetic fertilizers and chemical pesticides. There ensuring that the seeds saved can withstand harsh conditions and can grow at their natural full capability. Massawe, Mayes and Cheng (2016) assert that indigenous farming practices often promote crop diversity, contributing to biodiversity conservation. These practices help preserve indigenous crop varieties and the wild relatives of theses cultivated species, maintaining their genetic resources that are crucial in adapting to changing environmental conditions caused by climate change and human activities (Richard, 2023). Furthermore, indigenous farming practices are rooted in the cultural traditions of South African communities. Where they are often accompanied by rituals, ceremonies, and festivals that celebrate the planting and harvesting of crops such as the first haverst ceremony, where the first yield of the crops is taken the community leader as offering and a sign of the communities' good harvest (Ohaja and Anyim, 2021). These practices reinforce cultural identity and social cohesion within communities.

METHODS AND MATERIALS

The study was conducted at Mhinga Traditional Authority, 22° 45' 0" South, 30° 54' 0" East, at Vhembe district in the North of Limpopo province, South Africa. It is one of the rural settlements of Collins Chabane municipality approximately 41km from Malamulele (Stats SA, 2011; Sikhweni and Hassan, 2014). MTA is adjacent to the Kruger National Park (KNP), and it covers an area of 20 000 ha of cropping and grazing land (Chaminuka, McCrindle and Udo, 2012). It has approximately 7000 households and an estimate of 10 524 residents (Stats SA, 2022). The community is a home to Xitsonga speaking people and it is under the traditional leadership of Chief Shilungwa Mhinga. The study was conducted at nine villages under MTA namely Mabiligwe (Qaza), Josepha or Josefa, Mhinga Botsoleni, Matiyani, Maphophe, Mhinga Zone 1, Mhinga Zone 2 and Mhinga Zone 3.

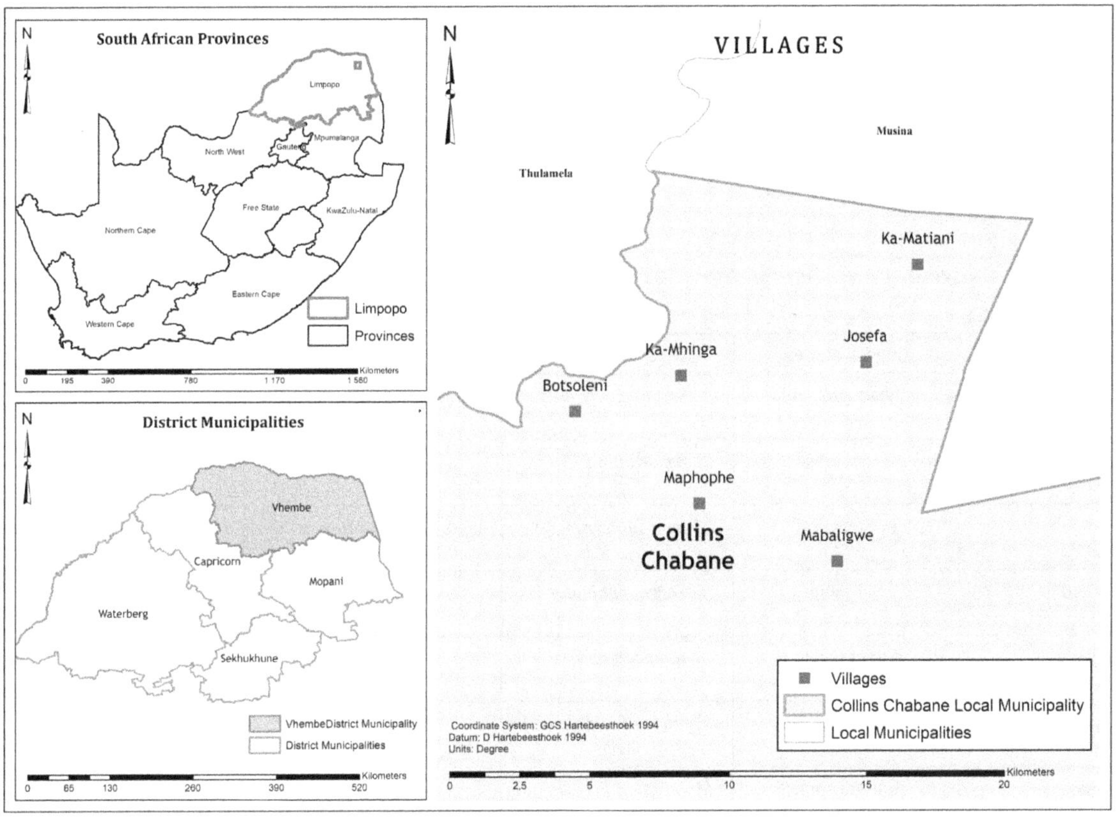

Methodology

A qualitative approach was used for an in-depth perspective and understanding of the documentation of indigenous crops utilized by Vatsonga households for food security in Mhinga Traditional Authority and its importance in day-to-day life of an average Mutsonga consumer. The approach is qualitative as participants will need to provide their experiences with these crops at free flow. The study is exploratory in nature and has provided background of in-depth information on how indigenous crops are being consumed by Vatsonga people in MTA. This method has helped in gaining insights on the significance indigenous crops in documentation and survival. Purposive sampling with a snowballing technic was used to acquire participants with the relevant knowledge to the study (Taherdoost, 2016; Etikan, et al 2016). This sampling was mainly giving preference to participants with the knowledge on indigenous crops and have practical knowledge of the knowledge. An indigenous paradigm lens was applied so that the study moves away from the western lens that has led to these crops being left in the margins of the knowledge economy and the local economy of indigenous communities. In-depth interviews with an in-depth interview guide were utilized to have open and prolonged discussions with the participants. The study uses data collection interviews. The interview was guided by interview questions that emerged from the research questions. The interviews were in-depth-face to face, focus groups and field walks.

The data was collected and analysed using the 6 steps of thematic analysis by Braun and Clarke (2012). The themes will be reported in this paper as finding of the study.

FINDINGS FROM THEMES THAT EMERGED

Table 1. Plant inventory of some indigenous crops used within Vatsonga households of Mhinga Traditional Authority

Xitsonga Names	English Names	Scientific Names	Parts used	Season planted/ grow
Tindluwa	Bambara groundnut	*Vigna subterranea*	Roots (Legumes contained in the roots)	Summer (Basu,et al, 2007)
Theka	Amaranthus	*Amaranthus spp*	Leaves	Summer (Trucco and Tranel, 2011)
Guxe	Jutes mallow	*Vigna subterranean*	Leaves and flowers	All year (Palada and Chang, 2003)
Handelo	Sorghum	*Sorghum bicolor*	Grains	Spring- Summer (Orr, et al, 2016)
Mihlanta	Sweet Potato	*Ipomoea batatas*	Roots	Summer (Mavhungu, et al, 2022)
Tinyawa	Cowpeas	*Vigna unguiculata (L.) Walp*	Legumes and Leaves	Summer (Asiwe,2009)
Nkaka	Balsam apple	*Momordica Balsamina*	Fruits and leaves	All year (Venter et al, 2021)
Tinyiya	Bird Plum or Brown Ivory	*Berchemia Discolor*	Legumes and leaves	Summer (Rampedi, 2010)
Tinhwembe Xiphaswu	Pumpkin and Pumpkin leaves	*C. Moschata*	Leaves, flowers and pumpkin	Summer (Kock, 2020)
Bangala	Cleome	*Cleome gynandraL*	Leaves and Flowers	Summer (Kock, 2020)
Manjanja	Millet	*Panicum miliaceum*	Grains	Winter
Makavatlha	Maakataan	*Citrullus lanatus,*	Fruit	Winter and beginning of spring
Timanga	Bambara groundnuts	*Arachis Hypogaea*	legumes	Summer
Mpfilwa	Common Medlar	*Mespilus germanica*	Fruit	Summer
Nkanyi or Makanyi	Marula	*Sclerocarya birrea*	Fruit	Summer
Ntoma	Common Medlar	*Mespilus germanica*	Fruit	Summer
Muxiji	Black-Jack	*Bidens pilosa L*	Leaves and Flowers	All year
Marhanga	Wild melon	*Lagenaria Sicerai Or Cucurbita pepo*	Fruit	Summer

Strategies Used By Vatsonga From MTA to Name Their Indigenous Crops

Indigenous crops have been the primary source of food for some participants since birth. These partipants grew up being taught the fundamentals of indigenous crop cultivation and the different types of indigenous crops that Vatsonga consume. For them the names of the crops of indigenous crops that they grew up eating is not just a symbol of food security and nutrition, but they have meanings embedded in them. These indigenous crops were named by the ancestors based on their characteristics and this knowledge has been passed on for generations. Some of the characteristics that influenced by the naming of some Vatsonga indigenous crops was a result of how they grow while for some it is because of the association that people with the indigenous crops. Participant 3 indicated that leafy vegetables such as Tinhwembe (*C. Moschata)* was named because of how one crop can occupy a large surface area and its ability to bear pumpkin. Unlike other indigenous leafy vegetables, *C. Moschata* can withstand the test of time. Moreover, because of its resistant abilities to withstand harsh weather conditions and being accessible all year around. The naming of this indigenous crops was also influenced by the be of men. With participant 8 stating that "*N'wanuna I nhwembe loko yi kula ya nava*" simply translating to a "A *man is like a C. Moschata when it grows, it spreads".* The moral behind this association of men to a *C. Moschata* is that men are naturally polygamous and as they grow, they start exhibiting traits and behaviours of having a man that has more than one woman. The very same phase is used to comfort the wife of the man by the elders when she starts raising her consents about the husband's behaviour or used during the Vatsonga bridal shower, were those who have years of marriage experience will use the indigenous crops as an example or a symbol resilience in marriage and understanding a man's behaviour.

Other names of the indigenous crops used by the people of MTA, were named because of how they taste. Those that taste bitter tend to have names that are gravitating towards their bitterness such as Bangala leafy vegetables that are known for their bitter taste. The name "Bangala- *Cleome gynandraL*" can easily be associated to "Ku biha- to be bitter". Other crops were named because of how they look, their shape and longevity. Indigenous crops such as Marhanga (*Lagenaria Sicerai*) were named because of their texture on the outside and on the inside in both states of when they are raw and cooked. While some were named because of the methods that is used during their processing for preservation. Handelo (*Sorghum bicolor)* was named after the process they do before washing and drying. The process of "Ku *Handa" "Scratch or run hands through something"* is usually done after harvesting, this process ensures that the unwanted parts of the sorghum plant are not mixed with the sorghum that will be washed, dried and grinded in sorghum flour.

Production and Utilization Methods of Indigenous Crops Within Households

Most the indigenous crops consumed by Vatsonga from MTA villages are crops that grow naturally. These crops are annual crops while some are seasonal. The indigenous crops that grow naturally are usually harvested with indigenous crops that are planted during the beginning of the harvest season. While some indigenous crops are harvest as their demand and use such as Guxe "*Vigna subterranean*". Participant 28 alluded that her and the neighbours prefer consuming Guxe while it is still fresh and because the indigenous crop of this species grow occasionally after it has rained, they hardly over harvest it for preservation. For other crops such as Tinyawa "*Vigna unguiculata (L.) Walp*" and Tindluwa "*Vigna subterranean*", these indigenous crops are only assessable through planting. They are also seasonal crops that thrive in the summer while in winter more resilient crops are planted such as Tinhwembe *(C. Moschata)*.

The indigenous crops are planted using timetable that is associated with season, weather and climate. They usually plant their crops after rain to increase the chances of growth. Moreover, they determine what to plant in that season based on the behaviour of the sky. They are always able to tell when to plant and what top plant after observing the behaviour of the sky during the day and at night. Their production methods also use planting methods such as mixed cropping as well as crop rotation. These indigenous crops are utilized in Vatsonga households for consumption, maintain food security, income generation, nutrition and some they can also be used for health purposes. In the utilization for food security and nutrition they use can use the tubers of the crop as a staple and the leafy parts as a relish or sauce that can be paired with the tubers for a delicious and nutritious meal. The crops are also used for income generation where those who manage to have surplus yield after harvesting and preserving can use indigenous to generate extra income. While some plant indigenous crops for the purpose of making profits and maintain their households such as paying for their children's school fees and their households basic needs.

Strategies Used By Vatsonga From MTA to Maintain Indigenous Crop Practices

The Vatsonga community residing in MTA has successfully employed indigenous farming practices to secure food for their households and the wider community. Their strong connection to the environment, rooted in traditional knowledge passed down through generations, has allowed them to innovate and adapt to the challenges posed by climate change. Despite external influences on food trends, the community remains committed to methods that are both familiar and environmentally sustainable, such as crop rotation, mixed cropping, organic soil fertilization, and intercropping. These strategies have not only helped maintain agricultural yields but have also reduced the community's expenses. Farmers actively share knowledge about these practices, ensuring that newcomers learn to save seeds from their best harvests. Fertilization relies on organic materials, such as leftover Tinhwembe "amaranth" leaves, which enrich the soil both by being buried and applied on top. Crop rotation is utilized primarily during the growing season of leafy vegetables, ensuring that indigenous crops like amaranth thrive with adequate nutrients. Meanwhile, mixed cropping facilitates the simultaneous planting of grains, legumes, and other vegetables, providing resilience against adverse weather conditions.

To address water scarcity, the community employs rainwater harvesting techniques, storing water in buckets and wells. Several farmers have also adapted their planting methods by digging deeper holes for better water absorption during rainfall, allowing for longer retention in the soil. They specifically water their crops at night to maximize absorption, a strategy well-suited to Limpopo's hot and humid climate. These indigenous practices together enhance the community's agricultural productivity and sustainability. According to participant 12 in MTA has been teaching the young on how to cultivate and plant indigenous crops. These skills are taught from every young age. Through oral orientations and practices. According to the norms of the community, the elders start teaching the young the importance of indigenous crops through the food they consume. They introduce delicacies cooked from indigenous crops to the young, so that they may start appreciating them through consumption. As they grow, they young a taught how to plant indigenous crops that do grow naturally.

In MTA the primary indigenous food producers are female. As a result, most of the females from MTA are the main producers of these food. This has influenced the indigenous crop maintenance norms. Thus, apart from passing on the knowledge from generation to generation, they have taught their young females how to plant, how to prepare the soil using natural compost or animal manure, how to mix crops using crops that can thrive when planted together. How to harvest as well as the harvesting procedures. Moreover, the myths and taboos associated with females planting and being in the fields. In MTA it is believed that when a woman is on her cycle she can go near the garden as her body heat can contribute towards leafy vegetable such as tinhwembe drying up "Tinhwembe (C. Moschata)". Moreover, the same can be associated with women that are expecting.

Economic Benefits From Indigenous Crops in MTA

Indigenous crops are currently considered unique delicacies and are often not accessible in conventional supermarkets. This limited availability creates an accessible market for those who want to reconnect with their roots and cherish their nostalgic upbringing. In today's world, where genetically modified foods dominate the diet, there is growing awareness of the benefits of indigenous crops. In MTA, the primary consumers of these crops are the elderly and those who produce them. Farmers who manage to grow surplus crops often sell them to community members who either cannot grow their own or lack access to these products. This practice significantly contributes to household income, especially for families with larger plots of land. The income generated from selling indigenous crops has had a positive impact on the community. Many Indigenous people in MTA have used the profits to invest in their households, fund their children's higher education, build decent homes, and establish new businesses. Moreover, for some participants, they do not only sell crops that they have planted but have managed to leverage what nature has given them. In MTA there are some community members that do not have huge plots of land by are still able to run an indigenous crop sale business only using those indigenous crops that grow naturally. With some selling fresh and dry jutes mallow to those people in urban areas who do not have access to these crops.

Prospects Of Indigenous Crops in MTA

Participant 6 highlighted a positive outlook on the future use of indigenous crops by participants, emphasizing their significance not only for food security but also for their medicinal and nutritional benefits. While acknowledging that some indigenous crops may not be as abundant as conventional produce, the participants believe these crops can contribute to the community's well-being and economic growth. Through selling their indigenous crops for profit. As this will allow them to generate wealth for their households. Moreover, be able to branch into the conventional market using the well packed indigenous crops products. This also allows household to manage their house bills and children's school needs. While some expressed a disconnect with conventional supermarket goods, sharing personal experiences of improved health after consuming indigenous foods, especially in relation to managing chronic illnesses like high blood pressure and diabetes. Indigenous crops are more affordable and accessible, organic, and often more affordable compared to store-bought organic produce.

The participants advocate for the integration of these crops into the conventional market, suggesting this could provide better access for individuals unable to afford organic options, particularly those with chronic health issues. Some even reported feeling that the consumption of indigenous crops has positively impacted their longevity compared to peers who have not participated in this dietary shift. The participants also recognized the unique value of each community's indigenous crops, believing that promoting these crops in larger markets could foster self-sufficiency and growth through shared agricultural knowledge. Those with larger plots of land aspire to package their indigenous produce effectively and sell it to major supermarkets, aiming to reintroduce these crops and their benefits to a wider audience. This further promotes financial growth within households. Moreover, the prospect of higher education institutions is important for them as documenting indigenous crops will help in preserving the knowledge as well ensuring that the knowledge safely kept in literature and continues in practices within various MTA communities.

Documenting these crops serves two essential purposes, the preservation and evidence of their economic potential for local communities like MTA. This documentation will ensure that future generations are aware of these crops, their benefits, and their potential for income generation and food security. Furthermore, it helps to maintain the cultural significance of these crops within the Vatsonga community of MTA, ensuring that the knowledge surrounding them is not lost. By preserving this knowledge, they also safeguard the community's heritage and protect against the commercial exploitation of their agricultural practices

CHALLENGES OF INDIGENOUS CROPS IN MTA

The challenges that MTA is facing is climate change. For the past few years MTA indigenous crop harvest has been inconsistent. With some planting seasons having more yield than other seasons. While some having a bad harvest due to the heavy rains and in most cases extreme heatwaves. Resulting some crops that are planted dying. Moreover, these weather conditions are also affecting those that grow naturally as they now grow in limited access. With some that for part of the community's delicacy being rare to find in areas where they used to be found in abundance. This not only affect those who use these crops for personal consumption but also those who rely on these crops for income generation. In addition, this also becomes a challenge in indigenous crop knowledge preservation. As these effects are making difficult to practice indigenous farming methods as well as passing them from one generation to the next. Lastly, it affects those households that supplement what they cannot afford for consumption with indigenous crops. As this has resulted in some of the indigenous households not having enough food as before that is nutritious and promotes their food security.

SENSE OF VUMUNHU (UBUNTU) IN MTA

The cultivation of indigenous crops within the MTA framework brings numerous economic benefits, primarily driven by a deep-rooted passion for preserving traditional farming practices. Participants in the study highlighted that for them, planting these crops is not just a source of income, but also a means of honouring their identity and promoting health. The accessibility of these crops either through planting or seasonal harvesting, enables farmers to assist those who are less privileged and food insecure. This communal spirit, rooted in the philosophy of Ubuntu, fosters a sense of togetherness, allowing households to share their yields and ensure that their neighbours do not go to bed hungry. The practice of sharing can take various forms, from providing crops at a reduced price to outright gifts, which further strengthens community bonds. Additionally, this focus on indigenous crops facilitates the preservation of traditional knowledge, norms, and practices, while also improving the livelihoods of farmers. Those cultivating these crops benefit from maximizing their profits and ensuring effective crop maintenance throughout each planting season. Moreover, the cultivation of indigenous crops encourages hard work since many of these plants require significant manpower for tasks like digging, weeding, and properly organizing planting areas. This not only promotes physical labour but also instils values of kindness and community support, enhancing the overall fabric of the community through shared efforts and resources.

DISCUSSION

There is an African proverb that states, "One who eats alone cannot discuss the taste of the food with others". For most indigenous people in Africa, this proverb embodies a way of life that promotes community food security and humility. Ensuring that no neighbour goes to bed hungry when there's plenty of food to share. This way of life can be traced back to the early history of people in Africa. Where indigenous people have been using indigenous farming practices as the basis of their community's survival. These skills have been passed on for generations through oral traditions and practice. While promoting food security and development with indigenous communities. With indigenous people in the Southern hemisphere of Africa having continued the use of indigenous crops even in the present time. In the same way the use of indigenous crops by the current generation of descendants can be traced far back to the history of the Bantu people who were in central Africa but migrated to the South in 1000 B.C.E (Ehret,1972). This practice not only ensures that Vatsonga communities under Mhinga traditional authority (MTA) have sufficient food but are also able to generate household income by leveraging their indigenous crops and environment. Moreover, maintaining their economic stability, food security, food availability and nutrition. Additionally, with Vatsonga from MTA having maintained their unique knowledge of indigenous crops and benefits. This has resulted in a supply of balanced meals within most MTA households. Resulting in communities under MTA consistently being able to maintain, utilize indigenous crops and their unique farming practises for food security as well as their surplus crops for income generation.

Simultaneously, for centuries the use of indigenous crops in MTA has been fundamental for their survival, uniqueness, norms, practices, traditions and knowledge preservation. As the world progressed, food preferences and the globe's palate changed and started to gravitate towards commercial food and conventional farming produce such as those that have GMO's that assists in stimulating the growth of these crops as a way of trying to meet the world food demands and supply. Influencing their accessibility in supermarkets, use of minimal manpower by producers, reduced cooking time by consumers, affordable prices, mass production creating a more reliable supply that is not affected by nature as well as climate change and Eurocentric food trends. While some residents of MTA continued with consuming, harvesting, planting, selling and preserving the knowledge of Vatsonga indigenous crops within their various communities. Promoting sustainable food security and economic development from household income generation in the area from the relationship they have with their natural surroundings, indigenous knowledge, practices and culture.

The use of indigenous crops in MTA ensures that is knowledge dissemination. As the young learn from the elders and the cycle of knowledge sharing continues as they the young grow up and have families of their own. However, this does not guarantee the effective system of knowledge sharing as this knowledge sharing is based on knowledge that is stored on people's minds. Thus, the saying "when an elder die is the same as a library burning down" similarly MTA's knowledge should also be documented in literature to ensure that those with the knowledge do not die with. With indigenous crops being embedded in the culture of the people, the loss of this knowledge will also result in an indigenous aspect of Vatsonga from MTA dying that's contributing to the slow extinction of indigenous knowledge of the people. With MTA having various communities under it. The use of indigenous crops has contributed to the community's development. As this has influenced a lot of the families financially. Thus, the documentation of the knowledge is important for the continuous use of indigenous crops for financial growth. In MTA, the growth of these households has influenced the development of communities because households being in a better financial position as this helps in influencing the growth and development of communities.

FUTURE RESEARCH DIRECTIONS

To document more of Vatsonga indigenous agricultural knowledge from the perspectives of the custodians of the knowledge. Promote the use of indigenous crops in the conventional market by small-scale farmers. Integrate western marketing strategies in creating awareness indigenous crops and affordability. Conduct studies on different sectors on investing in small community businesses with indigenous people who still see a future for indigenous farming practices and indigenous as this will help in creating awareness on issues of community and business collaborations while ensuring that there is growth and community benefits.

CONCLUSION

The integration of indigenous crops within the MTA is crucial for preserving community traditions and practices. This reliance on such crops demonstrates the resilience and value of indigenous agricultural knowledge, which plays a significant role in transforming lives and supporting households through improved nutrition and economic benefits. Despite facing challenges like climate change, the continued use of indigenous crops remains essential for those in the community who cultivate them. This study emphasized the importance of indigenous agricultural practices in promoting sustainability, knowledge preservation and an understanding of indigenous crops and the positive impacts that are embedded in the use of indigenous crops.

This book chapter highlighted and contributed to documenting indigenous crops from a Vatsonga perspective. Through correcting some literature discrepancies and adding knowledge of indigenous crops that Vatsonga from MTA use that is yet to be documented using community-based knowledge from MTA knowledge holders. It serves as a literature resource from a Vatsonga perspective in indigenous crop agriculture and the indigenous knowledge of crops and in community maintenance. While putting into perspective the importance of indigenous crops and their economic potentials from a view of MTA's day to day life. The cultivation of indigenous crops within the MTA framework brings numerous economic benefits, primarily driven by a deep-rooted passion for preserving traditional farming practices. Participants in the study highlighted that for them, planting these crops is not just a source of income, but also a means of honouring their identity and promoting health. The accessibility of these crops either through planting or seasonal harvesting, enables farmers to assist those who are less privileged and food insecure. This communal spirit, rooted in the philosophy of Ubuntu, fosters a sense of togetherness, allowing households to share their yields and ensure that their neighbours do not go to bed hungry. The practice of sharing can take various forms, from providing crops at a reduced price to outright gifts, which further strengthens community bonds. Additionally, this focus on indigenous crops facilitates the preservation of traditional knowledge, norms, and practices, while also improving the livelihoods of farmers. Those cultivating these crops benefit from maximizing their profits and ensuring effective crop maintenance throughout each planting season. Moreover, the cultivation of indigenous crops encourages hard work since many of these plants require significant manpower for tasks like digging, weeding, and properly organizing planting areas. This not only promotes physical labour but also instils values of kindness and community support, enhancing the overall fabric of the community through shared efforts and resources.

REFERENCES

Agrawal, A. (1995). Dismantling the divide between indigenous and scientific knowledge. *Development and Change*, *26*(3), 413–439. DOI: 10.1111/j.1467-7660.1995.tb00560.x

Akinola, R., Pereira, L. M., Mabhaudhi, T., De Bruin, F. M., & Rusch, L. (2020). A review of indigenous food crops in Africa and the implications for more sustainable and healthy food systems. *Sustainability*, *12*(8), 3493. DOI: 10.3390/su12083493 PMID: 33520291

Aliber, M., & Hall, R. (2012). Support for smallholder farmers in South Africa: Challenges of scale and strategy. *Development Southern Africa*, *29*(4), 548–562. DOI: 10.1080/0376835X.2012.715441

Ardhede, H. (2007). *Traditional Knowledge and the Patent System-Irreconcilable differences or a simple case of mistaken identity?* Asiwe, J.A.N. (2009). Needs assessment of cowpea production practices, constraints and utilization in South Africa. *African Journal of Biotechnology*, *8*(20).

Ayanwale, A. B., Amusan, C. A., Adeyemo, V. A., & Oyedele, D. J. (2016). Analysis of household demand for underutilized indigenous vegetables. *International Journal of Vegetable Science*, *22*(6), 570–577. DOI: 10.1080/19315260.2015.1103350

Balasubramanian, A. (2014). Classification of agricultural crops. Africa. *Agrekon*, *48*(4), 459–482.

Banerjee, A. V., & Duflo, E. (2005). Growth theory through the lens of development economics. *Handbook of economic growth*, 1,473 -552.

Basu, S., Roberts, J. A., Azam-Ali, S. N., & Mayes, S. (2007). Bambara groundnut. In *Pulses, Sugar and Tuber Crops* (pp. 159–173). Springer Berlin Heidelberg. DOI: 10.1007/978-3-540-34516-9_10

Bostoen, K. (2018). *The bantu expansion*. Oxford University Press.

Braun, V., & Clarke, V. (2012). *Thematic analysis*. American Psychological Association. DOI: 10.1037/13620-004

Brush, S. B. (2005). Protecting traditional agricultural knowledge. *Wash. UJL & Pol'y*, *17*, 59.

Bvenura, C., & Afolayan, A. J. (2015). The role of wild vegetables in household food security in South Africa: A review. Food Research International, 76, pp.1001-1011.change adaptation decisions to enhance food security in Sub-Saharan Africa? *Agricultural Systems*, *127*, 161–177.

Chibarabada, T. P., Modi, A. T., & Mabhaudhi, T. (2019). Water use of selected grain legumes in response to varying irrigation regimes. *Water S.A.*, *45*(1), 110–120.

Chikaire, J. Osuagwu, C.O. Ihenacho, R.A. Oguegbuchulam, M.N. Ejiogu-Okereke, N. & Obi, K.U.(2012). Indigenous knowledge system: The need for reform and the way forward. *Global advanced research journal of agricultural science*, 1(8),201-209.

Cleveland, D. A., & Murray, S. C. (1997). The world's crop genetic resources and the rights of indigenous farmers. *Current Anthropology*, *38*(4), 477–516. DOI: 10.1086/204641

Cloete, P. C., & Idsardi, E. F. (2013). Consumption of indigenous and traditional food crops: Perceptions and realities from South Africa. *Agroecology and Sustainable Food Systems, 37*(8), 902–914. DOI: 10.1080/21683565.2013.805179

Constitution of the Republic of South Africa. (1996). Pretoria. Government printer.2013 reprint.

Creswell, J. W., Hanson, W. E., Clark Plano, V. L., & Morales, A. (2007). Qualitative research designs: Selection and implementation. *The Counseling Psychologist, 35*(2), 236–264. DOI: 10.1177/0011000006287390

Creswell, J. W., & Poth, C. N. (2016). *Qualitative inquiry and research design: Choosing among five approaches.* Sage publications.

Deci, E. L., Olafsen, A. H., & Ryan, R. M. (2017). Self-determination theory in work organizations: The state of a science. *Annual Review of Organizational Psychology and Organizational Behavior, 4*(1), 19–43. DOI: 10.1146/annurev-orgpsych-032516-113108

Denison, J. A., & Wotshela, L. (2012). An overview of indigenous, indigenised and contemporary water harvesting and conservation practices in South Africa. *Irrigation and Drainage, 61*(S2), 7–23. DOI: 10.1002/ird.1689

Dury, J., Schaller, N., Garcia, F., Reynaud, A., & Bergez, J. E. (2012). Models to support cropping plan and crop rotation decisions. A review. *Agronomy for Sustainable Development, 32*(2), 567–580. DOI: 10.1007/s13593-011-0037-x

Ehret, C. (1972). Bantu origins and history: Critique and interpretation. *Transafrican Journal of History, 2*(1), 1–9.

Eliazer Nelson, A.R.L. Ravichandran, K. & Antony, U. (2019). The impact of the Green.

Elmore, R. W. (1996). Our relationship with the ecosystem and its impact on agriculture. *Journal of Production Agriculture, 9*(1), 42–45. DOI: 10.2134/jpa1996.0042

Emmanuel, O. C., & Babalola, O. O. (2022). Amaranth production and consumption in South Africa: The challenges of sustainability for food and nutrition security. *International Journal of Agricultural Sustainability, 20*(4), 449–460. DOI: 10.1080/14735903.2021.1940729

Etikan, I., Musa, S. A., & Alkassim, R. S. (2016). Comparison of convenience sampling and purposive sampling. *American Journal of Theoretical and Applied Statistics, 5*(1), 1–4. DOI: 10.11648/j.ajtas.20160501.11

Ferranti, P., Berry, E., & Jock, A. (2018). *Encyclopedia of food security and sustainability.* Elsevier.

Food and Agricultural Organization. (2002). *The state of Food insecurity in the World.* FOA.

Haumba, E. N., & Kaddu, S. (2017). Documenting and disseminating agricultural indigenous knowledge for sustainable food security in Uganda. *University of Dar es Salaam.Library Journal, 12*(1), 66–86.

Hiddink, G. A., Termorshuizen, A. J., & van Bruggen, A. H. (2010). Mixed cropping and suppression ofs oilborne diseases. *Genetic Engineering, biofertilisation, soil quality and organic farming,* pp.119-146.

Holden, C. J. (2017). The spread of Bantu languages, farming, and pastoralism in sub-equatorial Africa. In Mapping our ancestors (pp. 247-268). Routledge. indigenous ways of knowing. *The Journal of Pan African Studies*, *2*(4), 184–202.

Isern, N., & Fort, J. (2019). Assessing the importance of cultural diffusion in the Bantu spread into South Eastern Africa. *PLoS One*, *14*(5), e0215573. DOI: 10.1371/journal.pone.0215573 PMID: 31067220

Johnson, B. & Lundvall, B.A. (1994). The learning economy. *Journal of industry studies*, 1(2), 23- 42.

Kalland, A. (2000). Indigenous knowledge: prospects and limitations. *Indigenous environmental knowledge and its transformations: critical anthropological perspectives*, 316.

Kock, K. T. (2020). The influence of culture on the utilization of traditional leafy vegetables among VaTsonga and Ba-Pedi women of Bushbuckridge, Mpumalanga Province (Doctoral dissertation, North-West University (South Africa)).

Kristin, B., Kelly, S., Travis, H., Joseph, L., & Lori, C. (2017). Retail food environments, shopping experiences, First Nations and the provincial Norths. *Health Promotion and Chronic Disease Prevention in Canada : Research, Policy and Practice*, *37*(10), 333–341. DOI: 10.24095/hpcdp.37.10.03 PMID: 29043760

Kuhnlein, H., Eme, P., & Larrinoa, Y. D. (2019). Indigenous food systems: contributions to sustainable food systems and sustainable diets. In Sustainable diets: linking nutrition and food systems (pp. 64-78). Wallingford UK: CABI. (ASAP). DOI: 10.1079/9781786392848.0064

Lara, S. W., Tsiami, A., & Cross, P. (2023). Discovering and mapping colloquial terminologies describing underutilized and neglected food crops—A comprehensive review. *Foods*, *12*(12), 2428. DOI: 10.3390/foods12122428 PMID: 37372639

Lewis, J. L., & Sheppard, S. R. (2006). Culture and communication: Can landscape visualization improve forest management consultation with indigenous communities? *Landscape and Urban Planning*, *77*(3), 291–313. DOI: 10.1016/j.landurbplan.2005.04.004

Leyew, Z. (2024). Indigenous knowledge practices in danger of extinction due to cultural and environmental changes: The case of Gojjam, Amhara region. *Ethiopian Journal of Language. Culture and Communication*, *9*(2), 30–75.

Liu, J. (2019). The impact of Bantu migration on agriculture in sub-Saharan Africa. *International Journal of New Developments in Engineering and Society*, *3*(2), 28–34.

Luthria, D. L., & Pastor-Corrales, M. A. (2006). Phenolic acids content of fifteen dry edible bean (Phaseolus vulgaris L.) varieties. *Journal of Food Composition and Analysis*, *19*(2-3), 205–211. DOI: 10.1016/j.jfca.2005.09.003

Makuse, S. M., & Mbhenyane, X. G. (2011). Health benefits and omega-3-fatty acid content of selected indigenous foods in the Limpopo Province, South Africa. *Indilinga*, *10*(2), 182–194.

Mapiye, O., Chikwanha, O. C., Makombe, G., Dzama, K., & Mapiye, C. (2020). Livelihood, food, an nutrition security in Southern Africa: What role do indigenous cattle genetic resources play? *Diversity*, *12*(2), 74. DOI: 10.3390/d12020074

Marnewick, J. L., Venter, I., Rautenbach, F., & Piek, H. (2019). Rooibos herbal tea: An optimal cup an its consumers. *Health SA Gesondheid*, *24*(1), 1–9. PMID: 31934411

Maroyi, A. (2012). Use of traditional veterinary medicine in Nhema communal area of the Midlands province, Zimbabwe. *African Journal of Traditional, Complementary, and Alternative Medicines : AJTCAM*, *9*(3), 315–322. DOI: 10.4314/ajtcam.v9i3.3 PMID: 23983361

Massawe, F., Mayes, S., & Cheng, A. (2016). Crop diversity: An unexploited treasure trove for food security. *Trends in Plant Science*, *21*(5), 365–368. DOI: 10.1016/j.tplants.2016.02.006 PMID: 27131298

Mavengahama, S. (2013). The contribution of indigenous vegetables to food security and nutrition within selected sites in South Africa.

Mavhungu, T. J., Nesamvuni, A. E., Tshikolomo, K. A., Mpandeli, N. S., & van Niekerk, J. (2022). Productivity and profitability of sweet potato (ipomoea batatas l.), dry bean (Phaseolus Vulgaris) and maize (Zea mays l.) as selected field crops in irrigated smallholder agricultural enterprises (ISAEs) in Vhembe District, Limpopo Province, South Africa. *Technium Soc. Sci. J.*, *29*, 683–699. DOI: 10.47577/tssj.v29i1.5932

Mbhenyane, X. G. (2017). Indigenous foods and their contribution to nutrient requirements. *The South African Journal of Clinical Nutrition : SAJCN : the Official Journal of the South African Society of Parenteral and Enteral Nutrition and the Association for Dietetics in South Africa*, *30*(4), 5–7.

McDonough, S. (2013). Decolonizing Methodologies: Research and Indigenous Peoples by Linda Tuhiwai Smith. *Collaborative Anthropologies*, *6*(1), 458–464. DOI: 10.1353/cla.2013.0001

Mgwenya, L. I., Agholor, I. A., Ludidi, N., Morepje, M. T., Sithole, M. Z., Msweli, N. S., & Thabane, V. N. (2025). Unpacking the Multifaceted Benefits of Indigenous Crops for Food Security: A Review of Nutritional, Economic and Environmental Impacts in Southern Africa. *WORLD (Oakland, Calif.)*, *6*(1), 16. DOI: 10.3390/world6010016

Muchineripi, P.C. (2008). Feeding Five Thousand: The Case for Indigenous Crops, in Zimbabwe.

Murray, J. S. (1970). Social Factors in the Development of Agriculture in the Bantu Areas with Special Reference to Land Tenure Systems. *Agrekon*, *9*(1), 29–34. DOI: 10.1080/03031853.1970.9524422

National Research Council. Global Affairs, Board on Science, Technology for International Development and Ad Hoc Panel of the Advisory Committee on Technology Innovation, (1989). *Lost crops of the Incas: little-known plants of the Andes with promise for worldwide cultivation.* National Academies Press.

Ncube, B. (2018). Insights into indigenous knowledge strategies for coping and adapting to drought in agriculture: A Karoo scenario, South Africa. *Indilinga*, *17*(1), 92–108.

Nelson, R. (1997). How new is new growth theory? *Challenge*, *40*(5), 29–58. DOI: 10.1080/05775132.1997.11471992

Nesamvuni, C., Steyn, N. P., & Potgieter, M. J. (2001). Nutritional value of wild, leafy plants consumed by the Vhavenda. *South African Journal of Science*, *97*(1), 51–54.

Njeme, C., Goduka, N. I., & George, G. (2014). Indigenous leafy vegetables (imifino, morogo, muhuro) in South Africa: A rich and unexplored source of nutrients and antioxidants. *African Journal of Biotechnology, 13*(19), 1933–1942. DOI: 10.5897/AJB2013.13320

Nyota, S., & Mapara, J. (2008). *Shona Traditional children's games and play songs as Nyota, S., Mapara, J. and Centre for Advanced Studies of African Society, (2007). Language as indigenous knowledge.* Centre for Advanced Studies of African Society.

Ohaja, M., & Anyim, C. (2021). Rituals and embodied cultural practices at the beginning of life: African perspectives. *Religions, 12*(11), 1024. DOI: 10.3390/rel12111024

Olum, S., Okello-Uma, I., Tumuhimbise, G. A., Taylor, D., & Ongeng, D. (2017). The relationship between cultural norms and food security in the Karamoja Sub-Region of Uganda. *Journal of Food and Nutrition Research, 5*(6), 427–435. DOI: 10.12691/jfnr-5-6-10

Orr, A., Mwema, C., Gierend, A. & Nedumaran, S.(2016). Sorghum and millets in Eastern and Southern Africa: facts, trends and outlook

Philipson, D. W. (1977). *The later prehistory of the Eastern and Southern Africa.* Heinemann Educational Books.

Pinhasi, R., & Stock, J. T. (Eds.). (2011). *Human bioarchaeology of the transition to agriculture.* John Wiley & Sons. DOI: 10.1002/9780470670170

Pinstrup-Andersen, P. (2009). Food security: Definition and measurement. *Food Security, 1*(1), 5–7. DOI: 10.1007/s12571-008-0002-y

Poulton, C., Dorward, A., & Kydd, J. (1998). The revival of smallholder cash crops in Africa: Public and private roles in the provision of finance. Journal of International Development. *The Journal of the Development Studies Association, 10*(1), 85–103.

Puoane, T. Matwa, P. Hughes, G. & Bradley, H.A. (2006). Socio-cultural factors influencing food consumption patterns in the black African population in an urban township in South Africa.

Raimondo, D. (2009). *Red list of South African plants.* South African National Biodiversity Institute.

Rampedi, I. T. (2010). Indigenous plants in the Limpopo Province: *Potential for their commercial beverage production*. Doctoral dissertation, University of South Africa.

Reeve, J. (2002). *Self-determination theory applied to educational settings.* Revolution on indigenous crops of India. *Journal of Ethnic Foods, 6*(1), 1–10.

Richard, T.T. (2014). Indigenous farming methods in the Mambilla Plateau of Nigeria: the need for a sustainable maintenance culture policy. *Science Journal of Sociology and Anthropology*, 2014.

Rima, I. H. (2004). Increasing returns, new growth theory, and the classicals. *Journal of Post Keynesian Economics, 27*(1), 171–184.

Rindos, D. (2013). *The origins of agriculture: an evolutionary perspective.* Academic Press.

Rivera, D., Allkin, R., Obón, C., Alcaraz, F., Verpoorte, R., & Heinrich, M. (2014). What is in a name? The need for accurate scientific nomenclature for plants. *Journal of Ethnopharmacology*, *152*(3), 393–402. DOI: 10.1016/j.jep.2013.12.022 PMID: 24374235

Ruelle, M. (2015). *Human-plant ecology of an afromontane agricultural landscape: diversity, knowledge, and food sovereignty in Debark, Ethiopia.*

Schoenbrun, D. L. (1993). We Are What We Eat: Ancient Agriculture Between the Great Lakes1. *Journal of African History*, *34*(1), 1–31. DOI: 10.1017/S0021853700032989

Seko, J., Bain, E., & Maponya, P. (2021). Assessing the Impact of Indigenous Knowledge Systems on Sustainable Agriculture: A Case Study of the Selected Communities in the City of Tshwane Metropolitan, Gauteng Province, South Africa. Sustainable Bioeconomy: *Pathways to Sustainable Development Goals*, pp.183-208.

Smith, A. B. (2019). Ancient and traditional agriculture, pastoralism, and agricultural societies in sub-Saharan Africa. In Oxford Research Encyclopedia of Environmental Science.

Socha, T., Chambers, L., Zahaf, M., Abraham, R., & Fiddler, T. (2011). Food availability, food store management, and food pricing in a northern community First Nation community. *International Journal of Humanities and Social Science*, *1*(11), 49–61.

Solow, R. M. (1994). Perspectives on growth theory. *The Journal of Economic Perspectives*, *8*(1), 45–54. DOI: 10.1257/jep.8.1.45

Stilwell, C. (2010). Understanding indigenous knowledge: Bridging the knowledge gap through a knowledge creation model for agricultural development. *South African Journal of Information Management*, *12*(1), 1–8.

Taherdoost, H. (2016). Sampling methods in research methodology; How to choose a sampling technique for research. *SSRN*, *5*(2), 18–27. DOI: 10.2139/ssrn.3205035

Tirivangasi, H. M., & Tayengwa, D. (2017). Indigenous knowledge systems and food security in South Africa: Is land reform a prerequisite? *Journal of Human Ecology (Delhi, India)*, *57*(3), 118–124.

Trucco, F., & Tranel, P. J. (2011). Amaranthus. In Wild Crop Relatives: Genomic and Breeding Resources: Vegetables (pp. 11-21). Berlin, Heidelberg: Springer Berlin Heidelberg.Palada, M.C. and Chang, L.C. (2003). Suggested cultural practices for jute mallow. *International Cooperator Guide,* 2(14), 1-4

Tshidzumba, S. E. (2020). *Nutritional composition of traditional mixed-dishes eaten by elderly women in Vhembe District* (Doctoral dissertation).

Turner, N. J., & Turner, K. L. (2007). *Traditional food systems, erosion and renewal in Northwestern.* North America.

Uusiku, N. P., Oelofse, A., Duodu, K. G., Bester, M. J., & Faber, M. (2010). Nutritional value of leafy vegetables of sub-Saharan Africa and their potential contribution to human health: A review. *Journal of Food Composition and Analysis*, *23*(6), 499–509. DOI: 10.1016/j.jfca.2010.05.002

Vallerand, R.J. Pelletier, L.G. & Koestner, R. (2008). Reflections on self-determination theory.

Van der Merwe, J. D., Cloete, P. C., & Van der Hoeven, M. (2016). Promoting food security through indigenous and traditional food crops. *Agroecology and Sustainable Food Systems*, *40*(8), 830–847. DOI: 10.1080/21683565.2016.1159642

Vasconcellos, D., Parker, P. D., Hilland, T., Cinelli, R., Owen, K. B., Kapsal, N., Lee, J., Antczak, D., Ntoumanis, N., Ryan, R. M., & Lonsdale, C. (2020). Self-determination theory applied to physical education: A systematic review and meta-analysis. *Journal of Educational Psychology*, *112*(7), 1444–1469. DOI: 10.1037/edu0000420

Venter, P., Malemela, K., Mbazima, V., Mampuru, L. J., Muller, C. J., & Riedel, S. (2021). An RP-LC- UV-TWIMS-HRMS and Chemometric Approach to Differentiate between Momordica balsamina Chemotypes from Three Different Geographical Locations in Limpopo Province of South Africa. *Molecules (Basel, Switzerland)*, *26*(7), 1896. DOI: 10.3390/molecules26071896 PMID: 33801575

Vorster, H. J., Stevens, J. B., & Steyn, G. J. (2008). Production systems of traditional leafy vegetables: Challenges for research and extension. *South African Journal of Agricultural Extension*, *37*, 85–96.

Webber, H. Gaiser, T. & Ewert, F. (2014). What role can crop models play in supporting climate

Zadik, M. (1986). Teaching horticulture with a human perspective. *NACTA Journal*, *30*(1), 59.

Zondi, N. T. B., Ngidi, M. S. C., Ojo, T. O., & Hlatshwayo, S. I. (2022). Impact of market participation of indigenous crops on household food security of smallholder farmers of South Africa. *Sustainability*, *14*(22), 15194. DOI: 10.3390/su142215194 PMID: 39659967

ADDITIONAL READING

Kesselman, B., & Zukulu, S. (2025). Traditional foodways of the Amadiba: A struggle for indigenous food sovereignty in Mpondoland, South Africa. *Journal of Political Ecology*, *32*(1). Advance online publication. DOI: 10.2458/jpe.5933

Khumalo, T. A., Chakale, M. V., Asong, J. A., Aremu, A. O., & Amoo, S. O. (2025). Indigenous farming methods and crop management practices used by local farmers in Madibeng local municipality, South Africa. *Scientific Reports*, *15*(1), 8918. DOI: 10.1038/s41598-025-91210-w PMID: 40087342

Matloga, S. T. (2025). The Impact of Climate Change on Rural Communities of Makhado Local Municipality, Limpopo Province. *African Journal of Public Administration and Environmental Studies*, *4*(1), 223–243.

Ndhlovu, E. (2025). Agricultural Development, Innovation, and Mechanisation in Pre-Colonial Africa. In *Agriculture, Autonomous Development, and Prospects for Industrialization in Africa* (pp. 21–36). Springer Nature Switzerland. DOI: 10.1007/978-3-031-84992-3_2

Sahani, A. K., Gupta, G., Anand, S., Sharma, V. R., Singh, R., Kamil, A., & Anand, J. (2025). Indigenous Knowledge and Water Conservation Practices in South Africa: A Systematic Literature Review. *Journal of Environment and Earth Science*, *7*(02), •••.

KEY TERMS AND DEFINITIONS

Indigenous Knowledge: It is a complex set of knowledge that is acquired over generations by community members through interactions between human beings and the environment. It is embedded in technology, philosophy, culture, economics, governance systems and community learning.

Indigenous Community: It is a distinct cultural group with their own unique ways of life. Indigenous community often use indigenous language, and it has self-identification and identification by others as distinct cultural group members.

Food Security: It is a state that exists when people have access to food that is nutritionally balanced. This situation ensures that all people always have food that has sufficient nutritious and is safe, while meeting their dietary requirements and food preferences that are healthy.

Indigenous Farming Practices: Are the result of millennia of cultural and biological evolution and are a culmination of the experiences of indigenous farmers. Without using outside inputs, capital, or current scientific understanding, the farmers cultivate their crops by using their knowledge of seasonal variation and environmental circumstances.

Indigenous Crops: Are edible vegetations that originated from a specific region and are recognised as cultural or traditional food by the people of that specific region. These crops are high in nutrients, they are drought resistant and linked to the knowledge and lifestyle of the local people.

Economy: It is a system of making decisions pertaining to resources within a region as well as goods and services that a local place and country possesses that can be produced and utilized by the country's inhabitants.

Chapter 13
The Role of the Fourth Industrial Revolution (4IR) Technologies in Digital Preservation and Documentation of Indigenous Knowledge Systems

Xholiso Ennocent Khazamula
https://orcid.org/0000-0002-1796-2156
University of Pretoria, South Africa

ABSTRACT

The Fourth Industrial Revolution (4IR) heralds a new era of technological advancements that fundamentally alter how societies function, particularly in the realm of knowledge preservation and documentation. This study explores the application of 4IR technologies in the digital preservation and documentation of indigenous knowledge. Indigenous knowledge systems, rich in cultural heritage and traditional practices, face the risk of erosion due to globalization and technological dominance. Leveraging 4IR technologies such as artificial intelligence (AI), blockchain, Internet of Things (IoT), and big data analytics presents a promising solution to this challenge. This study aims to examine how these technologies can be effectively utilized to safeguard and perpetuate indigenous knowledge for future generations. This chapter concludes by emphasizing the transformative potential of 4IR technologies in preserving and documenting IK.

INTRODUCTION AND BACKGROUND

Indigenous knowledge encapsulates the wisdom, traditions, and cultural practices of Indigenous communities, often transmitted orally across generations. This knowledge is integral to the identity, heritage, and sustainable development of these communities (Battiste & Henderson, 2000). It represents the collective wisdom of communities developed over centuries through interaction with their natural

DOI: 10.4018/979-8-3693-7964-6.ch013

and cultural environments. Battiste and Henderson (2000) further noted that indigenous knowledge encompasses knowledge about agriculture, medicine, governance, and spiritual practices, often transmitted orally or through cultural rituals. However, the rapid pace of modernization, urban migration, digital homogenization, globalization and urbanization has placed IK at risk of erosion and extinction. The challenge lies in preserving these cultural reservoirs while ensuring they are still accessible to communities and researchers alike. Traditional methods of knowledge transmission are increasingly vulnerable to disruption, leading to a potential loss of invaluable cultural heritage (Nakata, 2002). The rapid pace of technological advancement and globalization also poses significant threats to the preservation of Indigenous knowledge. The Fourth Industrial Revolution (4IR) has ushered in transformative technological advancements, offering unprecedented opportunities for the preservation and documentation of Indigenous Knowledge (IK). The advent of the Fourth Industrial Revolution introduced a suite of advanced technologies that could transform the landscape of knowledge preservation. Technologies such as AI, blockchain, IoT, and big data analytics offer innovative solutions to digitize, store, and manage Indigenous knowledge systematically. These technologies not only enhance the accessibility and dissemination of Indigenous knowledge but also ensure its authenticity and integrity (Schwab, 2016).

The Fourth Industrial Revolution (4IR) characterized by the fusion of digital, biological, and physical innovations thus offers tools to address this challenge (Schwab, 2017). Emerging technologies such as AI, Blockchain, and IoT hold the potential to document, store, and disseminate IK in ways that were previously unimaginable. These tools not only safeguard knowledge but also empower communities to maintain ownership and control over their heritage. Despite the potential benefits, the application of 4IR technologies in preserving Indigenous knowledge remains underexplored. Challenges such as the digital divide, data privacy, and cultural sensitivity need careful consideration to ensure that these technologies are effective and respectful of Indigenous communities' values and rights (Grobler, 2018). This chapter explores how 4IR technologies such as Artificial Intelligence (AI), Blockchain, Big Data Analytics, Virtual Reality (VR), and the Internet of Things (IoT) can be leveraged to digitally safeguard Indigenous Knowledge.

PROBLEM STATEMENT

The preservation and documentation of Indigenous knowledge face significant challenges in the modern era due to the rapid advancement of technology and globalization. Indigenous knowledge, which encompasses traditional practices, cultural heritage, and ecological wisdom, is at risk of being eroded as younger generations become increasingly disconnected from their ancestral roots (Duarte & Belarde-Lewis, 2021). Traditional methods of knowledge transmission, primarily oral and community-based, are insufficient to withstand the pressures of a globalized world where digital communication and information storage dominate. Consequently, there is an urgent need to explore innovative solutions to ensure that this invaluable knowledge is preserved and accessible for future generations. The emergence of Fourth Industrial Revolution (4IR) technologies offers a potential solution to this pressing issue. Technologies such as artificial intelligence, blockchain, the Internet of Things (IoT), and big data analytics can revolutionize how Indigenous knowledge is digitized, authenticated, and disseminated (Tsai, 2020). However, the application of these technologies in preserving Indigenous knowledge is fraught with challenges. Issues such as data privacy, cultural sensitivity, and the digital divide between urban and rural or Indigenous communities must be carefully navigated to prevent further marginalization of Indigenous peoples

(Kwet, 2019). Moreover, there is a lack of comprehensive studies and practical frameworks on integrating 4IR technologies with Indigenous knowledge systems, highlighting a significant gap in current research.

This problem affects not only the Indigenous communities who are the custodians of this knowledge but also the broader global society. The loss of Indigenous knowledge would result in a significant cultural and intellectual deficit, depriving the world of diverse perspectives on sustainable living, ecological stewardship, and cultural heritage (Gibson, 2019). For Indigenous communities, the erosion of their traditional knowledge threatens their cultural identity, autonomy, and ability to maintain their unique ways of life. Addressing this issue is crucial for fostering inclusive technological development that respects and preserves cultural diversity, thus benefiting both indigenous and non-indigenous populations alike.

The chapter's objectives are as follows:

- To evaluate the potential of artificial intelligence in the digitization and analysis of Indigenous knowledge.
- To investigate the role of blockchain technology in ensuring the authenticity and integrity of digitized Indigenous knowledge.
- To assess the application of the Internet of Things (IoT) in enhancing the accessibility and real-time documentation of Indigenous knowledge.
- To explore the use of big data analytics in managing and disseminating indigenous knowledge effectively.

The research questions are as follows.

- How can artificial intelligence be utilized to digitize and analyze Indigenous knowledge effectively?
- In what ways can blockchain technology contribute to maintaining the authenticity and integrity of digitized Indigenous knowledge?
- What are the potential applications of the Internet of Things (IoT) in improving the accessibility and real-time documentation of Indigenous knowledge?
- How can big data analytics be employed to manage and disseminate indigenous knowledge efficiently?

THEORETICAL FRAMEWORK

The theory of Socio-Technical Systems (STS) was developed in the mid-20th century to address the interplay between social and technical components in organizational settings. Trist and Bamforth (1951), early proponents of STS, introduced the concept to emphasize the interdependence between humans (the social system) and technology (the technical system) in achieving optimal outcomes. The theory posits that neither social nor technical components can be studied or optimized in isolation, as their interactions shape the overall system's effectiveness. STS theory has since evolved, becoming a foundational framework for analyzing how technological systems interact with human elements in diverse contexts, including digital preservation and knowledge management.

A core tenet of STS is the principle of joint optimization, which emphasizes that social and technical subsystems must be aligned to achieve desired outcomes (Baxter & Sommerville, 2011). In the context of digital preservation using Fourth Industrial Revolution (4IR) technologies, this alignment ensures that the design and implementation of technologies such as Artificial Intelligence (AI), Blockchain, and Virtual Reality (VR) are sensitive to the needs, values, and practices of the communities involved. For example, the use of AI to preserve African traditional medicine requires not only advanced technical infrastructure but also the active participation of traditional healers, whose expertise informs the system's accuracy and cultural relevance (Omole & Olusegun, 2019). This illustrates how STS provides a framework for integrating Indigenous Knowledge into digital ecosystems without eroding its contextual integrity.

STS theory also highlights the dynamic and adaptive nature of systems. Technologies are not static artefacts but evolve alongside the social systems they interact with (Leonardi, 2012). This adaptability is particularly relevant to Indigenous Knowledge preservation, where traditional practices often change over time in response to environmental and cultural shifts. For instance, the use of Blockchain to secure knowledge ownership among India's tribal communities demonstrates how technology adapts to address specific challenges, such as intellectual property rights and benefit-sharing (Kumar et al., 2019). By embedding social considerations into technical design, STS ensures that digital preservation efforts remain responsive to the needs of Indigenous communities while respecting their autonomy and values. STS theory is equally concerned with power dynamics and stakeholder participation. Researchers have noted that technological systems often reflect the power structures of their designers, potentially marginalizing certain groups (Bijker & Law, 1992). In the study of digital preservation, this concern underscores the importance of participatory approaches that involve Indigenous communities in decision-making processes. For example, Virtual Reality (VR) projects documenting Amazonian tribal practices have succeeded when tribes actively contributed to content creation and storytelling, ensuring that the resulting representations align with their cultural narratives (Silva et al., 2020). Such practices reflect the STS principle of fostering equitable collaboration between technical and social actors. Linking STS theory to the study at hand, the preservation of Indigenous Knowledge using 4IR technologies exemplifies a socio-technical challenge. The technical systems AI, Blockchain, VR, and others serve as tools for documentation, analysis, and dissemination. However, their effectiveness hinges on their integration with social systems, including Indigenous practices, community governance structures, and intergenerational knowledge transfer. For instance, the success of AI-based cataloguing in African traditional medicine depends on building trust between researchers and Indigenous communities, ensuring that data collection respects cultural protocols and intellectual property rights (Chikezie et al., 2020). This alignment of technical capabilities with social dynamics embodies the core principles of STS.

Moreover, STS theory offers insights into addressing potential tensions between technological innovation and cultural preservation. Indigenous Knowledge systems are inherently social, characterized by contextuality, spirituality, and oral transmission. Digital technologies, on the other hand, tend to standardize and codify information, risking the loss of nuance and context (Smith, 2012). By applying STS principles, researchers can design socio-technical solutions that preserve the holistic nature of Indigenous Knowledge. For example, Blockchain's decentralized architecture allows Indigenous communities to retain control over their knowledge, mitigating concerns about cultural misappropriation (Hardy & Maurushat, 2021). These synergies demonstrate how STS theory informs the design of systems that respect and amplify Indigenous agency.

The theory of Socio-Technical Systems provides a robust framework for analyzing and addressing the challenges of preserving Indigenous Knowledge using 4IR technologies. Its emphasis on joint optimisation, adaptability, stakeholder participation, and equitable power dynamics aligns with the complexities of this study. By integrating the technical capabilities of AI, Blockchain, and VR with the social realities of Indigenous communities, STS theory ensures that digital preservation efforts are effective and culturally respectful. This theoretical foundation enhances the study's analytical depth and guides the development of inclusive, sustainable solutions for safeguarding Indigenous Knowledge.

OVERVIEW OF 4IR TECHNOLOGIES

The Fourth Industrial Revolution (4IR) represents a paradigm shift in technological advancement, characterized by the integration of digital, biological, and physical systems (Schwab, 2017). Central to 4IR are technologies such as Artificial Intelligence (AI), Blockchain, the Internet of Things (IoT), Virtual Reality (VR), and Big Data. AI involves the creation of systems capable of performing tasks that typically require human intelligence, such as language processing and pattern recognition. Blockchain, a decentralized ledger system, ensures secure and transparent record-keeping. IoT connects physical devices to the internet, enabling real-time data collection and monitoring. VR offers immersive simulations, while Big Data analytics extracts meaningful patterns from massive datasets. Together, these technologies are reshaping industries, including heritage preservation and cultural documentation.

In the realm of digital preservation, these technologies present unprecedented opportunities. AI, for example, can analyze complex patterns in Indigenous oral traditions, translating and categorizing them for broader understanding (Ongena et al., 2020). Blockchain facilitates the secure documentation of traditional knowledge, ensuring that ownership rights remain with Indigenous communities (Hardy & Maurushat, 2021). IoT-enabled devices can monitor ecological systems linked to Indigenous practices, providing real-time data that supports the preservation of environmental knowledge. Meanwhile, VR recreates cultural rituals, allowing users to experience them in an interactive format, enhancing understanding and empathy. Big Data plays a pivotal role in identifying trends and threats to Indigenous Knowledge systems, offering insights that drive preservation efforts. These applications demonstrate how 4IR technologies transcend mere documentation by actively engaging with the dynamic nature of Indigenous Knowledge. Unlike static archives, 4IR tools allow for the continuous updating and expansion of knowledge systems, preserving not only the content but also the processes by which this knowledge evolves. For example, digital storytelling platforms supported by AI and VR provide communities with tools to share their traditions interactively, ensuring that the younger generation connects with their heritage in meaningful ways. Such innovations reinforce the relevance of Indigenous Knowledge in contemporary contexts while addressing historical challenges of documentation and access. Despite their promise, the deployment of 4IR technologies in digital preservation is not without challenges. Issues of accessibility, data privacy, and cultural sensitivity need careful consideration. Indigenous Knowledge systems are deeply contextual, and the abstraction of knowledge into digital forms risks decontextualizing and misrepresenting it (Smith, 2012). Furthermore, the high cost and technical expertise required for 4IR technologies often exclude marginalized communities from fully benefiting from these innovations. Therefore, any implementation must prioritize inclusivity and collaboration, ensuring that Indigenous communities are active participants in the preservation process rather than passive subjects of it.

CHARACTERISTICS AND VULNERABILITIES OF INDIGENOUS KNOWLEDGE

Indigenous Knowledge (IK) is deeply rooted in the lived experiences, customs, and interactions of local communities with their natural and social environments (Battiste & Henderson, 2000). It is often transmitted orally, through storytelling, rituals, and practice, rather than written records. This oral tradition fosters a dynamic and contextual understanding of the world but also makes Indigenous Knowledge highly vulnerable to loss. Factors such as globalization, urban migration, and generational shifts threaten the continuity of these practices. With fewer community elders available to pass on their wisdom, traditional knowledge risks disappearing altogether. The loss of oral traditions has profound implications for both cultural identity and practical knowledge systems. For instance, in Africa, the use of herbal medicine derived from centuries of empirical understanding of local flora is rapidly declining as younger generations shift towards modern pharmaceuticals (Asfaw, 2001). Similarly, Indigenous Australian Dreamtime stories, which encode complex ecological and cosmological knowledge, are being lost as Western education systems displace traditional ways of learning (Rose, 1996). These examples illustrate the fragility of knowledge systems that rely on interpersonal transmission and highlight the urgent need for preservation efforts.

Beyond oral traditions, Indigenous Knowledge systems face systemic vulnerabilities due to their marginalization in dominant epistemologies. Mainstream knowledge systems often dismiss Indigenous Knowledge as anecdotal or unscientific, further undermining its survival (Smith, 2012). This epistemic injustice not only diminishes the value of Indigenous Knowledge but also hinders its integration into global solutions for pressing issues like climate change and sustainable development. The compartmentalization of Indigenous Knowledge as folklore or culture, rather than a valid body of science, exacerbates its vulnerabilities in an era of rapid modernization. Moreover, the commodification of Indigenous Knowledge, particularly in the pharmaceutical and agricultural industries, poses another significant threat. Without mechanisms to protect intellectual property rights, Indigenous communities risk losing ownership of their knowledge to corporations that exploit it for profit (Battiste, 2008). This exploitation further disempowers communities, reducing their ability to safeguard and transmit their heritage. Therefore, strategies for preserving Indigenous Knowledge must address not only the technical challenges of documentation but also the broader socio-political dynamics that renders it vulnerable.

SYNERGIES BETWEEN 4IR AND INDIGENOUS KNOWLEDGE

The convergence of 4IR technologies and Indigenous Knowledge presents transformative opportunities for digitization, storage, and accessibility. For instance, digital platforms supported by AI and Blockchain enable the systematic cataloguing of Indigenous Knowledge while ensuring authenticity and ownership. Digitization allows oral traditions to be transcribed, translated, and archived in formats accessible to both Indigenous communities and global audiences. These digital repositories can be updated dynamically, preserving the living nature of Indigenous Knowledge and ensuring its relevance to contemporary issues. The use of mobile technologies in reaching remote communities exemplifies the potential of 4IR to democratize access to digital preservation tools. Mobile phones, often the most accessible technology in underserved regions, can act as entry points for capturing and sharing Indigenous Knowledge. Apps designed for data collection and storytelling enable communities to document their practices in real-time, empowering them to control the narrative of their heritage (Ongena et al.,

2020). Additionally, mobile networks facilitate the dissemination of digital archives, ensuring that knowledge is accessible to diaspora communities and researchers alike. However, the synergy between 4IR and Indigenous Knowledge goes beyond mere preservation. It offers pathways for innovation and resilience. For example, AI can analyze patterns in traditional agricultural practices to optimize crop yields in changing climatic conditions, effectively blending traditional wisdom with modern science (Hardy & Maurushat, 2021). Similarly, Blockchain technology ensures that any commercialization of Indigenous Knowledge benefits the rightful owners, addressing historical injustices related to biopiracy. Such integrations reaffirm the value of Indigenous Knowledge in addressing global challenges while safeguarding its cultural integrity. Despite these opportunities, the application of 4IR technologies must be guided by principles of cultural sensitivity and inclusivity. Indigenous Knowledge systems are not merely databases to be digitized; they are lived practices imbued with spiritual and cultural significance. Ethical frameworks that prioritize community involvement and respect for local norms are essential to ensuring that digitization efforts enhance rather than undermine Indigenous heritage. Collaborative approaches, involving Indigenous leaders, technologists, and policymakers, are crucial for realizing the full potential of 4IR in the preservation of Indigenous Knowledge.

CASE STUDIES IN DIGITAL PRESERVATION USING 4IR TECHNOLOGIES

The Fourth Industrial Revolution (4IR) has introduced transformative technologies such as Artificial Intelligence (AI), Blockchain, the Internet of Things (IoT), and Virtual Reality (VR), which are reshaping how knowledge is documented, preserved, and shared. In the context of digital preservation, these technologies have proven instrumental in safeguarding cultural heritage and Indigenous Knowledge (IK) systems that are increasingly under threat due to globalization, modernization, and environmental changes. Scholars such as Marwala and Hurwitz (2017) emphasize that 4IR technologies not only enable the digitization of information but also provide platforms for authenticating, analyzing, and disseminating knowledge while preserving its cultural integrity. However, the application of these technologies is context-dependent, requiring adaptation to the unique cultural and social realities of Indigenous communities. Exploring case studies offers valuable insights into how different regions leverage 4IR tools for the digital preservation of their knowledge systems.

Case studies from Africa, Asia, and South America highlight the diverse applications of 4IR technologies in preserving Indigenous Knowledge. In Africa, AI has been employed to catalogue and analyze traditional medicinal practices, providing a robust framework for integrating local knowledge into global health solutions. In Asia, Blockchain has emerged as a critical tool for securing the authenticity and ownership of tribal knowledge, particularly in agricultural practices among India's Indigenous communities. Meanwhile, in South America, Virtual Reality has revolutionized the teaching of Indigenous history and the promotion of cultural heritage through immersive eco-tourism experiences, such as those with Amazonian tribes. These examples illustrate the potential of 4IR technologies to safeguard Indigenous Knowledge while simultaneously raising questions about accessibility, equity, and ethical considerations in their implementation. Through critically examining these case studies, this study seeks to understand the benefits and limitations of using 4IR tools for digital preservation in diverse cultural contexts.

Africa: Preserving Traditional Medicine With AI

Africa's rich repository of traditional medicinal knowledge passed down through generations, is under threat due to globalization and the decline of oral traditions. To combat this, Artificial Intelligence (AI) has emerged as a powerful tool for cataloguing and analyzing African medicinal plants, providing an innovative approach to preserving this crucial aspect of Indigenous Knowledge. AI algorithms, particularly those in machine learning and natural language processing, have been employed to document traditional remedies, identify bioactive compounds in plants, and establish links between traditional practices and modern pharmacological research (Chikezie et al., 2020). For instance, AI-driven databases are being developed to catalogue the properties, uses, and ecological requirements of medicinal plants, offering a digital repository accessible to both researchers and Indigenous communities. Collaborations between local communities and technology firms have been instrumental in these efforts. Community members, including traditional healers, contribute their expertise by providing detailed descriptions of medicinal plants, their preparation methods, and usage contexts. This data is then integrated into AI systems, which analyze patterns and generate insights that inform conservation and research (World Health Organization [WHO], 2013). For example, in Nigeria, partnerships between traditional medicine practitioners and AI researchers have led to the creation of a digital platform that identifies plants with high potential for treating diseases such as malaria and diabetes (Omole & Olusegun, 2019). These collaborations not only preserve Indigenous Knowledge but also validate its scientific relevance, fostering mutual respect between traditional and modern knowledge systems. However, challenges remain in ensuring that these initiatives benefit the source communities. Issues such as data sovereignty, intellectual property rights, and equitable benefit-sharing must be addressed. Many Indigenous communities are wary of exploitation, as historical precedents of biopiracy have eroded trust in collaborations with external entities. Thus, ethical frameworks that prioritize community consent, transparency, and fair compensation are essential to the success of AI-driven preservation projects. By addressing these challenges, Africa's traditional medicinal knowledge can be safeguarded as a living, evolving resource for future generations.

Asia: Blockchain for Knowledge Ownership

In Asia, blockchain technology is being used to address the critical issue of ownership and benefit-sharing in the documentation of Indigenous Knowledge. Blockchain's decentralized and tamper-proof ledger system provides a secure method for recording and protecting knowledge, ensuring that communities retain control over their intellectual property (Hardy & Maurushat, 2021). This is particularly relevant in India, where tribal communities have extensive agricultural practices rooted in centuries of ecological wisdom. For instance, the documentation of traditional crop varieties and farming techniques using blockchain ensures that these contributions are attributed to their rightful custodians.

One successful example involves the Santal community in Eastern India, which has collaborated with local NGOs and blockchain developers to record their knowledge of indigenous rice varieties. This initiative not only documents their practices but also safeguards their intellectual property from unauthorized use by corporations or researchers (Kumar et al., 2019). By creating blockchain-enabled smart contracts, these projects ensure that any commercial use of the knowledge such as the development of high-yield seeds results in equitable benefit-sharing. This approach empowers communities to monetize their knowledge while preserving their cultural heritage. Moreover, blockchain facilitates transparency and trust in transactions involving Indigenous Knowledge. The technology records every transaction or access request, enabling communities to track how their knowledge is used and ensuring accountability from external parties. This is particularly valuable in agricultural practices, where traditional methods are increasingly being recognized for their sustainability and climate resilience. For example, blockchain has been used to document organic farming techniques in India's Northeastern states, providing certifications that enhance market access for small-scale farmers (Sharma & Saha, 2020).

South America: Virtual Reality and Cultural Heritage

In South America, Virtual Reality (VR) is being utilized to create immersive experiences that preserve and promote Indigenous cultural heritage. VR technology allows users to explore digital recreations of traditional practices, rituals, and landscapes, offering a unique method of documenting and teaching Indigenous history. This is particularly significant in regions like the Amazon, where deforestation and cultural assimilation threaten the survival of Indigenous traditions (Santos et al., 2021). By capturing the sights, sounds, and narratives of Indigenous communities, VR preserves the sensory richness of cultural practices that are difficult to document through traditional means. One notable case involves the use of VR to promote eco-tourism and cultural education among Amazonian tribes. Collaborations between Indigenous leaders and tech companies have resulted in VR projects that simulate tribal ceremonies, storytelling sessions, and daily activities. These immersive experiences are used in museums and educational institutions worldwide, raising awareness about the Amazon's cultural and ecological diversity (Silva et al., 2020). At the same time, they generate revenue for the tribes, who retain control over the content and narratives presented in these projects. VR also plays a role in preserving endangered languages and oral traditions. By integrating audio recordings of Indigenous elders into VR environments, these projects ensure that the linguistic and narrative elements of cultural heritage are not lost. For example, a VR project developed with the Yawanawá tribe in Brazil has documented their myths and rituals, creating a resource for both academic research and community education. This approach highlights the potential of VR to bridge the gap between traditional and modern knowledge systems, making Indigenous Knowledge accessible to global audiences while respecting its original context. While VR offers powerful tools for cultural preservation, it is not without limitations. The high cost of VR equipment and the need for specialized technical skills can exclude many Indigenous communities from participating fully in these projects. Furthermore, there is a risk of commodifying cultural heritage for entertainment purposes, undermining its spiritual and cultural significance. To address these issues, VR initiatives must be guided by principles of cultural sensitivity and community ownership. Through prioritizing ethical considerations, VR can serve as a transformative medium for preserving South America's rich cultural heritage.

THE ROLE OF ARTIFICIAL INTELLIGENCE IN SAFEGUARDING INDIGENOUS KNOWLEDGE SYSTEMS IN RURAL AFRICAN COMMUNITIES

The safeguarding of Indigenous Knowledge (IK) systems is critical in preserving the cultural heritage, traditional practices, and ecological insights of rural African communities. As Indigenous Knowledge faces threats from globalization, cultural homogenization, and generational attrition, Artificial Intelligence (AI) has emerged as a promising tool for its documentation, analysis, and dissemination. Scholars such as Marwala and Hurwitz (2017) have extensively discussed the potential of AI in addressing these challenges, offering a foundation for understanding its applications and implications. However, while AI offers substantial benefits in safeguarding IK, it also presents notable limitations that must be critically evaluated.

Artificial Intelligence (AI) has emerged as a transformative tool for safeguarding Indigenous Knowledge (IK) systems in rural African communities, where traditional practices are deeply rooted in oral traditions and local cultural frameworks. AI technologies, particularly Natural Language Processing (NLP), have been instrumental in documenting Indigenous languages and oral histories, ensuring their preservation for future generations. As argued by Mushi et al. (2019), AI-powered transcription tools can effectively convert oral narratives into digital formats, thus protecting these traditions from the risk of generational attrition. Similarly, Mhlanga and Molokwane (2020) highlight the capacity of AI to bridge linguistic divides by developing algorithms capable of interpreting and standardizing Indigenous languages, which are often underrepresented in global linguistic repositories. However, the effectiveness of AI in this context depends on the availability of diverse datasets, which are often scarce in rural African settings.

AI also plays a significant role in enhancing the accessibility and dissemination of Indigenous Knowledge. For instance, applications of AI in agriculture, such as predictive analytics and automated advisory systems, have been tailored to incorporate traditional farming practices, enabling rural farmers to make informed decisions while preserving their cultural approaches to cultivation (Mutanga et al., 2018). Furthermore, Marwala and Hurwitz (2017) argue that AI can integrate Indigenous ecological knowledge into global climate models, thereby validating and amplifying its relevance in addressing environmental challenges. These systems empower Indigenous communities by providing real-time insights and extending the reach of their knowledge to a global audience. However, critiques by Juma and Agyei-Mensah (2021) caution against the potential commodification of IK, where external entities profit from these innovations without equitably compensating the communities that hold the knowledge. A crucial benefit of AI is its ability to connect Indigenous Knowledge systems with modern scientific frameworks, fostering innovation and sustainable development. Traditional medicine, for example, has benefited from AI systems capable of analyzing the chemical properties of medicinal plants and correlating them with biomedical data (Chikezie et al., 2020). Marwala and Hurwitz (2017) assert that such applications not only confirm traditional healing practices but also open pathways for collaborative research between Indigenous communities and pharmaceutical industries. However, the integration of IK into scientific models must be conducted with cultural sensitivity and respect for intellectual property rights, as highlighted by Battiste and Henderson (2000). Without proper safeguards, communities risk losing control over their heritage, which could lead to exploitation or cultural erasure. Despite its potential, the application of AI in safeguarding Indigenous Knowledge is not without challenges. One significant limitation is the digital divide, which disproportionately affects rural African communities. As Ndung'u et al. (2019) observe, inadequate access to electricity, internet infrastructure, and digital literacy poses significant barriers to the adoption of AI technologies. Moreover, Mushi et al. (2019) emphasize that the training of AI models often relies on datasets that inadequately represent Indigenous Knowledge, leading to biases or misinterpretations in AI-driven analyses. This highlights the importance of developing inclusive and community-driven approaches to AI implementation, where Indigenous people play active roles in data collection, algorithm design, and technology governance.

Ethical considerations remain at the forefront of debates surrounding AI and Indigenous Knowledge systems. The exploitation of digitized IK by external entities, such as corporations or academic institutions, is a growing concern. Marwala and Hurwitz (2017) acknowledge the need for legal frameworks that protect the ownership and distribution of IK, advocating for the use of AI technologies like Blockchain to secure intellectual property rights. Additionally, Juma and Agyei-Mensah (2021) argue that AI-driven solutions must prioritize community empowerment over mere technological advancement, ensuring that rural African communities derive tangible benefits from these innovations. Addressing these ethical challenges requires collaborative efforts among policymakers, technologists, and Indigenous communities to create equitable and sustainable systems for safeguarding IK.

THE POTENTIAL OF AI IN SAFEGUARDING INDIGENOUS KNOWLEDGE

AI technologies are transformative in digitizing and preserving Indigenous Knowledge. Tools such as machine learning and Natural Language Processing (NLP) have been used to transcribe oral histories, a cornerstone of many African Indigenous Knowledge systems, into digital formats. This process ensures the longevity of these narratives, many of which are at risk of being lost due to the decline of oral tradition in younger generations. For example, AI-powered transcription systems can process Indigenous languages with high precision, creating comprehensive databases that capture the cultural nuances of rural African communities (Chikezie et al., 2020). Tshilidzi Marwala, in his work on AI and African development, argues that machine learning can bridge the gap between traditional knowledge and contemporary science by identifying patterns and insights that are not immediately apparent (Marwala & Hurwitz, 2017). For instance, in traditional medicine, AI systems can analyze large datasets of herbal remedies, linking their chemical properties to potential biomedical applications. This not only validates the knowledge within scientific frameworks but also provides opportunities for the integration of IK into global health solutions. Marwala's perspective highlights AI's potential to elevate the status of Indigenous Knowledge, positioning it as a critical resource in addressing global challenges such as climate change and public health crises.

AI also enhances the accessibility of IK. Through AI-driven platforms, Indigenous practices can be disseminated widely, allowing both local and international audiences to engage with the knowledge. This accessibility fosters greater appreciation for African cultures while empowering rural communities to control the narrative of their heritage. For example, AI-enabled mobile applications have been deployed in rural areas to document and share knowledge about sustainable farming practices, bridging the gap between Indigenous wisdom and modern agricultural technologies (Omole & Olusegun, 2019). The benefits of AI in safeguarding Indigenous Knowledge are multifaceted. First, AI enables the systematic documentation of complex, orally transmitted knowledge, preserving it for future generations. By creating digital repositories, AI safeguards IK against the risks of cultural erosion and generational gaps. Second, AI fosters innovation by linking Indigenous practices to contemporary challenges. For example, traditional African agricultural techniques, analyzed through AI systems, have informed sustainable farming practices adapted to changing climatic conditions. Third, AI empowers communities by providing them with tools to document and disseminate their knowledge independently, reducing reliance on external actors. Moreover, AI's ability to process multilingual data ensures inclusivity in documenting the diversity of Indigenous Knowledge across Africa's numerous linguistic groups. This inclusivity strengthens the representation of marginalized voices, ensuring that all knowledge systems are preserved equitably.

SHORTFALLS OF AI IN INDIGENOUS KNOWLEDGE SAFEGUARDING

Despite its potential, AI's application in safeguarding Indigenous Knowledge faces significant challenges. One major critique of Marwala's work is its optimistic view of AI's neutrality and scalability. While he acknowledges the need for inclusivity, his analysis often underestimates the socio-cultural barriers to AI adoption in rural contexts. For example, AI systems require significant infrastructural support, including stable electricity and internet connectivity, which are often lacking in rural African communities (Marwala & Hurwitz, 2017). These infrastructural gaps limit the feasibility of large-scale AI implementation, leaving many communities unable to benefit from these technologies. AI systems are not inherently culturally sensitive. The development and training of AI models often rely on datasets that do not adequately reflect the complexity and diversity of Indigenous Knowledge. This limitation can lead to misrepresentation or oversimplification of IK, undermining its contextual richness. For instance, NLP systems may struggle with the nuances of Indigenous languages, leading to inaccuracies in transcription and interpretation.

Another significant issue is the potential for knowledge appropriation. Marwala's work acknowledges the importance of intellectual property rights but fails to fully address the ethical dilemmas posed by AI-driven documentation. Once digitized, Indigenous Knowledge may become vulnerable to exploitation by external entities, such as corporations or researchers, without adequate benefit-sharing agreements. This risk underscores the need for robust legal frameworks and community governance to protect IK from being commodified. AI systems can inadvertently perpetuate existing power imbalances. The design and control of AI technologies are often concentrated in the hands of external developers or organizations, sidelining the agency of Indigenous communities. This dynamic raise concerns about digital colonialism, where the digitization process becomes a means of extracting and profiting from Indigenous Knowledge without empowering the communities themselves.

AI holds immense potential to safeguard Indigenous Knowledge systems in rural African communities, offering innovative solutions for documentation, analysis, and dissemination. Scholars such as Tshilidzi Marwala have highlighted its transformative role in validating and integrating IK into contemporary frameworks. However, realizing this potential requires addressing the significant infrastructural, ethical, and cultural challenges associated with AI adoption. While Marwala's contributions provide a valuable foundation, his optimistic outlook must be tempered with critical attention to the socio-economic realities of rural African contexts. To maximize the benefits of AI while mitigating its shortfalls, a participatory approach is essential. Indigenous communities must play an active role in the design, implementation, and governance of AI systems to ensure that these technologies align with their values and priorities. Moreover, investments in infrastructure, legal protections, and culturally sensitive AI models are necessary to create fair and sustainable solutions for safeguarding Indigenous Knowledge.

THE ROLE OF 4IR IN SAFEGUARDING INDIGENOUS KNOWLEDGE SYSTEMS IN THE SOUTH AFRICAN GOVERNMENT CONTEXT

The integration of Fourth Industrial Revolution (4IR) technologies into governmental frameworks has created new opportunities for the preservation and promotion of Indigenous Knowledge Systems (IKS) in South Africa. 4IR technologies, including Artificial Intelligence (AI), Blockchain, and Big Data, have been recognized by policymakers as tools that can enhance the documentation and dissemination of Indigenous Knowledge, especially in rural and marginalized communities. According to Marwala and Hurwitz (2017), AI has the potential to digitize oral histories and traditional practices, ensuring their long-term preservation in accessible formats. Similarly, Ndlovu-Gatsheni (2018) emphasizes that South Africa's decolonial framework aligns with the adoption of 4IR tools to restore and amplify Indigenous voices within governance structures. These technologies not only safeguard cultural heritage but also empower communities by integrating their knowledge into national development plans. The South African government's policy landscape reflects a commitment to leveraging 4IR technologies for IKS preservation. The Protection, Promotion, Development, and Management of Indigenous Knowledge Act (2019) explicitly recognizes the role of technology in ensuring the intellectual property rights of Indigenous communities. Complementing this, Molefe and Mkhize (2021) argue that Blockchain can enhance the security of digital repositories, preventing unauthorized access or exploitation of Indigenous data. This legislative and technological synergy aims to mitigate the historical marginalization of Indigenous communities while fostering their active participation in the knowledge economy. However, as pointed out by Smith, Ndwandwe, and Dlamini (2020), challenges such as digital infrastructure disparities in rural areas must be addressed to realize the full potential of these initiatives. Another critical area of application is traditional medicine, where AI has been employed to catalogue medicinal plants and their uses. South Africa's wealth of biodiversity, combined with the extensive traditional knowledge of its Indigenous communities, positions the country as a global leader in biocultural research. According to Chinsembu (2016), digital platforms supported by AI can bridge the gap between Indigenous medicinal knowledge and modern pharmacological research, promoting the integration of local remedies into the healthcare system. Marwala and Hurwitz (2017) argue that machine learning algorithms can identify patterns and applications in traditional medicinal practices that were previously undocumented, thus enriching scientific understanding. While these advancements hold great promise, Molefe et al. (2021) caution against the commercialization of such knowledge without equitable benefit-sharing agreements that prioritize the rights and welfare of Indigenous communities.

The educational sector also plays a vital role in the government's approach to integrating 4IR into IKS preservation. Initiatives such as e-learning platforms and digital storytelling applications have been implemented to teach Indigenous languages and cultural practices to younger generations. Nkondo (2012) argues that these technologies can counter the decline of oral traditions by ensuring their accessibility in modern educational systems. Furthermore, the Department of Science and Innovation has launched programs to promote the digitization of Indigenous Knowledge repositories, enabling their incorporation into national curricula. However, Van der Merwe and Chitiga (2020) highlight that these efforts require substantial investments in capacity-building and digital literacy to ensure their sustainability and effectiveness, particularly in rural areas where access to technology remains limited.

Despite these advancements, the use of 4IR technologies in safeguarding IKS is not without its challenges. Ethical concerns around data ownership, privacy, and cultural sensitivity remain significant hurdles. For instance, Smith et al. (2020) argue that while Blockchain provides robust mechanisms for securing Indigenous data, its implementation must be guided by ethical principles that prioritize community consent and control. Additionally, questions about the sustainability of technology-based interventions arise, particularly considering South Africa's socioeconomic disparities. Without addressing these foundational issues, the application of 4IR technologies risks perpetuating existing inequalities rather than alleviating them. As such, a balanced approach that combines technological innovation with community-centered governance is essential for the long-term preservation and promotion of Indigenous Knowledge Systems in South Africa.

South Africa's government has undertaken several initiatives to integrate Fourth Industrial Revolution (4IR) technologies in safeguarding Indigenous Knowledge Systems (IKS), with the National Recordal System (NRS) being a flagship project. Launched by the Department of Science and Innovation (DSI) in 2013, the NRS aims to document, preserve, and protect Indigenous Knowledge by creating a centralized digital repository. This system leverages advanced technologies such as Artificial Intelligence (AI) and Big Data to catalog oral traditions, traditional medicinal practices, and other cultural assets. According to Tangwa and Tiebere (2019), the NRS represents a groundbreaking effort to prevent the erosion of Indigenous Knowledge by digitizing it in formats that are accessible and secure. The repository not only preserves this knowledge for future generations but also creates a framework for its ethical use in scientific research and commercialization. The use of 4IR technologies in the NRS extends beyond mere documentation. For instance, the system integrates Blockchain technology to secure data ownership and protect against unauthorized exploitation.

Molefe and Mkhize (2021) highlight that Blockchain ensures the authenticity of recorded information while maintaining transparency and trust between stakeholders. By involving local communities in the recording process, the NRS also promotes inclusive governance and ensures that Indigenous custodians retain control over their intellectual property. However, critics such as Smith, Ndwandwe, and Dlamini (2020) argue that while the system demonstrates technological sophistication, its implementation has faced logistical challenges, particularly in rural areas with limited digital infrastructure. Another significant project is the digitization of Indigenous languages to preserve linguistic heritage and promote multilingualism. The Pan South African Language Board (PanSALB), in collaboration with the Department of Arts and Culture, has utilized Natural Language Processing (NLP) and machine learning tools to develop digital dictionaries and language corpora for Indigenous languages such as isiZulu, Sepedi, and Xitsonga. Nkondo (2012) emphasizes that such initiatives are critical in safeguarding linguistic diversity, especially as younger generations increasingly adopt global languages over Indigenous ones. AI-driven language tools further enable the integration of these languages into modern communication technologies, such as voice assistants and translation software, thus fostering their everyday use. However, as Van der Merwe and Chitiga (2020) caution, these efforts require sustained investment and technical expertise to ensure that they remain relevant and widely adopted.

The government has also initiated projects to support traditional medicine through digital platforms. The South African Traditional Medicines Research Initiative (SATMRI), supported by the DSI, utilizes AI and Big Data to map and analyze the country's rich biodiversity of medicinal plants and the associated Indigenous Knowledge. According to Chinsembu (2016), these projects aim to integrate traditional medicine into the mainstream healthcare system by validating its efficacy and documenting its applications in secure, accessible formats. Marwala and Hurwitz (2017) argue that such initiatives position South Africa as a leader in biocultural innovation, creating opportunities for both economic development and health equity. However, ethical concerns regarding benefit-sharing and biopiracy remain unresolved, necessitating robust legal frameworks to protect community interests.

In addition to these efforts, partnerships between the government and academic institutions have played a vital role in advancing 4IR-driven IKS preservation. Universities such as the University of Venda and the University of KwaZulu-Natal have collaborated on projects to digitize traditional farming practices and Indigenous ecological knowledge. These initiatives align with the National Development Plan (NDP), which underscores the importance of leveraging technology to promote sustainable development and social inclusion. As Ndlovu-Gatsheni (2018) notes, these collaborative efforts reflect a broader commitment to decolonizing knowledge systems by embedding Indigenous perspectives into

South Africa's technological and policy frameworks. While these projects illustrate the transformative potential of 4IR technologies, challenges persist in their execution. Digital literacy, infrastructure disparities, and funding constraints continue to hinder the full realization of these initiatives, particularly in under-resourced rural areas. Furthermore, as Tangwa and Tiebere (2019) argue, the sustainability of these projects depends on fostering strong community engagement and ensuring that Indigenous Knowledge custodians are at the forefront of decision-making processes. By addressing these challenges, South Africa can set a global precedent for the ethical and effective use of 4IR technologies in preserving Indigenous Knowledge Systems.

METHODOLOGY

This study employed a secondary data methodology, focusing on the analysis of existing literature, case studies, and reports. Secondary data sources include academic journals, books, conference papers, and reputable online resources. The methodology involved a comprehensive review of the current state of research on 4IR technologies and their application in the digital preservation and documentation of indigenous knowledge, examination of specific instances where 4IR technologies have been implemented in preserving indigenous knowledge, analysis of their outcomes and lessons learned, identification of recurring themes, challenges, and best practices from the literature and case studies, comparison of different 4IR technologies to determine their relative effectiveness and suitability for various types of indigenous knowledge.

FINDINGS OF THE STUDY

Effective Utilization of Artificial Intelligence for Indigenous Knowledge Digitisation and Analysis

Artificial Intelligence (AI) presents transformative potential in the digitization and analysis of Indigenous Knowledge by automating the documentation process, enabling sophisticated analysis, and preserving the cultural essence of the knowledge. AI-driven tools such as Natural Language Processing (NLP) can be used to transcribe oral traditions, transforming Indigenous narratives into structured digital formats. For instance, AI applications have been employed to record and translate Indigenous languages, ensuring their preservation and accessibility for future generations (Chikezie et al., 2020). Machine learning algorithms can also classify and analyze large datasets of traditional medicinal practices, identifying patterns and connections that were previously undocumented. Furthermore, AI enhances the accessibility of Indigenous Knowledge by developing multilingual systems that cater to diverse linguistic groups. For example, AI has been used in Africa to catalogue traditional medicinal plants, integrating information about their uses, ecological requirements, and cultural contexts into digital repositories (Omole & Olusegun, 2019). By leveraging AI's analytical capabilities, researchers and communities can link traditional practices with modern scientific knowledge, validating and expanding the relevance of Indigenous Knowledge in contemporary contexts. However, ethical considerations remain paramount. AI systems must be designed in collaboration with Indigenous communities to ensure cultural sensitivity and accuracy. Additionally, mechanisms must be in place to protect the intellectual property rights of the communities, avoiding misappropriation or exploitation of their knowledge.

Ways Blockchain Technology Can Contribute to Maintaining the Authenticity and Integrity of Digitized Indigenous Knowledge

Blockchain technology provides a decentralized and secure framework for documenting Indigenous Knowledge, ensuring its authenticity and integrity. The immutability of blockchain records prevents unauthorized alterations, making it a trusted platform for storing sensitive cultural information. This is particularly critical for maintaining the authenticity of oral traditions and traditional practices that are prone to distortion when transferred to digital formats (Hardy & Maurushat, 2021). Blockchain also enables the implementation of smart contracts, which facilitate fair benefit-sharing agreements. Communities can establish conditions under which their knowledge may be accessed or utilized, ensuring transparency and accountability. For instance, blockchain has been used in India to protect Indigenous agricultural knowledge, preventing biopiracy by multinational corporations (Kumar et al., 2019). This approach not only safeguards the knowledge but also empowers communities to monetize their intellectual property without losing control. Moreover, blockchain's decentralized architecture allows for the creation of distributed knowledge repositories that multiple stakeholders can access while maintaining data sovereignty. Indigenous communities can retain ownership of their knowledge, deciding how and by whom it can be accessed. This aligns with ethical principles of equity and respect, addressing historical grievances related to the exploitation of Indigenous Knowledge.

The Potential Applications of the Internet of Things (IoT) in Improving the Accessibility and Real-Time Documentation of Indigenous Knowledge

The Internet of Things (IoT) offers innovative ways to enhance the accessibility and real-time documentation of Indigenous Knowledge by creating interconnected networks of devices that collect, store, and disseminate information. IoT sensors and devices can be deployed in remote areas to monitor environmental conditions and document traditional ecological knowledge in real-time. For example, IoT systems could record data on traditional agricultural practices, such as planting and harvesting cycles, weather patterns, and soil conditions, providing valuable insights for both local communities and researchers (Sundmaeker et al., 2010). IoT-enabled mobile applications also bridge the gap between rural Indigenous communities and global audiences. Communities can use IoT devices to share their knowledge directly, bypassing intermediaries and ensuring that the content remains authentic. For instance, wearable devices or mobile apps could record traditional ceremonies, songs, or stories, preserving their dynamic and performative aspects in digital formats. Additionally, IoT facilitates real-time collaboration between Indigenous communities and researchers. Environmental IoT systems, for instance, can alert communities to changes in biodiversity, enabling them to adapt their traditional practices accordingly. This integration of real-time data with Indigenous Knowledge not only enriches the knowledge base but also ensures its continued relevance in addressing contemporary challenges such as climate change and biodiversity loss (Wortmann & Flüchter, 2015).

Efficient Employment of Big Data Analytics to Manage and Disseminate Indigenous Knowledge Efficiently

Big data analytics plays a critical role in managing and disseminating Indigenous Knowledge by enabling the processing of vast amounts of information to identify patterns, trends, and insights. With Indigenous Knowledge systems often comprising diverse and complex datasets spanning oral traditions, environmental observations, and medicinal practices big data analytics provides the tools needed to structure and analyze this wealth of information systematically. For instance, big data analytics can integrate data from various sources, such as oral histories, environmental monitoring systems, and historical records, to create comprehensive digital repositories of Indigenous Knowledge (Laney, 2001). These repositories can then be used to generate visualizations and predictive models that support decision-making in areas such as sustainable agriculture, conservation, and community health. For example, by analyzing data on traditional medicinal plants, big data systems can identify potential new drugs and link traditional practices to scientific research findings (Chikezie et al., 2020).

Dissemination is another key application. Big data systems can segment knowledge into user-friendly formats tailored to different audiences, such as educational materials for schools, policy briefs for governments, or interactive platforms for the public. Furthermore, big data analytics can track how Indigenous Knowledge can be accessed and used, providing communities with valuable feedback and ensuring compliance with ethical standards. However, employing big data analytics requires addressing challenges such as data privacy, ownership, and consent. Indigenous communities must be involved in decisions about how their knowledge is collected, analyzed, and shared. This participatory approach ensures that the benefits of big data analytics are equitably distributed, aligning with the broader goals of digital preservation and cultural respect.

FUTURE RESEARCH DIRECTIONS

A valuable future research would be to explore the effectiveness and ethical implications of integrating Artificial Intelligence and Blockchain technologies in the curation and dissemination of Indigenous Knowledge Systems (IKS). While the chapter highlights the potential of 4IR tools in preserving and sharing IKS, more empirical studies are needed to assess how these technologies can be implemented in ways that respect community ownership, cultural sensitivities, and data sovereignty. Additionally, research could focus on developing inclusive digital platforms that are co-designed with Indigenous communities to ensure accessibility, cultural relevance, and long-term sustainability.

CONCLUSION AND RECOMMENDATIONS

This chapter has explored the potential of Fourth Industrial Revolution (4IR) technologies in the digital preservation of Indigenous Knowledge, highlighting their transformative role in documentation, analysis, and dissemination. Through a detailed examination of AI, Blockchain, IoT, and big data analytics, the study has demonstrated how these technologies can address the challenges of knowledge loss, ensure authenticity and integrity, and enhance accessibility for future generations. Artificial Intelligence has been identified as a powerful tool for digitizing oral traditions and analyzing complex datasets, while Blockchain ensures the security and ownership of digitized knowledge. The Internet of Things facilitates real-time documentation and extends the reach of preservation efforts to remote areas, and big data analytics enables efficient management and dissemination of Indigenous Knowledge. Each of these technologies, however, requires careful implementation guided by ethical considerations and active community participation. The findings underscore the importance of aligning technical innovations with the cultural and social realities of Indigenous communities. The application of 4IR technologies should not only aim to preserve knowledge but also empower communities, respect their autonomy, and safeguard their intellectual property. By adopting a socio-technical approach that emphasizes collaboration, equity, and sustainability, stakeholders can ensure that the digital preservation of Indigenous Knowledge becomes a transformative force for cultural resilience and innovation. This chapter contributes to the growing body of knowledge on the intersection of technology and culture, providing a roadmap for leveraging 4IR technologies in ways that respect and amplify the voices of Indigenous communities. Future research should focus on the long-term impacts of these technologies and the development of best practices for their implementation in diverse cultural contexts. Based on the findings of this study, several recommendations are proposed to enhance the role of Fourth Industrial Revolution (4IR) technologies in the digital preservation of Indigenous Knowledge:

Collaborative Development of AI Systems

AI systems should be developed in collaboration with Indigenous communities to ensure cultural sensitivity and relevance. Active community participation in data collection, algorithm design, and validation processes can help bridge the gap between technological innovation and cultural preservation. Governments and academic institutions should fund projects that prioritize co-design methodologies in AI applications for Indigenous Knowledge documentation.

Adoption of Blockchain for Knowledge Sovereignty

Blockchain technology should be adopted as a standard for ensuring the authenticity, integrity, and ownership of digitized Indigenous Knowledge. Policymakers should work with Indigenous communities to establish frameworks for benefit-sharing agreements facilitated by smart contracts. This will protect communities from exploitation while enabling equitable access to their cultural heritage.

Integration of IoT for Real-Time Knowledge Documentation

IoT systems should be deployed to capture real-time data on environmental practices, traditional ceremonies, and ecological knowledge. This can be achieved by distributing affordable IoT devices in remote areas, coupled with training programs to ensure their effective use by Indigenous communities.

Leveraging Big Data for Knowledge Management and Dissemination

Indigenous Knowledge systems should be integrated into large-scale data analytics platforms to enhance their accessibility and relevance. These platforms should employ user-friendly interfaces that cater to diverse audiences, including educators, policymakers, and researchers. Big data tools must also prioritize privacy, ensuring that all data usage aligns with the ethical principles of informed consent and community ownership.

Policy and Ethical Frameworks

Governments, NGOs, and international bodies should establish ethical and legal frameworks that protect Indigenous Knowledge in the digital domain. These frameworks should address issues of intellectual property, data sovereignty, and the equitable distribution of benefits derived from the use of Indigenous Knowledge.

Capacity Building and Education

Indigenous communities should be empowered through education and capacity-building programs to engage with 4IR technologies effectively. Training in AI, Blockchain, IoT, and big data analytics will enable these communities to become active stakeholders in the preservation of their cultural heritage.

REFERENCES

Asfaw, Z. (2001). The role of home gardens in the production and conservation of medicinal plants. Ethnobotany and Conservation of Medicinal Plants, 79–100.

Battiste, M. (2008). The decolonization of Aboriginal education: Dialogue, reflection, and action in Canada. Educational theories and practices from the majority world, 168-195.

Battiste, M., & Henderson, J. Y. (2000). *Protecting Indigenous Knowledge and Heritage: A Global Challenge*. Purich Publishing. DOI: 10.59962/9781895830439

Baxter, G., & Sommerville, I. (2011). Socio-technical systems: From design methods to systems engineering. *Interacting with Computers*, *23*(1), 4–17. DOI: 10.1016/j.intcom.2010.07.003

Bijker, W. E., Hughes, T. P., & Pinch, T. (1987). *The Social Construction of Technological Systems: New Directions in the Sociology and History of Technology*. MIT Press.

Bijker, W. E., & Law, J. (Eds.). (1992). *Shaping technology/building society: Studies in sociotechnical change*. MIT Press.

Chikezie, P. C., Ogidi, C. O., & Anosike, C. A. (2020). Artificial intelligence in traditional medicine research. *Journal of Ethnopharmacology*, *262*, 113254.

Chinsembu, K. C. (2016). Ethnobotanical study of medicinal plants used to treat sexually transmitted infections in Northern Namibia (Oshikoto region). *Journal of Ethnobiology and Ethnomedicine*, *12*(1), 1–12. DOI: 10.1186/s13002-016-0087-2

Duarte, M. E., & Belarde-Lewis, M. (2021). Imagining: Creating Spaces for Indigenous Ontologies. *The Journal of Documentation*, *77*(5), 1053–1066.

Gibson, K. (2019). Digital Preservation, Social Justice, and the Role of Information Professionals. *Library Trends*, *68*(2), 155–172.

Grobler, M. (2018). The ethical challenges of the Fourth Industrial Revolution. *South African Journal of Science*, *114*(1/2).

Hardy, K., & Maurushat, A. (2021). Blockchain and the protection of Indigenous Knowledge. Journal of Law. *Information, and Science*, *28*(1), 123–145.

Israel, B. A., Schulz, A. J., Parker, E. A., & Becker, A. B. (1998). Review of Community-Based Research: Assessing Partnership Approaches to Improve Public Health. *Annual Review of Public Health*, *19*(1), 173–202. DOI: 10.1146/annurev.publhealth.19.1.173 PMID: 9611617

Juma, K., & Agyei-Mensah, S. (2021). AI ethics and Indigenous Knowledge: Examining intellectual property challenges in Africa. *African Journal of Science, Technology, Innovation and Development*, *13*(3), 225–236.

Kumar, R., Singh, S., & Pradhan, R. (2019). Blockchain in agriculture: Preserving knowledge and innovation. *International Journal of Technology Management & Sustainable Development*, *18*(1), 78–95.

Kwet, M. (2019). Digital Colonialism: US Empire and the New Imperialism in the Global South. *Race & Class*, *60*(4), 3–26. DOI: 10.1177/0306396818823172

Leonardi, P. M. (2012). Materiality, sociomateriality, and socio-technical systems: What do these terms mean? How are they related? Do we need them? *Information and Organization*, *22*(3), 139–149.

Marwala, T., & Hurwitz, E. (2017). *Artificial intelligence and economic theory: skynet in the market* (Vol. 1). Springer International Publishing.

Mhlanga, S., & Molokwane, T. (2020). AI for Indigenous language preservation: A case study in Southern Africa. *SA Journal of Linguistics*, *12*(2), 45–58.

Molefe, T., & Mkhize, T. (2021). Blockchain technology as a solution for Indigenous Knowledge preservation in South Africa. *Journal of African Studies*, *47*(2), 101–116.

Mushi, G., Ndhlovu, T., & Makalima, K. (2019). Digital tools for cultural heritage preservation: Exploring AI's potential in Africa. *Journal of Cultural Heritage Studies*, *8*(4), 361–375.

Mutanga, C., Mlambo, M., & Sithole, P. (2018). Harnessing AI in traditional agriculture: A Zimbabwean perspective. *Agrárinformatika Folyóirat*, *9*(3), 14–22.

Nakata, M. (2002). Indigenous Knowledge and the Cultural Interface: Underlying Issues at the Intersection of Knowledge and Information Systems. *IFLA Journal*, *28*(5-6), 281–291. DOI: 10.1177/034003520202800513

Ndlovu-Gatsheni, S. J. (2018). *Epistemic freedom in Africa: Deprovincialization and decolonization*. Routledge. DOI: 10.4324/9780429492204

Ndung'u, N., Paul, S., & Waema, T. (2019). Bridging the digital divide in Africa: AI and the rural challenge. *The African Journal of Information Systems*, *11*(2), 65–83.

Nkondo, G. (2012). Indigenous Knowledge Systems in the 21st century: Perspectives on possibilities and challenges in South Africa. *South African Journal of Education*, *32*(2), 185–200.

Omole, O. H., & Olusegun, A. (2019). Preserving African traditional medicine using AI. *African Journal of Health Sciences*, *26*(3), 112–120.

Ongena, G., Hossain, F., & Steijn, C. (2020). Mobile technology for cultural heritage. *Digital Humanities Quarterly*, *14*(2), 1–20.

Santos, J., Silva, M. J., & Costa, P. (2021). Virtual reality as a tool for Indigenous knowledge preservation. *Digital Cultural Heritage*, *9*(2), 145–162.

Schwab, K. (2016). PDF The Fourth Industrial Revolution By.

Schwab, K. (2017). *The fourth industrial revolution*. Crown Currency.

Sharma, A., & Saha, B. (2020). Blockchain-enabled transparency in Indigenous agricultural practices. *Indian Journal of Sustainable Development*, *14*(3), 215–230.

Silva, M. J., Santos, J., & Rocha, L. (2020). Virtual heritage: Documenting Indigenous cultural practices through immersive technology. *Heritage Science*, *8*(1), 45–62.

Silva, M. J., Santos, J., & Rocha, L. (2020). Virtual heritage: Documenting Indigenous cultural practices through immersive technology. *Heritage Science*, *8*(1), 45–62.

Smith, A., Ndwandwe, S., & Dlamini, N. (2020). Ethical considerations in the use of Blockchain for Indigenous Knowledge preservation. *South African Journal of Information Management*, *22*(1), 1–10.

Smith, L. T. (2012). *Decolonizing Methodologies: Research and Indigenous Peoples*. Zed Books.

Tangwa, G., & Tiebere, F. (2019). Digital preservation strategies for Indigenous Knowledge systems in Africa. *International Journal of Cultural Policy*, *25*(4), 451–469.

Trist, E., & Bamforth, K. (1951). Some social and psychological consequences of the longwall method of coal-getting. *Human Relations*, *4*(1), 3–38. DOI: 10.1177/001872675100400101

Tsai, C. (2020). The Impact of the Fourth Industrial Revolution on Knowledge Management. *Journal of Knowledge Management*, *24*(2), 291–307.

Van der Merwe, A., & Chitiga, L. (2020). Technology, education, and the preservation of Indigenous Knowledge. *African Journal of Education and Technology*, *15*(3), 25–40.

World Health Organization (WHO). (2013). *Traditional Medicine Strategy 2014-2023*. WHO.

KEY TERMS AND DEFINITIONS

Fourth Industrial Revolution (4IR): It refers to the ongoing technological transformation characterized by the integration of advanced technologies such as Artificial Intelligence (AI), Blockchain, Internet of Things (IoT), Virtual Reality (VR), and Big Data into various sectors of society and the economy.

Indigenous Knowledge (IK): It refers to the traditional, local, and culturally specific knowledge systems that have been developed, sustained, and passed down over generations within Indigenous communities. It encompasses practices, beliefs, and skills related to areas such as medicine, agriculture, environmental management, and social structures, rooted in the deep understanding of local ecosystems and cultural heritage.

Digital Preservation: It is the process of maintaining, archiving, and ensuring the long-term accessibility of digital content, including texts, images, audio, and other forms of data. It involves a combination of strategies such as digitization, metadata creation, file format migration, and secure storage to safeguard information against technological obsolescence, physical degradation or accidental loss.

Internet of Things (IoT): It refers to a network of interconnected physical devices, vehicles, sensors, appliances, and other objects embedded with software, sensors, and connectivity features that enable them to collect, share, and analyse data in real time.

Blockchain: It is a decentralized, distributed ledger technology that records transactions in a secure, transparent, and tamper-resistant manner. Each transaction is stored in a block, and these blocks are linked chronologically to form a chain, creating an immutable record.

Chapter 14
Using Social Media Networks in the Management of Indigenous Knowledge in Selected Public Libraries in KwaZulu-Natal, South Africa

Petros Nhlavu Dlamini
https://orcid.org/0000-0002-0014-419X
University of Zululand, South Africa

Siphamandla Mgwengwe
University of Zululand, South Africa

ABSTRACT

The management of indigenous knowledge (IK) through social media networks is one of the most imperative subjects that need to be improved to ensure IK is captured, stored and disseminated effectively and efficiently. Indigenous knowledge is widely known as tacit knowledge and preserved in the human mind, especially in the memories of elders. It has been widely acknowledged that this knowledge is at risk and will become extinct if proper initiatives to capture, store and disseminate IK are not in place in organisations, institutions and information centres.

INTRODUCTION

The preservation, management, and sharing of Indigenous Knowledge (IK) are widely recognized as essential for social and economic development in Africa (Owiny, Mehta & Maretzki, 2014; Ajan et al., 2024). The significance of IK in sustaining human societies underscores the need for its preservation (Botangen, Vodanovich & Yu, 2017; Oosthuizen, Ungerer & Volschenk, 2023). As a crucial component of community, national, and global development, the documentation and preservation of IK have become priorities for information centers, libraries, and academic institutions (Yunnus, 2017; Gwerevende & Mthombeni, 2023). Public libraries play a key role in identifying, creating, storing, and managing IK.

DOI: 10.4018/979-8-3693-7964-6.ch014

The integration of social media has further enabled libraries to adapt to the evolving information landscape. Given their public mandate, libraries must be adequately resourced to meet the diverse information needs of the communities they serve (Mugwisi, Jiyane & Fombad, 2018). Dlamini and Ocholla (2018) and Zimu-Biyel (2016) highlight that ICT advancements have enabled knowledge digitisation for preservation and global access. Social media platforms like Facebook, Twitter, YouTube, and WhatsApp are valuable tools for managing Indigenous Knowledge (IK), facilitating its creation, storage, retrieval, and dissemination. These platforms, widely used for daily interactions, break geographical barriers and support spontaneous and formal exchanges (Amedie, 2015). Kaplan and Haenlein (2010) emphasise that social media fosters user-generated content, making it essential for IK management in public libraries. Public libraries serve as vital institutions for facilitating access to various forms of knowledge, including Indigenous Knowledge (IK), which is essential for cultural preservation and community development (Afful-Arthur et al., 2022; Chigwada & Chiparausha, 2020). According to the International Federation of Library Associations and Institutions (IFLA, 2002), Hangshing and Laloo (2021), public libraries have a critical role in promoting local knowledge, which contributes to achieving the Sustainable Development Goals (SDGs). Furthermore, UNESCO/IFLA (2002) highlights three key responsibilities of public libraries in relation to IK: fostering intercultural dialogue to promote cultural diversity, supporting oral traditions to preserve indigenous narratives, and ensuring community-wide access to relevant information.

With the advancement of digital technologies, public libraries are increasingly utilising social media networks as tools for storing and disseminating IK. Social media platforms offer an interactive and dynamic way to engage communities in sharing and preserving indigenous practices, histories, and knowledge. However, a gap exists in research regarding the role of social media networks in managing IK. Previous studies (Dlamini & Ocholla, 2018; Lwoga & Ngulube, 2008) have predominantly focused on the use of ICT networks in IK management, without adequately exploring the potential of social media in this context. Dlamini and Ocholla (2018) emphasise that social media platforms present effective opportunities for the management and preservation of IK for future generations. Given the widespread adoption of social media by various communities, public libraries have the potential to leverage these platforms to enhance their IK management strategies. This study, therefore, seeks to examine the extent to which public libraries are involved in managing IK through social media networks, exploring the challenges and opportunities presented by these digital tools. By investigating this area, the study aims to provide insights into how public libraries can effectively integrate social media into their IK management practices for the benefit of present and future generations.

RESEARCH PROBLEM

Public libraries have long been recognised as essential institutions for preserving and providing access to information and knowledge. Historically, in the 19th century, their primary role was limited to safeguarding printed materials. However, in contemporary society, their responsibilities have expanded significantly. Public libraries are now regarded as custodians of both knowledge and cultural heritage, encompassing a diverse range of materials such as drawings, paintings, manuscripts, records, books, and audio-visual resources (Isah, Bashorun & Omopupa, 2012). Beyond serving as repositories of written and printed knowledge, public libraries and other information centers have evolved into treasuries of indigenous experiences, traditional knowledge, and historical narratives. They play a critical role in preserving the cultural identity of communities by safeguarding indigenous knowledge (IK) and ensuring that valuable traditions, stories, and practices are documented and shared for future generations (Hangshing & Laloo, 2021; Masenya, 2024). In this digital era, libraries are leveraging modern technology to expand their preservation efforts, integrating digital repositories and social media platforms to enhance accessibility and engagement with IK.

Public libraries ensure free and equal access to information, adapting to the Fourth Industrial Revolution by integrating technology for managing oral information (Greyling, 2010). Modern libraries must provide access to diverse formats, including digital and oral records. Social media has become a valuable tool for preserving and sharing Indigenous Knowledge (IK) in public libraries. While global institutions like the Smithsonian and New York Public Library have long embraced technology for IK management, African libraries have lagged (Greyling, 2010). This study explores how South African public libraries manage IK, identifying challenges and recommending strategies for improved preservation and accessibility. Indigenous Knowledge (IK) is deeply rooted in Africa's cultural and ecological diversity, and has been used for centuries to address developmental and environmental challenges (Mohamedbhai, 2013). However, this knowledge is fading due to poor management and limited involvement of public libraries, which should serve as custodians of information. Mhlongo and Ngulube (2018) noted that the decline of IK in public libraries hampers access to vital information in indigenous communities. Additionally, the current generation's understanding of IK is dwindling as Western knowledge overshadows traditional practices. Jain (2014) emphasised the power of social media in preserving and sharing knowledge. However, limited research has been done in KwaZulu-Natal to assess public libraries' use of social media for managing IK. This study is crucial as it will evaluate how public libraries in the province utilise social media to manage IK, providing insights that can guide library management in decision-making and policy development. The study also aims to enhance collaboration between rural communities and public libraries by facilitating their participation in IK-related sessions through social media platforms

AIM AND RESEARCH QUESTIONS OF THE STUDY

The aim of the study was to investigate the use of social media networks for the management of indigenous knowledge by public libraries in the province of KwaZulu-Natal, South Africa. To achieve this objective, the study was guided by the following main research questions:

- What is the level of awareness of the use of social media networks by public libraries in the management of indigenous knowledge?
- What is the status of managing indigenous knowledge using social media networks by public libraries?
- What are the various types of social media networks used by public libraries to manage indigenous knowledge?

CONCEPTUALISATION/THEORETICAL MODEL

This study adopted Nonaka's (1995) SECI model, also known as the Knowledge Creation Theory, which consists of four key elements: Socialisation, Externalisation, Combination, and Internalisation (SECI). These elements work together organically and dynamically to generate knowledge (Stilwell,

2010). According to Dlamini and Ocholla (2018), the knowledge creation model is particularly effective in the generation, sharing, documentation, and transfer of knowledge through various knowledge management approaches. This study applied the SECI model to examine the use of social media networks in managing Indigenous Knowledge (IK) in public libraries. The table below illustrates the process of knowledge creation through these four key elements:

Table 1. Knowledge creation elements

Knowledge creation Model	Descriptions
Socialisation	Where tacit knowledge is shared through shared experiences, for example face-to-face conversations.
Externalisation	Where tacit knowledge is converted to explicit knowledge with the help of metaphors and analogies, for example, printed materials and rock paintings.
Combination	Where explicit knowledge is systemised and refined, for example, by utilising information and communication technologies and existing databases.
Internalisation	Where explicit knowledge is transferred to tacit knowledge, for example, learning by doing or translating theory into practice.

a. Socialisation

Nonaka (1994) defined socialisation as the process of creating common tacit knowledge through shared experiences. In order for socialisation to be effective, people need to build a "field" of interaction, where individuals share experiences and space at the same time, thereby creating common unarticulated beliefs or embodied skills. Bratianu (2010) opined that socialisation is the first knowledge transfer process considered by Nonaka, which reflects the tacit knowledge-tacit knowledge exchange. It is the process of bringing together tacit knowledge through shared experiences. However, since tacit knowledge is context-specific, it is important to note that people can share same experience through joint activities. According to Pescaru (2018), socialisation has as its main purpose the ability to obey the exigencies of social norms and ideals materialised in the cultural values of the group to which man belongs. The content of socialisation not only has a psychological significance (the maturing of the young person), but also a culturological (internalising social norms and values) and sociological (social role-playing and development of appropriate behaviours) significance. The socialisation process is based on a set of variables that include:

- Individual characteristics (age, gender, maturity, development, intelligence).
- Methods and forms of transmission (language, control mechanisms, rituals, child-raising practices and social integration, forms of imitation, identification, substitution, inhibition or reinforcement).
- Structures of attitudes, values, actions and behaviours (roles and social status, morality of social relations, ethics of work, political-civic orientation, performance, altruism, integration, conformation, etc.).

Pescaru (2018) was of the opinion that socialisation is an interactive communication process that involves both individual development and personal influences, namely the personal reception and interpretation of all social messages, but also the dynamics and content of social influences. Thus, socialisation is a complex process that is based on the theory of social learning - learning which a fundamental mechanism of assimilation of social experience is. Although retaining certain autonomy, man and society form an indissoluble unity, and society is generated by the relationship between its members. To note, Mosoti and Masheka (2010) emphasised that during socialisation, the conversion is tacit-to-tacit (for example, watching somebody doing something, and then copying it). It is widely known that the effectiveness of the socialisation process depends on the social structure and the balance between individuals and groups' cooperation. According to Nonaka and Konno (1998), in socialisation, tacit knowledge can only be shared if the self is freed to become the larger self that includes the tacit knowledge of the other. Dlamini (2017) emphasised that socialisation must go beyond the everyday dialogues and exchange of neutral phrases. However, it must stimulate deeper layers of experiences and stored knowledge (Bratianu & Orzea 2010; Holste & Fields 2010; Marra 2004). For example, this study supported the socialisation process because it involved social media networks which allow people to share and possibly manage IK by public libraries.

b. Externalisation

According to Nonaka and Konno (1998) and Clark (2004), the process of externalisation is supported by two key factors: the conversion of tacit knowledge into explicit knowledge that involves techniques to help express one's ideas or images as words, concepts, figurative language (e.g. metaphors, analogies or narratives) and visuals, and translating the tacit knowledge of customers or experts into readily understandable forms. A study by Scharmer (1996) regarded externalisation as the process of articulating tacit knowledge into explicit knowledge as concepts and/or diagrams, often using sketches. The advantage of externalisation, according to Marley (2012), is that it makes tacit knowledge transferable to others, and makes it useful and applicable in a wider context. The author further mentioned that tacit knowledge needs to be codified and converted into an understandable format, such as words or images. It is further indicated that some technologies, such as groupware, databases, radio, television, the Internet, flash drives and many others, assist in the externalisation of tacit knowledge. It is worth noting that this is an area where social media networks are applied to safeguard traditional knowledge for future use. Finally, the methods of externalisation entail capturing the personal, historical, social and material contexts of knowledge, including both the individual and collective work practices. It also involves being adaptable to, and reflecting, the dynamics of knowledge in work activities and in various organisational settings, in particular between communities and various stages of knowledge-management maturity. According to Jaleel and Verghis (2015), externalisation involves the conversion of tacit knowledge into explicit knowledge. This process attempts to rationalise tacit knowledge and articulate it into explicit concepts and formal models (e.g., to write instruction manuals). Externalisation is triggered by dialogue or collective reflection and is useful in the concept phase to articulate tacit knowledge; as Peltokorpi, Nonaka, and Kodama (2007) pointed out that "exposure to diverse ideas during the externalisation phase is important as every step in the innovation process is proposed to be about someone asking about imaginary possibilities, speculating about what would happen if, and reflecting on, yet unrealised and perhaps unrealisable solutions" (Peltokorpi, Nonaka, & Kodama, 2007:56).

c. Internalisation

Internalisation is the process where explicit knowledge is changed into tacit knowledge, which is connected to "learning by doing" (Nonaka, 1994:20). The explicit knowledge is captured in a form of mental models, skills or technical know-how possessed by individuals (Nonaka & Takeuchi, 1995; Nonaka, Umemoto, & Sasaki, 1998). Knowledge internalisation includes efforts to absorb accumulated organisational know-how, in which individuals strive to learn the recipe of "how to do it" (Kale & Singh, 2007). According to Jaleel and Verghis (2015), internalisation is a process of embodying explicit knowledge into tacit knowledge and internalising the individual experiences gained through the other models of knowledge creation in the form of shared mental models. Through internalisation, explicit knowledge created is shared through an online learning community and converted into tacit knowledge by individuals.

According to Tsai and Lee (2006), knowledge internalisation is the process of embodying explicit knowledge into tacit knowledge. Through internalisation, explicit knowledge created is shared throughout an organisation and converted into tacit knowledge by individuals. Determining the best method for enabling the widespread internalisation of explicit knowledge throughout an organisation is an issue of significant concern to firms and schools. The knowledge internalisation mentioned here refers primarily to ''knowledge applicative capability,'' or the ability to apply knowledge in real situations (Broadbent & Weill, 1993; Nonaka & Takeuchi, 1995; Duncan, 1995). However, in today's rapidly changing and fiercely competitive environment, firms' expectations regarding their employees are no longer passive in terms of acquiring and applying knowledge. The more complex jobs become, and the more rapidly the environment changes, the more necessary it is for employees to be able to adjust and update the knowledge they have learned, in order to go further and create better knowledge to perform their jobs, and thereby enhance the effectiveness of their work. For this reason, the internalisation discussed in this study comprises not only ''knowledge applicative capability'' but also the ''knowledge creation capability'' (Nonaka & Takeuchi, 1995).

Becerra-Fernandez (2001) noted that internalisation involves the conversion of explicit knowledge into internalised knowledge, i.e., individual or organisational learning. Internalisation requires that individuals identify explicit knowledge that is personally relevant, and create their own tacit knowledge based on it. During internalisation, individuals identify and acquire explicit knowledge that is created during the externalisation and combination phase. This knowledge is then explored by means of practice, following which individuals acquire and understand the tacit dimension of this knowledge (Nonaka & Konno, 1998; Marley, 2012). In addition, Bratianu and Orzea (2010) posited that the advantage of internalised knowledge is that it increases the level of individual understanding and absorptive capacity. However, this increases the knowledge capacity of an individual promoting absorption of information. Dlamini (2017) summarised that internalisation is the process of understanding and absorbing explicit knowledge into the tacit knowledge held by an individual. In the process of internalisation, the experiences of experienced workers in an organisation are transferred to the tacit knowledge of the individual. Additionally, knowledge in its tacit form is actionable by the owner while others are observing.

d. Combination

Combination integrates the newly-created explicit knowledge with the organisation's knowledge base to facilitate innovation (Shu, Page, Gao, & Jiang, 2012). As a result of knowledge integration, technological advancements can be interpreted in a new, innovative way, which is more likely to facilitate knowledge reconfiguration. Therefore, knowledge configuration facilitates translating new concepts to marketable products. Lee and Choi (2003) empirically supported the positive effect of combination on innovation and organisational creativity, which was defined as "the creation of valuable, useful, product, idea by individuals working together in a complex social system" (Lee & Choi, 2003:4). Furthermore, Schulze and Hoegl (2006) argued that combination has a positive effect on new product success at both the concept and the development phase (their study supported the effect during the development phase, while it failed to support the effect during the concept phase). Evidence also gained that combination helps in creating valuable knowledge from explicit knowledge to generate innovative product ideas (Crawford & Di Benedetto, 2008). Problem-solvers decompose, and recombine pieces of information in order to examine the cause–effect linkages and specify requirements for more useful and differentiated products (Corti & Storto, 2000).

Furthermore, combination embarks and builds on the organisation's current capability in order to verify the feasibility of the initial product idea. Hence, "combinative capability," as proposed by Kogut and Zander (1992), aims to synthesise and apply current and acquired knowledge from various domains to envision potential reconciliation between current limitations and technologies, in order to create product ideas with differentiated characteristics (e.g., applying technologies from the optical signal processing to mobile phones systems, adapted from Schulze and Hoegl (2006). In addition, combinative capability can use reports of previous unsuccessful product ideas to avoid falling into the same mistakes. Hence, combination facilitates NPD through quick idea generation within the specified timeframe and budget (Nonaka & Takeuchi, 1995; Schulze & Hoegl, 2006). Knowledge combined from different domains can help in realising new innovative solutions based on looking at challenges from different perspectives and integrating multiple dimensions into local solutions.

REASONS FOR USING KNOWLEDGE CREATION MODEL IN THIS STUDY

This study applied the Knowledge Creation Theory, which emphasises the possibility of social media networks in managing Indigenous Knowledge (IK). Socialisation, one of the key components of the SECI model, serves as a foundational method for capturing and sharing IK through social media platforms. By leveraging social media networks, public libraries can facilitate the process of socialisation, allowing community members to share their tacit knowledge with others, which is a crucial step in managing IK. Knowledge creation is a dynamic process where tacit and explicit knowledge are exchanged between individuals or groups within a predefined environment. In the case of public libraries, this environment is the digital space provided by social media networks. Through these platforms, community members can share their local knowledge, traditions, and cultural practices, making it accessible to a broader audience and ensuring its preservation for future generations. The researcher argues that public libraries in KwaZulu-Natal should take advantage of the widespread use of social media networks to manage IK effectively, as many community members are actively engaged on these platforms.

Dlamini and Ocholla (2018) highlighted that the effectiveness of any model is determined by its ability to achieve its intended goals. In this context, the Knowledge Creation Theory is relevant as it offers a framework for understanding how public libraries can use social media networks to facilitate the creation, sharing, and management of IK. By applying this theory, the study aims to demonstrate that social media networks can play a significant role in the management of IK in public libraries, especially in areas like KwaZulu-Natal, where these platforms are widely used by community members. The Knowledge Creation Model (KCM) is particularly relevant to this study due to its focus on managing tacit knowledge, which is central to Indigenous Knowledge (IK). Tacit knowledge, often embedded in local customs, practices, and experiences, is at risk of becoming extinct if not appropriately managed and preserved. The model's framework allows for the transfer of tacit knowledge to a more accessible form, making it crucial for the preservation of IK, which is often passed down through oral traditions and community practices. Baruah (2012) highlighted that social media platforms offer a wide range of tools for knowledge sharing, including blogs, microblogs, wikis, podcasts, photos, videos, and social bookmarking. These platforms facilitate the exchange of ideas, views, and information, making them ideal for managing and disseminating IK. Social media not only enables the sharing of content but also fosters collaboration and interaction among users from diverse backgrounds and geographical areas. Singh and Pandey (2017) further emphasised that social media brings together people from different parts of the world, providing a platform where individuals can share their emotions, ideas, and information freely. The numerous social networking sites such as Facebook, WhatsApp, Instagram, Twitter, and LinkedIn have revolutionized the way people communicate, allowing them to exchange knowledge instantaneously. With advancements in technology, social media users no longer have to wait for traditional dissemination processes. Instead, every social media user has become a potential source of information, contributing to the global flow of knowledge. In the context of this study, social media provides an ideal environment for managing and sharing IK, as it aligns with the Knowledge Creation Model's emphasis on capturing and disseminating tacit knowledge within a dynamic and interactive space. This makes social media networks not only valuable for managing indigenous knowledge but also instrumental in ensuring its preservation and global accessibility.

RESEARCH METHODOLOGY

The study adopted an interpretivism paradigm to explore participants' views, interpretations, and experiences related to the management of Indigenous Knowledge (IK) through social media networks in public libraries. The interpretivist paradigm emphasises understanding the subjective experiences and perspectives of participants, making it essential for the study to employ a qualitative research approach. Qualitative methods align well with interpretivism as they enable an in-depth exploration of participants' perspectives using various instruments, such as interviews. By adopting this paradigm, the study aimed to assess the extent to which public libraries in KwaZulu-Natal, South Africa, utilise social media networks for managing IK. The research questions were primarily open-ended, allowing for rich, detailed responses, with only a few closed-ended questions to provide additional context. Additionally, an in-depth literature review and document analysis were incorporated into the study, contributing to a qualitative content analysis that further supported the research findings and conclusions. This approach ensured a comprehensive understanding of the role social media plays in the management and preservation of IK within public libraries.

The study employed purposive sampling, a non-probability method that selects participants based on specific criteria or choice. The data collection method used was semi-structured interviews, which were deemed essential for gathering valid and relevant information for the study. According to Neuman (2014), an interview is a verbal conversation between two people aimed at collecting pertinent data for research purposes. Due to the large size of KwaZulu-Natal, purposive sampling was used to focus specifically on public libraries using social media platforms to manage Indigenous Knowledge (IK). As a result, the study sample was limited to the Pietermaritzburg Bessie Head Public Library, as other libraries in Durban were unable or unwilling to participate due to other commitments. The Bessie Head Public Library comprises various library staff, including eight data capture librarians, five knowledge manager librarians, two marketing librarians, two audio-visual librarians, one reference librarian, and one lending librarian, totaling 19 participants in the study. Using purposive sampling, the study interviewed these 19 librarians to gather qualitative data, which was then analysed using thematic analysis. This approach allowed for the identification of key themes and patterns within the data, providing insights into the management of IK through social media networks in public libraries.

Table 2. Demographic profile of participants

Variable	**PMB Public Library**	
Gender		
	Frequency	**Percentage (%)**
Female	10	52.6%
Male	09	47%
Total	19	**100%**
Position		
Knowledge Manager Librarian	05	23%
Marketing Librarian	02	10%
Audio-Visual Librarian	02	10%
Data Capturer Librarian	08	43%
Lending Librarian	01	05%
Reference Librarian	01	05%
Total	**19**	**100%**
Work experience		
1-5 years	13	68.4%
6-10 year	01	5.2%
11- and above	05	26.3%
Other	00	00
Total	**19**	**100%**

RESULTS AND DISCUSSIONS

Awareness Of Social Media Tools in the Management of Indigenous Knowledge

This question aimed to assess public libraries' awareness of social media tools for managing Indigenous Knowledge (IK). It sought to determine their familiarity with these platforms. Data was analysed using content analysis, and the results are presented in Table 3 below:

Table 3. Awareness of social media networks used for the management of indigenous knowledge (N =19)

No. Of participants	Position in the library	Response to the question
08	*Data Capturer Librarians*	***Yes***
05	*Knowledge Manager Librarians*	***Yes***
02	*Marketing Librarians*	***Yes***
02	*Audio-Visual Librarians*	***Yes***
01	*Lending Librarian*	***Yes***
01	*Reference Librarian*	***Yes***

The results indicated that librarians shared similar views and understanding regarding the use of social media networks for managing indigenous knowledge. Specifically, all participants were aware of these platforms' role in managing IK (see Table 3). In that light, the study results collaborated with Bullas (2014) who reported that librarians are aware of social media usage and its ease of access which makes it possibly got them to communicate directly for the purpose of exchanging and improving ideas and data, thereby advancing science. Gupta, Gautam and Khare (2014) also reported that social media applications are being used by many libraries to interact and build relationships with their communities by using wikis, flickr, blogs etc., to publish historical content and answering questions using applications like IM, SMS, and Twitter etc., as it is convenient for users. A study by Egbeyemi and Oniyide (2020) also highlighted that social network websites' usage has inevitably become a growing global phenomenon in information centres and is highly used in the management of information. Hence, Suraweera et al. (2011) also added that the use of online social networks by libraries and information organisations is also an increasingly prevalent and growing tool that is being used to communicate with more potential library users, as well as extending the services provided to individual libraries.

How Did You Become Aware of Social Media Networks for Managing Indigenous Knowledge?

This question aimed to explore how participants became aware of social media networks for managing IK. Understanding their awareness process was crucial for the study, as it revealed different ways participants learned about these platforms. Their responses are outlined below:

Table 4. How did you become aware of social media networks for managing indigenous knowledge?

Participant	Responses
Data Capturer Librarian 1:	*"From the American corner presentation on different social media platforms to promote libraries".*
Data Capturer Librarian 2:	"Through internet usage".
Data Capturer Librarian 3:	*"From Ulwazi programme".*
Data Capturer Librarian 4:	*"From the conferences and workshop offered by Municipality library".*
Data Capturer Librarian 5:	*"I become aware of this through studies"*
Data Capturer Librarian 6:	*"Conferences and workshops"*
Data Capturer Librarian 7:	*"When I was doing my honours degree at Unizulu".*
Data Capturer Librarian 8:	*"From the workshop I attended in Durban".*
Knowledge Manager Librarian 1:	*"From the workshop I attended in Durban".*
Knowledge Manager Librarian 2:	*"From social media discussions".*
Knowledge Manager Librarian 3:	*From workshops conducted in Pietermaritzburg".*
Knowledge Manager Librarian 4:	*"Through attending conferences 2 to 3 years ago".*
Knowledge Manager Librarian 5:	*"Through reading of literature".*
Marketing Librarian 1:	*"From a friend of mine from another public library".*
Marketing Librarian 2:	*"From the workshop I attended".*
Audio-Visual Librarian 1:	*From the American corner presentation on different social media platforms to promote libraries".*
Audio-Visual Librarian 2:	*Through Ulwazi programme in conference".*
Lending Librarian 1:	*"Refer to by a family member".*
Reference Librarian 1:	*"From a student who was conducting research on social media usage for promoting IK".*

The study findings revealed that participants learned about social media networks for managing IK through various sources. These included workshops, seminars, and conferences, as well as reading papers, browsing the Internet, engaging with the Ulwazi programme, studying, and learning from other researchers. These findings concurred with Priya and Rabindra (2010) who also reported that it is essential to propagate the use of indigenous traditional knowledge for human causes through certain activities, such as seminars, workshops, debates, lectures, and exhibitions in which such stories of indigenous traditional knowledge need to be reflected. In a nutshell, the study findings suggested that there are several means to promote the use of social media networks in support of the management of IK. These findings also concurred with Nonaka (1994) through knowledge creation where socialization is the only ingredient where people learn from others by observing and doing. In other words, Nonaka (1994) emphasized the importance of learning by doing where people copy good things from others and implement those things to improve their lives or working environment.

What is the Level of Awareness and Use of Social Media Networks By Public Libraries in the Management of Indigenous Knowledge in KwaZulu-Natal?

This question aimed to assess the level of awareness of social media networks among public libraries in KwaZulu-Natal and to observe whether these libraries were embracing technology to store and manage information. The results are presented as on Table 5 below as follows:

Table 5. What is the level of awareness and use of social media networks by public libraries in the management of indigenous knowledge in KwaZulu-Natal?

Participant	Responses
Data Capturer Librarian 1:	*"At the moment it is not promising".*
Data Capturer Librarian 2	*"Very poor in our province".*
Data Capturer Librarian 3:	*"There is a high level of awareness of social media tools like academic Edu, Google docs and Facebook etc. in our library but in the province as a whole, it is very poor".*
Data Capturer Librarian 4:	*"Libraries are now in a digital age, and they know a lot about social media networks so level must be high, but in our province, it is very poor".*
Data Capturer Librarian 5:	*"The young librarians are aware of the social media tools whereas the old generation finds it difficult to know and use them. I can say it is very low".*
Data Capturer Librarian 6:	*"The level is very low when it comes to the aged librarians, but the young ones are very much aware and active on these platforms to find information".*
Data Capturer Librarian 7:	*"It is poor at the moment and the management is doing nothing about it"*
Data Capturer Librarian 8:	*"The awareness is poor which requires young and energetic youth to ensure it grows".*
Knowledge Manager Librarian 1:	*"Not satisfactory because only few are aware of it".*
Knowledge Manager Librarian 2:	*"The awareness is very poor in our service and only the management team is aware of because they attend workshops and conferences".*
Knowledge Manager Librarian 3:	*"According to my view it is very low because we are still glued in books".*
Knowledge Manager Librarian 4:	*"It is somehow good because there are few social media tools used for advertisement on library services".*
Knowledge Manager Librarian 5:	*"Good in our service but not all libraries are aware of it"*
Marketing Librarian 1:	*"The level is good but not in all public libraries".*
Marketing Librarian 2:	*"Not promising in our province due to poor management".*
Audio-Visual Librarian 1:	*"The level is currently poor in our province".*
Audio-Visual Librarian 2:	*"It is very poor in our province".*
Lending Librarian 1:	*"I do not know because sometimes we are not being told of any new developments".*
Reference Librarian 1:	*"For now, it seems good because Facebook is currently used to promote library services".*

The study findings revealed that the level of awareness regarding the use of social media for managing IK in KwaZulu-Natal is generally low. This indicates that public libraries have not yet fully embraced online platforms to promote the information products and services they offer. The study findings agreed with Liu, Weng and Sanga (2019) who revealed that most librarians in the developing countries are not aware of social networking services' even the few that are aware are still struggling to discover the productive uses of these sites for library services. The study findings also concurred with Folarin

(2005) who reported that "most librarians in the developing countries are not aware of the impact and usefulness of social networking services, even the few that are aware are only aware of popular social media tools such as Facebook, WhatsApp, and Twitter." Wagner (2008) noted that aside from the lack of awareness which hamper librarians' use of communication and information technology, phobias for the use of the different Internet technologies is what drives many librarians to claim ignorance of the existence of such technologies with such functionalities. The study findings were affirmed by Plockey (2014) who argued that these are causing cultural gaps between many generations and denying the continent the power and knowledge that it has developed for centuries. While technological advancement constitutes a vital component of sustainable development, there is a conspicuous and persistent lack of African IK technological advancement in Africa, and a greater dependency on "western scientific" technologies. In that light, Chisita (2011) revealed that libraries in Southern Africa are a by-product of colonial rule and as such they have to grapple with reorienting and reengineering their services in line with the needs of the local people. These findings are supported by Owiny, Mehta and Maretzki (2014) whose views showed that many African libraries cannot afford the cost of documentation, preservation and dissemination of IK using technology.

What is the Current Status of Managing Indigenous Knowledge Using Social Media Networks in Your Library?

This question was about the current status of managing IK using social media networks in public libraries. This question assisted the current study in understanding the level of use of social media networks by individual public library. The following is the narration of the results:

Table 6. What is the current status of managing indigenous knowledge using social media networks in your library?

Participant	Responses
Data Capturer Librarian 1:	*"I can say it is not looking good because there are staff members who do not know that we use some of the social media networks to manage IK like only Facebook. All I can say is that the status is still poor".*
Data Capturer Librarian 2	*"It is promising even though we have a long way to go".*
Data Capturer Librarian 3:	*"It is poor at the moment because the library only utilises Facebook to try and manage IK".*
Data Capturer Librarian 4:	*"The management of indigenous knowledge is at the infant stage, and we have a lot to do in this digital age".*
Data Capturer Librarian 5:	*"The library only digitised information using computers and databases to disseminate using Facebook".*
Data Capturer Librarian 6:	*"In my view the status is poor in this library because we are hardly trained to use social media tools to manage indigenous knowledge".*
Data Capturer Librarian 7:	*"The current status is poor in the province and in our library".*
Data Capturer Librarian 8:	*"It is poor at the moment because the library only utilises Facebook, but the other platforms are not used".*
Knowledge Manager Librarian 1:	*"Only Facebook which shows the possibility of managing IK in our library".*

continued on following page

Table 6. Continued

Participant	Responses
Knowledge Manager Librarian 2:	*"The management of indigenous knowledge is better in our library because knowledge can be accessed through the social media Network and eBooks".*
Knowledge Manager Librarian 3:	*Not promising at the moment and seems low".*
Knowledge Manager Librarian 4:	*"Not good for now some employees don't believe in using social media tools prefer old formats".*
Knowledge Manager Librarian 5:	*"It is very bad because we do not have sections for using social media tools for indigenous knowledge".*
Marketing Librarian 1:	*"Quite poor in our service as they do not consider social media tools for indigenous knowledge".*
Marketing Librarian 2:	*"Our library pays little attention in the management of indigenous knowledge using social media networks".*
Audio-Visual Librarian 1:	*"Not promising as we are not trained to apply social media tools for managing IK and other subjects".*
Audio-Visual Librarian 2:	*"It is poor due to lack of training provision".*
Lending Librarian 1:	*"We are not being trained even when the information is available they do not share anything, so the current status is poor".*
Reference Librarian 2:	*"Some employees use social media tools not for managing indigenous knowledge but for entertainment purpose, so current the status is still very low".*

What Needs to be Done to Ensure That the Current Status of Managing Indigenous Knowledge Using Social Media Networks Improves in Public Libraries?

This question discussed ways of improving the current status of using social media networks to manage IK in public libraries. The study had to uncover various ways for improving the use of social media networks to manage IK. This question was meant to bring a solution to the low status of using social media networks in the management of IK by public libraries. The results are narrated on Table 7 below:

Table 7. What needs to be done to ensure that the current status of managing indigenous knowledge using social media networks improves in public libraries?

Participant	Responses
Data Capturer Librarian 1:	*"By being provided with social media networks relevant for the management of IK. Let us not use only one but let us explore because there are many of them".*
Data Capturer Librarian 2	*"By utilising several social media networks which can be used to create, preserve and disseminate of indigenous knowledge in public libraries".*
Data Capturer Librarian 3:	*"The management must allow the staff members to use social media networks and implement those that are suitable for managing IK".*
Data Capturer Librarian 4:	*"Public libraries need to be aware of social media networks and be orientated on how to use them"*

continued on following page

Table 7. Continued

Participant	Responses
Data Capturer Librarian 5:	*"Implementation of social media networks, training staff members to utilise these tools for the management of indigenous knowledge".*
Data Capturer Librarian 6:	*"It must be well promoted by teaching the library staff on how to use these platforms for retrieving and managing indigenous knowledge".*
Data Capturer Librarian 7:	*"Individual developments in terms of keeping up with new technological trends is crucial".*
Data Capturer Librarian 8:	*"The library needs to be updated with ICT facilities and librarians must have sufficient skills".*
Knowledge Manager Librarian 1:	*"The library staffs needs to be frequently trained in order to easily use these platforms"*
Knowledge Manager Librarian 2:	*"Provision of orientation regarding the use of social media networks for the management of indigenous knowledge".*
Knowledge Manager Librarian 3:	*"Certain programmes must be implemented by municipality to educate public libraries to use social media networks to manage indigenous knowledge".*
Knowledge Manager Librarian 4:	*"Introducing digital resources as to encourage employees to be familiar with social media networks".*
Knowledge Manager Librarian 5:	*"Certain standards and procedures must be implemented in order to use social media networks freely in public libraries".*
Marketing Librarian 1:	*"Conferences and workshops must be encouraged to library staff for learning and observation purposes".*
Marketing Librarian 2:	*"Library management must change the thinking capacity to meet the Morden world by engaging with researchers".*
Audio-Visual Librarian 1:	*"Improve library staff by hiring young and fresh graduates".*
Audio-Visual Librarian 2:	*"Orientation and training is required time to time in the usage of social media to manage IK".*
Lending Librarian 1:	*"Public libraries must hire young professionals to innovate new ideas and strategies".*
Reference Librarian 2:	*"Government needs to improve public libraries by offering new facilities and resources".*

The study found that the current status of managing IK among public libraries in KwaZulu-Natal is generally very low. This was demonstrated by many participants who agreed that they were not happy with the way public libraries in KwaZulu-Natal participated in the management of IK using social media networks. The study concurred with Gautam and Gupta (2014) who reported that library staff are also not adequately aware of the usefulness of social media applications since they are not properly knowledgeable nor trained on these applications. The study also corroborated with Carlson and Dreher (2018) who reported that the rapid rise of the use of social media as a means of social, cultural and political interaction among indigenous peoples and groups is an intriguing development. While this is not to suggest there is no digital inequality, it does counter any assumptions that indigenous peoples may have little interest in the possibilities of technology and the online environment. While recognising the significant challenges and costs for indigenous participation online, overall the collection highlights resistance to colonial discourses and policies, and the enormous scope for creativity, critique and imagining alternative futures as social media platforms enable connectivity beyond the colonial state. In that light, Reijerkerk and Nyitray (2023) recommended that librarians should acquire in-depth knowledge about social media so as to put it into more effective use as well as using it beyond personal or individual uses.

TYPES OF SOCIAL MEDIA NETWORKS USED BY YOUR PUBLIC LIBRARY FOR MANAGING INDIGENOUS KNOWLEDGE

This is the third objective of the study which discussed the types of social media networks used by public libraries to manage IK. It was crucial for the participants to spell out the types of social media networks being used in their public libraries to manage IK. As such, this section discusses several themes which emanated from this objective. They are discussed as follows:

Use of Social Media Networks to Manage Indigenous Knowledge in Public Libraries

The aim of this section was to uncover whether Bessie Head Public Library used social media networks for managing IK. It was essential for the study to ask participants to agree or disagree that they were using social media networks to manage IK. This question was meant to provide a clear view on the use of social media networks in managing IK. The participants' results are summarised below:

What Types of Social Media Networks are Used in Your Library for the Management of Indigenous Knowledge?

A follow-up question was asked to identify the specific social media networks used by public libraries to manage indigenous knowledge (IK). It's important to note that not all participants responded to this question, as some indicated they were not using social media networks for managing IK. As a result, twelve out of the nineteen participants answered this question. The responses are summarised in the table below:

Table 8. Types of social media networks for managing indigenous knowledge

Position in the library	*Types of social media*
Data Capturer Librarian 1:	*Facebook and YouTube*
Data Capturer Librarian 2:	*Facebook*
Data Capturer Librarian 3:	*Facebook*
Data Capturer Librarian 4:	*YouTube, Instagram and Facebook*
Data Capturer Librarian 5:	*YouTube and Facebook*
Data Capturer Librarian 6:	*Facebook, YouTube and Instagram*
Knowledge Manager Librarian 1:	*Facebook and YouTube*
Knowledge Manager Librarian 2:	*YouTube and Facebook*
Marketing Librarian ×2:	*Facebook YouTube*
Audio-Visual Librarian ×2:	*Facebook and YouTube*

Table 8 shows various social media networks that are used at Bessie Head Public Library for managing IK. The findings showed that the mostly commonly used social media network in the management of IK at Bessie Head Public library was Facebook, closely followed by YouTube and Twitter. The study findings were in line with Kaur and Manhas (2019) who reported that social media applications are widely accepted in various fields, and they are also making their way in the libraries. These study findings corroborated with Okuonghae (2018) whose study also reported that the wide use of popular social media enterprises such as Twitter and Facebook demonstrated the importance of this technology and the ways in which social media have changed the means of communication. Prabhakar and Rani (2017) also reported that social networking applications have emerged as a very interesting and powerful tool for libraries to deliver and meet the expectations of users; with increasing interest in visual platforms, such as YouTube, Flickr, Instagram and Pinterest (Ian McCallum, 2015). According to Bullas (2014), Gruzd and Staves (2011), libraries use social media networks such as twitter, Facebook, LinkedIn and WhatsApp.

Which Social Media Networks Do You Use for Capturing or Recording Indigenous Knowledge?

The aim of this question was to understand which social media tools were used by the participants for capturing or recording IK. It was important for the study to understand the types of social media networks currently used in Bessie Head Public Library for capturing and recording IK to provide a clear insight in the usage of social media networks. Content analyses were used to analyse the data collected. As such, this section discussed several themes, which emerged from this objective and are discussed follows:

Table 9. Social media networks used for capturing and recording IK

Position in the library	*Types of social media*
Data Capturer 1	*Facebook*
Data Capturer 2	*Facebook*
Data Capturer 3	*Facebook*
Data Capturer 4	*YouTube*
Data Capturer 5	*Facebook*
Data Capturer 6	*Facebook and YouTube*
Knowledge Manager Librarian 1	*Facebook and YouTube*
Knowledge Manager Librarian 2	*Twitter and Facebook*
Marketing Librarian×2	*Facebook, Instagram*
Audio-Visual Librarian×2	*Facebook and YouTube*

Table 9 shows the various social media networks that were being used at Bessie Head Public Library for capturing and recording indigenous knowledge. The findings showed that the mostly commonly used social media networks for capturing and recording of IK at Bessie Head Public Library were Facebook and YouTube. There was only one participant who also used Instagram. The study findings demonstrated that the participants agreed that they were using social media networks to manage IK in their public library. This was revealed by 12 of the participants who agreed that their library was using social media networks and only six (6) said they were not using social media networks. In supporting their views, the participants who agreed they were social media networks further mentioned different types of social media networks they were using to manage IK. The study showed that the social media networks used by the participants to manage IK were Facebook, YouTube and Instagram (see Table 5, in Chapter

Four). The study findings were in line with Kaur and Manhas (2019) who reported that social media applications are widely accepted in various fields, and they are also making their way in the libraries. These study findings corroborated with Okuonghae (2018) whose study also reported that the wide use of popular social media enterprises such as Twitter and Facebook demonstrated the importance of this technology and the ways in which social media have changed the means of communication. Prabhakar and Rani (2017) also reported that social networking applications have emerged as a very interesting and powerful tool for libraries to deliver and meet the expectations of users; with increasing interest in visual platforms, such as YouTube, Flickr, Instagram and Pinterest (McCallum, 2015). According to Bullas (2014) and Gruzd and Staves (2011), libraries use social media networks such as twitter, Facebook, LinkedIn and WhatsApp.

What is the Effectiveness of Social Media Networks in Disseminating of Indigenous Knowledge?

The main purpose of the question was to uncover the effectiveness of social media networks in disseminating IK. It must be noted that this question was only responded to by those who agreed that they were using social media networks in Bessie Head Public Library. This question helped to understand the effectiveness of social media networks in disseminating IK. The participants' results were narrated as follows:

Table 10. What is the effectiveness of social media networks in disseminating of indigenous knowledge?

Participant	Responses
Data Capturer Librarian 1:	*"They are effective because there is huge percentage of users which are easily reached by one post".*
Data Capturer Librarian 2	*"It is very good because it is easily shared by many users or patrons".*
Data Capturer Librarian 3:	*"Excellent because many young people are on these networks which can promote the effectiveness of the service".*
Data Capturer Librarian 4:	*"Very good as it able to reach a number of users in a short space of time".*
Data Capturer Librarian 5:	*"Poor some users cannot access share contents due internet data and poor network coverage".*
Data Capturer Librarian 6:	*"Disseminated content can provide knowledge to a wide audience".*
Data Capturer Librarian 7:	*"Disseminated contents can stay for a very long time and it very effective".*
Data Capturer Librarian 8:	*"All contents are disseminated as quickly as possible".*
Knowledge Manager Librarian 1:	*"It makes reaching widely to audience is possible with social media, very effective".*
Knowledge Manager Librarian 2:	*"Facebook and other networks provide unlimited range of dissemination of information".*
Knowledge Manager Librarian 3:	*"Conversion of content can sometimes be challenging as some format don't support the other format, which make it hard to disseminates, less effective".*
Knowledge Manager Librarian 4:	*"Some content lose quality when disseminated due to poor devices".*

The study demonstrated the social media networks used, were Facebook, YouTube, Google Docs, Instagram and Twitter - and these findings suggested that multiple social media networks are effective and reliable in disseminating IK. These findings agreed with those of Lwoga (2009) who also found that ICTs are effective networks for disseminating exogenous knowledge in the communities. These findings also agreed with the findings of Le Roux (2003) and Tihapi (2004) who reported that the use of the Internet in disseminating IK is crucial and does not only disseminate but also communicates and preserves IK. In that light, Lor (2003) also affirmed that Internet connectivity spread throughout communities can promote the awareness and appreciation of IK nationwide. Other authors who supported the current study are Maina (2012); Owiny, Mehta and Maretzki (2014). They emphasised that the Internet has removed physical (geographical) barriers, thus, there's a significant increase in cultural interaction. The researcher suggests that the availability of these networks depended largely on funds to acquire them. The study findings also demonstrated that there is hope for the survival of IK through the use of social media networks in managing IK. This is in line with the findings of Ngulube (2003) who reported that it is possible to manage IK using knowledge management approaches and new technologies.

FUTURE RESEARCH DIRECTIONS

Given the findings of this study, future research should be conducted in the following areas:

a) **Nationwide Assessment of Social Media Use in Indigenous Knowledge Management**
   - o A broader study should be conducted across all provinces in South Africa to evaluate the extent to which public libraries utilize social media networks in managing indigenous knowledge. This would provide comparative insights and highlight regional disparities in technology adoption.

b) **Effectiveness of Different Social Media Platforms**
   - o Further research should examine which social media platforms are most effective for preserving and disseminating indigenous knowledge. This could include an analysis of engagement levels, user preferences, and content reach among different platforms.

c) **Challenges and Barriers to Digital IK Management**
   - o A study should explore the challenges that libraries face in integrating social media for indigenous knowledge management, such as lack of digital skills, funding constraints, technological infrastructure, and ethical considerations.

d) **Collaboration Between Libraries and Indigenous Knowledge Holders**
   - o Research should investigate how public libraries can establish stronger partnerships with indigenous knowledge holders to ensure authentic, respectful, and community-driven knowledge-sharing through digital platforms.

e) **Impact of Social Media on Indigenous Knowledge Preservation**
   - o A longitudinal study should assess the long-term impact of social media use in managing and preserving indigenous knowledge, particularly in terms of knowledge retention, community engagement, and intergenerational transmission.

These future research areas would help in developing evidence-based strategies and policies to enhance the role of social media networks in safeguarding and promoting indigenous knowledge within public libraries.

CONCLUSION AND RECOMMENDATIONS

The study concluded that librarians at Bessie Head Public Library are aware of social media platforms for managing Indigenous Knowledge Systems (IKS). The participants demonstrated a good understanding of the application of social media networks for this purpose. They became aware of these platforms through various channels such as workshops, seminars, conferences, reading academic papers, browsing the internet, engaging with the Ulwazi programme, and studying the work of other researchers. However, the study found that, despite these awareness-raising activities, the overall level of awareness regarding the use of social media networks for managing IK in public libraries in KwaZulu-Natal was alarmingly low. This suggests that the methods employed to raise awareness have not been sufficiently effective in advancing the educational and practical use of these technologies in libraries. The study also identified various social media networks known to the participants. These included Facebook, Twitter, Google Docs, YouTube, Instagram, MySpace, NING, and WhatsApp. These platforms are commonly recognised as useful for managing and sharing IK in public libraries. Despite the availability of these tools, the study revealed a significant gap between knowledge of these platforms and their practical application in managing IK in KwaZulu-Natal's public libraries. This gap is particularly concerning given that the province is rich in indigenous knowledge practices, with many forms of IK still actively practiced and preserved by local communities.

Given these findings, the study emphasises the urgent need for librarians to engage more with research and professional development programs that will expose them to the latest trends in Library and Information Science (LIS). As the world continues to evolve in the context of the 4th Industrial Revolution, it is crucial that librarians are equipped with the necessary skills and awareness to effectively leverage social media platforms in their work. This knowledge and skillset could greatly benefit the management of IK in public libraries and enhance the services these libraries offer to their communities. Therefore, the study recommends that public libraries in KwaZulu-Natal invest in targeted training and professional development to improve librarians' awareness and utilisation of social media platforms for managing indigenous knowledge.

REFERENCES

Adebayo, C., Oluwaseye, J., & Adeyemo Cln, A. A. (2017). Documentation and dissemination of indigenous knowledge by library personnel in selected research institutes in Nigeria. *Library Philosophy and Practice (e-journal)*, 16-28.

Afful-Arthur, P., Kwafoa, P. N. Y., Ampah-Johnston, M., & Mensah, V. (2022). Managing and accessing indigenous knowledge for national development: The role of academic libraries in Ghana. *Information Development*, *38*(4), 535–548. DOI: 10.1177/02666669211009916

Ajani, Y. A., Oladokun, B. D., Olarongbe, S. A., Amaechi, M. N., Rabiu, N., & Bashorun, M. T. (2024). Revitalizing indigenous knowledge systems via digital media technologies for sustainability of indigenous languages. *De Gruyter*, *53*(1), 35–44. DOI: 10.1515/pdtc-2023-0051

Amedie, J. (2015). *The impact of social media on society. Advanced writing: Pop Culture Intersections, 2*. Retrieved April, 19, 2021, from https://scholarcommons.scu.edu/engl_176/2/

Baruah, T. D. (2012). Effectiveness of social media as a tool of communication and its potential for technology enabled connections: A micro-level study. *International Journal of Scientific and Research Publications*, *2*(5), 1–10.

Becerra-Fernandez, R. S. (2001). Organizational knowledge management: A contingency perspective. *Journal of Management Information Systems*, *18*(1), 23–55. DOI: 10.1080/07421222.2001.11045676

Botangen, K. A., Vodanovich, S., & Yu, J. (2017). Preservation of indigenous culture among indigenous migrants through social media: the Igorot peoples. *Proceedings of the 50th Hawaii International Conference on System Sciences*, 2303–2312. Retrieved May 27, 2022, from: https://doi.org/DOI: 10.24251/HICSS.2017.278

Bratianu, C. (2010). A critical analysis of Nonaka's model of knowledge dynamics. *Electronic Journal of Knowledge Management*, *8*(2), 193–200.

Bratianu, C., & Orzea, I. (2010). Organizational knowledge creation. *Management. Marketing Challenges for Knowledge Society*, *5*(3), 41–62.

Broadbent, M., & Weill, P. (2003). Improving business and information strategy alignment: Learning from the banking industry. *IBM Systems Journal*, *32*(1), 162–179. DOI: 10.1147/sj.321.0162

Bullas, J. (2014). An exploratory study of Indian University students' use of social networking websites: Implications for the workplace. *Business Communication 32 Quarterly*, 72(1), 105-110.

Carlson, B., & Dreher, T. (2018). Introduction: Indigenous innovation in social media. *Media International Australia, Incorporating Culture & Policy*, *169*(1), 16–20. DOI: 10.1177/1329878X18803798

Chaputula, A. H., Abdullah, H., & Mwale, B. (2020). Proliferation of social media in academic libraries: Use of WhatsApp as a platform for providing library services. *Library Management*, *41*(8/9), 717–729. DOI: 10.1108/LM-04-2020-0075

Chigwada, J., & Chiparausha, B. (2020). Efficacy of culture houses and centres in the acquisition, preservation, and dissemination of indigenous knowledge in Zimbabwe. In: Information Resources Management Association (ed.) *Indigenous Studies: Breakthroughs in Research and Practice*. Hershey, PA: IGI Global, 498–518. DOI: 10.4018/978-1-7998-0423-9.ch027

Chisita, C. T. (2011) Role of libraries in promoting the dis-semination and documentation of indigenous agricultural information: case study of Zimbabwe. *In: World Library and Information Congress: 77th IFLA General Conference and Assembly*. Retrieved May 10, 2023, from http://conference. ifla. org/past-wlic/2011/78-chisita-en

Clark, D. (2004). Apprentice and applied theoretical knowledge. *Educational Philosophy and Theory*, *36*(5), 509–521. DOI: 10.1111/j.1469-5812.2004.087_1.x

Clark, K. B., & Fujimoto, T. (1991). *Product development performance: Strategy, organization, and management in the world auto industry*. Harvard Business Press.

Corti, E., & Storto, C. (2000). Knowledge creation in small manufacturing firms during product innovation: An empirical analysis of cause–effect relationships among its determinants. *Enterprise and Innovation Management Studies*, *1*(3), 245–263. DOI: 10.1080/14632440010023190

Crawford, C. M., & Di Benedetto, A. (2008). *New products management* (9th ed.). McGraw-Hill Irwin.

Dlamini, P., & Ocholla, D. N. (2018). Information and communication technology tools for managing indigenous knowledge in KwaZulu-Natal Province, South Africa. *African Journal of Library Archives and Information Science*, *28*(2), 137–153.

Dlamini, P. N. (2017). Use of information and communication technology tools to capture, store and disseminate indigenous knowledge: A literature review. In Ngulube, P. (Ed.), *Handbook of Research on Theoretical Perspectives on Indigenous Knowledge Systems in Developing Countries*. IGI Global Publication. DOI: 10.4018/978-1-5225-0833-5.ch010

Duncan, N. B. (1995). Capturing of information technology infrastructure: A study of resource characteristics and their measure. *Journal of Management Information Systems*, *12*(2), 37–57. DOI: 10.1080/07421222.1995.11518080

Folarin, J. (2005). The impact of knowledge sharing, organizational capacity and partnership quality on IS outsourcing success. *Information & Management*, *38*(5), 323–335.

Greyling, B. (2010). *A model for community participation in African libraries to preserve indigenous knowledge*. Retrieved May 13, 2023, from https://www.inst.at/trans/17Nr/8-2/8-2_greyling.htm

Greyling, E., & Zulu, S. (2010). Content development in an indigenous digital library: A case study in community participation. *IFLA Journal*, *36*(1), 30–39. DOI: 10.1177/0340035209359570

Gruzd, A., & Staves, K. (2011). Trends in scholarly use of online social media. Position paper presented at the workshop on changing dynamics of scientific collaboration, *the 44th Annual Hawaii International Conference on System Sciences (HICSS)*. Retrieved August 24, 2022, from: https://dalspace.library.dal.ca/handle/10222/14427

Gupta, R. K., Gautam, J. N., & Khare, V. P. (2014). Awareness and use of social media applications among library staff of power sector organizations. [ALIS]. *Annals of Library and Information Studies, 61*(4), 320–331.

Gwerevende, S., & Mthombeni, Z. M. (2023). Safeguarding intangible cultural heritage: Exploring the synergies in the transmission of indigenous languages, dance and music practices in Southern Africa. *International Journal of Heritage Studies, 29*(5), 398–412. DOI: 10.1080/13527258.2023.2193902

Hangshing, J. & Laloo, B. (2021). Role of libraries in preservation and accessibility of indigenous knowledge: A study of University Libraries in Northeast India. *Library Philosophy and Practice (e-journal)*. 5254.

Holste, J. S., & Fields, D. (2010). Trust and tacit knowledge sharing and use. *Journal of Knowledge Management, 14*(1), 128–140. DOI: 10.1108/13673271011015615

IFLA. (2002). *IFLA Statement on indigenous traditional knowledge.* Retrieved June 10, 2024, from https://www.ifla.org/publications/ifla-statement-on-indigenous-traditional-knowledge

Isah, A., Bashorun, M. T., & Omopupa, K. T. (2012). Libraries and preservation of indigenous knowledge in developing countries: The Nigeria experience. In *Library and Information Science in Developing Countries* (pp. 96–106). Contemporary Issues. DOI: 10.4018/978-1-61350-335-5.ch008

Jain, P. (2014). Indigenous knowledge management in Botswana using ICT applications. In *Concepts and Advances in Information Knowledge Management* (167-191). DOI: 10.1533/9781780634357.2.167

Jaleel, S., & Verghis, A. M. (2015). Knowledge creation in constructivist learning. *Universal Journal of Educational Research, 3*(1), 8–12. DOI: 10.13189/ujer.2015.030102

John, H. C., Egbeyemi, T. A., & Oniyide, F. D. (2020). Social networking technologies in libraries and the role of librarians as agent for repositioning Nigerian education. *International Journal of Library and Information Studies, 10*(2), 75–84.

Kale, P., & Singh, H. (2007). Building firm capabilities through learning: The role of the alliance learning process in alliance capability and firm-level alliance success. *Strategic Management Journal, 28*(10), 981–1000. DOI: 10.1002/smj.616

Kaplan, A. M., & Haenlein, M. (2010). Users of the world, unite! the challenges and opportunities of social media. *Business Horizons, 53*(1), 59–68. DOI: 10.1016/j.bushor.2009.09.003

Kogut, B., & Zander, U. (1992). Knowledge of the firm, combinative capabilities, and the replication of technology. *Organization Science, 3*(3), 383–397. DOI: 10.1287/orsc.3.3.383

Le Roux, C. J. B. (2003). Tapping indigenous knowledge on the world-wide web. *Indilinga, 2*(1), 107–113.

Lee, H., & Choi, B. (2003). Knowledge management enablers, processes, and organizational performance: An integrative view and empirical examination. *Journal of Management Information Systems, 20*(1), 179–228. DOI: 10.1080/07421222.2003.11045756

Liu, Y., Weng, C., Li, H., & Sanga, G. (2019). Social media-based digital storytelling for preserving oral tradition: A case study in Tibetan Culture. *Digital Scholarship in the Humanities, 34*(1), 165–179.

Lor, P. J. (2003). National libraries and the digital divide. *Mousaion*, *21*, 62–78.

Lwoga, E. D., Ngulube, P., & Stilwell, C. (2017). Indigenous knowledge management practices in indigenous organizations in South Africa and Tanzania. In: Ngulube P (ed), *Handbook of research on social, cultural, and educational considerations of indigenous knowledge in developing countries*. IGI Global, 181–201. Retrieved October 14, 2021, from: https://doi.org/DOI: 10.4018/978-1-5225-0838-0.ch010

Lwoga, E. T., & Ngulube, P. (2008). Managing indigenous and exogenous knowledge through information and communication technologies for agricultural development and achievement of the UN Millennium Development Goals, In B. Njobvu S. Koopman (Eds.), *Libraries and information services towards the attainment of the UN Millennium Development Goals*. Berlin: Walter de Gruyter.

Lwoga, E. T., Ngulube, P., & Stilwell, C. (2010). Managing indigenous knowledge for sustainable agricultural development in developing countries: Knowledge management approaches in the social context. *The International Information & Library Review*, *42*(3), 174–185. DOI: 10.1080/10572317.2010.10762862

Maina, C. K. (2012). Traditional knowledge management and preservation: Intersections with Library and Information Science. *The International Information & Library Review*, *44*(1), 13–27. DOI: 10.1080/10572317.2012.10762911

Marley, I. R. (2012). Investigating the appropriateness of the theory of organisational knowledge creation as a management model for practice-led research. *Journal of Literacy Criticism. Comparative Linguistic and Literacy Studies*, *31*(1), 1–10.

Marra, M. (2004). The contribution of evaluation to socialization and externalization of tacit knowledge: The case of the World Bank. *Evaluation*, *10*(3), 263–283. DOI: 10.1177/1356389004048278

Masekoameng, C. M. (2022). *Digital preservation and access to the South African documentary national heritage at the National Library of South Africa. Doctoral dissertation*, University of Limpopo. http://ulspace.ul.ac.za/handle/10386/3898

Masenya, T. M. (2024). Revitalization and digital preservation of indigenous knowledge systems for sustainable development of indigenous communities in South Africa. *The Serials Librarian*, •••, 1–17. DOI: 10.1080/0361526X.2023.2277962

McCallum, I. (2015). Use of social media by the library: Current practices and future opportunities. A White Paper from Taylor and Francis. *The Australian Library Journal*, *64*(2), 161–162. DOI: 10.1080/00049670.2015.1040364

Mhlongo, M., & Ngulube, P. (2018). Public libraries as conduits for indigenous knowledge in South Africa. *Library Philosophy and Practice*, (e-journal). Retrieved June 23, 2020, from: https://digitalcommons.unl.edu/libphilprac/1757/

Mohamedbhai, G. (2013). Indigenous knowledge must be harvested for development. *Traditional African Clinic*, *10*(262), 12–13.

Mosoti, Z., & Masheka, B. (2010). Knowledge management: The case for Kenya. *The Journal of Language. Technology and Entrepreneurship in Africa*, *2*(1), 107–133.

Mugwisi, T., Jiyane, G. V., & Fombad, M. C. (2018). Public libraries as facilitators of information services: A case study of selected libraries in KwaZulu-Natal. *Information Development*, *34*(1), 31–43. DOI: 10.1177/0266666916672718

Neuman, W. (2014). *Social research methods: Qualitative and quantitative approaches* (7th ed.). Pearson Education.

Ngenge, G. A. (2003). *Gender equity and new information and communication technologies*. Retrieved May 07, 2022, from: www.sdnp.undp.org/sdncmr/fawecam/ngenge.htm

Ngulube, P. (2003). Using the SECI knowledge management model and other networks to communicate and manage tacit indigenous knowledge. *Innovation*, *27*(1), 21–30.

Nonaka, I. (1994). A dynamic theory of organizational knowledge creation. *Organization Science*, *5*(1), 14–37. DOI: 10.1287/orsc.5.1.14

Nonaka, I., & Konno, N. (1998). The concept of "Ba": Building a foundation for knowledge creation. *California Management Review*, *3*(40), 40–54. DOI: 10.2307/41165942

Nonaka, I., & Takeuchi, H. (1995). *The knowledge-creating company: How Japanese companies create the dynamics of innovation*. Oxford University Press. DOI: 10.1093/oso/9780195092691.001.0001

Nonaka, I., Umemoto, K., & Sasaki, K. (1998). Three tales of knowledge-creating companies. In von Krogh, G., Roos, J., & Kleine, D. (Eds.), *Knowing in firms. Understanding, managing and measuring knowledge* (pp. 146–172). Sage. DOI: 10.4135/9781446280256.n7

Nonaka, I., Umemoto, K., & Sasaki, K. (1998). Three tales of knowledge-creating companies. In Von Krogh, G., Roos, J., & Kline, D. (Eds.), *Knowing in Firms. Understanding, Managing and Measuring Knowledge*. Sage. DOI: 10.4135/9781446280256.n7

Nonaka, I., & Von Krogh, G. (2009). Perspective-tacit knowledge and knowledge conversion: Controversy and advancement in organizational knowledge creation theory. *Organization Science*, *20*(3), 635–652. DOI: 10.1287/orsc.1080.0412

Okuonghae, O. (2018). Librarians' awareness of social media usage for informal scientific communication in university libraries in south-south. *Library Philosophy and Practice. (e-journal)*, 1–19. Retrieved May 14, 2022, from: https://www.thefreelibrary.com/Librarians%27+Use+Of+Social+Media+For+Informal+Scientific+Communication...-a0539387789

Oosthuizen, J. H., Ungerer, M., & Volschenk, J. (2023). A fourth industrial revolution integrated intelligence taxonomy for top management. *Journal of Contemporary Management*, *20*(1), 404–443. DOI: 10.35683/jcm1018.202

Owiny, S. A., Mehta, K., & Maretzki, A. N. (2014). The use of social media technologies to create, preserve, and disseminate Indigenous knowledge and skills to communities in East Africa. *International Journal of Communication*, *8*(10), 234–247. Retrieved March 05, 2020, from https://ijoc.org/index.php/ijoc/article/view/1667

Peltokorpi, V., Nonaka, I., & Kodama, M. (2007). NTT DoCoMo's launch of I-mode in the Japanese mobile phone market: A knowledge creation perspective. *Journal of Management Studies*, *44*(1), 50–72. DOI: 10.1111/j.1467-6486.2007.00664.x

Pescaru, M. (2018). The importance of the socialization process for the integration of the child in the society. *Revista Universitară de Sociologie*, *14*(2), 18–26.

Pham, L. (2018). A review of key paradigms: Positivism, interpretivism and critical inquiry. *School of Education, Med Program,* 1–7. Retrieved July 20, 2021, from: DOI: 10.13140/RG.2.2.13995.54569

Plockey, F. D. (2014). The role of Ghana public libraries in the digitization of indigenous knowledge: Issues and prospects. *The Journal of Pan African Studies*, *6*(10), 20–36.

Plockey, F. D. D. (2015). Indigenous knowledge production, digital media and academic libraries in Ghana. *The Journal of Pan African Studies*, *8*(4), 32–44.

Prabhakar, S. V. R., & Rani, S. M. (2017). Influence of social networking sites on library and information centers. [IJLIS]. *International Journal of Library and Information Science*, *6*(1), 83–87.

Pretty, J., & Vodouhe, S. (1997). Using rapid or participatory rural appraisal. In Swanson, B., Bentz, R. and Sofranko, A. (eds.) 1997. *Improving agricultural extension, 47-55*: A reference manual. Food and agriculture organisation of the United Nations, Rome. Retrieved May 20, 2022, from: https://www.fao.org/docrep/W5830E/w5830e08.htm

Priya, R. M., & Rabindra, K. M. (2010). Documenting indigenous traditional knowledge in Odish*a. Orissa Review*, •••, 99–103. Retrieved June 24, 2022, from https://www.academia.edu/76494585/Documenting_Indigenous_Traditional_Knowledge_In_Odisa

Reijerkerk, D., & Nyitray, K. J. (2023). Moving indigenous knowledge boundaries: A Review of Library and Archival Collection Literature since the Enactment of the Native American Graves Protection and Repatriation Act (NAGPRA). *Collection Management*, *48*(1), 22–47. DOI: 10.1080/01462679.2022.2033144

Schulze, A., & Hoegl, M. (2006). Knowledge creation in new product development projects. *Journal of Management*, *32*(2), 210–236. DOI: 10.1177/0149206305280102

Shu, C., Page, A. L., Gao, S., & Jiang, X. (2012). Managerial ties and firm innovation: Is knowledge creation a missing link? *Journal of Product Innovation Management*, *29*(1), 125–143. DOI: 10.1111/j.1540-5885.2011.00883.x

Singh, G., & Pandey, N. (2017). Role and impact of media on society: A sociological approach with respect to demonetisation. *IMPACT: International Journal of Research in Humanities. Arts and Literature*, *5*(10), 127–136.

Stilwell, C. (2010). Understanding indigenous knowledge: Bridging the knowledge gap through a knowledge creation model for agricultural development. *South African Journal of Information Management*, *12*(1), 1–8.

Tihapi, W. R. (2004). *The use of intellectual property as a form of information warfare with specific reference to indigenous knowledge*: A critical analysis and proposed solution. (PhD Thesis), University of Pretoria, South Africa.

Tsai, M., & Lee, K. (2006). A study of knowledge internalization: From the perspective of learning cycle theory. *Journal of Knowledge Management, 10*(3), 57–71. DOI: 10.1108/13673270610670858

Yunnus, F. (2017). *Preservation of indigenous knowledge (IK) by public libraries in Westcliff, Chatsworth, Durban*. Retrieved July 20, 2020, from https://core.ac.uk/display/154920975

Zimu-Biyela, A. N. (2016). *The management and preservation of indigenous knowledge in Dlangubo village in KwaZulu-Natal, South Africa*. (Doctoral dissertation, University of South Africa), South Africa.

KEY TERMS AND DEFINITIONS

Fourth Industrial Revolution: It is the current era of rapid technological advancements marked by the fusion of technologies that are transforming industries, economies, and societies.

Indigenous Knowledge: It refers to practices, and beliefs developed by indigenous communities through their long-standing interaction with their environment and cultural traditions.

Knowledge Creation: It is the process of generating new ideas and insights often through collaboration, interaction, and the transformation of existing knowledge.

Public Library: It is a library that is accessible by general public and provides access to knowledge, information, and resources often free of charge.

Social Media Networks: Are online platforms that facilitate social interaction and communication, allowing users and organizations to connect and share knowledge.

Chapter 15
Integrating Indigenous Knowledge Systems in the Fourth Industrial Revolution (4IR):
Preserving Indigenous Knowledge Through 4IR

Margaret Chauke
https://orcid.org/0000-0002-6402-9801
University of South Africa, South Africa

ABSTRACT

This chapter examines the integration of Indigenous Knowledge Systems (IKS) within the Fourth Industrial Revolution (4IR), highlighting the synergy between advanced technologies and traditional knowledge for preserving cultural heritage. Using a multidisciplinary approach and case studies, it demonstrates how technologies like Artificial Intelligence (AI), blockchain, and the Internet of Things (IoT) can safeguard Indigenous knowledge. Specific applications include AI for preserving endangered languages, blockchain for ensuring the authenticity of cultural artifacts, and IoT for monitoring sacred sites. The chapter also explores ethical considerations such as data sovereignty, intellectual property rights, and cultural commodification risks. By addressing these issues, it provides a thorough analysis of the challenges and opportunities at the intersection of 4IR technologies and Indigenous perspectives, contributing to discussions on sustainable development and cultural preservation for a more inclusive technological future.

INTRODUCTION

The Fourth Industrial Revolution (4IR) represents a transformative era driven by the convergence of digital, biological, and physical systems, reshaping industries, economies, and societies (Schwab, 2018). Technologies such as artificial intelligence (AI), blockchain, and the Internet of Things (IoT)

DOI: 10.4018/979-8-3693-7964-6.ch015

have enhanced efficiency, productivity, and interconnectivity (Xu, David & Kim, 2018). However, these advancements also pose challenges in preserving and promoting indigenous knowledge systems (IKSs) (Donkor & Mearns, 2022). IKS encompass the accumulated knowledge, practices, and worldviews of indigenous communities, forming the foundation of cultural identity, environmental stewardship, and sustainable resource management (Dudgeon & Bray, 2019). Yet, globalisation and rapid technological evolution threaten their survival, raising concerns about cultural erosion, data sovereignty, and heritage commodification (Mazzocchi, 2022). This chapter explores the interplay between 4IR technologies and IKS, assessing both risks and opportunities. While AI supports the revitalisation of endangered languages, blockchain secures indigenous intellectual property, and IoT monitors holy sites, their implementation often reflects Western-centric approaches misaligned with indigenous epistemologies (Bird & Yibarbuk, 2024; Giannossa et al., 2020). Decolonising technology applications is essential to ensure indigenous communities maintain control over their knowledge systems (Ofosu-Asare, 2024). Ethical concerns surrounding digitalisation and indigenous data sovereignty remain central to this discourse. While digitisation enhances accessibility, it also risks exploitation and misappropriation without proper legal frameworks and ethical safeguards (Tsosie, 2021). This chapter examines mechanisms for equitable knowledge governance, such as community-led initiatives and participatory technology design. Governments and organisations such as UNESCO advocate for inclusive digital heritage policies, yet meaningful indigenous representation in decision-making remains limited (UNESCO, 2020). This chapter critiques existing frameworks and highlights successful case studies where indigenous communities have co-designed technological solutions aligned with their values (Labadi, 2024). Finally, fostering inclusive dialogues between 4IR proponents and indigenous communities is crucial. Rather than imposing technology, participatory approaches should empower indigenous groups to shape technological advancements that protect their heritage (Mawere & Mukonza, 2024). This chapter contributes to scholarly discussions on cultural preservation, social justice, and sustainability, advocating for a paradigm where indigenous epistemologies inform technological development.

CHAPTER OBJECTIVES

This chapter aims to critically examine the integration of indigenous knowledge systems into the technological landscape, ensuring that innovation aligns with cultural preservation, community empowerment, and sustainable development. The following objectives guide this exploration:

- Preservation of indigenous knowledge through 4IR technologies: to assess how Artificial Intelligence (AI), Blockchain, and Internet of Things (IoT) can safeguard indigenous knowledge, cultural practices, and heritage in the digital era.
- Ethical considerations in digital integration: to examine challenges such as data sovereignty, intellectual property rights, and cultural commodification, emphasising responsible innovation to prevent misappropriation.
- Artificial intelligence and Blockchain for language and cultural heritage: to investigate AI-driven tools for documenting endangered languages and blockchain's role in authenticating and protecting cultural artefacts.
- IoT applications in safeguarding indigenous sites: to explore IoT-based environmental monitoring and security systems to protect holy sites from ecological risks and unauthorised access.

- IKS and 4IR for sustainable development: to evaluate how integrating 4IR technologies with indigenous knowledge contributes to cultural resilience and global sustainability.

THEORETICAL FRAMEWORK

The integration of indigenous knowledge systems (IKS) within the fourth industrial revolution (4IR) requires a theoretical framework that ensures ethical, equitable, and culturally sensitive data management. Indigenous data sovereignty (IDS) serves as the foundation, affirming the rights of indigenous communities to govern their data, aligning practices with their values, priorities, and self-determination (Bowman & Bremner, 2024). This chapter's theoretical framework is underpinned by three key IDS principles:

Control Over the Data Ecosystem

Indigenous data sovereignty asserts that indigenous communities must retain full authority over data throughout its lifecycle creation, management, analysis, and dissemination. This control safeguards the integrity and cultural significance of indigenous knowledge while preventing misappropriation and exploitation. Self-determined data governance enables communities to protect intellectual property, assert rights over traditional knowledge, and uphold ethical data practices (Wingara, 2018). Indigenous communities use data for governance, cultural preservation, and advocacy. IDS empowers them to collect and analyse data on land use, climate, and biodiversity, ensuring culturally relevant decision-making. Technologies such as blockchain and AI-driven repositories strengthen long-term data stewardship by securing indigenous knowledge from unauthorised access while ensuring accurate representation (Kukutai & Taylor, 2019).

Contextual and Disaggregated Data

IDS emphasise the importance of data that reflects the socio-cultural, economic, and environmental realities of specific indigenous groups. Disaggregated data enables tailored policies and interventions, ensuring equitable resource allocation in key sectors such as healthcare, education, and environmental management. By integrating culturally responsive methodologies, IDS reinforce self-determination and supports sustainable development (Hudson et al., 2023).

Empowerment Through Data

A core principle of IDS is empowering indigenous communities by ensuring full ownership and control over their data. This autonomy fosters informed decision-making in economic development, land rights, and heritage preservation while preventing external exploitation. IDS also serve as a tool for resisting colonial data frameworks, promoting data justice, and ensuring indigenous knowledge remains protected and community-driven (Oguamanam, 2020).

RESEARCH METHODOLOGY

Qualitative Case Study Design

This chapter employs a qualitative case study design to explore the integration of indigenous knowledge systems (IKS) with fourth industrial revolution (4IR) technologies, specifically artificial intelligence (AI), blockchain, and the Internet of Things (IoT). A case study is well-suited for investigating complex phenomena in real-life contexts, particularly when conceptual boundaries are fluid (Yin, 2017). Qualitative case studies align with constructivist paradigms, which recognise multiple realities and emphasise participants' perspectives (Guba & Lincoln, 1994). This approach is especially relevant to IKSs, given their deep cultural, social, and historical roots. By examining how 4IR technologies can support the preservation of indigenous knowledge, this chapter also addresses ethical considerations, including data sovereignty, intellectual property rights, and equitable knowledge-sharing (Guba & Lincoln, 1994; Yin, 2017). The chapter adopts a comparative case study approach, drawing on examples from South Africa, Australia and New Zealand to assess variations in technology adoption and its implications for indigenous communities (Rule & Yazan, 2024). This comparison provides a nuanced understanding of how AI, blockchain, and IoT are being adapted to safeguard and promote indigenous knowledge in different sociopolitical and technological environments.

Multidisciplinary Approach

The study integrates insights from multiple disciplines, including anthropology, computer science, and ethics, to provide a comprehensive analysis of IKSs in the context of 4IR technologies. Given the complex interplay between traditional knowledge and technological innovations, a multidisciplinary perspective is essential for addressing key concerns such as data sovereignty and intellectual property rights (Flick, 2017). By synthesising contributions from both the humanities and technology-related fields, the study offers a holistic framework for understanding how modern innovations can support the safeguarding of indigenous knowledge (Oguamanam, 2022).

Community-Led Research

A community-led research approach was adopted to ensure Indigenous scholars and practitioners played a central role in the research process. This method prioritises cultural sensitivity and aligns with the values and priorities of Indigenous communities, fostering trust and long-term sustainability in technology integration (Williams & Shipley, 2023). By centering Indigenous participation, the study ensures that IKS are respected and that technological advancements complement rather than replace traditional knowledge systems (Chilisa, 2019). In addition, this approach reinforces indigenous ownership of knowledge and enhances the ethical development of digital platforms designed to preserve cultural heritage (Smyth, Martin & Downing, 2025).

ETHICAL INTEGRATION OF 4IR TECHNOLOGIES IN PRESERVING INDIGENOUS KNOWLEDGE

The emergence of fourth industrial revolution (4IR) technologies, including artificial intelligence (AI), blockchain, and the Internet of Things (IoT), presents both opportunities and challenges for preserving indigenous knowledge systems (IKS). These technologies offer potential benefits such as cultural documentation, artefact authentication, and environmental monitoring. However, their implementation often disregards indigenous perspectives, raising concerns about data sovereignty, intellectual property rights, and cultural autonomy (Donkor & Mearns, 2022; Mazzocchi, 2022). A key issue is the lack of indigenous control over how their knowledge is digitised, stored, and utilised. External entities frequently appropriate indigenous knowledge without consent or equitable benefit-sharing, leading to cultural commodification and exclusion from decision-making (Kukutai & Taylor, 2019). AI-driven language preservation, blockchain-based artefact verification, and IoT-enabled environmental monitoring require governance frameworks that respect indigenous values and epistemologies to prevent exploitation and ensure equitable participation (Bird & Yibarbuk, 2024). Without ethical and inclusive integration, 4IR technologies risk deepening historical marginalisation rather than fostering empowerment. This chapter explores strategies for leveraging emerging technologies to safeguard and revitalise IKSs while promoting indigenous data sovereignty, cultural sustainability, and community-led technological adaptation (Guarda, Portela & Diaz-Nafria, 2023).

CASE STUDIES FROM AUSTRALIA, NEW ZEALAND AND SOUTH AFRICA

The integration of fourth industrial revolution (4IR) technologies has significantly contributed to the preservation and promotion of indigenous knowledge systems (IKSs). From language revitalisation to the protection of cultural artefacts and holy sites, technological advancements empower indigenous communities to safeguard their heritage while navigating the digital era. However, the digitisation of indigenous knowledge raises concerns about ownership, commercialisation, and cultural appropriation. Who controls the digitalised knowledge, and how can indigenous communities ensure equitable benefits from technological innovations? This section presents three case studies illustrating the transformative impact of artificial intelligence (AI), blockchain, and the Internet of Things (IoT) in South Africa. The first explores AI-driven initiatives, such as the Bantu Language Technology Project (BLTP), which supports the revitalisation of indigenous languages. The second highlights Blockchain's role in authenticating and protecting traditional crafts, ensuring fair recognition and economic benefits for indigenous artists. The third examines how IoT technologies help monitor and preserve holy sites, reinforcing indigenous control over cultural landscapes. These cases also integrate successful models from Australia to draw comparisons and highlight best practices

Case Study 1: Artificial Intelligence in Language Preservation

In South Africa, the Bantu Language Technology Project (BLTP), led by the University of Pretoria, utilises AI to preserve and revitalise endangered indigenous languages such as isiXhosa, isiZulu, and Sesotho. By developing speech recognition and language-processing tools, the project facilitates the digitisation and documentation of these languages, making them accessible through educational resources and mobile applications (Mlambo & Matfunjwa, 2024). The initiative aligns with global efforts such as Canada's *First Voices* and Google's *AI for Indigenous Languages* project, which similarly aim to prevent digital disenfranchisement by ensuring indigenous ownership of digital language data. Despite these advances, indigenous language preservation faces challenges such as:

- Dominance of English in education and media – English remains the primary language of instruction in many South African schools, limiting the formal development of indigenous languages.
- Limited digital presence – The absence of indigenous languages in mainstream digital spaces reduces their visibility.
- Urbanisation and globalisation – Migration to urban areas promotes the use of English for economic and social mobility.
- Lack of standardisation and formal recognition – Many indigenous languages struggle with standardised writing systems and institutional support.
- Intergenerational language shift – Parents prioritise English over indigenous languages, affecting transmission to younger generations.

The Bantu Language Technology Project (BLTP), plays a crucial role in reversing these trends by enhancing the digital presence of indigenous languages, ensuring their continued relevance in modern society. In New Zealand, Te Hiku Media developed an automatic speech recognition (ASR) model for the Māori language, Te Reo, achieving a 92% accuracy rate surpassing international tech giants. By ensuring Indigenous ownership of the dataset, this model protects against digital exploitation, setting an ethical standard for similar AI-driven language preservation projects (Kukutai et al., 2024; Soylu & Şahin, 2024). Policy recommendations:

- Community-owned digital platforms – indigenous communities must retain control over AI-generated linguistic data.
- Free, Prior, and Informed Consent (FPIC) – AI projects should ensure transparency and ethical use of indigenous language data.
- Funding for indigenous-led tech projects – Investment in indigenous-driven AI initiatives is crucial for sustainable linguistic preservation.

Case Study 2: Blockchain and Cultural Artefacts

South Africa has embraced blockchain technology for the protection and authentication of indigenous cultural artifacts. The Custodianship Blockchain Initiative (CBI), launched in 2021, uses blockchain to authenticate traditional crafts such as beadwork, pottery, and textile designs by the Ndebele and Zulu communities (Masenya, 2023). The system prevents counterfeit products from entering the market and ensures fair economic returns for Indigenous artists. However, the commercialisation of indigenous crafts raises concerns about cultural appropriation, as companies may exploit traditional designs without returning benefits to indigenous creators. Australia's First Nations Blockchain Pty Ltd has developed a global blockchain hub to authenticate indigenous art, preventing exploitation while ensuring provenance and fair trade. Their patented *tag/track/trace/trade* platform upholds transparency and protects Indigenous cultural intellectual property (The Australian Jewish News, 2024).

Challenges and Considerations

- Exploitation risks: Companies may profit from Indigenous art without returning benefits to the communities.
- Ownership and control: Who governs the blockchain authentication process?
- Equitable benefit distribution: Ensuring that economic benefits flow back to Indigenous artisans rather than external stakeholders.

Policy Recommendations

- Adaptation of intellectual property rights (IPR) systems – Legal frameworks should protect collective Indigenous knowledge from exploitation.
- Ethical technology standards – Blockchain initiatives should comply with ethical guidelines such as FPIC.
- Indigenous co-creation labs – Collaboration between indigenous artists and blockchain developers is essential for fair and transparent implementation.

Case Study 3: Internet of Things in Holy Site Protection

In South Africa, the Richtersveld World Heritage Site, a holy cultural landscape of the Nama people, utilises IoT technology to monitor environmental changes and prevent unauthorised access to holy areas. IoT sensors detect ecological disturbances and human activity, allowing for real-time alerts and swift community intervention. This integration ensures indigenous autonomy over holy sites, mitigating risks of cultural encroachment (Department Science and Technology, 2011; UNESCO, 2007). New Zealand has successfully applied similar technologies to protect the holy *Maunga* (mountains) of the Māori people. IoT-based environmental monitoring ensures that spiritual sites remain undisturbed while allowing indigenous communities to manage conservation efforts (Michaela, 2024).

Challenges and Considerations

- Technological accessibility: How can indigenous communities directly manage and benefit from IoT?
- Data sovereignty: Who owns and controls the environmental data collected by IoT devices?
- Long-term sustainability: How can these technologies remain accessible without dependence on external funding?

Policy Recommendations

- Indigenous hackathons and co-creation labs – Indigenous groups should collaborate with technology developers to create community-driven digital solutions.
- Community-Owned Digital Platforms – Data sovereignty must be prioritised, ensuring Indigenous control over environmental data.
- Sustainable funding for IoT projects – Investment in indigenous-led technological initiatives is essential for long-term success.

Future perspectives and recommendations for action - The integration of 4IR technologies presents both opportunities and challenges for indigenous communities. While AI, Blockchain, and IoT can empower indigenous knowledge preservation, ethical concerns regarding ownership, commercialisation, and technological accessibility must be addressed.

Key Recommendations

- Development of ethical technology standards – Any technological application involving Indigenous knowledge must include FPIC and ensure community-led governance.
- Community-owned digital platforms – Indigenous communities should retain control over digitised cultural and linguistic data.
- Adaptation of IPR systems – Legal frameworks must protect collective Indigenous knowledge from corporate exploitation.
- Funding for community-driven tech projects – Governments and private sectors should support indigenous-led innovations.
- Education and digital participation – Indigenous groups must actively shape the 4IR through indigenous coding schools and tech scholarships.

By centering indigenous agency in technological advancements, South Africa and Australia can set a global precedent for ethical, sustainable, and inclusive digital innovation.

ETHICAL CONSIDERATIONS TO PREVENT THE EXPLOITATION AND MARGINALISATION OF INDIGENOUS COMMUNITIES

The integration of fourth industrial revolution (4IR) technologies with indigenous knowledge systems (IKS) presents transformative opportunities for cultural preservation and sustainability. However, these advancements also introduce ethical challenges that must be carefully managed to prevent the exploitation and marginalisation of indigenous communities. Key concerns include data sovereignty, intellectual property rights, and cultural commodification, all of which require collaborative and community-led approaches to ensure ethical and sustainable implementation.

Data Sovereignty and Indigenous Control

Indigenous data sovereignty asserts the right of indigenous communities to control the collection, storage, and dissemination of their knowledge and cultural data. Ensuring that digital governance frameworks respect these rights is critical to preventing the unauthorised exploitation of indigenous intellectual property. Policies enforcing Free, Prior, and Informed Consent (FPIC) play a crucial role in protecting indigenous knowledge from misappropriation by external entities (Bowman & Bremner, 2024). To foster ethical technological integration, 4IR solutions must align with indigenous governance structures, allowing communities to retain ownership over their data. Embedding data sovereignty principles within AI, blockchain, and IoT applications ensures that these technologies serve as tools for empowerment rather than exploitation. For example, blockchain technology can facilitate transparent benefit-sharing mechanisms, ensuring that indigenous groups receive recognition and fair compensation for the use of their knowledge (Smyth, Martin & Downing, 2025).

Preventing Cultural Commodification

Cultural commodification occurs when indigenous traditions, artefacts, and knowledge are commercialised without proper consent or compensation. The rise of AI-generated indigenous art, digital tourism experiences, and virtual reconstructions of cultural heritage sites risks misrepresenting and decontextualising indigenous traditions if developed without meaningful indigenous participation. To safeguard cultural integrity, ethical frameworks must prioritise indigenous-led technology development. AI-driven platforms that translate indigenous languages or create digital replicas of traditional practices must be guided by cultural custodians to ensure authenticity and avoid distortion. Similarly, blockchain-based intellectual property protections can prevent the unauthorised reproduction and sale of indigenous cultural artefacts, ensuring that communities retain control over their heritage. While the integration of 4IR technologies with IKS presents immense potential for language revitalisation, cultural preservation, and environmental sustainability, ethical challenges must be proactively addressed. Ensuring data sovereignty, intellectual property protection, and fair representation is crucial to preventing the marginalisation of indigenous knowledge holders. Sustainable technological adoption must be driven by collaborative approaches that place indigenous communities at the centre of development and decision-making processes. By embedding ethical considerations into digital innovations, these technologies can serve as powerful tools for cultural resilience, self-determination, and sustainable development.

APPLICATION OF INDIGENOUS DATA SOVEREIGNTY PRINCIPLES IN FOURTH INDUSTRIAL TECHNOLOGIES

a) Artificial Intelligence

AI plays a crucial role in documenting, preserving, and revitalising indigenous languages and cultural practices. Indigenous data sovereignty principles ensure that AI-driven data collection is conducted with indigenous consent, preventing misuse or appropriation of knowledge. Ethical AI frameworks must incorporate indigenous perspectives to avoid bias and ensure that AI applications serve community interests (Soylu & Şahin, 2024).

b) Blockchain Technology

Blockchain technology offers a transformative approach to safeguarding indigenous intellectual property rights by providing transparent and immutable records. This ensures the authenticity and protection of cultural artefacts, land claims, and traditional ecological knowledge, effectively preventing unauthorised commercial exploitation. By leveraging blockchain's decentralised and tamper-proof ledger system, communities can securely document and verify ownership and provenance, thereby preserving their cultural heritage (Qizi & Kamalovich, 2025).

c) Internet of Things (IoT)

Internet of Things devices offer valuable applications for monitoring holy sites, tracking environmental changes, and managing natural resources. However, their implementation must adhere to IDS principles to prevent unauthorised surveillance or exploitation of indigenous lands and knowledge. Indigenous-led governance structures should oversee data collection and usage to ensure alignment with community values (Torre, Chennamaneni, & Rodriguez, 2023). By embedding indigenous data sovereignty principles within fourth industrial revolution technologies, this framework ensures that technological advancements serve to ethically preserve and promote indigenous knowledge systems. This approach upholds indigenous sovereignty, enhances self-determination, and fosters equitable participation in the digital age.

DEFINITION OF THE FOURTH INDUSTRIAL REVOLUTION

The fourth industrial revolution (4IR) refers to the rapid transformation of industries through advanced technologies that integrate the physical, digital, and biological spheres. Characterised by artificial intelligence (AI), blockchain, the Internet of Things (IoT), robotics, and biotechnology, 4IR enables automation, advanced data analysis, and seamless communication, reshaping economies and societies (Schwab, 2018; Xu et al., 2018). Unlike previous revolutions, 4IR is defined by exponential technological advancements and global disruptions.

Artificial Intelligence

Artificial intelligence (AI) encompasses computer systems designed to perform tasks requiring human intelligence, such as speech recognition, decision-making, and language translation (Russell & Norvig, 2016). AI plays a crucial role in automating processes, analysing large datasets, and enhancing decision-making across industries. In the context of indigenous knowledge systems (IKSs), AI aids in language preservation, knowledge documentation, and cultural heritage digitisation. AI-driven tools support indigenous language revitalisation by developing digital dictionaries, automated translation tools, and speech recognition systems. Natural Language Processing (NLP) applications facilitate the documentation and transmission of endangered languages through interactive platforms (Bird & Yibarbuk, 2024). Initiatives such as Google AI's Endangered Languages Project exemplify AI's role in sustaining linguistic diversity.

Artificial intelligence in Cultural Heritage Digitisation

AI enhances the digitisation and preservation of indigenous cultural heritage through advanced image recognition, cataloguing, and virtual museum experiences. AI-powered image recognition identifies and classifies indigenous artefacts, ensuring accurate representation and storage in digital repositories (Soylu & Şahin, 2024). Museums and cultural institutions leverage AI to curate and manage artefact collections ethically and sustainably (Osagie & Oladokun, 2024). In addition, computer vision technology reduces mislabelling risks and enhances accessibility, allowing global audiences to engage with indigenous heritage through virtual exhibits (Akyol & Avcı, 2023).

DEFINITION OF INDIGENOUS KNOWLEDGE SYSTEMS

Indigenous knowledge systems (IKS) encompass the knowledge, skills, and practices developed by indigenous communities over generations. Rooted in cultural, social, and environmental contexts, IKSs contribute to resource management, healthcare, agriculture, and cultural preservation. These dynamic systems continuously adapt to social and environmental changes while maintaining core values (Sen & Nagendra, 2019; Takazira & Dzingirayi, 2024). In addition, this knowledge system is used daily in various activities such as subsistence farming, food preparation, health care, decision-making, education, marital practices, risk management, and natural resource management (Zidny, Sjöström & Eilks, 2020).

Scope of Indigenous Knowledge Systems

Indigenous knowledge system covers various domains, including:

- Cultural heritage and identity: IKSs preserve traditions, rituals, and craftsmanship, reinforcing indigenous identity and ancestral connections (Molebatsi, 2023; Takazira & Dzingirayi, 2024).
- Environmental stewardship: Indigenous ecological knowledge informs sustainable resource management, biodiversity conservation, and climate adaptation (Sen & Nagendra, 2019).
- Health and medicine: Traditional medicinal practices, including the use of plants and natural remedies, complement modern healthcare, especially in remote areas (Takazira & Dzingirayi, 2024; WHO, 2023).
- Knowledge transmission: Oral traditions, storytelling, and mentorship ensure the intergenerational transfer of indigenous wisdom (Malapane, Chanza & Musakwa, 2024).
- Integration with modern technology: The incorporation of AI, IoT, and blockchain into IKSs supports knowledge documentation, protection, and dissemination while addressing intellectual property rights concerns (Alharbi, Attiah & Alghazzawi, 2022; Saeed, 2024).

Indigenous knowledge systems (IKSs) contribute to global sustainability, climate resilience, and cultural preservation through traditional environmental management, biodiversity conservation, and community-driven solutions (Alharbi, Attiah & Alghazzawi, 2022). IKSs enhance sustainable agriculture, water management, and disaster risk reduction, complementing modern science. However, challenges such as data sovereignty, ethical commodification, and misappropriation threaten indigenous rights and self-determination. Digitisation further raises concerns about intellectual property, consent, and equitable access. Addressing these issues requires policies that uphold indigenous agency, promote ethical knowledge-sharing, and ensure fair compensation.

BLOCKCHAIN TECHNOLOGY IN PRESERVING INDIGENOUS CULTURAL HERITAGE

Blockchain technology, initially recognised for its application in cryptocurrencies, has emerged as a pivotal tool in safeguarding indigenous knowledge systems (IKS). Its decentralised and immutable nature ensures transparent documentation, traceability, and protection of indigenous intellectual property, addressing concerns related to data sovereignty and equitable knowledge-sharing (Connolly, Nam & Goodman,

2022). Smart contracts facilitate fair benefit-sharing mechanisms, enabling indigenous communities to retain control over their traditional knowledge while preventing misappropriation. However, challenges such as accessibility, technological infrastructure, and ethical considerations must be addressed to ensure the effective integration of blockchain in protecting IKSs (UNESCO, 2023; Wagner & de Clippele, 2023). In South Africa, blockchain has the potential to authenticate traditional art forms, such as Zulu beadwork and Ndebele house painting, ensuring artists receive fair recognition and protection against unauthorised commercialisation (Masenya, 2023; Soylu & Şahin, 2024). By creating a decentralised record of these art forms, blockchain guarantees authenticity and ownership, thereby preserving cultural heritage and providing fair rewards to creators. In addition, blockchain can secure traditional medicinal knowledge, ensuring communities such as the San and Khoi Khoi benefit from the commercial use of indigenous plants like Rooibos and Hoodia (Mangla et al., 2022). By recording and tracking the use of these plants on a blockchain, these communities can protect their intellectual property rights and ensure they receive appropriate benefits from commercialisation.

In Australia, blockchain technology is utilised to authenticate indigenous art and artefacts, protecting cultural intellectual property and ensuring transparency in the art market (Whitaker et, al, 2021). For instance, platforms like IndigiLedger use blockchain and secure 'smart labels' to verify Aboriginal and Torres Strait Islander artworks, combating the proliferation of counterfeit pieces and ensuring artists receive due recognition and compensation. Furthermore, blockchain can document and regulate access to holy sites, integrating with IoT sensors to monitor environmental conditions and visitor impact, thus ensuring ethical preservation and sustainable management (Trček, 2022). This approach aligns with the need to protect cultural heritage sites from over-tourism and environmental degradation. In addition, blockchain records traditional fire management practices such as "cultural burning," promoting their ethical application in conservation efforts (Brown et al., 2022). By leveraging blockchain technology, indigenous communities can preserve their cultural heritage, protect intellectual property rights, and ensure sustainable management of their resources. Bottom of Form

FOURTH INDUSTRIAL REVOLUTION TECHNOLOGIES IN PRESERVING INDIGENOUS KNOWLEDGE SYSTEMS

Artificial Intelligence and Language Preservation

Artificial Intelligence-driven tools, particularly natural language processing (NLP) and machine learning, are crucial for preserving endangered indigenous languages. These technologies analyse linguistic patterns, generate real-time translations, and create digital archives to document oral traditions (Bird & Yibarbuk, 2024). AI-powered chatbots and voice recognition software enhance language revitalization by enabling interactive learning (Mazzocchi, 2021). However, challenges related to Indigenous data sovereignty and ethical AI development necessitate collaboration with Indigenous communities (Tsosie, 2021).

Blockchain for Cultural Artefact Authentication

Blockchain provides decentralised, tamper-proof records to verify the authenticity and provenance of indigenous cultural artifacts Smart contracts enable indigenous-controlled intellectual property rights, preventing unauthorised commercialisation of traditional knowledge (Ajani et. al, 2024; Iredale, 2021).

Case studies from Australia and South Africa illustrate blockchain's role in securing community-held knowledge while promoting equitable benefit-sharing (Donkor & Mearns, 2022). However, digital literacy and infrastructure limitations must be addressed to maximise blockchain's impact in Indigenous contexts.

Internet of Things for Environmental and Cultural Site Protection

Internet of Things technologies facilitate real-time monitoring of holy sites, helping mitigate ecological degradation and unauthorised access (Ajani et. al, 2024; Iredale, 2021). Sensors track climate conditions, pollution, and land encroachments, enabling data-driven conservation strategies (Dudgeon & Bray, 2019). However, concerns regarding data privacy and indigenous governance over IoT-generated information highlight the need for policies aligned with indigenous data sovereignty (IDS) principles (Bowman & Bremner, 2024).

ENSURING INDIGENOUS COMMUNITIES' ACCESS TO AND UTILISATION OF DIGITAL TECHNOLOGIES

The technologies such as artificial intelligence (AI), blockchain, and the internet of Things (IoT) are instrumental in preserving indigenous knowledge and languages, it is essential to ensure that indigenous communities can access and effectively utilise these digital resources (Soylu & Şahin, 2024). To bridge the digital divide and empower current generations, several key strategies should be implemented:

- Community-led digital initiatives: Empowering indigenous communities to spearhead digital projects ensures that technological solutions align with their cultural contexts and needs. A notable example is Te Hiku Media in New Zealand, which developed an automatic speech recognition (ASR) model for the Māori language, achieving a 92% accuracy rate. This success highlights the importance of indigenous ownership and control over digital data to prevent digital disenfranchisement (Agbeyangi & Jere, 2024).
- Capacity building and digital literacy: Investing in digital literacy and technical training within indigenous communities enables active participation in technology development and utilisation. Initiatives such as indigenous coding workshops and digital skills training can empower community members to create and manage digital platforms that preserve and disseminate their knowledge. For instance, Te Hiku Media's Papa Reo project focuses on nurturing and proliferating the Māori language unique to the haukāinga of Te Hiku o Te Ika, highlighting the role of community-driven digital initiatives in language preservation (Jones, Mahelona, Duncan & Leoni, 2023).
- Ethical frameworks and data sovereignty: Establishing ethical guidelines that respect indigenous data sovereignty ensures that communities maintain control over their digital assets. The CARE Principles (Collective Benefit, Authority to Control, Responsibility, and Ethics) for indigenous data governance provide a framework that emphasises the rights of indigenous peoples to govern the collection, ownership, and application of their data. Collaborative efforts between technologists and indigenous leaders can develop frameworks that govern the collection, storage, and sharing of indigenous knowledge, aligning with cultural values and priorities (Kelsey, Russo, Martinez & McElroy, 2023).

By implementing these strategies, present-day indigenous communities can create opportunities to access digital resources, apply advanced technologies for cultural preservation, reinforce their traditions through digital storytelling and archival systems, and empower future generations with knowledge sovereignty. This proactive engagement ensures that indigenous knowledge is not only safeguarded but also shared, adapted, and revitalised in ways that strengthen cultural identity and promote intergenerational learning.

SYNERGISTIC INTEGRATION OF 4IR TECHNOLOGIES AND INDIGENOUS KNOWLEDGE SYSTEMS

The convergence of fourth industrial revolution (4IR) technologies with indigenous knowledge systems (IKSs) presents transformative opportunities for preserving, safeguarding, and disseminating traditional knowledge in the digital era. Artificial intelligence (AI) facilitates the classification and management of extensive indigenous knowledge repositories, enhancing accessibility and organisation. AI-powered translation tools further support the preservation and global sharing of indigenous languages (Ajani et. al, 2024). Blockchain technology plays a crucial role in securing intellectual property rights, preventing unauthorised exploitation, and enabling fair benefit-sharing mechanisms that compensate indigenous communities for their knowledge contributions (Anderson, 2015). The Internet of Things (IoT) bridges traditional ecological practices with modern sensor technology, empowering indigenous farmers to optimise agricultural techniques while maintaining environmental sustainability (Simelane, Hansen & Munghemezulu, 2021). Immersive technologies such as virtual reality (VR) and augmented reality (AR) digitally reconstruct historical events, cultural ceremonies, and oral traditions, fostering deeper engagement and cross-cultural appreciation of IKS (Xiao, Jeremy, Jin 2019; Oladokun, Abdulahi, Adeyinka, 2024). Big data analytics enhance indigenous ecosystem management and climate resilience by identifying environmental patterns that complement traditional ecological knowledge, ensuring sustainable resource use (Dorji et. al, 2024). In addition, drones and geographic information systems (GIS) provide real-time data for land and resource management, enabling indigenous communities to monitor and protect their territories effectively (Opedes et. al, 2023).

IMPORTANCE OF SUSTAINABLE DEVELOPMENT AND CULTURAL PRESERVATION

Sustainable development and cultural preservation are intrinsically linked, ensuring that progress does not come at the expense of identity, heritage, and community well-being. The United Nations (2020) defines sustainable development as meeting present needs without compromising the ability of future generations to meet their own. While often framed in economic and environmental terms, sustainability also encompasses cultural dimensions, with indigenous knowledge systems (IKS) playing a critical role in fostering resilience and adaptability. For centuries, indigenous communities have practiced sustainable living through environmental stewardship and social equity, aligning closely with the principles of sustainable development (Department Science and Technology, 2011; Pisolkar, 2024). Recognising and protecting IKS is essential for biodiversity conservation, equitable resource distribution, and ethical governance, particularly in the face of rapid technological advancements associated with the fourth industrial revolution (4IR). However, the risk of commodification and exploitation of indigenous knowledge highlights the need for robust protective measures

Cultural preservation is not merely about safeguarding historical artefacts or traditions; it is a dynamic force that shapes collective identity, fosters social cohesion, and strengthens intergenerational continuity. Development strategies that disregard cultural heritage risk eroding the social fabric of communities, leading to fragmentation and loss of traditional values. The pressures of globalisation further threaten indigenous cultures, making their protection crucial for inclusive and sustainable development (Mawere & Walter Tshamano, 2023; UNESCO, 2021). The United Nations Sustainable Development Goals (SDGs) acknowledge this interconnection, particularly in SDG 11, which emphasises the importance of creating inclusive, resilient cities that safeguard cultural and natural heritage (United Nations, 2021). This recognition highlights cultural preservation as an essential pillar of sustainability, ensuring that communities retain their identities while participating meaningfully in global progress.

TECHNOLOGICAL APPLICATIONS IN CULTURAL PRESERVATION

The integration of 4IR technologies with IKS presents ground-breaking opportunities to protect, enhance, and sustain cultural heritage. Advanced digital tools such as artificial intelligence (AI), blockchain, and the Internet of Things (IoT) are increasingly leveraged to address critical aspects of cultural preservation, including language revitalisation, artefact protection, and the safeguarding of holy sites. AI plays a transformative role in language preservation, aiding in the documentation and revitalisation of endangered indigenous languages through automated transcription, translation, and speech recognition technologies (Jafari, 2023). These tools ensure that linguistic heritage remains accessible to future generations while fostering intergenerational learning and engagement (Wanjiku, 2024). Blockchain technology, meanwhile, provides a secure and transparent framework for safeguarding indigenous intellectual property rights. By ensuring authenticity, secure ownership, and fair attribution, blockchain prevents the unauthorised exploitation of cultural artefacts and traditional knowledge (Anderson, 2015). This strengthens indigenous communities' ability to control their cultural assets and benefit from their knowledge systems.

IoT technologies further contribute to cultural preservation by enabling real-time monitoring and management of holy and heritage sites. These systems support indigenous communities in protecting their environments while ensuring that traditional practices remain intact (Makananise & Madima, 2025; Trček, 2022) By integrating digital surveillance and environmental sensors, IoT helps communities maintain sustainable interaction with their cultural landscapes. Beyond preservation, these technological innovations create new socio-economic opportunities for indigenous populations, enabling them to leverage their knowledge while maintaining sovereignty over their cultural resources. By providing platforms for economic empowerment, skill development, and digital inclusion, fourth industrial revolution technologies ensure that cultural heritage remains a living, evolving force rather than a relic of the past (Salas-Pilco, 2025).

FINDINGS AND DISCUSSION

The integration of fourth industrial revolution (4IR) technologies specifically Artificial Intelligence (AI), Blockchain, and the Internet of Things (IoT) into indigenous knowledge systems (IKSs) presents significant opportunities for cultural preservation, knowledge transmission, and sustainable development (Connolly, Nam & Goodman, 2022). These advancements, however, introduce ethical, cultural, and practical challenges that require careful consideration. This section reports on the role of each technology in safeguarding and revitalising indigenous heritage while ensuring community agency and control.

Artificial Intelligence and Language Preservation

AI has become a vital tool in the preservation of indigenous languages, many of which face endangerment due to declining speakers and limited documentation. AI-powered applications such as automatic translation tools, speech recognition systems, and text analysis platforms have facilitated the digitisation and revitalisation of these languages. These technologies support the development of digital repositories, interactive language-learning platforms, and real-time translation services, contributing to intergenerational linguistic continuity. By digitising indigenous languages, AI enhances accessibility and ensures their survival in the digital era. AI-driven projects, such as Google's Indigenous Language Translation initiative, have successfully documented and translated low-resource languages, including Owens Valley Paiute, showcasing AI's potential in linguistic revival. However, concerns regarding data ownership, linguistic accuracy, and community consent persist. To maintain cultural integrity and linguistic authenticity, indigenous communities must be actively involved in the development and deployment of AI language models.

Blockchain for Cultural Artefacts and Intellectual Property Protection

Blockchain technology provides a secure and transparent approach to safeguarding indigenous cultural artefacts, intellectual property, and traditional knowledge. Its decentralised and immutable nature enables the creation of digital certificates of authenticity, ownership records, and benefit-sharing mechanisms. This is particularly beneficial for indigenous communities aiming to maintain sovereignty over their cultural heritage and prevent the unauthorised exploitation of their artefacts and traditional knowledge. Several initiatives have successfully leveraged blockchain for cultural preservation. For instance, First Nations Blockchain has been used to authenticate indigenous art and artefacts, ensuring provenance while protecting cultural intellectual property. In addition, blockchain has been applied to validate provenance claims in the native foods and botanicals supply chain, promoting economic empowerment and ecological sustainability for indigenous communities. Despite these advantages, challenges such as accessibility, technical literacy, and ensuring indigenous ownership over blockchain platforms remain. To fully realise the potential of blockchain, indigenous communities require training, policy support, and collaborative frameworks that prioritise their interests and knowledge sovereignty (Connolly, Nam & Goodman, 2022).

Internet of Things for Monitoring Holy Sites and Environmental Conservation

The IoT has emerged as a crucial tool for monitoring and protecting holy sites by providing real-time data on environmental conditions, visitor access, and potential threats. IoT-enabled sensors support environmental conservation by detecting climate changes, soil degradation, and human interference, enabling indigenous communities to retain control over their holy landscapes (Alharbi, Attiah & Alghazzawi, 2022). By integrating traditional ecological knowledge with IoT-based monitoring, indigenous communities can manage and protect their ancestral lands while addressing modern environmental challenges. In Australia, for example, IoT technology has been deployed to monitor Indigenous Protected Areas (IPAs), allowing Aboriginal and Torres Strait Islander communities to track ecological changes and implement conservation strategies rooted in traditional knowledge systems.

However, while IoT enhances environmental monitoring, issues of data sovereignty, cybersecurity, and long-term accessibility remain critical. Indigenous communities must be equipped with the necessary skills and infrastructure to manage IoT systems autonomously, ensuring that data collection aligns with their needs rather than external interests. The findings indicate that the integration of fourth industrial revolution technologies with indigenous knowledge systems must be approached with cultural sensitivity, ethical governance, and indigenous-led innovation. AI, blockchain, and IoT offer unprecedented opportunities to protect and revitalise indigenous knowledge, yet their success depends on ensuring community agency, equitable access, and the protection of intellectual property rights. Moving forward, the development of policies and frameworks must empower indigenous groups to co-create technological solutions that align with their cultural values, traditions, and aspirations (Bawack et. al., 2025).

FUTURE RESEARCH DIRECTIONS

Future research should focus on the sustainable integration of fourth industrial revolution (4IR) technologies into indigenous knowledge systems (IKSs) to ensure ethical, culturally appropriate, and community-driven applications. One critical area for further exploration is the long-term effectiveness of AI-powered language preservation tools in promoting intergenerational linguistic transmission. Studies should assess whether these technologies enhance linguistic diversity while respecting indigenous cultural autonomy and traditional knowledge transmission methods. In addition, empirical research is needed to evaluate the efficacy of blockchain-based systems in protecting indigenous intellectual property rights and preventing cultural misappropriation. Future studies should examine how blockchain can be incorporated into indigenous governance structures, ensuring that communities maintain sovereignty over their cultural and artistic expressions. Furthermore, research should investigate the scalability, affordability, and adaptability of IoT applications for cultural and environmental conservation, particularly in mitigating climate change impacts on culturally significant landscapes. It is essential to explore how IoT can be leveraged to safeguard holy sites and biodiversity while ensuring that indigenous knowledge remains central to conservation strategies. Ethical considerations also require deeper inquiry, particularly in establishing frameworks that guarantee the equitable participation of indigenous communities in technology design and decision-making processes. The intersection of Free, Prior, and Informed Consent (FPIC) with digital governance models is an important area of study, especially in shaping ethical AI, blockchain, and IoT applications. Policy development and implementation must be a priority in future research, with a focus on upholding indigenous data sovereignty, intellectual property rights, and ethical technological use. Comparative policy analyses across different regions could provide insights into best practices for integrating indigenous knowledge with 4IR innovations while safeguarding indigenous rights. Addressing these research gaps will contribute to the responsible and sustainable integration of advanced technologies with indigenous knowledge systems, ensuring that digital innovation aligns with cultural preservation, self-determination, and community empowerment.

CONCLUSION

The integration of fourth industrial revolution (4IR) technologies into indigenous knowledge systems (IKSs) presents both opportunities and challenges, necessitating ethical, sustainable, and culturally appropriate implementation. While AI, blockchain, and IoT have the potential to preserve and promote indigenous languages, protect intellectual property, and support environmental conservation, their application must be guided by principles of indigenous data sovereignty, ethical governance, and equitable participation. Ensuring that indigenous communities maintain control over their digital heritage and actively engage in the development of these technologies is essential for fostering self-determination and cultural sustainability. Future research should prioritise participatory methodologies that center indigenous voices in technological innovation, ensuring that AI-driven language tools are co-developed with linguistic experts to prevent algorithmic bias. Similarly, blockchain should be explored as a mechanism for securing ownership rights over cultural artefacts and traditional knowledge, preventing unauthorised commercialisation and cultural misappropriation. The role of IoT in environmental and cultural conservation must be further examined, particularly in mitigating climate change threats to holy sites and ecologically significant landscapes. However, the implementation of these technologies must align with indigenous conservation principles to ensure ethical use and long-term sustainability.

Furthermore, strengthening legal protections through national and international policy frameworks, such as the United Nations Declaration on the Rights of Indigenous Peoples (UNDRIP), is critical in safeguarding indigenous rights in the digital age. Governments, policymakers, and researchers must collaborate with indigenous communities to formulate policies that uphold ethical technological use, protect intellectual property, and promote digital literacy. By fostering cross-disciplinary partnerships between indigenous scholars, technologists, and policymakers, the responsible and sustainable integration of 4IR technologies can be achieved. Finally, digital innovation should serve as a tool for cultural empowerment rather than an instrument of exploitation. By adopting ethical guidelines, reinforcing indigenous governance structures, and ensuring equitable participation in technology development, 4IR technologies can support the preservation and revitalisation of indigenous knowledge systems. Moving forward, a holistic, community-centered approach is essential in ensuring that digital advancements align with indigenous traditions, values, and self-determined aspirations.

RECOMMENDATIONS

To ensure that fourth industrial revolution (4IR) technologies contribute meaningfully to the preservation and promotion of indigenous knowledge Systems (IKSs), several key recommendations must be implemented. Strengthening indigenous data sovereignty is essential, as policymakers and technology developers should integrate indigenous governance frameworks into digital initiatives. This would ensure that indigenous communities retain control over their data and digital heritage, preventing external exploitation and safeguarding their intellectual property. In addition, capacity building and digital literacy must be prioritised to equip indigenous communities with the technical skills and resources necessary to actively participate in the development and management of AI, blockchain, and IoT technologies. Tailoring these digital tools to their cultural needs will promote self-reliance and sustainable innovation. In the development of AI-driven language tools, ethical considerations must be at the forefront. AI tools for language preservation should be co-developed with indigenous language experts to ensure linguistic accuracy while preventing algorithmic biases that may misrepresent or distort indigenous languages. Blockchain technology also presents an opportunity for cultural protection, as governments and indigenous organisations should explore the adoption of blockchain-based registries to secure ownership rights for cultural artefacts, indigenous art, and traditional knowledge. This would prevent unauthorised commercialisation and ensure that cultural heritage remains under the stewardship of indigenous communities. Furthermore, IoT applications should be leveraged for environmental and cultural conservation. indigenous communities should be supported in implementing IoT-based monitoring systems to safeguard holy sites and mitigate environmental threats.

However, it is crucial that these technologies align with indigenous conservation principles to ensure ethical and culturally respectful implementation. Strengthening policy formulation and legal protections is also necessary. National and international policies should be reinforced to uphold indigenous rights in the digital age, ensuring compliance with frameworks such as the United Nations Declaration on the Rights of Indigenous Peoples (UNDRIP). Finally, collaborative research and development should be prioritised to ensure indigenous perspectives are at the center of technological advancements. Future research should adopt participatory methodologies that engage indigenous scholars, technologists, and policymakers in co-creating solutions. By fostering these partnerships, 4IR technologies can serve as tools for cultural empowerment rather than instruments of exploitation. Sustainable and ethical technological integration will enable indigenous communities to safeguard their heritage while benefiting from digital advancements in ways that respect their traditions and values.

REFERENCES

Agbeyangi, A., & Jere, N. (2024). IsiXhosa in the Digital Age: Navigating Language Preservation and Innovation (A Systematic Scoping Review). *IEEE Access*. https://gsarpublishers.com/journal-gjahss-home/

Ajani, Y. A., Baloyi, N. C., Chibike, J. J., Elisifa, Z., Lubambo, R. J., Mabika, M., & Xaba, B. B. B. M. (2024). *Digital Media and the Preservation of Indigenous Languages in Africa: Toward a Digitalized and Sustainable Society*. Rowman & Littlefield.

Ajani, Y. A., Oladokun, B. D., Olarongbe, S. A., Amaechi, M. N., Rabiu, N., & Bashorun, M. T. (2024). Revitalizing indigenous knowledge systems via digital media technologies for sustainability of indigenous languages. *Preservation. Digital Technology & Culture*, *53*(1), 35–44. DOI: 10.1515/pdtc-2023-0051

Akyol, G., & Avcı, A. B. (2023). Ai Applications in Cultural Heritage Preservation: Technological Advancements for the Conservation. https://hdl.handle.net/11147/13725

Alharbi, S., Attiah, A., & Alghazzawi, D. (2022). Integrating blockchain with artificial intelligence to secure IoT networks: Future trends. *Sustainability (Basel)*, *14*(23), 16002. DOI: 10.3390/su142316002

Anderson, J. E. (2015). Indigenous knowledge and intellectual property rights. In *International Encyclopaedia of the Social & Behavioural Sciences* (2nd ed., pp. 769–778). Elsevier Inc., DOI: 10.1016/B978-0-08-097086-8.64078-3

Banda, L. O. L., Banda, C. V., Banda, J. T., & Singini, T. (2024). Preserving cultural heritage: A community-centric approach to safeguarding the Khulubvi Traditional Temple Malawi. *Heliyon*, *10*(18), e37610. Advance online publication. DOI: 10.1016/j.heliyon.2024.e37610 PMID: 39347398

Bawack, R., Roderick, S., Badhrus, A., Dennehy, D., & Corbett, J. (2025). Indigenous knowledge and information technology for sustainable development. *Information Technology for Development*, •••, 1–18. DOI: 10.1080/02681102.2025.2472495

Bird, S., & Yibarbuk, D. (2024). Cntering the speech community. In *Proceedings of the 18th Conference of the European Chapter of the Association for Computational Linguistics* (Vol. 1, pp. 826–839). Long Papers., https://aclanthology.org/2024.eacl-long.50/

Bowman, D. N., & Bremner, L. (2024). Indigenous Data Sovereignty: Applying It By, With, For, and Through Indigenous Evaluators and Evaluations. *The Canadian Journal of Program Evaluation*, *39*(2), 265–287. DOI: 10.3138/cjpe-2024-0039

Brown, P., Smith, J., & Hunter, L. (2022). Blockchain and cultural burning: Managing Indigenous fire practices. *Journal of Australian Environmental Management*, *29*(2), 134–147.

Carroll, S. R., Garba, I., Plevel, R., Small-Rodriguez, D., Hiratsuka, V. Y., Hudson, M., & Garrison, N. A. (2022). Using indigenous standards to implement the CARE principles: Setting expectations through tribal research codes. *Frontiers in Genetics*, *13*, 823309. DOI: 10.3389/fgene.2022.823309 PMID: 35386282

Carroll, S. R., Rodriguez-Lonebear, D., & Martinez, A. (2019). Indigenous data governance: Strategies from United States native nations. *Data Science Journal*, *18*(1), 31. DOI: 10.5334/dsj-2019-031 PMID: 34764990

Chilisa, B. (2019). *Indigenous research methodologies* (2nd ed.). Sage publications.

Connolly, D., Nam, S., & Goodman, K. (2022). Solving old problems or making new ones? Blockchain technology for the protection of refugees and migrants. *Journal of Human Rights, 22*(2), 109–134. DOI: 10.1080/14754835.2022.2100984

Copeland, A. (2025). Engagement for Sustainable Communities: Leading, Collaborating and Supporting Roles for Librarians. *Public Library Quarterly*, •••, 1–6. DOI: 10.1080/01616846.2025.2464515

Creswell, J. W., & Poth, C. N. (2018). *Qualitative inquiry and research design: Choosing among five approaches* (4th ed.). Sage publications.

Department Science and Technology. (RSA). (2011). Indigenous Knowledge System Policy. https://www.dsti.gov.za/images/pdfs/IKS_Policy%20PDF.pdf

Donkor, F. K., & Mearns, K. (2022). Harnessing Indigenous knowledge systems for enhanced climate change adaptation and governance: Perspectives from Sub-Saharan Africa. In *Indigenous knowledge and climate governance: A Sub-Saharan African perspective* (pp. 181–191). Springer International Publishing. DOI: 10.1007/978-3-030-99411-2_14

Dorji, T., Rinchen, K., Morrison-Saunders, A., Blake, D., Banham, V., & Pelden, S. (2024). Understanding How Indigenous Knowledge Contributes to Climate Change Adaptation and Resilience: A Systematic Literature Review. *Environmental Management, 74*(6), 1101–1123. DOI: 10.1007/s00267-024-02032-x PMID: 39215837

Flick, U. (2017). *The Sage handbook of qualitative data collection*. Sage publications.

Giannossa, L. C., Muntoni, I. M., Laviano, R., & Mangone, A. (2020). Building a step-by-step result in archaeometry. Raw materials, provenance and production technology of Apulian Red Figure pottery. *Journal of Cultural Heritage, 43*, 242–248. DOI: 10.1016/j.culher.2019.11.002

Guarda, T., Portela, F., & Diaz-Nafria, J. M. (Eds.). (2023). *Advanced Research in Technologies, Information, Innovation and Sustainability: Third International Conference, ARTIIS 2023, Madrid, Spain, October 18–20, 2023, Proceedings, Part I.* Springer Nature

Guba, E. G., & Lincoln, Y. S. (1994). Competing paradigms in qualitative research. *Handbook of qualitative research, 2*(163-194), 105.

Gupta, A. D. (2012). Way to study indigenous knowledge and indigenous knowledge system. *Research Journal of Recent Sciences. ISSN, 2277*, 2502. Available online at: www.isca.in, www.isca.me

Hu, X., Ng, J., & Lee, J. H. (2019). VR creation experience in cultural heritage education: A preliminary exploration. *Proceedings of the Association for Information Science and Technology, 56*(1), 422–426.

Hudson, M., Carroll, S. R., Anderson, J., Blackwater, D., Cordova-Marks, F. M., Cummins, J., David-Chavez, D., Fernandez, A., Garba, I., Hiraldo, D., Jäger, M. B., Jennings, L. L., Martinez, A., Sterling, R., Walker, J. D., & Rowe, R. K. (2023). Indigenous Peoples' rights in data: A contribution toward Indigenous Research Sovereignty. *Frontiers in Research Metrics and Analytics, 8*, 1173805. DOI: 10.3389/frma.2023.1173805 PMID: 37215248

Iredale, G. W. Y. N. E. T. H. (2021). What are the different types of blockchain technology? 101Blockchains. Retrieved from https://101blockchains.com/types-of-blockchain

Jafari, Z. (2023). The Role of AI in Supporting Indigenous Languages.https://www.researchgate.net/publication/379040413_The_Role_of_AI_in_Supporting_Indigenous_Languages

Jones, P. L., Mahelona, K., Duncan, S., & Leoni, G. (2023). Kia tangata whenua: Artificial intelligence that grows from the land and people.

Kelsey Leonard (Shinnecock Nation), Stephanie Russo (Native Village of Kluti-Kaah), Andrew Martinez (O'odham, Diegueño, and Yoeme), Lauren McElroy (2023). Our Common Agenda Global Digital Compact March 2023: CARE Statement for Indigenous Data Sovereignty (Scottish Canadian)

Kukutai, Tahu; McIntosh, Tracey; Boulton, Amohia; Durie, Meihana; Foster, Meika; Hutchings, Jessica; et al. (2024). Te Pūtahitanga: A Tiriti-Led Science-Policy Approach for Aotearoa New Zealand. AgResearch. Report. **https://doi.org/**DOI: 10.57935/AGR.26001496.v1

Kukutai, T., & Taylor, J. (2019). *Indigenous data sovereignty: Toward an agenda.* ANU Press.

Labadi, S. (2024). The World Heritage Convention at 50: Management, credibility and sustainable development. *Journal of Cultural Heritage Management and Sustainable Development, 14*(4), 750–764. DOI: 10.1108/JCHMSD-05-2022-0077

Lubinga, S. N., Maramura, T. C., & Masiya, T. (2023). Adoption of Fourth Industrial Revolution: challenges in South African higher education institutions. http://hdl.handle.net/2263/94027

Makananise, F. O., & Madima, S. E. (Eds.). (2025). *Decolonising digital media and indigenisation of participatory epistemologies.* Routledge.

Malapane, O. L., Chanza, N., & Musakwa, W. (2024). Transmission of indigenous knowledge systems under changing landscapes within the Vhavenda community, South Africa. *Environmental Science & Policy, 161*, 103861. DOI: 10.1016/j.envsci.2024.103861

Mangla, S. K., Kazançoğlu, Y., Yıldızbaşı, A., Öztürk, C., & Çalık, A. (2022). A conceptual framework for blockchain-based sustainable supply chain and evaluating implementation barriers: A case of the tea supply chain. *Business Strategy and the Environment, 31*(8), 3693–3716. DOI: 10.1002/bse.3027

Masenya, T. M. (2023). Revitalization and Digital Preservation of Indigenous Knowledge Systems for Sustainable Development of Indigenous Communities in South Africa. *The Serials Librarian, 84*(5–8), 86–102. DOI: 10.1080/0361526X.2023.2277962

Mawere, J., & Mukonza, R. M. (2024). Empowering marginalised communities: Leveraging indigenous knowledge for sustainable energy development in South Africa. *Sustainable Development.* Advance online publication. DOI: 10.1002/sd.3243

Mawere, J., & Walter Tshamano, N. (2023). *The Role of African Indigenous Knowledge Systems in Achieving Restorative Justice among the Vhavenda of South Africa.* IntechOpen., DOI: 10.5772/intechopen.1002677

Mazzocchi, F. (2022). Diving deeper into the concept of 'cultural heritage'and its relationship with epistemic diversity. *Social Epistemology*, *36*(3), 393–406. DOI: 10.1080/02691728.2021.2023682

Michaela, R. (2024). Restoring Taranaki Maunga: 'Religion' and Conservation in Aotearoa New Zealand. Open Access Te Herenga Waka-Victoria Unziversity of Wellington. Thesis. https://doi.org/DOI: 10.26686/wgtn.26860153

Mlambo, R., & Matfunjwa, M. (2024). The use of technology to preserve indigenous languages of South Africa. *Literator-Journal of Literary Criticism, Comparative Linguistics and Literary Studies, 45*(1), 2007. https://hdl.handle.net/10520/ejc-literat_v45_n1_a2007

Molebatsi, J. K. (2023). *African identity in architecture: Guiding principles for the architectural design of traditional health practices in Gauteng, South Africa* (Doctoral dissertation, University of Pretoria (South Africa).

Nakamoto, S. (2008). Bitcoin: A peer-to-peer electronic cash system. Retrieved from. https://bitcoin .org/bitcoin.pdf

Oguamanam, C. (2020). Indigenous peoples, data sovereignty, and self-determination: Current realities and imperatives. *The African Journal of Information and Communication*, *26*(26), 1–20. DOI: 10.23962/10539/30360

Oguamanam, C. (2022). Transition to the Fourth Industrial Revolution: Africa's Science, Technology and Innovation Framework and Indigenous Knowledge Systems. *African Journal of Legal Studies*, *15*(1), 1–37. DOI: 10.1163/17087384-bja10058

Oladokun, B. D., Abdulahi, F., & Adeyinka, T. (2024). Bridging the Digital Divide: Empowering Nigerian Universities through Technological Advancements in Academic Libraries Conference: https://www.researchgate.net/publication/382947230_Bridging_the_Digital_Divide_Empowering_Nigerian_Universities_through_Technological_Advancements_in_Academic_Libraries

Opedes, H., Nedala, S., Mücher, C. A., Baartman, J. E., & Mugagga, F. (2023). How can drones uncover land degradation hotspots and restoration hopespots? An integrated approach in the Mount Elgon region with community perceptions. *Land (Basel)*, *13*(1), 1. https://www.mdpi.com/2073-445X/13/1/1#. DOI: 10.3390/land13010001

Osagie, O., & Oladokun, B. (2024). Usefulness of artificial intelligence to safeguard records in libraries: A new trend. *Southern African Journal of Security, 2*, 13-pages. https://doi.org/DOI: 10.25159/3005-4222/16803

Pansoni, S., Tiribelli, S., Paolanti, M., Di Stefano, F., Frontoni, E., Malinverni, E. S., & Giovanola, B. (2023). Artificial intelligence and cultural heritage: Design and assessment of an ethical framework. *The International Archives of the Photogrammetry, Remote Sensing and Spatial Information Sciences*, *48*, 1149–1155. DOI: 10.5194/isprs-archives-XLVIII-M-2-2023-1149-2023

Philbeck, T., & Davis, N. (2018). The fourth industrial revolution. *Journal of International Affairs*, *72*(1), 17–22. https://www.jstor.org/stable/26588339

Pisolkar, Y. (2024). Cultural Heritage Management and Sustainable Development: Major Themes and Research Trajectories, *Journal of Electrical Science*. DOI: 10.52783/jes.3224

Qizi, L. R. M., & Kamalovich, B. K. (2025). Blockchain Technology Usage on Intellectual Property Rights. *International Journal for the Semiotics of Law*, *38*(2), 363–380. DOI: 10.1007/s11196-024-10224-1

Rule, P., & Yazan, B. (2024). Knowledge and knowing in case study research: towards a relational epistemology. In *Handbook of Case Study Research in the Social Sciences* (pp. 31–50). Edward Elgar Publishing. DOI: 10.4337/9781803920320.00012

Russell, S. J., & Norvig, P. (2016). Artificial intelligence: a modern approach. *Pearson.* Ofosu-Asare, Y. (2024). Cognitive imperialism in artificial intelligence: counteracting bias with indigenous epistemologies. *AI & Society*, •••, 1–17. DOI: 10.1007/s00146-024-02065-0

Saeed, J. (2024). Guardians of the Data: Government Use of AI and IoT in the Digital Age.

Saffer, C. (2024). Secure from exploitation. Cultural artwork protected by blockchain. The Australian Jewish News, 2024).

Salas-Pilco, S. Z. (2025). Information technology (IT) and indigenous knowledge (IK) in Latin America: Pro-indigenous, para-indigenous, and per-indigenous IT–IK integration for development. *Information Technology for Development*, •••, 1–24. DOI: 10.1080/02681102.2025.2462122

Schwab, K. (2018). "The Fourth Industrial Revolution" Encyclopaedia Britannica Anniversary Edition: 250 Years of Excellence (1768–2018).

Sen, A., & Nagendra, H. (2019). The role of environmental placemaking in shaping contemporary environmentalism and understanding land change. *Journal of Land Use Science*, *14*(4-6), 410–424. DOI: 10.1080/1747423X.2020.1720841

Simelane, S. P., Hansen, C., & Munghemezulu, C. (2021). The use of remote sensing and GIS for land use and land cover mapping in Eswatini: A review. *South African Journal of Geomatics*, *10*(2), 181–206. DOI: 10.4314/sajg.v10i2.13

Smith, L. T. (2021). *Decolonising methodologies: Research and Indigenous peoples* (3rd ed.). Zed Books. DOI: 10.5040/9781350225282

Smyth, B. M., Martin, M. A., & Downing, M. (Eds.). (2025). *The Routledge handbook of human research ethics and integrity in Australia*. Routledge, Taylor & Francis Group.

Soylu, D., & Şahin, A. (2024). The Role of AI in Supporting Indigenous Languages. *AI and Tech in Behavioral and Social Sciences*, *2*(4), 11–18. DOI: 10.61838/kman.aitech.2.4.2

Takazira, L., & Dzingirayi, P. (2024). The Use of African Indigenous Knowledge Systems in Climate Change adaptation in Zimbabwe. *East African Journal of Education and Social Sciences*, *5*(6), 90–97. DOI: 10.46606/eajess2024v05i06.0417

Torre, D., Chennamaneni, A., & Rodriguez, A. (2023). Privacy-preservation techniques for IoT devices: A systematic mapping study. *IEEE Access : Practical Innovations, Open Solutions*, *11*, 16323–16345. DOI: 10.1109/ACCESS.2023.3245524

Trček, D. (2022). Cultural heritage preservation by using blockchain technologies. *Heritage Science, 10*(6), 6. Advance online publication. DOI: 10.1186/s40494-021-00643-9

Treleaven, P., Brown, R. G., & Yang, D. (2017). Blockchain technology in finance. *Computer, 50*(9), 14–17. DOI: 10.1109/MC.2017.3571047

Tsosie, R. (2021). Indigenous data sovereignty and digital inclusion. *Indigenous Law Journal at the University of Toronto Faculty of Law, 20*(2), 45–68. DOI: 10.1016/j.culher.2019.11.002

Udayakumar, P., & Anandan, R. (2024). Get Started with IoT. In *Design and Deploy Microsoft Defender for IoT*. Appress., DOI: 10.1007/979-8-8688-0239-3_1

UNESCO. 2021. World Health Convention. World Heritage and Indigenous People.

UNESCO. (n.d.). *Richtersveld Cultural and Botanical Landscape*. UNESCO World Heritage Centre. Retrieved from https://whc.unesco.org/en/list/1265

United Nations. (2020). Sustainable Development Goals. https://sdgs.un.org/goals

United Nations. (2021). Scientific and Cultural Organisation (UNESCO). Cultural heritage and sustainable development. https://unesdoc.unesco.org/

Wagner, A., & de Clippele, M.-S. (2023). Safeguarding Cultural Heritage in the Digital Era – A Critical Challenge. *International Journal for the Semiotics of Law - Revue internationale de Sémiotique juridique*, 36, 1915–1923. https://doi.org/SpringerLinkDOI: 10.1007/s11196-023-10040-z

Wanjiku, J. (2024). The Impact of Blockchain Technology on ART Authentication and Provenance Kiu Publication Extension. Kampala International University, Ugandahttps://www.researchgate.net/publication/382878671_The_Impact_of_Blockchain_Technology_on_ART_Authentication_and_Provenance

Whitaker, A., Bracegirdle, A., de Menil, S., Gitlitz, M. A., & Saltos, L. (2020). Art, antiquities, and blockchain: New approaches to the restitution of cultural heritage. *International Journal of Cultural Policy, 27*(3), 312–329. DOI: 10.1080/10286632.2020.1765163

Williams, D. H., & Shipley, G. P. (2023). Indigenous research methodologies: Challenges and opportunities for broader recognition and acceptance. *Open Journal of Social Sciences, 11*(5), 467–500. DOI: 10.4236/jss.2023.115030

Wingara. M.N, (2018). Indigenous Data Sovereignty Communique Indigenous Data Sovereignty Summit 20th June 2018, Canberra, ACT. Retrieved from Maiam Nayri Wingara.(n.d.) Principles.

World Health Organisation. (WHO) (2023). Traditional medicine has a long history of contributing to conventional medicine and continues to hold promise. https://www.who.int/news-room/feature-stories/detail/traditional-medicine-has-a-long-history-of-contributing-to-conventional-medicine-and-continues-to-hold-promise

Xu, M., David, J. M., & Kim, S. H. (2018). The fourth industrial revolution: Opportunities and challenges. *International journal of financial research, 9*(2), 90-95. URL: https://doi.org/DOI: 10.5430/ijfr.v9n2p90

Yin, R. K. (2017). *Case study research and applications: Design and methods*. Sage publications.

Zidny, R., Sjöström, J., & Eilks, I. A. (2020). Multi-Perspective Reflection on How Indigenous Knowledge and Related Ideas Can Improve Science Education for Sustainability. *Science & Education*, *29*(1), 145–185. DOI: 10.1007/s11191-019-00100-x

Zoannos, N., Chourdaki, P., & Assimakopoulos, N.UNESCO. (2023). Can UNESCO Use Blockchain to Ensure the Intangible Cultural Heritage of Humanity? *Heritage*, *6*(3), 3232–3255. DOI: 10.3390/heritage6030171

ADDITIONAL READINGS

Global Indigenous Data Alliance. (n.d.). *CARE Principles*. Retrieved from: Global Indigenous Data Alliance (24 March 2025).

Maiam nayri Wingara (2018). Indigenous Data Sovereignty Communique. Canberra, Australia: Maiam Nayri Wingaara. Available at: https://static1.squarespace.com/static/5b3043afb40b9d20411f3512/t/5b6c0f9a0e2e725e9cabf4a6/1533808545167/ Communique%2B-%2BIndigenous%2BData%2BSovereignty%2BSummit.pdf (Accessed 2 April 2025).

National Indigenous Australians Agency. (2024). *Framework for Governance of Indigenous Data*. Retrieved from: NIAA (24 March 2025).

Nations Institute. (n.d.). *Indigenous Data Sovereignty & Governance*. Retrieved from: Native Nations Institute (24 March 2025).

United Nations Educational, Scientific and Cultural Organization (UNESCO), "Charter on the Preservation of the Digital Heritage" (United Nations, 2003), https://unesdoc.unesco.org/ark:/48223/pf0000179529 (Assessed, 24 March 2025).

US Indigenous Data Sovereignty Network. (2024). *Indigenous Data Governance Brief.* Retrieved from: U.S. Indigenous Data Sovereignty Network (24 March 2025).

KEY TERMS AND DEFINITIONS

Artificial Intelligence (AI): It is a technology that enables computers and machines to perform tasks that typically require human intelligence.

Blockchain: It is a decentralised digital ledger that securely records transactions across a distributed network of computers. It ensures transparency, immutability, and resistance to tampering by linking "blocks" of data in a chronological "chain".

Cultural Preservation: It refers to efforts made to maintain and protect a culture's unique practices, beliefs, and artefacts from being lost or diminished, particularly in the face of external pressures such as globalisation.

Fourth Industrial Revolution (4IR): It is the current era of technological advancement, characterised by a fusion of technologies that blur the lines between the physical, digital, and biological spheres.

Indigenous Data: It refers to information or knowledge, in any format or medium, that pertains to and may affect Indigenous peoples both collectively and individually.

Indigenous Data Governance: It is the right of indigenous peoples to autonomously decide what, how, and why indigenous data is collected, accessed, and used. It ensures that data about indigenous peoples reflects their priorities, values, cultures, worldviews, and diversity.

Indigenous Data Sovereignty: It is the right of Indigenous peoples to exercise ownership over Indigenous data. This ownership is expressed through the creation, collection, access, analysis, interpretation, management, dissemination, and reuse of Indigenous data.

Indigenous Knowledge Systems (IKSs): It refers to the body of knowledge developed by a group of people over generations through close interaction with nature. This knowledge is passed down orally and through cultural practices, encompassing language, classification systems, resource use practices, social interactions, rituals, and spirituality.

Internet of Things (IoT): It is a network of physical objects, or "things," embedded with sensors, software, and other technologies that enable them to connect and exchange data with other devices and systems over the Internet or other communication networks.

Chapter 16
Harnessing 4IR Technologies:
Innovating the Digital Preservation of Indigenous Knowledge

Muhammad Usman Tariq
https://orcid.org/0000-0002-7605-3040
Abu Dhabi University, UAE & University College Cork, Ireland

ABSTRACT

The revolutionary potential of Fourth Industrial Revolution (4IR) technologies in conserving and advancing indigenous knowledge systems is examined in this chapter. In the modern world indigenous knowledge – which is frequently transmitted orally and through cultural practices – is at serious risk of eroding especially as technological advancements and globalization change cultural landscapes. In order to preserve these knowledge systems for upcoming generations this chapter looks at the function of 4IR technologies such as blockchain artificial intelligence (AI) and digital archiving tools. The chapter stresses how crucial it is to respect cultural sensitivities integrate indigenous knowledge into the digital age and make sure that digital preservation techniques are consistent with traditional values. It discusses the difficulties and possibilities that digital technologies bring to the development of sustainable culturally relevant and easily accessible platforms for the documentation of indigenous knowledge.

INTRODUCTION

The Fourth Industrial Revolution also known as the 4IR is the age, which includes the integration of digital, physical and biological components. AI and others including big data, blockchain, Internet of things (IoT), advanced robotics etc. have acted as key enabling and bring radical transformation across sectors and societies. As given that these technologies tend to reinforce our global society, they also foster distinctive possibilities and complications regarding the preservation of culture (Adigun & Enakrire, 2024). Embedded in conceptual systems and practices of indigenous people over generations of interactions with their environments, indigenous knowledge, therefore, occupies the crossroads of this revolution. However, this type of knowing is at risk of degradation due to globalization, urbanization and the fast advance in the complexity of technologies. Therefore, the use of 4IR technologies to support digital preservation of indigenous knowledge is a necessity (Adigun & Enakrire, 2024). And it is indisputable that indigenous knowledge systems are still of immense importance to the people who nurture them and

DOI: 10.4018/979-8-3693-7964-6.ch016

the society all over the world. They represent commitment to the conservation of the environment, and proper use of natural resources and diverse culture. For example, practices in skills of African agronomic cultures that were drawn from conventional knowledge have been effective in the fight against climate volatility. However, these knowledge systems are mainly available in an oral mode thus constituting a challenge especially with the increase in technological dissemination. The documentation of the same remains important, not as an anthropological heritage, but as a way of inspiring and providing solutions to modern day issues including climate change, loss of biodiversity and development (Rambe & Maime, 2023). The 4IR technologies present innovative means for encoding, storing and sharing indigenous facts about the world. Some technologies, such as blockchain, can put traditional practices that are shared in the public domain through social platform beyond the reach of personalization or commercialization. In a similar fashion, artificial intelligence can turn large collections of oral history, traditional cultural expressions, and language usage into databases that are easy to store. These digital archives can minimize generational gaps that could make the younger generation lack touch with their past in ways that a modern format appeals because they are always in touch with computers and the internet than books and maps (Adigun & Enakrire, 2024). In adopting such technologies, however, there is one cavity that is encountered by every investor. Questions of equality or inequality in the digital age, the ownership and protection of data, privacy, and ownership all require scrutiny. The purpose of this chapter is to examine how knowledge of 4IR technologies can be used to transform the preservation services offered by digital libraries in the preservation of indigenous knowledge. What is more, while considering these opportunities, it tries to consider the possible drawbacks of such advancements.

The chapter will also evaluate the impact of digitizing indigenous knowledge from the ethical, cultural and technological viewpoint. For instance, although, the impact of digitization mainly entails preservation, it has the pitfall of stripping knowledge of its culture thereby making it saleable or mis understandably assessable. There is always the line between pure preservation and integration of the new (Rambe & Maime, 2023). These dynamics will be illustrated with examples below: collaborative projects that include the use of AI in New Zealand, and the research shows how endangered languages can be rejuvenated through digitization of Māori language resources. On the other hand, conflicts in trying to patent indigenous knowledge on medicinal practices question the propriety of commercialization and methods that results to cultural misrepresentation. It also highlights the need for indigenous people's full participation as stakeholders respecting and incorporating their knowledge in preservation (Adigun & Enakrire, 2024).

BACKGROUND

Indigenous knowledge refers to the sum of the philosophy, beliefs, technology, practices, and impulses of indigenous people all over the globe. These knowledge systems carry extensive longitudinal, cultural, and ecological epistemology of the world, and therefore are highly valuable in today's globalized world when facing complex challenges that the world poses in the form of climate change, loss of species' habitats or biodiversity or sustaining development (Ndzendze et al., 2024). But, to some extent this cultural richness is already endangered primarily because of globalization, loss of native languages, and the exclusion of indigenous peoples. Due to the realization that there is a need to conserve this knowledge that held by indigenous people these new innovations brought about by the Fourth industrial revolution has brought the power of transformation to preserve and renew the indigenous systems in the contemporary world

(Suryadi & Firdaus, 2022). The Fourth Industrial Revolution, particularly AI, block chain, IoT and big data analytics have changed the way knowledge is developed, shared and stored. Indigenous knowledge may now be preserved, shared, and documented in previously unthinkable ways because to this technology. AI-powered technologies, for example, may be used to interpret and transcribe oral histories, helping to preserve endangered languages. Like this, blockchain technology provides a transparent and safe way to preserve intellectual property rights while documenting and disseminating indigenous customs. These instruments' combination with digital platforms has created new opportunities to preserve indigenous cultures' intangible elements, like rituals, oral traditions, and ecological knowledge, in addition to their physical elements, such objects and literature (Padeli & Dwita, 2025).

Notwithstanding the potential of 4IR technologies, cultural, ethical, and practical considerations must be carefully considered when incorporating these instruments into the preservation of indigenous knowledge. Because indigenous knowledge systems are intricately entwined with particular social, spiritual, and ecological settings, maintaining them is a challenging task. To prevent preservation attempts from unintentionally decontextualizing or commercializing the knowledge, technological solutions must take these subtleties into account. In addition, issues pertaining to intellectual property, data sovereignty, and fair access to technology continue to be major obstacles. The necessity for participatory techniques that give priority to the voices and views of indigenous people in the development of digital preservation efforts is underscored by the requirement that these groups have agency over the digitization, storage, and use of their knowledge (Suryadi & Firdaus, 2022). In addition to being a technological undertaking, the digital preservation of indigenous knowledge is also a moral necessity to save humanity's intellectual legacy and cultural variety. Collaboration amongst many stakeholders, such as governments, academic institutions, indigenous people, and technology suppliers, is necessary for successful preservation (Ndzendze et al., 2024).

The global movement to combine cultural preservation and technical innovation is shown by programs such as the UNESCO Indigenous Knowledge Program and the World Intellectual Property Organization's attempts to create frameworks for the protection of traditional knowledge. To meet the requirements of various indigenous populations, these initiatives must be expanded and modified (Padeli & Dwita, 2025). Offering insights into the prospects and problems in this ever-evolving subject, this chapter examines the relationship between 4IR technologies and the digital preservation of indigenous knowledge. The chapter seeks to provide readers a thorough grasp of how digital technologies might be used to preserve and honor indigenous peoples' wisdom by looking at international efforts, technology applications, and community-led tactics. It emphasizes the necessity for a future where indigenous knowledge flourishes in harmony with contemporary achievements by highlighting the need of cultural sensitivity and inclusion while utilizing technology for cultural preservation through critical analysis and case studies (Suryadi & Firdaus, 2022).

GLOBAL INITIATIVES FOR DIGITAL PRESERVATION

International concerns of preserving indigenous knowledge soils have increased as its exceptional cultural, environmental and scientific values are acknowledged. Various organizations, governments and societies of the world have embarked on undertaking efforts to preserve these knowledge systems given threats of disappearance. The main purpose is to guard and promote the indigenous people's knowledge assets to continue to occur in present day in the interest of the indigenous peoples and of the global society (Nkosi et al., 2020). Using these technologies, such modern endeavors have started to close the challenging chasm between conventional interventions and their modern-day counterparts. But this means that the path to efficient preservation is strewn with problems, and their analysis and improvement are subjects to scientific scrutiny (Uwaoma & Eloghosa, 2023). One of the examples of such initiatives it is the WIPO that has developed Indigenous peoples' cultural heritage framework and Guidelines for the endorsement of digital archives by indigenous peoples. Traditional knowledge documentation toolkit developed by WIPO gives proper insights into how to digitize knowledge while preserving culture. Another interesting project is the program called Memorial of the World of UNESCO that is aimed at conservation of documentary heritage in the perspective of using new technologies, such as indigenous manuscripts and oral traditions (Nkosi et al., 2020). The suggested programs have stressed the significance of cooperation, in the preservation efforts and with indigenous peoples and communities around the world (Olaitan & Mapanga, 2024; Jane, 2005).

Several case studies based on successful projects indicate the potential and the challenges of digital preservation. Another product is the "Mukurtu CMS" designed together with indigenous peoples which is a digital archive tool designed about their culture. For example, the community owners can restrict who has access to some forms of content, which makes sacred or restricted knowledge remain sacrosanct. Likewise, the "Endangered Languages Project" sponsored by Google has develop a catalogue of threatened languages which include sounds and videos, texts, and few others to save the language culture of further generations. These examples explain how even local communities can use advanced technology to take charge of their own cultural assets (Uwaoma & Eloghosa, 2023). Conventional patterns, nonetheless, put into light primary accomplishments, experiences from other countries indicate essential queries in molding. Several projects have been confronted with problems of access to technology, use of technology or inadequate funding. For instance, several good-meaning initiatives failed: it was impossible for communities to maintain digital archives by including appropriate facilities; or people failed to make proper use of the projects appropriately due to insufficient means on how to search and analyze the information (Nyagadza & Hove, 2022). Also, concerns involving copyright or property rights are also realized mainly due to portrayal of the external organizations taking over the digitalized cultures facilitating conversion of cultural heritage into commercial commodities. These challenges pose the importance of strengthening community capacity, coupled with fair collaborations with communities where indigenous stakeholders are involved (Olaitan & Mapanga, 2024). A detailed evaluation of the work done on the international level shows that there is always a conflict of interest when it comes to extending the media to the international market on the one hand and at the same time maintaining each culture of the world on the other hand. On the advantage side of digitization, it facilitates wide sharing of indigenous knowledge with others while on the disadvantage side, the digitization placed the indigenous knowledge out of the cultural setting that defines it takes away its cultural value (Nyagadza & Hove, 2022).

Further, the increased sales of the indigenous knowledge, for instance use of traditional medicines has led to controversy on biopiracy and distortion. These problems describe why ethical rules and Plains culture principles combined with involving indigenous peoples into every step of the preservation process should be elaborated and acted (Uwaoma & Eloghosa, 2023). The urge towards digital preservation globally is evidence of acknowledgement of the importance of indigenous knowledge system. But the road map to this output requires regulation and appropriate technological advancement without infringing the cultural rights of people. Therefore, by analyzing the progress of successful initiatives and the problems in their implementation, the subsequent projects can protect indigenous knowledge across the world in preserving cultural heritage and making it applicable in the contemporary global environment (Nyagadza & Hove, 2022; Tariq, 2024).

CHALLENGES AND OPPORTUNITIES IN DIGITAL PRESERVATION

Digital preservation of indigenous stand entails many problems despite the promising directions provided by technological advances. Another worthy cause is the inhibitory factor associated with the sensitivity of indigenous knowledge systems that are in most cases embodied in traditions and passed from one generation to another through word of mouth (Niyitunga, 2024). The procedure of capturing such knowledge electronically is unproductive and disruptive because it detaches practices from their contexts or even changes them. For example, traditional dancing or chanting or traditional forms of expressing or gaining healing might reduce or minimize its meaning when conveyed through technology. This is made worse by the fact most indigenous communities are lacking in infrastructure, technical knowhow, and capital that is instrumental to meaningful contribution towards preservation (Nyagadza & Hove, 2022). There is also substantial concern about the unethical nature of digital preservation as a major challenge. They are usually exposed to bio-piracy in a way that their knowledge is harvested or used for commercial gain without permission. Studies including the biopiracy of traditional medicine practices showed that even priceless cultural resources could be stolen by business entities or scholars. Socially, these practices put into doubt the ownership of other assets such as intellectual property, data and sharing of the benefits that may be got from the indigenous knowledge. In addition, since most digitization processes are done in various vernacular languages, marginalized groups like women, the poor, and the disabled cannot get the various digital jobs hence they also suffer from cultural decay (Ojo & Ndzendze, 2025).

However, the 4IR technologies present immense opportunities in handling these problems as is discussed below. For instance, blockchain helps legitimate digital records' sources and guard against piracy. Explaining decentralized, private structures, blockchain can assist communities in protecting the possession of their know ledge, and achieve better consent (Niyitunga, 2024). AI and machine learning tools are ideal for the fast digitization of various oral histories, languages, and traditions, which greatly convert massive collections of cultural information into digital form. For instance, it is possible to use the services of artificial intelligence-based transcription to save endangered languages by translating the spoken words into the written ones, for instance in the Cherokee Language Program in the United States of America (Nyagadza & Hove, 2022). For optimally utilizing these opportunities, a mechanism to address technical, ethical and cultural challenges should be planned. Indigenous people must be encouraged to engage in preservation as capacity building initiatives that entails offering digital literacy and preservation training enhance the involvement of indigenous people. Community-based interventions through supporting shared ownership and leadership in the creation of technology-mediated initiatives are likely to produce technologies that are culturally attuned and acceptable. WIPO for instance has guidelines that could be used to promote equitable use of indigenous knowledge in the social media platform (Ojo & Ndzendze, 2025). Thus, this chapter has centered on innovative fourth industrial revolution technology's advantages in not only preserving but reinventing indigenous knowledge systems. Through positive solutions to these challenges, indigenous knowledge can be correctly kept as a best practice for supporting sustainable and constructive cultural transmission in a rapidly advancing digital world (Niyitunga, 2024).

TOOLS, SYSTEMS AND TECHNOLOGIES FOR DIGITAL PRESERVATION

New technologies have made it easier to document and preserve the indigenous knowledge through enhanced tools and system for documentation, storage and dissemination of cultural materials. These technologies aid in documentation and preservation of the oral histories / relics/ languages/ and other culturally/spiritual practices etc. which otherwise are at risk of being lost in the future. At the heart of these developments is digitization of both archives and their respective metadata as well as the collaborative work environments where it is done (Ajani & Dlamini, 2024). Application like Mukurtu CMS which have been designed for indigenous people are actual examples of CSMs that are responsive to cultural requirements and them-based sharing options. This way, the platforms allow communities to be responsible for the preservation of their data and thus promote a form of preservation that is culturally sensitive (Zhou & Chibaya, 2024). AI has a critical role in the modern development of digital preservation. Machine learning in its basic and complex forms can guarantee the processes like transcription, translation and preliminary organizing of IK information. For instance, during Linguistics Documentation and description, machine learning has been utilized in transforming spoken languages into analyzable and searchable data bases. A good example is the Living Tongues Institute for Endangered Languages that employs its AI to develop digital dictionaries and record how various words and phrases are pronounced to help save on Extinct language (Ogur, 2024).

In the same way, it is possible to recognize and recover originals of manuscripts and other historical items and electrons them for their prolonged preservation. These technologies' analysis highlights the make-or-break nature of inclusion and participation. All design and deployment of digital preservation programs should include indigenous participation in decision-making processes and use the technology (Ajani & Dlamini, 2024). In addition, using the open-source instruments and cooperation with non-profit organizations can solve problems concerning the question of resources to introduce the newest systems. In this way it will be possible to incorporate the tools and technologies as they develop into the toolbox of the indigenous scholars in ways that are appropriately culturally sensitive; conversely, as the current presentation of tools and technologies shows, it is easy to incorporate a tool or technology which is not culturally appropriate into the toolbox of indigenous scholars in a way which nullifies the potential benefits of the tool or technology to indigenous knowledge (Zhou & Chibaya, 2024).

INDIGENOUS KNOWLEDGE SHARING TECHNIQUES AND STRATEGIES

Indigenous systems of knowledge transfer have always utilized oral means of communication, storytelling, rites and passage, and apprenticeship through which the knowledge bearers transferred knowledge from one generation to the other. They make practices continuous, the knowledge is then associated with its culture, with the community. For instance, the Inuit in the Arctic pass information on survival and living sustainably in the Arctic through storytelling. These practices, however, are run into some challenges in the current society and context such as urbanization, new generation and the decrease in number of knowledge bearers (Oguamanam, 2021). New technologies have brought in fresh ways you can share and preserve indigenous knowledge. Technology has improved the practice in that digital platforms, video documentation, and online repositories of women documentation have expanded. For example, the Indigenous Knowledge Centers that incorporate Multimedia in developing libraries or museum disseminate oral narratives, music and visual arts. These technologies potentially hold ways of getting to the younger people since many of these generations have daily interaction with digital content. However, issues to do with its appropriation of culture and loss of context make it necessary to develop these tools hand in hand with the appropriate cultures (Ogur, 2024). Implementation of such tools with indigenous methods constitutes one of the main challenges that should be carefully planned. One strategy enables the production of digital assets in collaboration with community stakeholders; this shall guarantee cultural resonance and accuracy (Adigun & Enakrire, 2024). For example, applications of augmented reality can also replicate concrete ceremonies or arts and crafts, giving educational value for its tourists, but at the same time following cultural norms. Accessibility remains also a concern; this is because transformed strategies must also consider offering interfaces in languages that include indigenous ones. Applications for usage in supporting indigenous plants or shared digital narratives, which integrate oral histories with innovative technologies also preserve and encourage both (Oguamanam, 2021).

It is also shown in case that well-designed initiatives are indeed possible. The Our Languages website of Australia has catalogued more than sixty indigenous languages and provides tutorials and tools for experimenting with those languages. Likewise, the Zuni Elders and Youth Program in the United States gets video recordings and online portals present to tie youths with original teachings and pass on knowledge from one generation to another. Such initiatives aim to stress flexibility as knowledge sharing develops in a world grounded in tradition (Adigun & Enakrire, 2024). A critical evaluation of the knowledge-sharing literature shows that the merely considered efforts depend on the strategic fitness with the community culture. Thus, the potential of digital tools for growth and innovative solutions should not overshadow the fact that in their absence the authors discreetly commodify or decontextualize ritual practices. There is a focus on its continued longevity, to involve the people and to ensure respect for their culture and countries' sovereignty. Allowing the communities to set the boundaries regarding the SH capabilities guarantees that the strategies bolstered by innovation uphold the cultural altitude of continuity in their traditions for posterity (Oguamanam, 2021; Tariq, 2025).

DECOLONIZATION AND INDIGENIZATION OF KNOWLEDGE SYSTEMS

Self-empowerment of indigenous knowledge systems is critical to the formulation of decolonization and regain of culture and identity. From the dawn of Colonialism, indigenous epistemologies were downplayed and erased and replaced by a colonial knowledge system. This suppression distorted the practice of knowledge management and deprived communities of their idea legacy. To decolonize such systems is to attempt to undo this type of epistemological violence, to reclaim the primacy of the documented indigenous ways of knowing. They entail eradicating colonial structures that prioritize western forms of knowing and asking that indigenous peoples head the processes of preserving and reimaging their culture according to indigenous dictate (Agbehadji & Ngowi, 2021). A significant approach to decolonization is always the return of the authority of the knowledge systems to the indigenous people. This entails ownership of digital archives, control over access and rights to decide how knowledge is to be recorded and disseminated. The Mukurtu CMS platform is an example of this approach by adopting collaborative projects. For indigenous peoples, Mukurtu offers filters according to their specific customs so that the communities themselves decide the limits and parameters for information access so that certain knowledge is protected. Their undertakings call into question the extractive methods of conventional anthropology and instead rebuild the power relations system. Foreignization of digital preservation is inversely proportion with indigenization and Africanization as they are intrinsically connected to Decolonization. Some of these processes involve the assimilation of the indigenous peoples' knowledge base, their language and culture into the means and approaches to conservation. For example, in Africa, such as the African Digital Heritage initiative, it is stated methods honoring cultural norms for preserving the traditions. This has entailed the use of local languages in digital archives, the display of oral historiography including the participation of indigenous people, and the use of participation design in architecture and design. From the use of digital storytelling in Nigeria, the fight to preserve the Yoruba folklore preserves local narratives using technology while embracing older methods (Agbehadji & Ngowi, 2021). Examples are given to illustrate that when decolonization and indigenization is applied, notable changes occur. In Canada, First Nations principles of Ownership, Control, Access and Possession (OCAP) are used to guide digital preservation. These principles mean that the communities where the research is conducted have complete authority over their data as a way of overcoming biased relationships behind the regular academic or government surveys. In the same way, the Maori "Te Reo Irirangi o Te Hiku o Te Ika" project in New Zealand brings back the dying Maori language by developing AI applications to teach and preserve the language that lays stake on Maori cultural principles (Tariq, 2024). However, critical analysis brings out the fact that the whole process of decolonization and indigenization has its difficulties. Their achievements are sometimes hampered by issues of structure, funding, and digital instrument availabilities.

However, with the act of incorporation of the modern technology it becomes quite challenging to protect the cultural integrity of the knowledge being stored without compromising it with modern techniques of knowledge management. For example, although AI can help protect languages, biases in data can result in preposterous or stereotyped representations or language and complicate what are very well-organized systems. This underscores the need for indigenous participation in every stage of the process of designing technology and its implementation (Agbehadji & Ngowi, 2021). Therefore, the decolonial turns and Indigenous epistemological frameworks are not only about protecting, asserting knowledge sovereignty but guarding the victims', Indigenous peoples, and minoritarians' justice. In asserting the sovereignty of indigenous peoples for technology-mediated archival endeavors on behalf of the survivors of colonialization, these initiatives underscore the sustained significance of indigenous knowledge, as well as redress justice. As such, to achieve the dreams of a post-colonial world where the indigenous knowledge systems have, they dominated the modern world, it will be very important for the nations to continuously embrace partnerships, cultural sensitivity and digitalization.

INDIGENOUS KNOWLEDGE IN HEALTHCARE AND AGRICULTURE

Traditional knowledge is valuable for healthcare procedures because it provides information regarding surrounding environment and organism, local plants with curative properties and balanced attitudes to health. In many indigenous cultures there are people commonly referred to as "medicine men and women," who have insight into the uses of natural resources for healing. For instance, in central South America and specifically in the rainforest, indigenous peoples have continued to use plant material such as the Quinine's barks from the cinchona trees, which is known to cure Malaria. Likewise, in Africa, traditional information about plant uses for treating different diseases, for instance, the bitter kola nut, is applied. Such knowledge is usually taken through generations, and thus it complements current practices in the health sector. Nevertheless, this knowledge is under threat of overshadowing by globalization as well as gradual encroachment of industrialization. Researchers have been able to depict an example in the use of digital databases that archive indigenous related health care information like the traditional medicines and healing. Nevertheless, which questions are in question is a debate on the ethical use of this knowledge, arguments over the intellectual property rights and cultural appropriation (Awuzie & Moghayedi, 2024). Thus, agricultural indigenous knowledge is also important in sustainable farming practices. For thousands of years indigenous farmers have adapted methods for growing crops, for maintaining soil fertility, and conserving biodiversity of the immediate ecosystems. Most traditional practices of farming are practiced round the globe like polyculture and agroforestry that teach that to survive that is spearheaded by indigenous people, it is important to cooperate with nature and not exploit it. For example, the milpa pattern of cropping practiced by native people of Meso America includes several crops in the same field and it prevents soil erosion and keeps off synthetic fertilizers from the field. They help succeed the current state of ecological balance, are integrated into climate change resilience frameworks, and are beneficial to the overall function of supporting prolonged food security. Presently, many modern farming processes incorporate these techniques; the Task Force of the Food and Agriculture Organization (FAO) global, supports the incorporation of indigenous knowledge in the present world farming procedures. However, the question seems to arise in a way today's industrial farming system can adopt these practices while the process is organized for efficiency and quick returns (Ogedengbe & Eboigbe, 2023). As a result, digitization of indigenous healthcare and agriculture systems knowledge is vital to ensure Contin adage of these practices. Traditional knowledge can be documented and database archived by digital platforms and hence made for public use. For instance, the "Digital Preservation of Indigenous Knowledge in Agriculture" project that is happening in India is to archive ma- agricultural practices from the various indigenous peoples, practices that encourage diversity and sustainable use of resources. Likewise, the World Health Organization has endorsed the effort to index traditional medicine to make sure that indigenous information is incorporated into healthcare. However, there are always emerging issues regarding data sovereignty, which is the ability of an indigenous community to own or share the data fairly taking benefit of the knowledge gained from it. The difficulty lies in navigating between preservation of the online content and a no less important task of preserving cultures from external invasion. Nonetheless, digital media reveal potential routes to extend and disseminate indigenous health and farming knowledge for better climatic health and environmental fitness worldwide (Awuzie & Moghayedi, 2024).

ROLE OF MEMORY INSTITUTIONS AND CULTURAL HERITAGE AGENCIES IN PRESERVING INDIGENOUS KNOWLEDGE

Libraries, archives as well as museums are among memory institutions that are significant for preservation of indigenous knowledge, particularly holding such knowledge as cultural memory. Even in today's technological advance world these institutions have all the might, skills and facilities to obtain gather, document and store artifacts, manuscripts and oral history which are the bulwark of indigenous knowledge systems. For instance, history and language specially, may be documented in scripts by librarians and archivists; cultural and family history, may be stored in old records; object culture such as clothing, farming utensils, paintings can be conserved in museums. Some of these institutions also have archives of oral tradition and histories, which are ingredients of indigenous knowledge, thus provide a picture of indigenous peoples' life and perceptions (Koopman & Koopman, 2023). Yet the interaction between memory institutions and indigenous communities is a contested one in most cases partly due to ownership and management disputes. To many indigenous peoples, the practice of indigenous cultural materials being collected, archived, and catalogued by and for non-indigenous institutions continues to be a project of extraction: with little reciprocity or agency from the communities whose knowledge has been so aggregated. This has raised the need for a more holistic and honorable way of protecting indigenous knowledge base (Chisita & Ngoaketsi, 2021).

Decision making process by CH agencies and indigenous organizations is crucial on how best indigenous knowledge can be archived in a manner that has a positive impact on those communities. Such a partnership entails a change of paradigm from the archival role of memory institutions to the approved knowledge-sharing roles. Cultural agency must consult native peoples to close the circles and guarantee the native stuff is collected in the proper ways that will protect it from distortion or misuse. For instance, in New Zealand, materials owned by the National Library of the country have been involved in digitization of Maoris' oral history so that it is not only retrievable but also represents the Maoris. Likewise, AIATSIS provides guidelines for archiving and disseminating Aboriginal and Torres Strait Islander's materials using their consent and participation in the dissemination of the knowledge that adheres to cross-cultural practices (Francis & Hagenmeier, 2022). Positive cultural heritage agency- indigenous people partnership was deemed effective to offer lessons on similar partnerships in future. A very good example is the relationship that has been formed between the Smithsonian Institution's National Museum of the American Indian and indigenous peoples of the USA. This relationship has caused the return of artefacts and has made indigenous groups to have a say in how the arts that culturally represent them are curated (Koopman & Koopman, 2023). The UNESCO sponsored the Memory of the World in the same context with a project, aimed at the digitization of indigenous languages and traditional culture. These initiatives show that to develop effective partnerships, it is crucial to understand each other, as well as being essentially open and willing to collaborate based on the decision-making powers of indigenous people to shield their information and their culture (Chisita & Ngoaketsi, 2021). But analysis of the case studies suggests that some issues persist when entering these partnerships such as lack of sustainability and adverse tendencies toward the commodification of indigenous knowledge. Though these collaborations are a step in the right direction, the effectiveness of those contracts will only be confirmed by further work to check the exploitation of minorities and indigenous people and by their ability to control production of their cultural artifacts. Furthermore, to ensure that indigenous languages and traditions continue to be alive, dynamic systems of knowledge for future generations, it is necessary to broaden the scope of these projects to include not just knowledge preservation but also their revival (Francis & Hagenmeier, 2022).

IMPACT OF POLICIES AND PROTOCOLS

It is an important reason why indigenous knowledge should be protected through policies and protocols to enhance integrity and cultural value of indigenous people. In the last few decades, nation states and other international systems, and indigenous people have come up with various policies and frameworks on the recognition of indigenous knowledge. UN Declaration on the Rights of Indigenous Peoples (UNDRIP) which is a standard framework, protects indigenous people's rights to preserve and develop their cultural identity and heritage. In a statement that underlines the call for EPS member group sovereignty on the use of its knowledge, it demands that indigenous knowledge systems should be respected alongside scientific systems of knowledge. National policies in Canadian, Australian, and

New Zealand countries also indicate these principles recognizing indigenous people's certain rights to cultural goods and advocating self-determination concerning knowledge assets management (Kruger, 2024). Such big and important policies or protocols that are formulated are still not very effective and productive at large and in practice. For instance, although UNDRIP sets out frameworks for the protection of indigenous knowledge, there is no directive means of implementation. Likewise, in specific countries native populations may have legal ownership of their knowledge heritage but the policies do not offer adequate capital or technical assistance to apply the protection. There may be favorable laws in one region but then there may be loopholes in another and since it is not global jurisdiction, these loopholes leave indigenous knowledge open for exploitation.

Probably one of the largest topics of discussion is the problem of Intellectual Property Rights where indigenous knowledge is recognized neither under the intellectual property laws as they are formulated with euro American ownership systems in mind (Kruger & Steyn, 2024). Due to this kind of policy exclusion, indigenous knowledge can be used by commercial users in so far as policy makers and governing agencies do not intervene, thus areas such as the use of indigenous knowledge in the production of pharmaceuticals, agriculture, and tourism may experience commercial exploitation of indigenous knowledge. When contemplating technological and organizational changes needed to support long term preservation of digital resources, these policies have far-reaching implications. On the one hand, new digital technologies provide a unique chance to record, store and make accessible indigenous knowledge which, in turn, might help avoid its loss (Kruger, 2024). But interestingly, the digital world is not risk-free either. If there are not enough policies in place, there is a possibility that whoever tries to engage with indigenous knowledge will just do so in ways that are exploitative, misrepresentative, or extractive without goes through proper process/policy to first consult with those indigenous peoples whose knowledge is sought for. Digital preservation therefore should not be entered in a blind and mechanical manner but with precaution to make certain that indigenous communities are being protected. For this reason, there is need to develop policies that deal with ethical usage of indigenous knowledge especially in the online platform to prevent misuse of the same. For example, protocols for decision-making and communities' agreement, data sovereignty, and the specification of rules for access and utilization are critical so that interoperable digital curation complies with the indigenous persons' rights and values systems (Kruger & Steyn, 2024). The specific suggestions for the enhancement of relevant policy are as follows: it is suggested to work on the improvement of legal frameworks that would give enough protection to indigenous knowledge. That is; one of the recommendations is development of special IP laws that will prevent cloning of indigenous knowledge but at the same time provide means for proper access and share. Also, there is a need for policies that would promote the engagement of the indigenous people in all processes in the documentation and digitization of indigenous knowledge, in the design of the access right policies on digitized materials, among others. Governments and international organizations should also fund capacity development activities which will enable the indigenous people and organs to have adequate capacity to deal with the protection of knowledge in the digital environment (Kruger, 2024). Last but not the least; policies can be further improved, consolidated in a way that the indigenous knowledge not only is saved but also is promoted and incorporated into the international land national debates on science, culture and even education. Therefore, the paper concludes by arguing to the effect that while there has been tremendous progress on the part of policy makers in putting into place policies and protocols that protect indigenous knowledge, the job is far from done. Not only does indigenous knowledge require legal protection and adherence to ethical norms; indigenous people can and should play a proactive role in protecting their own knowledge in the digital environment. Otherwise, the chances of exploitation and erasure persist that is why indigenous communities struggle to reclaim their assets (Kruger & Steyn, 2024).

INDIGENOUS KNOWLEDGE AND SUSTAINABLE DEVELOPMENT

Indigenous knowledge contributes to the satisfactions of the SDGs whenever it delivers simple, culturally appropriate solutions to the problems facing people in their sustainable development endeavors in environmental, food, health, and cultural aspects. CKW and indigenous knowledge systems are holistic in approach and have been developed over time and therefore are in harmony with UNESCO's global sustainable development goals, including climate action (SDG 13), water sustainable use (SDG 6), and conservation of terrestrial ecosystems (SDG 15). For years, the UN has shifted to reframing the perception of indigenous knowledge as a severe input towards sustainable development concepts that are inclusive, effective in providing resilience, and equitable (Ogedengbe & Eboigbe, 2023). This chapter looking at indigenous knowledge, bring out a case which involve sustainability, that is, traditional agriculture. The indigenous people have been practicing agriculture in ways that have ensured soil productivity, increased on the variety in the production system and increased production especially in changing climatic conditions. Toes like those of the Andes, indigenous growers employ terracing methods to counter erosion and increase production throughout the slope. All of these are in line with the support of the SDG 2, also known as Zero Hunger and promoting sustainable use of Soil for Agriculture. In the same way that water harvesting enhances the efficiency of water use to support SDG 6, so too do traditional methods, such as India's stepwells and Peruvian amunas systems. These present ways show that indigenous knowledge has solutions to paramount global problems without overexploiting the environment (Mafikeng, 2022).

Approaches to utilizing indigenous knowledge in sustainable development involve partnership, capacity enhancement and technology adoption. Engaging indigenous people in the development projects that will be implemented in their midst will guarantee that whatever is done whether developmental or corrective is appropriate. For instance, harnessing participatory mapping strategies, indigenous populations in the Amazon region have been enabled to implement own mapping systems of their occupation of the forests, as a means of protection of trees and at the same time as a source of livelihoods. This approach helps in the implementation of SDG 13 with regards to climate response via community led forest management. Further, whenever incorporation of indigenous knowledge in national education systems is practiced, this encourages the grown as well as sustainable practice of these optimal knowledge as rooted in indigenous practices. Programs such as the integration of Aboriginal ecological knowledge into Australian school syllabi reveal that education remains important in the transition between the indigenous elders' and any scientific knowledge to bring out developmentalism propped on the fourth sustainable development goal, quality education (Malgwi & Sadiq, 2023). Nevertheless, integrating indigenous knowledge into sustainable development has considerable challenges. One of the weaknesses is lack of representation of indigenous people in decision-making concerning policies and resources. It is regrettable that most governments and institutions today, continue to adopt western thought systems and paradigms, while they disregard native frameworks. Moreover, another drawback is that lack of secure tenure right and external intervene to control resources undermine indigenous organizations. As for these barriers, what is needed is legal and institutional changes that would respect and protect indigenous peoples' rights and recognize them as useful and active partners to harvest the Sustainable Development Goals (Mafikeng, 2022).

Successful examples of indigenous knowledge application to a sustainable development agenda are discussed by cases. In Kenya, Maasai people use the latest methods in grazing that preserves soil fertility and avoids excessive use and as such, concurrently achieving SDG 15 of 'Life on Land.' In a similar way, the Inuit people of Canada are also involved in the fight against climate change since their traditional but effective way of studying sea ice provides scientific researchers working in the country with knowledge on climate change in accordance with SDG 13. These cases show how indigenous knowledge can be proactive in providing localized and efficient solutions that are also feasible (Malgwi & Sadiq, 2023). Therefore, indigenous knowledge is a critical resource for the realization of the SDGs, using system-based, contextual and sensitive approaches to development. However, for this potential to be realized to the optimum, there are important steps that must be taken when embracing indigenous knowledge, these include acknowledge and respect of indigenous people's rights, create inclusive systems and structures together with establish partnerships that blend traditional EC and modern technology. Community-based knowledge system promotes that sustainable development strategies advocate for a sustainable and equitable future (Ogedengbe & Eboigbe, 2023).

4IR TECHNOLOGIES IN DIGITAL PRESERVATION

The emergence of the fourth industrial revolution (4IR) technologies have advanced the nature, techniques, tools and systems of digital preservation to document, protect and share knowledge. In this respect, these technologies offer ways of preserving the indigenous knowledge so that they will remain useful to future generations. AI, IoT and blockchain have been more effective in addressing the requirements, using innovative strategies to protect the wealth of the indigenous people's oral history, ecology and spirituality (Makazhe & Maramura, 2023). To support indigenous knowledge, AI has a critical role since it can document and process indigenous knowledge automatically. Recent advances in machine learning can also translate endangered languages, transcribe oral history and assess the style and pattern into traditional art. For example, AI transcribe technologies have been applied in the process of collecting and storing of the oral history genres of the Maoris in New Zealand. Moreover, natural language processing helps to build digital dictionaries for indigenous languages and, therefore, to revive language heritage (Niyitunga, 2024).

The IoT fosters the preservation process by adding sensor-based systems used in the protection of cultural objects and indigenous natural assets (Tariq, 2025). For instance, the use devices of the Internet of Things IoT make it is possible to monitor changes in the environment of indigenous sacred territories, especially in relation to rights as the data is relayed to the groups in the real-time basis. Such systems make it possible to encompass conservation not only of cultural items but also of environments supporting indigenous practices. By the uses of blockchain technology, it is possible to create a safe and transparent means of handling issues to do with ownership of indigenous peoples' know-how. Through the development of decentralized ledgers of ownership and consent, blockchain guarantees sovereignty over knowledge assets and knowledge sharing and utilization. For example, in Australia, it has already been used to log the origin of the indigenous arts to combat counterfeits and guarantee that proceeds end up with the artists (Makazhe & Maramura, 2023). Examples are made to show that such technologies have revolutionary characteristics. First Nations Digital Archive in Canada incorporated AI and blockchain to validate KE and provide means to indigenous peoples to engagement in climatic change discussion using validated fundamental ecological knowledge. Likewise, in India, through implementation of Documentation and tracking using IoT systems of the traditional Agriculture systems has helped in the improvement of food security and agricultural biodiversity. However, despite these developments' antecedents remain. Implementation costs, technological reach, and issues of data sovereignty continue to be issues for many indigenous peoples. There is a need for mutually beneficial partnership, which has recognized the inclusion of indigenous peoples and respect their sovereignty as the key to overcoming these challenges. If all the 4IR technologies are used wisely then digital preservation can be an important way to honor and preserve the difference in human culture (Niyitunga, 2024).

MODELS AND FRAMEWORKS FOR DIGITAL PRESERVATION

Digital preservation therefore involves the use of models and frameworks that will organize a way of preserving cultural, historical and indigenous knowledge in a digital format. These models dictate how documents are created, used and accessed, and how they are maintained in the long run so that valuable information is not archived for future use or gets wiped away by progression. In the past, preservation models such as O_AI still provided the high-level conceptual architecture for preservation. The OAIS model, used in archives and libraries, is focused on the ingest process, preservation, and delivery of information products and its features concern with the authenticities and integrated aspects. Nevertheless, due to advances in technology and the rising needs of complexity and inclusiveness in digital preservation, new frameworks implementing 4IR technology become available (Mawere, 2024). Current models, although quite sound, lack some basic features that may be essential in the case of indigenous knowledge, including context awareness and information diffusion using spoken form. This gap has prompted development of frameworks that deploy new generation technologies, such as artificial intelligence (AI) technology, blockchain, and the Internet of Things (IoT). For example, AI frameworks can bring algorithmic solutions for classification and speech-to-text component that will keep the qualities of effectiveness of the oral histories. Likewise, in cases of indigenous peoples, blockchain-based systems offer firm and indisputable record of ownership and consent, which helps safeguarding of such people's intellectual property rights (Nyagadza & Hove, 2022).

To create new frameworks for management, close attention should be paid to the issues of encompassing representatives of Indigenous peoples and their active participation in the process of its creation. For instance, the FAIR data principles were adopted to identify the ethical principles to be used in indigenous contexts to incorporate the community-driven methods. These adaptations help make sure that frameworks protect cultural values and sovereignty so that everyone can trust how preservation really works. About these evolving frameworks contemporary cases provide insights. The Mukurtu system based and created and for Australian Aboriginal peoples incorporates Traditional Ecological Knowledge frameworks with regards to usage and access of content. It provides paradigm of how pro-technology frameworks can be used to address the social aspects of cultures. Likewise, Digital Himalaya Project, which attempts to record the culture of Himalayan people, apply modular architectures that includes artificial intelligence analysis of the data collected, while the outcome is made open accessible (Nyagadza & Hove, 2022).

As much has been detailed and critiqued elsewhere there are difficulties associated with the application of these frameworks, namely cost, technology and cultural sensitivity. For example, thereby limiting accessibility to communities without technical support or experience with the tools that are used. To solve these problems, there are possibilities to use partially traditional and digital methods that unite possibilities and help to expand using. Thus, the models and the frameworks directing and enabling digital preservation must also transform and aspire to do so with the help of 4IR technologies while being as flexible as to avoid disrespecting the cultures of its target sources. Intervention strategies that identify possibilities for preservation of cultural values and human assets require a jointly appointed and interprofessional approach (Mawere, 2024).

WEB-ARCHIVING OF INDIGENOUS KNOWLEDGE SYSTEMS

Consequently, Web-archiving helps in preservation of indigenous knowledge systems as it avails an opportunity to correct encyclopedias in the digital age controlled by online platforms. Globalization and technology threaten to erode traditional knowledge handed down for generations from one generation to the other. Web-archiving opens an opportunity to preserve these traditions by archiving and cataloguing the digital content concerning language, rites, ecology, and history. This way it also guarantees accessibility to future generations and performs educational and advocacy functions (Mkhize, 2024). As highlighted below, web-archiving depends on the use of sophisticated tools as well as the best practices. Web crawlers like that used in the Wayback Machine owned by the Internet Archive play a key role in archiving indigenous knowledge Websites. For instance, endangered language or traditional medicinal practices archive may employ these tools to gather information from community enforced projects or government organized and implemented projects. Another important method to is metadata creation as it facilitates locating the materials stored in the archives as well as providing them with some background (Tariq, 2025). Applications like Archive-It, a web application for the organizations involved in cultural heritage, comprise the ways in which advanced communities set the priorities regarding the archiving process and organize their content in a culturally appropriate manner (Mlambo et al., 2024).

Other successful cases of web-archiving projects are presented here to show the possibility of this approach. For example, the Endangered Languages Project primarily employs web-archiving as a means of documenting World's linguistic diversity by collecting the relevant videos, audio, and text resources. It does not only collect endangered languages but also serves as a basis for language revitalization. Likewise, the Native Land Digital project collects and preserves mapping and other materials that illustrate indigenous lands and backgrounds so that it stays relevant to support advocacy and education cues (Mkhize, 2024). However, web-archiving indigenous knowledge has advantages, yet it also has its drawbacks. Concerns about ethics of the ownership and responsiveness of the archived content are in place, indigenous communities can be exploited or misrepresented. Also, technical factors which have impounded participation in archiving activities include lack of reliable internet to access both voice and data, especially in the rural areas. To solve these problems, it is necessary to use models that imply active cooperation with the community and its agreement. It is important to recognize the gaps in communications between researchers and indigenous peoples Particularly, it appears that to fill these gaps, it is necessary and useful to develop fissiparous instruments, firstly to find partners and establish cooperation with indigenous organizations, secondly to address the creation of appropriate instruments, which would be accurate in consideration of the culture of local populations (Mlambo et al., 2024). Alternatively, web-archiving proves to be a potentially rich technique for preserving and documenting indigenous knowledge systems for posterity. Thus, the presented web-archiving approach may be used both as a protective shield and as a tool to promote cultural minorities' activities. Successful realization of such efforts depends on application of technology and complementarily of the moral imperative to have indigenous people continue to remain an active part of the evolving global identity (Mkhize, 2024).

FUTURE RESEARCH AND DIRECTIONS

Digubybo, the practice of preserving indigenous knowledge through modern technology is very dynamic due to the changing nature of trends and technological innovations that exist with the potential of improving both the availability, reliability and longevity of cultural assets. This is one of the biggest trends as more and more AI technologies are implemented to transcribe and translate indigenous languages and traditions, as well as create catalogues for them. Some of the applications of artificial intelligence include using natural language recognition and understanding to transcribe oral histories and oral literatures that are crucial components of indigenous knowledge, important segment of every souring culture that is frequently at risk of dying out. For instance, AI has applied to reduce the visual and oral written form of indigenous languages like Cherokee and Māori and help in their reclamation by digitization to the learners (Nkosi et al., 2020). Another developing trend has been the use of blockchain technology in a way that will protect the ownership of intellectual property for Indigenous people and the sharing of such property ethically. Blockchain brings transparency and immutability to the history of ownership and consent required to preserve traditional knowledge from infringement. Several indigenous communities now use blockchain to control the assets that contain this valuable indigenous knowledge for the good of their communities and not for third parties. It is a most promising trend for the establishment of an efficient and safe system for sharing of knowledge (Zhou & Chibaya, 2024).

This chapter has also identified that the future trends in integration and use of technologies in preservation of digital content would involve the sophistication of reader friendly tools that integrate traditional knowledge-based practices into the digital space. One of them is the continuation of the use of virtual and augmented reality and practices in which signs are created in digital space and users can interact with them in simulations involving historical and cultural contexts to support indigenous knowledge. These technologies could also assist in the struggle to connect native knowledge with digital databases for the present by providing a simulation of the indigenous actuality with realistic interaction for the districts community and the public (Nkosi et al., 2020). Further, digital preservation as a field is expected to have a stronger comparative of disciplinary cooperation between computer science specialists, anthropologists, indigenous knowledge bearers, and legal scholars. It may also mean the development of better frameworks whereby indigenous peoples' rights, dignity and sovereignty are protected as well as cultural appropriation of the techniques taken into preservative digital methodologies. One of the important directions for further investigation is the creation of expanded and liberalized technologies that could address inflicted-based communities with no or poor Web access and restricted gadgetry options, and to avoid isolation of various communities from the digital preservation procedures (Mogajane, 2022; Baporikar, 2022).

In this chapter, the authors have sought to understand how the application of new preservation principles in digital environments has unfolded in relation to indigenous knowledge with specific regard to Fourth Industrial Revolution (4IR) technologies. We started with issues of the efficacy of the process of indigenous knowledge as a resource that is of useful not only for the communities that possess it but for society. The issues indigenous knowledge experiences in the digital realm including misrepresentation and respect for cultural protocols were discussed alongside Opportunities offered by technology including artificial intelligence (AI) blockchain and virtual reality (VR). These technologies provide new approaches to classical questions regarding accessibility, documentation, and ethical and effective protection, for better development of more inclusive, secure, and transparent systems for preservation and use of indigenous knowledge (Zhou & Chibaya, 2024). The key debates included exploring the current instruments and approaches to digital preservation, and the novel paradigms embracing the 4IR technologies. AI for language documentation, which is a key area of focus of this research, blockchain for protection of intellectual property right, and VR to make a cultural experience was listed as the potential advancements towards the digital preservation platform.

Signatures including the use of AI for transcription and translation of first nations languages or Mukurtu content management system that provides indigenous people with a control of how their knowledge is shared online, show how these advanced technologies can yield such results. However, these tools also have limitations; some of them are technological, ownership issues and cultural considerations when implementing the above tools (Nkosi et al., 2020). Finally, the future of indigenous knowledge preservation remains in methodological innovative and synergy between technology professionals, indigenous people, and authorities. It can therefore be seen that there is a definite need for other similar tools to be designed that reach out to other out of the way places so as not to leave the technologically challenged behind (Mogajane, 2022). Also, thinking about ethics and how this is an ethical practice—how issues like indigenous rights and fears on exploitation are being considered at every layer of this process. Thus, the application of 4IR technologies for the conservation of indigenous knowledge remains a promising Avenue but it calls for proper incorporation into the Indigenous teleology. It is possible to save and disseminate the indigenous population's knowledge that would have otherwise disappeared; indigenous people can successfully exist in the information society. However, the achievement of these objectives is premised on sustained invention, integrity in the business practices, and inter-sectoral cooperation. Let it not only the indigenous communities are being represented but also, they are the actors in this kind of digital preservation. Such a synergy will be beneficial in guaranteeing that indigenous knowledge is one of the ways that humanity is enriched in the future (Zhou & Chibaya, 2024). The data of this chapter has both a practical and a theoretical orientation. It will explore approaches to digital preservation, highlight specific successful practices, and outline models for ethical and inclusive partnership. Thus, the present chapter aims to call for methodological approaches and works that interconnect and integrate technology and culture to make indigenous knowledge not only sustainable but also dominant in the modern world (Tariq, 2025).

CONCLUSION

This chapter has examined how the digital preservation of indigenous knowledge is changing, especially as it relates to Fourth Industrial Revolution (4IR) technology. We started out by talking about how crucial it is to preserve indigenous knowledge since it is valuable to both the communities that possess it and all of mankind. Along with the potential presented by technology like artificial intelligence (AI), blockchain, and virtual reality (VR), the difficulties indigenous knowledge confronts in the digital age such as cultural sensitivity and misrepresentation were also addressed. These technologies enable more inclusive, safe, and transparent methods for sharing and conserving indigenous knowledge by providing creative answers to persistent problems with accessibility, documentation, and ethical protection. The several frameworks and methods now employed in digital preservation, as well as the new models that integrate 4IR technology, were the subject of one of the crucial talks. Among the promising approaches to improve digital preservation efforts were the use of blockchain for intellectual property administration, AI for language preservation, and virtual reality for cultural immersion. The concrete advantages of these technologies are illustrated by examples like the Mukurtu Content Management System, which gives indigenous people control over access to their digital information, and the application of AI for the transcription and translation of indigenous languages. But there are drawbacks to these tools as well, such as the requirement for culturally appropriate implementation, worries about data ownership, and technological obstacles. Going forward, the preservation of indigenous knowledge depends on ongoing innovation and cooperation between governments, indigenous people, and technology specialists. To ensure that isolated or underprivileged populations are not left behind, further research is obviously needed to create more accessible and inclusive solutions that close the technology divide. Furthermore, it is crucial to make sure that the ethical ramifications of digital preservation such as safeguarding cultural sovereignty and avoiding exploitation are considered at every stage.

In conclusion, there is a lot of promise for using 4IR technology to preserve indigenous knowledge, but careful culturally sensitive integration is needed. Indigenous cultures may flourish in the digital era by using these tools to communicate and preserve information that might otherwise be lost. However, sustained innovation, moral behavior, and cross-sector cooperation are necessary for these programs to succeed. Together, we must make sure that indigenous people are actively involved in determining how their knowledge is recorded, disseminated, and safeguarded, in addition to being the focus of digital preservation. To guarantee that indigenous knowledge continues to benefit mankind for many generations to come, cooperation will be essential.

REFERENCES

Adigun, G. O., Ajani, Y. A., & Enakrire, R. T. (2024). The intelligent libraries: Innovation for a sustainable knowledge system in the fifth (5th) Industrial Revolution. *Libri*, *74*(3), 211–223. DOI: 10.1515/libri-2023-0111

Agbehadji, I. E., Awuzie, B. O., & Ngowi, A. B. (2021). COVID-19 pandemic waves: 4IR technology utilisation in multi-sector economy. *Sustainability (Basel)*, *13*(18), 10168. DOI: 10.3390/su131810168

Ajani, Y. A., Tella, A., & Dlamini, N. P. (2024). Indigenous language preservation and promotion through digital media technology in the Fourth Industrial Revolution. In *Digital Media and the Preservation of Indigenous Languages in Africa* (p. 49). Toward a Digitalized and Sustainable Society.

Awuzie, B., & Moghayedi, A. (2024). Harnessing 4IR/5IR technologies for improved procurement and delivery. In *Routledge Handbook of Construction Project Procurement and Delivery* (pp. 371–390). Routledge.

Baporikar, N. (2022). Strategies to manage and preserve indigenous knowledge. In *Handbook of research on protecting and managing global indigenous knowledge systems* (pp. 207–222). IGI Global. DOI: 10.4018/978-1-7998-7492-8.ch012

Chisita, C. T., Enakrire, R. T., Durodolu, O. O., Tsabedze, V. W., & Ngoaketsi, J. M. (Eds.). (2021). *Handbook of research on records and information management strategies for enhanced knowledge coordination*. IGI Global. DOI: 10.4018/978-1-7998-6618-3

Francis, J., Nyamukondiwa, P., & Hagenmeier, C. (2022). How can rural-based universities harness the Fourth Industrial Revolution to strengthen community engagement beyond national borders? *Journal of Community Engagement and Scholarship*, *13*(2). Advance online publication. DOI: 10.54656/CFHX2494

Hunter, J. (2005). The Role of Information Technologies in Indigenous Knowledge Management. *Australian Academic and Research Libraries*, *36*(2), 109–124. DOI: 10.1080/00048623.2005.10721252

Kgomoeswana, K. B., & Makgahlela, L. (2024). Key challenges faced when preserving records in traditional councils during the 4th Industrial Revolution. In *Multidisciplinary Approach to Information Technology in Library and Information Science* (pp. 62–80). IGI Global.

Koomen, O., & Koopman, K. J. (2023). How the technological epoch and COVID-19 silenced indigenous knowledge in the university curriculum. In *Decolonizing the South African University: Towards Curriculum as Self Authentication* (pp. 57–75). Springer Nature Switzerland.

Kruger, S. (2024, June). Transcultural leadership and sustainable development in the digital era: Navigating the 4IR in South Africa. In *International Conference on Society 5.0* (pp. 207–217). Cham: Springer Nature Switzerland.

Kruger, S., & Steyn, A. A. (2024). Leveraging technology adoption to navigate the 4IR towards a future-ready business: A systematic literature review. *Engineering Reports*, *6*(5), e12762. DOI: 10.1002/eng2.12762

Makazhe, E. C., & Maramura, T. C. (2023, September). The role of Fourth Industrial Revolution (4IR) technologies in achieving sustainable development goals. In *The Global Conference on Entrepreneurship and the Economy in an Era of Uncertainty* (pp. 985–998). Springer Nature Singapore.

Malgwi, P. G., Otubelu, B. N., & Sadiq, H. (2023). Enhancing library services in the Fourth Industrial Revolution (4IR): The innovative role of emerging technologies.

Mawere, J. (2024). Green energy solutions in Africa: Rediscovering indigenous knowledge for the Fourth Industrial Revolution. In *Contributions of Africa's Indigenous Knowledge to the Wave of Digital Technology: Decolonial Perspectives* (pp. 159–177). IGI Global.

Mkhize, P. (2024, September). Navigating the Fourth Industrial Revolution: Generational differences, COVID-19 impact, and knowledge sharing in the modern workplace. In *European Conference on Innovation and Entrepreneurship* (Vol. 19, No. 1, pp. 515–522). DOI: 10.34190/ecie.19.1.2783

Mlambo, V. H., Thusi, X., Shoba, M., Mlambo, H., Mnguni, H., & Mbongwa, L. (2024). Harnessing digital technology for economic development in Africa. In *Contributions of Africa's Indigenous Knowledge to the Wave of Digital Technology: Decolonial Perspectives* (pp. 232–259). IGI Global. DOI: 10.4018/978-1-6684-7851-6.ch009

Mogajane, N. M. (2022). Digital technology impact in the access, preservation, and curation of indigenous artefacts at Mafikeng Museum (Doctoral dissertation, North-West University, South Africa).

Nakaziba, S., & Ngulube, P. (2024). Harnessing digital power for relevance: Status of digital transformation in selected university libraries in Uganda. *Collection and Curation*, *43*(2), 33–44. DOI: 10.1108/CC-11-2023-0034

Ndzendze, B., Singh, A., Timm, S., Meyers, C., Malatji, M. J., Malatji, M. J., ... & Manyana, Z. (2024). The 4IR and the humanities in South Africa: Perspectives on innovation, power and potentialities.

Niyitunga, E. B. (Ed.). (2024). *Contributions of Africa's Indigenous Knowledge to the Wave of Digital Technology: Decolonial Perspectives*. IGI Global. DOI: 10.4018/978-1-6684-7851-6

Nkosi, T., Aboginije, A., Mashwama, N., & Thwala, W. (2020, March). Harnessing Fourth Industrial Revolution (4IR) for improving poor universities infrastructure in developing countries: A review. In *Proceedings of the International Conference on Industrial Engineering and Operations Management Dubai,* UAE (pp. 10–12).

Ogedengbe, D. E., James, O. O., Afolabi, J. O. A., Olatoye, F. O., & Eboigbe, E. O.Damilola Emmanuel Ogedengbe1Oladapo Olakunle JamesJennifer Osayawe Atu AfolabiFunmilola Olatundun OlatoyeEmmanuel Osamuyimen Eboigbe. (2023). Human resources in the era of the Fourth Industrial Revolution (4IR): Strategies and innovations in the global south. *Engineering Science & Technology Journal*, *4*(5), 308–322. DOI: 10.51594/estj.v4i5.617

Ogur, E. O. (2024, September). The Fourth Industrial Revolution: Opportunities and challenges for Kenya. In *Proceedings of the Sustainable Research and Innovation Conference* (pp. 76–87).

Ojo, T. A., & Ndzendze, B. (2025). Women and the 4IR in Africa. In *African Women in the Fourth Industrial Revolution* (pp. 8–35). Routledge.

Padeli, W., Mustafa, W. A., Pangil, F., Abd Kadir, K., & Dwita, V. (2025). Knowledge management and the Fourth Industrial Revolution (4IR): A recent systematic review. *Journal of Advanced Research in Applied Sciences and Engineering Technology*, *51*(2), 18–33. DOI: 10.37934/araset.51.2.1833

Rambe, P., & Maime, R. (2023). The impact of 4IR technologies on venture creation and technology commercialisation: Insights and exemplars from an emerging economy context. In *Small Business and Entrepreneurial Development in Africa: A Route to Sustained Economic Development* (pp. 149–178). Springer Nature Switzerland. DOI: 10.1007/978-3-031-37675-7_7

Suryadi, B., Larasati, A. S., Verniastika, R., & Firdaus, S. N. (2022). Harnessing Fourth Industrial Revolution (4IR) technologies for climate change mitigation in the Asia-Pacific region.

Tariq, M. U. (2025). Leveraging Libraries and Information Centers for Startup Innovation and Growth. In Gupta, V., & Gupta, C. (Eds.), *Navigating Strategic Partnerships for Sustainable Startup Growth* (pp. 1–28). IGI Global Scientific Publishing., DOI: 10.4018/979-8-3693-4066-0.ch001

Tariq, M. U. (2025). Bridging Generations: Digital Transformation and Its Multifaceted Impact Across Societal Sectors. In Anshari, M., Almunawar, M., & Ordóñez de Pablos, P. (Eds.), *Impacts of Digital Technologies Across Generations* (pp. 1–22). IGI Global Scientific Publishing., DOI: 10.4018/979-8-3693-6366-9.ch001

Tariq, M. U. (2025). Empowering Learning Through Networked and Connected Education: Case Studies in Digital Engagement. In Gierhart, A. (Ed.), *Cases on Enhancing P-16 Student Engagement With Digital Technologies* (pp. 169–198). IGI Global Scientific Publishing., DOI: 10.4018/979-8-3693-5633-3.ch007

Tariq, M. U. (2025). The Belmont Report: Guiding Ethical Principles in Human Research. In Throne, R. (Ed.), *IRB, Human Research Protections, and Data Ethics for Researchers* (pp. 245–268). IGI Global Scientific Publishing., DOI: 10.4018/979-8-3693-3848-3.ch010

Uwaoma, P. U., Eboigbe, E. O., Kaggwa, S., Akinwolemiwa, D. I., & Eloghosa, S. O. (2023). Ecological economics in the age of 4IR: Spotlight on sustainability initiatives in the global south. *International Journal of Advanced Economics*, *5*(9), 271–284. DOI: 10.51594/ijae.v5i9.647

Zhou, Z., & Chibaya, T. (2024). The opportunities and challenges of harnessing digital technologies. In *Tourism and Hospitality for Sustainable Development: Technological Innovations and Development Realities* (1), 111.

KEY TERMS AND DEFINITIONS

Blockchain: A decentralized digital ledger technology used to securely store data, ensuring transparency and immutability.

Decolonization: The process of challenging and dismantling colonial frameworks and empowering local knowledge systems to reclaim their authenticity and representation.

Digital Preservation: The process of storing and safeguarding digital content and assets to ensure their long-term accessibility and usability.

Fourth Industrial Revolution (4IR): A technological paradigm shift characterized by the integration of advanced technologies like AI, blockchain, and IoT into various domains.

Indigenous Knowledge: Traditional knowledge and cultural practices developed over generations within specific communities.

Chapter 17
Indigenous Knowledge and Environmental Sustainability:
A Technology Perspective

Princess Blose
https://orcid.org/0000-0001-9825-8090
University of South Africa, South Africa

Mishack Thiza Gumbo
https://orcid.org/0000-0001-6760-4341
University of South Africa, South Africa

ABSTRACT

This qualitative case study explores the essence of indigenous knowledge (IK) and environmental sustainability from a Technology perspective. IK is linked to the natural environment because it is framed within cultural practices. However, IK is sidelined in education and seeking the environment's sustainable solutions, thus disregarding its critical contribution to preserving the environment. Eight Senior Phase Technology teachers and a Curriculum Subject Advisor (CSA) were purposively selected and interviewed individually. Document analysis was also conducted on the CAPS Technology, Sasol Inzalo textbooks, and Annual Teaching Plans (ATPs). Data was analysed following the inductive thematic analysis framework. It was learned that teachers understood indigenous technology and articulated the importance of teaching about education for sustainable development (ESD). However, they did not integrate it into their teaching despite the opportunities in the analysed documents. The chapter contributes a more inclusive and culturally informed framework for teaching about SD and ESD.

INTRODUCTION AND BACKGROUND

The nonrecognition of Indigenous and Local Knowledge (ILK) negatively affects the quality of education in African centers, as learners' lived experiences in their socio-cultural environments are not contextualized to enhance the relatability of their learning subjects (Mbah & Ezegwu, 2024). From this perspective, we draw our chapter from the decolonization of IK in educational systems within the Botho/Ubuntu philosophy for sustainability teaching in the Technology curriculum. David (2024) espouses

DOI: 10.4018/979-8-3693-7964-6.ch017

that the subordination of the African way of life within the education system has restricted the potential contribution of African IK in the Anthropogenic Global Warming (AGW) discourse. Nonetheless, its decoloniality has strong roots in Botho's philosophy, which is based on finding truth through mutual sharing and interaction among members is easier (Omodan & Diko, 2021). IK is a dynamic, place-based system of understandings, skills, and philosophies developed through long histories of interaction between indigenous communities and their natural surroundings (Petzold et al., 2020). It encompasses ecological, cultural, spiritual, and practical insights deeply tied to cultural identity and environmental stewardship. Passed through generations via oral traditions, observations, and practices, IK informs decision-making on fundamental aspects of life, fostering sustainability, adaptation, and resilience, particularly in climate change (Petzold et al., 2020; Jessen et al., 2022). Indigenous Knowledge Systems (IKS) emphasize relational perspectives that connect people, organisms, and ecosystems, offering a holistic understanding of socioecological systems (Jessen, 2022; Littletree et al., 2020). While distinct from Western scientific knowledge, they are increasingly recognized as complementary, contributing valuable insights to biodiversity conservation, environmental management, and sustainable development (SD) (Jessen et al., 2022; McLean et al., 2023). Efforts to integrate IK prioritize intercultural dialogue, community participation, and curricular adaptation to preserve cultural heritage and promote sustainability, reflecting its transformative potential for fostering equitable and inclusive approaches to global challenges (Da Silva et al., 2024; McLean et al., 2023). Integrating IK into educational settings is crucial for fostering a holistic understanding of the environment, sustainability, and the interconnectedness of humans and nature. IK systems emphasize experiential learning and intergenerational knowledge transfer, which can enrich curricula by offering alternative perspectives on environmental and cultural sustainability.

Incorporating IK into education can promote cultural diversity, respect for indigenous worldviews, and recognition of alternative approaches to problem-solving, particularly in areas like climate change, biodiversity conservation, and sustainable resource management (Mishra et al., 2024; Brondízio et al., 2021). This integration counters cultural erosion and knowledge loss while empowering indigenous communities by validating their contributions formally. Educational institutions can serve as platforms for collaborative knowledge production, blending indigenous wisdom with scientific inquiry to create innovative, context-specific solutions to global challenges. Integrating IK into modern education can foster respect, inclusivity, and a deeper appreciation for diverse ways of knowing, which is essential for addressing the complexities of a rapidly changing world (Brondízio et al., 2021). Considering this, the study was conducted at six schools in the Ehlanzeni district of Mpumalanga Province. A case study was used as a research design to study the essence of Indigenous Knowledge (IK) and environmental sustainability from a Technological perspective. Eight Senior Phase Technology teachers and a Curriculum Subject Advisor (CSA) were the target participants of the study.

STATEMENT OF THE PROBLEM

IK is a dynamic, place-based system of understandings, skills, and philosophies developed through long histories of interaction between indigenous communities and their natural surroundings (Petzold et al., 2020). Despite its richness and vibrancy, IK has often been marginalized within formal education and decision-making processes due to the dominance of Western epistemologies (Da Silva et al., 2024). This marginalization leads to the loss of valuable knowledge that could significantly contribute to sustainable development. One of the critical Science, Technology, Engineering and Mathematics subjects in which IK is underplayed is Technology Education. For this reason, the study aims to address this gap by exploring how IK can be integrated into the Technology (as Technology Education is called in South Africa) curriculum to promote environmental sustainability and enhance the educational experience for students. The problem leads to the research questions and objectives in the below sections.

RESEARCH QUESTIONS

- What is the essence of Indigenous Knowledge (IK) in the context of environmental sustainability and technology?
- How do Senior Phase Technology teachers and Curriculum Subject Advisors understand and apply IK in their teaching practices?
- How can IK be effectively incorporated into the Technology curriculum to promote sustainable development education?

RESEARCH OBJECTIVES

- To explore the essence of IK in the context of environmental sustainability and technology.
- To examine Senior Phase Technology teachers' and Curriculum Subject Advisors' understanding and application of IK in their teaching practices.
- To explore IK's effective incorporation into the Technology curriculum to promote sustainable development education.

THEORETICAL/CONCEPTUAL FRAMEWORK

Within the Botho philosophy, we found the Cultural Historical Activity Theory (CHAT) the most relevant framework guiding this chapter. Both frameworks highlight the significance of community and collaborative learning. In educational settings, this means fostering environments where learners learn through social interactions and community involvement. CHAT is a sociocultural framework that examines human activities as complex, socially situated phenomena, emphasizing the role of cultural and historical context in shaping human actions and interactions. Botho is an African philosophy that emphasizes the interconnectedness of individuals with their communities and the environment. It is based on "I am because we are," highlighting the importance of community, mutual respect, and shared humanity (David, 2024). Botho values collective well-being over individualism and promotes the idea that individuals find their identity and purpose through their relationships with others (Omodan & Diko, 2021). Similarly, CHAT integrates tools, community, rules, and division of labor to understand how individuals and groups engage in goal-directed activities (Foot, 2014; Yamagata-Lynch, 2007). An activity system in CHAT consists of six interconnected elements: 1) subject (actor), 2) object (goal or desired outcome), 3) tools (resources), 4) community (people involved), 5) rules (regulations), and 6) division of labor (roles and responsibilities) (Foot, 2014; Mathieson et al., 2024).

CHAT identifies contradictions within and between these elements as sources of development and innovation, including primary contradictions (tensions within each component), secondary contradictions (tensions between different components), tertiary contradictions (tensions from new elements), and quaternary contradictions (tensions between activity systems) (Foot, 2014; Yamagata-Lynch, 2007). The theory introduces the concept of expansive learning, which involves a cycle of learning actions questioning, analyzing, modeling, examining, implementing, reflecting, and consolidating that help map the development of activity systems and identify opportunities for change (Mathieson et al., 2024; Yamagata-Lynch, 2007). CHAT provides a comprehensive framework for analyzing and understanding the dynamic interactions within human activities, making it valuable for research and practice in various fields, including education, social work, and organizational development (Foot, 2014; Mathieson et al., 2024). The application of CHAT in the chapter provides a comprehensive framework for integrating IK into modern educational practices, particularly within the Technology curriculum. CHAT's holistic approach, which includes analyzing the interconnected elements of an activity system—subject, object, tools, community, rules, and division of labor—aligns well with the chapter's emphasis on the interconnectedness of IK with cultural, ecological, and social contexts (Foot, 2014; Yamagata-Lynch, 2007). We, therefore, believe that by identifying and addressing contradictions within and between activity systems, CHAT helps surface tensions between IK practices and modern educational frameworks, facilitating the development of strategies that respect and preserve indigenous practices while enhancing modern education (Mathieson et al., 2024). The concept of expansive learning in CHAT, involving a cycle of questioning, analyzing, modeling, examining, implementing, reflecting, and consolidating, encourages continuous reflection and adaptation, leading to transformative changes in teaching practices and educational policies (Mathieson et al., 2024; Yamagata-Lynch, 2007). Furthermore, CHAT emphasizes the role of community and collaborative learning, which aligns with the chapter's focus on the importance of community involvement and the principles of Botho and Community of Practice (CoP) in promoting indigenous technology and sustainable practices (Mathieson et al., 2024). Thus, balancing indigenous and modern tools and methods, CHAT supports the chapter's argument that indigenous and modern technologies can complement each other, leading to innovative and culturally appropriate solutions for environmental sustainability (Foot, 2014; Mathieson et al., 2024). Additionally, incorporating IK into the curriculum through CHAT helps strengthen cultural identity and promote SD, aligning with the chapter's emphasis on the importance of cultural identity in environmental management and the preservation of indigenous cultural heritage (Foot, 2014; Mathieson et al., 2024). This integration fosters a deeper understanding of sustainability and cultural diversity, leading to transformative changes in education and aligning with the goals outlined in the chapter (Blose & Gumbo, 2024; Brondízio et al., 2021; Jessen et al., 2022; McLean et al., 2023; Da Silva et al., 2024).

LITERATURE REVIEW

Pedagogical Approaches for Integrating IK

In the transition to integrating IK with the current curriculum, Bothogogy (Ubuntugogy) emerges as a pedagogy that supports teaching and learning by considering the cultures and values of the community, specifically the learners in this context. Bothogogy emphasizes people's values, empowering individuals to recognize their history beyond colonization and seeking to address Afro-European epistemological differences (Omodan & Diko, 2021). The pedagogy of Bothogogy is based on African-ness, African values, beliefs, and ideas (Omodan & Diko, 2021). Additionally, Bothogogy relies on a social interpretation and acknowledgement of human histories and their unique attributes, which are crucial to learners' success (Donohue, 2020). From this perspective, Traditional Ecological Knowledge (TEK), Reciprocity and Stewardship, and Community-Based Resource Management are essential for teaching sustainability in the Technology subject.

In the CAPS Technology Grades 7–9, integrating IK is a practical and transformative approach (Department of Basic Education [DBE], 2011). Moreover, CAPS is guided by transformational principles such as valuing Indigenous Knowledge Systems (IKS), human rights, social justice, and inclusivity. The Technology-specific curriculum highlights the importance of understanding how indigenous communities historically developed and utilized technology to solve problems with locally available materials and sustainable methods. This understanding should extend to current developments in IK rather than being confined to historical contexts. IK helps learners appreciate the interplay between technology, society, and the environment, fostering awareness of diverse knowledge systems. Notably, IK provides real-world contexts rooted in environmental and social justice. For instance, learners explore indigenous building techniques and materials for constructing emergency shelters, which promotes sustainability and cultural relevance. This approach deepens their understanding of the practical application of technology while showcasing environmentally friendly and cost-effective solutions (DBE, 2011). Therefore, investigation activities should not be limited to urban settings dominated by conventional technology. The curriculum also tackles biases in conventional technology by advocating for recognizing indigenous intellectual property rights (DBE, 2011). It challenges the predominance of Western paradigms in education by highlighting the importance of indigenous practices in addressing contemporary societal challenges. By incorporating IK, the curriculum fosters inclusivity, equity, and respect for cultural diversity while preparing learners for practical problem-solving in a globally interconnected world (DBE, 2011). Nonetheless, integrating IK into Technology can be perceived as superficial and confusing for teachers, particularly in educational contexts where Western paradigms are predominant. For example, a study by Blose and Gumbo (2024) revealed that IK is underexplored in Technology teachings. Thus, this chapter investigates the essence of IK and environmental sustainability from a Technology perspective, aiming to advance environmental sustainability and enhance learners' understanding of sustainable development (SD). We contend that addressing issues such as IK preservation, environmental degradation, resource depletion, gaps in SD education, the tension between tradition and modernity, cultural identity, and sustainability will strengthen the teaching of IK and environmental sustainability in Technology. The chapter is grounded in the literature on SD and education for SD, exploring the intersection of IK and environmental sustainability.

Preservation of Indigenous Knowledge

One of the main concerns is the potential loss of IK, given that modernity and globalization increasingly dominate cultural landscapes. According to Iloka (2016), the richness of this knowledge is not well understood in reducing disaster risks, as policymakers rely on conventional scientific mitigation strategies. Although much is written about IK, Africa's disaster risk reduction and climate change documentaries are limited (Iloka, 2016). We, therefore, believe that the erosion of indigenous customs and wisdom threatens sustainable life, especially in communities relying on indigenous customs for generations. Iloka (2016) also claims that protecting natural resources for sustainability is another area where IK is invaluable. Therefore, the chapter emphasizes the relevance and applicability of IK to contemporary environmental challenges and advocates its preservation and revitalization role. It explores integrating this knowledge into educational systems, particularly in the Technology curriculum, to ensure its dissemination to future generations.

Msuya (2007) discusses the importance of IK as a vital resource for economic development, cultural preservation, political transformation, and poverty reduction. IK is characterized by its local specificity, oral transmission, adaptability, and communal ownership. Msuya (2007) highlights the practices of the Maasai pastoralists and the Sambaa and Zigua indigenous medicine men in Tanzania, who possess extensive environmental and medicinal knowledge passed down through generations. However, IK faces threats of extinction due to a lack of documentation, marginalization in modern education systems, and insufficient integration into the global knowledge system. Efforts to protect and preserve IK in Tanzania include the development of databases like the ESRF database, which aims to capture, store, and disseminate IK. Ethical issues such as ownership of research findings and inclusivity in research methods are also discussed. Msuya (2007) emphasizes the need for appropriate policies, research, and the establishment of IK resource centers to ensure IK's survival and continued relevance. This chapter responds to the needed research about this phenomenon. Collaborative efforts with other developing countries are recommended to advocate for IK at international forums. Overall, preserving IK is crucial for maintaining its cultural significance and practical applications in daily life.

According to Zimu-Biyela (2016), preserving IK involves blending indigenous and modern methods to ensure its continuity and accessibility. Indigenous methods include oral traditions, where knowledge is passed down through storytelling and oral history, and apprenticeship, where skills are transferred through hands-on experience. Artefacts and rituals also play a crucial role in maintaining the cultural context of IK. Modern methods involve documentation, codification, and digitization, which convert IK into written or digital formats for long-term preservation and global sharing. The use of information and communication technologies (ICTs) and databases further aids in storing and disseminating IK widely. However, challenges such as balancing indigenous and modern methods, declining knowledge-sharing activities, and limited adoption of modern tools due to low literacy levels and resource constraints persist. Policies and legislation, like the National Indigenous Knowledge Systems Office (NIKSO) and intellectual property laws, protect IK from misuse and ensure communities benefit from their knowledge. Community involvement through social structures and communities of practice (CoPs) is vital for the socialization and transmission of IK (Zimu-Biyela, 2016). Educational programs are also necessary to raise awareness about the importance of preserving IK and the tools available to preserve it. A holistic approach that respects indigenous practices while leveraging modern technologies and legal frameworks is crucial for effectively preserving IK. The preservation of IK in South Africa involves significant digitization and policy development, primarily through the National Recordal System (NRS) initiative, which focuses on digitizing indigenous medicine, plants, and food to protect this knowledge from biopiracy and misappropriation. Despite policies like the National Policy on Digitization of Heritage Resources and the IKS Policy, challenges such as inadequate digital preservation policies at institutional levels, insufficient infrastructure, and limited awareness among staff persist. Measures like metadata management, regular backups, and disaster preparedness plans are in place but need standardization and uniform implementation. Ensuring the authenticity and trustworthiness of digitized IK through trusted digital repositories and compliance with standards like the Open Archival Information System (OAIS) is crucial. Gumbo (2023) cautions about digitizing IK in ways that do not promote inclusivity, beneficiation and sustainability.

Correspondingly, Tsetselelane et al. (2021) write that preserving IK through digital libraries is crucial for ensuring the intergenerational transmission of culturally unique knowledge, which is at risk of extinction due to modernization and changing lifestyles. Digital libraries make IK accessible to a broader audience, including younger tech-savvy generations, and protect it from intellectual theft and misappropriation (Tsetselelane et al., 2021). This modern approach integrates IK with contemporary scientific knowledge, creating a hybrid that respects indigenous practices while embracing technological advancements. Community involvement ensures accurate representation and preservation according to indigenous protocols, while ethical considerations address the collective nature of IK and prevent misrepresentation. Overall, we concur with (Tsetselelane et al., 2021) that digital libraries offer a sustainable, accessible, and ethically sound method to protect and share valuable cultural heritage in the modern world. We, however, warn that indigenous stakeholders should be actively involved in the digitalizing project considering the right to knowledge, sifting between the disclosure and non-disclosure of knowledge such as sacred knowledge, collective ownership of knowledge within a community, etc.

As much as we advocate for the digitization of IK, we are aware of the unauthorized use and commercialization of Indigenous cultural heritage, such as traditional knowledge, artefacts, and sacred symbols may be realized. Hence, we encourage a combination of Transformative Learning Theory (TLT) with Botho philosophy. TLT emphasizes profound shifts in individuals' values and perspectives, particularly regarding their relationships with the environment. On the other hand, Botho, with its principles of reciprocity, harmony, and solidarity, provides a transformative pathway to rethink and respond to environmental crises such as climate change (David, 2024). Therefore, a Both-inspired TLT can foster a deeper understanding of the interconnectedness of all beings and promote ethical and responsive interactions with nature. In this way, the IK will be continually passed to generations within the context of Botho through education systems that appreciate it. In the next section, we make a case about the critical contribution of IK and SD in addressing environmental degradation and resource depletion.

IK and SD: A Panacea for Environmental Degradation and Resource Depletion

The fundamental assumption of ecological restoration is that human beings are responsible for degrading the natural environment and that they are responsible for repairing it (Egan et al., 2011). Rapid environmental degradation and unsustainable use of natural resources are pressing global concerns. The chapter discusses how modern development practices often conflict with indigenous sustainable methods used by indigenous communities. Thus, by presenting indigenous communities' environmentally friendly practices, the chapter argues for their integration into modern development strategies. It aims to prove that indigenous methods can provide sustainable resource management and environmental protection alternatives. Evans et al. (2011) believe that humans are part of nature and play an essential role in consciously or otherwise determining the condition of the environment in which they live. For instance, Ali et al. (2021) explored the interplay between natural resource depletion, renewable energy consumption, and environmental degradation, comparing developed and developing nations. It highlights that resource depletion significantly contributes to environmental degradation in developed countries, while in developing nations, the impact is less pronounced due to external exploitation by wealthier nations. Fossil fuel consumption is identified as a significant driver of environmental harm globally, with a more substantial impact in developing regions. Contrastingly, renewable energy reduces environmental degradation in developed countries due to efficient systems, but in developing nations, inefficiencies in renewable energy utilization can ironically exacerbate environmental issues. Urbanization and trade positively influence environmental outcomes; urbanization promotes sustainable practices and incentivizes environmentally friendly production. Economic growth, however, remains a double-edged sword, driving development at the cost of environmental degradation. The study calls for developed nations to prioritize reducing resource depletion and improving renewable energy systems while developing countries should focus on enhancing energy efficiency and reducing fossil fuel dependency. Therefore, globally, policies fostering urbanization and sustainable trade practices are essential for mitigating environmental harm while supporting economic growth.

Similarly, Muhammad et al. (2021) investigated the complex interactions among environmental-related technologies, energy intensity, renewable energy production, natural resource depletion, and other factors impacting environmental sustainability in Canada from 1989 to 2020. They employed a dynamic ARDL simulation model to analyze these relationships. The study finds that environmental-related technologies significantly reduce environmental degradation in the short and long term. At the same time, factors such as energy intensity, financial development, natural resource depletion, renewable energy production, and temperature exacerbate environmental degradation. The study underscores the need to adopt and enhance environmental technologies, promote renewable energy efficiency, and reduce dependency on natural resource exploitation to achieve SD in Canada. It also highlights the role of financial investment and policy interventions in fostering a sustainable environment. We argue that IK would also have a great impact if considered.

In the same light, Rozi and Taufik (2020) discuss the significant impact of human activities on environmental degradation and resource depletion, highlighting issues such as illegal logging, fishing, mining, pollution, and environmental crimes driven by commercial profit. They emphasize the importance of local wisdom and IK in protecting and preserving the environment, using examples from Indonesian communities like the Mentawai tribe, the people living on the slopes of Mount Merapi, and the Balinese. These communities utilize their IK to manage natural resources and respond to environmental challenges. The Mentawai tribe maintains forest conservation through the arat sabulungan tradition, the Mount Merapi community relies on cosmological teachings and rituals for disaster management, and the Balinese use the Wariga calendar to manage agricultural activities and climate change. The article also addresses these communities' pressures, including modernization and government policies that often marginalize IK. It recommends adapting and integrating IK with modern scientific approaches to address contemporary environmental issues collaboratively and sustainably. The next section discusses the gaps in SD education, thus suggesting the need for this integration of IK.

Gaps in SD Education

Contemporary educational systems, especially in African contexts, are increasingly concerned that they do not sufficiently emphasize the importance of SD and do not adequately integrate IK into the curriculum. After all, when educating African children, foreign colonial forces ignore African knowledge that lay dormant in African families (Mji et al., 2017). This leads to a disconnect between learners and their cultural heritage and a lack of understanding of sustainable practices. The chapter addresses this gap by proposing strategies to integrate IK into the educational framework through the Technology curriculum. It emphasizes bridging these gaps and promotes a sense of environmental responsibility in the curriculum by teaching principles of SD based on indigenous practices. A few studies have identified gaps in education for sustainable development (ESD) and education for ESD goals (ESDGs). Firstly, Kopnina's (2020) study critiqued the current framework for conflating economic growth with sustainability, exacerbating inequalities and environmental degradation. The study highlighted the anthropocentric focus of ESDGs, prioritizing human needs and economic development over ecological integrity, and pointed out the lack of critical reflection on the contradictions within the ESDGs. It also addressed the issue of cognitive imperialism, where Western educational models devalue IK and practices. Furthermore, the study noted the neglect of alternative economic models like degrowth and steady-state economy, which emphasize reducing consumption and maintaining ecological balance. The overemphasis on social and economic aspects often comes at the expense of ecological considerations, treating nature instrumentally rather than recognizing its intrinsic value. The potential for ESDGs to become a form of indoctrination is also discussed, as it promotes a narrow, economy-centered view of sustainability without encouraging critical thinking. Lastly, the study criticized ESDGs for not adequately addressing population growth and overconsumption, major drivers of environmental degradation. The author advocated a more ecocentric and critically reflective approach to SD education, incorporating IK, alternative economic models, and a deeper appreciation for ecological integrity.

Secondly, Acosta Castellanos and Queiruga-Dios (2022) argued that despite the formal commitment to ESD, significant research and practice remained rooted in environmental education (EE), indicating a slow and uneven transition. Geographical disparities existed, with Europe leading in ESD research while regions like Latin America focused more on EE. There was a lack of comprehensive integration of ESD in higher education curricula, particularly in engineering programs, and insufficient governmental policies and institutional support hindered widespread adoption. Economic disparities and cultural influences also affected the acceptance and implementation of ESD. Additionally, there was confusion between EE and ESD, with some studies treating them as synonymous, which could lead to unclear educational objectives. Addressing these gaps can help higher education institutions better align with ESDGs and effectively prepare learners to tackle environmental challenges. These gaps are also realized at schools (Blose, 2023), particularly in the Technology curriculum.

Thirdly, Glavič (2020) proposes IK's integration into educational developmental initiatives. This includes the need for modern and effective content to keep pace with rapid technological and environmental changes and a more holistic approach that integrates environmental, social, and economic dimensions. There is also a significant need for increased stakeholder awareness and participation, capacity building for teachers, and transformative teaching methodologies. Additionally, the development and use of metrics to measure ESD effectiveness, stronger policy and institutional support, and a shift from growth-centric economic and social paradigms to more sustainable models are essential. Adequate funding and enhanced global and regional cooperation are critical to support ESD initiatives. Addressing these gaps can help integrate ESD more effectively into educational systems and contribute to broader SD goals. Despite the suggestion made by Glavič (2020), we contend that IK has a strong background in teaching ESD and environmental sustainability, hence the need to consider balancing IK practices and modernity.

The Balance Between IK Practices and Modernity

The tension between maintaining indigenous practices and adapting to the advances of modern technology is another problem. IK is often misunderstood and incompatible with modernity, leading to its marginalization. This chapter explores the possibility of integrating IK and modern technology, demonstrating that the two approaches are complementary and not mutually exclusive. It shows how indigenous practices can strengthen modern innovations and create culturally and environmentally appropriate sustainable solutions. For instance, balancing tradition and modernity in Technology, as outlined in the CAPS for Grades 7–9, involves integrating foundational knowledge and skills with contemporary technological advancements and methodologies. This balance can be achieved by incorporating IKS, valuing indigenous technologies, and exploring their modern applications through culturally relevant case studies and projects. The curriculum emphasizes the design process (Investigate, Design, Make, Evaluate, Communicate), encouraging learners to engage in systematic problem-solving that blends indigenous and modern methods. Teaching methodologies focus on active and critical learning, suppressing rote memorization and favoring dynamic approaches that include critical thinking, problem-solving and collaborative learning through group projects. Modern tools and resources, such as computer-aided design (CAD) software and three-dimensional (3D) printing, are integrated alongside indigenous tools, ensuring learners can access both. Assessment strategies include formative and summative assessments, with Mini-Practical Assessment Tasks (Mini-PATs) designed to evaluate practical skills and integrate indigenous and modern knowledge. The curriculum covers a wide range of topics, ensuring a comprehensive understanding of indigenous and modern technologies, and lessons are designed to be relevant to learners' lives and communities. Ethical and environmental considerations are also emphasized, encouraging learners to consider the environmental impact of technologies and explore sustainable solutions that blend indigenous practices with modern innovations while discussing the ethical implications of technology use, including issues of bias and inclusivity (DBE, 2011). The following section raises the importance of cultural identity and sustainability.

Cultural Identity and Sustainability

The loss of cultural identity due to the loss of indigenous practices is a significant problem, as it can weaken the relationship between communities and the environment. This disconnect can result in unsustainable practices that further damage the environment. The chapter highlights the importance of cultural identity in promoting SD. It argues that the conservation and teaching of IK are crucial to environmental management and preserving indigenous cultural heritage and identity. Mapara (2022) emphasizes that disciplines must incorporate an epistemic, cognitive, or thought change and recognizes that IK is necessary for curriculum development. Zhang et al. (2020) examined how cultural identity influences the sustainability of heritage tourism, using Suzhou Kunqu Opera as a case study. It highlighted that cultural identity, a sense of belonging, and psychological commitment to a particular culture significantly impacts tourists' engagement and consumption intentions. The study employed the Theory of Planned Behavior (TPB) to show that positive attitudes, social influences, and perceived ease of engaging with heritage sites enhance tourists' willingness to participate in cultural tourism. The findings suggested that fostering cultural identity can drive sustainable tourism development by attracting tourists, generating economic benefits, and promoting the preservation of intangible cultural heritage. The study provided practical implications for tourism management, emphasizing the need for supportive policies, effective marketing strategies, and enhanced cultural experiences to achieve the long-term sustainability of heritage sites.

Conversely, Li et al. (2021) examined the relationship between cultural identity and sustainability in Jeju Island's tourism development. They highlighted the importance of preserving the island's unique cultural heritage through effective collaboration between tourism developers and local communities. They identified several barriers to successful collaboration, including power disparities, lack of institutional support, and residents' reluctance to participate. While tourism can stimulate cultural exchange and increase cultural activities, it can also lead to cultural intrusions and environmental damage, threatening the local cultural identity. The authors recommended empowering local communities, providing institutional support, and adopting sustainable tourism practices that respect and integrate local cultural values to ensure long-term cultural sustainability. Moreover, Clayton et al. (2021) examined the concept of environmental identity (EID) and its significance for sustainability across different cultural contexts. The revised EID scale, designed to be more inclusive of diverse populations and urban experiences, was tested in five countries (USA, Russia, Switzerland, Taiwan and Peru). The study found strong internal consistency and significant correlations between EID, pro-environmental behavior, and environmental concern across all locations. This suggests that environmental identity is a meaningful construct globally, influencing sustainable behaviors. The findings highlighted the importance of cultural context in promoting environmental sustainability and provided a reliable tool for assessing environmental identity in various cultural settings. Their study focused more on sustainability in the tourism sector, not in the educational sector. So, we believe the gaps or concerns raised would help build a strong foundation for environmental sustainability in the education sector, particularly in the Technology curriculum. To present that foundation, we collected and analyzed data as detailed in the following sections.

METHODOLOGY

The research methodology employed in this chapter was designed to explore the integration of IK into the Technology curriculum to promote environmental sustainability and preserve cultural heritage. The study was conducted using a qualitative case study approach in the Ehlanzeni District of Mpumalanga Province, South Africa. This approach was chosen to gain an in-depth understanding of the experiences and perspectives of teachers regarding incorporating IK into their teaching practices.

Research Design

A qualitative case study approach was selected to provide a detailed and contextualized understanding of the research problem. This approach allowed for exploring complex phenomena within their real-life context, making it particularly suitable for studying the integration of IK in education. According to Mesec (1998), a case study involves describing and analyzing an individual matter or case to identify variables, structures, forms, and order of interaction between participants.

Sample Selection

Participants were selected purposively to ensure they had relevant experience and knowledge about the Technology curriculum and the integration of IK. The sample consisted of eight Senior Phase Technology teachers from six different schools in the Ehlanzeni District and one Curriculum Subject Advisor (CSA).

The purposive sampling method was used to select participants who could provide rich, detailed insights into the research questions. The teachers were pseudonymized as Teacher A, Teacher B, Teacher C, and so on to maintain confidentiality.

Data Collection

Data was collected through individual face-to-face semi-structured interviews and document analysis.

Semi-Structured Interviews

The eight Senior Phase Technology teachers and one CSA were interviewed. The interviews were designed to be flexible, allowing in-depth exploration of emerging themes and issues. The interview guide included questions about the participants' understanding of indigenous technology, sustainable development (SD), and education for sustainable development (ESD). It also explored their experiences with collaborative learning and teamwork approaches for promoting indigenous technology. The interviews were conducted conversationally, allowing participants to freely share their experiences and perspectives. This approach ensured that the interviews were natural and focused on the unique context of each participant.

Document Analysis

CAPS Technology curriculum, Sasol Inzalo textbooks, and Annual Teaching Plans (ATPs) were analyzed. The document analysis aimed to identify how IK and environmental sustainability are addressed in the Technology curriculum and teaching materials. This analysis provided a comprehensive understanding of the existing opportunities and gaps in the curriculum for integrating IK and promoting sustainable practices.

Data Analysis

Data was analyzed using inductive thematic analysis. The thematic analysis involved identifying, analyzing, and reporting patterns (themes) within the data. The steps in Table 1 by Braun and Clarke (2006) were followed in the data analysis process:

Table 1. Thematic steps (adapted from Braun & Clarke, 2006)

Familiarization with the data: **We transcribed the interviews and thoroughly read the transcripts and documents to familiarize ourselves with the content.**
Generating initial codes: We generated initial codes by identifying significant phrases, sentences, or data sections relevant to the research questions. These codes were used to label and organize the data.
Searching for themes: We grouped the initial codes into broader themes that captured the key patterns and insights from the data. We looked for themes related to the context of indigenous technology, teachers' understanding of SD, collaborative and teamwork approaches, and the integration of IK in the curriculum.
Reviewing themes: We reviewed and refined the identified themes to ensure they accurately represented the data. We checked for coherence and consistency within each theme and across the entire dataset.
Defining and naming themes: We defined and named the final themes to convey their meaning and relevance to the research questions. Each theme was described in detail, highlighting the key findings and supporting evidence from the data.
Writing the report: We synthesized the findings and wrote the report, presenting the themes and their implications for integrating IK into the Technology curriculum. The report included direct participant quotes to illustrate the themes and provide rich, contextualized insights.

Ethical Considerations

The study adhered to ethical guidelines to ensure the protection and respect of the participants. Informed consent was obtained from all participants, and the researchers ensured that the data were securely stored and used only for the study.

Trustworthiness

Data triangulation was employed by using multiple data sources (interviews and document analysis) to ensure the study's trustworthiness. This approach helped to validate the findings and provide a comprehensive understanding of the research problem.

FINDINGS AND DISCUSSION

The findings are presented under the themes developed following the analysis steps described under data analysis process above.

Context of Indigenous Technology in Technology Curriculum

Teachers generally understood indigenous technology as the knowledge and methods used by their ancestors to solve everyday problems before the advent of modern technology. Teacher H stated that *it is the technology that was done, that was used by our forefathers in solving problems before the new technology came into existence.* Similarly, the CSA said: *When we deal with indigenous technology, we're looking at how things were done before and how they've improved now.* This includes indigenous food preservation methods and using natural materials for various purposes. Teacher D stated thus:

Especially when it comes to processing because there are some things like grinding mealies, grinding peanuts that I still do like using the old ways of using a stone. Teachers believed that the Technology curriculum incorporates elements of IK, particularly in topics like food preservation, mechanical systems and control, electrical systems and control, and the impact of technology on the environment. For example, Teacher F claimed that *back in the days, whenever a tyre was flat, they used to jack a car with bricks or manpower. A jack was then introduced, then it means it minimized the number of people and even the time. The indigenous way gave someone an idea that you can develop something that can better this one.* However, the integration is limited and not fully exploited in day-to-day teaching. Teacher G, in this case, indicated: *There is still a gap to cover this indigenous knowledge.* The teachers and the CSA further stated that the TE curriculum provides opportunities to integrate indigenous technology, especially in areas like recycling, processing, mechanical systems and control, electrical systems, and understanding the impact of technology on society and the environment. The CSA said: *You know, we're dealing with electricity. We don't only deal with the content part but how does it affect the environment, what are the dangers?* It is, therefore, the perspective that teachers can use these opportunities to teach about sustainable practices rooted in IK.

The ATPs for Grades 7–9 address indigenous technology, particularly in specific terms. For example, the Grade 7 ATP focuses on indigenous technology and investigation skills in Term 4, where learners investigate materials and building techniques used by indigenous people for constructing housing in rural South Africa. Similarly, the Grade 8 ATP focuses on the impact of technology and IK in Term 3, where learners explore the environmental impact of mining and indigenous mining practices. The Grade 9 ATP addresses processing indigenous technology and recycling in Term 4, where learners explore indigenous methods of preserving metals and food. However, the integration is limited and not fully exploited in day-to-day teaching, with IK and environmental sustainability primarily addressed in specific terms and limited integration across other terms. Moreover, the Sasol Inzalo textbooks touch on environmental sustainability and indigenous practices but do not explicitly address IK. For instance, the Grade 7 textbook emphasizes the environmental impact of product design but lacks explicit incorporation of indigenous practices and wisdom into the curriculum. Including IK could provide learners with a broader perspective on sustainable practices and the value of indigenous methods. While environmental sustainability is mentioned, it is not consistently integrated across all topics.

Overall, the ATPs provide opportunities for teachers to integrate IK, particularly in topics like food preservation, mechanical systems, and the impact of technology on the environment. For example, the Grade 7 ATP includes activities where learners investigate indigenous building techniques, and the Grade 8 ATP covers the environmental impact of mining and indigenous mining practices. However, these opportunities are often underutilized due to a lack of detailed guidance and practical examples. The textbooks provide content that explains indigenous methods and their modern applications, but there is a gap in explicitly addressing IK. For example, the Grade 7 textbook's emphasis on the environmental impact of product design can help teachers understand and teach the importance of sustainable resource use, drawing parallels with indigenous practices.

Teachers' Knowledge of the Concept of SD and ESD

Teachers and the CSA are aware of the concept of SD which emphasizes caring for the environment. Teacher B, for instance, said: *Something that you will continue using and helping the community positively.* This was supported by Teacher E: *Sustainable development is the way we make use of resources to strengthen our economy or our country, using various factors e.g., use of electric transports to try to avoid emitting dangerous and harmful fumes and causing air pollution*. Teacher C added to say: *So, it's about teaching learners how to save the environment.* While teachers are aware of SD and ESD, there is a gap in how to apply the concepts in the classroom. For example, Teacher B showed limited understanding, which questions its practical implementation in their teaching. It refers to environmental sustainability as education is to learn about real-life situations, not to learn, *only to pass to the next grade*. Nonetheless, CAPS is a foundation for integrating IK and provides many opportunities for environmental sustainability. Thus, CAPS mention including IK but does not provide detailed guidance on its implementation. Therefore, there is a need for a framework to guide the incorporation of ESD in the Technology curriculum. Such a framework should detail the aspects of SD and provide practical examples of integrating IK into teaching.

On the other hand, we realized that the ATPs include SD topics, such as the environmental impact of technology and recycling practices. However, there is a gap in detailed guidance on ESD, and the integration of these principles is limited to specific terms. For example, the Grade 7 ATP addresses indigenous technology and investigation skills in Term 4, but ESD principles are less integrated throughout the year. The textbooks cover environmental sustainability in several chapters, but there is a need for more comprehensive coverage of ESD topics. For instance, the Grade 8 textbook discusses recycling and energy efficiency, but there is limited integration of IK. Including IK could give learners a broader perspective on sustainable practices and the value of indigenous methods. Therefore, we believe the suggested framework in the conclusion section would be helpful.

Collaborative and Teamwork Approaches for Indigenous Technology

The teachers and CSA believed that collaborative learning and teamwork are effective methods for promoting indigenous technology in the classroom. For instance, the CSA said: ... *collaboration, because at the end of the day, they have to work together to make sure that the assessment all the aspects of the assessment are covered, and they know that they also rely on one another*. These approaches allow learners to share knowledge from their diverse cultural backgrounds, which can enhance their understanding of sustainable practices. Moreover, design and problem-solving activities in the Technology curriculum can help learners relate to their cultural contexts and co-create technological knowledge. Teacher C said: *When I design a practical assessment task for learners, I will ensure that there is always a section where the learners are supposed to investigate the environmental impact of whatever they are planning to do*. These activities can be geared towards recognizing and incorporating ESD principles. The teachers and the CSA believed that the principles of Botho and the CoP are important for the successful integration of ESD in Technology. These principles emphasize collaboration, respect, and the involvement of community stakeholders in the educational process. Teacher H said: *Yes, Ubuntu is very important, tolerance, forgiveness, understanding that you're not alone.* Similarly, Teacher A said CoP is considered in their teaching and *yes, we do. In technology kids, they must work together. We have activities that include group work.*

Correspondingly, the ATPs encourage collaborative projects and group work, which can facilitate the sharing of IK among learners. For example, the Grade 7 ATP includes activities where learners investigate indigenous building techniques in groups, promoting teamwork and collaborative learning. However, these approaches are not consistently applied across all grades and subjects. In the same light, the textbooks include activities that promote teamwork and problem-solving, but there is a gap in explicitly addressing IK. For example, the Grade 8 textbook includes practical activities where learners make new objects from recycled materials but does not explicitly incorporate indigenous practices.

The findings highlight the potential for integrating IK into the Technology curriculum to promote environmental sustainability. However, they also highlight the need for more detailed guidance and a structured framework to help teachers effectively incorporate these concepts into their teaching practices.

DISCUSSION

Integrating IKS into Technology is essential for promoting SD and preserving cultural heritage. This discussion highlights the significance of IK in Technology. The study reveals that teachers generally understand indigenous technology as methods used by ancestors to solve everyday problems. This understanding is crucial as it forms the basis for integrating IK into the curriculum. However, the practical application of this knowledge in the classroom is limited. We, therefore, believe that CHAT provides alternative knowledge and pedagogical forms aligned to indigenous technology that can balance IK and modern educational practices to offset the limited practical application of IK in the classroom. Nonetheless, teachers recognize the presence of IK in topics like food preservation and the impact of technology on the environment but note that its integration is not fully exploited.

The ATPs and textbooks could emphasize promoting IK through specific projects and investigations. For example, the Grade 7 ATP includes activities where learners investigate indigenous building techniques, highlighting the value of preserving these practices. However, there is a gap in consistently integrating these themes throughout the year. Including IK in the curriculum can give learners a broader perspective on sustainable practices and the value of indigenous methods (Zidny et al., 2023). Moreover, the ATPs address the environmental impact of technology and resource management, linking these to indigenous practices that promote sustainability. For example, the Grade 8 ATP focuses on the environmental impact of mining and indigenous mining practices. However, there is a gap in consistently integrating these themes throughout the year. The textbooks discuss the consequences of environmental degradation and the role of IK in mitigating these effects, but there is a gap in explicitly addressing IK (Khumalo & Mapotse, 2023). Nonetheless, the ATPs identify gaps in the integration of ESD and propose strategies to fill these gaps using IK. For example, the Grade 7 ATP addresses indigenous technology and investigation skills in Term 4, but ESD principles are less integrated throughout the year. Therefore, there is a need for more comprehensive coverage of ESD topics throughout the year. The textbooks address sustainability in certain chapters, suggesting a need for more comprehensive coverage of ESD topics (Sivapalan & Speight, 2019).

Teachers know SD concepts and ESD, emphasising environmental care and resource use. However, a more practical application of these concepts within the classroom is missing. The CAPS document provides a foundation for integrating IK but needs more detailed guidance on implementing ESD. Detailed guidance on IK integration with its added content could expand CAPS's scope to cover ESD. In this instance, CHAT can help identify the systemic issues that prevent effective integration of ESD principles, such as lack of detailed guidance and support. CHAT could unpack the indigenous historical sociocultural realities and activities of a technological nature, to make sense of CAPS and teach the content from the learners' cultural backgrounds. The study highlights the importance of contextualizing SD within Africa. Indigenous practices offer time-tested methods for sustainable resource management and environmental protection. Teaching and learning about IK can help learners suggest solutions to the harsh climate change and conditions in their design ideas. Integrating these practices into the curriculum can enhance learners' understanding of sustainability and promote culturally relevant education. It can also help them graduate into future designers of optimum solutions to environmental issues informed by indigenous perspectives.

Teachers can use the ATPs and textbooks to emphasize the importance of preserving IK through specific projects and investigations. Including IK in the curriculum can provide teachers with a broader perspective on sustainable practices and the value of indigenous methods (Zidny et al., 2023). Furthermore, teachers can use the ATPs to address the environmental impact of technology and resource management, linking these to indigenous practices that promote sustainability. For example, the Grade 8 ATP focuses on the environmental impact of mining and indigenous mining practices. However, there is a gap in consistently integrating these themes throughout the year. The textbooks discuss the consequences of environmental degradation and the role of indigenous knowledge in mitigating these effects, but there is a gap in explicitly addressing IK (Khumalo & Mapotse, 2023). Teachers can use the ATPs to identify gaps in the integration of ESD and propose strategies to fill these gaps using IK, such as the Grade 7 ATP, which addresses indigenous technology and investigation skills in Term 4. Currently, there is limited integration of ESD principles, creating "hunger" for a more comprehensive coverage of ESD topics throughout the year. While we acknowledge that the textbooks address sustainability in certain chapters, there is a need for more comprehensive coverage of ESD topics (Sivapalan & Speight, 2019).

Collaborative learning and teamwork are essential for integrating indigenous technology into Technology. These approaches allow learners to share knowledge from diverse cultural backgrounds, enhancing their understanding of sustainable practices. The principles of Botho and CoP are important for the successful integration of ESD in Technology. Correspondingly, the CHAT emphasizes collaborative efforts. Design and problem-solving activities in CAPS can help learners relate what they learn to their cultural contexts and co-create technological knowledge. These activities should be geared towards recognizing and incorporating ESD principles. The ATPs would balance indigenous practices with modern technology by integrating both approaches into the curriculum. The textbooks show the evolution of indigenous methods into modern applications, but there is a gap in explicitly addressing the balance between indigenous and modern versions (Khumalo & Mapotse, 2023). Moreover, the ATPs promote cultural identity by including IK and sustainable practices. For example, the Grade 7 ATP includes activities where learners investigate indigenous building techniques, highlighting the cultural significance of these practices. However, there is again a gap in consistently integrating these themes throughout the year. The textbooks discuss the cultural significance of indigenous practices and their contribution to sustainability, but there is a gap in explicitly addressing cultural identity (Payne & O'Neil 2019).

CONCLUSION

Integrating IKS into Technology is vital for promoting SD and preserving cultural heritage. However, this integration faces several challenges, including limited practical application, lack of detailed guidance, and insufficient teacher professional development. Addressing these challenges requires a comprehensive framework (described below) for incorporating IK into the curriculum to promote collaborative learning and involve the IK holders. In implementing these strategies, Technology can effectively balance IK, modernity, and sustainability, preparing learners for the challenges of the modern world while respecting and valuing their cultural heritage. We therefore suggest the framework below. More research should be conducted, exploring the pedagogies that integrate indigenous knowledge. As a more practical subject, Technology can be promoted and supported to be sensitive to society and the environment, directly implicating teaching for sustainable development. Researchers should, therefore, develop an interest in advancing the indigenous technological perspective to the content and teaching. Further research should be conducted into the proposed model below to validate and/or improve it.

Model for Teaching Technology

Teaching technology model is suggested to balance IK, modernity and sustainability in Technology for Grades 7–9, as outlined below.

a) Curriculum design
   - Integrate IK: Include topics that explore indigenous technologies and their evolution. Use case studies to show how modern technology can adapt indigenous practices.
   - Emphasize design process: Structure lessons around the investigate, design, make, evaluate, communicate (IDMEC) framework. Ensure learners apply indigenous and modern methods in their projects by drawing from CHAT. However, the design process should be implemented flexibly, being open to how indigenous people approach design in their technological activities.
   - Diversify the design scenarios to cover indigenous alternatives instead of conventional ones, such as telephone towers and the headgear of the crane for mining.

b) Teaching methodologies
   - Active and critical learning: Encourage learners to think critically and problem-solve rather than promote rote memorization. Use real-world problems, especially from the learners' contexts, to make learning relevant. Furthermore, teachers can use CHAT to identify how the subject (learner) engages with the object (learning outcomes) using tools (educational resources), including those within the community (classroom) under specific rules (curriculum guidelines) and division of labor (learner roles). They should consider involving IK holders from the communities.
   - Collaborative learning: Facilitate group projects where learners can share diverse perspectives and knowledge bases, blending indigenous and modern insights. CHAT can be used to analyze how the activity system elements facilitate collaborative projects.

c) Resource utilization
   - Modern and indigenous tools/techniques: Provide access to modern tools (e.g., CAD software, 3D printers) and indigenous tools (e.g., hand tools, indigenous materials). Ensure learners learn to use both effectively. Using CHAT would ensure that learners understand the value of indigenous and modern technologies and how they can be integrated to solve real-world problems.
   - Resource accessibility: Ensure all learners access necessary materials and tools, promoting equity in learning opportunities.

d) Assessment strategies
   - Balanced assessment: Mix of formative and summative assessments to evaluate learners' understanding and application of indigenous and modern technologies. Capitalize on oral presentations and storytelling as alternative methods dominant in IK-sharing circles.

- Mini-Practical Assessment Tasks (Mini-PATs): Design these tasks to assess practical skills and the integration of indigenous and modern knowledge. Ensure they cover diverse topics and skills.

e) Curriculum content
   - Diverse topics: Cover many topics, including structures, mechanical systems, electrical systems, and processing. Ensure content is comprehensive and integrates both indigenous and modern technologies.
   - Contextual learning: Design lessons relevant to learners' lives and communities, making learning more meaningful and engaging.

f) Ethical and environmental considerations
   - Sustainability: Encourage learners to consider the environmental impact of technologies. Promote sustainable solutions that blend indigenous practices with modern innovations.
   - Ethical awareness: Discuss the ethical implications of technology use, including issues of bias and inclusivity. Ensure learners understand the importance of ethical considerations in technology. CHAT would ensure that learners understand the importance of ethical considerations in technology and address any contradictions related to ethical awareness.

g) Project-based learning
   - Real-world projects: Assign projects requiring learners to solve real-world problems using indigenous and modern technologies, e.g. design a sustainable shelter using indigenous materials and modern construction techniques.
   - Interdisciplinary approach: Encourage projects integrating knowledge from other subjects, such as science and social studies, to provide holistic learning experience.

h) Professional development for teachers
   - Continuous training: Provide ongoing professional development for teachers to stay updated on indigenous and modern technologies. Encourage them to integrate new teaching methodologies and tools into their lessons.
   - Collaborative networks: Create networks for teachers to share best practices and resources, fostering a community of continuous learning and improvement. CHAT may foster a community of continuous learning and improvement.

By implementing this model, Technology can effectively balance IK, modernity, and sustainability, preparing learners for the challenges of the modern world while respecting and valuing their cultural heritage. In addition, we recommend that Technology teachers plan their teaching in a way that integrates the IK aspects for ESD recognition. Further research using mixed methods on IK and sustainability teaching in the Technology curriculum is also recommended.

REFERENCES

Acosta Castellanos, P. M., & Queiruga-Dios, A. (2022). From environmental education to education for sustainable development in higher education: A systematic review. *International Journal of Sustainability in Higher Education*, *23*(3), 622–644. DOI: 10.1108/IJSHE-04-2021-0167

Adeoye-Olatunde, O. A., & Olenik, N. L. (2021). Research and scholarly methods: Semi-structured interviews. *Journal of the American College of Clinical Pharmacy: JAACP*, *4*(10), 1358–1367. DOI: 10.1002/jac5.1441

Ali, A., Audi, M., & Roussel, Y. (2021). Natural resources depletion, renewable energy consumption and environmental degradation: A comparative analysis of developed and developing world. *International Journal of Energy Economics and Policy*, *11*(3), 251–260. DOI: 10.32479/ijeep.11008

Blose, P. (2023). *The incorporation of education for sustainable development in Technology Education for an indigenous context* (Doctoral Thesis). https://uir.unisa.ac.za/handle/10500/30606

Blose, P., & Gumbo, M. T (2024). Developing a framework for integrating IKS indigenous knowledge systems in Technology Education with sustainable development principles. *African Journal of Research in Mathematics, Science and Technology Education*, 1-13. https://doi.org/DOI: 10.1080/18117295.2024.2356416

Braun, V., & Clarke, V. (2006). Using thematic analysis in psychology. *Qualitative Research in Psychology*, *3*(2), 77–101. DOI: 10.1191/1478088706qp063oa

Brondízio, E. S., Aumeeruddy-Thomas, Y., Bates, P., Carino, J., Fernández-Llamazares, Á., Ferrari, M. F., Galvin, K., Reyes-García, V., McElwee, P., Molnár, Z., Samakov, A., & Shrestha, U. B. (2021). Locally based, regionally manifested, and globally relevant: Indigenous and local knowledge, values, and practices for nature. *Annual Review of Environment and Resources*, *46*(1), 481–509. DOI: 10.1146/annurev-environ-012220-012127

Clayton, S., Czellar, S., Nartova-Bochaver, S., Skibins, J. C., Salazar, G., Tseng, Y. C., Irkhin, B., & Monge-Rodriguez, F. S. (2021). Cross-cultural validation of a revised environmental identity scale. *Sustainability (Basel)*, *13*(4), 1–12. DOI: 10.3390/su13042387

Da Silva, C., Pereira, F., & Amorim, J. P. (2024). The integration of indigenous knowledge in school: A systematic review. *Compare: A Journal of Comparative Education*, *54*(7), 1210–1228. DOI: 10.1080/03057925.2023.2184200

David, J. O. (2024). Decolonizing climate change response: African indigenous knowledge and sustainable development. *Frontiers in Sociology*, *9*, 1456871. DOI: 10.3389/fsoc.2024.1456871 PMID: 39640009

Dawadi, S. (2020). Thematic analysis approach: A step-by-step guide for ELT research practitioners. *Journal of NELTA*, *25*(1-2), 62–71. DOI: 10.3126/nelta.v25i1-2.49731

Department of Basic Education. (2011). *National Curriculum Statement: Curriculum and Assessment Policy Statement (CAPS) Technology Grades 7 – 9*. Government Printers.

Donohue, D. (2020). Culture, cognition, and college: How do cultural values and theories of intelligence predict students' intrinsic value for learning? *Journal of Culture and Values in Education*, *4*(1), 1–14. DOI: 10.46303/jcve.2020.3

Egan, D., Hjerpe, E. E., & Abrams, J. (Eds.). (2011). *Why people matter in ecological restoration*. Island Press. DOI: 10.5822/978-1-61091-039-2_1

Foot, K. A. (2014). Cultural-historical activity theory: Exploring a theory to inform practice and research. *Journal of Human Behavior in the Social Environment*, *24*(3), 329–347. DOI: 10.1080/10911359.2013.831011

Glavič, P. (2020). Identifying key issues of education for sustainable development. *Sustainability (Basel)*, *12*(16), 1–18. DOI: 10.3390/su12166500

Gumbo, M. T. (2023). Digitisation of higher education and research: Raising inclusivity and equity issues for indigenous students. *South African Computer Journal = Suid-Afrikaanse Rekenaartydskrif*, *35*(1), 149–163. DOI: 10.18489/sacj.v35i1.1107

Hignett, S., & McDermott, H. (2013). Qualitative methodology for Ergonomics. In Wilson, J. R., & Sharples, S. (Eds.), *Evaluation of Human Work. A practical ergonomics methodology* (4th ed.). CRC Press.

Iloka, N. G. (2016). Indigenous knowledge for disaster risk reduction: An African perspective. *Jàambá*, *8*(1), 1–7. DOI: 10.4102/jamba.v8i1.272 PMID: 29955294

Jessen, T. D., Ban, N. C., Claxton, N. X., & Darimont, C. T. (2022). Contributions of indigenous knowledge to ecological and evolutionary understanding. *Frontiers in Ecology and the Environment*, *20*(2), 93–101. DOI: 10.1002/fee.2435

Khumalo, S. B., & Mapotse, T. A. (2023). Building modern technology innovation on indigenous knowledge systems in Technology Education. In Gumbo, M. T., & Williams, P. J. (Eds.), *Indigenous Technology Knowledge Systems: Contemporary issues in Technology Education*. Springer., DOI: 10.1007/978-981-99-1396-1_4

Kopnina, H. (2020). Education for the future? Critical evaluation of education for sustainable development goals. *The Journal of Environmental Education*, *51*(4), 280–291. DOI: 10.1080/00958964.2019.1710444

Li, X., Kim, J. S., & Lee, T. J. (2021). Collaboration for community-based cultural sustainability in island tourism development: A case in Korea. *Sustainability (Basel)*, *13*(13), 1–17. DOI: 10.3390/su13137306

Littletree, S., Belarde-Lewis, M., & Duarte, M. (2020). Centering relationality: A conceptual model to advance indigenous knowledge organization practices. *Knowledge Organization*, *47*(5), 410–426. DOI: 10.5771/0943-7444-2020-5-410

Mapara, J. (2022). Indigenous knowledge across the curriculum: An epistemic and cognitive shifting process. In Muyambo, T., Mapangisana-Hlatywayo, A., & Sithole, P. (Eds.), *Re-imagining indigenous knowledge and practices in 21st century Africa: Debunking myths and misconceptions for conviviality and sustainability* (pp. 41–68). Langaa Publishers.

Mathieson, S., Black, K., Allin, L., Hooper, H., Penlington, R., Mcinnes, L., & Anderson, E. (2023). New academics' experiences of induction to teaching: Using Cultural Historical Activity Theory (CHAT) to understand and improve induction experiences. *The International Journal for Academic Development*, •••, 1–15.

Mbah, M., Ajaps, S., & Molthan-Hill, P. (2021). A systematic review of the deployment of indigenous knowledge systems towards climate change adaptation in developing world contexts: Implications for climate change education. *Sustainability (Basel), 13*(9), 4811. DOI: 10.3390/su13094811

McLean, M., Warner, B., Markham, R., Fischer, M., Walker, J., Klein, C., Hoeberechts, M., & Dunn, D. C. (2023). Connecting conservation & culture: The importance of indigenous knowledge in conservation decision-making and resource management of migratory marine species. *Marine Policy, 155*, 105582. DOI: 10.1016/j.marpol.2023.105582

Mesec, B. (1998). *Uvod v kvalitativno raziskovanje v socialnem delu.* Ljubljana: Visoka šola za socialno delo.

Mishra, M. P., Joshi, B., Agarwal, P., & Varshney, I. (2024). Sustainable development goals–Role of indigenous knowledge. *Educational Administration: Theory and Practice, 30*(5), 4521–4527.

Mji, G., Kalenga, R., Ned, L., Alperstein, M., & Banda, D. (2017). Indigenous knowledge exclusion in education systems of Africans: Impact of beingness and becoming an African. In Ngulube, P. (Ed.), *Handbook of research on social, cultural, and educational considerations of indigenous knowledge in developing countries* (pp. 36–59)., DOI: 10.4018/978-1-5225-0838-0.ch003

Mokuku, T., & Janse van Rensburg, E. (1997). Reflecting on socially transformative environmental literacy for Lesotho. *Southern African Journal of Environmental Education, 17*, 31–39.

Morgan, H. (2022). Conducting a qualitative document analysis. *The Qualitative Report, 27*(1), 64–77. DOI: 10.46743/2160-3715/2022.5044

Msuya, J. (2007). Challenges and opportunities in the protection and preservation of indigenous knowledge in Africa. *International Journal of Information Ethics, 7*, 346–353. DOI: 10.29173/irie38

Omodan, B. I., & Diko, N. (2021). A Conceptualisation of Ubuntugogy as a Decolonial Pedagogy in Africa. *Journal of Culture and Values in Education, 4*(2), 95–104. DOI: 10.46303/jcve.2021.8

Payne, L., & O'Neil, J. K. (2019). Cultural sustainability in higher education. In Leal Filho, W. (Ed.), *Encyclopedia of sustainability in higher education.* Springer., DOI: 10.1007/978-3-319-63951-2_109-1

Petzold, J., Andrews, N., Ford, J. D., Hedemann, C., & Postigo, J. C. (2020). Indigenous knowledge on climate change adaptation: A global evidence map of academic literature. *Environmental Research Letters, 15*(11), 113007. DOI: 10.1088/1748-9326/abb330

Rozi, S., & Taufik, Z. (2020). Adaptation of religion and local wisdom in global environmental issues in Indonesia. *Religious: Jurnal Studi Agama-Agama Dan Lintas Budaya, 4*(3), 191–203. DOI: 10.15575/rjsalb.v4i3.9593

Senanayake, S. G. J. N. (2006). Indigenous knowledge as a key to sustainable development. *Journal of Agricultural Sciences–Sri Lanka, 2*(1), 87–94. DOI: 10.4038/jas.v2i1.8117

Sivapalan, S., & Speight, S. (2019). Role of education for sustainable development. In Leal Filho, W. (Ed.), *Encyclopedia of sustainability in higher education*. Springer., DOI: 10.1007/978-3-030-11352-0_28

Tsetselelane, M. D., Sekgothe, M., Stewart, K. L., & Vuma, L. (2021). Knowledge management: Preserving, managing and sharing indigenous knowledge through digital library. *Theological Studies*, *77*(2).

Yamagata-Lynch, L. C. (2007). Confronting analytical dilemmas for understanding complex human interactions in design-based research from a cultural—Historical activity theory (CHAT) framework. *Journal of the Learning Sciences*, *16*(4), 451–484. DOI: 10.1080/10508400701524777

Zhang, G., Chen, X., Law, R., & Zhang, M. (2020). Sustainability of heritage tourism: A structural perspective from cultural identity and consumption intention. *Sustainability (Basel)*, *12*(21), 1–17. DOI: 10.3390/su12219199

Zidny, R., Sjöström, J., & Eilks, I. (2023). Indigenous Knowledge and Science and Technology Education. In Akpan, B., Cavas, B., & Kennedy, T. (Eds.), *Contemporary Issues in Science and Technology Education. Contemporary Trends and Issues in Science Education* (Vol. 56). Springer., DOI: 10.1007/978-3-031-24259-5_12

Zimu-Biyela, A. N. (2016). *The management and preservation of indigenous knowledge in Dlangubo Village in KwaZulu-Natal, South Africa* (Doctoral dissertation, University of South Africa).

ADDITIONAL READING

Agrawal, A. (1995). Dismantling the divide between indigenous and western knowledge. *Development and Change*, *26*(3), 413–439. DOI: 10.1111/j.1467-7660.1995.tb00560.x

Burgos-Ayala, A., Jiménez-Aceituno, A., Torres-Torres, A. M., Rozas-Vásquez, D., & Lam, D. P. M. (2020). Indigenous and local knowledge in environmental management for human-nature connectedness: A leverage points perspective. *Ecosystems and People (Abingdon, England)*, *16*(1), 290–303. DOI: 10.1080/26395916.2020.1817152

Carrin, M. (2024). Indigenous Knowledge and Sustainability. *The Oriental Anthropologist: A Bi-annual International Journal of the Science of Man, 24*(1), 13-31. https://doi.org/DOI: 10.1177/0972558X2412278

Kanene, K. M. (2016). Indigenous practices of environmental sustainability in the Tonga community of southern Zambia. *Jàambá*, *8*(1), a331. DOI: 10.4102/jamba.v8i1.331 PMID: 29955300

Masoga, M. A., & Shokane, A. L. (2019). Viewpoint: Indigenous knowledge systems and environmental social work education: Towards environmental sustainability. *Southern African Journal of Environmetal Education*, *35*(1), 1–11. DOI: 10.4314/sajee.v35i1.14

Obiora, A. C., & Emeka, E. E. (2015). African indigenous knowledge system and environmental sustainability. *International Journal of Environmental Protection and Policy*, *3*(4), 88–96. DOI: 10.11648/j.ijepp.20150304.12

Ubale, S., & Badade, K. (2024). Cultural roots of conservation: Indigenous knowledge in environmental sustainability. *Indian Journal of Integrated Research in Law*, *IV*(VI), 414–432.

KEY TERMS AND DEFINITIONS

Indigenous Knowledge: Also viewed as traditional or local knowledge, refers to knowledge, skills, and beliefs unique to a specific community. Its custodians, indigenous people, have sustained it from generation to generation. It is mainly taught and learned through observation and experience and orally.

Environmental Sustainability: Saving the (natural) environment from degradative activities and human behavior by balancing it and conserving resources to support the well-being of the current and future generations.

Technology: The use of knowledge, skills and resources to manipulate the environment for humans' intended purposes while ensuring respectful treatment of the environment.

Botho: A set of values to interconnect people with their communities and the environment by expressing compassion, dignity, reciprocity, and mutuality.

Sustainable Development: A development that meets people's current needs while ensuring that future generations' needs are uncompromised; it focuses on the economic, social and environmental dimensions.

Curriculum: A standardized and planned experience to enable learners to practice and achieve the outcomes in the packaged content and applied learning skills.

Chapter 18
Bridging Heritage and Enterprise to Address Socio-Economic Issues:
A Policy Analysis of Indigenous Knowledge System Businesses in South Africa

Mpho Gift Makgamatha
https://orcid.org/0000-0001-6799-4554
University of Limpopo, South Africa

ABSTRACT

This research examines the role of government policies in supporting the establishment and growth of Indigenous Knowledge Systems (IKS) in South Africa. It evaluates key legislative frameworks, such as the Indigenous Knowledge Systems Policy (2004) and the Protection, Promotion, Development, and Management of Indigenous Knowledge Act (2019), and provincial programs aimed at fostering IKS enterprises. The study also examines the effectiveness of these policies in promoting entrepreneurship, safeguarding Indigenous knowledge, and ensuring equitable benefit-sharing. It also examines the impact of these policies on marginalized communities, particularly in rural areas. The research also highlights the need for more inclusive, adaptive, and context-specific policies that integrate Indigenous knowledge into mainstream economic frameworks. The study advocates for policies that promote IKS businesses and contribute to a more sustainable, equitable, and culturally rich South Africa.

INTRODUCTION

In South Africa, the intersection of IKS and business ventures offers a unique approach to addressing pressing socio-economic challenges. IKS, which encompass traditional cultural practices, natural resource management, agricultural techniques, and medicinal knowledge, have been passed down through generations and are deeply intertwined with the identities of indigenous communities (Al-Mansoori & Hamdan, 2023). However, the modern economy has often neglected these valuable resources, resulting in underutilization and economic marginalization of communities that possess this knowledge. Bridging the gap between heritage and enterprise is not only a means of preserving cultural identity but also an

DOI: 10.4018/979-8-3693-7964-6.ch018

opportunity to foster sustainable economic growth, particularly in rural and historically disadvantaged areas (Lerario, 2022). This policy analysis examines how the incorporation of indigenous knowledge into business models can create opportunities for socio-economic development, focusing on South Africa's rich cultural heritage. South Africa, with its diverse array of indigenous communities, presents a fertile ground for integrating IKS into contemporary business ventures (Al-Mansoori & Hamdan, 2023). The country is home to more than 50 ethnic groups, each with distinct practices, languages, and knowledge systems. Yet, despite this wealth of cultural heritage, many communities continue to experience poverty, unemployment, and limited access to education. For example, the unemployment rate among South Africa's youth remains alarmingly high, with recent statistics from the South African Statistics Office indicating a youth unemployment rate of over 60% (Mncayi, 2020). This situation calls for innovative economic strategies that leverage the indigenous knowledge base. By promoting indigenous businesses, whether in agriculture, eco-tourism, natural product industries (Herbal Remedies, Plant-Based Fertilizers, Neem-Based Products), or crafts, policymakers can create new avenues for employment and business growth. The integration of IKS into business models has shown promise in several sectors. For instance, the growing global demand for organic and sustainable products provides an opportunity for businesses that use indigenous plant knowledge to develop herbal products, medicines, and beauty treatments (Bareetseng, 2022). Indigenous knowledge on local plants and natural remedies is an untapped market, with several small businesses and cooperatives already capitalizing on these opportunities.

The rise of community-based eco-tourism businesses, where tourists experience indigenous cultures and natural landscapes, also demonstrates the potential for sustainable business ventures that both preserve heritage and create employment (Bareetseng, 2022). These businesses are particularly impactful in rural areas, where opportunities for formal employment are scarce. Despite these successes, the path to scaling IKS businesses faces significant challenges. There are gaps in policy support, infrastructure, and access to capital that hinder the growth of these enterprises. Indigenous businesses often struggle to navigate formal markets due to a lack of formal recognition of traditional knowledge, intellectual property rights, and business skills (Jones, Seet, Acker & Whittle, 2021). Furthermore, there is a need for educational initiatives that equip indigenous entrepreneurs with the necessary tools for business development while respecting cultural norms. This policy analysis will explore how targeted interventions can bridge these gaps and enhance the potential of indigenous knowledge systems to address socio-economic issues in South Africa, fostering a more inclusive and sustainable economy. The objectives formulated for this chapter were to:

- Determine the role of indigenous knowledge systems in South Africa.
- Identify socio-economic challenges addressed by IKS in South Africa.
- Examine the policy frameworks and initiatives to support indigenous knowledge system businesses in South Africa.
- Determine the role of government and policy in promoting indigenous knowledge businesses in South Africa.

HISTORICAL CONTEXT: THE EVOLUTION OF IKS IN SOUTH AFRICA

Indigenous knowledge has long been integral to South African communities, encompassing a wide range of traditional practices, agricultural techniques, medicinal knowledge, and cultural crafts (Bhuda & Marumo, 2020). Historically, this knowledge was passed down orally, embedded in everyday life and the customs of various ethnic groups. The early economies of South Africa were shaped by these practices, which not only facilitated the survival and prosperity of communities but also fostered sustainable interactions with the environment (Lam et al., 2020). For example, the Zulu people's use of medicinal plants, like the *Aloe vera*, and the Xhosa's traditional farming methods, such as the use of organic fertilizers, were foundational to local economies (Cock & Van Vuuren, 2020; Bese, Zwane & Cheteni, 2021). However, during the colonial era, indigenous knowledge was often undermined or dismissed by Western colonial powers in favour of imported European knowledge systems (Fernández-Llamazares, Lepofsky, Lertzman, Armstrong, Brondizio, Gavin, Lyver, Nicholas, Pascua, Reo & Reyes-García, 2021). This disjunction created a disconnect between traditional practices and Western economic structures. In many cases, colonial policies led to the displacement of indigenous people and the erosion of their traditional economies (Fernández-Llamazares et al., 2021). For instance, land dispossession and the forced migration of communities disrupted agricultural practices that had been passed down for generations, significantly limiting the potential for indigenous knowledge to evolve into business ventures (Mawere, 2023). The post-apartheid era, however, has witnessed a resurgence of interest in indigenous knowledge systems, particularly in the context of addressing socio-economic challenges (Odora Hoppers, 2021). Following the end of apartheid, there was a growing recognition of the value of indigenous knowledge not only in preserving cultural heritage but also in creating viable business opportunities. The South African government began to place greater emphasis on inclusive economic growth, acknowledging the potential of indigenous businesses to contribute to national development (Balogun & Kalusopa, 2021). For example, the rise of indigenous craft businesses, such as beadwork and weaving, has created significant employment in rural areas, especially for women. According to the Department of Trade and Industry (DTI), the South African craft sector, which includes a variety of indigenous cultural products, employs over 100,000 people, with a majority coming from previously marginalized communities (Ramaano, 2023). Furthermore, government institutions, such as the National Heritage Council's efforts to promote cultural tourism, have provided avenues for indigenous knowledge businesses to gain access to broader markets, both locally and internationally (Ramaano, 2023). These efforts include providing funding for community heritage projects that focus on research and documentation of diverse cultural practices, rituals, and traditional arts and crafts. By financially supporting such projects, the NHC helps preserve indigenous knowledge and facilitates its integration into the cultural tourism sector, thereby opening avenues for local and international market exposure. This shift in the policy landscape is an acknowledgment of how indigenous knowledge can be transformed into profitable ventures, bridging the gap between heritage preservation and enterprise.

THEORETICAL FRAMEWORK

This analysis draws upon the Cultural Capital Theory and Sustainable Livelihoods Approach to examine the intersection of heritage and enterprise in IKS businesses in South Africa (Manama, Goldman & Thomas, 2024). Cultural Capital Theory, as articulated by Pierre Bourdieu, underscores the value of indigenous knowledge as a form of intangible cultural capital that can be leveraged to foster social and economic development (Daraz, 2023). It emphasizes the potential of IKS to generate economic value while preserving cultural identity. Complementing this, the Sustainable Livelihoods Approach focuses on enhancing the capacity of communities to achieve long-term socio-economic well-being by utilizing their natural, social, and cultural assets. By integrating these theories, the framework explores how policies can empower IKS businesses to address poverty, unemployment, and inequality while ensuring heritage is preserved and sustainably utilized.

RESEARCH METHODOLOGY

This is a desktop study, which relied on qualitative data to collect secondary sources and analyze the role of IKS businesses in addressing socio-economic issues in South Africa. A comprehensive review of existing literature, policy documents, government reports, and case studies was undertaken to understand the current landscape of IKS enterprises and their socio-economic contributions. The research focuses on identifying patterns and opportunities in the intersection of cultural heritage and enterprise development. This approach provides a broad understanding of the policy frameworks and institutional mechanisms that support or hinder the growth of IKS businesses (Barman, 2024). Qualitative data was drawn from secondary sources such as academic articles, policy briefs, and sector-specific reports, enabling a nuanced analysis of the socio-economic impact of IKS businesses. Key themes explored included poverty alleviation, employment generation, and cultural preservation. The study prioritized secondary data to provide a holistic and contextually relevant perspective, avoiding the logistical challenges and resource constraints associated with primary data collection. The data was systematically analyzed through thematic analysis, identifying recurring patterns and insights across different sources. By synthesizing findings from multiple references, the study established a robust foundation for evaluating past policy interventions and proposing strategies to strengthen the role of IKS enterprises in sustainable development.

THE ROLE OF INDIGENOUS KNOWLEDGE SYSTEMS IN SOUTH AFRICA

IKS in South Africa play a pivotal role in shaping both cultural and economic landscapes, particularly in rural areas. These knowledge systems are deeply embedded in the traditions and practices of local communities, reflecting a profound understanding of the environment, agriculture, healthcare, and social organization (Lam, Hinz, Lang, Tengö, von Wehrden & Martín-López, 2020). Historically, IKS have been passed down orally through generations, encompassing a wide range of practices, from traditional medicine to sustainable farming methods. With the growing interest in sustainable development and eco-friendly solutions, Indigenous Knowledge is gaining recognition for its relevance in addressing modern-day challenges such as climate change, biodiversity loss, food insecurity, water scarcity, and sustainable healthcare (Lam et al., 2020). For instance, many South African farmers rely on Indigenous farming techniques such as intercropping, crop rotation, and the use of organic fertilizers (Malobane, Nciizah, Mudau & Wakindiki, 2020). These practices are not only cost-effective but also environmentally sustainable (Patel, Sharma & Singh, 2020). According to a study by the South African Department of Science and Innovation (DSI), indigenous agricultural knowledge contributes to the resilience of farming communities, especially in the face of climate change. In rural areas, these practices help farmers manage water scarcity, soil degradation, and unpredictable weather patterns, thereby securing food production and improving livelihoods.

In the field of healthcare, traditional healers in South Africa are integral to community health systems, especially in remote areas where access to modern healthcare is limited (Bhuda & Marumo, 2020). A significant percentage of the population, especially in rural settings, still relies on traditional medicine for primary health care. The South African Health Department estimates that approximately 60-80% of the rural population uses traditional healing practices (Cock & Van Vuuren, 2020). Many indigenous plants, such as the Aloe Vera plant used in treating wounds and skin conditions, are known for their medicinal properties, which have been passed down through generations of indigenous knowledge (Cock & Van Vuuren, 2020). Furthermore, IKS are integral to the tourism sector, particularly in cultural and eco-tourism. For example, the Khoisan communities, known for their ancient cultural practices, attract tourists who are keen to explore indigenous lifestyles, traditions, and local wisdom (Ntlhabo, 2022). The inclusion of these communities in the tourism sector has opened up new business opportunities, allowing them to leverage their cultural heritage for economic gain. This has also led to a rise in artisanal craft markets where traditional knowledge is showcased and sold to tourists (Ntlhabo, 2022).

The potential of IKS in driving South Africa's economy is closely tied to the effectiveness of policies such as the National Indigenous Knowledge Systems Policy Framework, which aims to recognize, preserve, and promote these knowledge systems (Balogun & Kalusopa, 2021). While IKS businesses can create jobs, bridge socio-economic gaps, and preserve cultural heritage, their economic impact remains underutilized due to challenges like limited commercialization, inadequate policy support, and lack of integration into mainstream industries. The success of these policies depends on fostering collaboration between government, communities, and businesses to ensure that indigenous knowledge is not only preserved but also innovatively applied in modern enterprises (Bhuda & Marumo, 2020). By strengthening policy implementation and aligning IKS with contemporary economic practices, South Africa can unlock its full potential, empowering marginalized communities and contributing to sustainable economic growth.

SOCIO-ECONOMIC CHALLENGES ADDRESSED BY IKS IN SOUTH AFRICA

IKS in South Africa has become a crucial component in addressing various socio-economic challenges. The IKS businesses, rooted in local cultural practices, traditional knowledge, and Indigenous skills, can tackle a range of issues from unemployment and poverty to social exclusion and environmental degradation (Akolgo-Azupogo, Rubens & Bardy, 2021). One of the most pressing challenges in South Africa is the high unemployment rate, particularly among young people and rural communities. According to Statistics South Africa, the unemployment rate for youth (ages 15–34) stands at a staggering 63.9%, a statistic that reflects the difficulties faced in securing stable, meaningful employment in urban areas (Mama, 2020). However, IKS, particularly those in rural areas, offer alternative employment opportunities by utilizing traditional craftsmanship, herbal medicine, agriculture, and tourism to create jobs. For instance, businesses that focus on traditional herbal remedies and natural products, such as those that manufacture and sell indigenous teas or oils derived from local plants like rooibos or marula, employ local communities in production, harvesting, and packaging processes (Cock et al., 2021). These businesses not only create jobs but also support the preservation of indigenous knowledge and local biodiversity. A well-known example is the rooibos tea industry, which has expanded into international markets, providing employment to thousands of people in rural communities in the Western Cape (Sousa and Raizada, 2020). In 2020, the South African Rooibos Council reported that the industry employed over 8,000 people directly, many of whom are from disadvantaged rural areas (Sousa & Raizada, 2020).

Moreover, IKS contribute to social inclusion by empowering women and marginalized groups. In rural communities, where gender inequality remains prevalent, indigenous knowledge businesses offer women the opportunity to participate in income-generating activities (Gwandure & Lukhele-Olorunju, 2023). For instance, women's cooperatives that produce traditional crafts such as beadwork, basket weaving, and clothing are becoming more prominent (Gwandure & Lukhele-Olorunju, 2023). These cooperatives are not only creating employment for women but also helping to preserve and promote South African cultural heritage. One example is the women's cooperatives in Limpopo, where indigenous knowledge of beadwork is being used to create both traditional and contemporary designs that are marketed locally and internationally (Ngomane, 2023).

Environmental sustainability is another area where IKS play a key role in addressing socio-economic challenges. Many indigenous knowledge enterprises operate in ways that are in harmony with the environment, using natural resources in a sustainable manner. For example, small-scale organic farming initiatives that rely on indigenous agricultural knowledge help combat food insecurity while preserving the environment (Cheteni & Umejesi, 2023). These businesses, which often utilize age-old practices of crop rotation and water conservation, provide fresh produce for local markets, contributing to food security in rural areas. According to the South African Food Sovereignty Campaign, small-scale farmers can help reduce the reliance on large-scale commercial farming that often depletes soil quality and exacerbates climate change (Cheteni & Umejesi, 2023).

POLICY FRAMEWORKS AND INITIATIVES TO SUPPORT INDIGENOUS KNOWLEDGE SYSTEM BUSINESSES

a. National Development Plan (NDP) 2030

The National Development Plan (NDP) envisions a South Africa where IKS businesses are leveraged for inclusive economic growth and development (Letshaba, 2024). The NDP highlights the importance of promoting cultural heritage, traditional practices, and local innovation as contributors to job creation and poverty alleviation. By integrating indigenous knowledge into the mainstream economy, the NDP seeks to empower marginalized communities while preserving traditional knowledge systems (Letshaba, 2024). Support for such businesses includes promoting access to resources, funding, and training to ensure sustainability and competitiveness in local and global markets.

b. Indigenous Knowledge Systems Policy (IKSP) 2004

The Indigenous Knowledge Systems Policy (IKSP) specifically aims to recognize, protect, and promote the use of indigenous knowledge in South Africa (Dansu, 2021). This policy provides a framework for the documentation, preservation, and commercialization of traditional knowledge in sectors like agriculture, healthcare, and arts. It also seeks to empower indigenous communities by ensuring they benefit equitably from the commercial use of their knowledge. The IKSP supports capacity-building initiatives, enabling indigenous entrepreneurs to participate effectively in value chains while ensuring their intellectual property rights are safeguarded (Dansu, 2021).

c. Protection, Promotion, Development, and Management of Indigenous Knowledge Act, 2019 (IK Act)

The IK Act provides legal mechanisms to protect indigenous knowledge and its holders from exploitation while facilitating its sustainable use for economic purposes (Fernández-Llamazares, Lepofsky, Lertzman, Armstrong, Brondizio, Gavin, Lyver, Nicholas, Pascua, Reo & Reyes-García, 2021). This legislation ensures that communities retain ownership and control over their knowledge, with mechanisms to share benefits from its commercialization. Indigenous knowledge system businesses are encouraged to register their knowledge under this act to secure their intellectual property rights, enabling them to negotiate fair partnerships and gain market recognition for their unique contributions (Fernández-Llamazares et al., 2021).

d. Small Enterprise Development Agency (SEDA) and Support Programs

SEDA provides tailored support to IKS businesses through mentorship, business development training, and access to funding. Programs under SEDA focus on equipping entrepreneurs with the skills to commercialize traditional knowledge in areas such as herbal medicine, crafts, and food production (Iwara & Kilonzo, 2022). By connecting these businesses with local and international markets, SEDA plays a crucial role in scaling indigenous enterprises and integrating them into the broader economy (Dansu, 2021).

The DTIC actively supports IKS businesses through funding and market access initiatives. Programs such as the Industrial Development Corporation (IDC) and the Export Marketing and Investment Assistance Scheme (EMIA) prioritize traditional and indigenous enterprises (Stephenson, Hamid, Peter, Sauvant, Seric & Tajoli, 2021). These initiatives help indigenous businesses expand their reach, improve production capabilities, and gain visibility in global markets. The DTIC also fosters partnerships between indigenous businesses and mainstream industries to enhance competitiveness and innovation.

e. Agri-Parks Program

Indigenous agricultural practices and knowledge are central to the Agri-Parks program, which promotes rural development and food security (Khan, 2021). By incorporating indigenous farming methods, the program supports traditional farmers in improving productivity while maintaining ecological sustainability. The Agri-Parks program facilitates access to infrastructure, resources, and value chains, enabling indigenous farmers to commercialize their knowledge and contribute to local economic development (Khan, 2021).

f. Cultural and Creative Industries Development Framework (CCIDF)

The CCIDF recognizes the potential of cultural heritage and traditional arts in fostering economic growth (Santoro, Bresciani & Papa, 2020). This framework supports indigenous artisans, musicians, and other creative professionals in developing sustainable businesses. It provides grants, training, and platforms to showcase their work, promoting the commercialization of indigenous cultural products. By linking cultural industries with tourism and export markets, the CCIDF ensures that IKS businesses gain visibility and profitability (Santoro et al., 2020).

g. Integrating Indigenous Knowledge with Modern Economic Strategies

Integrating Indigenous Knowledge with modern economic strategies creates a powerful synergy that fosters sustainable development, cultural preservation, and economic inclusion (Lerario, 2022). By leveraging traditional practices, such as sustainable harvesting and water-efficient agriculture, alongside modern technologies and global markets, communities can unlock new economic opportunities. Examples like the Marula and Rooibos industries in South Africa demonstrate how this approach can preserve biodiversity, generate income, and empower marginalized groups (Sousa & Raizada, 2020). Such integration not only addresses socio-economic challenges but also enriches innovation by drawing on the deep ecological and cultural wisdom of Indigenous communities.

THE ROLE OF GOVERNMENT AND POLICY IN PROMOTING INDIGENOUS KNOWLEDGE BUSINESSES

The government of South Africa plays a crucial role in promoting indigenous knowledge-based businesses through supportive policies and initiatives aimed at integrating traditional knowledge into the mainstream economy. Policies such as the Indigenous Knowledge Systems Policy (IKSP) and Protection, Promotion, Development and Management of Indigenous Knowledge Act, 2019 (Act No. 6 of 2019) was promulgated in Government Gazette No. 42647, published on August 19, 2019, provide legal frameworks to safeguard indigenous knowledge while enabling its commercialization (Masenya, 2022). Government agencies like the Department of Trade, Industry and Competition (DTIC) and the Small Enterprise Development Agency (SEDA) support these businesses by offering funding, mentorship, and market access programs (Balogun & Kalusopa, 2021; Fernández-Llamazares et al., 2021; Iwara & Kilonzo, 2022). Moreover, infrastructure development and tax incentives are implemented to create a conducive environment for indigenous entrepreneurs (Malatji and Makgamatha, 2024). By fostering partnerships between communities, private sector players, and global markets, the government ensures that indigenous businesses not only preserve cultural heritage but also contribute to economic development, job creation, and poverty alleviation in rural and marginalized areas.

CASE STUDIES: INDIGENOUS KNOWLEDGE BUSINESSES IN SOUTH AFRICA

a. The Rooibos Tea Industry

The Rooibos tea industry is a prime example of an IKS business in South Africa that has achieved global success. Rooibos, a herbal tea derived from a plant native to the Western Cape, was traditionally used by the Khoi and San communities for its medicinal properties (Docrat, Abdul & Marnewick, 2024). By combining traditional knowledge with modern commercialization, South Africa has established Rooibos as a valuable export product. The industry contributes significantly to the local economy, providing income and employment to rural communities (Docrat et al., 2024). Additionally, benefit-sharing agreements have been introduced to ensure that the Khoi and San communities are compensated for their contributions to this indigenous knowledge.

b. Marula Oil Production

The production of marula oil, derived from the fruit of the marula tree, highlights the integration of traditional practices into the modern cosmetic and skincare industry (Hlangwani, Hal, Moganedi & Dlamini, 2023). Indigenous communities have long used marula oil for its moisturizing and healing properties. Today, marula oil is a key ingredient in international beauty products. Businesses like Eudafano Women's Cooperative in Namibia and others in South Africa empower rural women by involving them in the harvesting, processing, and selling of marula oil (Nakanyete, Matengu & Diez, 2024). These enterprises not only preserve traditional knowledge but also provide sustainable livelihoods for local communities.

c. Imvelo Eswatini (Traditional Herbal Products)

Imvelo Eswatini produces a range of traditional herbal remedies and wellness products rooted in indigenous knowledge (Moyo & Moyo, 2021). Drawing on traditional practices, the company sources medicinal plants from rural communities in a sustainable manner. Imvelo Eswatini's products, such as herbal teas, tinctures, and essential oils, have gained popularity for their authenticity and health benefits (Moyo & Moyo, 2021). By promoting local harvesting practices and knowledge systems, the business empowers indigenous communities and supports biodiversity conservation.

d. Craft and Beadwork Enterprises

Traditional craft and beadwork enterprises showcase how indigenous knowledge and artistry can thrive in global markets. Businesses such as Monkeybiz, based in Cape Town, collaborate with local artisans to create unique, handmade beadwork (De Villiers, 2020). These enterprises draw on traditional designs while innovating for contemporary aesthetics. Through these ventures, artisans are able to earn a living while preserving cultural traditions and transferring skills to younger generations (De Villiers, 2020). Moreover, such businesses contribute to the tourism industry by offering culturally authentic products.

e. Eco-Tourism Initiatives

Eco-tourism ventures that incorporate indigenous knowledge, such as those in Limpopo and the Eastern Cape, demonstrate how traditional practices can be integrated into sustainable tourism. These businesses provide visitors with immersive cultural experiences, including guided tours of traditional villages, storytelling, and demonstrations of indigenous crafts and cuisine (Kostauli, 2024). The ventures not only preserve indigenous knowledge but also generate income for local communities. For instance, the Leshiba Wilderness in Limpopo combines eco-tourism with cultural heritage preservation, offering tourists a unique experience while supporting rural development (Ramaano, 2023). Mr William Motjila Samora Kgoale, a prominent individual from Botlokwa Village (Mphakane) in Limpopo Province, South Africa, was widely recognized for his leadership as the former chairperson of the Batlokwa Ikageng movement (PressReader, 2021). He has advocated for the integration of traditional knowledge with modern entrepreneurship to foster sustainable livelihoods while safeguarding cultural heritage. The Botlokwa Monument, situated in the centre of Botlokwa Village (Mphakane) in Limpopo Province, South Africa, was established through his efforts and serves as a landmark cultural and historical monument

(PressReader, 2021). The monument stands as a tribute to their struggles, traditions, and contributions to the region's development. Moreover, due to its historical significance, the Botlokwa Monument plays a central role in promoting tourism, education, and community pride. It is frequently visited during cultural festivals and serves as an inspiration for initiatives that blend cultural preservation with modern economic opportunities, contributing to the broader socio-economic development of the region (Ngoepe & Bhebhe, 2023). The picture depicts Mr William Motjila Samora Kgoale and the Botlokwa Monument located in Mphakane Village.

(https://www.google.co.za/url?sa=t&rct=j&q=&esrc=s&source=web&cd=&cad=rja&uact=8&ved=2ahUKEwjYvorA0O2KAxXOSUEAHaAiEREQFnoECBoQAQ&url=https%3A%2F%2Fwww.pressreader.)

f. Baobab Fruit Products

Baobab fruit products, derived from the iconic baobab tree, demonstrate the commercial potential of indigenous knowledge in food and health industries (Mwangi, 2023). Rich in nutrients, the fruit has been traditionally used by indigenous communities for its health benefits. Companies like EcoProducts have harnessed this knowledge to create a range of baobab-based items, including powders and oils, for global markets (Wang & Tian, 2023). These businesses not only generate income for rural communities but also promote sustainable harvesting practices to protect baobab trees for future generations.

g. Sangoma Herbal Remedies

Traditional healers, known as sangomas in South Africa, have long utilized indigenous plants for healing and wellness (Maluleka, 2020). Businesses inspired by this heritage, such as African Remedies, have emerged to commercialize traditional herbal medicines. These businesses focus on creating accessible products while respecting the cultural origins of their practices. Sangoma herbal remedies have gained recognition for their efficacy, and efforts are being made to standardize and certify products for broader acceptance in healthcare markets (Maluleka, 2020).

h. Indigenous Clothing and Fashion

Businesses like MaXhosa Africa, founded by Laduma Ngxokolo, merge traditional Xhosa patterns and designs with modern fashion. Inspired by Xhosa beadwork and cultural symbolism, MaXhosa Africa has gained international acclaim (Hlengiwe, 2023). This success story demonstrates how indigenous knowledge can be reimagined in the global fashion industry while promoting cultural pride and creating economic opportunities for artisans. The Indigenous business fashion collaboration between Bavarian Motor Works (BMW) and the renowned Ndebele artist and "gogo" (grandmother), Dr Esther Mahlangu, stands as a testament to the global appeal of Indigenous art. Esther Mahlangu, a celebrated Ndebele painter, became the first woman to create artwork for BMW's Art Car series in 1991 (Bhuda, 2021; Mhlambi, 2021). Using traditional Ndebele geometric patterns and vivid colours, she painted the exterior of a BMW 525i, merging traditional heritage with modern automotive design. This partnership catapulted Mahlangu and Ndebele art onto the international stage, showcasing the beauty and cultural significance of Indigenous knowledge in a contemporary context. Beyond the Art Car, Mahlangu has worked with BMW on various projects, including customizing luxury vehicle interiors with her iconic designs (Mhlambi, 2021). These collaborations not only highlight the adaptability of traditional art in modern industries but also provide a platform for preserving and celebrating Ndebele cultural heritage. Esther Mahlangu's work has inspired a growing appreciation for Indigenous art in global fashion, design, and luxury industries, demonstrating how traditional knowledge can thrive in a globalized economy while empowering local communities (Bhuda, 2021). Below is the image of Dr. Esther Mahlangu:

(https://www.press.bmwgroup.com/southafrica/article/detail/T0439118EN/bmw-celebrates-artistic-journey)

i. Traditional Beer Brewing Enterprises

Traditional beer brewing businesses, such as those producing umqombothi (sorghum beer), highlight how indigenous food and drink practices can be commercialized (Hlangwani, Adebiyi, Doorsamy & Adebo, 2020). These enterprises cater to both local and international markets, celebrating traditional brewing techniques. By integrating modern production standards, such businesses maintain cultural authenticity while ensuring product quality and safety (Hlangwani et al., 2020).

CONCLUSION

Bridging heritage and enterprise through IKS businesses in South Africa offers a powerful strategy to address socio-economic challenges such as unemployment, inequality, and environmental degradation. Policies that foster collaboration between Indigenous communities and modern industries can unlock immense economic potential while preserving cultural heritage. By implementing frameworks that ensure equitable benefit sharing, intellectual property protection, and sustainable practices, South Africa can position IKS-based enterprises as drivers of inclusive growth. The success of ventures in sectors like traditional medicine, agriculture, and eco-tourism highlights the transformative potential of integrating heritage with innovation. A deliberate policy focus on this synergy can pave the way for sustainable development, cultural preservation, and economic empowerment across the nation.

SOLUTIONS AND RECOMMENDATIONS

a. Establish Legal and Policy Frameworks to Protect Indigenous Knowledge Rights

To foster IKS businesses, the government should strengthen legal frameworks like the Intellectual Property Laws Amendment Act. These frameworks must ensure that communities retain ownership of their knowledge while benefiting from commercial partnerships. Clear guidelines on benefit-sharing agreements can prevent exploitation and encourage trust between stakeholders.

b. Provide Financial and Technical Support for IKS-Based Enterprises

IKS businesses often lack access to funding and modern tools. Establishing targeted grant programs, microfinance opportunities, and incubation centres can help these businesses thrive. For example, integrating traditional healers into value chains for herbal medicine production could increase their market reach and incomes.

c. Facilitate Skills Development and Knowledge Exchange

Government and private sectors can initiate training programs that blend Indigenous practices with modern business skills. For instance, workshops on product branding and digital marketing for IKS-based products like Marula oil or Rooibos tea can enhance their competitiveness in global markets.

d. Promote Collaborative Research and Development

Encouraging partnerships between local communities, universities, and industries can lead to innovative products and sustainable practices. Research projects on traditional agricultural methods or medicinal plants should prioritize co-ownership of intellectual property to benefit Indigenous communities directly.

e. Integrate IKS into National Development Plans

Incorporating IKS into broader socio-economic and environmental policies ensures their recognition as valuable assets. This integration can highlight the role of IKS in achieving Sustainable Development Goals, such as poverty reduction and biodiversity preservation, fostering a cohesive national development approach.

FUTURE RESEARCH DIRECTIONS

Future research should explore the intersection of heritage and enterprise by analyzing policies that govern Indigenous Knowledge System (IKS) businesses in South Africa and their impact on socio-economic development. Investigating the effectiveness of existing regulatory frameworks, funding mechanisms, and capacity-building initiatives can provide insights into how IKS businesses contribute to job creation, poverty alleviation, and sustainable economic growth. Furthermore, comparative studies with other countries that have successfully integrated indigenous knowledge into mainstream economies could offer valuable policy lessons. Emphasis should also be placed on understanding the challenges faced by IKS entrepreneurs, such as intellectual property protection, market access, and technological adaptation, to formulate inclusive policies that support their growth. By bridging heritage and enterprise, future research can help policymakers develop strategies that enhance the viability of IKS businesses while preserving cultural heritage and fostering economic resilience.

REFERENCES

Akolgo-Azupogo, H., Rubens, A., & Bardy, R. (2021). Soft skills in developing economies: An African view on the hidden linkage between indigenous knowledge and business perspectives. *Journal of African Studies and Development, 13*(3), 47–58. DOI: 10.5897/JASD2021.0625

Al-Mansoori, F., & Hamdan, A. (2023). Integrating indigenous knowledge systems into environmental education for biodiversity conservation: A study of sociocultural perspectives and ecological outcomes. *AI. IoT and the Fourth Industrial Revolution Review, 13*(7), 61–74.

Balogun, T., & Kalusopa, T. (2021). A framework for digital preservation of Indigenous knowledge system (IKS) in repositories in South Africa. *Records Management Journal, 31*(2), 176–196. DOI: 10.1108/RMJ-12-2020-0042

Balogun, T., & Kalusopa, T. (2021). A framework for digital preservation of Indigenous knowledge system (IKS) in repositories in South Africa. *Records Management Journal, 31*(2), 176–196. DOI: 10.1108/RMJ-12-2020-0042

Bareetseng, S. (2022). The worldwide herbal market: Trends and opportunities. *Journal of Biomedical Research & Environmental Sciences, 2766*(5), 2276. DOI: 10.37871/jbres1482

Barman, P. (2024). The Ethical Frontier: Integrating IKS for Sustainable Economic Development. *MSW Management Journal, 34*(1), 228–238.

Bese, D., Zwane, E., & Cheteni, P. (2021). The use of sustainable agricultural methods amongst small-holder farmers in the Eastern Cape province, South Africa. *African Journal of Science, Technology, Innovation and Development, 13*(3), 261–271. DOI: 10.1080/20421338.2020.1724388

Bhuda, M.T. (2021). The role of art in sustaining the livelihoods of Amandebele women in South Africa.

Bhuda, M.T. & Marumo, P. (2020). African traditional medicine and healing in South Africa: Challenges and prospects before and during COVID 19. *Ife Centre for Psychological Studies/Services, 18*, 16710

Bhuda, M. T., & Marumo, P. (2020). African traditional medicine and healing in South Africa: Challenges and prospects before and during COVID 19. *Ife Centre for Psychological Studies/Services. PO Box 1548. Ile-Ife Osun State, Nigeria, 18*(4), 16710.

Cheteni, P., & Umejesi, I. (2023). Application of indigenous knowledge and sustainable agriculture by small-scale farmer households and entrepreneurs in South Africa. *International Journal of Development and Sustainability, 12*(1), 1–18.

Cindi, L. (2021). Incorporating African Indigenous Knowledge Systems into the Basic Education curriculum: Experiences from two schools in the Gauteng and KwaZulu-Natal Provinces, South Africa. University of the Free state

Cock, I. E., & Van Vuuren, S. F. (2020). The traditional use of southern African medicinal plants in the treatment of viral respiratory diseases: A review of the ethnobotany and scientific evaluations. *Journal of Ethnopharmacology, 262*, 113194. DOI: 10.1016/j.jep.2020.113194 PMID: 32730880

Dansu, T. (2021). Impact of Indigenous Knowledge System and Practices (IKSP) on Secondary School Students' Comprehension and Entrepreneurial Skill in Chemistry Concepts. *Available atSSRN* 4348210.

Daraz, B. (2023). *Cultural capital and academic advising for international students* (Doctoral dissertation, Fordham University).

De Villiers, K. (2020). *Expanding the Product Range of Reaboka Foundation Through Action-based Research*. University of Johannesburg.

Docrat, T. F., Abdul, N. S., & Marnewick, J. L. (2024). The phytotherapeutic potential of commercial South African medicinal plants: Current knowledge and future prospects. *Physical Sciences Reviews*, *9*(1), 271–289. DOI: 10.1515/psr-2022-0136

Fernández-Llamazares, Á., Lepofsky, D., Lertzman, K., Armstrong, C. G., Brondizio, E. S., Gavin, M. C., Lyver, P. O. B., Nicholas, G. P., Pascua, P. A., Reo, N. J., & Reyes-García, V. (2021). Scientists' warning to humanity on threats to indigenous and local knowledge systems. *Journal of Ethnobiology*, *41*(2), 144–169. DOI: 10.2993/0278-0771-41.2.144

Fernández-Llamazares, Á., Lepofsky, D., Lertzman, K., Armstrong, C. G., Brondizio, E. S., Gavin, M. C., Lyver, P. O. B., Nicholas, G. P., Pascua, P. A., Reo, N. J., & Reyes-García, V. (2021). Scientists' warning to humanity on threats to indigenous and local knowledge systems. *Journal of Ethnobiology*, *41*(2), 144–169. DOI: 10.2993/0278-0771-41.2.144

Gwandure, C., & Lukhele-Olorunju, P. (2023). Women's use of indigenous knowledge in Africa. *Social Sciences & Humanities Open*, *8*(1), 100741. DOI: 10.1016/j.ssaho.2023.100741

Hlangwani, E., Adebiyi, J. A., Doorsamy, W., & Adebo, O. A. (2020). Processing, characteristics and composition of umqombothi (a South African traditional beer). *Processes (Basel, Switzerland)*, *8*(11), 1451–1481. DOI: 10.3390/pr8111451

Hlangwani, E., Hal, P. H. V., Moganedi, K. L., & Dlamini, B. C. (2023). The future of African wild fruits–a drive towards responsible production and consumption of the marula fruit. *Frontiers in Sustainable Food Systems*, *7*, 1294437. DOI: 10.3389/fsufs.2023.1294437

Hlengiwe, M. (2023). *From Beadwork to Africanfuturism: Exploring Maxhosa Africa's SS22 Collection* (Master's thesis, University of Pretoria (South Africa) Indigenous knowledge system (IKS) in repositories in South Africa. *Records Management Journal, 31*(2), 176-196

Iwara, I. O., & Kilonzo, B. M. (2022). Towards a conceptual understanding of an effective rural-based entrepreneurial University in South Africa. *Social Sciences (Basel, Switzerland)*, *11*(9), 388. DOI: 10.3390/socsci11090388

Jones, J., Seet, P. S., Acker, T., & Whittle, M. (2021). Barriers to grassroots innovation: The phenomenon of social-commercial-cultural trilemmas in remote indigenous art centres. *Technological Forecasting and Social Change*, *164*, 119583. DOI: 10.1016/j.techfore.2019.02.003

Kandolo, K. M., & Ngibe, M. (2024). Exploring financial knowledge influences on the economic empowerment of women entrepreneurs in Kwazulu-Natal, South Africa. *International Journal of Research in Business & Social Science*, *13*(8), 46–61. DOI: 10.20525/ijrbs.v13i8.3683

Khan, S. (2021). Rural land reform and local economic development through agri-parks: A case study of Harry Gwala Municipality in KwaZulu-Natal. *The New Political Economy of Land Reform in South Africa*, 179-197.

Kostauli, M. R. (2024). *Towards the co-management of natural resources in protected areas in South Africa: a study of the Silaka nature reserve, in Eastern Cape province* (Doctoral dissertation, University of the Witwatersrand, Johannesburg).

Lam, D. P., Hinz, E., Lang, D. J., Tengö, M., von Wehrden, H., & Martín-López, B. (2020). Indigenous and local knowledge in sustainability transformations research: A literature review. *Ecology and Society*, *25*(1), 23–45. DOI: 10.5751/ES-11305-250103

Lerario, A. (2022). The role of built heritage for sustainable development goals: From statement to action. *Heritage*, *5*(3), 2444–2463. DOI: 10.3390/heritage5030127

Letshaba, R. K. (2024). *Developing a framework for the implementation of knowledge management within skills development providers in South Africa* (Doctoral dissertation, North-West University (South Africa).

Malatji, T., & Makgamatha, M. G. (2024). An Assessment on The Effect of Immigrant Entrepreneurship on Local Economic Development in South Africa: A Literature Analysis. [Public Administration Journal]. *Jurnal Administrasi Publik*, *14*(2), 92–103.

Malobane, M. E., Nciizah, A. D., Mudau, F. N., & Wakindiki, I. I. (2020). Tillage, crop rotation and crop residue management effects on nutrient availability in a sweet sorghum-based cropping system in marginal soils of South Africa. *Agronomy (Basel)*, *10*(6), 776–886. DOI: 10.3390/agronomy10060776

Maluleka, R. J. (2020). The status of traditional healing in the Limpopo province of South Africa. *Hervormde Teologiese Studies*, *76*(4), 36–43. DOI: 10.4102/hts.v76i4.6103

Mama, M. L. (2020). *Promoting agriculture amongst the black youth of South Africa: the role the sector has to play in creating employment, alleviating poverty, and achieving equitable economic growth* (Doctoral dissertation, Stellenbosch: Stellenbosch University).

Manyaga, M. B., Goldman, G. A., & Thomas, P. (2024). Sustaining SMEs through Indigenous Knowledge Systems: Exploring opportunities and challenges. *Southern African Journal of Entrepreneurship and Small Business Management*, *16*(1), 1–12. DOI: 10.4102/sajesbm.v16i1.882

Mawere, J. (2023). *Exploring The Efficacy of Customary Tenure as A Tool for Rural Development and Food Security in Vhembe District of South Africa: An Indigenous Knowledge System Perspective* (Doctoral dissertation)

Mncayi, N. P. (2020). *An analysis of underemployment amongst young people in South Africa: The case of university graduates* (Doctoral dissertation, North-West University (South Africa).

Moyo, O.N. & Moyo, O.N. (2021). Performing Africanity Southern African Immigrants' Perspectives on Ubuntu. *Africanity and Ubuntu as Decolonizing Discourse*, 151-189.

Mwangi, A. W. (2023). *Development and Quality Evaluation of a Functional Food From Cultured Milk Supplemented With Baobab (Adansonia Digitata L.) Fruit Pulp* (Doctoral dissertation, University of Nairobi).

Nakanyete, N.F., Matengu, K.K. & Diez, J.R. (2024). 6 Requirements for community-company partnerships in non-timber forest product trade: The case of San communities in northern Namibia. *Indigenous Knowledge for Sustainable Livelihoods: Evaluating Non-Timber Forest Product Value Chains for San Communities in Northern Namibia*, 84.

Ngoepe, M., & Bhebhe, S. (2023). *Indigenous Archives in Postcolonial Contexts: Recalling the Past in Africa*. Taylor & Francis. DOI: 10.4324/9781003277989

Ngomane, T.S. (2023). Fostering women entrepreneurship in the tourism industry: a look into obtacles and opportunities of women selling arts and crafts around the Kruger national park. *International Journal of Research in Business and Social Science (2147-4478), 12*(10), 175-181.

Ntlhabo, M. A. (2022). *Navigating between the sacred and the profane: Mohokare sacred sites, spiritual tourism and the challenges in the formal heritage sector* (Doctoral dissertation, University of the Free State).

Odora Hoppers, C. (2021). Research on Indigenous knowledge systems: The search for cognitive justice. *International Journal of Lifelong Education*, *40*(4), 310–327. DOI: 10.1080/02601370.2021.1966109

Patel, S. K., Sharma, A., & Singh, G. S. (2020). Traditional agricultural practices in India: An approach for environmental sustainability and food security. *Energy, Ecology & Environment*, *5*(4), 253–271. DOI: 10.1007/s40974-020-00158-2

Ramaano, A. I. (2023). *The prospects of using the tourism industry to advance community livelihoods in Musina Municipality, Limpopo Province* (Doctoral dissertation).

Santoro, G., Bresciani, S., & Papa, A. (2020). Collaborative modes with cultural and creative industries and innovation performance: The moderating role of heterogeneous sources of knowledge and absorptive capacity. *Technovation*, *92*, 102040. DOI: 10.1016/j.technovation.2018.06.003

Sousa, E. C., & Raizada, M. N. (2020). Contributions of African crops to American culture and beyond: The slave trade and other journeys of resilient peoples and crops. *Frontiers in Sustainable Food Systems*, *4*, 586340. DOI: 10.3389/fsufs.2020.586340

Stephenson, M., Hamid, M. F. S., Peter, A., Sauvant, K. P., Seric, A., & Tajoli, L. (2021). More and better investment now! How unlocking sustainable and digital investment flows can help achieve the SDGs. *Journal of International Business Policy*, *4*(1), 152–167. DOI: 10.1057/s42214-020-00094-2

Wang, J., & Tian, G. (2023). Sustainability of Forest Eco-Products: Comprehensive Analysis and Future Research Directions. *Forests*, *14*(10), 2008–2220. DOI: 10.3390/f14102008

Yumagulova, L., Parsons, M., Yellow Old Woman-Munro, D., Dicken, E., Lambert, S., Vergustina, N., Scott, J. C., Michell, P., & Black, W. (2023). Indigenous perspectives on climate mobility justice and displacement-mobility-immobility continuum. *Climate and Development*, •••, 1–18. DOI: 10.1080/17565529.2023.2227158

ADDITIONAL READINGS

Cheteni, P., & Umejesi, I. (2024). Indigenous Knowledge in Entrepreneurship and Cultural Tourism in the Rural Areas. *Comparative Sociology*, *23*(3), 340–359. DOI: 10.1163/15691330-bja10110

Mdhluli, T. D., Mokgoatšana, S., Kugara, S. L., & Vuma, L. (2021). Knowledge management: Preserving, managing and sharing indigenous knowledge through digital library. *Hervormde Teologiese Studies*, *77*(2). Advance online publication. DOI: 10.4102/hts.v77i2.6795

Ntuli, L. F. (2024). Governance Legacy and Digitisation in Contextualising Systems Approaches for Local Knowledge and Values Preservation. A Case study of South Africa. *Governance: An International Journal of Policy, Administration and Institutions*, *26*(2).

Adeola, O. (2023). Advancing Indigenous Knowledge and Building Sustainable Business Structures in Africa. In *Casebook of Indigenous Business Practices in Africa: Apprenticeship, Craft, and Healthcare–Volume 1* (pp. 239-257). Emerald Publishing Limited.

KEY TERMS AND DEFINITIONS

Indigenous Knowledge Systems (IKS): A body of knowledge, innovations, and practices developed by indigenous communities over generations, often transmitted orally and deeply rooted in cultural traditions, local ecosystems, and social practices.

IKS Businesses: Businesses that leverage indigenous knowledge to create products and services in sectors such as traditional medicine, agriculture, crafts, cosmetics, and tourism, contributing to economic sustainability and cultural preservation.

Socio-Economic Development: A process of improving the well-being of individuals and communities through economic activities, job creation, and social empowerment, often linked to poverty alleviation and sustainable livelihoods.

Policy Analysis: The systematic evaluation of government policies, regulations, and initiatives aimed at fostering the growth and integration of Indigenous Knowledge Systems within the formal economy.

Cultural Preservation: Efforts to safeguard and promote indigenous traditions, languages, practices, and knowledge, ensuring they remain integral to community identity and economic activities.

Intellectual Property Rights (IPR) in IKS: Legal mechanisms, including patents and trademarks, used to protect indigenous knowledge from exploitation while ensuring fair compensation for knowledge holders.

Traditional Medicine: Health practices, remedies, and healing systems based on indigenous knowledge, often using natural resources like medicinal plants and spiritual healing methods.

Commercialization of IKS: The process of transforming indigenous knowledge into marketable goods and services while maintaining ethical and cultural integrity.

Chapter 19

The Commercialization and Intellectual Property Protection of Indigenous Knowledge:

A Look Into Ndebele Artefacts Entrepreneurs

Monicca Thulisile Bhuda
https://orcid.org/0000-0002-8506-9562
University of Mpumalanga, South Africa

ABSTRACT

Ndebele women have been creating murals and beading as a part of their cultural identity for decades and they have passed this expertise down to next generation. To create sustainable livelihoods, they have been traveling the world to showcase their skills and sell crafts to businesses and tourists. However, an existing challenge today is that most entrepreneurs are exploited of their knowledge due to lack of education on policies and poor government outreach initiatives on protection of Indigenous Knowledge (IK). Therefore, it is critical that IK entrepreneurs comprehend their rights and financial power over their expertise with government assistance. This study, which employed qualitative methods, was carried out in the KwaMhlanga area in Mpumalanga province, with Ndebele women who are artefacts entrepreneurs. It sought to examine existing IK policies that these entrepreneurs can use to protect their artefacts. The study also discussed government initiatives that guarantee entrepreneurs recognize their sovereignty over their expertise and receive a fair economic share.

INTRODUCTION

Previously, Indigenous Knowledge (IK) was regarded as mysterious, outmoded, unpopular, and unimportant, particularly in terms of developing-country development (Mkapa, 2004). As a result of this attitude, IK has been marginalized, and African rural people have been suffering in extreme poverty as a result of food insecurity (Ponge, 2011; Sharma, 2021). Women in many African rural areas have limited or no access to productive resources, limiting their ability to participate in or benefit from economic or social growth. As a result, in African rural communities, women are poorer than men (Nyaleka, et al.,

DOI: 10.4018/979-8-3693-7964-6.ch019

2019). For instance, according to Statistics South Africa (2017), there are considerable variations in rural poverty experiences between men and women. Poverty in male-headed households climbed from 31.7 percent in 2011 to 33.0 percent in 2015, while poverty in female-headed households increased from 47.8 percent in 2011 to 50 percent in 2015. Thus, according to Statistics South Africa (2017), poverty in South Africa is on the rise, with over 30 million people living in poverty, with women being the most vulnerable to growing poverty.

The country's poverty level shows that it is critical for women to engage in sustainable income-generating livelihood strategies such as IK-based artefacts in order to raise and improve their income and, eventually, their household food security. As per Poverty Report Statistics South Africa (2017), the percentage of people living in rural poverty was 77 percent in 2011 and 81.3 percent in 2015. The poverty rate increased from 35.6 percent to 40.7 percent, widening the divide between rural and urban regions. Therefore, people who live in rural areas experience more poverty than in urban areas making people use their IK for income generation (Francis & Webster, 2019).

With the intentions of reducing poverty and creating jobs using IK the Indigenous Knowledge Systems (IKS) Policy of 2004 encouraged knowledge custodians to use their expertise for wealth creation, employment, and economic growth under the 2004 IKS policy. This would also require ensuring that communities receive fair and consistent recognition for their knowledge, as well as financial remuneration when appropriate. Women's participation is essential from the start and at all levels. The IKS Act of 2019 was introduced in South Africa with the goal of protecting, promoting, developing, and managing IK. It also intends to provide for the administration of IK communities' rights (Bhuda, 2019).

PROBLEM STATEMENT

Ndebele women who are custodians of artefacts have gained interests commercialization them which is one method to preserve this cultural practice. The commercialization of Ndebele artefacts by women targets selling them to local and international customers. However, there are challenges of selling these artefacts as Ndebele women are exploited of by companies and individual who are interested in this market. In the past, Ndebele women have faced unauthorized extraction and commercialization of their artefacts without adequate compensation or recognition for the people who possess it. Like other indigenous communities, Ndebele women lack the means and authority to profit from the commercialization of their knowledge, whereas multinational businesses and other organizations do. According to studies by Amechi (2018) and Kugedera et al. (2021), indigenous people's involvement in research pertaining to their knowledge frequently results in their exploitation by capitalist people and organizations, which supports the concerns expressed in this study. Accordingly, Rajah's (2019) study suggested that community engagements relevant to IK and development are necessary to be conducted. However, there is insufficient academic proof that South African indigenous communities engage in community activities that emphasize the commercialization of IK, agreements for Access and Benefit Sharing (ABS), and legal protection against exploitation. Based on the identified gaps, the study aims to investigate the commercialization of Ndebele artefacts by women and the IP protection rights of this knowledge. Government initiatives in protecting Ndebele artefacts and policies in place will also be discussed in this study. The objectives of this study are as follows:

- To examine the commercialization of artefacts by Ndebele women.

- To investigate the government initiatives on the protection and preservation of Ndebele artefacts.
- To investigate challenges encountered by Ndebele women in commercializing their artefacts.

The above objectives assisted in asking the research questions below as follows:

- How do Ndebele women commercialize their artefacts?
- Which initiatives do government use to protect and preserve Ndebele artefacts?
- What challenges encountered by Ndebele women in the commercializing their artefacts?

INDIGENOUS KNOWLEDGE

IK is frequently considered as historical and ancient African peoples' customs, which is a flawed viewpoint. The term indigenous has frequently been applied to specific groups of people defined by ancestral lands, collective cultural configurations, and historical locales (Angioni, 2003; Dei, 2002). Bhuda and Koitsiwe (2022) stated that IK can be defined as a multidimensional body of knowledge, practices, and representations preserved and produced by peoples with long histories of intimate involvement with the local natural surroundings.

As a result, the word indigenous suggests that the knowledge is unique and belongs to peoples from certain locations who share cultural and social ties (Thompson, et al., 2020). Thus, IK is a process of learning and sharing about each cultural group's social existence, histories, identities, economic, and political activities. Magni (2017) argued that this reflects the distinctiveness of how specific civilizations make sense of the world and how such types of knowledge address context-specific issues and solutions. IK is handed down from one generation to another through symbols, art, oral narratives, proverbs, and performance such as songs, storytelling, wise sayings, riddles, and dances (Arsenault et al., 2019; Owusu-Ansah & Mji, 2013).

Grenier (1998: 1) defines IK succinctly as "…the unique, traditional knowledge existing within and developed around specific conditions of women and men indigenous to a particular geographic area". What makes the knowledge indigenous is its inalienable link to the native people or aborigines of a particular locality: "It is knowledge peculiar to the cultural system of such communities in a given locale" (Grenier, 1998: 1). It has been found to be structured and systematic, with gender and age-specific training taking place, "is stored in peoples' memories and activities and is expressed in stories, songs, folklore, proverbs, dances, myths, cultural values, beliefs, rituals, community laws, local language and taxonomy, agricultural practices, equipment, materials, plant species and animal breeds. IK is shared and communicated orally, by specific example, and through culture. Indigenous forms of communication and organisation are vital to local-level decision-making processes and the preservation, development and spread of indigenous knowledge" (Grenier, 1998: 2).

IK is analogous to intellectual capital in that it is dynamic and has the potential to transform production and provide competitive advantage (Bhuda, 2024). However, measuring knowledge-based assets is difficult across cultures, and IK is frequently laden with values that go beyond monetary value (Briggs, 2005). IK, for example, is frequently associated with a deep understanding of and relationship to a specific local environment, which has social, spiritual, and ecological elements. Thus, commercialization of IK can be a means of sustaining livelihoods, stimulating economic activity in indigenous communities, and supporting the preservation and growth of traditional practices (Bhuda, 2019).

NDEBELE WOMEN AND ART FORMS

An unconventional form or medium through which impulses seen to be creative may be conveyed is referred to as an art form. An art form can be further seen as a creative or expressive action. Painting, sculpture, literature, architecture, theatre, film, and music are among the seven types of art forms. An art form therefore is a practice or product that can be categorized as a form of artistic expression (Arslanbek et al., 2022). Ndebele women have had the right and responsibility to represent society through their art forms for millennia, according to tradition. Traditional Ndebele sensitivity to form and colour, as well as geometric symmetry preferences (Schneider, 1985), have survived the early stages of modernization, thanks in part to their annual initiation process, which is ritually performed by elder Ndebele women with their female adolescents. Initiation serves as a medium for transmitting the customary skills, customs, and attitudes required to be a good Ndebele woman (Bhuda & Saurombe, 2022).

Recognizing as Ndebele contributes to a secure feeling of self for young Ndebele women whose exposure to the western world, its images, lifestyles, and value systems exceeds that of their parents (Bhuda, 2019). Bhuda and Koitsiwe (2021) stated that this initiation process, which restores balance to the life equation in these South African provinces, is based on Ndebele art. Though not the only ethnic example, South African Ndebele women artists show the constant minute alterations in consciousness required to know oneself at all times. Given the arrival of foreign cultural influence, with its plethora of aesthetic options and creative and economic motives, these women artists' interpretations remain astonishingly true to the image of things Ndebele. Despite exposure with Global capitalism, their ancestral visual memory appears to be intact. Bhuda and Koitsiwe (2022) argued that Ndebele art today functions as an expressive cultural identity and an economic driver for custodians and their families.

AN OVERVIEW ON THE PROTECTION AND PRESERVATION OF INDIGENOUS KNOWLEDGE FOR COMMERCIALIZATION

IK is protected under Canadian law as Aboriginal and treaty rights (Battiste, 2005). Under section 35, IK is inextricably tied to Aboriginal and treaty rights (1). As a result, the Supreme Court of Canada has established that a substantive Aboriginal right will typically include the incidental right to impart such a practice, custom, and tradition to a younger generation in order to maintain the continuity of Aboriginal customs and traditions. In treaty interpretation, similar reasonable incident rights exist that would apply to education provisions. Federal and provincial educational laws, regulations, and practices have yet to be implemented or reconciled with constitutional rights to teach IK. Sections 35 and 52 of the 1982 Constitution Act provide the best protection for IK (Oguamanam, 2004).

As per Oguamanam (2004), the Trade Related Aspects of Intellectual Property Rights (TRIPs) agreement puts emphasis on the endeavour to integrate IK -protection protocols into the IP discourse. TRIPs is an attempt to globalize IP, but it does not affect the status of IP as a subject of constitutional authority. The effort to incorporate indigenous and local communities' knowledge-protection regimes into the IP legal structure could have global repercussions. National governments would be acting within their legitimate rights in enacting legal protection for traditional knowledge, both under the TRIPs Agreement and their commitments to other international instruments on indigenous peoples' protection. "The need to integrate indigenous knowledge-protection protocols into IP discourse is a consequence of an indigenous renaissance and resistance that has yielded a logic of epistemological pluralism. Under this pluralism, indigenous holistic and subjective understanding of phenomena or ways of knowing compete for a space within knowledge-protection jurisprudence" (Oguamanam, 2004: 5).

Upreti (2021) argued that the modern understanding of IP is built on the notion of protection. This principle implies exclusive rights, in addition to other legal measures that limit third-party use of the protected material or impose conditions for their authorized use, such as imposing compensation or patenting (Makgamatha & Moikanyane, 2024). IP can also be protected as confidential information against unauthorized disclosure. The principle's main goals are to assure honest business dealing by prohibiting unfair competition, to inspire people to produce new items and come up with new ideas, and to improve transaction and innovation safety. These goals, however, cannot always be tailored to the context of traditional knowledge. The possible clash between IP law and traditional knowledge necessitates an examination of the latter from an IP standpoint (Botha, 2013; Hart, 2007).

The original purpose of IP law was to protect business transactions. It is intended to stimulate commercial and industrial development. TRIPs were incorporated into the General Agreement on Tariffs and Trade (GATT) in 1994, resulting in increased commercialization. Commercialization limits IP law's ability to identify and defend indigenous communities' IP rights (Yassine, 2018). Mbatha (2013) argued that the property is part of indigenous communities' heritage, which is recreated by groups in reaction to their environment. It gives them a sense of belonging and continuity, encourages cultural diversity and human ingenuity, and secures their survival. As a result, it should be modified to the changing environment, and its IP rights should reflect these special considerations (Du Bois, 2013; Yassine, 2018).

IP rights are granted by society to individuals or organizations through governments and mandated international entities over their creative endeavors manifested in inventions, literary and artistic works, musical performance, symbols, names, images, and designs utilized for commerce. For a limited time, IP rights grant the creator the exclusive right to prevent others from making unlawful use of their creation (Marinova & Raven, 2006). Industrial property (functional inventions) and artistic and literary property are the two types of IP law (cultural creations).

South Africa's regime for the protection of IK has four primary features: bioprospecting regulation and economic development; an innovative IK documentation system for defensive protection; positive protection through an as-yet unimplemented sui generis initiative; and linkages with the IP system. Prominent examples of ABS arrangements are scarce, but they appear to be growing, aided in part by organizations such as the Council for Scientific and Industrial Research (CSIR), which has been assisting interested IK -holder communities in translating their knowledge into economic development engines (Bagley, 2018). Knowledge custodians who want to protect their IK and commercialize from it have to register their IP using Laws of copyright, designs, trademark and geographical indications. The Indigenous Knowledge Systems Documentation Centre (IKSDC), the Department of Science and Innovation (DSI) and the National Indigenous Knowledge Systems Office (NIKSO) can be invited to directly come and spend the day with the community whereby they can share those ways in which they can assist in documenting and protecting their knowledge (Balogun & Kalusopa, 2021).

METHODOLOGY

Firstly, a qualitative document analysis was utilized to examine legal documents pertaining to the protection of Ndebele artefacts and other indigenous works. The researcher carried out desk research in order to identify documents which were aligned with this study. The documents that have been analyzed are as follows:

- Patent Act of 1978
- The 2003 Convention for the Safeguarding of the Intangible Cultural Heritage
- Indigenous Knowledge Systems Policy of 2004
- The Intellectual Property Laws Amendment Act 28. 2013 (the IPLA Act)
- The South African Intellectual Property Policy Act of 2018
- Indigenous Knowledge Act of 2019

The study further used qualitative research method and an exploratory research design whereby the researcher wanted to explore the commercialization of the Ndebele art by custodians of the Ndebele artefacts. In qualitative research, exploratory research design is a tool for examining topics or under-studied subjects. Before providing recommendations, it is frequently utilized to create baseline data and investigate a variety of choices (Mainardes et al., 2010).

The target population for this study were ten women from KwaMhlanga area and only five people were sampled using expert purposive sampling procedure. The participants' ages range from 50 to 75, and their demographic information is shared below on table 1. In-depth (face to face) interviews were used as a guide to collect data from the participants. A voice recorder, note book and a camera were used to store data. Themes were obtained using thematic analysis which in social sciences, thematic analysis is a qualitative research technique that is used to find and display recurrent themes or patterns in data. It entails closely reading and analyzing the content in order to derive meaning and comprehend various topics and interpretations (Clarke & Braun, 2017).

TRIANGULATION

Triangulation was used to ensure the accuracy and consistency of the data gathered for this article. To increase the validity and trustworthiness of the results, triangulation entails using several data sources, approaches, or viewpoints. Triangulation was used in this study's setting to improve the qualitative document analysis's overall caliber. The researcher gained access to a range of scholarly work on IK, IP, policies and commercialization by using a variety of sources, including Google Scholar, JSTOR, ResearchGate, government policy documents and university repositories. Additionally, field notes and observations made throughout the data collecting period with Ndebele women who are artefacts entrepreneurs were used for triangulation (Denzin, 2012).

Table 1. Demographic information of participants

Feature	Frequency (n)	Percentage (%)
Age group		
50–59	2	20
60–69	4	40
70 and above	4	40
Type of entrepreneur		
Mural artist	3	30
Beadwork artist	7	70
Gender		
Male	0	0
Female	10	100
Marital status		
Widower/widow	4	40
Single	2	20
Married	4	40
Divorced	0	0

RESULTS

The section of the study focuses on documents that protect, promote and disseminate IK, cultural knowledge, traditional art forms, living heritage and other cultural expressions. Qualitative document analysis was carried out using the following processes: (1) ready materials, (2) extract data, (3) analyze data and (4) distil the findings. Qualitative document analysis is a process of examining documents to determine their meaning and obtain new perspectives. This kind of qualitative document analysis (QDA) looks for themes and patterns in non-numerical data. Qualitative document analysis, like other analytical techniques in qualitative research, necessitates the examination and interpretation of data in order to extract meaning, comprehend, and generate empirical knowledge (Bowen, 2009). Documents that have been evaluated are presented below.

FINDINGS FROM QUALITATIVE DOCUMENT ANALYSIS

The Importance of the Patent Act of 1978

According to the Patents Act of 1978, a patent may be granted for any novel invention that contains an innovative step and is capable of being used or used in trade, industry, or agriculture. As a result, a patent is a negative right that grants inventors the exclusive right "to restrict others from creating, selling, distributing, importing, or using their invention for a specified period of time, without permission or authorization" (Flint & William, 1989: 159-162; Jacob & Alexander, 1993: 21-33). TRIPS requires a minimum of twenty years of protection from the date of filing, and in South Africa, the life of a patent, unless otherwise provided, is twenty years from the date of application (Ss 46 (1); Burrell, 1999: 547). In exchange, society expects the patent applicant to disclose the innovation in a way that allows others to put it into practice (Ss 32).

The Financial Compensation of Indigenous Knowledge Custodians

The IKS Policy, enacted in November 2004, was the result of a multi-departmental effort to create a framework for the recognition, understanding, integration, and promotion of South Africa's substantial IK resources. The participation of IK practitioners and holders has been critical. The IKS Policy of 2004 established a framework for encouraging and enhancing IK's contribution to South Africa's social and economic development.

Several areas of action are identified in the strategy, including the protection of IK (such as artifacts) and the people who keep it from exploitation. This would also require ensuring that communities receive fair and consistent recognition for their knowledge, as well as financial remuneration when appropriate. The Cabinet authorized the IKS Policy (2004), with the overarching purpose of facilitating the acknowledgment, affirmation, promotion, protection, and growth of IK in South Africa. The IKS Policy encourages women to participate in the protection and maintenance of IKS. It goes on to say that women should take part in creating wealth and job possibilities through IKS.

The Protection, Promotion, Development and Management of Indigenous Knowledge

According to the Department of Science and Innovation (DSI), the IK Act of 2019 is mostly about reparations and bringing IK into the mainstream. It was also a game changer since it addressed how indigenous peoples may contribute to and integrate into the mainstream economy by leveraging their indigenous wisdom. One of its goals was governance, with the notion of forming an advisory council that would be diverse in terms of gender and youth. Prior learning was recognized, making the transition between formal and prior IK easier, and an outcomes-based approach was used. The IK Act of 2019 protects, promotes, develops, and manages IK. Furthermore, it addresses the administration of IK communities' rights.

The Viability of the Intangible Cultural Heritage

The Convention for the Safeguarding of the Intangible Cultural Heritage is a UNESCO treaty adopted by the UNESCO General Conference on 17 October 2003 (Keitumetse, 2006). Safeguarding, according to the 2003 UNESCO Convention, refers to "actions aimed at ensuring the viability of the intangible cultural heritage, including the identification, documentation, research, preservation, protection, promotion, enhancement, transmission, particularly through formal and non-informal education, as well as the revitalization of the various aspects of such heritage" (The draft of the National Policy on Living Heritage in South Africa, 2009:14).

In South African history, the state's responsibility is to assist in the recognition and celebration of the importance of living legacy traditions by ensuring that:

- Processes are created to help people to record their heritage in order to identify and safeguard its value to communities
- Common interests are protected within a democratic state
- People contribute innovatively to industries (in areas such as medicine) and to tourism, using IKS for innovation
- People contribute to sustainable economic development and social cohesion and communities continue to celebrate their heritage
- People are afforded proper coordination of living heritage in South Africa and in relation with other countries
- Community participation is integral to the management of living heritage

The value of community involvement in the protection of living heritage cannot be emphasized. Community participation is recognized as a vital component of heritage management in the 2003 Convention. This policy framework recognizes the inextricable link between living heritage and the communities in which it resides.

The Databases for Indigenous Knowledge

According to the Intellectual Property Laws Amendment Act of 2013 (the IPLA Act), databases for IK must be kept in the prescribed manner at the offices of the registrars of patents, copyright, trademarks, and designs, as part of the existing IP registers where applicable, in the National data base. Subsection (1) envisions databases that are subsets of existing IP registers, with separate sections for recording information on diverse manifestations of indigenous cultural expressions or knowledge, as described in subsection (2). (3). No right in a derivative indigenous work permitted by the Intellectual Property Laws Amendment Act, 2013, shall be eligible for registration unless an ABS agreement has been struck between the applicant and the relevant authorities or indigenous community. If an indigenous community has established a community protocol, the protocol must be taken into account when making contact with the indigenous community envisioned in subsection (4).

The Recognition and Protection of Certain Manifestations of Indigenous Knowledge

The South African Intellectual Property Policy Act of 2018 seeks to recognize and protect specific expressions of IK as IP. A concerted approach is required to raise IP awareness among South Africans in order to conserve indigenous resources, traditional innovation, and traditional knowledge. The construction of a framework for protecting traditional knowledge that will secure it from misuse and abuse while also encouraging additional research and development into traditional knowledge-based products and services.

There is also legislation that protects aspects of IK, such as the National Heritage Resources Act 25 of 1999, the World Heritage Convention Act 1999, the National Environment Management: Biodiversity Act (no 10 of 2004), and Sui Generis systems (which protect Plant Breeders' Rights and Database Protection). Finally, the most recent policy framework, The Protection of IK through the IP System, contained the Patents Amendment Act of 2005 and the Intellectual Property Laws Amendment Bill of 2007. These laws will be addressed in depth in the chapter on IP laws in South Africa.

INTRODUCTION TO DATA OBTAINED FROM IN-DEPTH INTERVIEWS

This section of the study focuses on the themes that were generated when data was collected using in-depth (face to face) interviews. The themes for this section are presented as follows: current status of commercializing Ndebele artefacts; Practical government initiatives on the protection and preservation of Ndebele artefacts; Challenges encountered during the commercialization Ndebele artefacts.

Current status of Commercializing Ndebele Artefacts

According to the study's findings, there is a considerable demand for Ndebele artefacts from both domestic and international purchasers. The Ndebele art custodians are pleased with the number of people who come to buy their artefacts and pleased that artefacts are being used for fashion purposes, and they are the main producers since people prefer to buy from the original manufacturers of the artefacts. Participant 4 mentioned that she has entered art competitions and won the Innibos National artefacts Awards 2020 in the Beading Category. She went on to say that she is grateful to her people's IK for helping her become internationally famous for her artistic creations. Figure 1 depicts the winning bicycles in the beading category.

Figure 1. The participant with the award-winning beaded artefacts

Participant 7 reported that she sells beadwork and murals to tourists from other countries. She even promotes her business on social media in order to attract a broader audience. She said that the proceeds from her business allowed her to send her children to school and even build a house. She combines modern ideas that appeal to young people with cultural designs for cultural ceremonies. She stated that the economic position of Ndebele women has greatly improved, and their businesses are thriving as they continue to find new ways to attract customers and learn to use digital platforms to do so.

Participant 6 stated that selling artefacts such as beadwork and mural art is another approach to preserve and safeguard the Ndebele IK which is an ongoing tradition as seen in figure 2. She remarked that more young people are being involved in and encouraged to develop artefacts not only for cultural but also for commercial purposes. She stated that elderly women transmit their knowledge to younger women, who later join the commercialization of the artefacts after perfecting their skills. Younger women are collaborating with the elder generation to advertise using social media and the internet in order to reach a broader audience. This approach was particularly effective during COVID-19, when many old entrepreneurs complained about their enterprises collapsing due to national lockdown laws, which included the closure of some businesses and the suspension of international travel, among other things.

Figure 2. Mural art and beaded artefacts sold by participants

Practical Government Initiatives on the Protection and Preservation of Ndebele Artefacts

The study's findings show that the government has not been implementing community outreach initiatives or programs that educate people how to protect their artefacts by using the law. The Ndebele artists indicated that they are unaware of any existing laws that safeguard their expertise, which is why anyone is allowed to manufacture and sell them even though they do not own them. The participants emphasized that they are exploited and that there is no ABS agreement between them as caregivers and others who wish to profit from the artefacts as a result of the lack of community educational activities. As a result, the Ndebele people are not aware of or employ any legally enforceable agreements.

Participant 3 claimed that if there are any laws, they are not taught to the people, and the government takes an up to bottom approach rather than a bottom-up approach. The participant further claims that the country's laws are not as active, which is why they do not know how to preserve their expertise and benefit from them. The participant expressed concern about how policies may be in place without adequate educational initiatives to put them into practice in local areas. Due to the government's continued failure to support indigenous communities, people remain exploited without an ABS agreement in place. Due to their lack of formal education, it is incredibly difficult for them to even learn about legal regulations, and they must rely completely on government to protect them from exploiters, a strategy that has utterly failed.

The participants stated that they only see the government's initiatives during other events that are not really specifically for Ndebele people. Ndebele women agree that they do participate in national activities and showcase their skills, yet they do not attend any workshops or meet with officials that talk about the protection of their artefacts. They have also mentioned that the local municipalities have no clue on how the IP laws work so they do not even attempt to engage them when they encounter any challenges or want to ask questions. Their hope is provincial and national government to educate them which thus far has not happened.

Participant 2 indicated that she was one of those who started this adventure of traveling all over the world to display her skills, and she was exploited by people because she was unaware of her rights. She added that there will be those who buy and sell twice or three times the amount in their home nations. Others will claim to want to be managers but will take larger percentages when revenues are shared. As a result, the Ndebele women will return home with pennies while their labor continues to produce a fortune, particularly in industrialized countries. In her own words, she further elaborated that:

> "We have been taken advantage of before, and there were never any disclosures of the money we used to make. We could only be thankful that we were able to travel and stay for free. I believe that if I had been treated fairly, I would have become a millionaire by now. Even today, one of our managers lives in the Johannesburg South suburbs. He made use of all the advantages he gained from our skills and talents. I am heartbroken to realize that so much money was made that most Ndebele women did not profit from."

Participant 1 stated that the government cannot be blamed if traditional leaders, who are local authorities, do not work hard to ensure that people understand the policies protecting their artefacts. As a result, traditional leaders must take steps to conserve Ndebele artifacts that serve as symbols of cultural identity and expression. The researcher explained to the participants that there is an existing panel of IKS/ intangible heritage experts that advise the minister of Sport, Arts and culture on the preservation and documenting of IK. The panel is also responsible for community outreach and making sure that the communities understand the protection of their knowledge and how to further commercialize it. As a member of the panel, the researcher promised to submit their concerns and allow the department to undertake the appropriate measures to resolve challenges regarding IK protection and commercialization.

Challenges Encountered During the Commercialization of Ndebele Artefacts

The Ndebele women mentioned that there is too much competition even from people who are not Ndebele and there is nothing that can be done at the moment because the laws of South Africa are showing to be lenient and not practical for people who are not knowledge holders while practicing or using other people's knowledge. Participants also complained about not having stable selling markets for their customers. They mentioned that there is no proper business market, and they have to resort to sell in the comfort of their own homes or sell in the Pretoria CBD at a place called Marabastad which is unhygienic and more likely international; tourists do not get there. In her words, participant 8 stated that:

> "Marabastad is a good marketplace for local potential customers, but for foreign tourists we need a legitimate market. This is not to say that local consumers do not deserve to buy from such a site, but because Marabastad is a CBD for many local people, it makes it easier for them to approach us".

Furthermore, participants have also complained about heritage sites that have been neglected by government such as Mapoch (Mabhoko) village in Klipgat, and Erholweni, Mapoch (Mabhoko) caves. They believe that if there was enough budget and strategies in place, they would be the ones who get the tender to paint the buildings of the heritage sites and sell their artefacts (beaded and painted) to international tourists. They feel neglected by the government and that everything depends on their own capacity. They wish to open business closer to the tourism attraction places especially in Mpumalanga province where there is safari and the Kruger international airport.

Moreover, they think that most international tourist travel to Mbombela city and surrounding areas to be in contact with nature, relax, experience the local culture and buy cultural items. Therefore, Ndebele artists hope to team up with the airport and accommodation places in order to sell their artefacts to international travelers. Even though their business is booming because travelers do get to KwaMhlanga and surrounding, the artists believe that there is a market to explore within the city of Mbombela and surrounding areas because over the years it has been proven that these areas are tourism attraction places.

DISCUSSIONS

The study has provided data about elderly Ndebele women who are entrepreneurs and use their IK of artefacts to sustain themselves and to grow their local economy. This has been supported by the IKS Policy of 2004 which supports the use of IKS for wealth creation and rural economic development. However, the Ndebele women raised concerns that they do not know any policies that protect their rights and there are no current initiatives by the government to ensure that people know about their rights and how they can benefit from their knowledge. There is a current existing ABS agreement signed into South African law which state that IK custodians need to benefit from their knowledge which is not always the case and custodians lack knowledge about it. (Bhuda, 2019; Bhuda & Kgwadi, 2023).

According to Pandor (2015), in order to get access to IKS for commercial or industrial purposes, an ABS agreement must be signed between the community and the third party. The advantage of the ABS agreement must include terms that assure fair compensation for such access and the equitable distribution of any benefits derived from it. The NIKSO may help indigenous communities reach ABS agreements with outside parties. Commercialization of IK / Traditional Knowledge (TK) requires a strong legal framework and institutions that assure TK holders of their TK's protection - a system that ensures fairness even to non-TK holders while not exposing TK holders to exploitation.

Indigenous people need to understand their IP rights and that can be done through continuous engagements which policy makers in local communities. Odora-Hoppers (2002) stated that IP rights were often regarded as a threat to developing countries' young economy, where counterfeiting was common for a variety of reasons related to failure to register. This is especially true in the context of traditional knowledge, as indigenous communities are usually found in underdeveloped countries. As a result, they perceive IP law as a threat, primarily because it reflects the Western perspective of traditional knowledge; as a result, indigenous tribes criticize the legislation.

Milius (2009) claimed in a book on IP and traditional knowledge that the fundamental principles of IP law are provided by specific legal instruments with the same breadth of protection. When the broad aspects of IP regulations are assessed with the particularities of traditional knowledge, such protection can extend to include traditional knowledge in specific contexts where IP principles may apply. A separate treaty protecting traditional knowledge does not exist in IP law. As a result, it is not expressly protected, although it is not outside the scope of the WIPO's mission. A legislative instrument on traditional knowledge is being considered by the Intergovernmental Committee on Genetic Resources, Traditional Knowledge, and Folklore (Milius, 2009).

Agan (2013) did similar research, stating that the current accessible options for protecting IP in the Republic of South Africa (RSA) are Patents, Trade Secrets, Copyrights, Trademarks, and Industrial Design. Parties whose rights have been violated can also seek common law remedies. The laws controlling these IP regimes were enacted at various times, some before South Africa became a republic in 1963, some after, and yet others after apartheid was abolished in 1990.

The RSA was required to alter or abolish legislation passed prior to the TRIPs in 1994 and establish TRIPs compatible legislation. However, due to difficulties that cannot be handled by an international IP regime, one area of IP for Indigenous people, also known as Traditional Knowledge (TK), has not been fully safeguarded. This has resulted in poor or insufficient commercialization of TK. TRIPs also do not cover TK, further marginalizing it. This paper's scope is limited to the commercialization of TK. However, it should be noted that commercialization cannot occur in a vacuum. Thus, protecting TK is a must for commercialization.

North-West University, South Africa has encouraged community engagements in partnership with a government department, the Department of Science and Innovation (DSI), in order for indigenous people to understand their IP rights, and their most recent engagements were in members of the Khoi and San communities in 2020 (North-West University, 2020). The Ndebele people would also benefit from such initiatives in which the government and higher education institutions that understand can conduct outreach programmes. Concerning the Ndebele people and their artefacts, meaningful engagements are required so that they may show how their knowledge cannot be further exploited by capitalists and non-knowledge bearers (Bhuda, 2019).

FUTURE RESEARCH DIRECTIONS

The findings of this investigation on the commercialization of Ndebele artefacts as the IK of the Ndebele people will help further research on both commercialization and IK. Most studies on IK and commercialization focus on traditional plants and little research exists about other indigenous practices. This study is therefore important to provide researchers with a different lens and perspective on IK and commercialization. Research on the Sustainable Development Goals in relation to IK and poverty reduction is further advanced by this work and can be used for guidance by researchers. Furthermore, comparable research on South Africa's IKS and IP rights can be guided by the policies discussed in this paper. This research is beneficial to indigenous people and can be used as a case study by government entities who are working towards elevating indigenous communities.

CONCLUSION AND RECOMMENDATIONS

The study focused on the IP protection of Ndebele artefacts in order to promote long-term economic prosperity. It also looked into government initiatives to protect Ndebele artefacts and ensure that communities of practice benefit from their knowledge. Policies that foster and protect IK forms and elements have also been prioritized. According to the study's findings, Ndebele women are entrepreneurs use their artefacts to sustain themselves and grow the local economy. They have produced wealth and jobs for themselves, as encouraged by the IKS strategy of 2004. Elderly women have passed on their knowledge to younger generations and have recently begun to collaborate in order to reach a larger audience through social media platforms and the internet. They have, however, been victims of exploitation and governmental neglect. Their main issue is that no outreach programs or campaigns are in place to educate them about their IP rights and ABS agreements. They claim that policymakers stay in their offices, develop policies, and never visit communities to ensure that they are aware of existing policies and how they can be implemented. This research looked at policy documents that protect IK, traditional knowledge, and heritage. Unfortunately, the communities who need to be taught about such policies are left vulnerable and exploited because knowledge holders in the instance of Ndebele women are primarily elderly people who never attended any formal schooling and lack education. As per the study, there should be an ABS agreement between Ndebele women artefacts entrepreneurs and people interested in purchasing (usually in bulk) and collaborating with the women for commercialization purposes. A fair ABS agreement can be achieved only if government and traditional leaders work with Ndebele women to prevent further exploitation. The study recommends the following:

- Community outreach programs by the government focus on IK, commercialization and IP rights.
- Appointment of relevant individuals in provincial and local government to oversee the exchange of IK goods between big enterprises and entrepreneurs.
- Proper registration of small businesses which belong to IK holders who are entrepreneurs.
- Involvement of more young people to participate in IK commercialization which assists to sustain livelihoods and also preserve IK.
- Annual visit to indigenous communities by NIKSO with an aim of documenting knowledge through IKSDCs.

REFERENCES

Agan, W. (2013). Commercialization of traditional knowledge in South Africa: whether the existing intellectual property framework encourages commercialization. Cape Town: University of Cape Town. (Dissertation-PhD).

Amechi, E. P. (2018). Bio-economy, patents, and the commercialization of traditional knowledge on the medicinal uses of plants in South Africa. *SA Mercantile Law Journal, 30*(2), 251–279.

Angioni, G. (2003). Indigenous Knowledge: Subordination and Localism. In G. Sanga & G. Ortalli (Ed), *Nature, Knowledge: Ethnoscience, cognition, and utility* (pp. 287- 296). New York: Oxford.

Arsenault, R., Bourassa, C., Diver, S., McGregor, D., & Witham, A. (2019). Including indigenous knowledge systems in environmental assessments: Restructuring the process. *Global Environmental Politics, 19*(3), 120–132. DOI: 10.1162/glep_a_00519

Arslanbek, A., Malhotra, B., & Kaimal, G. (2022). Indigenous and traditional arts in art therapy: Value, meaning, and clinical implications. *The Arts in Psychotherapy, 77*, 101879. DOI: 10.1016/j.aip.2021.101879

Bagley, M. (2018). *Toward an effective indigenous knowledge protection regime case study of South Africa*. https://www.cigionline.org/search?search_api_fulltext=Bagley%20&sort_by=created

Balogun, T., & Kalusopa, T. (2021). A framework for digital preservation of Indigenous knowledge system (IKS) in repositories in South Africa. *Records Management Journal, 31*(2), 176–196. DOI: 10.1108/RMJ-12-2020-0042

Battiste, M. (2005). Indigenous knowledge: Foundations for first nations. *WINHEC: International Journal of Indigenous Education Scholarship*, (1), 1–17.

Bhuda, M., & Saurombe, T. (2022). Mechanisms of Ethnomathematical Knowledge Transmission amongst AmaNdebele women: Ways of Transmitting Indigenous Knowledge. *Journal of Indigenous Social Development, 11*(1), 97–119.

Bhuda, M. T. (2019). *The role of ethnomathematics in the cultural life of AmaNdebele women at Ekosini village in Mpumalanga Province. Mafikeng: North-West University*. Dessertation-M.IKS.

Bhuda, M. T. (2024). Decolonising and indigenising African Indigenous research: Why Indigenous ways of knowing, being and living matter in research. In *Higher Education Transformation in Africa* (pp. 50–69). Routledge. DOI: 10.4324/9781003528319-4

Bhuda, M. T., & Kgwadi, B. (2023). The economic measures established to rebuild the South African cultural heritage tourism after the COVID-19 impact. *Gender & Behaviour, 21*(1), 21398–21412.

Bhuda, M. T., & Koitsiwe, M. (2021). Practicing rites of passage by Southern Ndzundza Ndebele during Covid-19: Is it safe for male initiation schools to operate during Covid-19? *Gender & Behaviour, 19*(1), 17280–17293.

Bhuda, M. T., & Koitsiwe, M. (2022). The Covid-19 Pandemic and The Challenges Encountered by The Southern Ndebele Crafts Entrepreneurs Due to Closure of The Tourism Sector. *Gender & Behaviour, 20*(1), 19188–19202.

Botha, C. (2013). Patent protection of traditional knowledge relating to biological resources in South Africa. *Responsa Meridiana*.

Bowen, G. A. (2009). Document analysis as a qualitative research method. *Qualitative Research Journal*, *9*(2), 27–40. DOI: 10.3316/QRJ0902027

Briggs, J. (2005). The use of indigenous knowledge in development: Problems and challenges. *Progress in Development Studies*, *5*(2), 99–114. DOI: 10.1191/1464993405ps105oa

Burrel, T. (1999) *Burrell's South African Patent and Design Law*. Durban: Butterworths

Clarke, V., & Braun, V. (2017). Thematic analysis. *The Journal of Positive Psychology*, *12*(3), 297–298. DOI: 10.1080/17439760.2016.1262613

Dei, G. J. S. (2002). Rethinking the role of inthgenous knowledges in the acad-emy. *International Journal of Inclusive Education*, *4*(2), 111–132. DOI: 10.1080/136031100284849

Denzin, N. K. (2012). Triangulation 2.0. *Journal of Mixed Methods Research*, *6*(2), 80–88. DOI: 10.1177/1558689812437186

Du Bois, M. (2013). Protection of intellectual property rights as human rights in international law. *South African Yearbook of International Law*, *38*(1), 92–122.

Flint, M. T., & William, A. (1989). *Intellectual Property the new law*. Butterworths.

Francis, D., & Webster, E. (2019). Poverty and inequality in South Africa: Critical reflections. *Development Southern Africa*, *36*(6), 788–802. DOI: 10.1080/0376835X.2019.1666703

Grenier, L. (1998). *Working with indigenous knowledge: a guide for researchers*. International Development Research Centre Books.

Hart, M. (2007). Indigenous knowledge and research: The míkiwáhp as a symbol for reclaiming our knowledge and ways of knowing. *First Peoples Child & Family Review: A Journal on Innovation and Best Practices in Aboriginal Child Welfare Administration, Research. Policy & Practice*, *3*(1), 83–90.

Jacob, R., & Alexander, D. (1993). *A guide to intellectual property*. Sweet & Maxwell.

Keitumetse, S. (2006). UNESCO 2003 convention on intangible heritage: Practical implications for heritage management approaches in Africa. *South African Archaeological Bulletin*, *61*(184), 166–171. DOI: 10.2307/20474924

Kugedera, A. T., Sakadzo, N., Museva, T., Kokerai, L. K., Muridzi, G., & Musara, A. (2021). Role and commercialisation of indigenous knowledge in the control of Covid-19 pandemic. *The Dyke*, *15*(2), 43–57.

Magni, G. (2017). Indigenous knowledge and implications for the sustainable development agenda. *European Journal of Education*, *52*(4), 437–447. DOI: 10.1111/ejed.12238

Mainardes, E. W., Alves, H., & Raposo, M. (2010). An exploratory research on the stakeholders of a university. *Journal of Management and strategy*, *1*(1), 76.

Makgamatha, M. G., & Moikanyane, K. K. (2024). The role of community development projects in tackling socio-economic issues in South Africa. *Management, Business, Administration and Legal*, 256.

Marinova, D., & Raven, M. (2006). Indigenous knowledge and intellectual property: A sustainability agenda. *Journal of Economic Surveys*, *20*(4), 587–605. DOI: 10.1111/j.1467-6419.2006.00260.x

Mbatha, B. (2013). Unpacking four sequential modes of knowledge conversion in managing indigenous knowledge. *Indilinga*, *12*(2), 171–187.

Milius, D. (2009). Justifying intellectual property in traditional knowledge. *Intellectual Property Quarterly*, *2*, 185–216.

Mkapa, B. (2004). Indigenous knowledge-a local pathway to Global Development: In World Bank, Indigenous knowledge. Local pathways to Global Development. Marking Five Years of the World Bank Indigenous Knowledge for Development Program, World Bank. http://www. worldbank. org/afr/ik/ikcomplete. pdf

North-West University. (2020). NWU partners with department to promote Indigenous Knowledge Act. https://news.nwu.ac.za/create-news

Nyeleka, S., Taruvinga, A., Zhou, L., & Mopipi, K. (2019). Income and Food Security Status Among rural Women Indigenous Knowledge Based Crafters and Non-crafters: Evidence from Rural South Africa. *International Journal of Food Engineering*, *5*(1).

Odora Hoppers, C. A. (2002). Indigenous knowledge systems, sustainable livelihoods and the intellectual property system: A peace action perspective. *Journal of Peacebuilding & Development*, *1*(1), 106–112. DOI: 10.1080/15423166.2002.373049356048

Oguamanam, C. (2004). Protecting indigenous knowledge in international law: Solidarity beyond the nation-state. *Law Text Culture*, *8*(8), 191–230. DOI: 10.14453/ltc.858

Owusu-Ansah, F. E., & Mji, G. (2013). African indigenous knowledge and research. *African Journal of Disability*, *2*(1), 1–5. DOI: 10.4102/ajod.v2i1.30 PMID: 28729984

Pandor, N. (2015). Protection, Promotion, Development and Management of Indigenous Knowledge Systems Bill, 2014. (Bill Notice 243 of 2015). *Government gazette,* 38574: 20 March.

Pongel, A. (2011). Integrating Indigenous Knowledge for Food Security: Perspectives from the Millennium Village Project at in Nyanza Province in Kenya. *International Conference on Enhancing Food Security in the Eastern and Horn of Africa regions.* Kampala, Uganda.

Rajah, S. S. (2019). Conceptualising community engagement through the lens of African indigenous education. *Perspectives in Education*, *37*(1), 1–14.

Schneider, E. A. (1985). Ndebele mural art. *African Arts*, *18*(3), 60–67. DOI: 10.2307/3336357

Sharma, A. (2021). Decolonizing international relations: Confronting erasures through Indigenous knowledge systems. *International Studies*, *58*(1), 25–40. DOI: 10.1177/0020881720981209

Statistics South Africa. (2017). *Poverty trends in South Africa, No. 03-10–06.* Statistics South Africa.

Thompson, K. L., Lantz, T., & Ban, N. (2020). A review of Indigenous knowledge and participation in environmental monitoring. *Ecology and Society*, *25*(2), art10. DOI: 10.5751/ES-11503-250210

Unesco. (2009). The draft of the National Policy on Living Heritage in South Africa. https://www.maropeng.co.za/uploads/files/National_Policy_on_South_African_Living_Heritage__ICH.pdf

Upreti, P. N. (2021). The Role of National and International Intellectual Property Law and Policy in Reconceptualising the Definition of Investment. *IIC; International Review of Industrial Property and Copyright Law*, *52*(2), 103–136. DOI: 10.1007/s40319-020-01009-7

Yassine, J. (2018). IP Rights and Indigenous Rights: Between Commercialization and Humanization of Traditional Knowledge. *San Diego Int'l LJ*, *20*, 71.

ADDITIONAL READING

Agan, W. (2013). *Commercialisation of traditional knowledge in South Africa: whether the existing intellectual property framework encourages commercialisation* (Doctoral dissertation).

Ao, M. (2020). Branding and Commercialisation of Traditional Knowledge and Traditional Cultural Expressions: Customary Law of North East Vis-à-Vis Contemporary Law. *Indigenous Peoples'. JL Culture & Resistance*, *6*(1), 75. DOI: 10.5070/P661051238

Dountio, J. (2011). The protection of traditional knowledge: Challenges and possibilities arising from the protection of biodiversity in South Africa. *South African Journal of Art History*, *26*(1), 10–22.

Janke, T. (2018). From smokebush to spinifex: Towards recognition of Indigenous knowledge in the commercialisation of plants. *International Journal of Regional. Rural and Remote Law and Policy*, (1), 1–27.

Leakey, R. R., Tientcheu Avana, M. L., Awazi, N. P., Assogbadjo, A. E., Mabhaudhi, T., Hendre, P. S., Degrande, A., Hlahla, S., & Manda, L. (2022). The future of food: Domestication and commercialization of indigenous food crops in Africa over the third decade (2012–2021). *Sustainability (Basel)*, *14*(4), 2355. DOI: 10.3390/su14042355

Ndhlovu, P. T., Omotayo, A. O., Otang-Mbeng, W., & Aremu, A. O. (2021). Commercialization potential of six selected medicinal plants commonly used for childhood diseases in South Africa: A review. *Sustainability (Basel)*, *14*(1), 177. DOI: 10.3390/su14010177

Torri, M. C. (2011). Bioprospecting and commercialisation of biological resources by indigenous communities in India: Moving towards a new paradigm? *Science, Technology & Society*, *16*(2), 123–146. DOI: 10.1177/097172181001600201

KEY TERMS AND DEFINITIONS

Artefacts: An artefact is a human-made object, usually one with historical or cultural significance.

Commercialization: Commercialization is a process of launching new goods or services for profit

Indigenous Knowledge: Indigenous knowledge is the term used to describe the distinct perspectives, abilities, and beliefs that local cultures have acquired via extensive engagement with their surroundings.

Intellectual Property: refers to the legal protection of intangible works of art, literature, inventions, designs, and symbols to enable creators to make money off their concepts.

Economic Development: Economic development includes methods to raise per capita income, raise living standards, and encourage sustainable growth with the goal of bettering the economic and social circumstances of emerging nations.

Entrepreneurs: An entrepreneur is a person who launches and manages a company, assuming the risks and benefits, frequently with a novel good or service.

Ndebele Women: Ndebele women are part of the Ndebele nation and are renowned for their elaborate mural art, vivid traditional clothing, and vibrant beadwork, all of which show their cultural identity and social standing.

Chapter 20
The Role of Government in the Commercialization of Indigenous Knowledge-Based Businesses in South Africa:
Opportunities and Challenges

Mpho Gift Makgamatha
https://orcid.org/0000-0001-6799-4554
University of Limpopo, South Africa

ABSTRACT

The commercialization of indigenous Knowledge-based Businesses in South Africa is a crucial step towards economic empowerment and cultural heritage preservation. The government plays a crucial role in facilitating this process, providing financial support, capacity-building programs, and market access through government-led partnerships. However, challenges such as bureaucratic inefficiencies, limited awareness of government programs, and inadequate infrastructure hinder the full realisation of these opportunities. The tension between modern commercialisation demands and the preservation of traditional practices also complicates the process. The study underscores the need for more robust, culturally sensitive, and inclusive strategies to support indigenous businesses in South Africa.

INTRODUCTION

The commercialization of indigenous Knowledge-based businesses in South Africa, represents a significant opportunity for economic growth, cultural preservation, and social empowerment (Nkosi et al., 2020). Indigenous Knowledge- based businesses often serve as lifelines for rural communities, providing employment opportunities and preserving traditional knowledge (Bhuda, 2023). However, the successful integration of these enterprises into the mainstream economy requires substantial government involvement. Approximately 60% of small and medium enterprises (SMEs) in South Africa cite limited access to funding and resources as a primary barrier to growth (Makgamatha, 2022). This challenge is even more pronounced among indigenous businesses, which often lack formal structures and access to

DOI: 10.4018/979-8-3693-7964-6.ch020

institutional support (Nkosi, 2020). Government initiatives aimed at fostering indigenous knowledge-based businesses development have made notable strides. For instance, 48% of government-funded enterprise development programs have incorporated aspects of cultural and rural business support (Matubatuba, 2022). These programs offer training, financial assistance, and market access to promote sustainability (Makgamatha, 2022). However, challenges persist, including bureaucratic inefficiencies, policy misalignment, and the lack of a cohesive framework to integrate traditional practices with modern commercial systems (Satika, 2024). Moreover, commercialization efforts face resistance from global market standards that often marginalize indigenous products. For example, while 70% of South Africans report using traditional medicine, the global herbal product market remains underexplored by indigenous businesses like Mpesu (Bhuda & Marumo, 2020; Adeola, 2023). Effective commercialization would require the government to address these systemic barriers, implement robust intellectual property protections for indigenous knowledge, and foster partnerships between traditional practitioners and modern enterprises.

STATEMENT OF THE PROBLEM

In South Africa, indigenous knowledge systems (IKS) represent a vast reservoir of cultural, medicinal, agricultural, and ecological wisdom developed over generations. Despite their rich potential to contribute to economic development and social empowerment, the commercialization of indigenous knowledge-based businesses remains limited and fragmented. The government has introduced various policies and strategies such as the Indigenous Knowledge Systems Policy and the Protection, Promotion, Development and Management of Indigenous Knowledge Act (2019) to support the recognition and commercialization of indigenous knowledge. However, there is a significant gap between policy intent and practical implementation. Many indigenous knowledge holders, particularly in rural communities, lack access to the technical, legal, and financial resources required to transition their knowledge into viable business enterprises. This disconnect raises critical questions about the effectiveness of government interventions and the enabling environment created for such businesses to flourish. Furthermore, the commercialization of indigenous knowledge-based businesses faces complex challenges such as intellectual property rights protection, cultural sensitivity, inadequate infrastructure, limited market access, and a lack of coherent institutional support. While government agencies may offer support mechanisms, such as funding, training, and incubation, these are often fragmented or inaccessible to grassroots innovators. Additionally, the lack of collaboration between traditional knowledge holders, academic institutions and industry limits innovation and sustainable commercialization. This research seeks to investigate the role of government in bridging these gaps, evaluating both the opportunities and systemic barriers that influence the growth and sustainability of indigenous knowledge-based businesses in South Africa. Understanding these dynamics is essential for formulating effective, inclusive, and culturally respectful policies that promote equitable economic participation and the preservation of indigenous heritage. The objectives of the chapter are as follows:

- Examine the current landscape of indigenous knowledge-based business in South Africa.
- Assess government policies and programs supporting indigenous knowledge-based. businesses in South Africa.
- Determine the opportunities for indigenous knowledge-based businesses in South Africa.
- Establish the challenges faced by indigenous knowledge-based businesses in South Africa.

THE HISTORICAL CONTEXT OF INDIGENOUS KNOWLEDGE-BASED BUSINESSES IN SOUTH AFRICA

Indigenous knowledge-based businesses in South Africa have deep historical roots, tracing back to pre-colonial times when local communities relied on traditional knowledge systems for survival, trade, and economic sustainability (Masenya, 2023). Indigenous knowledge (IK) was passed down through generations, shaping industries such as traditional medicine, agriculture, craft-making, and spiritual practices (Cindi, 2021). These businesses played a crucial role in fostering self-sufficiency and economic resilience among indigenous communities. However, with the advent of colonialism and later apartheid, indigenous industries were marginalized, and their contributions were either overlooked or suppressed in favour of Western economic models. This led to a decline in traditional practices as many indigenous entrepreneurs were excluded from formal economic participation, and their knowledge was often exploited without recognition or benefit to the original knowledge holders (Akahome & Wilson, 2024). Despite these historical setbacks, indigenous knowledge-based businesses have experienced a revival in post-apartheid South Africa, driven by policy reforms, cultural reclamation efforts, and global interest in traditional healing, organic products, and sustainable practices. The government has acknowledged the value of IK in fostering inclusive economic development and has introduced initiatives to support the commercialization of indigenous businesses (Ntlangani, 2021).

THEORETICAL FRAMEWORK

Innovation Systems Theory emphasizes the role of government in fostering knowledge-based economic activities by creating an enabling environment for innovation and commercialization (Asheim et al., 2011). In the context of indigenous knowledge-based businesses in South Africa, the government plays a critical role in providing research and development (R&D) support, establishing policies that protect indigenous intellectual property, and facilitating access to markets. Through initiatives such as funding programs, technology transfer offices, and partnerships between traditional knowledge holders and formal enterprises, the state can drive commercialization while ensuring the sustainability of indigenous knowledge. However, challenges such as bureaucratic inefficiencies, inadequate infrastructure, and weak enforcement of intellectual property rights often hinder the full realization of these opportunities. Public-Private Partnership (PPP) Theory also highlights the collaborative role of government and private entities in economic development, particularly in sectors requiring extensive investment and expertise (Cheng et al., 2021). In the commercialization of indigenous knowledge-based businesses, PPPs provide a framework where the government can facilitate knowledge valorization through policy support, funding mechanisms, and business incubation while leveraging private sector efficiency, market access, and innovation capabilities. In South Africa, successful commercialization depends on balancing government interventions with private sector dynamism, ensuring that indigenous communities retain control over their intellectual assets while benefiting from commercialization.

RESEARCH METHODOLOGY

This study employed a qualitative research approach to explore the role of government in the commercialization of indigenous knowledge-based businesses in South Africa, with a specific focus on identifying the opportunities and challenges inherent in this process. A qualitative methodology was deemed appropriate due to its strength in capturing complex, context-specific and socially embedded phenomena such as indigenous knowledge systems (IKS). This approach allowed for a deeper understanding of the nuanced relationship between government policy, indigenous innovation, and socio-economic development. Rather than relying on numerical data or quantifiable variables, the study emphasized meaning, interpretation and the lived experiences of indigenous knowledge holders, policy actors and other stakeholders within the commercialization ecosystem. The study relied exclusively on secondary data sources which included government policy documents, legislative frameworks, strategic plans, academic literature, published case studies, evaluation reports and official publications from relevant institutions such as the Department of Science and Innovation, the National Indigenous Knowledge Systems Office (NIKSO) and the Department of Trade, Industry and Competition (DTIC). These documents were selected based on their relevance to the themes of indigenous knowledge, commercialization government intervention and enterprise development.

Secondary data analysis provided a rich foundation for understanding the historical evolution, current initiatives and institutional structures surrounding indigenous knowledge commercialization in the country. It also allowed for the triangulation of multiple perspectives to enhance the reliability and depth of the findings. To analyze the collected data, the study utilized thematic analysis, a widely used method in qualitative research that involves identifying, organizing, and interpreting patterns of meaning (themes) across the dataset. The thematic analysis enabled the researcher to systematically examine policy content, trace recurring issues and distill insights related to the opportunities and constraints in the government's role. The process involved coding relevant textual data, grouping similar codes into overarching themes, and interpreting these themes about the research objectives. Key themes that emerged included policy alignment and coherence, capacity building, intellectual property rights, access to funding and markets, and stakeholder collaboration. This analytical approach not only facilitated a comprehensive understanding of existing government interventions but also illuminated the gaps and systemic challenges that persist in the promotion of indigenous knowledge-based entrepreneurship.

THE CURRENT LANDSCAPE OF INDIGENOUS KNOWLEDGE-BASED BUSINESS

In South Africa

Indigenous knowledge-based businesses in South Africa, while diverse in nature, play a crucial role in the country's socio-economic fabric. These businesses are primarily situated in rural and peri-urban areas and are often SMEs, with a strong focus on community-oriented and culture-based industries (Johnson, 2019). According to Statistics South Africa, approximately 47% of South Africa's population identifies as Black African, a large portion of which includes indigenous groups that have historically been marginalized in the formal economy (Evans, 2023). Despite the immense potential these businesses have, they face numerous barriers to growth, most notably limited access to capital, lack of market access, and socio-economic disadvantages such as low levels of education and skills (Morris et al., 2020). A significant portion of indigenous knowledge-based businesses operates in sectors such as agriculture, arts and crafts, eco-tourism, and traditional medicine. In the agricultural sector, for instance, indigenous farmers and producers, often practicing subsistence farming, face challenges related to low yields and outdated techniques. The South African government's Department of Agriculture, Forestry and Fisheries reported in 2022 that around 70% of agricultural production is concentrated in the hands of large-scale commercial farmers, leaving a meagre share of the market for small-scale, indigenous farmers (Mdletshe, 2023). This disparity is exacerbated by poor infrastructure, such as limited access to transportation, storage facilities, and technology, which hinders the growth of indigenous farming businesses.

In the arts and crafts sector, indigenous artisans often struggle with market integration, both domestically and internationally (Adebola, 2020). The global market for ethnic and traditional crafts is substantial, yet indigenous entrepreneurs frequently find themselves competing against mass-produced goods that flood the market at lower prices. South Africa's Department of Trade, Industry, and Competition (2022/23) reported in 2023 that the small businesses sector, which includes many indigenous-run businesses, contributed only 28% to the country's GDP, a figure that lags behind more industrialized sectors. These businesses lack the financial resources to engage in effective marketing and distribution strategies, which makes it difficult to reach a broader customer base. Despite these challenges, the government has recognized the importance of supporting indigenous knowledge-based businesses through various initiatives. The 2021/2022 report from the Small Enterprise Development Agency (SEDA) indicated that over 25,000 indigenous knowledge-based businesses had received support in the form of funding, mentorship, and training (Simango, 2022). These support programs, however, remain underfunded and often fail to reach all eligible businesses, particularly those in remote areas.

The broader socio-economic context of indigenous knowledge-based businesses cannot be separated from the legacy of apartheid, which resulted in deep disparities in wealth, education, and infrastructure (Tshishonga, 2019). Indigenous people, especially in rural areas, continue to face barriers to formal employment, which compounds the challenges of entrepreneurship. The 2021 Census data revealed that unemployment among Black South Africans remains stubbornly high, with the rate for youth unemployment exceeding 60% (Magoro, 2021). This situation is mirrored in indigenous people, where informal and survivalist businesses are often the only source of income. According to the National Youth Development Agency 2020-2030, more than 50% of young indigenous entrepreneurs do not have access to formal business training or financial management skills, thus limiting their ability to scale and expand their businesses. Despite these barriers, there is a growing recognition of the value that indigenous businesses bring to the national economy, especially through their connection to cultural heritage and sustainable practices. The government has increasingly looked to sectors such as eco-tourism and the preservation of Indigenous Knowledge Systems (IKS) as potential growth areas (Barman, 2024). Yet, these businesses remain vulnerable to broader economic trends, such as the fluctuating global interest in travel and tourism, which means that long-term sustainability depends on both domestic and international market dynamics.

Government Policies and Programs Supporting Indigenous Knowledge-Based Businesses in South Africa

a. The Protection, Promotion, Development, and Management of Indigenous Knowledge Act (2019)

The Protection, Promotion, Development, and Management of Indigenous Knowledge Act (2019) is a landmark legislation in South Africa aimed at safeguarding indigenous knowledge systems while enabling their commercialization in a way that benefits traditional knowledge holders (Masenya, 2022). The Act provides a comprehensive legal framework for the recognition, recording, and protection of indigenous knowledge, ensuring that local communities receive fair economic returns from their contributions. A key feature of the Act is the establishment of the National Indigenous Knowledge Management System (NIKMAS), a database that documents and safeguards indigenous knowledge from misappropriation (Zondi, 2021). This ensures that traditional healers, artisans, and other indigenous entrepreneurs maintain control over their intellectual property and can commercialize their knowledge legally. According to government reports, South Africa's bioprospecting industry, which largely relies on indigenous medicinal plants, is valued at approximately R2.9 billion annually, with plants like Rooibos and Hoodia being prime examples of indigenous knowledge-based commercial successes (Mugari, 2021).

The Act also mandates benefit-sharing agreements, meaning that if a company wishes to use indigenous knowledge for product development, a portion of the profits must go back to the knowledge holders. For instance, the San and Khoi communities successfully negotiated a benefit-sharing agreement with the Rooibos tea industry, securing financial compensation for their historical contribution to the knowledge of Rooibos plant benefits (Schroeder et al., 2020). Despite these advancements, challenges remain, particularly in enforcing the Act and ensuring that indigenous communities are adequately informed about their rights. The government has introduced educational programs and legal assistance initiatives to help indigenous entrepreneurs navigate the complexities of intellectual property protection (Masenya, 2022). However, limited funding and bureaucratic inefficiencies often slow down the registration process, making it difficult for many indigenous businesses to fully benefit from the law (Bushe, 2019). Strengthening implementation mechanisms and increasing awareness among indigenous entrepreneurs will be crucial in ensuring that the Act fulfils its intended purpose of protecting and promoting indigenous knowledge while driving economic development.

b. The Indigenous Knowledge Systems (IKS) Policy (2004)

The IKS Policy, introduced by the South African government in 2004, aims to recognize, protect, and promote indigenous knowledge as a valuable asset for economic and cultural development (Balogun & Kalusopa, 2021). This policy led to the establishment of the National Indigenous Knowledge Systems Office (NIKSO), which focuses on integrating indigenous knowledge into various sectors, including agriculture, health, and environmental management (Balogun & Kalusopa, 2021). One significant initiative under this policy is the Indigenous Knowledge Registration System (IKRS), launched in March 2022, which has documented over 20,000 claims of indigenous knowledge across various provinces (Department of Science and Innovation, 2022). The IKRS serves as a secure digital repository, preserving traditional knowledge and facilitating benefit-sharing agreements between communities and commercial entities. Additionally, the policy has fostered collaborations between indigenous practitioners and scientific researchers, leading to the development of products such as traditional medicines and nutraceuticals.

For example, South Africa's rich biodiversity includes approximately 24,000 plant species, with around 4,000 used in traditional medicine (Muleba, 2020). The IKS Policy has been instrumental in harnessing this biodiversity for economic growth, ensuring that indigenous communities benefit from the commercialization of their traditional knowledge.

c. Council for Scientific and Industrial Research (CSIR)

The Council for Scientific and Industrial Research (CSIR) plays a crucial role in supporting indigenous knowledge-based businesses in South Africa by leveraging scientific research and technological innovation to enhance the commercialization of indigenous knowledge-based products (Kumari, Pandey, Giri, Chauhan, Mishra, Verma, Tripathi, Singh, Bajpai & Mishra, 2022). The CSIR works closely with traditional knowledge holders, entrepreneurs, and research institutions to develop sustainable solutions that bridge the gap between indigenous practices and modern industry standards. One of its key initiatives is the Bioprospecting and Indigenous Knowledge-Based Technology Innovation Unit, which focuses on validating, refining, and commercializing indigenous products in sectors such as traditional medicine, cosmetics, and agriculture (Kumari et al., 2022). For instance, the CSIR has conducted extensive research on Aspalathus linearis (Rooibos) and Sutherlandia frutescens (Cancer Bush), two indigenous plants with significant medicinal properties, ensuring that they meet global regulatory standards for export (Docrat, Abdul & Marnewick, 2024).

Furthermore, the CSIR contributes to the National Recordal System (NRS), which documents indigenous knowledge to prevent biopiracy and unauthorized commercialization (Amechi, 2018). Through its Biomanufacturing Industry Development Centre (BIDC), the CSIR has assisted small indigenous businesses in developing high-value products such as herbal teas, plant-based cosmetics, and nutritional supplements (Mudau, Chimonyo, Modi & Mabhaudhi, 2022). According to CSIR reports, the BIDC has supported over 20 small enterprises in scaling their operations, creating over 100 jobs in the process (Kumari et al., 2022). The organization also collaborates with government entities like the Department of Science and Innovation (DSI) to provide funding and technical expertise to indigenous entrepreneurs. In 2021, the CSIR played a pivotal role in developing an indigenous herbal remedy for immune support, incorporating traditional knowledge with scientific validation to ensure safety and efficacy (Kumari et al., 2022). By providing state-of-the-art research facilities, business incubation, and market access support, the CSIR continues to be a driving force in transforming indigenous knowledge into economically viable businesses, contributing to both job creation and cultural preservation in South Africa.

Opportunities For Indigenous Knowledge-Based Businesses In South Africa

Indigenous knowledge-based businesses in South Africa have significant opportunities in sectors that align with the country's rich cultural heritage and growing global trends towards sustainability and cultural tourism (Nkosi et al., 2020). One of the most prominent sectors offering opportunities is eco-tourism. South Africa, with its diverse landscapes and indigenous cultures, has seen increasing interest in sustainable travel experiences (Litheko, 2021). Indigenous communities, especially those in rural and remote areas, are increasingly capitalizing on eco-tourism by offering cultural tours, wildlife experiences, and heritage-focused activities (Bhuda & Marumo, 2021). For example, the Xhosa and Zulu communities have developed cultural villages and heritage tours, which not only preserve their traditions but also provide an important source of income (Spiegel, 2020). The government's support for eco-tourism, such as grants from the Tourism Incentive Program, has played a role in helping indigenous businesses tap into the global tourism market, though challenges related to infrastructure and marketing remain (Bhuda & Kgwadi, 2023; Ntlangani, 2021). Another area of opportunity for indigenous businesses is in agriculture. South Africa's indigenous farming knowledge, including traditional crops and sustainable farming practices, is gaining attention both locally and internationally. The Department of Agriculture, Forestry and Fisheries (DAFF) has reported a rise in interest in organic and heritage crops, many of which are rooted in indigenous farming traditions (Viljoen, 2020). For example, millet, sorghum, and teff – traditional grains used by indigenous farmers are experiencing growing demand in both local and international markets due to their health benefits and environmental sustainability (Taylor & Duodu, 2017). In 2021, exports of organic agricultural products from South Africa increased by 14%, with indigenous crops like these contributing to the market share (Amelework et al., 2021). Indigenous farmers who use environmentally sustainable techniques and traditional knowledge are also well-positioned to benefit from the growing global trend towards organic and sustainably sourced food. However, government support in the form of training, funding, and market access is still crucial to enable these businesses to scale up and compete with larger commercial farms (Mokgomo et al., 2022). The IKS sector offers another growing opportunity for indigenous businesses. The South African government has increasingly recognized the value of IKS, which includes knowledge related to medicine, agriculture, and crafts, and has started to support initiatives that preserve and commercialize this knowledge (Bhuda & Marumo, 2020). In 2022, the Department of Science and Innovation (DSI) launched the IKS Policy Framework, which is designed to foster innovation and economic growth by promoting the commercialization of indigenous knowledge (Oguamanam, 2022). The market for herbal and traditional medicines is expanding globally, with an estimated value of over USD 100 billion per year (Gebashe et al., 2022). In South Africa, indigenous-run businesses that focus on traditional healing practices or natural cosmetics are tapping into this market. For example, the African Potato and Baobab fruit industries, which rely on indigenous knowledge of plants and healing, have grown significantly in recent years (Van Wyk, 2011; Welford et al., 2015). Indigenous businesses in this space have benefited from both domestic interest and the growing global demand for natural, plant-based products. However, the commercialization of IKS is still hindered by intellectual property issues and regulatory barriers, which the government is working to address.

Challenges Faced by Indigenous Knowledge-Based Businesses In South Africa

Indigenous knowledge-based businesses in South Africa encounter several structural and systemic challenges that hinder their growth and sustainability. One of the primary obstacles is limited access to finance, with indigenous entrepreneurs often unable to secure loans or venture capital due to the perceived high risk and lack of collateral (Tshikovhi, 2021). According to the South African Reserve Bank, only 5% of small and medium-sized enterprises (SMEs) owned by indigenous entrepreneurs were able to access formal financing in 2020 (Sibiya et al., 2023). This gap is further exacerbated by a lack of financial literacy, as most indigenous entrepreneurs are not familiar with the formal banking system or available funding mechanisms. The government's financial support programs, such as the National Empowerment Fund (NEF), while designed to assist historically disadvantaged groups, have been critiqued for their complex application processes and bureaucratic hurdles (Vilakazi & Moore, 2023). As a result, most indigenous businesses in rural and underserved areas remain informal or struggle to scale up, relying instead on small-scale, self-financing models that limit their potential for expansion (Simatele & Kabange, 2022).

Another significant challenge faced by indigenous entrepreneurs is market access, particularly in competitive industries like agriculture, arts, and tourism (Scheyvens et al., 2021). They face significant challenges in market access due to limited resources, networks, and infrastructure, as well as discrimination, lack of formal certifications, and lack of exposure to global markets, making it difficult to compete with established businesses (Scheyvens et al., 2021). Despite the global demand for organic, sustainable products, indigenous farmers often struggle to compete with large commercial producers who benefit from economies of scale and better access to distribution networks. In the agriculture sector, for example, the Department of Agriculture, Forestry and Fisheries reported in 2022 that 70% of agricultural production is dominated by large-scale commercial farmers, leaving indigenous smallholder farmers with limited market opportunities (Dube et al., 2023). Indigenous-run cultural and eco-tourism ventures often lack resources such as marketing expertise, technology, professional networks, and funding to effectively promote their products. They may also face challenges in building brand recognition and securing access to international travel platforms. The lack of knowledge stems from limited exposure to formal tourism training and business development programs, as well as a lack of experience in navigating the complexities of global tourism markets. Moreover, most indigenous entrepreneurs may not have the resources to hire marketing professionals or access marketing tools, which hinders their ability to compete with established tourism providers. A 2021 report by the Tourism Business Council of South Africa indicated that 60% of indigenous tourism businesses struggled to reach global markets due to inadequate digital marketing skills and limited exposure (Sixaba & Rogerson, 2023). These challenges highlight the need for more targeted government support to help indigenous businesses overcome barriers to market entry and compete effectively within the broader economy. The South African government often struggles to support or regulate certain indigenous knowledge-based businesses, such as mpesu (traditional medicines and practices), due to a lack of clear legal frameworks, regulatory inconsistencies, and concerns over the protection of intellectual property (Ekene et al., 2023). While the government has made strides in promoting IKS, the absence of standardized regulations and formal recognition of traditional practices hampers the ability of businesses like mpesu to access funding, formal markets, and legal protection (Ekene et al., 2023). Moreover, there is often insufficient coordination between different government departments, resulting in policies that fail to address the unique challenges faced by these businesses, such as cultural sensitivity and the need for specialized training in both traditional practices and modern business management.

CONCLUSION

In conclusion, the government plays a critical role in shaping the future of indigenous knowledge-based businesses in South Africa by creating an enabling environment that addresses the unique challenges these businesses face. While opportunities exist in sectors such as eco-tourism, agriculture, and Indigenous Knowledge Systems, these businesses often struggle due to barriers like limited access to finance, inadequate infrastructure, and complex regulatory frameworks. By improving access to capital, providing targeted skills development, streamlining policies, and promoting market access, the government can unlock the full potential of indigenous enterprises, driving economic growth and fostering cultural preservation. Ultimately, a more coordinated and supportive approach from the government is essential for ensuring the long-term sustainability and success of indigenous knowledge-based businesses in South Africa.

SOLUTIONS AND RECOMMENDATIONS

a. Improved Access to Finance and Credit

The Department of Small Business Development (DSBD) should create targeted financial products for indigenous businesses, such as low-interest loans and microfinance opportunities, tailored to the specific needs of these enterprises. Financial institutions should be incentivized to offer favorable terms to indigenous entrepreneurs, especially in rural areas where access to capital is limited. Expanding the scope of funding programs such as the National Empowerment Fund (NEF) to reach smaller, community-based businesses would help address this critical barrier.

b. Capacity-Building and Skills Development

There is a need for government-led initiatives to provide comprehensive business training for indigenous entrepreneurs. Department of Small Business Development (DHET), in collaboration with institutions like the Sector Education and Training Authorities (SETAs), should provide entrepreneurship training programs to equip indigenous entrepreneurs with the necessary skills to grow their businesses. This should include training in financial literacy, digital marketing, supply chain management, and leadership. Partnerships between government bodies, universities, and business development organizations can foster skill-building programs designed specifically for indigenous communities. Empowering indigenous entrepreneurs with modern business skills will improve their ability to scale and compete in national and global markets.

c. Streamlining Regulatory Frameworks and Policies

The Department of Trade, Industry, and Competition (DTIC) oversees the Companies and Intellectual Property Commission (CIPC), which facilitates business registration and intellectual property protection, crucial for indigenous knowledge-based businesses (IKBs). The government should simplify the regulatory processes that indigenous knowledge-based businesses must navigate, making it easier for them to formalize their operations and access market opportunities. This includes creating a clear legal framework that recognizes the commercial value of IKS and protects traditional practices. Establishing a dedicated regulatory body for indigenous businesses could ensure that policies are more coordinated and better tailored to the unique needs of these enterprises.

d. Market Access and Promotion

The Department of Small Business Development (DSBD), through SEDA, should assist indigenous knowledge-based businesses in improving their branding, digital marketing, and access to e-commerce platforms. To help indigenous knowledge-based businesses reach wider markets, the government should invest in marketing and branding initiatives that promote indigenous products and services both locally and internationally. Supporting indigenous knowledge-based businesses in participating in trade fairs, expos, and digital platforms can significantly enhance their visibility. Government-funded campaigns that highlight the cultural and environmental value of indigenous products could also increase consumer demand, particularly in growing sectors like eco-tourism, organic agriculture, and IKS-based products.

e. Public-Private Partnerships

The government should foster stronger partnerships between indigenous knowledge-based businesses and larger corporations, as well as with international businesses, to enhance market access and encourage investment. By facilitating collaborations and joint ventures, the government can help indigenous businesses gain access to more sophisticated technologies, larger distribution channels, and expertise in business operations. Public-private partnerships could also provide indigenous entrepreneurs with access to more resources, networks, and knowledge-sharing platforms, allowing them to expand and modernize their businesses.

f. Protection and Commercialization of Indigenous Knowledge

The Department of Science and Innovation (DSI), which has led efforts to safeguard indigenous knowledge through policies like the Protection, Promotion, Development, and Management of Indigenous Knowledge Act (2019). Therefore, the government should prioritize the protection and commercialization of indigenous knowledge-based businesses, particularly in fields such as traditional medicine, crafts, and agriculture. This can be achieved by strengthening intellectual property laws that specifically address the needs of indigenous communities, ensuring they benefit financially from the commercialization of their traditional knowledge. Creating mechanisms for communities to register and protect their cultural practices and traditional knowledge will prevent exploitation and encourage sustainable business models.

FUTURE RESEARCH DIRECTIONS

Future research on the role of government in the commercialization of Indigenous Knowledge-based businesses in South Africa should explore both the opportunities and challenges that shape this sector. Key areas of investigation could include policy frameworks that support or hinder commercialization, the effectiveness of government funding and incubation programs, and the role of intellectual property rights in protecting Indigenous Knowledge while fostering innovation. Moreover, research could examine how public-private partnerships can enhance the sustainability and scalability of Indigenous Knowledge-based enterprises. Challenges such as bureaucratic barriers, market access limitations, and the potential exploitation of Indigenous communities should also be critically analyzed. A comparative analysis with other countries that have successfully integrated Indigenous Knowledge into their economies could provide valuable insights. Ultimately, such research can inform policies that balance economic development with cultural preservation, ensuring that Indigenous communities benefit equitably from the commercialization of their knowledge.

REFERENCES

Adebola, A. O. (2020). Protection of Craft Products Embodied in Cultural and Creative Industries in South Africa. In Sustainable Organizations-Models, Applications and New Perspectives. *IntechOpen.*

Adeola, O. (2023). Introduction: A Casebook of Indigenous Business Practices in Africa. I*n Casebook of Indigenous Business Practices in Africa* [Emerald Publishing Limited.]. *Apprenticeship, Craft, and Healthcare, 1*, 1–14.

Akahome, J. E., & Wilson, E. F. (2024). Indigenous Entrepreneurship as a Pathway for Nollywood's Global Relevance and SDG 8 Intervention. In *Indigenous Entrepreneurship in Sub-Saharan Africa: Implications for Theory and Practice* (pp. 185–199). Springer Nature Switzerland. DOI: 10.1007/978-3-031-64924-0_8

Amechi, E. P. (2018). Bio-economy, patents, and the commercialisation of traditional knowledge on the medicinal uses of plants in South Africa. *SA Mercantile Law Journal, 30*(2), 251–279.

Amelework, A. B., Bairu, M. W., Maema, O., Venter, S. L., & Laing, M. (2021). Adoption and promotion of resilient crops for climate risk mitigation and import substitution: A case analysis of cassava for South African agriculture. *Frontiers in Sustainable Food Systems, 5*, 617783. DOI: 10.3389/fsufs.2021.617783

Asheim, B. T., Smith, H. L., & Oughton, C. (2011). Regional innovation systems: Theory, empirics and policy. *Regional Studies, 45*(7), 875–891. DOI: 10.1080/00343404.2011.596701

Balogun, T., & Kalusopa, T. (2021). A framework for digital preservation of Indigenous knowledge system (IKS) in repositories in South Africa. *Records Management Journal, 31*(2), 176–196. DOI: 10.1108/RMJ-12-2020-0042

Barman, P. (2024). The Ethical Frontier: Integrating IKS for Sustainable Economic Development. *MSW Management Journal, 34*(1), 228–238.

Bhuda, M. T. (2023). The process towards cannabis commercialization by indigenous communities for medicinal purposes in South Africa. In *Rapid innovation and development in the global cannabis market* (pp. 231–245). IGI Global. DOI: 10.4018/978-1-6684-6346-8.ch013

Bhuda, M. T., & Kgwadi, B. (2023). The economic measures established to rebuild the South African cultural heritage tourism after the COVID-19 impact. *Gender & Behaviour, 21*(1), 21398–21412.

Bhuda, M.T. & Marumo, P. (2020). African traditional medicine and healing in South Africa: Challenges and prospects before and during COVID 19. *Ife Centre for Psychological Studies/Services,* 18, 16710.

Bhuda, M. T., & Marumo, P. (2021). The role of art in sustaining the livelihoods of Amandebele women in South Africa. *In: Proceedings of the conference on the implications of Covid-19 on gender and behaviour in Africa*. Johannesburg: The Havilah Glo Marketing.

Bushe, B. (2019). The causes and impact of business failure among small to micro and medium enterprises in South Africa. Africa's public service delivery and performance review, 7(1), 1-26.

Cheng, Z., Wang, H., Xiong, W., Zhu, D., & Cheng, L. (2021). Public–private partnership as a driver of sustainable development: Toward a conceptual framework of sustainability-oriented PPP. *Environment, Development and Sustainability*, *23*(1), 1043–1063. DOI: 10.1007/s10668-019-00576-1

Cindi, L. (2021). *Incorporating African indigenous knowledge systems into the Basic Education curriculum: Experiences from two schools in the Gauteng and KwaZulu-Natal provinces, South Africa.* (Master's dissertation). University of Free State, Bloemfontein.

Docrat, T. F., Abdul, N. S., & Marnewick, J. L. (2024). The Phyto therapeutic potential of commercial South African medicinal plants: Current knowledge and future prospects. *Physical Sciences Reviews*, *9*(1), 271–289. DOI: 10.1515/psr-2022-0136

Dube, I., Makola, Z. S., Ntoyanto, N., & Msiza, C. (2023). Investigating Small-scale Farmers and Barriers to Entry in Cannabis Cultivation in South Africa. *Politeia*, *42*(1), 277–290. DOI: 10.25159/2663-6689/15002

Ekene, A. K., Lee, K. S., Decide, M. T., & Thobejane, T. D. (2023). Placing Indigenous Knowledge at the Core of Production and Marketing of Medicinal Plant (Mpesu) in Zimbabwe and South Africa. *In Casebook of Indigenous Business Practices in Africa* [Emerald Publishing Limited.]. *Apprenticeship, Craft, and Healthcare*, *1*, 167–185.

Evans, I. (2023). *Bureaucracy and race: Native administration in South Africa* (Vol. 53). Univ of California Press. DOI: 10.2307/jj.5973059

Gebashe, F. C., Naidoo, D., Amoo, S. O., & Masondo, N. A. (2022). Cosmeceuticals: A newly expanding industry in South Africa. *Cosmetics*, *9*(4), 77–89. DOI: 10.3390/cosmetics9040077

Johnson, T. G. (2019). Existing conceptual models of arts and culture: an inventory. *The Rural Policy Research Institute (RUPRI)–University of Iowa, Iowa City*. Dostupno na: http://www. rupri. org/wp-content/uploads/Conceptual-Mo- dels-of-Arts-and-Culture. pdf (pristupljeno: 20. srpnja 2021).

Jongwe, A. I., Moroz, P. W., Gordon, M., & Anderson, R. B. (2020). Strategic alliances in firm-centric and collective contexts: Implications for indigenous entrepreneurship. *Economies*, *8*(2), 31–57. DOI: 10.3390/economies8020031

Kumari, M., Pandey, S., Giri, V. P., Chauhan, P., Mishra, N., Verma, P., Tripathi, A., Singh, S. P., Bajpai, R., & Mishra, A. (2022). Integrated approach for technology transfer awareness of traditional knowledge for upliftment of circular bioeconomy. In *Biomass, Biofuels, Biochemicals* (pp. 613–636). Elsevier. DOI: 10.1016/B978-0-323-89855-3.00005-4

Litheko, A. (2021). Small tourism business growth through eco-tourism in Mahikeng, South Africa. *African Journal of Hospitality, Tourism and Leisure*, *10*(1), 256–269. DOI: 10.46222/ajhtl.19770720-100

Magoro, B. K. (2021). *Implications of youth unemployment on socio-economic development in South Africa: A case of Mankweng Township in Limpopo Province* (Doctoral dissertation).

Masenya, T. M. (2022). Decolonization of indigenous knowledge systems in South Africa: Impact of policy and protocols. [IJKM]. *International Journal of Knowledge Management*, *18*(1), 1–22. DOI: 10.4018/IJKM.310005

Masenya, T. M. (2023). Revitalization and digital preservation of Indigenous Knowledge Systems for sustainable development of Indigenous communities in South Africa. *The Serials Librarian, 84*(5-8), 86–102. DOI: 10.1080/0361526X.2023.2277962

Matubatuba, M. G. (2022). *The contribution of small, medium and micro enterprises to rural development in the Matatiele local municipality, South Africa* (Doctoral dissertation, North-West University (South Africa).

Mdletshe, S. T. C. (2023). *Viability of Government-Funded Broiler Production: Lessons from Northern KwaZulu-Natal, South Africa* (Doctoral dissertation, University of Fort Hare).

Mokgomo, M. N., Chagwiza, C., & Tshilowa, P. F. (2022). The impact of government agricultural development support on agricultural income, production and food security of beneficiary small-scale farmers in South Africa. *Agriculture, 12*(11), 1760. DOI: 10.3390/agriculture12111760

Morris, M. H., Santos, S. C., & Neumeyer, X. (2020). Entrepreneurship as a solution to poverty in developed economies. *Business Horizons, 63*(3), 377–390. DOI: 10.1016/j.bushor.2020.01.010

Mudau, F. N., Chimonyo, V. G. P., Modi, A. T., & Mabhaudhi, T. (2022). Neglected and underutilized crops: A systematic review of their potential as food and herbal medicinal crops in South Africa. *Frontiers in Pharmacology, 12*, 809866. DOI: 10.3389/fphar.2021.809866 PMID: 35126143

Mugari, D. (2021). *Bioprospecting of indigenous species berchemia discolor and its socio-economic prospects in Vhembe District Municipality* (Doctoral dissertation).

Muleba, J. L. I. N. (2020). *Understanding the use patterns of medicinal plants: investigating three ethnobotanical hypotheses in the Mpumalanga Province, South Africa*. University of Johannesburg.

National Youth Development Agency. (2020-2030)._https://www.nyda.gov.za/Portals/0/downloads/National%20Youth%20Policy%202030.pdf. (Accessed 6/10/2024)

Nkosi, N. N., Mostert, T. H. C., Dzikiti, S., & Ntuli, N. R. (2020). Prioritization of indigenous fruit tree species with domestication and commercialization potential in KwaZulu-Natal, South Africa. *Genetic Resources and Crop Evolution, 67*(6), 1567–1575. DOI: 10.1007/s10722-020-00932-5

Ntlangani, B. S. (2021). *An evaluation of community tourism organisations' role in small, medium and micro enterprises (SMME) tourism development in the eThekwini region* (Doctoral dissertation).

Oguamanam, C. (2022). Transition to the Fourth Industrial Revolution: Africa's Science, Technology and Innovation Framework and Indigenous Knowledge Systems. *African Journal of Legal Studies, 15*(1), 1–37. DOI: 10.1163/17087384-bja10058

Satika, R., Salahuddin, , Kasman, , & Suraya, . (2024). Implementation of Public Service Reform In Enhancing The Welfare Of Bima District Community: Challenges And Opportunities. *KOMUNITAS, 15*(1), 91–106. DOI: 10.20414/komunitas.v15i1.9989

Scheyvens, R., Carr, A., Movono, A., Hughes, E., Higgins-Desbiolles, F., & Mika, J. P. (2021). Indigenous tourism and the sustainable development goals. *Annals of Tourism Research, 90*, 103260. DOI: 10.1016/j.annals.2021.103260

Schroeder, D., Chennells, R., Louw, C., Snyders, L., & Hodges, T. (2020). The Rooibos benefit sharing agreement–Breaking new ground with respect, honesty, fairness, and care. *Cambridge Quarterly of Healthcare Ethics*, *29*(2), 285–301. DOI: 10.1017/S0963180119001075 PMID: 31685054

Sibiya, A., van der Westhuizen, J. & Sibiya, B. (2023). Challenges Experienced by SMMEs and Interventions by the South African National and Provincial Government: A Literature Review. *African Journal of Inter/Multidisciplinary Studies,* 5(1), 1-11.

Sibiya, P. (2023). *Predicting Loan Defaults in Development Financing in South Africa.* (Master's dissertation). University of Cape Town, Cape Town.

Simango, S. S. (2022). *Examining the impact of business incubation on the growth of tourism SMMEs: the case of Pilanesberg business Incubator Programme*. PBIP.

Simatele, M., & Kabange, M. (2022). Financial inclusion and Intersectionality: A case of business funding in the South African informal sector. *Journal of Risk and Financial Management*, *15*(9), 380. DOI: 10.3390/jrfm15090380

Sixaba, Z., & Rogerson, C. M. (2023). 'Transformation' and Tourism Small Firm Development in South Africa: Evidence from Eastern Cape Province. *Studia Periegetica*, *41*(1), 83–108. DOI: 10.58683/sp.378

Small Enterprise Development Agency. (2020). s://www.seda.org.za/Publications/Publications/SEDA%20Integrated%20report%20-%20DSBD%2024%20Feb%2020.pdf. (Accessed 15/11/2024)

Spiegel, A. D. (2020). Struggling with tradition in South Africa: the multivocality of images of the past. In *Social Construction of the Past* (pp. 185–202). Routledge. DOI: 10.4324/9780203754412-17

State of Tourism Report. (2022/21). https://www.tourism.gov.za/AboutNDT/Publications/Annual%20State%20of%20Tourism%20Report%202020-21.pdf (Accessed 17/11/2024).

Taylor, J. R., & Duodu, K. G. (2017). Sorghum and millets: Grain-quality characteristics and management of quality requirements. In *Cereal grains* (pp. 317–351). Woodhead Publishing. DOI: 10.1016/B978-0-08-100719-8.00013-9

Tshikovhi, N. (2021). A theory of indigenous entrepreneurship in the modern African state: The case of South Africa and Zimbabwe. *International Journal of African Renaissance Studies-Multi-. Inter-and Transdisciplinary*, *16*(2), 21–60.

Tshishonga, N. (2019). The legacy of apartheid on democracy and citizenship in post-apartheid South Africa: An inclusionary and exclusionary binary? *African Journal of Development Studies*, *9*(1), 167–191. DOI: 10.31920/2075-6534/2019/9n1a8

Van Wyk, B. E. (2011). The potential of South African plants in the development of new medicinal products. *South African Journal of Botany*, *77*(4), 812–829. DOI: 10.1016/j.sajb.2011.08.011

Vilakazi, M., & Moore, C. (2023). The Black Industrialist Programme's prospects to the advancement of developmental state in South Africa. *Africa's Public Service Delivery and Performance Review*, *11*(1), 602. DOI: 10.4102/apsdpr.v11i1.602

Viljoen, D. (2020). *Business framework for managing the sustainability of emerging farmers in South Africa* (Doctoral dissertation, North-West University (South Africa).

Welford, L., Venter, S., Dohse, C., & Chibaya, I. (2015). Harvesting from the Tree of Life: responsible commercialization of baobab in South Africa and Malawi. In *Ecological sustainability for non-timber Forest products* (pp. 90–102). Routledge.

ADDITIONAL READING

Brondízio, E. S., Aumeeruddy-Thomas, Y., Bates, P., Carino, J., Fernández-Llamazares, Á., Ferrari, M. F., Galvin, K., Reyes-García, V., McElwee, P., Molnár, Z., Samakov, A., & Shrestha, U. B. (2021). Locally based, regionally manifested, and globally relevant: Indigenous and local knowledge, values, and practices for nature. *Annual Review of Environment and Resources*, *46*(1), 481–509. DOI: 10.1146/annurev-environ-012220-012127

Rosli, A., & Cacciolatti, L. (2022). The role of universities in the development of the local knowledge base: Supporting innovation ecosystems through skills development and entrepreneurship. *International Journal of Intellectual Property Management*, *12*(1), 64–87. DOI: 10.1504/IJIPM.2022.121018

Seko, J., Bain, E. & Maponya, P. (2021). Assessing the impact of indigenous knowledge systems on sustainable agriculture: a case study of the selected communities in the city of Tshwane Metropolitan, Gauteng province, South Africa. *Sustainable Bioeconomy: Pathways to Sustainable Development Goals*,183-208.

Zhang, M., & Dannenberg, P. (2022). Opportunities and challenges of indigenous food plant farmers in integrating into Agri-food value chains in Cape Town. *Land (Basel)*, *11*(12), 2267. DOI: 10.3390/land11122267

KEY TERMS AND DEFINITIONS

Indigenous Knowledge (IK): It is the traditional knowledge, innovations, and practices developed by indigenous communities over generations, often passed down orally and closely tied to cultural heritage.

Indigenous Knowledge-Based Businesses (IKBBs): Are enterprises that leverage indigenous knowledge for commercial purposes, including traditional medicine, crafts, agriculture, and cosmetics.

Commercialization: It is the process of transforming indigenous knowledge into marketable products or services for economic gain while ensuring sustainability and benefit-sharing.

Government Intervention: Are policies, programs, and regulations implemented by the government to support, regulate, and promote indigenous knowledge-based businesses.

Sustainable Development: It is the approach to economic growth that ensures indigenous knowledge-based businesses operate in an environmentally, socially, and economically viable manner.

Funding Mechanisms: It refers to government and private-sector financial support systems, such as grants, loans, and incentives, for indigenous entrepreneurs.

Market Access: It is the ability of indigenous knowledge-based businesses to enter and compete in local and global markets.

Chapter 21
Indigenous Digital Sovereignty:
Data Cooperatives as the New Standard for an Inclusive Digital Economy?

Silvio Andrae
https://orcid.org/0000-0002-8586-7812
Independent Researcher, Germany

ABSTRACT

This research emphasizes the importance of Indigenous data sovereignty (IDSov). It foregrounds the rights of Indigenous peoples to exercise control over their data while resisting data colonialism and its myriad harms. It argues that including Indigenous knowledge in the design, development, and implementation of data-based technologies is essential. The CARE Principles for Indigenous Data Governance are applied to underpin this. The four case studies from Canada, New Zealand, Australia, and the US illustrate the links between Indigenous peoples' rights and data and ecosystem protection and show how responsible data management can benefit people and the environment. Such an understanding of data sovereignty illuminates the exploration of data cooperatives. The focus is on data cooperatives as new institutional models enabling Indigenous communities to exercise digital sovereignty. This term refers to active control and ownership of digital assets and data. It then examines whether the data and sovereignty principles from the Indigenous context can be transferred to other contexts.

INTRODUCTION

The increasing digitalization and datafication of the global economy have brought the value and importance of data to the fore. Datafication is the process of transforming various aspects of life and society into quantifiable data. In this context, the concept of data sovereignty is of growing importance. Data sovereignty refers to the right of individuals, communities, or states to control the collection, use, and sharing of their data by the laws and protocols of the nation-state where the information resides (Kukutai & Taylor, 2016). Unlike traditional data use, which companies and governments often control without the involvement of affected communities, data sovereignty ensures that this control is exercised by those whose data is processed. It is essential for Indigenous peoples, whose data is often used without regard to their cultural rights and traditions (Toki, 2024). Interest in Indigenous knowledge is growing, focusing more on relationships than acquisition. However, turning to Indigenous knowledge and bridg-

DOI: 10.4018/979-8-3693-7964-6.ch021

ing knowledge systems carries the risk of reinforcing the racist hierarchy and the supposed assertion of sovereignty that already underpins the prevailing system of resource regulation and knowledge. However, Indigenous peoples' knowledge systems predate today's nation-states and have their own rules. They capture how peoples, lands, nature, history, and knowledge should be represented, who has the right to use information, and for what purpose (Carroll et al., 2020). It is crucial to support and understand the importance of community self-determination in the context of data sovereignty. This support is reinforced by Article 31 of the United Nations Declaration on the Rights of Indigenous Peoples (UNDRIP) (United Nations, 2008), which directly addresses Indigenous peoples' control over data and information. The principles of the UNDRIP express the importance of geographic and cultural context and community self-determination through meaningful participation (Lovett et al., 2019). The relevance of data sovereignty in the context of Indigenous communities goes beyond the protection of privacy and data sovereignty. Indigenous knowledge holders are under pressure from non-indigenous scientists to work together, for example, to address environmental issues, while the open data movement is pressuring them to make their data public (Cannon et al., 2024). Indigenous data sovereignty (IDSov) refers to the collective right of Indigenous peoples to regulate the collection, storage, and use of data about their communities.

> 'This lens on data sovereignty not only exceeds narrow notions of sovereignty as data localization and jurisdictional rights but also upends the assumption that the nation-state is the legitimate locus of power. IDSov has thus become an important catalyst for broader conversations about what Indigenous sovereignty means in a digital world and how some measure of self-determination can be achieved under the weight of Big Tech dominance'. (Kukutai, 2023)

In the digital world, protecting Indigenous rights is paramount in the face of historical and ongoing marginalization. The CARE (Collective Benefit, Authority to Control, Responsibility, Ethics) Principles for Indigenous Data Governance serve as a crucial ethical framework. These principles, developed by Indigenous communities, aim to integrate Indigenous rights and values into data use, ensuring Indigenous peoples' self-determination, cultural integrity, and well-being. They emphasize the collective benefit of data use, the authority of Indigenous communities to control their data, the responsibility of data users to respect Indigenous values and traditions, and the ethical considerations that should guide data use (Carroll et al., 2020).

This study delves into the concepts and challenges of Indigenous digital sovereignty, with a specific focus on data cooperatives as a potential solution. The rising importance of digital sovereignty, especially for Indigenous communities, who are demanding control over their data and its usage, is a key area of exploration. While large tech companies and state actors often use data as a resource for economic and administrative purposes, the question remains: to what extent can Indigenous data cooperatives help redistribute data power and enable a fairer data economy? This understanding of data sovereignty sheds light on research into data cooperatives. It underscores the importance of cultural sensitivity, inclusivity, and ethical responsibility in data governance practices. The focus is on data cooperatives as innovative institutional models that actively empower Indigenous communities to exercise their digital sovereignty. These cooperatives, with their robust democratic governance, allow data decisions to be made by the members themselves or by representatives acting on their behalf. Digital sovereignty means more than purely legal interpretation and rhetorical use in political discourses. Data cooperatives are the vehicle to interpret digital sovereignty 'from within' as an ongoing series of negotiations, practices, struggles, and controversies embedded in infrastructures, their creation, development, and maintenance (Musiani, 2025).

They act as data intermediaries and provide a user-owned alternative for data management. This personal responsibility of users for their data brings economic and social advantages. It promotes a decentralized responsibility for data that balances collective power with individuals' or groups' autonomy of data ownership (Miller, 2021). This study aims to illuminate the role of data cooperatives in this context and analyze their potential and challenges. It will examine the extent to which they can empower Indigenous communities in their self-determination and decision-making authority over data. In addition, the economic viability of such cooperatives will be considered, especially in competition with well-capitalized companies. The central research questions are:

- To what extent do data cooperatives offer a sustainable alternative to the data-driven economy of large technology companies?
- What are Indigenous communities' structural, technical, and legal challenges in implementing such models?
- What political and regulatory measures could promote the development and scalability of data cooperatives?

METHODOLOGICAL APPROACH

The study is based on a multidimensional methodological approach that combines theoretical analysis, case studies, and empirical evaluations. This combination of methods provides a holistic overview that allows for the development of practice-oriented recommendations for policymakers, businesses, and Indigenous organizations.

Theoretical analysis: The first part of the study examines the concept of digital sovereignty from a theoretical perspective. It links it to related concepts such as data protection, self-determination, and algorithmic fairness. The principles of CARE (Collective Benefit, Authority to Control, Responsibility, Ethics) are considered normative frameworks to define the requirements for Indigenous data practices. Data cooperatives play a crucial role in establishing a technical, legal, and governance structure for the collective management and dissemination of data on behalf of the cooperative members. This cooperative structure allows members to retain control over their data while benefiting from collective resources, knowledge, and experience (Bühler et al., 2023; Kapoor et al., 2024; Ruije, 2021; Fink, 2024). It is important to note that data trusts, as fiduciaries or administrators, go beyond the business model of cooperatives. They also manage data on behalf of data producers and ensure that the agreed terms for data use are met. Data trusts are beneficial when dealing with sensitive data requiring high control and transparency (Zarkadakis, 2020). Despite these differences, data cooperatives, data trusts, decentralized autonomous organizations, and data associations have in common that they view data as a form of collective value, the benefits of which are passed on to their members. These intermediaries work to align the social and economic value of aggregated data with the normative interests of the individuals described in it (Duncan, 2023).

Empirical case studies: Four case studies from Indigenous contexts are analyzed, including the First Nations Information Governance Centre (FNIGC), Te Hiku Media and the Māori Data Sovereignty Network, the Open Data Institute (ODI) and Indigenous Data Network, and the Indigenous Data Sovereignty Collective (IDSC). The case studies shed light on the extent to which these cooperatives meet the requirements of Indigenous sovereignty. Subsequently, the study examines whether the data and sovereignty principles from the Indigenous context can be transferred to other contexts.

Comparative assessment: The study examines existing data cooperatives and assesses their social, economic, and technical viability. Particular emphasis is placed on financing, governance, and data protection challenges. The analysis of the case studies shows that data cooperatives offer significant potential for strengthening Indigenous digital sovereignty. The following aspects are particularly noteworthy. Indigenous data cooperatives enable communities to define access rights and terms of use for their members' data. It strengthens their independence and protects against external data use without consent. While some data cooperatives have successfully established economic models, challenges remain in terms of long-term financing, technical infrastructure, and regulatory uncertainties. In particular, building and maintaining data stores requires financial resources that are not always available. A lack of legal frameworks makes it difficult for Indigenous communities to enforce their data sovereignty rights effectively. Existing data protection laws are often not geared towards collective data rights and primarily focus on individual data protection regulations. Data cooperatives promote the exchange of knowledge and the development of shared digital infrastructures that contribute to the long-term strengthening of Indigenous self-determination.

The study shows that Indigenous data cooperatives can make a decisive contribution to Indigenous communities' self-determination in the digital age. Legal adjustments, targeted financial support, and interdisciplinary cooperation could make such models more sustainable. However, long-term success requires a fundamental reform of the global data economy to ensure that Indigenous communities are not just passive data suppliers but active shapers of their digital future. Developing and promoting sustainable, ethical, and fair data models is not only a technical challenge but also a socio-political one that requires collective solutions.

STRUCTURAL BARRIERS TO FAIR DIGITAL PARTICIPATION FOR INDIGENOUS COMMUNITIES

The shift from Indigenous data sovereignty to more comprehensive digital sovereignty is of utmost importance, as it comes with specific challenges that are deeply rooted in the colonial past of Indigenous peoples and current digital inequalities. Indigenous communities are faced with controlling not only their data but also the underlying networks and digital infrastructures. It requires understanding the historical, legal, and technological barriers to full digital sovereignty (Walter & Carroll, 2021). The enduring impact of the colonial past on Indigenous peoples' current digital standing is a testament to their resilience. Despite the historical processes of dispossession and marginalization, Indigenous communities have shown remarkable strength and determination. This resilience is evident in their continued fight for digital equality, a testament to their unwavering spirit. Urgent action is needed to address the paramount issue of digital inequality. It manifests itself in several forms, from insufficient access to technological infrastructure to a lack of legal frameworks that could protect the digital rights of Indigenous peoples. While large parts of the world benefit from high-speed internet and cloud services, many Indigenous communities in rural and remote areas are cut off from these developments. This digital divide prevents

Indigenous peoples from managing their data and information self-determinedly or asserting their rights in the digital space, underscoring the need for immediate and concerted action.

Furthermore, the unequal distribution of power in the digital world means that global technology companies and governments often have control over Indigenous peoples' data. These colonial dynamics continue in the digital sphere, where Indigenous communities are often excluded from the benefits of datafication, and their data is used for external purposes without their sharing of the resulting profits (Milan & Treré, 2019). The colonial past of Indigenous peoples thus reinforces existing digital inequalities and makes it difficult for Indigenous communities to achieve true digital sovereignty. The importance of Indigenous data sovereignty cannot be overstated, as it is a fundamental aspect of their rights and autonomy in the digital age. Another central problem in the context of Indigenous digital sovereignty is the protection of cultural knowledge and intellectual property. Indigenous communities have deeply rooted knowledge systems often based on centuries-old traditions and practices. This knowledge be it related to medicine, agriculture, ecology, or spirituality is often seen as a collective heritage that is shared and passed on within the community (Robbins et al., 2023; Artita, 2025)

However, in the digital age, Indigenous communities are under immediate threat. Their cultural knowledge risks being digitized and exploited for commercial gain without consent. This form of knowledge appropriation, known as 'digital colonization' or 'data colonialism,' is a dispossession that integrates traditional knowledge into the global data economy, stripping the affected communities of control over its use. It can lead to the commercialization of knowledge that is sacred to Indigenous peoples or holds special cultural significance (Mejias & Couldry, 2024; Couldry & Mejias, 2019b). Based on a study in West, Papua, Sianturi et al. (2025) argue that when technology is disconnected from Indigenous knowledge systems (IKS), it risks exacerbating cultural disengagement and dependence on external frameworks, potentially leading to abuse. Conversely, a lack of technology adaptation to the cultural context creates a significant gap in engaging younger generations and sustaining Indigenous cultural practices in a rapidly digitizing world. The authors underscore the need for a balanced synergy between Indigenous ways of knowing, being, and doing and technology to ensure culturally responsive, sustainable, and empowering educational practices in Indigenous settings.

Protecting intellectual property is a challenge for Indigenous communities in the digital space. Traditional concepts of intellectual property rights are based on Western notions of individual rights and property. Indigenous communities need to recognize collective knowledge systems based on community and tradition. Therefore, Indigenous communities are often unable to protect their cultural heritage in a way that does justice to their collective claims. In the context of digital sovereignty, developing mechanisms to protect cultural knowledge and intellectual property is paramount. These mechanisms must ensure that Indigenous communities retain complete control over their data and knowledge. One such initiative that provides a sense of reassurance is the CARE Principles (Collective Benefit, Authority to Control, Responsibility, Ethics). These principles safeguard Indigenous rights and values in the digital space. They underscore the necessity for cultural knowledge and intellectual property data to benefit the communities and be rooted in their traditions and norms. While the shift towards Indigenous digital sovereignty is a significant development, it has challenges. This shift enables Indigenous communities to assert control over their data, cultural knowledge, and digital infrastructures in alignment with their rights and values. However, the colonial past and current inequalities in the digital space complicate this process. They necessitate the development of innovative strategies to empower Indigenous communities to shape their digital future self-determinedly.

Principles of Indigenous Digital Sovereignty

Most IDSov networks have developed their principles, guidelines, and tools for data sovereignty during digitalization. These principles preserve and strengthen Indigenous peoples' sovereignty in the digital space. They are based on deeply rooted values such as self-determination, the protection of cultural values, and community-based approaches. These approaches enable Indigenous peoples to control their data and information and involve them in the process, fostering a sense of engagement and ownership in their digital journey.

The First Nations Information Governance Centre (2014) in Canada was the first network to lay the foundation for IDSov with the OCAP principles that ensure ownership, control, access, and possession of First Nations data.

- Indigenous communities own their information, empowering them like an individual would own their personal information.
- Indigenous communities and nations can seek control over research data and its management at all research cycle stages.
- Indigenous communities and nations must be able to access data about themselves and have the right to make or be involved in decisions about access to the data.
- Physical data control should reside with the First Nation or Indigenous-controlled custodian or an external data custodian (designated by the Indigenous community or Nation) who ensures that ownership, control, and access principles are respected.

A central principle of Indigenous digital sovereignty is the urgent need for self-determination over data. This principle emphasizes that Indigenous peoples must completely control their data collection, storage, management, and use. Indigenous communities decide what data is collected, how it is used, and who can access it. This aspect of self-determination over data goes far beyond traditional data protection, which usually focuses on individual rights. Instead, data sovereignty is seen as a collective right that affects entire communities. For Indigenous communities, self-determination over data means protecting their data and controlling information related to their land, resources, and cultural knowledge. With this right, Indigenous peoples can avoid having their data used for other people's purposes, as was often the case when colonial and state institutions collected data without consent.

The GIDA CARE Principles for Indigenous Data Governance address the inherent tensions in open data at the forefront. Data justice movements typically focus on protecting individual data rights (WEF, 2024). IDSov, on the other hand, focuses on the protection, preservation, and obligations of collective rights. The CARE Principles, which were developed specifically for the protection of Indigenous data, emphasize the authority of Indigenous communities to make self-determined decisions about the use of their data and require that data activities are consistent with their values and rights (Carroll et al., 2020; Roberts & Montoya, 2023). The abbreviation 'CARE' stands for collective benefit, authority to control, responsibility, and ethics.

- Collective benefit means that data activities should be consistent with Indigenous values and bring recognizable benefits to the communities from which the data originate, emphasizing these activities' communal value and impact.

- The power of control affirms Indigenous peoples' right to control information about their people, traditions, and territories.
- Responsibility is to consult and involve Indigenous communities in data processes and uses.
- Ethics requires respect for the values and rights of Indigenous peoples in all data practices. Recognizing these principles acknowledges the pluralistic nature of data sovereignty and emphasizes that it must always be rooted in the local context and respect local rights and traditions.

In Aotearoa, the Māori Data Governance Model (https://www.kahuiraraunga.io/maoridatagovernance) was developed in 2023 by Māori data experts for use by public agencies as part of a contractual relationship agreement between the National Tribal Forum and Stats NZ. The model is based on the vision of 'data for self-determination.' It states that the collection of Māori data should prioritize the information needs of Māoriand tribes, uphold community values of respect and dignity, and enshrine the principle of free, prior, and informed consent. However, the model has yet to be fully implemented, and in the absence of IDS legislation, there is concern that authorities will do little to change poor data practices. For now, the algorithms currently in use in Aotearoa have the potential to harm Māori. Therefore, Māori must actively and meaningfully participate in the provision, development, and use of algorithms applied to them. Brown et al. (2023) developed the idea of MASov: the underlying is an extension of MDSov principles based on Te Ao Māori's values and Te Tiriti's principles.

The Local Contexts initiative (2021) has created labels and cues that make it possible to recognize the embeddedness of published data in Indigenous worldviews and contexts. Although data governance rules vary widely between Indigenous groups and most initiatives are context-specific, core principles for IDSov include the right to Indigenous ownership of knowledge about its use, Indigenous authority to control and access knowledge, prioritization of collective over individual benefit, recognition of context specificity, a values-based approach that goes beyond consent, and consideration of future generations. A unique perspective on data underpins Indigenous digital sovereignty. For many Indigenous communities, data is a resource and a profound expression of their cultural identity. This perspective underscores the paramount importance of protecting data that encapsulates their cultural knowledge and traditional practices. Indigenous peoples' cultural heritage, such as oral histories, traditional medical knowledge, or spiritual rituals, is intricately linked to their collective identity. This data cannot be passed on or used commercially without the consent of the communities.

The Māori principles of data sovereignty go a step further and state that in some data contexts, collective Māori rights take precedence over the rights of individuals. Data with a precise collective dimension includes DNA and genomic data, tribal nation data, and traditional knowledge. The Māori Data Sovereignty Network is a beacon of these principles, based on the belief that 'Data is a living tāonga.' It upholds the guiding principles of *Rangatiratanga*, *Kotahitanga* (unity), *Mana Akitanga* (generosity), and *Kaiti Akitanga* (guardianship) with unwavering commitment. The Charter recognizes the need for both governance and operational levels. In particular, data can support the expression of Rangatiratanga (sovereignty/autonomy). The Charter emphasizes the potential contribution of ethical data use and identifies as an essential consideration the need to appoint appropriate data guardians and the principles by which they will operate (Te Mana Rarunga, 2018).

Indigenous digital sovereignty plays a proactive role in preventing the exploitation of Indigenous knowledge. It encompasses the fight against the commodification of Indigenous knowledge. In this historical issue, Indigenous peoples' cultural and spiritual practices have been adopted, commercialized, and integrated into Western knowledge systems without benefiting the communities. By seeking to prevent such practices, Indigenous digital sovereignty ensures that communities' cultural knowledge remains protected by their norms and traditions. It includes legal and technical measures to regulate access to cultural data and prevent its misuse. Community-based approaches form the backbone of Indigenous digital sovereignty. Indigenous peoples often emphasize the collective nature of their decisions and values, meaning that data handling is regulated collectively. Instead of individual approaches to data use, the focus is on collective responsibility and shared benefit for the entire community. It reflects Indigenous peoples' social and political structures, traditionally based on collective decisions and the good of the community. Community-based approaches to data management often involve the formation of data cooperatives in which Indigenous communities work together to control the collection, management, and use of their data. These cooperatives allow communities to establish collective decision-making processes and ensure that data activities align with their cultural values. An important goal of such cooperatives is to distribute the economic benefits arising from the use of Indigenous data equitably within the community and to ensure that communities are involved in decisions about the use of their data (Flemmer & Schilling-Vacaflor, 2015; Merino, 2018; Austin et al., 2020). Another critical aspect of community-based approaches is integrating local knowledge systems and technologies. Rather than relying on external technologies or companies, Indigenous communities increasingly develop digital infrastructures that reflect their cultural and social needs. These local solutions, often facilitated through partnerships or local technology workshops, offer communities the opportunity to shape and secure their digital sovereignty. Community-led environmental monitoring projects, for example, apply IDSov principles. The Prey Lang Network (Cambodia) has developed a forest monitoring application to monitor illegal deforestation (Brofeldt et al., 2018). The network decides which data to collect, and the collected data is owned and managed by the users. The digital toolset MAPEO (www.digital-democracy.org/mapeo), an offline-first application jointly developed by Digital Democracy and Indigenous peoples worldwide, allows users to map their land and collect evidence of environmental and human rights violations. MAPEO ensures local data sovereignty and sovereignty via a peer-to-peer database, enabling local data sharing without a centralized external server. Communities can also customize the forms, categories, symbols, and maps to remain culturally relevant (Reyes-García et al., 2022).

In Brazil, national research and non-profit organizations have worked together to develop the Tô no Mapa application (https://tonomapa.org.br). 'Tô no Mapa' is an interactive digital map project that makes traditional communities and family farms visible. It supports Indigenous peoples, *quilombolas* (descendants of enslaved Africans in Brazil who live in self-organized communities known as quilombos), small farmers, and other marginalized groups in documenting and promoting their habitats. A map allows users to view the locations and information of these communities. It increases awareness and legal protection of their territories and resources. The application has a feature that allows communities to choose whether to allow current or future disclosure of registered area data, with the option to make this information public. In summary, Indigenous digital sovereignty is based on the principles of self-determination, protecting cultural values, and community-oriented approaches. The voluntary and legally non-binding nature of many IDSov principles creates a space where Indigenous peoples can assert their sovereignty over data without seeking permission from nation-states. However, this approach also means that there are no mechanisms of public accountability to ensure compliance with the IDSov principles (Reyes-García et al., 2022).

CASE STUDIES

In recent years, Indigenous communities worldwide have launched various projects based on data and digital sovereignty. These initiatives show how Indigenous peoples are putting their rights to data control into practice through innovative and community-oriented approaches. In the following, some outstanding examples are reflected on and compared with each other in order to illustrate the different approaches and successes in the implementation of Indigenous digital sovereignty.

First Nations Information Governance Center (FNIGC)-Canada

The FNIGC is a First Nations organization in Canada that protects Indigenous communities' data sovereignty. The Center developed the First Nations Regional Health Survey (RHS), Canada's most significant Indigenous health database. The FNIGC is best known for developing the First Nations Regional Health Survey, which was initiated, administered, and analyzed by Indigenous communities. At the heart of the FNIGC's work are the OCAP (Ownership, Control, Access, and Possession) principles, which state that First Nations retain complete control over their data. The OCAP principles reflect the Indigenous values of self-determination and collective ownership and safeguard the communities' rights to their data.

The FNIGC project is a beacon of hope, demonstrating how data sovereignty in the health sector can be realized by giving Indigenous communities control over the collection and use of health data (Engstrom et al., 2024). It is a successful model for how Indigenous communities can use data to improve health care and social services while maintaining community autonomy over their data. The data supports First Nations' policy decisions and health strategies, enabling them to make informed decisions and better address their needs. The FNIGC works with various First Nations, health organizations, and government institutions to improve data collection and management without compromising Indigenous communities' autonomy. The data cooperative model implements the principle of responsibility and control over data.

Te Hiku Media-New Zealand

Te Hiku Media, a New Zealand data cooperative, is a testament to the power of community-based approaches in preserving Māori cultural heritage through digital media and technology. The project operates a digital archive and radio programs promoting Māori language and culture. It developed a speech recognition platform for Te Reo Māori to promote language literacy and preservation. A significant project of Te Hiku Media was the development of a corpus of the Māori language through the use of language recordings and data. Te Hiku Media's focus on keeping control of the language data in the hands of the Maori community ensures that external actors do not commercialize it, providing a sense of reassurance about the preservation of Māori cultural heritage.

Unlike many significant language platforms such as Open AI, Google, or Amazon, which often use language data without consulting communities, Te Hiku ensures that Māori themselves decide how their data is used. Te Hiku Media protects Māori cultural knowledge by ensuring that no commercial use occurs without community consent. The project places particular emphasis on protecting intellectual property and upholding traditional values. Technological innovation: By utilizing advanced technologies such as artificial intelligence and machine learning, Te Hiku Media is helping to preserve the Māori language and make it accessible for future generations. It also promotes network sovereignty as communities control their technologies and infrastructure.

In software engineering, Bowen and Hinze (2022) examine the ethical use of data, including cultural aspects and data ownership rights. They propose an extension of typical user-centered design processes. They call this 'participatory data design.' The design will include frameworks and tools that focus specifically on managing data within the cultural context from which it originates. They will apply the proposal to a data collection project of Māori workers in the New Zealand forestry industry. The authors suggest that a cultural analysis of the data collected in the final product is conducted as part of the initial requirements elicitation. Finally, it is essential to develop solutions that collect and use the data culturally sensitively. It requires a mechanism to understand data within different cultural frameworks. In the context of Aotearoa, there is no such framework for data collection. The data collected in the Haki-Turi system can be categorized according to dimensions, such as personal, contextual, or business-related. Bowen and Hinze (2022) point out that the goal of Indigenous data sovereignty is not automatically achieved through the proposed software engineering design process. It relies on software engineers fully engaging in the process and not just playing a tokenistic role.

Open Data Institute (ODI) and Indigenous Data Network-Australia

The Indigenous Data Network (IDN) in Australia is a shining example of a successful data cooperative. The IDN, in partnership with the Open Data Institute (ODI), is dedicated to developing technical solutions and governance models that empower Indigenous communities to manage their data. This collaborative approach underscores the network's commitment to supporting and strengthening Indigenous data sovereignty (Rana & Azeez, 2025).

The IDN supports Indigenous communities in Australia in building the capacity to collect, analyze, and use data. It provides training and technical support to strengthen control over Indigenous data. The network also facilitates the sharing of best practices between different Indigenous communities to improve data management and maximize the collective benefits of data. Finally, by working with the ODI and other international partners, the IDN advocates for more robust legal frameworks and technical solutions supporting Indigenous communities' data sovereignty.

The IDN promotes self-determination over data and knowledge building within Indigenous communities. It supports creating an infrastructure enabling communities to manage their data securely and responsibly while maximizing collective benefits.

Indigenous Data Sovereignty Collective (IDSC)-USA

The Indigenous Data Sovereignty Collective in the USA is a remarkable example of implementing Indigenous data sovereignty. This network of Indigenous scientists and activists is a strong advocate for Indigenous communities' rights to their data. The collective has developed guidelines to help Indigenous communities control access to their data, particularly in environmental, resource management, and research, thereby promoting political engagement and data sovereignty. Compared to FNIGC and Te Hiku Media, IDSC takes a more proactive role in advocating for broader legal protections for Indigenous data sovereignty. While FNIGC focuses on health data and Te Hiku on language data and cultural rights, IDSC takes a broader approach by emphasizing and lobbying for data protection in many areas of the Indigenous population's lives. Although all three projects are based on the principle of Indigenous digital sovereignty, there are essential differences in their focus and implementation. FNIGC focuses strongly on the health sector and shows how collecting and using health data can achieve positive outcomes for Indigenous communities. It provides a structured model with clearly defined principles (OCAP) for data ownership and control. Te Hiku Media, on the other hand, is dedicated to protecting and promoting cultural and linguistic identity. The project shows how language data can be protected in the digital space and how cultural values can be preserved in data-driven projects. It relies on community-based approaches to ensure Māori control over their language data.

Finally, the Indigenous Data Sovereignty Collective offers a broader framework that is not limited to a specific area such as health or language. It encompasses the full range of data affecting Indigenous communities. The focus here is more on the legal and political enforcement of data sovereignty and developing guidelines that promote the general protection of Indigenous data. The projects mentioned above show that Indigenous digital sovereignty can be implemented in various ways, depending on the specific needs and priorities of the communities (Table 1). FNIGC shows how Indigenous rights can be strengthened through clear principles and data infrastructures in the health sector. Te Hiku Media demonstrates how protecting cultural values is possible through controlling language data. At the same time, IDSC takes a broader political and legal approach to defending the rights of Indigenous communities in different areas.

Table 1. Comparison of the four Indigenous data cooperatives

Criterion	First Nations Information Governance Centre (FNIGC)	Te Hiku Media	Indigenous Data Network (IDN)	Indigenous Data Sovereignty Collective (IDSC)
Location	Canada	New Zealand	Australia	USA
Target audience	Indigenous First Nations communities	Māori communities	Australian indigenous communities	Indigenous communities in the USA
Focus	Health and wellness data	Language and culture	Data governance and technical support	Broad legal and policy data sovereignty
Founding Year	1997 (approx.)	2012	Unknown	2016
Primary goal	Control over health data, aligning with OCAP principles	Preserve Māori language and culture	Strengthen data self-management	Promote policy and legal support for data sovereignty
Data focus	Health-related data	Language daat	General data (context-specific)	Various sectors, including environment and resources
Key activities	Health surveys, policy support, advocacy	Digital media archiving, AI for Māori language	Training, legal support, data exchange	Legal lobbying, policy development
Technology use	Health data infrastructure	AI, machine learning for language preservation	Technical training and infrastructure setup	Legal frameworks and data governance
Cultural orientatino	OCAP (Ownership, Control, Access, Possession) principles	Māori cultural values	Cultural sensitivity in data usage	Emphasis on legal-cultural frameworks

Source: Own representation

These initiatives have one thing in common, however: They leave control over Indigenous data in the hands of the communities, thereby contributing to strengthening Indigenous self-determination in the digital age. They show that Indigenous communities can develop digital infrastructures and models to manage their data using their values and traditions.

SHARING DATA: THE POTENTIAL OF DATA COOPERATIVES

What are Data Cooperatives?

Data cooperatives, an innovative model for collective data management, have shown significant potential in preserving digital sovereignty for communities, particularly Indigenous peoples. This key benefit of the model not only reassures the audience about its effectiveness but also inspires hope for the empowerment of Indigenous communities. This study will abstract from the specific Indigenous context and examine whether the data cooperative model is also suitable as an organizational form for other contexts without compromising the principles (Bühler et al., 2023; Fink, 2024). At its core, a data cooperative is an organizational form in which members jointly manage, control, and use their data to achieve collective benefits. In contrast to traditional data management models, where individual actors or companies have control over data, a data cooperative is based on the principles of community ownership, democratic co-determination, and the promise of collective benefit (Parra-Moyano, 2024).

A data cooperative is a collaborative organization in which multiple stakeholders—individuals, communities, or organizations-democratically pool their data and make collective decisions about its collection, storage, and use. This model ensures that the cooperative members retain control over their data and that decisions are made according to the community's shared goals and values. Therefore, a data cooperative combines aspects of data sovereignty, such as self-determination and collaborative management, with a cooperative organizational principle, making the audience feel included and part of a collective decision-making process (Bühler et al., 2023; Andrae, 2025). The International Cooperative Alliance (ICA) proposes seven core principles that are shared by the cooperative movement worldwide (Cheney et al., 2023):

- Voluntary membership
- Democratic member control
- Economic participation of the members
- Autonomy and independence
- Education, training, and information
- Cooperation between cooperatives
- Concern for the community

Data cooperatives are a particular type of cooperative. They act as intermediary organizations (also known as data brokers). They collect data from their members and trade it to create value for those members. This value can be in the form of insights, services, or financial returns. The data collected can be diverse, including personal, business, or community data. The producers usually own existing data cooperatives; those who provide data and benefit from its trade are the 'producers.' Membership in data cooperatives for individuals may require providing data and other resources, such as a monetary fee (Cheney et al., 2023). The functioning of a data cooperative is rooted in the democratic participation of its members. In such a cooperative, the data is collectively owned, and decisions about its use are made democratically. This approach ensures that data is not seen as an individual resource but as a collective good managed for the entire community's benefit. Each person or group contributing data has a say in decision-making and stands to benefit from the value generated from the data (Fink, 2024) (Figure 1).

Figure 1. Organizational structure of a data cooperative

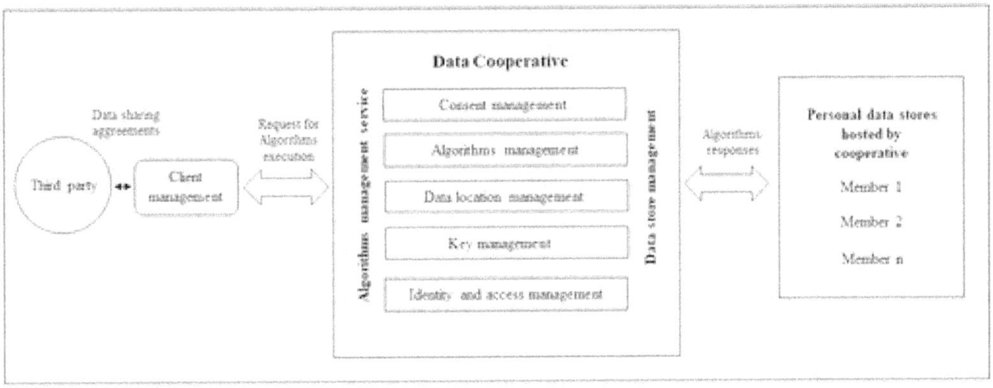

Source: Own representation

The members of a data cooperative work together to define,

- Which data is collected
- How this data is used and shared
- Who has access to the data
- Which ethical and legal standards must be observed

Data cooperatives can pursue different goals. Some focus on the commercial exploitation of data, with the proceeds being used collectively, while others focus on the protection and sustainable use of community data. A vital advantage of this model is that it balances out the power asymmetries that often favor large companies or government institutions in the conventional data economy. Instead of centralizing data power, a cooperative distributes control and ensures that the communities from which the data originate share in the decisions and potential profits, offering the audience an optimistic view of the future of data governance (Bühler et al., 2023; Fink, 2024).

Once managed in a cooperative, data can be used, managed, and shared for many purposes. The largest and most successful data cooperatives are health data cooperatives, where individuals upload and share health data while retaining detailed control over their data. So instead of giving consent at the time of data collection and the data disappearing into databases that normal users cannot access, Dealer Management Systems allow users to change access permissions at any time and understand which groups and for what purpose their data is being used. Individuals thus join data cooperatives to gain more power and control over data sharing, but also for data altruism—the ability to share personal data collectively and consensually for altruistic purposes such as pharmaceutical research (Bietti et al., 2021). Some cooperatives may be structured and require little member involvement, while others may involve significant involvement in individual data management and governance (Hafen et al., 2014; Hardjono & Pentland, 2019; Micheli et al., 2023).

In addition, data cooperatives are often based on transparent governance structures. It gives members insight into how their data is used and managed. In the case of Indigenous communities, data cooperatives can also be supplemented by culturally and ethically sensitive guidelines that consider the communities' specific needs and traditions. It could include, for example, the protection of cultural knowledge or compliance with Indigenous values and norms. Overall, data cooperatives are a promising model for strengthening Indigenous digital sovereignty. They offer a platform for collective self-determination, promote the protection of cultural values, and, at the same time, enable the fair distribution of the benefits created by data (Bühler et al., 2023; Fink, 2024).

ROLES AND FUNCTIONS OF DATA COOPERATIVES

Data cooperatives fulfill various essential roles and functions that cover the entire data lifecycle from collection and management to analysis and use to knowledge creation and transfer. These processes enable members to control and utilize data effectively. Importantly, they ensure that decision-making is collective and transparent, reflecting the democratic nature of data cooperatives (Fink, 2024; Hardjono & Pentland, 2019). The following section describes the central roles of data cooperatives and the associated processes. The first essential role of a data cooperative is organizing and coordinating data collection and management. Data collection and management involve ensuring that data is collected from members in a way that meets the cooperative's ethical standards and respects members' individual and collective rights. Rights can be particularly important for Indigenous communities as they often involve sensitive information, cultural knowledge, or data on natural resources.

In practice, this means that the members jointly decide which data is essential for the cooperative, who has access to the data collection, and how the data collection is organized. A particular focus is protecting data sovereignty, which Abbas et al. (2024) define as comprising protection, participation, and provision (the three Ps). Protection emphasizes fundamental rights such as data ownership, participation ensures a clear division of tasks, and provision includes control, security, and compliance mechanisms to ensure that fundamental rights are respected during and after data sharing. For instance, when personal data is shared, data privacy becomes a crucial protection facet, leading to challenges in terms of ownership between data providers and data subjects, compliance requirements, and control enforcement. The concept presented by Abbas et al. (2024) sets the stage for a comprehensive and coherent theory development regarding data sovereignty as a complex, multifaceted construct. It could be effectively tested in data cooperatives, providing a secure environment for data sharing.

Data cooperatives often develop specific policies to ensure that data collection aligns with the community's needs and values. For example, a community governance model that ensures integrity and privacy can oversee digital tools and platforms for data collection (Fink, 2024). Another critical aspect of management is storing the collected data. Here, data cooperatives play a central role by creating secure infrastructures that enable members to store and access their data securely. Technical aspects, such as encryption and data backup, as well as social and cultural aspects, such as the protection of traditional knowledge, are considered. The second important role of data cooperatives lies in data analysis and the responsible use of the collected data. Once the data has been collected and securely stored, the analysis process gains valuable information and insights. Here, cooperatives play a crucial role by ensuring that the analysis processes align with the interests of the community and respect the values of its members.

Data cooperatives often tap into collective knowledge systems that blend traditional and modern analysis methods. It is particularly significant for Indigenous communities, which possess profound knowledge of their environment and resources (Reyes-Garciá et al., 2022). By fusing local wisdom with modern analytical tools, cooperatives can unearth unique insights that pique the community's interest. Another crucial aspect of data usage is its potential economic, political, or scientific exploitation. Data cooperatives present an avenue to distribute the economic benefits of data usage equitably. It could be achieved by providing data for research or economic projects, with the cooperative guaranteeing that the community receives a fair share of the profits. Simultaneously, it's imperative to ensure that sensitive information remains protected and is only used in line with shared values (Fink, 2024). As a central hub for building and transferring knowledge within the community, data cooperatives empower all members by providing access to specialized knowledge and skills. This aspect is crucial, as data and digital technologies often require expertise that is not equally accessible to all. By acting as a platform for exchanging knowledge and skills, cooperatives ensure that all members can participate in decision-making, fostering empowerment and inspiration (Fink, 2024). One example is training members in data management, protection, and analysis. By offering educational programs and technical training, cooperatives enable members better to understand data management's technical and legal aspects. Building technical knowledge is crucial to maintaining digital sovereignty, especially in Indigenous communities, which have historically often been excluded from technological developments. Furthermore, data cooperatives serve as a bridge for the transfer of knowledge beyond the boundaries of their community. It can be achieved by exchanging best practices with Indigenous communities or collaborating with academic or civil society organizations. In doing so, cooperatives contribute to raising global awareness of Indigenous digital sovereignty and ensure that Indigenous voices are heard in global discussions on data management, fostering a sense of connection and belonging.

FUTURE RESEARCH DIRECTIONS

Future research should investigate how Indigenous data sovereignty can be effectively integrated into global data governance frameworks. It includes exploring policy mechanisms ensuring Indigenous communities retain data control while still participating in data-driven economies. The importance of interdisciplinary research cannot be overstated in this context, as it is needed to assess the technological advancements required to support Indigenous-led data governance, such as decentralized data storage, encryption methods, and AI systems designed with Indigenous knowledge principles in mind. Comparative studies between Indigenous data governance models and existing legal frameworks in different jurisdictions could offer valuable insights into best practices and potential regulatory reforms, highlighting the need for collaboration across disciplines.

Another crucial research avenue is the socio-economic impact of Indigenous data cooperatives. Studies could examine how these cooperatives influence economic development, community resilience, and digital inclusion. There is also a need for empirical research on the scalability of Indigenous data governance models beyond local contexts—potentially informing global debates on ethical data stewardship. Moreover, future research could analyze how emerging technologies like blockchain and federated learning can be leveraged to strengthen Indigenous data sovereignty without reinforcing digital colonialism, offering exciting new possibilities for the future of data governance. Research addressing these challenges can contribute to more equitable, culturally sensitive, and sustainable data ecosystems.

CONCLUSION

This study discusses the growing concept of Indigenous digital sovereignty and presents data co-operatives as a promising model for promoting this sovereignty. The article begins with a definition of IDSov. It emphasizes its importance in the context of Indigenous peoples seeking to regain historical and contemporary forms of control over their data and knowledge. The discussion covers the theoretical foundations of data sovereignty, its links to related concepts such as self-determination and data privacy, and the practical implications for Indigenous communities, particularly in an increasingly data-driven world. Various unanswered questions accompany the concept of IDSov, such as how Indigenous contexts related to privacy and security reveal third-party information sharing for industrial computation (Carroll et al., 2024). In this respect, broadening the focus from data to infrastructure is essential. This broadening is expressed in Indigenous digital sovereignty, which includes data protection and control over networks and infrastructures. Indigenous communities face specific challenges, such as the colonial past, which exacerbates digital inequality and the misuse of cultural knowledge, as well as the protection of intellectual property. The study explains the principles of Indigenous digital sovereignty, including self-determination over data, the protection of cultural values, and the importance of community-based approaches.

The concept of Indigenous digital sovereignty emphasizes the fundamental coloniality of data. One could also include algorithmic discrimination here (Bonini & Treré, 2024). Mumford (2022) agrees that colonialism is primarily only used as a motif in Couldry and Mejias (2019a). The prevailing approach to data also shapes our knowledge systems (Quijano, 2000). The universalism practiced and the problematic notion of objectivity and epistemology imposed on historically colonized communities by the Global North and increasingly by China must be explored further (Ricaurte, 2019; Couldry & Mejias, 2023; Milan & Treré, 2019). For example, agricultural development initiatives aimed at more comprehensive data collection and disclosure can be observed, extracting information from local contexts in ways that ultimately benefit multinational agribusinesses rather than local farms (Bronson, 2022). Data cooperatives play a crucial role in promoting Indigenous digital sovereignty. These cooperatives, where communities control data collection, management, and use, provide a practical model for protecting data rights and strengthening self-determination. By examining successful case studies such as the First Nations Information Governance Centre (FNIGC) in Canada, Te Hiku Media in New Zealand, and the Indigenous Data Network (IDN) in Australia, the study demonstrates how these cooperatives are implemented in practice, the challenges they overcome, and the successes they achieve. These successes, such as the FNIGC's empowerment of First Nations communities in Canada, instill a sense of inspiration and hope. Such Indigenous data networks favor community-driven organizational models instead of the typical hierarchical data structures. By defining data sovereignty as a collective right, Indigenous movements challenge the neoliberal data economy, prioritizing individual data ownership over community sovereignty.

The synergies between data cooperatives and the principles of Indigenous digital sovereignty are unmistakable. Data cooperatives provide a platform enabling communities to gain control over their data, protect their cultural values, and make decisions according to their traditional governance structures. By promoting self-determination and the protection of cultural values, data cooperatives not only support the advancement of Indigenous digital sovereignty but also provide a model tailored to meet these communities' specific needs, instilling a sense of reassurance and confidence. Graeber and Wengrow's (2021) exploration of mutual aid as a social force resonates with the Indigenous approach to knowledge sharing and data stewardship. The concept of the data cooperative, in particular, is a practical application of their ideas. It is a system where communities come together to manage their data and share resources for mutual benefit, not for profit. This cooperative approach effectively translates Graeber and Wengrow's theories into action, creating a knowledge economy that serves the common good. Based on the results of the analysis, several recommendations for action can be derived:

a) Strengthening the legal framework: National and international legislators should develop clear rules for Indigenous data rights. These include laws that regulate the use of Indigenous data by third parties and enshrine the need for free, prior and informed consent (FPIC).
b) Promote financial and technical infrastructure: Indigenous communities need access to financial resources to build technical infrastructure for data storage and analysis. Funding programs and collaborations with universities or non-profit organizations could help here.
c) Introduction of education and training programs: To strengthen digital sovereignty, training courses on data analysis, data protection, and ethical AI development should be developed specifically for Indigenous communities. This would enable Indigenous actors to manage their own data structures in the long term.
d) Promote interdisciplinary collaboration: Social scientists, legal scholars, and technology developers must collaborate to develop innovative and practical solutions for Indigenous data cooperatives. Cooperation between Indigenous organizations and data policy institutions is crucial.
e) Establishing a global network for Indigenous data cooperatives is a crucial step: An international platform could consolidate best practices, legal developments, and technical innovations, thereby fostering knowledge exchange between Indigenous communities worldwide. This global network could serve as a hub for sharing experiences, learning from each other's challenges and successes, and collectively advancing the cause of Indigenous digital sovereignty.

As technologies like AI increasingly dominate human activity, the need to reclaim human agency becomes more urgent. It's crucial to remember that humans are an integral part of a complex ecosystem and a fragile web of life. Data impacts all organic forms on the planet, presenting both unprecedented opportunities and existential risks to our species. The long-standing colonial doctrines of nature as separate from human life and destined for limitless appropriation are at the heart of today's data-intensive practices of 'knowledge extraction.' This datafication and the resulting production of new data relations abstract and reorganize social life and alter human relations to nature. Overall, the study clarifies that promoting Indigenous digital sovereignty depends on close cooperation between communities, governments, organizations, and individuals (Garba et al., 2023). The Indigenous communities are demanding recognition as 'interest groups' in digital policy and as key players with their own legal and cultural sovereignty. A more inclusive and equitable digital economy can emerge by establishing data cooperatives and recognizing Indigenous rights in the digital space. According to Hudson et al. (2023), whether with or without data cooperatives, the rights of Indigenous peoples are to be anchored in data and represent a concrete step towards the operationalization of Indigenous digital sovereignty.

REFERENCES

Abbas, A. E., van Velzen, T., Ofe, H., van de Kaa, G., Zuiderwijk, A., & de Reuver, M. (2024). Beyond control over data: Conceptualizing data sovereignty from a social contract perspective. *Electronic Markets*, *34*(1), 20. DOI: 10.1007/s12525-024-00695-2

Andrae, S. (2025). Data cooperatives to strengthen digital citizenship: Opportunities and risks. In Hawamdeh, M. Kh. M. (Ed.), *Digital citizenship and the future of AI engagement, ethics, and privacy* (pp. 533–568). IGI Global Publishing., DOI: 10.4018/979-8-3693-9015-3.ch018

Arita, M. (2025). Data sovereignty and open sharing: Reconceiving benefit-sharing and governance of digital sequence information. *Data Science Journal*, *24*(9), 1–5. DOI: 10.5334/dsj-2025-009

Austin, B. J., Robinson, C. J., Mathews, D., Oades, D., Wiggin, A., Dobbs, R. J., Lincoln, G., & Garnett, S. T. (2019). An Indigenous-led approach for regional knowledge partnerships in the Kimberley region of Australia. *Human Ecology: an Interdisciplinary Journal*, *47*(4), 577–588. DOI: 10.1007/s10745-019-00085-9

Bietti, E., Etxeberria, A., Mannan, M., & Wong, J. (2021). Data cooperatives in Europe: A legal and empirical investigation. The New School's Platform Cooperativism Consortium. https://cyber.harvard.edu/sites/default/files/2022-02/Data_Cooperatives_Europe-group2.pdf

Bonini, T., & Treré, E. (2024). *The everyday fight against platform power*. The MIT Press., DOI: 10.7551/mitpress/14329.001.0001

Bowen, J., & Hinze, A. (2022). Participatory data design: Managing data sovereignty in IoT solutions. *Interacting with Computers*, *34*(2), 60–71. DOI: 10.1093/iwc/iwac031

Brofeldt, S., Argyriou, D., Turreira-García, N., Meilby, H., Danielsen, F., & Theilade, I. (2018). Community-based monitoring of tropical forest crimes and forest resources using information and communication technology – Experiences from Prey Lang, Cambodia. *Citizen Science: Theory and Practice*, *3*(2), 1–14. DOI: 10.5334/cstp.129

Bronson, K. (2022). *The immaculate conception of data: Agribusiness, activists, and their shared politics of the future*. McGill-Queen's University Press.

Brown, P. T., Wilson, D., West, K., Escott, K.-R., Basabas, K., Ritchie, B., ... (2023). Māori algorithmic sovereignty: Idea, principles, and use. *arXiv*. https://doi.org//arXiv.2311.15473DOI: 10.48550

Bühler, M. M., Calzada, I., Cane, I., Jelinek, T., Kapoor, A., Mannan, M., Mehta, S., Mookerje, V., Nübel, K., Pentland, A., Scholz, T., Siddarth, D., Tait, J., Vaitla, B., & Zhu, J. (2023). Unlocking the power of digital commons: Data cooperatives as a pathway for data sovereign, innovative and equitable digital communities. *Digital*, *3*(3), 146–171. DOI: 10.3390/digital3030011

Cannon, S. E., Moore, J. W., Adams, M. S., Degai, T., Griggs, E., Griggs, J., Marsden, T., Reid, A. J., Sainsbury, N., Stirling, K. M., Barnes, A. A. Y. S., Benson, R., Burrows, D., Chamberlin, G. R., Charley, B., Dick, D., Duncan, A. T., Liddle, K. K. M., Paul, M., & Wilson, K. B. (2024). Taking care of knowledge, taking care of salmon: Towards Indigenous data sovereignty in an era of climate change and cumulative effects. *Facets*, *9*, 1–21. DOI: 10.1139/facets-2023-0135

Carroll, S. R., Duarte, M., & Liboiron, M. (2024). Indigenous data sovereignty. *Data & Society. 4*, 207–223. https://datasociety.net/wp-content/uploads/2024/04/Keywords_Indigenous_Data_Sovereignty_Carroll_Duarte_Liboiron_04242024.pdf

Carroll, S. R., Garba, I., Figueroa-Rodríguez, O. L., Holbrook, J., Lovett, R., Materechera, S., Parsons, M., Raseroka, K., Rodriguez-Lonebear, D., Rowe, R., Sara, R., Walker, J. D., Anderson, J., & Hudson, M. (2020). The CARE principles for Indigenous data governance. *Data Science Journal, 19*(43), 1–12. DOI: 10.5334/dsj-2020-043

Carroll, S. R., Herczog, E., Hudson, M., Russell, K., & Stall, S. (2021). Operationalizing the CARE and FAIR principles for Indigenous data futures. *Scientific Data, 8*(1), 108. DOI: 10.1038/s41597-021-00892-0 PMID: 33863927

Cheney, G., Noyes, M., Do, E., Vieta, M., Azkarraga, J., & Michel, C. (2023). *Cooperatives at work.* Emerald Publishing Limited., DOI: 10.1108/9781838678258

Couldry, N., & Mejias, U. A. (2019a). *The costs of connection: How data is colonizing human life and appropriating it for capitalism.* Stanford University Press.

Couldry, N., & Mejias, U. A. (2019b). Data colonialism: Rethinking big data's relation to the contemporary subject. *Television & New Media, 20*(4), 336–349. DOI: 10.1177/1527476418796632

Couldry, N., & Mejias, U. A. (2023). The decolonial turn in data and technology research: What is at stake and where is it heading? *Information Communication and Society, 26*(4), 786–802. DOI: 10.1080/1369118X.2021.1986102

Duncan, J. (2023). Data protection beyond data rights: Governing data production through collective intermediaries. *Internet Policy Review, 12*(3). Advance online publication. DOI: 10.14763/2023.3.1722

Engstrom, T., Lobo, E. H., Watego, K., Nelson, C., Wang, J., Wong, H., Kim, S. L., Oh, S. I., Lawley, M., Gorse, A.-D., Ward, J., & Sullivan, C. (2024). Indigenous data governance approaches applied in research using routinely collected health data: A scoping review. *npj. Digital Medicine, 7*(1), 68. DOI: 10.1038/s41746-024-01070-3 PMID: 38491156

Fink, A. (2024). Data cooperative. *Internet Policy Review, 13*(2). Advance online publication. DOI: 10.14763/2024.2.1752

First Nations Information Governance Centre. (2014). Ownership, control, access, and possession (OCAP): The path to First Nations information governance. First Nations Information Governance Centre. https://achh.ca/wp-content/uploads/2018/07/OCAP_FNIGC.pdf

Flemmer, R., & Schilling-Vacaflor, A. (2015). Unfulfilled promises of the consultation approach: The limits to effective Indigenous participation in Bolivia's and Peru's extractive industries. *Third World Quarterly, 37*(1), 172–188. DOI: 10.1080/01436597.2015.1092867

Garba, I., Sterling, R., Plevel, R., Carson, W., Cordova-Marks, F. M., Cummins, J., Curley, C., David-Chavez, D., Fernandez, A., Hiraldo, D., Hiratsuka, V., Hudson, M., Jäger, M. B., Jennings, L. L., Martinez, A., Yracheta, J., Garrison, N. A., & Carroll, S. R. (2023). Indigenous peoples and research: Self-determination in research governance. *Frontiers in Research Metrics and Analytics, 8*, 1272318. Advance online publication. DOI: 10.3389/frma.2023.1272318 PMID: 38033627

Graeber, D., & Wengrow, D. (2021). *The dawn of everything: A new history of humanity*. Macmillan.

Hardjono, T., & Pentland, A. (2019). Data cooperatives: Towards a foundation for decentralized personal data management. *arXiv*. https://doi.org//arXiv.1905.08819DOI: 10.48550

Hudson, M., Russo Carroll, S., Anderson, J., Blackwater, D., Cordova-Marks, F. M., & Cummins, J.. (2023). Indigenous peoples' rights in data: A contribution toward Indigenous research sovereignty. *Frontiers in Research Metrics and Analytics, 8*, 1173805. DOI: 10.3389/frma.2023.1173805 PMID: 37215248

Kapoor, A., Tait, J., & Vaitl, B. (2024). How to build a data cooperative: A practitioner's handbook. https://data2x.org/wp-content/uploads/2024/06/Data-Cooperative-Handbook.pdf

Kukutai, T. (2023). Indigenous data sovereignty—A new take on an old theme. *Science, 382*(6674), eadl4664. Advance online publication. DOI: 10.1126/science.adl4664 PMID: 38033063

Kukutai, T., & Taylor, J. (2016). Indigenous data sovereignty. In Kukutai, T., & Taylor, J. (Eds.), *Indigenous data sovereignty: Towards on agenda* (pp. 1–22). ANU Press., DOI: 10.22459/CAEPR38.11.2016.01

Lim, J. B. Y. (2025). *Digital media interventions in Southeast Asia*. Palgrave Macmillan. DOI: 10.1007/978-3-031-78211-4

Local Contexts. (2021). Grounding Indigenous rights. Local Contexts. https://localcontexts.org

Lovett, R., Lee, V., Kukutai, T., Cormack, D., Rainie, S., & Walker, J. (2019). Good Data Practices for Indigenous Data Sovereignty and Governance. In Daly, A., Devitt, S. K., & Mann, M. (Eds.), *Good Data* (pp. 26–36). Institute of Network Cultures.

Mejias, U. A., & Couldry, N. (2024). *Data grab: The new colonialism of Big Tech and how to fight back*. University of Chicago Press. DOI: 10.7208/chicago/9780226832319.001.0001

Merino, R. (2018). Re-politicizing participation or reframing environmental governance? Beyond Indigenous' prior consultation and citizen participation. *World Development, 111*, 75–83. DOI: 10.1016/j.worlddev.2018.06.025

Micheli, M., Farell, E., Carballa-Smichowski, B., Posada-Sánchez, M., Signorelli, S., & Vespe, M. (2023). Mapping the landscape of data intermediaries: Emerging models for more inclusive data governance. [Report]. Publications Office of the European Union. https://data.europa.eu/doi/10.2760/8943

Milan, S., & Treré, E. (2019). Big data from the south(s): Beyond data universalism. *Television & New Media, 20*(4), 319–335. DOI: 10.1177/1527476419837739

Miller, K. (2021). Radical proposal: Data cooperatives could give us more power over our data. Stanford University Human-Centered Artificial Intelligence, Oct 20, 2021. https://hai.stanford.edu/news/radical-proposal-data-cooperatives-could-give-us-more-power-over-our-data

Mumford, D. (2021). Data colonialism: Compelling and useful, but whither epistemes? *Information Communication and Society*, *25*(10), 1511–1516. DOI: 10.1080/1369118X.2021.1986103

Musiani, F. (2025). Reassessing "infrastructuring digital sovereignty": Digital self-determination as a set of infrastructure-embedded practices. *Frontiers in Communication*, *10*, 1562072. Advance online publication. DOI: 10.3389/fcomm.2025.1562072

Parra-Moyano, J. (2024). Data entrepreneurs of the world, unite! How business leaders should react to the emergence of data cooperatives. *The Capco Institute Journal of Financial Transformation*, *59*, 20–25. https://www.capco.com/Capco-Institute/Journal-59-innovation-and-control

Quijano, A. (2000). Coloniality of power and eurocentrism in Latin America. *International Sociology*, *15*(2), 215–232. DOI: 10.1177/0268580900015002005

Rana, V., & Azeez, G. K. (2025). Indigenous data sovereignty in Australian higher education: Paving the way for First Nations' self-determination. *Journal of Higher Education Policy and Management*, 1–9. Advance online publication. DOI: 10.1080/1360080X.2025.2469920

Reyes-Garciá, V., Tofighi-Niaki, A., Austin, B. J., Benyei, P., Danielsen, F., Fernández-Llamazares, Á., Sharma, A., Soleymani-Fard, R., & Tengö, M. (2022). Data sovereignty in community-based environmental monitoring: Toward equitable environmental data governance. *Bioscience*, *72*(8), 714–717. DOI: 10.1093/biosci/biac048 PMID: 35923191

Ricaurte, P. (2019). Data epistemologies, the coloniality of power, and resistance. *Television & New Media*, *20*(4), 350–365. DOI: 10.1177/1527476419831640

Robbins, P., Hunt, H. H., Pelegri, F., & Gilbert, J. (2023). Sovereign genes: Wildlife conservation, genetic preservation, and Indigenous data sovereignty. *Frontiers in Conservation Science*, *4*, 1099562. Advance online publication. DOI: 10.3389/fcosc.2023.1099562

Roberts, J. S., & Montoya, L. N. (2022). Decolonization, global data law, and Indigenous data sovereignty. *arXiv*. https://doi.org//arXiv.2208.04700DOI: 10.48550

Roberts, J. S., & Montoya, L. N. (2023). In consideration of Indigenous data sovereignty: Data mining as a colonial practice. *arXiv*. https://doi.org//arXiv.2309.10215DOI: 10.1007/978-3-031-47451-4_13

Ruijer, E. (2021). Designing and implementing data collaboratives: A governance perspective. *Government Information Quarterly*, *38*(4), 101612. DOI: 10.1016/j.giq.2021.101612

Sianturi, M., Lee, J.-S., & Cumming, T. M. (2025). The synergy between Indigenous ways of knowing, being, and doing and technology in culturally responsive home-school partnerships. *Information Technology for Development*, 1–24. Advance online publication. DOI: 10.1080/02681102.2025.2465435

Te Mana Rarunga. (2018). Principles of Māori data sovereignty. https://www.otago.ac.nz/__data/assets/pdf_file/0014/321044/tmr-maori-data-sovereignty-principles-october-2018-832194.pdf

Toki, V. (2024). *Indigenous rights, climate change and governance: Measuring success and data*. Edward Elgar., DOI: 10.4337/9781803924984

United Nations. (2008). United Nations declaration on the rights of Indigenous peoples. chrome- https://www.un.org/development/desa/indigenouspeoples/wp-content/uploads/sites/19/2018/11/UNDRIP_E_web.pdf

Walter, M., & Carroll, S. R. (2021). Indigenous data sovereignty, governance and the link to Indigenous policy. In Walter, M., Kukutai, T., Carroll, S. R., & Rodriguez-Lonebear, D. (Eds.), *Indigenous data sovereignty and policy* (pp. 1–20). Routledge., DOI: 10.4324/9780429273957

World Economic Forum (WEF). (2024). Advancing data equity: An action-oriented framework. September 2024. https://www3.weforum.org/docs/WEF_Advancing_Data_Equity_2024.pdf

Zarkadakis, G. (2020). 'Data trusts' could be the key to better AI. Harvard Business Review. November 10, 2020. https://hbr.org/2020/11/data-trusts-could-be-the-key-to-better-ai

KEY TERMS AND DEFINITIONS

Data Colonialism: The exploitation and control of data by powerful actors, often to the detriment of marginalized groups, similar to the historical patterns of colonialism.

Data Commons: Shared data use and management by a community to maintain and protect data as a public good.

Data Cooperative: An organization where members jointly manage and control their data to maximize collective benefits.

Data Sovereignty: The right of individuals or states to determine how their data is used and managed.

CARE Principle: A framework for handling the data of Indigenous communities that emphasizes collective benefit, authority to control, responsibility, and ethics.

Digital Sovereignty: The ability of a state or community to independently control digital technologies, data, and infrastructure.

Indigenous Data Sovereignty: The right of Indigenous peoples to manage their data by cultural values and political rights of self-determination.

Chapter 22
Homogenizing Traditional Knowledge and Biodiversity Conservation:
A Study Intersecting IP Rights and Sustainable Development

Tasnim Jahan
https://orcid.org/0009-0004-1269-0122
M.S. Ramaiah University of Applied Sciences, India

Shashikant Saurav
https://orcid.org/0000-0002-9349-7452
M.S. Ramaiah University of Applied Sciences, India

ABSTRACT

This chapter explores the nexus between biodiversity conservation and the preservation of traditional knowledge, focusing on the need for an integrated approach that is in line with sustainable development goals. The chapter examines the mechanisms that exist for the protection of traditional knowledge and biodiversity by critically appraising the international frameworks such as the Convention on Biological Diversity and the Nagoya Protocol along with national legislations such as India's Biological Diversity Act, 2002. The emphasis here is on "biocultural heritage" and its integration into conservation frameworks that create traditional knowledge systems uniform. Fitted together with case studies and law contexts, this study highlights the important role that local communities are also assuming in custodianship over biodiversity and traditional knowledge. This research contributes to the broader discourse on environmental law and governance, including policy recommendations to harmonize biodiversity conservation with protection of traditional knowledge.

DOI: 10.4018/979-8-3693-7964-6.ch022

INTRODUCTION

Traditional knowledge (TK) refers to the living repository of cultural wisdom, encompassed within practices, innovations, or insights that indigenous communities come to develop over generations. Such knowledge is often entrenched in intimate ecological understandings and forms the heart of sustainable living for many around the world. There is, for example, guidance of crop rotation techniques given through TK systems, practice for herbal medicinal purposes, water conservation techniques, such as Dong Bund irrigation in India. These practices not only promote ecological harmony but also safeguard cultural identities, enriching human heritage. TK plays a crucial role in biodiversity conservation. Aboriginal communities, who are usually the custodians of TK, live within areas of high biological richness. Their lifestyles are examples of sustainable resource use, which makes living ecosystems resilient to environmental shifts. As noted in CBD (Convention on Biological Diversity, 1992), TK is essential for the achievement of conservation and sustainable exploitation of biodiversity. However, it is threatened by globalization processes, commercialization, as well as loss of intergeneration transmission. Biodiversity is a web of life that includes the diversity of ecosystems, species, and genetic diversity. This supports human survival and well-being through essential services such as food security, climate regulation, and disease prevention that serve as the foundation for the United Nations Sustainable Development Goals (SDGs). However, biodiversity is being attacked by habitat destruction, overexploitation, pollution, and climate change. Sustainable development, according to the Brundt land Commission (World Commission on Environment and Development, 1987), is one that satisfies the needs of the present without compromising the future generations' ability to meet their own needs. Bringing biodiversity conservation into the forefront of sustainable development frameworks forms a great basis. Tradition-based knowledge systems, represented by TK, provide sustainability alternatives in the approach to more conventional ways of development. The intersection of Intellectual Property Rights (IPRs) and TK has gained increasing value in legal and policy discourse. IPRs were designed with the objective of providing individual rights to creators to encourage innovation, but TK poses a challenge to this paradigm. TK is community-owned, intergenerational, and context-specific. This nature of TK disallows commodification tendencies within traditional IP regimes.

Efforts to protect TK through IPR frameworks have gained momentum. Instruments such as patents, Geographical Indications and trademarks are now being adapted to recognize that TK systems are collective. Thus, for instance, when India registered GIs on turmeric and basmati rice, it is protecting the cultural identity as well as the economic interest of communities associated with the resources. However, these frameworks often fail to address the nuances of TK, leading to sui generis approaches specially tailored to TK systems. The interaction between TK, IPRs, and biodiversity conservation poses challenges. Misappropriation of TK, often termed as "biopiracy," where innovations based on TK are patented without the knowledge or consent of indigenous communities, poses a key issue. The high-profile cases of neem and turmeric patents (Shiva, 1997) are just examples of how the failure of proper legal safeguards could lead to unequal outcomes. A more difficult issue is the non-commercial nature of TK's substance as juxtaposed to the market orientation of IPRs. Wherein TK is often embedded within community life, commercialization will then taint its value and reduce cultural importance. Fragmentation of international and national legal frameworks also creates ambiguity as TK holders are exposed to exploitation by businesses that aim to profiteer from pillage.

Objective of the Chapter

This chapter is written using doctrinal approach exploring ways through which TK and biodiversity are intersecting with each other affecting indigenous people. It explains how the intersection can be harmonized with respect to intellectual property management, particularly with sustainable development principles. Taking a closer look into various legal instruments such as CBD, Nagoya Protocol and India's Biological Diversity Act, 2002, the study aspires to bridge gaps across those frameworks. The ultimate object in the paper is to strategize homogenization and protection of TK, conservation of biodiversity, and equitable benefits while ensuring sustainable development.

RESEARCH METHODOLOGY

This chapter adopted a multidisciplinary and qualitative approach in its pursuit to study the relationship between IP rights, biodiversity conservation, and traditional knowledge. It formed the theoretical and contextual framework for this research based on an all-rounded review of primary and secondary literature, such as international treaties, national laws, case studies, and scholarly articles. This is a doctrinal legal analysis of the legislative frameworks at the global and national levels to comment on effectiveness and lacunas in protection of TK. Case studies drawn from India and other jurisdictions have been undertaken to determine challenges such as biopiracy and models that are successfully providing benefit sharing to the rightful benefactors. However, this study was not based on any empirical research but engaged with stakeholders such as local communities and policymakers who could help in providing insights from the field. The comparison analysis further contrasts legal frameworks and implementation strategies across countries to derive best practices. Synthesizing these findings, the study proposed actionable policy recommendations that align IP rights with biodiversity conservation and sustainable development goals. This integrative methodology thus allows for a robust examination of the interplay between TK and biodiversity while advocating for equitable and sustainable solutions to the management of biodiversity.

TRADITIONAL KNOWLEDGE AND BIODIVERSITY CONSERVATION

TK and biodiversity conservation interlink complex challenges, traditional knowledge, IPRs, and biodiversity conservation. The most critical challenge in this area probably lies in the fact that there is no harmonized legal framework that could adequately support the needs of TK holders, which are mostly indigenous or local communities. International agreements, such as the CBD (Convention on Biological Diversity, 1992) and its Nagoya Protocol (Joseph, 2010), have established equitable benefit sharing of TK, yet the practice differs considerably between jurisdictions. This inconsistency has led to a highly fragmented legal landscape that creates a problem for communities as they navigate overlapping systems. Lacking uniformity, much more exploitation occurs whereby opportunistic entities exploit the loopholes in laws to misappropriate TK or circumvent the obligations of benefit-sharing. These gaps must therefore be bridged through the development of coherent global frameworks that respect the cultural uniqueness of TK systems and integrate local realities. The crucial issue is that there is an inherent incompatibility between the communitarian nature of TK and the individualistic approach maintained by traditional IPR regimes. TK is always collective and develops over generations as a communal cultural property. On the other hand, conventional IPRs- patents and copyrights- have been framed to favor individual creativity and confer time-bound monopolies. These differences create doctrinal and practical problems in using IP tools for the protection of TK. For example, it becomes hard to determine legitimate claimants for common knowledge or practices, especially when that knowledge cuts across several geographical or ethnic areas.

In this regard, the finite nature of IPRs often conflicts directly with the intergenerational durability of TK and thus calls for sui generis systems of law. Intended for the nature of TK, sui generis models can, on paper, work better in satisfying the requirements of TK in terms of attaining suitable protection; but to actually make these systems functional and meaningful, there must be an international acceptance of such. One of the biggest difficulties in gaining fair protection of TK is still the lack of global acceptability of sui generis models (Kothari, 1999). What arises next are ethical and doctrinal questions regarding the commercialization of TK. While one can expect from IPR frameworks that economic incentive to innovation is provided, a system that puts TK up for grabs commodifies cultural heritage in ways the indigenous cultures would consider inconsistent with practices other than perhaps some indirect value added as wealth and in the marketplace. Often, that sacred or sensitive knowledge misappropriated remains unprotected as commercial exploitation of TK prevails.

Therefore, the ethical aspects of TK protection need community consent and cultural integrity to guide the administration of legal systems. In this respect, the customary laws and practices of the communities holding TK should be brought within the statutory framework. This reinforces the legitimacy of legal mechanisms: The recognition of customary norms brings legal mechanisms to the values and aspirations of the TK holders. Access and Benefit Sharing mechanisms, therefore, open up dealing with such challenges under the CBD (Convention on Biological Diversity, 1992) and the Nagoya Protocol (Joseph, 2010) by ensuring equitable sharing of benefits arising from the use of genetic resources and associated TK. Yet problems afflict many while initiating ABS mechanisms. The old trouble is the imbalance of power negotiating between indigenous communities and extra-player transnational corporations. It is but natural that the natives always lag at such venues for want of legal and technical competence and hence enter into poor terms most of which are averse to the holder of TK. Thus, it would be institutional empowerment and support for the holder of TK (Bhat, 2013). The institutionalization of community participation in ABS negotiations is the case of Biodiversity Management Committees constituted under India's Biological Diversity Act 2002. However, these committees can only be meaningful if they receive adequate financial support, have capacity development, and acquire legal empowering. Otherwise, ABS frameworks may, in practice, end up exacerbating existing inequalities rather than redressing them. Another problem also comes up at the implementation level. Even within jurisdictions that do have legal provision to protect TK, the enforcement mechanism is weak or at best underdeveloped. Patenting of neem and turmeric without permission is merely one example of how such a weak system of enforcement could be exploited for TK (Gadgil, 1991). Another considerable aspect is Agreement on Trade-Related Aspects of Intellectual Property Rights (TRIPS): This agreement, part of the World Trade Organization (WTO) framework, sets minimum standards for intellectual property protection, including patents. The TRIPs Agreement impacts several principles of the Convention on Biological Diversity (CBD) in the following ways:

(a) It requires developing countries to change their intellectual property laws to align more closely with those of industrialized nations.

(b) It calls for broad patent protection, ensuring patents are available for inventions whether they are products or processes in all fields of technology. This includes extending patent protection to certain life forms that were previously not patentable in most countries.
(c) When it comes to plant varieties, the TRIPs Agreement allows protection to be provided either through patents, a specialized sui generis system, or a combination of both.

In response to these international treaties, many countries such as Costa Rica, India, Mexico, Peru, and the Philippines are working on new laws or other measures to balance intellectual property rights (IPRs) with biodiversity concerns. International cooperation and monitoring systems must be there to trace the use of genetic resources and TK to counter this problem. Evidence of prior art in the form of comprehensive databases such as India's Traditional Knowledge Digital Library (CSIR, 2002) can be helpful here in this regard. Those databases are a barrier against wrongful grant of patents but also should be supplemented by other proactive measures so that holders of traditional knowledge benefit when their knowledge is commercially applied (Ghosh, 2006). Noteworthy in this regard is the role played by Intellectual Property Asset Management or the IPAM (WIPO, 2002) in dealing with challenges imposed by doctrines. Thus, IPAM (WIPO, 2002) represents a strategic utilization of IPs towards conservation or economic development. It can support TK in the process of structuring communal knowledge as an asset and shape the management. For instance, the development of portfolios of IPs concerning innovations based on TK will strengthen the bargaining hand of a community in ABS agreements. In so doing, however, it does require careful implementation to avoid commodifying TK or diminishing its cultural values. The IPAM (WIPO, 2002) strategies have to converge at the point of equitable benefit distribution, transparency in decision-making, and actual participation by holders of TK (Khor, 2002). Another dimension extends it towards an inter-generational equity question-the ethical considerations of protecting TK are also because indigenous communities owning TK, are actually trustees providing preservation for subsequent generations.

This has to be supported by a legal framework that is able to provide long-term protection, reflecting the long-term nature of TK. Because of their finite character, such systems are poorly suited to this task. Indefinite protection that could improve a sui generis system. The benefits sharing arrangement has to incorporate elements that require community participation in the establishment of projects for community development. Such activities will be developed over time to build capacities for the sustainability and usage of TK (May, 2006). In fact, these developments have added complexities to the doctrinal plane through globalization and new technologies. Digital diffusion of TK thus provides a mass scale recognition opportunity but poses risks in terms of the probable misuse. Open access databases and digital platforms, open access database and digital platforms should strictly adhere to the protocols regarding the appropriate use of TK under custodians' consent and in international legal frames, awareness regarding the cross-border issues wherein the case of misuse of TK is going to arise in jurisdictions wherein protections for TK are weak or nonexistent. Increased cooperation among nations will strengthen the nurturing of a global environment that would enable equitable protection of TK. At their deepest, the issues in TK, biodiversity conservation, and IPRs indicate deeper tensions between economic, cultural, and ecological priorities, which legal frameworks therefore must navigate in fostering synergies rather than conflict (Ostrom, 1990). Hence, a shift of perspective from the market-centric view to a more holistic view respecting the intrinsic value of TK is sought. Then, systems may be designed so that TK holders have the empowerment to enhance global biodiversity conservation efforts along with their legal protections such that they be aligned toward equity, sustainability, and cultural respect. The entire effort would be successful if all parties, including indigenous people, governments, international organisations, and corporate institutions, actively participated in the development and implementation of legal and policy frameworks (Mishra, 2024).

INTERSECTION OF TRADITIONAL KNOWLEDGE AND INTELLECTUAL PROPERTY RIGHTS

Taking about the history and origin of the intersection of the biodiversity vis-a vis Intellectual Property Rights, for centuries, human communities have depended on biodiversity for their food, medicine, and cultural life, their knowledge being freely shared across generations. Indigenous communities and traditional farming populations created diverse crop varieties and healing medicines through collective knowledge and not through individual property (Allison, 2023). But the industrial revolution and globalization saw a change of attitude—resources that were once considered part of our common heritage turned into commodities, and arguments over ownership and rights ensued. Intellectual property laws, which were first conceived for inventions and mechanical innovations, started to find their place in biological materials in the 20th century. 1930 U.S. Plant Patent Act was one of the earliest cases to subject forms of life to IPR protection and give exclusive control over new plant varieties bred by someone. The movement picked up speed with the establishment of the Union for the Protection of New Varieties of Plants (UPOV) in 1961, further solidifying breeders' rights. The 1980 U.S. Supreme Court decision in Diamond v. Chakrabarty which opened the door to the patenting of genetically engineered microorganisms established a precedent for more expansive claims over genetic resources (Amani, 2024).

Alongside globalization, the 1994 Trade-Related Aspects of Intellectual Property Rights (TRIPS) Agreement in the World Trade Organization (WTO) obliged nations to implement robust IPR protections, which extended to biological innovations. Coincidentally, the fear of biopiracy and pilferage of traditional knowledge also prompted the signature of the Convention on Biological Diversity (CBD) in 1992. It acknowledged sovereign rights over biodiversity and aimed to institute equitable benefit-sharing with local peoples (Dutfield, 2000). The constant tug of war between TRIPS and the CBD is an indication of the difficulty of reconciling commercial goals with ecological and ethical concerns. Such instances as the patenting of turmeric, neem, and basmati rice bring into question the need for systems that accord due respect to indigenous contributions while encouraging innovation. Therefore, the controversy surrounding biodiversity and IPR is not merely legal but an issue of justice, conservation, and sustainable development (Gepts, 2004).

Traditional knowledge, IPRs, and biodiversity conservation do interact in the context of complex challenges. One of the most significant challenges here may probably be that no harmonized legal frameworks exist that could effectively support the needs of TK holders- mainly indigenous or local communities. International agreements, for example, the CBD (Convention on Biological Diversity, 1992) and its Nagoya Protocol (Joseph, 2010), provide for equitable benefit sharing of TK, yet implementation largely differs between jurisdictions. This inconsistency has resulted in a very fragmented legal landscape, creating a problem for communities as they navigate overlapping systems. Lacking uniformity, much more exploitation occurs whereby opportunistic entities exploit the loopholes in laws to misappropriate TK or circumvent the obligations of benefit-sharing. These gaps must therefore be bridged through the development of coherent global frameworks that respect the cultural uniqueness of TK systems and integrate local realities. An important issue pertains to the inherent incompatibility between the communitarian nature of TK and the individualistic approach that traditional IPR regimes maintain. TK is often collective and evolves over generations as a communal cultural asset (Gupta, 2002). On the other hand, conventional IPRs, like patents and copyrights, are designed to favor individual creativity and confer time-limited monopolies. These differences pose doctrinal and practical difficulties in using IP tools to protect TK. It may be hard, for instance, to determine legitimate claimants for common knowledge or practices, especially when such knowledge spans multiple geographic or ethnic areas.

Moreover, the finite nature of IPRs frequently clashes with the intergenerational durability of TK, thereby requiring alternative legal frameworks such as sui generis systems. Designed to fit the nature of TK, sui generis models can better accommodate the needs of TK for adequate protection; however, for these systems to function meaningfully, international acceptance must exist (Kamau, 2010). The failure to achieve worldwide acceptability of sui generis models is a key obstacle in securing equitable protection for TK. This also leads to further ethical concerns about the commercialization of TK. Although IPR frameworks are expected to provide economic incentives for innovation, its application to TK risks commodifying cultural heritage. For many indigenous communities, TK is closely interlocked with spiritual and social practices that have purposes other than economic value (Jain, 1991). Commercial exploitation of TK without proper safeguards can easily denude the cultural significance attached to it and result in misappropriation of sacred or sensitive knowledge. Ethical aspects of TK protection thus call for the primacy of community consent and cultural integrity in the dispensation of legal systems. Under such circumstances, customary laws and practices of the communities holding TK should be included within the statutory framework. It strengthens legitimacy for legal mechanisms: By recognizing customary norms, legal mechanisms are integrated to the values and aspirations of TK holders (Mooney, 1996).

ABS mechanisms, therefore, open up dealing with such challenges under the CBD (Convention on Biological Diversity, 1992) and the Nagoya Protocol (Joseph, 2010) by ensuring equitable sharing of benefits arising from the use of genetic resources and associated TK. Yet problems afflict many while initiating ABS mechanisms (Ncube, 2024). The old trouble is the imbalance of power negotiating between indigenous communities and extra-player transnational corporations. It stands to reason that the natives always lag at such venues behind for lack of legal and technical competence and thus make poor terms most which are unfavorable to the holder of TK. Thus, it would be institutional empowerment and support for the holder of TK. Biodiversity Management Committees, as constituted under the Biological Diversity Act, 2002 of India, are an example of institutionalizing community participation in ABS negotiations. However, for these committees to be effective, there should be adequate funding, capacity building, and legal empowerment. Otherwise, ABS frameworks run the risk of reinforcing already existing inequities instead of redressing them (Posey, 1990) (Posey & Dutfield, 1996)

Another issue comes up in the area of enforcement. Even under jurisdictions with legal provisions for protection of TK, enforcement mechanisms remain weak or underdeveloped. Cases like patenting of 'neem' and turmeric without permission exemplify how weak enforcement facilitates the exploitation of TK. Monitoring systems and international cooperation are necessary for the tracking of the use of genetic resources and TK to counter this problem (Rao, 2003). Evidence of prior art in the form of comprehensive databases such as India's Traditional Knowledge Digital Library (CSIR, 2002) can be helpful here in this regard. Those databases are a barrier against wrongful grant of patents but also should be supplemented by other proactive measures so that holders of traditional knowledge benefit when their knowledge is commercially applied (Singh, 2000).

THE ROLE OF INTELLECTUAL PROPERTY ASSET MANAGEMENT

Worth mentioning in this regard is the role of Intellectual Property Rights Asset Management or IPAM (WIPO, 2002) in tackling challenges. IPAM (WIPO, 2002), therefore is strategic use of IP for attainment of objectives related to preservation or economic development. For TK, it would help structure the management of communal knowledge as an asset. For instance, developing portfolios of IP for innovations based on TK can enhance the bargaining hand of a community in ABS agreements. In so doing, though, this requires careful implementation to avoid commodifying TK or diminishing its cultural values. IPAM (WIPO, 2002) strategies must converge at the point of equitable benefit distribution, transparency in decision-making, and actual participation by holders of TK. From another side, the ethical aspects of protection of TK also stretch toward an intergenerational equity question, as indigenous communities, having ownership of TK, are actually trustees that would provide preservation to subsequent generations. 'IPAM (WIPO, 2002)' refers to systematic identification and structuring of data to transform it into valuable information that contributes to an organization's knowledge base and eventually becomes a critical asset. 'IP Assets' consist of various forms of intellectual property such as patents, trademarks, copyrights, industrial designs, and geographical indications, which are strategically selected for business value. Effective management enhances the value of these assets, thereby generating wealth (Rose, 2003). Thus, there are many examples of 20th-century organizations using their IP in a more effective way to facilitate financial growth. For example, IBM has shown the value of sound IP management by securing about $1.8 billion per year in royalties from its arsenal of more than 25,000 patents.

IPAM (WIPO, 2002) is relevant to increasing intangible assets and their effects on strategic markets and shareholder value. It helps both larger corporations and small and medium-sized enterprises (SMEs) and also universities. For instance, universities in Canada and the United States finance their budgets and support education and research initiatives by licensing their patented inventions. For example, Stanford University collected $49.5 million in gross royalties in the fiscal year of 2004-2005 by licensing around 1,100 inventions globally (Singh, 2000). Similarly, developing countries have started realizing the scope for indigenous innovation and accumulation of IP assets through technology transfer agreements, joint ventures, and regional market expansions. Mergers and acquisitions also increase corporate value because different IP assets are put into one entity, thereby enhancing the actual and perceived worth in the market. WIPO reports that IPAM (WIPO, 2002) activities include among others, IP assessment and planning, which include IP audit, strategic IP plan, clustering of target areas, human capital development, incentive structure, policies to address brain drain, and initiatives to support SMEs and research institutions. Regional cooperation and market integration further strengthen IPAM (WIPO, 2002) efforts (Subramanian, 2010).

A critical first step in IP management is conducting an 'IP audit' which is a systematic review of all intellectual assets owned, used, or acquired by a company. The objective of an IP audit is to inventory such assets, identify underutilized IP, address potential threats, and devise strategies to enhance the company's market position. For instance, an example of how effective an IP regime could be was when in 1995, the 'Council for Scientific and Industrial Research (CSIR)' of India defeated and revoked a U.S. patent for the medicinal use of turmeric. CSIR provided over 30 historical references in Sanskrit, Urdu, and Hindi which stated that Indians already knew the healing properties of turmeric long ago. The Patent and Trademark Office would cancel the patent for novelty (Swanson, 1995). This landmark case led to the establishment of the 'Traditional Knowledge Digital Library (CSIR, 2002) (TKDL)' and the inclusion of traditional knowledge in the 'International Patent Classification System'. It also inspired amendments to India's Patent Act of 1970 and the enactment of the Geographical Indications (Registration of Goods) Act, 1999. Similar cases, such as the Basmati case, demonstrate the importance of careful IP management in protecting traditional knowledge and increasing global awareness of indigenous intellectual property (Wynberg, 2009).

THE ROLE OF LEGAL FRAMEWORK

The role has to be supported by the legal framework, providing long-term protection, thus reflecting the long-term nature of TK. The conventional IP systems are not very well designed for this because of their finite nature. Indefinite protections may, on the other hand, be more satisfactory with sui generis systems. Further, the benefit-sharing agreement has to incorporate provisions involving the community in the establishment of community development initiatives to build future capacities for the sustainability and utilization of TK (Agrawal, 1995). The United Nations Declaration on the Rights of Indigenous Peoples (UNDRIP) confirms indigenous peoples' rights to preserve, control, and protect their traditional knowledge, cultural heritage, and biological resources. This is in line with the Convention on Biological Diversity (CBD) in promoting equitable benefit-sharing and prior informed consent in genetic resource access. Nevertheless, the tensions between TRIPS and these guidelines are evidence of ongoing challenges to ensure intellectual property rights do not violate indigenous sovereignty over their knowledge regimes (Zhang, 2017).

There has been an added layer of complexity with globalization and new technologies in the plane. The diffusion of TK digitally provides opportunities for wide-scale recognition but adds risks to possible misappropriation. Open access databases and digital platforms should adhere strictly to protocols in ensuring proper use of TK after obtaining custodians' consent. On the other hand, international legal frameworks need to be aware of cross-border issues involving the misuse of TK in jurisdictions where such protections are either weak or nonexistent. Increased cooperation among nations can help foster a global environment that accommodates the equitable protection of TK (Zoll, 2003). Ultimately, the issues in TK, biodiversity conservation, and IPRs reveal deeper tensions between economic, cultural, and ecological priorities. Such tensions must be therefore navigated by legal frameworks within fostering synergies rather than conflict. Hence, a shift of view from the market-centric towards a more holistic one, recognizing the intrinsic value of TK is necessary. Systems can be accordingly crafted to empower TK holders to further global biodiversity conservation efforts along with their legal protections so that these are aligned toward equity, sustainability, and cultural respect. Success would come only through active participation of all stakeholders like the indigenous communities, governments, the international organizations, and private entities involved, in shaping and implementation processes of legal and policy frameworks (Laird, 2018).

Understanding the Gap Between the Protection Over Traditional Knowledge and Its Access Within the Aboriginal Communities: A Case Study of Kolli Hills, Tamil Nadu, India

An empirical study was conducted in the Kolli Hills of Tamil Nadu to assess the gap in awareness among indigenous communities regarding the provisions of Access and Benefit Sharing (ABS) under the Biological Diversity Act, 2002. Kolli Hills was chosen as the study area because of its rich repository of Traditional Ecological Knowledge (TEK), particularly in sustainable agriculture and medicinal plant usage. The region is inhabited by the Malayali tribes. It has a long history of farming traditional millet varieties and, thus, indigenous medicinal knowledge related to healthcare. This unique knowledge system that has been passed down through generations plays an essential role in the conservation of biodiversity and sustainable development in this region. However, the knowledge about IP laws in these communities is relatively scarce, making the protection and fair sharing of knowledge by these communities challenging.

A questionnaire was prepared and translated into Tamil to make it better understandable, and responses were recorded to gauge their awareness level. A sample size of 72 respondents was chosen. The respondents were males as the women were excluded traditionally from the Village Knowledge Committees, which administer and transmit TEK in the area. The youngest respondent was 28 years old (Manikandan), and the oldest was 86 years old (Ramasamy). These were local committee members who were responsible for maintaining and managing the local traditional knowledge of agriculture and medical practice.

Currently, there are 13 such committees serving 95 villages in Kolli Hills, benefiting approximately 36,468 people. Considering the vast geographical area and limited time and resources, only three committees—numbers 3, 4, and 8 were considered for the study. The findings try to shed the light on the lacuna of awareness about IP laws and the necessity of targeted interventions to empower communities to protect their traditional knowledge through legal mechanisms. Since all the respondents belonged to rural areas, all interviews and responses to questionnaires were done in Tamil so that there is clarity and engagement. In a conversation with Ramasamy, a farmer, and advisor to the No. 3 Village Knowledge Committee in Kolli Hills, he recalled how their TEK has transformed their lives. He described it as a gift from their ancestors that enabled them to thrive in the otherwise challenging terrain of the region. According to him, the traditional irrigation practices and medicinal plant knowledge not only supported agriculture but also fulfilled household needs such as cooking, maintaining cattle, and addressing minor health issues with natural remedies. He pointed out how the water channels they conventionally managed led to agricultural productivity, household utility, and ecological sustainability. Flora and plants flourished in regions irrigated with the water channels, and also attracted diverse species of birds and animals that relied upon the water sources. According to Ramasamy, formerly, there was an era when collections of medicinal flora and agricultural exploitation were rampant that jeopardized resource sustainability. However, the Village Committees have put limitations over these practices for preserving biodiversity that is specifically found in Kolli Hills. TEK also has its economic benefits. According to Ramasamy, traditional millet varieties farmed using this knowledge have ensured food security and made financial stability for a family so that it can sell surplus harvests in local markets. He remembered how the collaborative management of resources created harmony among the communities in Kolli Hills, brought them together to work for the greater good. He regretted, however, that the youth do not find agriculture interesting anymore. Most of the youth have moved to urban centers in search of employment, while the middle-aged and elderly members find it challenging to maintain and fix the traditional water channels and irrigation systems. This trend, he feared, imperils the future of TEK.

A number of respondents expressed similar concerns. Ramasamy also noted that even though Kolli Hills has great biodiversity and the indigenous people possess traditional knowledge, the government has not been responsive to the local communities. The agricultural extension office, which was once functional, is now a derelict, and local issues are seldom brought to resolution in a timely manner. He showed that he was angered by other government-led projects like constructing dams and irrigation channels that fail. According to him, he said they don't consult local communities nor do they make provision for their region's own specific circumstances, such as the migration of water streams. After having his questionnaire, he said that he did not know much about BDA, 2002 nor how it might be applicable to his locality. He emphasized that all their appeals to the administration for a permanent concrete irrigation system to replace the present bamboo and stone infrastructures had fallen on deaf ears. Although the locals thought that a concrete system would lessen the annual repair effort, Bipin, an expert from a local biodiversity NGO who joined the researcher, disagreed with this. According to Bipin, concretization may seem useful in terms of reducing the seepage of water, but it would be detrimental to biodiversity. The water's natural seepage supports a range of flora and fauna, which could perish without it.

The researcher observed a strong sense of unity among the people during their stay, often seeing villagers gather around bonfires at night to discuss community issues. Meetings to allocate water resources were held regularly, but logistical challenges prevented attending one. This sense community revealed a profound relationship with the understanding of TEK, seemingly insensitive to the general implications associated with biodiversity conservation and the development and promotion of biological resources. Community campaigns by State Biodiversity Boards and Biodiversity Management Committees must be brought into existence for educating society of sustainable practice and the implication and incorporation of TEK in a productive manner. The study highlights the importance of engaging local communities in resource management, increasing their awareness of IP laws, and bridging the gap between government policies and indigenous knowledge systems. The integration of TEK into biodiversity management strategies could offer valuable lessons for sustainable development in Kolli Hills and beyond.

Graphical Representation

Figure 1. Community awareness and engagement with biodiversity

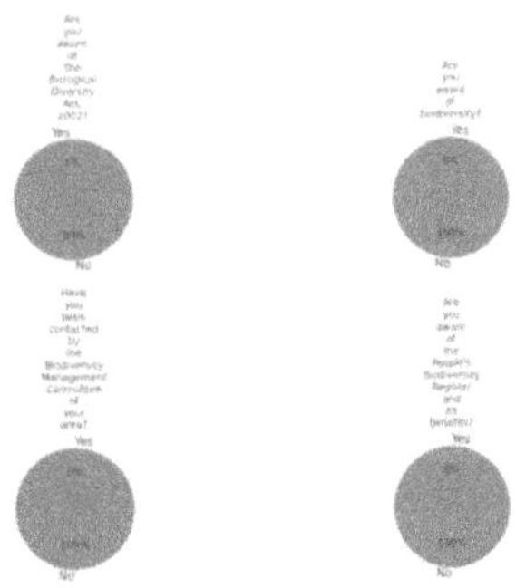

(Tasnim & Saurav, 2025)

The pie charts paint a dismal picture of community awareness and engagement with biodiversity conservation and the legal frameworks surrounding it, such as the Biological Diversity Act, 2002, and related mechanisms like BMCs and PBRs. These findings are critical to understanding the gap between legislative intent and on-ground implementation in the context of biodiversity conservation and the protection of traditional knowledge (TK).

Key Observations

Awareness of the Biological Diversity Act, 2002- Only 1% of the population of the surveyed respondents know about the Biological Diversity Act, 2002. This shows nearly total lack of awareness for a critically important legal instrument meant for conserving biodiversity and for protecting TK, that is essentially a part of the act. The Act focuses upon sustainable use and equitable sharing of biological resources; however, without community awareness, its prospects are largely unexplored.

Public Awareness of Biodiversity- Alarmingly, 100% of respondents were not aware of biodiversity. This is a rather significant educational and informational gap regarding the most essential environmental concepts. Without understanding biodiversity, communities cannot appreciate its role in sustaining ecosystems or know the importance of conserving traditional knowledge systems dependent on these ecosystems.

Interaction with Biodiversity Management Committees- No respondent reported having been approached by a BMC. BMCs are local-level institutions under the Biological Diversity Act that involve communities in biodiversity management and ensure equitable benefit-sharing. The lack of outreach indicates that where these committees' function, they might not be reaching out or may be idle in many regions. This calls into question the on-ground functioning of policies aimed at decentralization of biodiversity governance.

Awareness of the People's Biodiversity Register (PBR) and Its Gains- No respondent has knowledge of the PBR. The PBR aims at recording knowledge about biological resources and their uses, mainly among indigenous and local communities. Lack of awareness on PBR is an opportunity to save and promote TK as part and parcel of biodiversity conservation.

The study indicates the existence of a huge knowledge gap between the policy structures and the local people who are core custodians of TK. This gap affects the potential protection of TK in the context of the IPR framework and challenges the purposes served by biodiversity conservation legislation. The lack of awareness and action leads to loss of Traditional Knowledge. The lack of knowledge on legal mechanisms makes communities prone to bio-piracy, in which multinational corporations take advantage of TK without adequate benefit sharing. This normally results in loss of ownership and recognition by indigenous knowledge holders. Another Implication would be ineffective implementation of Policy. The lack of community participation undermines the effectiveness of the Biological Diversity Act and its related mechanisms. Community participation is essential in developing robust databases such as PBRs, which form the basis of IPR claims and equitable benefit-sharing arrangements.

Last but not the least Thwarted Sustainable Development is major concern as biodiversity conservation demands collective action and local stewardship. In the absence of community awareness, sustainable development goals remain unattained, affecting both ecological balance and socio-economic well-being. However, these gaps can be bridged by following targeted interventions such as Education Initiatives, Strengthening BMCs, and Popularize PBRs. Educating people concerning biodiversity, the Biological Diversity Act and the mechanisms associated with it will help them understand the consequences attached therein. These can be done by organizing workshops/seminars and carrying out multimedia campaigns in local languages. BMCs need to be empowered and made accountable to make the local communities function effectively. Develop PBRs in easy steps; involve them actively and impress communities of the benefits of preserving TK. Integration with IPR Education and highlight the role of biodiversity and TK in the global IPR regime, showing communities the potential for recognition and economic benefits. All things considered, the data points toward a critical need for implementers of policy to put more emphasis on grassroots-level awareness and engagement. Empowering communities with knowledge and tools to participate in biodiversity governance will transform them into active stakeholders in conservation and traditional knowledge protection, aligning with broader objectives for sustainable development and intellectual property rights frameworks.

FUTURE RESEARCH DIRECTIONS

- ***Enhancing Indigenous and Local Community Rights*** – There is a call for legal arrangements that specifically safeguard the rights of indigenous and local communities to their traditional knowledge and biological resources. This entails prior informed consent and equitable benefit-sharing within national and international settings.
- ***Reforming Intellectual Property Regimes*** – The TRIPS Agreement needs to be modified so that it affords more flexibility for nations to exclude life forms from patentability and implement sui generis systems that ensure conservation and equitable access to knowledge on biodiversity.
- ***Harmonizing TRIPS with CBD and UNDRIP*** – An effort should be made to harmonize intellectual property rights with biodiversity conservation and indigenous peoples' rights so that the goals of the Convention on Biological Diversity (CBD) and the United Nations Declaration on the Rights of Indigenous Peoples (UNDRIP) are not compromised.
- ***Developing Alternative Protection Models*** – Community-based intellectual property regimes, traditional resource rights, and defensive patenting strategies should be explored to safeguard biodiversity and prevent biopiracy. Mechanisms such as geographical indications and customary laws can be integrated into formal legal systems.
- ***Guaranteeing Transparent and Inclusive Decision-Making*** – International negotiations concerning biodiversity and intellectual property must be more transparent with meaningful participation of indigenous communities, local stakeholders, and civil society organizations to ensure equitable policy results.
- ***Enhancing Compliance and Enforcement*** – While the WTO enforces TRIPS in the form of trade sanctions, there is no corresponding enforcement process for biodiversity. Developing binding legal commitments for benefit-sharing agreement compliance and indigenous people's rights can balance the powers.
- ***Encouraging Public Awareness and Capacity Building*** – Creating awareness among policymakers, researchers, and local communities regarding the effects of IPR on biodiversity and indigenous knowledge is critical. Capacity-building must target empowering indigenous peoples to negotiate fair agreements.
- ***Utilizing Global Coalitions and Advocacy*** – Developing nations, indigenous groups, and environmental organizations should work together at the international level to advocate for greater legal protections, challenge unjust patents, and promote reforms in global trade and intellectual property policies.

CONCLUSION AND RECOMMENDATIONS

Traditional knowledge forms the keystone of biodiversity conservation because time-tested practices and models for sustainable use are highly consistent with ecological preservation. Erosion of TK, coupled with inadequacy in legal protection, however, poses serious challenges not only to cultural heritage but to environmental sustainability. This paper considers the important intersection of IP rights and biodiversity conservation, pressing home the urgency of mainstreaming TK within a structured and equitable framework. One of the core findings is the inadequacy of the international and national mechanisms to deal with the TK complexities in protection. Instruments such as the CBD (Convention on Biological Diversity, 1992) and the Nagoya Protocol (Joseph, 2010), though progressive, find themselves wanting because of loose definitions, minimum community engagement, and not enough or effective enforcement mechanism.

At the national level, India's Biological Diversity Act of 2002 is a very good model but turns out to be an issue at the grassroots where it fails to empower the Biodiversity Management Committees and activate the People's Biodiversity Registers. The study shows biopiracy as a serious threat to the TK, exploiting this for commercial gains without adequate benefit-sharing. The high-profile cases demonstrate exploitation of indigenous knowledge, and there is a need for robust legal frameworks to redress such injustices. At the same time, examples of successful community-driven models underline the potential for equitable benefit-sharing and the role of TK in promoting sustainable practices. A multipronged approach must be made for the harmonization of IP rights and biodiversity conservation. Building up community awareness and participation must be the bedrock of an effective conservation strategy. The biodiversity custodians-the local communities-must be empowered through programs that enhance their capacities, legal literacy, and their involvement in decision-making. This can be achieved through infusing traditional knowledge systems into national and international IP frameworks. This means a redefinition of patent laws to accommodate TK as collective property rather than individual property. Coordination between governments, NGOs, and other international organizations may advance collaborative policymaking in regard to TK holders' rights and remunerations. More broadly, the relevance of this research is for a biocultural heritage approach: an approach which upholds the inseparability of culture and biodiversity. That is not only safe-keeping traditional practices but, conversely, contributes towards a richer biodiversity conservation due to indigenous wisdom. Achieving a balance between IP rights, biodiversity conservation, and TK protection is a challenge both legally and morally. It is against this backdrop that law, policy, and grassroots action are fashioned synergistically to meet the objectives of environmental sustainability and socio-economic equity. By pushing for a future in which it not only saves but promotes traditional knowledge as the fulcrum of biodiversity conservation, this study adds to growing literature on sustainable development.

REFERENCES

Agrawal, A. (1995). Indigenous and scientific knowledge: Some critical comments. *Indigenous Knowledge and Development Monitor*, *3*(3), 3–6.

Allison-Cassin, S., Bateman, R., Bhardwaj, R. K., & Chisita, C. T. (2023). Introduction. *The Serials Librarian*, *84*(5–8), 62–65. DOI: 10.1080/0361526X.2024.2313400

Amani, B., Ncube, C. B., & Rimmer, M. (2024). Introduction: 'the people's agenda': a history of intellectual property and sustainable development. In *The Elgar Companion to Intellectual Property and the Sustainable Development Goals* (pp. 1–36). Edward Elgar Publishing. DOI: 10.4337/9781803925233.00007

Kothari, A., & Anuradha, R. V. (1999). Biodiversity and Intellectual Property Rights: Can the Two Co-Exist? *J. Int'l Wildlife L. & Pol'y*, *2*, 252.

Bhat, S. (2013). *Natural Resources Conservation Law*. Eastern Book Company.

Chisita, C. T. (2024). The Library Amid an Epistemic Storm: Research Descriptions on Indigenous Knowledge Services. In Evaluating Indigenous African Tradition for Cultural Reconstruction and Mind Decolonization (pp. 26-44). IGI Global.

Convention on Biological Diversity. (1992). *Text of the Convention*. Retrieved from https://www.cbd.int/convention/text/

Dutfield, G. (2000). *Intellectual Property Rights, Trade, and Biodiversity*. Earthscan. DOI: 10.4324/9781849776233

Gadgil, M., & Berkes, F. (1991). Traditional resource management systems. *Resource Management and Optimization*, *8*(3), 127–141.

Gepts, P. (2004). Who owns biodiversity, and how should the owners be compensated? *Plant Physiology*, *134*(4), 1295–1307. DOI: 10.1104/pp.103.038885 PMID: 15084724

Ghosh, P. K. (2006). Biodiversity conservation in India: A legal perspective. *Indian Journal of Environmental Law*, *7*(1), 5–18.

Gupta, A. (2000). *Rewarding traditional knowledge and contemporary grassroots creativity*. Indian Institute of Management.

Gupta, A. K. (2002). Value addition to local resources and institutions. *Electronic Journal of Sustainable Development*, *1*(1), 43–55.

International Union for the Protection of New Varieties of Plants. (1991). *UPOV Convention*. Retrieved from https://www.upov.int/

Jain, S. K. (1991). Ethnobotany and research on medicinal plants in India. *Ethnobotany: Principles and Applications*, 185–203.

Joseph, R. K. (2010). The Nagoya Protocol (Joseph, 2010) on Access and Benefit Sharing: International treaty on genetic resources comes into force. *Current Science*, *100*(10), 1441–1442.

Kamau, E. C., Fedder, B., & Winter, G. (2010). The Nagoya Protocol (Joseph, 2010) on Access to Genetic Resources and Benefit Sharing: What is new and what are the implications for provider and user countries and the scientific community? *Law, Environment & Development Journal*, *6*(3), 246–262.

Khor, M. (2002). *Intellectual Property, Biodiversity, and Sustainable Development*. Zed Books.

Laird, S. A., & Wynberg, R. P. (2018). A fair share? Sharing the benefits of botanical knowledge. *Nature Plants*, *4*(9), 605–609.

May, C. (2006). *The Global Political Economy of Intellectual Property Rights*. Routledge.

Ministry of Environment. (2002). Forest and Climate Change. Biological Diversity Act, 2002. Retrieved from https://www.moef.gov.in/

Mishra, M. P., Joshi, B., Agarwal, P., & Varshney, I. (2024). Sustainable Development Goals–Role Of Indigenous Knowledge. *Educational Administration: Theory and Practice*, *30*(5), 4521–4527.

Mooney, P. (1996). The parts of life: Agricultural biodiversity, indigenous knowledge, and the role of the third system. *Development Dialogue*, *1*(96), 73–94.

Ncube, C. B. (2024). Access to science, technology, and innovation: intellectual property, human rights, and sustainable development. In The Elgar Companion to Intellectual Property and the Sustainable DOI: 10.4337/9781803925233.00030

Ostrom, E. (1990). *Governing the Commons: The Evolution of Institutions for Collective Action*. Cambridge University Press. DOI: 10.1017/CBO9780511807763

Posey, D. A. (1990). Intellectual property rights: And just compensation for indigenous knowledge. *Anthropology Today*, *6*(4), 13–16. DOI: 10.2307/3032735

Posey, D. A., & Dutfield, G. (1996). *Beyond Intellectual Property: Toward Traditional Resource Rights for Indigenous Peoples and Local Communities*. IDRC.

Rao, M. (2003). *Biodiversity and Sustainable Development: Policies and Practices*. National Book Trust.

Rathore, S., & Rai, S. (2024). Interplay of Intellectual Property Rights, Technological Innovation, and Sustainable Food Security: A Comprehensive Analysis. In *E-Financial Strategies for Advancing Sustainable Development: Fostering Financial Inclusion and Alleviating Poverty* (pp. 351–361). Springer Nature Switzerland. DOI: 10.1007/978-3-031-67523-2_21

Rose, C. M. (2003). Property as the keystone right? Some reflections on the IP/biodiversity interface. *Arizona Law Review*, *45*, 445–470.

Shiva, V. (1997). *Biopiracy: The Plunder of Nature and Knowledge*. South End Press.

Singh, R. B. (2000). Indigenous knowledge and sustainable development in the Indian Himalayan Region. *Mountain Research and Development*, *20*(2), 156–160.

Subramanian, S. M., & Pisupati, B. (2010). *Traditional Knowledge in Policy and Practice: Approaches to Development and Human Well-being*. United Nations University Press.

Swanson, T. (1995). The reliance of northern economies on southern biodiversity: Biodiversity as information. *Ecological Economics*, *17*(1), 1–8. DOI: 10.1016/0921-8009(95)00101-8

Tobin, B. (2004). *The Role of Customary Law in Access and Benefit-sharing and Traditional Knowledge Governance: Perspectives from Andean and Pacific Island Countries*. UNU-IAS.

UNDP. (2004). *Biodiversity and Development: A Global Assessment of National Biodiversity Strategies and Action Plans*. Retrieved from https://www.undp.org/

Vermeylen, S. (2007). Contextualizing 'fair' and 'equitable': The San, Hoodia, and benefit-sharing. *Local Environment*, *12*(4), 423–436. DOI: 10.1080/13549830701495252

WIPO. (2002). *Traditional Knowledge and Intellectual Property*. World Intellectual Property Organization. Retrieved from https://www.wipo.int/

World Bank. (2002). *Indigenous Knowledge: Local Pathways to Global Development*. Retrieved from https://www.worldbank.org/

Wynberg, R., & Chennells, R. (2009). Indigenous peoples, consent, and benefit sharing: Lessons from the San-Hoodia case. *Environmental Sciences*, *6*(2), 139–156.

Zhang, Y., & Li, Y. (2017). Traditional ecological knowledge and sustainable development. *Sustainability*, *9*(2), 222.

Zoll, M. (2003). Intellectual property rights and bioprospecting: Policies for conserving biodiversity. *Environmental Policy and Governance*, *13*(5), 367–380.

KEY TERMS AND DEFINITIONS

Biodiversity: The variety of life forms, including species, ecosystems, and genetic diversity, on Earth.

Traditional Knowledge (TK): Indigenous and local knowledge systems developed over generations within a community.

Intellectual Property (IP): Legal rights granting creators protection over their innovations, inventions, or artistic works.

Bio-Piracy: The exploitation of biological resources and traditional knowledge without fair compensation to the indigenous communities.

Nagoya Protocol: A supplementary agreement to the CBD (Convention on Biological Diversity, 1992) focusing on access to genetic resources and fair benefit-sharing.

Sustainable Development: Development that meets present needs without compromising future generations' ability to meet their own needs.

Chapter 23
Indigenous Knowledge and Sustainable Development:
The Indian Story

Aparna Rao Yerramilli
https://orcid.org/0009-0005-0262-9712
GVP College of Engineering, India

ABSTRACT

This chapter explores the profound relationship between Indigenous knowledge and sustainable development in India. It highlights how traditional ecological knowledge, cultural practices, and community-driven initiatives of Indigenous populations contribute to environmental sustainability and socio-economic development. It delves into traditional ecological practices and community-driven initiatives that have sustained the environment and supported socio-economic development for centuries. By examining case studies from diverse regions of India, the chapter illustrates the successful integration of Indigenous knowledge with modern development practices. It also addresses the challenges and opportunities in leveraging Indigenous knowledge for sustainable development and offers policy recommendations to enhance its inclusion in development frameworks. This chapter aims to highlight the invaluable contributions of Indigenous communities to sustainable development and provide insights for policymakers, researchers, and development practitioners.

INTRODUCTION

> "Indigenous knowledge and wisdom are deeply intertwined with nature, offering solutions to sustainable development through localized practices, seed preservation, and ecological balance." – Dr. Vandana Shiva, Environmental Activist and Scholar.
>
> "For sustainable development, we must draw on the wisdom of our indigenous knowledge systems that have coexisted with nature for millennia, ensuring the harmonious survival of both." – Dr. A.P.J. Abdul Kalam, Former Indian President.

Indigenous Knowledge[1] Systems (IKS), also referred to as traditional knowledge (TK), is a body of knowledge entrenched deep within the cultural practices, belief systems, and livelihoods of Indigenous communities. IKS is holistic in nature and takes into consideration, various domains such as agriculture,

DOI: 10.4018/979-8-3693-7964-6.ch023

medicine, weather forecasting, and natural resource management. Since this knowledge is age old, it is transmitted orally and is deeply rooted in rituals, folklore, and every day practices of indigenous communities. Contrary to the modern development practices prioritizing rapid growth and industrialization, Indigenous Knowledge emphasizes balance, conservation, and long-term sustainability. Sustainable development is the development that meets the needs of the present generations without compromising on the ability of future generations to meet their needs. As such, Indigenous Knowledge provides a significant reason for rethinking these modern development models. Today, as the world grapples with the adverse effects of climate change and environmental degradation, there is a emergent the need to revisit and revitalize these traditional practices for holistic existence. India is a country which is endowed with rich cultural heritage. India is also home to diverse Indian Indigenous Knowledge systems, which have been cascaded down through generations through oral transmission. These systems are intricately connected with nature, thus forming the very foundation of sustainable living practices promoting ecological balance, resource conservation, as well as community well-being. There are several Indigenous communities in India, which have practiced sustainable methods in agriculture, water management, healthcare, and biodiversity conservation. The Indian story of Indigenous Knowledge and sustainable development is one of resilience, adaptability, and deep respect for nature.

METHODOLOGY

The study aims to explore how traditional practices and wisdom contribute to sustainability, assess their relevance in modern development paradigms, and propose ways to integrate IK into contemporary sustainable development strategies. The research employs literature reviews to establish a primary understanding of Indigenous Knowledge and its relationship with sustainable development. The following sources were consulted: Academic journals, books, reports, and articles focusing on Indigenous Knowledge systems, sustainable development practices in India, and case studies of specific Indigenous communities.

REVIEW OF LITERATURE

Agrawal (1995) critiques the rigid dichotomy between Indigenous and scientific knowledge systems, arguing that they should be seen as complementary rather than oppositional. He challenges the dominant narrative that portrays Indigenous knowledge as static and unscientific, emphasizing instead its dynamic nature and adaptability. His work is foundational in understanding how knowledge classification affects policy and practice, influencing debates on the integration of Indigenous perspectives into mainstream development and conservation efforts. Agrawal (2002) expands on his earlier arguments by analyzing the political and institutional factors that shape the categorization of Indigenous knowledge. He examines how the classification of knowledge systems is not merely a conceptual exercise but is deeply intertwined with power dynamics, governance structures, and resource management policies. By critically assessing how Indigenous knowledge is often marginalized or selectively appropriated, he provides valuable insights into the broader implications for local development initiatives and environmental sustainability. Altieri and Nicholls (2017) explore agroecology as an interdisciplinary and participatory field, emphasizing the role of Indigenous farming practices in sustainable agriculture. They argue that traditional agricultural knowledge, rooted in centuries of experimentation and adaptation, offers resilient and ecologically sound alternatives to industrial farming. Their work highlights the significance of Indigenous crop diversity, soil management techniques, and water conservation strategies in fostering food security and environmental sustainability.

Berkes (2012, 2018) presents a comprehensive view of traditional ecological knowledge (TEK) and its role in biodiversity conservation and environmental management. His research underscores the value of Indigenous knowledge in ecosystem stewardship, advocating for its integration into mainstream conservation efforts. Berkes (2012) demonstrates how Indigenous communities possess intricate understandings of ecological processes, which can complement scientific approaches to habitat restoration, climate resilience, and resource management. Boelens, Perreault, and Vos (2016) delve into issues of water governance and justice, shedding light on the ways Indigenous water rights are often marginalized despite the ecological wisdom embedded in Indigenous water management practices. They critically examine legal frameworks, policy decisions, and institutional barriers that restrict Indigenous access to water resources. Their work calls for a more equitable and inclusive approach to water governance, one that acknowledges Indigenous contributions and ensures participatory decision-making processes. Ens et al. (2015) focus on Australian Indigenous knowledge systems for land and biodiversity management, showcasing collaborative approaches between Indigenous and non-Indigenous stakeholders. Their study illustrates successful co-management models where Indigenous ecological knowledge is actively integrated into conservation strategies. They highlight case studies where Indigenous-led initiatives have enhanced biodiversity conservation, fire management, and habitat restoration, demonstrating the efficacy of knowledge-sharing partnerships. Ford, Smit, and Wandel (2020) analyze climate change adaptation strategies in Indigenous communities, emphasizing resilience, local knowledge, and policy integration. They explore how Indigenous communities across different geographical regions have developed adaptive strategies to cope with climate variability. Their research highlights the importance of including Indigenous perspectives in climate policies, as these communities possess unique insights into environmental changes and sustainable adaptation methods.

Gadgil, Berkes, and Folke (1993) discuss Indigenous contributions to biodiversity conservation, presenting community-based conservation as a viable alternative to Western conservation approaches. They argue that Indigenous communities, through their deep ecological knowledge and sustainable resource use practices, have played a crucial role in maintaining biodiversity. Their work advocates for a shift away from exclusionary conservation policies towards models that empower Indigenous stewardship. The Government of India (2023) outlines policies aimed at improving tribal development, acknowledging the role of Indigenous knowledge in sustainable development. The policy document recognizes Indigenous contributions in fields such as agriculture, forest management, and healthcare. It also highlights government initiatives that aim to protect Indigenous rights, promote cultural preservation, and ensure economic empowerment while integrating traditional knowledge into national development programs. IPCC (2019) addresses the impact of climate change on land and ecosystems, emphasizing Indigenous contributions to mitigation and adaptation strategies. The report underscores the role of Indigenous knowledge in enhancing climate resilience, particularly in managing forests, agriculture, and water resources. It calls for greater inclusion of Indigenous perspectives in global climate governance and adaptation frameworks, recognizing their deep-rooted connections to the environment.

Jain (1991) compiles ethnobotanical knowledge from Indian communities, providing a valuable resource on traditional medicine and plant-based remedies. His extensive documentation of Indigenous medicinal practices highlights the vast pharmacological potential of traditional plant-based treatments. His work serves as an important reference for researchers, conservationists, and policymakers interested in preserving and integrating Indigenous medical knowledge into modern healthcare systems. Kumar and Kerr (2013) explore the Recognition of Forest Rights Act (FRA) in India, illustrating the legal and social dynamics of Indigenous land rights. Their analysis highlights the struggles of Indigenous communities in asserting their land tenure rights and resisting encroachments by state and corporate actors. They argue that despite the FRA's progressive framework, challenges remain in its implementation, requiring stronger institutional support and community participation to ensure effective enforcement. Nakashima, Krupnik, and Rubis (2012) discuss the role of Indigenous knowledge in climate change assessment and adaptation, demonstrating its integration into global climate policies. Their work emphasizes that Indigenous knowledge systems offer valuable, place-based insights that complement scientific climate models. By advocating for a co-production of knowledge approach, they highlight how Indigenous and scientific knowledge can work together to address climate challenges more effectively.

Schwartzman, Zimmerman, and Kormos (2019) analyze how Indigenous lands and protected areas contribute to climate change mitigation by reducing deforestation and carbon emissions. They provide empirical evidence showing that Indigenous-managed forests and conservation areas are more effective at carbon sequestration than conventional protected areas. Their research strengthens the argument for recognizing Indigenous land tenure as a critical component of global climate mitigation strategies. Tiwari (2013) examines Indigenous knowledge systems in Northeast India, emphasizing their importance in sustainable development and ecological management. His work explores how Indigenous communities in this region have developed sustainable agricultural, forestry, and water management practices suited to their local ecosystems. Tiwari calls for greater policy support and recognition of these knowledge systems to ensure their preservation and continued contribution to environmental sustainability.

INDIGENOUS KNOWLEDGE AND SUSTAINABLE DEVELOPMENT ACROSS THE GLOBE

Sustainable development has emerged as a critical global concern, emphasizing the need for economic growth, environmental conservation, and social inclusion. Indigenous knowledge (IK), which encompasses the traditional practices, beliefs, and wisdom of Indigenous communities, plays a crucial role in advancing sustainable development goals (SDGs). Across the globe, Indigenous communities have maintained a harmonious relationship with nature, utilizing their knowledge systems to manage natural resources sustainably. IK systems are diverse, reflecting the ecological and cultural diversity of Indigenous groups worldwide. They encompass various domains, including agriculture, water management, medicine, and biodiversity conservation. For example, Indigenous agricultural practices, such as the Andean terrace farming and the Zai pit system in Africa, have demonstrated remarkable resilience to climate variability (Altieri & Nicholls, 2017).

Indigenous Knowledge and Environmental Conservation

One of the most significant contributions of Indigenous knowledge to sustainable development is environmental conservation. Indigenous communities possess deep ecological understanding and have developed sustainable resource management strategies that align with ecosystem dynamics. Forests, which play a critical role in carbon sequestration and biodiversity preservation, are often managed sustainably by Indigenous groups. The Kayapo people in Brazil, for example, protect large tracts of the

Amazon rainforest using traditional land management practices (Schwartzman et al., 2019). Similarly, the fire management techniques of Australian Aboriginal communities, such as controlled burning, help prevent catastrophic wildfires while promoting biodiversity (Ens et al., 2015). Moreover, Indigenous water management practices have proven effective in addressing water scarcity and quality issues. The Aymara people of the Andes have developed sophisticated irrigation systems that optimize water use while maintaining soil fertility (Boelens et al., 2016). Such practices highlight the potential of IK to complement modern environmental conservation efforts.

Climate Change Mitigation and Adaptation

Indigenous knowledge plays a crucial role in climate change mitigation and adaptation. Many Indigenous communities possess detailed knowledge of local climate patterns and have developed adaptive strategies to cope with environmental changes (Ford et al., 2020). For example, the Inuit in the Arctic rely on traditional weather forecasting techniques, such as cloud formation and animal behavior observations, to anticipate and respond to climate variability (Weatherhead et al., 2010). Similarly, Pacific Island communities utilize traditional navigation and coastal management strategies to protect against rising sea levels and extreme weather events (Nunn et al., 2017). By integrating Indigenous knowledge with scientific research, policymakers can develop more effective climate adaptation policies. The Intergovernmental Panel on Climate Change (IPCC) has recognized the value of Indigenous knowledge in enhancing climate resilience, particularly in vulnerable regions (IPCC, 2019).

Indigenous Knowledge and Sustainable Agriculture

Sustainable agriculture is another domain where Indigenous knowledge has made significant contributions. Traditional farming systems, such as agroforestry, intercropping, and rotational farming, have demonstrated superior sustainability compared to industrial agricultural methods (Altieri & Nicholls, 2017). For instance, the traditional Mayan milpa system, which combines maize, beans, and squash, enhances soil fertility, reduces pests, and increases crop yields without synthetic fertilizers (Gliessman, 2015). Similarly, rice-fish farming in Southeast Asia, where fish are raised in flooded rice paddies, promotes biodiversity and improves food security (Xie et al., 2019). Such Indigenous agricultural practices align with SDG 2 (Zero Hunger) and SDG 15 (Life on Land) by promoting food security, reducing deforestation, and enhancing soil health. Integrating these practices into modern agricultural policies can contribute to global sustainability efforts.

Challenges and Threats to Indigenous Knowledge

Despite its importance, Indigenous knowledge faces numerous challenges and threats. Globalization, climate change, land dispossession, and cultural assimilation have led to the erosion of traditional knowledge systems (Sillitoe, 2017). The expansion of industrial agriculture, deforestation, and extractive industries further marginalize Indigenous communities and undermine their sustainable practices. Additionally, Indigenous knowledge is often undervalued in mainstream policy-making and scientific discourse. Western development models tend to prioritize technological solutions, overlooking the potential of traditional practices (Agrawal, 2002). Recognizing and integrating Indigenous knowledge into formal education, research, and policy frameworks is essential to preserving its relevance and efficacy.

Towards a More Inclusive Sustainable Development Approach

To fully harness the potential of Indigenous knowledge for sustainable development, a more inclusive and participatory approach is necessary. Governments, international organizations, and researchers should collaborate with Indigenous communities to document, protect, and apply traditional knowledge. Legal frameworks, such as the United Nations Declaration on the Rights of Indigenous Peoples (UNDRIP), emphasize the importance of respecting Indigenous rights, knowledge, and cultural heritage (UN, 2007). Recognizing Indigenous land rights and ensuring their participation in environmental governance can empower Indigenous communities and enhance sustainability efforts. Furthermore, integrating Indigenous knowledge with scientific research through co-management models can lead to innovative solutions. For example, community-based conservation initiatives, such as those in the Amazon and Canada's boreal forests, have demonstrated success in protecting biodiversity while ensuring Indigenous autonomy (Berkes, 2018).

Indigenous knowledge is an invaluable resource for sustainable development across the globe. From environmental conservation and climate resilience to sustainable agriculture and cultural preservation, Indigenous knowledge offers holistic and time-tested solutions to pressing global challenges. However, ensuring its survival and integration into mainstream policies requires collaborative efforts that respect Indigenous rights and promote knowledge exchange. By embracing Indigenous knowledge alongside scientific advancements, the global community can move towards a more sustainable and inclusive future.

COMPARATIVE ANALYSIS OF INDIGENOUS KNOWLEDGE IN SUSTAINABLE DEVELOPMENT: A GLOBAL PERSPECTIVE

Indigenous knowledge, often referred to as Traditional Ecological Knowledge (TEK), has evolved through centuries of interaction between Indigenous communities and their environments (Berkes, 2018). This knowledge is deeply rooted in local ecosystems and encompasses resource management, agricultural practices, and conservation strategies that have proven effective over time. Across different continents, Indigenous communities have demonstrated sustainable ways of living that align with modern environmental and developmental goals. However, the marginalization of Indigenous voices and knowledge systems remains a challenge in global sustainability discourse (Whyte, 2020). This paper compares Indigenous knowledge applications from various regions, focusing on successful integration into modern sustainability initiatives. Indigenous knowledge has been integral to sustainable development across the world, guiding environmental conservation, resource management, and climate adaptation. This paper examines Indigenous knowledge systems in India, Australia, Canada, the Amazonian region, and Africa, highlighting their contributions to sustainability. Case studies include the Bishnoi community's wildlife conservation in India, Aboriginal fire management in Australia, the Inuit's climate adaptation strategies in Canada, Amazonian agroforestry practices, and African pastoralist techniques. The analysis underscores the importance of recognizing, integrating, and protecting Indigenous wisdom in global sustainability efforts while ensuring ethical collaboration and benefit-sharing.

Comparative Case Studies

India: Bishnoi Community and Wildlife Conservation

The Bishnoi community of Rajasthan has a long-standing tradition of environmental conservation rooted in religious and cultural beliefs. Their commitment to protecting trees and wildlife, exemplified by the 1730 Khejarli Massacre, has resulted in Bishnoi villages becoming safe havens for endangered species such as the blackbuck antelope (Sinha, Bhatia, & Vishwakarma, 2018). This conservation ethic aligns with modern wildlife protection strategies and demonstrates how Indigenous values can contribute to biodiversity conservation.

Australia: Aboriginal Fire Management

In Australia, Aboriginal communities have historically used controlled burning, known as "cultural burning" or "firestick farming," to manage landscapes and prevent wildfires (Bowman et al., 2020). This practice involves strategically setting small, low-intensity fires to reduce undergrowth, protect biodiversity, and maintain healthy ecosystems. Recent wildfires in Australia have reignited interest in integrating Aboriginal fire management into national fire prevention strategies (Steffensen, 2020).

Canada: Inuit Climate Adaptation Strategies

The Inuit people of the Arctic have developed intricate climate adaptation strategies based on their extensive understanding of sea ice, weather patterns, and wildlife behavior. Their traditional navigation and hunting techniques are crucial for survival in extreme environments. However, climate change has significantly altered their habitat, leading to collaborative projects between Inuit communities and climate scientists to integrate Indigenous knowledge into climate adaptation policies (Ford et al., 2016).

Amazon: Indigenous Agroforestry Practices

Indigenous communities in the Amazon rainforest, such as the Kayapó and Asháninka, practice agroforestry systems that enhance biodiversity and carbon sequestration (Levis et al., 2017). These communities cultivate diverse plant species in multilayered forest gardens, promoting soil fertility and resilience against deforestation. Unlike industrial agriculture, Indigenous agroforestry aligns with sustainable land management principles, offering valuable insights for combating deforestation and climate change.

Africa: Pastoralist Knowledge and Land Management

African pastoralist communities, including the Maasai in Kenya and Tanzania, have developed sophisticated land management systems that ensure sustainable grazing while maintaining biodiversity (Reid et al., 2014). Their rotational grazing techniques prevent overgrazing and allow ecosystems to regenerate. However, land privatization and climate change pose significant threats to their traditional practices, necessitating policies that recognize and integrate Indigenous pastoralist knowledge into modern land-use planning.

Comparative Analysis and Discussion

Despite regional differences, Indigenous knowledge systems share common themes of environmental stewardship, resilience, and resource efficiency. Table 1 summarizes the key characteristics of these Indigenous practices:

Table 1. Indigenous practices and their characteristics

Region	Indigenous Group	Sustainable Practice	Key Benefits
India	Bishnoi	Wildlife conservation	Biodiversity protection, ecological balance
Australia	Aboriginal Australians	Cultural burning	Fire prevention, ecosystem health
Canada	Inuit	Climate adaptation	Resilience to climate change, sustainable hunting
Amazon	Kayapó, Asháninka	Agroforestry	Biodiversity conservation, carbon sequestration
Africa	Maasai	Pastoralist land management	Sustainable grazing, ecosystem regeneration

The integration of Indigenous knowledge with modern science has shown promising results in sustainability efforts. For example, the collaboration between Aboriginal fire managers and Australian authorities has reduced the severity of wildfires (Steffensen, 2020). Similarly, Indigenous agroforestry in the Amazon has influenced international conservation strategies by demonstrating alternatives to deforestation-driven agriculture (Levis et al., 2017). However, challenges persist in incorporating Indigenous knowledge into mainstream sustainability policies. Many Indigenous communities face land dispossession, exploitation, and a lack of recognition for their intellectual contributions (Whyte, 2020). Ethical frameworks that ensure fair benefit-sharing and intellectual property rights are essential to avoid biopiracy and knowledge appropriation (Ford et al., 2016). Indigenous knowledge systems offer time-tested, sustainable solutions to global environmental challenges. The case studies presented in this paper illustrate how Indigenous practices contribute to biodiversity conservation, climate adaptation, and sustainable resource management. Recognizing and integrating Indigenous wisdom into modern sustainability efforts requires not only scientific collaboration but also policies that respect Indigenous sovereignty and ensure equitable partnerships. Moving forward, fostering dialogue between Indigenous communities and policymakers will be crucial for achieving truly sustainable development.

MAJOR INDIGENOUS COMMUNITIES IN INDIA

India's Indigenous communities are known as Adivasis or Scheduled Tribes[2]. They constitute a significant portion of the country's population. These communities are characterized by their distinct cultural practices, languages, and innate connection to the land. For many centuries, these indigenous communities have lived in absolute harmony with their environment. They have been instrumental in developing many traditional practices and have also contributed to holistic sustainable development. The Indigenous knowledge systems of the Mundas, Todas, Bodos, Gonds, Santhals, Bhils, Mizos, Nagaland tribes, and the Khasis, Jaintias, and Garos offer invaluable insights into sustainable living, resource management, and community governance. These systems are deeply rooted in ecological wisdom and cultural heritage, contributing to biodiversity conservation, resilience in the face of environmental challenges, and the preservation of cultural identity. As the world grapples with issues of sustainability and environmental degradation, there is much to learn from the time-tested practices and philosophies of these Indigenous communities. Recognizing and integrating their knowledge into modern development policies can pave the way for more inclusive and sustainable solutions to global challenges.

The Gonds: The Gonds are one of India's largest Indigenous communities, primarily found in the states of Madhya Pradesh, Chhattisgarh, Maharashtra, and Odisha. Their Indigenous knowledge system is interwoven with their cultural identity and natural surroundings.

a. **Sustainable Agriculture:** The Gonds practice shifting cultivation (slash-and-burn), allowing forest lands to regenerate after a period of use. This system is guided by traditional knowledge regarding soil fertility and crop rotation. They also cultivate a wide range of indigenous crops like millets and pulses, ensuring food security and biodiversity.
b. **Forest Resource Management:** The Gonds have traditional systems for managing forest resources, such as community-controlled harvesting of timber and non-timber forest products. Sacred groves, revered as homes of deities, are preserved and play a significant role in biodiversity conservation.
c. **Traditional Healing Practices:** Gond ethnomedicine involves the use of local herbs and plants for treating common ailments. Traditional healers possess knowledge of various plants' medicinal properties, which are used for holistic healing, addressing both physical and spiritual well-being.

The Santhals: The Santhals, predominantly residing in Jharkhand, Odisha, West Bengal, and Bihar, have a rich Indigenous knowledge system that centers on agriculture, forest resource management, and social customs.

a. **Agricultural Practices:** The Santhals rely on traditional agricultural practices like organic farming and mixed cropping. Their knowledge of seed preservation and pest management without synthetic chemicals supports sustainable farming. They cultivate staple crops such as rice, maize, and millets, using environmentally friendly methods.
b. **Forest and Biodiversity Conservation:** Santhals have a profound relationship with the forests, which provide them with food, medicinal herbs, and materials for livelihood. Their sacred groves, known as *jaher than*, are protected areas where biodiversity thrives. These groves are central to their religious and cultural rituals, emphasizing conservation.
c. **Community Governance and Social Harmony:** Santhal villages have well-organized governance systems that manage community resources and maintain social order. Decisions are made collectively, promoting a sense of unity and cooperation within the community.

The Bhils: The Bhils are one of the oldest tribal groups in India, primarily inhabiting Rajasthan, Gujarat, Madhya Pradesh, and Maharashtra. Their Indigenous knowledge systems are centered around natural resource management, agriculture, and cultural practices.

a. **Traditional Agriculture:** The Bhils practice subsistence farming, focusing on crops like millets, maize, and pulses. They have developed techniques for conserving water in arid regions, such as building check dams and small reservoirs to store rainwater. Their agricultural practices are adaptive to the challenging environments they inhabit.

b. **Forest Resource Utilization:** Bhils are skilled in the sustainable use of forest resources, including the collection of minor forest produce like tendu leaves, honey, and medicinal plants. They have community norms that regulate forest use, preventing overexploitation and ensuring resource availability for future generations.
c. **Ethno-medicine:** Traditional Bhil healers, known as *bhoopas*, have deep knowledge of medicinal plants and their therapeutic uses. Their healing practices are based on holistic approaches that consider the physical, spiritual, and environmental aspects of health.

The Mizos: The Mizos, residing predominantly in Mizoram, have an Indigenous knowledge system that is well-suited to the hilly terrain and heavy rainfall of their region.

a. **Shifting Cultivation (Jhum):** Mizos practice jhum cultivation, which involves clearing small patches of forest for crop cultivation and allowing the land to regenerate after use. This system, guided by traditional ecological knowledge, promotes soil fertility and prevents land degradation. The cycle of jhum is carefully managed to ensure sustainability.
b. **Community Resource Management:** Mizo villages are characterized by strong community governance, where decisions about resource use, land distribution, and forest management are made collectively. This communal approach ensures equitable access to resources and fosters social cohesion.
c. **Traditional Medicine:** Mizos have a rich tradition of using local herbs and plants for treating illnesses. Their knowledge of ethnobotany includes remedies for common ailments, as well as practices for spiritual healing. This knowledge is preserved and passed down within families and communities.

The Nagaland Tribes: Nagaland is a north eastern state in India, which is home to several tribes, including the Angami, Ao, and Konyak, each with unique Indigenous knowledge systems closely tied to their environment and cultural practices.

a. **Terrace Farming:** The tribes of Nagaland have perfected the art of terrace farming, creating stepped fields on hilly terrain to prevent soil erosion and retain water. This method allows them to cultivate rice, maize, and vegetables sustainably, despite the challenging geography.
b. **Forest Conservation:** Nagaland tribes have a tradition of community-owned forests, where rules are strictly enforced to prevent overexploitation. They follow traditional laws and rituals that govern hunting, logging, and resource extraction, helping preserve biodiversity and ecological balance.
c. **Craftsmanship and Cultural Knowledge:** Nagaland tribes are renowned for their craftsmanship in weaving, wood carving, and beadwork. These skills, passed down through generations, are not just economic activities but also a means of preserving cultural identity and transmitting traditional knowledge.

The Khasis, Jaintias, and Garos: The Khasis, Jaintias, and Garos inhabit the state of Meghalaya and are known for their matrilineal society and sustainable practices.

a. **Agriculture and Livelihoods:** The Khasis and Jaintias are skilled in wet rice cultivation, using traditional irrigation systems made of bamboo to channel water to terraced fields. The Garos, living in the hilly regions, practice jhum cultivation, growing diverse crops suited to their environment.

b. **Sacred Groves:** The sacred groves maintained by the Khasis and Jaintias are examples of Indigenous conservation practices. These groves, protected for religious reasons, are untouched by human activity and serve as sanctuaries for numerous plant and animal species.
c. **Community Governance and Social Structure:** The governance of Khasi, Jaintia, and Garo communities is based on traditional councils, known as *dorbar shnongs*. These councils manage land, resources, and social issues through consensus-based decision-making. The matrilineal system, where inheritance and lineage are traced through the female line, is a distinctive feature.

Todas: The Todas are a small pastoral community living in the Nilgiri Hills of Tamil Nadu. Their Indigenous knowledge system is primarily centered on buffalo herding, sustainable land use, and distinct cultural practices.

a. **Pastoralism and Buffalo Herding:** The Toda economy and culture are based on the herding of the indigenous Toda buffalo. This unique breed is central to their livelihood, providing milk, which is used in various rituals and ceremonies. The Todas have in-depth knowledge of buffalo management, grazing patterns, and sustainable ways of using pastures. They practice rotational grazing, allowing the grasslands to regenerate, which maintains the ecological balance of the Nilgiri Hills.
b. **Sacred Landscapes and Conservation:** The Todas consider the grasslands and specific hill areas as sacred, leading to their conservation over centuries. Toda settlements, known as *munds*, are located within these landscapes and consist of huts with a distinctive curved architecture. The sacredness attributed to certain areas has resulted in the preservation of biodiversity in these regions.
c. **Traditional Craftsmanship and Architecture:** The Todas are known for their traditional huts, which are semi-barrel shaped and made from bamboo, rattan, and thatch. The construction of these huts reflects their Indigenous knowledge of locally available materials and the environment. Toda embroidery, characterized by intricate geometric patterns, is another example of their Indigenous craftsmanship.
d. **Rituals and Social Structure:** The Toda community follows a set of rituals linked to their buffalo herding. The Toda religion involves nature worship and has specific rituals for the protection of their herds and the natural environment. The community's social structure is tightly knit, with decision-making processes guided by elders who possess extensive traditional knowledge.

The Bodos: The Bodos are one of the largest Indigenous communities in Assam and have a rich knowledge system related to agriculture, forest management, traditional healing, and cultural practices.

a. **Agriculture and Livelihoods:** The Bodos are primarily agriculturalists, with rice being their staple crop. They have developed traditional methods of rice cultivation that suit the flood-prone and alluvial plains of Assam. They also grow mustard, jute, and pulses. The Bodos use organic farming techniques, including composting and crop rotation, ensuring soil fertility and pest control without relying on chemicals.
b. **Water Management:** In their agricultural practices, the Bodos use traditional irrigation methods, including small canals and water channels that they construct to manage floodwaters. They also have knowledge of the seasonal rhythms of the rivers and use this understanding to plan their planting and harvesting cycles.

c. **Forest Resource Management:** The Bodos have a tradition of sustainable forest management, which involves the collection of medicinal herbs, bamboo, and other forest produce. They practice *jhumming* (shifting cultivation) in forested areas, which is adapted to the local ecology, allowing time for regeneration between cycles.
d. **Traditional Healing and Medicine:** Bodo communities have extensive knowledge of medicinal plants and herbs, which are used to treat common ailments and promote overall health. Traditional healers, known as *ojhas*, use a combination of herbal remedies and spiritual practices in their healing methods.
e. **Cultural Practices and Folklore:** The Bodos have a vibrant culture that includes folk songs, dances, and oral literature. The *Bagurumba* dance is a traditional dance form performed by Bodo women, symbolizing harmony with nature. Their folktales and oral histories are rich in knowledge about local flora, fauna, and environmental conservation.

The Munda Community is one of the prominent Indigenous groups in India, primarily residing in the states of Jharkhand, Odisha, West Bengal, and Bihar. The Munda knowledge systems encompass a wide range of practices and beliefs that are closely linked to their environment, agriculture, forest management, community organization, and cultural heritage. These systems have been passed down through generations and are rooted in the sustainable use of natural resources and strong community governance.

Agriculture and Land Management

- **Traditional Farming Practices:** The Mundas traditionally practice subsistence agriculture, primarily cultivating rice, millets, pulses, and vegetables. They rely on rain-fed farming and have developed techniques for mixed cropping and crop rotation, which help maintain soil fertility and reduce pest infestations. Their knowledge of local crops, seeds, and organic farming practices ensures sustainability and resilience against climate variability.
- **Shifting Cultivation (Jhum):** In certain areas, Mundas practice shifting cultivation, where land is cultivated for a few years and then left fallow to allow for natural regeneration. This cyclical use of land is based on in-depth knowledge of soil conditions, crop cycles, and forest regeneration.

Forest Resource Management

- **Community-Based Forestry:** The Munda community has traditionally relied on forests for their livelihood, gathering fruits, herbs, firewood, and other forest products. They have a community-based approach to forest management, where access and usage rights are regulated by customary laws. Sacred groves, which are small patches of protected forests, are maintained and respected as places of worship, contributing to the preservation of biodiversity.
- **Ethnobotanical Knowledge:** The Mundas possess extensive knowledge of medicinal plants and herbs, which they use for treating common ailments. Traditional healers, known as *ojhas*, play a key role in this system, combining herbal remedies with spiritual practices.

Water Conservation and Management

- **Indigenous Irrigation Techniques**: In areas with limited water resources, the Mundas have developed traditional irrigation systems, such as small earthen dams and canals, to conserve rainwater for agricultural use. They also build *dobhas* (small ponds) to collect rainwater, which is crucial for irrigation and drinking water during dry spells.
- **Water Harvesting Practices:** Their water conservation practices are based on the understanding of local hydrology and rainfall patterns. These techniques are simple, low-cost, and highly effective in maintaining water security for their communities.

Community Governance and Social Organization

- **Khutkatti System:** The Mundas have a unique land ownership system known as the *Khutkatti* system, where land is owned and managed collectively by the community. This system ensures equitable distribution of resources and prevents land alienation to outsiders. Community leaders, often called *Mankis* or *Pahans*, play a key role in managing land, resolving disputes, and performing religious rituals.
- **Adivasi Panchayats:** The traditional governance structure of the Mundas involves *panchayats* (village councils) that are responsible for decision-making, resource management, and social justice. These councils follow customary laws and practices that are designed to ensure harmony within the community and with nature.

Cultural and Spiritual Knowledge

- Sarhul and Karma Festivals: The Mundas celebrate festivals like *Sarhul* (worship of the sacred grove and nature) and *Karma* (festival of harvest), which highlight their deep connection to nature and agricultural cycles. These festivals are not just cultural events but also reinforce the community's relationship with the environment.
- Oral Tradition and Folklore: The Mundas have a rich tradition of oral literature, including songs, stories, and folklore that carry wisdom about the environment, history, and social norms. These stories often contain moral lessons related to ecological conservation, communal living, and resilience in the face of challenges.

Craftsmanship and Traditional Knowledge

- **Wood and Stone Carving:** The Mundas are skilled in wood and stone carving, creating traditional tools, household items, and ritual objects. Their craftsmanship reflects their deep understanding of natural materials and their ability to create functional and artistic objects using locally available resources.
- **Tattooing and Weaving:** Munda women traditionally practice tattooing (known as *godna*) and weaving, which are integral to their cultural identity. These crafts are passed down through generations and are imbued with cultural significance and symbolism.

SUCCESSFUL INTEGRATION OF INDIGENOUS KNOWLEDGE IN SUSTAINABLE DEVELOPMENT PROJECTS IN INDIA

Indigenous knowledge systems have historically contributed to sustainable development through ecologically sound and socially inclusive practices. This paper explores successful examples of Indigenous communities in India integrating their traditional knowledge with contemporary sustainability initiatives. Highlighted case studies include the Bishnoi community's commitment to wildlife conservation, the Apatani tribe's paddy-cum-fish cultivation, and the Kani tribe's medicinal plant conservation. Additional examples include the Dongria Kondh tribe's forest conservation efforts, the Khasi tribe's construction of living root bridges, and the Maldhari pastoralists' sustainable livestock management. Other cases examined are the Toda tribe's sustainable dairy farming, the Warli tribe's eco-friendly housing techniques, and the Siddi community's traditional agroforestry. These instances demonstrate how Indigenous wisdom, when acknowledged and integrated with modern scientific approaches, can significantly contribute to biodiversity conservation, food security, and socio-economic development. Indigenous knowledge, developed over generations through close interaction with the environment, provides valuable insights into sustainable development (Ramakrishnan, 2007). In India, various Indigenous communities employ traditional methods that align with contemporary sustainability goals. Their practices emphasize ecological balance, resource efficiency, and cultural continuity. This paper examines several successful cases where Indigenous knowledge has been effectively incorporated into modern sustainability initiatives.

Bishnoi Community and Wildlife Conservation

The Bishnoi community of Rajasthan has a deep-rooted commitment to environmental conservation, embedded in their religious and cultural beliefs (Sinha, Bhatia, & Vishwakarma, 2018). The 1730 Khejarli Massacre, where 363 Bishnois sacrificed their lives to protect sacred Khejri trees, exemplifies their unwavering dedication to conservation. Today, Bishnoi villages serve as safe havens for endangered species such as the blackbuck antelope, illustrating how Indigenous traditions can foster sustainable environmental stewardship (Roy & Basu, 2020).

Apatani Tribe and Paddy-Cum-Fish Cultivation

The Apatani people of Arunachal Pradesh practice an innovative agro-ecological system that integrates rice farming with fish rearing. This symbiotic method enhances soil fertility, naturally controls pests, and ensures food security (Food and Agriculture Organization [FAO], 2021). Recognized as a Globally Important Agricultural Heritage System (GIAHS) by the FAO, this Indigenous practice exemplifies sustainable agriculture by maintaining ecological balance while maximizing productivity.

Kani Tribe and Medicinal Plant Conservation

The Kani tribe of Kerala possesses extensive knowledge of medicinal plants and has collaborated with scientists from the Tropical Botanic Garden and Research Institute (TBGRI) to develop *Jeevani*, a herbal product derived from the *Arogyapacha* plant (Gadgil, Berkes, & Folke, 1993). This partnership resulted in financial benefits for the tribe and highlighted the potential of integrating Indigenous knowledge with modern scientific research. However, the case also raised concerns about biopiracy and the need for equitable benefit-sharing mechanisms.

Dongria Kondh Tribe and Forest Conservation

The Dongria Kondh tribe of Odisha has played a crucial role in protecting the Niyamgiri Hills, a biodiversity hotspot. Their resistance against mining activities in the region underscores the role of Indigenous communities in forest conservation (Ramakrishnan, 2007). Their sustainable land-use practices, including rotational farming and sacred grove preservation, contribute to ecological balance while ensuring their livelihoods.

Khasi Tribe and Living Root Bridges

In Meghalaya, the Khasi tribe has developed an ingenious method of constructing "living root bridges" by guiding the roots of rubber fig trees across rivers and streams. These natural bridges, which can last for centuries, provide a sustainable alternative to conventional infrastructure that deteriorates over time. This practice exemplifies Indigenous engineering that enhances resilience to climate change and promotes environmental sustainability (Roy & Basu, 2020).

Maldhari Pastoralists and Sustainable Livestock Management

The Maldhari pastoralists of Gujarat, residing in and around the Gir Forest, practice sustainable livestock management that harmonizes with wildlife conservation. Their traditional grazing patterns ensure that their livestock does not deplete local flora, thus maintaining ecological balance. Additionally, their coexistence with Asiatic lions in the Gir National Park highlights their role in biodiversity conservation (Sinha et al., 2018).

Toda Tribe and Sustainable Dairy Farming

The Toda tribe of the Nilgiris in Tamil Nadu is known for its sustainable dairy farming practices. Their buffaloes graze on naturally regenerating grasslands, which prevents overgrazing and helps maintain biodiversity. Their traditional dairy system follows a community-based approach that respects natural cycles and avoids excessive resource exploitation (Gadgil et al., 1993).

Warli Tribe and Eco-Friendly Housing Techniques

The Warli tribe of Maharashtra employs eco-friendly housing techniques using locally available materials such as mud, bamboo, and cow dung. Their traditional architecture is climate-responsive, ensuring natural cooling and insulation. These methods offer a model for sustainable rural housing that reduces dependency on non-renewable resources (Roy & Basu, 2020).

Siddi Community and Traditional Agroforestry

The Siddi community in Karnataka and Gujarat practices traditional agroforestry, integrating trees, crops, and livestock into a single farming system. This method enhances biodiversity, improves soil health, and contributes to climate resilience (FAO, 2021). Their agroforestry model aligns with modern sustainable agriculture principles and provides insights into combating deforestation and land degradation. Indigenous communities in India have long demonstrated sustainable practices that align with contemporary environmental and development goals. The case studies discussed illustrate how traditional knowledge can be effectively integrated into modern sustainability initiatives. Recognizing and respecting Indigenous wisdom is crucial for fostering holistic development that benefits both local communities and global sustainability efforts. Policies that support Indigenous rights, equitable benefit-sharing, and collaboration between Indigenous knowledge holders and scientific researchers can further enhance the successful integration of Indigenous practices into sustainable development frameworks.

INTEGRATING INDIGENOUS KNOWLEDGE INTO SUSTAINABLE DEVELOPMENT: A PATH TO INCLUSIVE AND EFFECTIVE SOLUTIONS

Indigenous knowledge has been an essential component of environmental stewardship, biodiversity conservation, and sustainable development for centuries. However, it has often been marginalized in favor of Western scientific approaches. This paper explores the necessity of recognizing Indigenous knowledge as equal to scientific knowledge, ensuring the participatory involvement of Indigenous communities, building trust-based relationships, protecting intellectual property rights, and integrating traditional and scientific knowledge. The study highlights the importance of strengthening Indigenous self-governance, promoting educational exchange programs, adapting development projects to local contexts, ensuring long-term financial support, and securing political and legal recognition. Through a critical analysis of these principles, the paper argues that sustainable development efforts can achieve greater success when Indigenous perspectives are fully integrated into planning and decision-making processes. Indigenous communities worldwide possess a wealth of knowledge regarding sustainable resource management, environmental conservation, and adaptation to climatic changes (Berkes, 2018). This knowledge, accumulated over generations, is deeply embedded in their cultural traditions and way of life. However, mainstream development and scientific discourse have historically marginalized Indigenous perspectives, often treating them as supplementary rather than equal to Western scientific approaches (Whyte, 2020). In recent years, there has been growing recognition of the value of Indigenous knowledge in addressing global challenges such as climate change, biodiversity loss, and food security.

Recognition and Respect for Indigenous Knowledge

For Indigenous knowledge to contribute meaningfully to sustainability, it must be recognized as equal to scientific knowledge (Smith, 2012). Traditionally, Indigenous wisdom has been viewed as anecdotal or informal, leading to its exclusion from formal policy and decision-making processes (Nakata, 2007). However, Indigenous knowledge systems are empirical, observational, and deeply intertwined with ecological processes, making them highly valuable in areas such as biodiversity conservation and climate adaptation (Mistry & Berardi, 2016). Recognition also requires shifting the perspective from seeing

Indigenous knowledge as a mere supplement to Western science. Instead, development frameworks should integrate both knowledge systems, creating a holistic approach to sustainability (Agrawal, 2002). For example, in Australia, Indigenous fire management techniques have been incorporated into national fire prevention policies, demonstrating the effectiveness of traditional practices in modern environmental management (Bowman et al., 2020).

Participatory Involvement of Indigenous Communities

Indigenous communities must be actively engaged in the planning, implementation, and evaluation of development projects from the outset. Historically, many projects have imposed external solutions without considering Indigenous perspectives, leading to cultural conflicts and unsustainable outcomes (David-Chavez & Gavin, 2018). Participation ensures that Indigenous values, priorities, and needs are fully integrated into decision-making processes, leading to greater acceptance and success. A notable example is the co-management of national parks in Canada, where Indigenous communities collaborate with government agencies in conservation efforts (Reid et al., 2014). This participatory model not only empowers Indigenous peoples but also enhances environmental outcomes by leveraging their traditional ecological knowledge.

Building Relationships Based on Trust

Long-term relationships built on mutual respect and trust are crucial for successful collaboration between Indigenous communities and external stakeholders (Whyte, 2020). Many Indigenous groups have experienced historical injustices, including land dispossession, forced assimilation, and exploitation of their resources. As a result, skepticism towards external interventions remains prevalent (Smith, 2012). Building trust requires open communication, transparency in decision-making, and the protection of Indigenous rights. A key example is New Zealand's Treaty of Waitangi, which provides a legal foundation for Indigenous participation in governance and resource management (Williams, 2018). Similar frameworks should be adopted worldwide to foster respectful and equitable partnerships.

Protecting Indigenous Intellectual Property

Indigenous knowledge is often passed down through generations and is closely linked to cultural identity. However, the commercialization and exploitation of Indigenous wisdom without fair compensation often referred to as biopiracy pose significant ethical and legal challenges (Posey & Dutfield, 1996). For instance, the commercialization of Indigenous medicinal plants by pharmaceutical companies has led to disputes over benefit-sharing (Bannister, 2018). To address these concerns, legal frameworks such as the Nagoya Protocol under the Convention on Biological Diversity (CBD) have been established to ensure that Indigenous communities retain control over their knowledge and benefit from its use (UNEP, 2011). Strengthening such mechanisms can help protect Indigenous intellectual property while fostering ethical collaboration.

Combining Traditional and Scientific Knowledge

Integrating traditional and scientific knowledge can lead to innovative and sustainable solutions. Many Indigenous agricultural practices, such as agroforestry, crop rotation, and polyculture, have been recognized as ecologically beneficial (Altieri, 2009). For example, the Apatani tribe in India practices paddy-cum-fish farming, an Indigenous agricultural method that enhances biodiversity and soil fertility while ensuring food security (FAO, 2021). Similarly, Indigenous knowledge of medicinal plants has contributed to modern pharmacology. The collaboration between the Kani tribe in India and scientists from the Tropical Botanic Garden and Research Institute (TBGRI) led to the development of Jeevani, a herbal drug derived from the Arogyapacha plant (Gadgil et al., 1993). These examples demonstrate the potential of blending Indigenous wisdom with scientific research.

Strengthening Indigenous Self-Governance and Decision-Making Powers

Empowering Indigenous communities to govern their lands and resources is fundamental to achieving sustainable development. Many Indigenous groups have their own governance structures that regulate resource use based on principles of sustainability (Cornell & Kalt, 2007). Strengthening these institutions allows Indigenous peoples to lead development initiatives on their own terms. The recognition of Indigenous land rights in countries like Australia, Canada, and Bolivia has improved environmental governance and conservation outcomes (Larson & Pulhin, 2012). Governments must continue to support Indigenous self-determination through policies that protect land tenure and resource rights.

Promoting Education and Knowledge Exchange Programs

Mutual learning between Indigenous communities and external stakeholders can enhance appreciation for Indigenous knowledge. Educational programs that incorporate Indigenous perspectives into curricula can promote cross-cultural understanding and knowledge-sharing (Battiste, 2013). For example, Indigenous ranger programs in Australia train Indigenous youth in conservation techniques while preserving traditional ecological knowledge (Hill et al., 2012). Such initiatives demonstrate the value of knowledge exchange in strengthening Indigenous leadership in environmental management.

Adaptation to Local Contexts

Sustainability initiatives must be adapted to the ecological, social, and cultural contexts of Indigenous communities. Standardized development models often fail because they do not account for local conditions (Escobar, 2011). Indigenous groups have in-depth knowledge of their ecosystems and can provide valuable insights into climate adaptation and resource management. For instance, Inuit communities in the Arctic have developed climate adaptation strategies based on their deep understanding of sea ice patterns and wildlife behavior (Ford et al., 2016). Collaborating with Indigenous communities ensures that solutions are locally relevant and sustainable.

Long-Term Support and Financing

Sustainable development requires long-term investment in Indigenous-led initiatives. Many Indigenous practices have been undermined by insufficient funding and short-term projects (Graham & Fortier, 2005). Governments and international organizations should establish financing models that provide stable resources for Indigenous communities to maintain and expand their traditional practices.

Political and Legal Recognition

Legal frameworks must protect Indigenous knowledge and promote its integration into national and international sustainability strategies. The UN Declaration on the Rights of Indigenous Peoples (UNDRIP) recognizes the rights of Indigenous peoples to their knowledge and resources (UN, 2007). Governments should align their policies with UNDRIP to ensure Indigenous participation in decision-making. Indigenous knowledge is a vital resource for sustainable development. Recognizing its value, ensuring participatory involvement, building trust, protecting intellectual property, and integrating traditional and scientific knowledge are essential steps towards inclusive and effective solutions. Strengthening Indigenous governance, promoting education, adapting projects to local contexts, ensuring long-term funding, and securing political recognition will further enhance the contributions of Indigenous communities to global sustainability. By respecting and integrating Indigenous wisdom, we can create a more just and sustainable future.

BEST PRACTICES AND LESSONS TO BE LEARN FROM OTHER COUNTRIES

Legal Recognition and Protection of Indigenous Knowledge

One of the most significant steps India can take is to establish strong legal frameworks that recognize and protect Indigenous knowledge. Countries like Canada and New Zealand have integrated Indigenous rights into their legal systems. The United Nations Declaration on the Rights of Indigenous Peoples (UNDRIP) (United Nations, 2007) has been instrumental in shaping these policies. In New Zealand, the Treaty of Waitangi provides a legal framework that acknowledges the rights of the Māori people. It ensures their active participation in environmental management and decision-making (Williams, 2018). India can adopt similar legal measures to recognize Indigenous knowledge and grant Indigenous communities' greater control over their resources.

Participatory Governance and Indigenous Leadership

Australia's Indigenous land management programs, such as the Indigenous Ranger Program, involve Indigenous communities in conservation projects and environmental governance (Hill et al., 2012). These programs empower Indigenous people by providing employment and integrating their knowledge into national conservation efforts. India can replicate this approach by involving Indigenous communities in forest conservation, water management, and climate adaptation projects. Strengthening Indigenous self-governance structures can lead to more effective and inclusive sustainable development initiatives.

Ethical Bioprospecting and Intellectual Property Rights

Many Indigenous communities worldwide have faced challenges regarding biopiracy, where corporations exploit traditional knowledge without fair compensation. The Nagoya Protocol (United Nations Environment Programme [UNEP], 2011) established an international framework for access and benefit-sharing related to genetic resources. India can strengthen its legal framework by implementing stricter policies to prevent biopiracy and ensure fair compensation for Indigenous communities. Learning from Canada's ethical biotrade models can help India develop fairer benefit-sharing mechanisms.

Integration of Indigenous Knowledge into Education and Research

Countries such as Canada and the United States have integrated Indigenous knowledge into their educational systems. Indigenous science programs and traditional ecological knowledge courses are incorporated into university curricula (David-Chavez & Gavin, 2018). India can introduce similar initiatives in schools and universities to foster a greater appreciation for Indigenous knowledge. Encouraging collaboration between Indigenous knowledge holders and scientific researchers can lead to innovative sustainability solutions.

Sustainable Livelihoods and Economic Support

Incorporating Indigenous knowledge into sustainable business models can improve economic opportunities for Indigenous communities. The Sami people of Scandinavia have successfully developed sustainable reindeer herding industries, supported by government policies and Indigenous cooperatives (Reid et al., 2014). India can support Indigenous economic development by promoting Indigenous-led ecotourism, sustainable handicrafts, and organic farming enterprises. Providing financial support and market access to Indigenous entrepreneurs can enhance their economic resilience.

CHALLENGES AND OPPORTUNITIES OF INDIGENOUS KNOWLEDGE IN INDIA

Challenges and Threats

By learning from international examples, India can adopt legal, institutional, and economic measures to strengthen Indigenous knowledge systems. Recognizing Indigenous knowledge as equal to scientific knowledge, ensuring participatory governance, and protecting Indigenous intellectual property are essential steps. Additionally, integrating Indigenous knowledge into education, research, and economic programs can foster long-term sustainability. Ultimately, a collaborative approach that respects, preserves, and promotes Indigenous knowledge will enable India to achieve sustainable development while empowering its Indigenous communities.

Modernization and Loss of Traditional Knowledge

As modernization spreads rapidly across the globe, many Indigenous communities are losing their traditional knowledge. Younger generations, in particular, are increasingly drawn toward urban lifestyles and modern education, which often devalue or disregard Indigenous practices. This generational disconnect leads to the gradual disappearance of sustainable practices that have been honed over centuries. In the face of globalization, traditional agricultural methods, spiritual connections to nature, and community-based environmental stewardship are being replaced by modern, industrialized systems, which often have detrimental effects on the environment.

Land Rights and Displacement Issues

Indigenous communities worldwide face significant challenges regarding land rights. Development projects, such as large-scale mining, hydropower dams, and infrastructure projects, often lead to the displacement of Indigenous peoples from their ancestral lands. This displacement severs their deep connection with the environment and disrupts their ability to practice sustainable land management. Land tenure insecurity also makes Indigenous people vulnerable to exploitation and resource extraction by external actors, further threatening their sustainable practices and livelihoods.

Climate Change Impacts on Indigenous Communities

Climate change poses a significant threat to Indigenous communities, particularly those whose livelihoods depend on natural ecosystems. Rising temperatures, changing precipitation patterns, and extreme weather events impact traditional agricultural practices, biodiversity, and water sources. For instance, Indigenous communities in coastal regions or small island nations are disproportionately affected by sea-level rise and coastal erosion, while those in arid or semi-arid regions face increased water scarcity and drought. These climate impacts not only threaten food security but also the cultural and spiritual identities of Indigenous communities, which are intricately linked to their environment.

OPPORTUNITIES AND STRATEGIES

Policy Recommendations for Integrating Indigenous Knowledge into Sustainable Development

Governments and international organizations should prioritize the integration of Indigenous knowledge into sustainable development policies. This could involve creating legal frameworks that recognize and protect Indigenous land rights, ensuring equitable participation of Indigenous people in policy-making, and fostering partnerships between Indigenous communities and scientific researchers. Indigenous knowledge systems should be formalized as part of national environmental and conservation strategies to ensure that traditional practices continue to inform sustainable development efforts.

Role of Education and Capacity Building

Education plays a vital role in preserving Indigenous knowledge and promoting its integration into sustainable development. By incorporating Indigenous environmental practices into school curricula, both Indigenous and non-Indigenous students can learn about the value of traditional ecological knowledge. Capacity-building initiatives, such as leadership and technical training, can empower Indigenous people to participate more effectively in sustainable development projects and advocate for their rights in broader policy forums.

Collaborations between Indigenous Communities, Government and NGOs

Collaboration between Indigenous communities, government agencies, and NGOs can create synergies for sustainable development. Governments can offer legal and financial support, while NGOs can facilitate capacity-building and create platforms for knowledge-sharing. Additionally, NGOs can serve as intermediaries, helping to mediate conflicts between Indigenous groups and external stakeholders, such as corporations or government agencies, ensuring that development projects respect the rights and contributions of Indigenous communities.

FUTURE RESEARCH DIRECTIONS

Future research should focus on documenting and preserving Indigenous knowledge, exploring ways to scale up successful practices, and identifying mechanisms for equitable benefit-sharing. Policymakers should prioritize the protection of Indigenous land rights and create inclusive platforms where Indigenous people can lead decision-making processes. Efforts to bridge the gap between Indigenous and scientific knowledge systems will be the key to achieving long-term sustainable development goals.

CONCLUSION

Indigenous knowledge plays a crucial role in sustainable development, offering time-tested solutions for biodiversity conservation, agriculture, healthcare, and climate adaptation. In India, Indigenous communities have demonstrated remarkable innovations in sustainable practices. However, their contributions often remain underrecognized in mainstream development policies. Indigenous knowledge systems provide sustainable solutions to contemporary challenges, from biodiversity conservation and food security to climate resilience. Case studies of the Bishnoi, Apatani, and Kani communities illustrate how traditional practices can be successfully integrated into sustainable development projects. However, these knowledge systems face numerous threats, including modernization, land displacement, and the impacts of climate change.

REFERENCES

Agrawal, A. (1995). Dismantling the divide between Indigenous and scientific knowledge. *Development and Change*, *26*(3), 413–439. DOI: 10.1111/j.1467-7660.1995.tb00560.x

Agrawal, A. (2002). Indigenous knowledge and the politics of classification. *International Social Science Journal*, *54*(173), 287–297. DOI: 10.1111/1468-2451.00382

Altieri, M. A. (2009). Agroecology, small farms, and food sovereignty. *Monthly Review (New York, N.Y.)*, *61*(3), 102–113. DOI: 10.14452/MR-061-03-2009-07_8

Altieri, M. A., & Nicholls, C. I. (2017). *Agroecology: A transdisciplinary, participatory and action-oriented approach*. CRC Press.

Bannister, K. (2018). Biopiracy or fair biotrade? Ethical dilemmas in biodiversity conservation and commercial use of traditional knowledge. *Ethics in Science and Environmental Politics*, *18*(2), 123–140. DOI: 10.3354/esep00182

Battiste, M. (2013). *Decolonizing education: Nourishing the learning spirit*. Purich Publishing.

Berkes, F. (2012). *Sacred ecology*. Routledge. DOI: 10.4324/9780203123843

Berkes, F. (2018). *Sacred ecology* (4th ed.). Routledge.

Boelens, R., Perreault, T., & Vos, J. (2016). *Water justice*. Cambridge University Press.

Bowman, D. M., O'Brien, J. A., & Goldammer, J. G. (2020). Indigenous use of fire: The social and environmental benefits of traditional knowledge. *Proceedings of the National Academy of Sciences of the United States of America*, *117*(34), 20475–20484. DOI: 10.1073/pnas.2008734117

Cornell, S., & Kalt, J. P. (2007). *The state of the Native nations: Conditions under U.S. policies of self-determination*. Oxford University Press.

David-Chavez, D. M., & Gavin, M. C. (2018). A global assessment of Indigenous community engagement in climate research. *Environmental Research Letters*, *13*(12), 123005. DOI: 10.1088/1748-9326/aaf300

Ens, E. J., Finlayson, M., Preuss, K., Jackson, S., & Holcombe, S. (2015). Australian approaches for managing 'country' using Indigenous and non-Indigenous knowledge. *Ecological Management & Restoration*, *16*(2), 84–92. DOI: 10.1111/emr.12120

Escobar, A. (2011). *Encountering development: The making and unmaking of the Third World*. Princeton University Press.

Food and Agriculture Organization (FAO). (2021). *Globally Important Agricultural Heritage Systems (GIAHS)*. Retrieved from https://www.fao.org/giahs/en/

Ford, J. D., Cameron, L., Rubis, J., Maillet, M., Nakashima, D., Willox, A. C., & Pearce, T. (2016). Including Indigenous knowledge and experience in IPCC assessment reports. *Nature Climate Change*, *6*(4), 349–353. DOI: 10.1038/nclimate2954

Ford, J. D., Maillet, M., Pouliot, V., Meredith, T., Cavanaugh, A., & Labbe, J. (2016). Adaptation and Indigenous peoples in the Arctic: An analysis of best practices in climate adaptation planning. *Environmental Research Letters, 11*(9), 093004. DOI: 10.1088/1748-9326/11/9/093004

Ford, J. D., Smit, B., & Wandel, J. (2020). Climate change adaptation in Indigenous communities. *Environmental Research Letters, 15*(8), 083001. DOI: 10.1088/1748-9326/ab9b90

Gadgil, M., Berkes, F., & Folke, C. (1993). Indigenous knowledge for biodiversity conservation. *Ambio, 22*(2-3), 151–156. PMID: 33566330

Government of India. (2023). *National policy on tribal development.* Ministry of Tribal Affairs.

Graham, J., & Fortier, J. (2005). *Indigenous governance: Emerging trends and ongoing challenges.* Institute on Governance. https://iog.ca/docs/IndigenousGovernance.pdf

Hill, R., Grant, C., George, M., Robinson, C. J., & Sparrow, A. (2012). A typology of Indigenous engagement in Australian environmental management: Implications for knowledge integration and social-ecological system sustainability. *Ecology and Society, 17*(1), 23. DOI: 10.5751/ES-04587-170123

Intergovernmental Panel on Climate Change (IPCC). (2019). *Special report on climate change and land.* https://www.ipcc.ch/srccl/

Jain, S. K. (1991). *Dictionary of Indian folk medicine and ethnobotany.* Deep Publications.

Kumar, M., & Kerr, J. M. (2013). Democratic assertions: The making of India's Recognition of Forest Rights Act. *Development and Change, 44*(3), 433–455.

Larson, A. M., & Pulhin, J. M. (2012). Enhancing forest tenure reforms through more responsive governance. *Conservation & Society, 10*(2), 103–113. DOI: 10.4103/0972-4923.97482

Levis, C., Costa, F. R. C., Bongers, F., Peña-Claros, M., Clement, C. R., & Ter Steege, H. (2017). Persistent effects of pre-Columbian plant domestication on Amazonian forest composition. *Science, 355*(6328), 925–931. DOI: 10.1126/science.aal0157 PMID: 28254935

Mistry, J., & Berardi, A. (2016). Bridging Indigenous and scientific knowledge. *Science, 352*(6291), 1274–1275. DOI: 10.1126/science.aaf1160 PMID: 27284180

Nakashima, D., Krupnik, I., & Rubis, J. (2012). *Indigenous knowledge for climate change assessment and adaptation.* Cambridge University Press.

Nakata, M. (2007). *Disciplining the savages: Savaging the disciplines.* Aboriginal Studies Press.

Posey, D. A., & Dutfield, G. (1996). *Beyond intellectual property: Toward traditional resource rights for Indigenous peoples and local communities.* International Development Research Centre.

Ramakrishnan, P. S. (2007). Traditional forest knowledge and sustainable forestry: A north-east India perspective. *Forest Ecology and Management, 249*(1-2), 91–99. DOI: 10.1016/j.foreco.2007.04.001

Reid, A. J., Eckert, L. E., Lane, J. F., Young, N., Hinch, S. G., Darimont, C. T., Cooke, S. J., Ban, N. C., & Marshall, A. (2014). Two-eyed seeing: An Indigenous framework to transform fisheries research and management. *Fish and Fisheries, 22*(1), 243–261. DOI: 10.1111/faf.12516

Reid, R. S., Fernández-Giménez, M. E., & Galvin, K. A. (2014). Dynamics and resilience of pastoralist societies. *Annual Review of Environment and Resources*, *39*, 217–242. DOI: 10.1146/annurev-environ-020713-163329

Roy, M., & Basu, R. (2020). Indigenous environmental knowledge and climate adaptation in India. *Current Science*, *119*(3), 467–475.

Schwartzman, S., Zimmerman, B., & Kormos, C. (2019). Indigenous lands, protected areas, and slowing climate change. *Conservation Biology*, *33*(4), 700–711. DOI: 10.1111/cobi.13329

Sinha, R. K., Bhatia, S., & Vishwakarma, R. (2018). Indigenous knowledge systems and sustainable development: A case study of the Bishnoi community. *Indian Journal of Traditional Knowledge*, *17*(2), 298–305.

Tiwari, B. K. (2013). Indigenous knowledge and sustainable development. *Journal of North East India Studies*, *3*(2), 1–8.

United Nations (UN). (2007). *United Nations Declaration on the Rights of Indigenous Peoples (UN-DRIP)*.https://www.un.org/development/desa/indigenouspeoples/declaration-on-the-rights-of-indigenous-peoples.html

KEY TERMS AND DEFINITIONS

Agroforestry: A land-use system that integrates trees and shrubs into agricultural landscapes, combining ecological benefits with sustainable livelihoods.

Folk Medicine: Herbal and natural remedies used in Indigenous healthcare systems, including Ayurveda, Siddha, and Unani medicine.

Indigenous Agricultural Practices: Traditional farming techniques such as organic farming, seed conservation, and natural pest management, used by Indian rural communities.

Indigenous Knowledge (IK): Traditional, local knowledge that has evolved over centuries within a particular community, passed down through generations. It includes agricultural practices, medicinal systems, ecological management, and cultural wisdom.

Sacred Groves: Protected forest patches conserved by Indigenous communities for religious and cultural reasons, playing a vital role in biodiversity conservation.

Sustainable Development: Development that meets present needs without compromising the ability of future generations to meet their own needs, incorporating economic, social, and environmental sustainability.

Traditional Ecological Knowledge (TEK): A subset of Indigenous Knowledge that focuses on environmental conservation, land management, and biodiversity preservation through practices developed over centuries.

Jal Shakti (Water Power): Traditional water management systems in India, including stepwells, baolis, and tank irrigation, which have been crucial for water conservation.

ENDNOTES

1 The word **"indigenous"** refers to people, cultures, or things that are native to a particular place or region and have a long-standing connection to it. Indigenous people (or communities) are often recognized for their deep spiritual connection to the land and for preserving traditional ways of life that have been passed down through generations.

2 The term "Adivasis," derived from the Hindi words "Adi" (original) and "Vasi" (inhabitants), refers to the Indigenous or tribal communities in India who have historically resided in forested and hilly regions across the country. These communities, officially recognized as Scheduled Tribes under the Indian Constitution, are distinguished by their unique cultural practices, languages, and traditional ways of life.

Chapter 24
In the Transit of Indigenous Knowledge Systems Through Digital Transformation Framework

Lindiwe Carol Mthethwa
https://orcid.org/0000-0001-9170-369X
University of Zululand, South Africa

ABSTRACT

This chapter explores the transition of Indigenous Knowledge Systems (IKS) through a digital transformation framework, emphasizing the need for culturally sensitive approaches that honour integrity and context of indigenous knowledge. This widens the potential of digital tools to enhance accessibility, foster innovation, and bridge generational knowledge gaps, while addressing critical issues such as intellectual property rights, data sovereignty, and ethical representation. Through a problem-based pedagogy, strategies for integrating IKS into modern digital ecosystems, promoting inclusive knowledge practices and global sustainability are outlined. The objectives propose a digital framework for transitioning IKS using digital platforms; to identify the role of digital transformation in preserving, transmitting, and revitalizing IKS; and to discuss ethical considerations, challenges and practical implementation in Higher Education (HE). This sparked debate about a redesigned role for digital framework education and authenticity. Indigenous Knowledge Systems Digital Transformation Framework (IKSDTF)is seen as having the ability to preserve, transmit, and revitalise IKS.

INTRODUCTION

Indigenous Knowledge Systems (IKS) represent a vast repository of cultural heritage, ecological wisdom, and traditional practices that have sustained communities for generations. However, the rapid advancement of digital technologies presents both opportunities and challenges for the preservation, dissemination, and application of these knowledge systems. Lecturers in Higher Education (HE) are in the transit of incorporating indigenous knowledge system in their teachings. Mindfulness of Digital Transformation Framework (DTF) is suggested as the main effort to rediscover, realign and revamp (3R)

DOI: 10.4018/979-8-3693-7964-6.ch024

the new HE curriculum. This includes the co-developing of culturally sensitive and sustainable digital systems for Indigenous Knowledge Systems. It aligns with values of inclusivity, respect, social justice and innovation. Presented in this chapter are the guidelines for utilising Indigenous Knowledge in Indigenous Higher Education contexts. In fulfilling this there is increasing need to improve Indigenous Knowledge systems. The DTF serves as a guide for embracing innovation and adjusting to the changing needs of education in the digital age. The chapter explores the shipment of indigenous knowledge systems through digital transformation framework. Changes in the society have teachings in a networked society not to be restricted in a specific place but can happen across time, space and technology. The escalating rising of people with gadgets implied the fast tracking of accumulating new skills of using technology so as to teach any time. Indigenous knowledge represents the accumulated experience, wisdom, and practices of indigenous peoples worldwide. Often localized and orally transmitted, this knowledge faces erosion due to urbanization, loss of native languages, and socio-economic pressures. Simultaneously, digital technologies offer new avenues for documenting, sharing, and revitalizing such knowledge. This chapter argues for a reimagined role of digital framework education as a vehicle to bridge indigenous knowledge with the demands of the digital age, without compromising its essence or authenticity.

The place-based pedagogies are elaborated to enhance culturally relevant philosophy. The term "Indigenous lecturer" here suggests to the historically marginalized group in South Africa, who have endured levels of subordination and persecution, even within their own communities. The experiences and indigenous ties of indigenous lecturer are highlighted in this chapter, which adds to the ontologies and epistemologies of how knowledge is created. Through a close cultural interpretation of the chapter, it offers greater insights into their distinct customs and experiences. The great stance is how lecturers draw their indigenous student in the production, documentation, and digitisation of Indigenous Knowledge as important Indigenous People (IP) (Winschiers-Theophilus, 2022). The importance of documentation is part of a comprehensive plan for IK prevention. Furthermore, IK documentation can be in textual registries and files, video, photographs, and audio recordings; it can be in indigenous or foreign languages; it can use current or older technology and be in any combination of these formats (Kovach, 2018).

This chapter aims to establish the transit of Indigenous Knowledge systems through the Digital Transformation Framework to bring together higher education lecturers, researchers, students, and practitioners (Mdhluli, Mokgoatsana, Kugara & Vuma, 2021). Lecturer's open spaces for student-led initiatives ranging from multidisciplinary and interdisciplinary. Intention is to explore the important link between innovation, ethics, and inclusive excellence in higher education teaching and learning. In this regard, there will be a promotion of debate and sharing of creative practices. That aim to improve equity and learning in response to the digital transformation of higher education. It is done through Indigenous Knowledge systems digitalised new technologies, and student and lecturer's diversity. Analysing traditional ecological methods improves critical thinking skills and problem-solving ability. The need for indigenisation of knowledge in African educational institutions is an acknowledgement of the rich knowledge found in African cultures.

BACKGROUND OF THE STUDY

South Africa is frequently portrayed in history as a nation that emerged with the advent of European settlers (Seroto, 2012). Best digital preservation procedures and archival standards should be followed when creating a dependable digital preservation system in a localised forms of knowledge that are unique to a given society and community (Buthelezi, Ocholla & Dlamini, 2024). Gumbo (2023) insisted that higher education research and education digitisation can be revolutionized if indigenous knowledge is not compromised. Thus, there has been an important call of promoting academics' active participation in the production, recording, and sharing of indigenous knowledge (Buthelezi, et al, 2024) This will involve having a system in place for mapping and auditing university IK research and instruction, curbing inferiorisation and distortion (Cakata & Ramose, 2023). It is time for adopting theories, methodologies and historical guides that are 'going transnational…because we live and have lived' (Badenoch & Follmer, 2018). I turn to differ from scholarly argument that claims that the meaning of life is infrequent in the African tradition, hence what is claimed to be common is the meaning in life. IKS policies should be developed in collaboration with a diverse group of stakeholders, including local non-governmental organisations and indigenous people. Global issues need to be addressed such as climate change, biodiversity loss, agricultural sustainability, and ecological stability by protecting and disseminating indigenous knowledge. Thus, incorporating IKS into mainstream procedures is critical. IKS will be used to explain the daily living environment. The Indigenous Knowledge Digital Transformation Framework (IKDTF) is used in this chapter as mind map to be used by lecturers. The challenges are presented to caution lecturers. Indigenous knowledge is the combined experience, wisdom, and traditions of indigenous peoples worldwide. Urbanisation, loss of native languages, and socioeconomic pressures are all threatening to erode this information, which is frequently localised and passed down orally. Simultaneously, digital technologies provide new opportunities to document, share, and revitalise such knowledge. The objectives of this chapter are as follows:

- To identify the role of digital transformation in preserving, transmitting, and revitalizing IKS.
- To discuss ethical considerations, challenges and practical implementation in Higher Education (HE).
- To propose a digital framework for transitioning IKS using digital platforms.

DEFINITION OF CONCEPTS

a) What is Indigenous Knowledge?

Indigenous knowledge is a word used to describe specialised information which is specific to a given culture and group with specialised knowledge types exclusive to a particular community or society (Buthelezi, et al, 2024). According to Odora Hoppers (2002), the word indigenous refers to the inherent nature of a particular setting or culture. Hence in this chapter, IKS is about Local Indigenous Knowledge systems and the strategies for incorporating them into mainstream formal education (Gumbo, 2023, Msila, 2021). Indigenous knowledge is the unique and traditional knowledge (Raseroka, 2008) that embodies the values, wisdom, and expectations of the community and the broader society. The digitisation of indigenous knowledge is widely documented in African countries, namely Ghana (Plockery, 2015); Tanzania (Mbilinyi & Mwabungulu, 2020); and Namibia (Chamunorwa, Winschiers-Theophilus, & Zaman, 2018). In this chapter African refer to Sub-Saharan countries, however my usage of African joins the scholarly argument of identifying what has been salient in the large Sub-Saharan countries and globally (Metz, 2020).

Keikelame and Swartx (2019) used the concept 'Indigeneity' as a concept that extends the significance and power of identifying oneself as an agent of continuity and change. Indigeneity refers to the cultural identity politics of the First Peoples who inhabit a geographic location, island, or nation. Thus naming, whether a person, or a political movement, makes one real. In this study, the current angle will be empirically fruitful answers to what is, when it was and how it will translate rich ecology (Badenoch & Follmer, 2018). There must be audiences that can trace, translate and transform for fruitful benefits, as the 'experience machine' (Metz, 2020).

b) Place-Based Pedagogy

Place-based education is an approach to curriculum development and instruction that starts with the local community and environment. Local culture, history, art, and other resources form the foundation for all the stake holders in teaching and learning, i.e. lecturers and students. Environmental sustainability and community interaction are crucial components of place-based pedagogy. Lecturers groom their students to frequently help the community tackle social or ecological challenges, which fosters a sense of civic responsibility and respect for the natural world. Place-based pedagogy has origins and draws inspiration from indigenous societies and educational systems. In many ways, place-based pedagogy is a reaction to colonisation, development, and globalisation processes that have damaged local environments and communities (Hayes, 2022). In the application of place-based education lecturers will be "recognising IKS in education (a)s a step towards indigenisation, which aims to recapture African ways of doing, living, and being in education" (Bhuda, 2021:69). The place-based pedagogy in this study do not need the foreign context to clarify education as it has been contested that "African philosophers have not sought to answer that meta-ethical question systematically in English-speaking works" (Metz, 2020:114)

c) Cultural-Relevant Philosophy

Culturally relevant philosophy in the context of Indigenous Knowledge Systems (IKS) and digital transformation refers to frameworks or philosophies that respect, incorporate, and adapt to the cultural, historical, and contextual realities of the communities they serve. Lecturers are having wide space of supporting government efforts (Cakata, 2015) in eradicating what Europe has done when distorting peoples' ways of knowing and relating to themselves (Cakata & Ramose, 2023). Indigenous people have often been the first people who to oppose policy implementation strategies of knowledge preservation. Cakata and Ramose (2023:488) claimed that the responsibility is thus upon the colonised indigenous people to demand their role as educators of their children Such philosophies aim to ensure that technological interventions and academic analyses are not only aligned with universal principles but are also deeply rooted in the lived experiences, values, and worldviews of indigenous communities. The communal nature of indigenous knowledge, where knowledge is collectively held and shared, is revisited. Advocates for collaboration in digital transformation projects, ensuring all stakeholders are valued. Emphasizes community, human dignity, and the interconnectedness of all beings. "I am because we are" (communal interconnectedness). Digitisation projects should respect and incorporate historical context, ensuring knowledge is not stripped of its roots. Akan people of Ghana, symbolised by the Sankofa bird, which looks back while moving forward. Encourages learning from the past to inform the future (Magoro, 2024). Digitisation projects should respect and incorporate historical context, ensuring knowledge is not stripped of its roots. This is a call of attention to the ways that colonial knowledge systems use truth-telling to silence, delegitimize, and diminish knowers and methods of knowing, being, and doing (Maddox & Ninomiya, 2025).

At minimum, research using Indigenous methodologies "will honour the protocol of introduction that speaks to the researcher's story, the motivations for research, and how the researcher is situated in relationship to community" (Kovach, 2018:400). Further, this claims for acknowledging the existence of 'cross-cultural engagement' in academia (Metz, 2020). In the study, IKS is transmitted through a variety of livelihoods, traditional rituals, principles, norms, standards, taboos, poetry, proverbs, practices, and teachings, all of which contribute to the community's rich indigenous knowledge and culture. The preservation of IKS is important to preventing extinction (Ajani, Oladokun, Olarongbe, Amaechi, Rabiu & Bashorun, 2024). As a result, the government must develop effective policies for protecting and advancing indigenous knowledge systems. These policies should be created in consultation with a wide range of stakeholders, including local non-governmental organisations and indigenous people. By preserving and spreading IKS, we can solve global concerns like climate change, biodiversity loss, agricultural sustainability, and ecological stability. Thus, integrating IKS into mainstream procedures is crucial. In education the inclusion of IKS is used to explain the daily life environment.

THE ROLE OF DIGITAL TRANSFORMATION IN PRESERVING, TRANSMITTING, AND REVITALIZING IKS

This chapter opened the discussion for a reimagined role of digital framework education as a vehicle to bridge indigenous knowledge with the demands of the digital age, without compromising its essence or authenticity. IKDTF is viewed having the abilities to preserve, transmit and revitalise IKS.

Theoretical Foundations

The Place-Based Spaces for Networked Learning (PBSNL) (Carvalho, Goodyear & de Laat, 2016) viewed the creation of spaces for network learning, how places and traces activity outwards into the digital networks that extend each place. They managed to identify what is digital versus what is material, the virtual as opposed to actual. However, as the paradigm shifts, archive science is not transitioning from its established principles and procedures to a new "normal" science. Archival science is transitioning from relative stability to continuous change. The Place-Based Spaces (PBS) is chosen due to its potential of improving 'capacity and connectedness of communities' (Chand, 2018:161). As, a result, the changing nature of the increasingly complex digital environment makes the problem for archives extremely tough; it is a challenge that can never be totally solved. Masenya (2024) proposed a framework for preservation of indigenous knowledge which was made of four components, enabling factors, anticolonial strategies, collaboration and partnerships and relevant stakeholders entrusted with safeguarding and preserving indigenous knowledge systems in South Africa. She further indicated that relevant stakeholders are expected to utilise and improve their resources based on human needs and improving sustainability. It is therefore important to clarify that the argument is not for the place-based spaces in an essentialist manner, but this chapter open a new dialogue on the new ways to preserve IKS using abstract digital place-based approaches.

The Indigenous Knowledge Systems (IKS) is defined as dynamic, context-specific understandings that influence local practices in agriculture, medicine, and culture. Indigenous theory principles in this discussion mean Indigenous teachings as it includes philosophy, values and practices (Kovach, 2018). There is a respect of indigenous laws, ethics, protocols and knowledge discourses. Incorporating Indigenous Knowledge Systems (IKS) into curricula allows students to engage with real-world, context-specific challenges. One individual in Africa is not complete without the other. Both combat and friendship that are ethically necessary (Mboti, 2015). Knowledge is accumulated "by observing human behaviour, patterns .... and socio-ecological relationships" (Chand, 2018:160). This theory refuted the usage of research due to its Western Imperialism but promoting respect for community and individuals, reciprocity and responsibility, connection between mind and heart, self-awareness, and subjectivity (Kovach, 2018).

While the Digital Framework Education (DFE) ensure encompassing the use of technology and digital literacy in teaching and learning, it emphasises on accessibility, inclusion, and adaptability. African communities have unique ways of knowing and doing, therefore Western pedagogy and methodologies are not always applicable (Cakata, 2023). To respond to the changing digital landscape and new technology, the contemporary generation must understand how society makes and shares records, be willing to use modern technologies in new archival practices. The nation-state paradigm of cultural memory is changing due to digitisation processes and the connectedness of digital technologies. It is now a transnational framework that permits both the worldwide exchange of cultural memories and the study of the dynamics of this exchange (Thylstrup, 2018).

Intersection of Digital Frameworks and Indigenous Knowledge

Literature confirms that, globally, technology components have made it possible to preserve, promote, and disseminate both tangible and intangible cultural heritage (Mendoza, De La Hoz Franco & Gómez Gómez, 2023; Chamunorwa et al, 2018; Du, 2017). DFE offers three key opportunities to sustain indigenous knowledge, namely documentation and archiving, dissemination and accessibility, rejuvenation through innovation (Ajani et al., 2024). The inventive use of digital tools is the applications to document, archive, and digitise indigenous knowledge. This is an extension of the types of information and communication technology tools that are currently used in the management of indigenous knowledge (Dlamini & Ochola, 2018). It also requires developing interactive and engaging language learning programs and online tools, which are critical components of the preservation toolkit (Ajani, et al, 2024). Indigenous people's skills in art, music, and oral knowledge transmission are said to be ideally suited to the digital environment that is enhanced with text, graphics, audio, and video files (Du, 2017. Communication networks have significantly reduced the distances to major cities, markets, and the global landscape, thereby enhancing the accessibility of all human being to the forces of modernization and transformation (Faldi, Fisher & Moretto, 2021). The combination of policies, strategies and actions will give birth to effective intersection of digital framework and IKS (Masenya & Ngulube, 2021). Intersection of digital and IKS is argued by Clammer that:

> We can be imaginative in the arts and in technology, but rarely in society and in conceiving of our desired social arrangements. Yet it is this failure of social imagination that lies at the root of many of our contemporary problems: a failure to see that what is does not necessarily define what can be. But

the question then arises as to what resources we can call on to promote and educate this imagination (Clammer, 2014:4).

There are challenges of digital communication like cost of internet, isolation, low computer skills, illiterate and language (Du, 2017), which pushes the digital framework outside these challenges (Madsen, 2018). There is a need to draw native inhabitant to become core-creators of their digital services (Magoro, 2024). This chapter argues for a redefined role of digital framework education as a means of connecting indigenous knowledge to the demands of the digital era while preserving its integrity and authenticity.

Documentation and Archiving

Digital tools can record oral histories, rituals, and traditional practices is (Masenya & Ngulube, 2021). Platforms such as interactive databases and multimedia storytelling can store these for educational and cultural purposes (Mhlongo & Ngulube, 2020). A transnational framework permits both the worldwide circulation of cultural memories and research on the dynamics of this circulation is replacing the nation-state paradigm of cultural memory (Madsen, 2018). This could be re-shaped by digitisation processes and the connectedness of digital technologies (Metz, 2021).

Dissemination and Accessibility

Online platforms, mobile apps, and open-source repositories can enable global and local access to indigenous knowledge. Tailored interfaces in native languages ensure inclusivity.

Rejuvenation Through Innovation

Virtual reality (VR) and augmented reality (AR) can recreate immersive experiences of cultural practices. This approach resonates with younger generations, blending tradition with technology. VR and AR allow learners to practise endangered languages by simulating real-life scenarios like conversations with native speakers or virtual field trips to language-speaking regions (Sweller et al., 2019). These technologies provide an interactive and engaging learning environment. VR technology offers a viable way to create immersive traditional experiences. Virtual reality allows users to interact with native speakers, improve language skills, and experience cultural situations. VR technology can digitally preserve cultural heritage places, traditional practices, and storytelling, enabling future generations to engage with their past and knowledge.

ETHICAL CONSIDERATIONS, CHALLENGES AND PRACTICAL IMPLEMENTATION IN HIGHER EDUCATION (HE)

The Role of Indigenous Education in the Digital Ecosystem

Indigenous knowledge systems, frequently passed down through generations, are critical for the preservation of unique identities and the sustainable management of ecosystems. Bhuda (2021) used her study to highlight the significance of indigenising education in South Africa and to reform the educational system while linking students to classroom subjects beyond the Covid-19 pandemic. Indigenous knowledge is currently endangered, with its risking the loss of vital knowledge. Leveraging digital media technology is a viable way to revitalise and preserve indigenous knowledge systems (Ajani et al, 2024).

Capacity Building

NIKSOPORTAL (2024) has exciting theme which is "ensuring quality knowledge for our children's future." Indigenous lecturers and youth must be equipped with digital skills to become custodians of their knowledge. Yasseri, Gildersleve, and David (2022) referred to the agency of collective memory in a digital age. The role of education in the digital ecosystem capacitates learning through culture, not about culture. IKS includes various skills, practices, attitudes, and ideas connected to the environment, agriculture, health, spirituality, and social organisation (Ajani et al, 2024). Indigenous knowledge is profoundly embedded in people's cultures and traditions, passed down orally from generation to generation. Preserving indigenous knowledge is crucial for cultural continuity and has contemporary relevance (Masenya, 2024). Above all indigenous knowledge need to be shared, to have knowledge integrated. Carvalho et al (2016, p.1) insisted that "people sharpen their skills, have new insights, and achieve deeper understandings of themselves and ...their use creates and modulates connections between people, things, ideas, and experiences".

Policy Integration

Governments and institutions should embed indigenous knowledge in curricula using digital frameworks to ensure both relevance and longevity. Indigenous people should have control over how their traditional practices are represented in digital spaces, ensuring cultural protocols and ethical issues are followed (Ajani et al, 2024). In the anticolonial strategies proposed by Masenya (2024) policy integration should improve decolonisation of indigenous knowledge systems, research methods, education and curriculum, policy, procedures and protocols.

Collaborative Platforms

Universities, NGOs, and tech companies can collaborate to develop culturally sensitive tools for knowledge preservation. This brought the flip ways how education happen, which is mainly collaborative (Ajani, et al, 2024). This aims at producing lifelong learners, which will never stop learning.

Challenges in Application of Indigenous Knowledge Digital Transformation Framework

Despite the good promise and goodness of employing IKDTF, implementing digital frameworks for indigenous knowledge faces barriers. The nation-state paradigm of cultural memory is changing due to digitisation processes and the connectedness of digital technologies. It is now a transnational framework (Badenoch. & Follmer, 2018) permits both the worldwide exchange of cultural memories and the study of the dynamics of this exchange (Metz, 2020). Below are few that are mentioned:

Digital Divide

Limited access to internet and technology in remote areas. History has confirmed that indigenous people has been victimised (Castellino, 2024).

Cultural Sensitivities

Ensuring the authenticity of knowledge while respecting indigenous intellectual property rights. Higher education institutions in South Africa should incorporate both Western and African-centred teaching and research paradigms. Africa's Indigenous Knowledge (IK) is on the verge of extinction due to antiquated knowledge management practices, with indigenous people anxious about Western exploitation. Nonetheless, African IK is vital for the survival of communities, as most of the food is generated utilising traditional farming practices (Mdhluli et al, 2021).

Sustainability

Long-term maintenance of digital archives and tools requires continuous investment and community engagement. There must be organisations that must develop systems that enable authenticity to be assessed depending on how reliable the digital entities' source is (Masenya & Ngulube, 2021). To conserve them in a sustainable manner, human beings must use the required tactics and technologies. In attaining sustainability society should be conceived as an epiphenomenon, container or structured modelling of psychological, spiritual and a primary focus on agency (Clammer, 2014:4).

THE PRACTICAL IMPLEMENTATION AND CONCEPTUAL MODEL

The empirical side of the model involves "observing, listening and asking those who live, work and use the particular space" that knowledge. Digital art has means of providing people with new ways for arts interaction, rather than just viewing and incorporating audience reactions (Nyoni, 2025). This chapter introduces a model with three pillars:

a) Preservation

Preservation is the main pillar for leveraging digital tools for archiving practices. Protocols for ethical research with Indigenous peoples and communities include respecting Indigenous knowledge. This includes multiple Indigenous voices (especially in urban contexts) and negotiating ownership with appropriate representatives. Use Indigenous research advisory committees in urban contexts for data control. There must be establishment of co-authorship agreements, and the demonstration of reciprocity by sharing research and creating community capacity (Kovach, 2018). Digital technologies have significantly influenced both the study and formation of collective memory in several ways. Firstly, the advent of the Internet and web-based information and communication technologies has transformed how collective memories are created, shared, and preserved. This transformation is characterized by the emergence of what is referred to as "Digitalized Collective Memory," which represents a new form of remembrance facilitated by digital tools (Yasseri, Gildersleve & David, 2022).

b) Transmission

Embedding indigenous knowledge in digital education curricula. Mboti (2015) viewed goodwill as equivalent to good sense. Good sense is like good citizenship. Good citizenship contributes to ethical living. Good citizenship is therefore necessary to keep society.

c) Innovation

Enhancing relevance through modern technological applications. 'Ubuntu' is embodied with an ethical understanding "of what it is to be human and what is necessary for human beings to grow and find fulfilment." (Mboti, 2015, p.135). Exposure to storytelling, oral traditions, and indigenous art forms inspires innovative thinking and alternative perspectives. Archivists are actively defining, categorising, and disseminating documents in innovative ways, demonstrating an open-minded approach. Therefore, a new trend in which any stored information can be considered current, as data mining algorithms and predictive analytic techniques can generate meaningful new patterns from old data. One of the key impacts of digital technologies is the ability to trace and study collective memory in ways that were previously not possible. Digital tools allow researchers to analyse vast amounts of trace data, which can provide insights into how collective memories are formed both within digital spaces and in the physical world (Yasseri et al, 2022). This capability opens new avenues for understanding the dynamics of memory construction, as it enables the examination of interactions, discussions, and negotiations that shape collective remembrance.

INDIGENOUS KNOWLEDGE DIGITAL TRANSFORMATION FRAMEWORK

New models, framework and policies are in demand lately to cope with the tests associated with maintenance of indigenous knowledge systems (Masenya, 2024). This chapter suggests guidelines for utilising Indigenous Knowledge in Indigenous Higher Education contexts. This will be done to improve Indigenous Knowledge systems, the chapter discusses the Digital Transformation Framework, which serves as a guide for embracing innovation and adjusting to the changing needs of education in the digital age. This adds to the framework suggested by Mhlongo 2021; FAIR and CARE framework (Ngulube, 2024); The anticipation is the improvement of teaching and learning results since the Framework of Digital Transformation identified students as the primary pillar. Digital transformations are not a one-and-done project. The concept 'indigenous' in this chapter refers to what is local, original and native geographically. In Africa, it is used as a basis for identity, and it provides a basis for problem-solving (Masenya, 2024). In short Indigenous multimodal literacy practices are shaped by cultural identities (Mills, Davis-Warra, Sewell, & Anderson, 2016). Technologies are used to include Indigenous Knowledge Holders in Digital Cultural Heritage Preservation.

THE DEMAND FOR INDIGENOUS KNOWLEDGE DIGITAL TRANSFORMATION FRAMEWORK

In South Africa there is the National IK Registration System formerly known as the National Record System and the National IK Management System (NIKMAS). This represents an Information and Communications Technology (ICT) platform for the management of Indigenous Knowledge (IK) as governed by the Protection, Promotion, Development and Management of Indigenous Knowledge Act No 6 of 2019. The IK Registration System is the technology platform that is connected to IKS Documentation Centres that support communities to record their indigenous knowledge. The IK information accessed and published through the national IK Registration System has ownership vested in communities (NIKSO-PORTAL, 2024). The shortage of resources to fund IKS is the major challenge in education (Mhlongo & Ngulube, 2020). The demand exceeds the parameters of geographical landscape while curation of digital cultural heritage materials into a significant cultural force, made information exchange and connecting possible (McCarthy, 2007).

The demand for IKDTF is joining the struggle that was "conceived out of the need to centre African epistemologies deliberately and unapologetically into scholarship" (Cakata, 2023). While there is no one-size-fits-all solution, numerous collecting institutions in sub-Saharan Africa faced challenges while converting oral indigenous assets into digital objects and making them available. These institutions appeared to find managing metadata difficult. Since IK and data were not included into pre-existing collections there has been a lack of standardised metadata, since they existed in silos. There is a chance that IK holders' knowledge may have been misrepresented because many attempts to develop metadata for indigenous data and knowledge were carried out without their active participation. To choose the best taxonomy for IK data, extensive collaboration with indigenous knowledge holders is required (Ngulube, 2024). IKDTF aims at fast tracking the involvement of students to craft numerous archives with reliable data that would be verified by students themselves.

In a scoping review done by Masenya (2024), two themes of findings were identified concerning the demand of digital transformation. They were factors affecting the effective sharing and digital preservation of indigenous knowledge systems in indigenous communities and anticolonial strategies for recovery and promotion of indigenous knowledge systems. For knowledge sharing, two barriers were identified, namely the individual and organisational barriers. Under individual barriers lectures need to teach students to guard against fear, lack of time, low level of awareness, selfishness, lack of interpersonal skills and poor communication. Then under organisational barriers, institutions of higher education need to improvise organisational culture, guard against inadequate infrastructure, insufficient funding, improvise institutional resources plus training on how to incorporate IKS into current curriculum with relevant technical support.

THE ROLE OF LECTURERS AS CUSTODIANS OF INDIGENOUS KNOWLEDGE SYSTEMS (IKS) IN HIGHER EDUCATION

Ownership is the best custodian of IKS (Ntuli, 2002). Literature indicated that "knowledge management and sharing systems will be implemented and adopted by indigenous knowledge owners and their communities to preserve and share indigenous knowledge for the use of future generations (Masenya, 2024). To choose the best taxonomy for IK data, extensive collaboration with indigenous knowledge holders is required (Ngulube 2024). Tutu (1999) viewed lecturers as being 'caught up' in their ubuntu ontologies and epistemologies since they are dependent (kind of) humanity. In this regard, IKS has ever since prioritised the inclusion of students, as the lecturers' involved students in knowledge generation. Tutu (1999) identified that there are two aspects of being as a one-ness and an indivisible wholeness. Lecturers use their research subconscious mind the identification of who owns, strategizes, interprets, accounts on and eventually receive paybacks from the investigative process and product (Nyoni, 2025). The 'new digital art' (Özdemir, 2022) pushes the promotion of visibility of the epistemologies of the Global South, as interrogation means of decoloniality in higher education curriculum (Nyoni, 2025). Lecturers are categorised by Cakatha (2015) as the people who are having power to challenge the colonial practices that are engrained in the education system. However, lecturers encounter problems that inhibit effective curricular decolonisation, including a lack of information and skills in curriculum construction (Masenya, 2024).

Indigenous knowledge bearers have occasionally accepted copyright agreements without being made aware of the terms governing the future use of their information. Their authority and control over the knowledge were weakened as a result, due to being the carries of limited digital skills. This chapter push the agenda of seismic shift from institutional e-learning formats to open learning movements led by humans (Traxler, 2019). Proposed means to cover the gap would be an inclusion scholar from the Global South as the knowledge custodians to bridge the gap between non-indigenous knowledge and IK's knowledge management perspective (Nyoni, 2025, Traxler, 2019). Ngulube (2024) indicated the importance of balancing the FAIR and CARE principles while developing digital IK and data gathering. Lecturers have a responsibility to teach humanity as a goodwill gift, since Ubuntu emphasise communitarianism and harmony bound collectivism (Mboti, 2015) The technologies involved in networked activity are changing very fast, which demand lecturers to be ahead of their students in their usage. The curriculum should allow diverse epistemologies, with the guidance from lecturers as to how new knowledge should be merged. Concerning the digitisation of IK in Africa, Balogun (2023, p.99) recommended that "academic institutions should be more involved in the data capturing and ingesting of IK in their institutional repositories". Indigeneity covers indigenous and inclusive of those in the developing countries who struggle to survive and who still rely on traditional forms of knowledge whether they be Indigenous within developed and developing nation-states (Nakata, 2003:21). Indigenising education involves assessing the extent to which existing material and pedagogy acknowledge Indigenous/Aboriginal peoples' existence and the significance of Indigenous knowledge (Bhuda, 2021).

THE DIGITAL FACTORY

The ongoing learning cycle in Western education indicates that there is an ultimate end which is getting a certificate and being awarded but in indigenous there is no end moment of learning. Pluriversity approaches of digital art involve influences of the subconscious mind (Nyoni, 2025). Institutions of higher learning become hubs for documenting and preserving IKS, ensuring its transmission to future generations. This is an extension of the types of information and communication technology tools that are currently used in the management of indigenous knowledge (Dlamini & Ochola, 2018). Using a quantum theory, Ntuli (2002) advised Africans to reclaim their culture and understand the world in the way they perceive it. Time has passed for putting people in boxes, walling culture, no exchanging, promoting collaboration, strategies of crafting a better future, new perspectives must be allowed to describe people's content and context (Odora Hoppers, 2002). Nevertheless, decolonisation must start within academia. It will be when there will be a clear identification of what is valuable knowledge and decentralised knowledge? Accepting the baseline, notion of decolonisation implied the movement of the urgency of change. The African Renaissance provides a unique chance for Africans to define themselves and their agenda based on their own reality and the world around them (Makgopa, 2022).

Africa is still lagging in preserving its own IKS epistemologies and research methodologies in education that can be adopted by the rest of the continent (Balogun, 2023) The book "Reclaiming Our Space and Centring Our Knowledge" provides clear and detailed explanations of difficulties, as well as effective corrective and healing solutions (Cakata, 2023). Using South Africa's shared languages and cultural traditions, this chapter has proposed a methodology that could also benefit the most oppressed cultural groups through a series of chapters, in addition to giving the culturally ingrained knowledge required. No wonder Odora Hopper (2002) viewed IKS as forming the backbone of the social, economic, scientific and technological identity of people. Cakata (2015) contested that African communities have unique ways of knowing and doing, therefore Western pedagogy and methodologies are not always applicable. Documentation of knowledge, capture, documenting and dissemination of the knowledge. Knowledge that has been undermined, from Indigenous communities are revitalised and captured. This enabled knowing that knowledge is not created at one part, by one person and on one time. Preserving and documenting Indigenous Knowledge must not be government's duty alone but academics and students must play a vital role (Balogun, 2023). The long digital factory struggle enables Africans to decentering spaces we knew that they came but the strategies of decentering were not clear. The question how to decenter knowledge is equivalent to the intentional on how to change knowledge? Ajani et al (2024) in their findings recommended the strategies for effective preservative strategies. Decentering will be exploring the origins of the awakening of African values (Makgoba, 2022). An important realisation among lecturers is that they do not have IK themselves thus the notion of bringing in student's families and communities or other knowledgeable persons as resources will be a major pillar of gaining knowledge. Further lecturer's willingness to give up the monopoly on knowledge and, rather, to assume the role of a facilitator of lecture room discussion. Odora Hopper's (2002) has recommended the integration as a major critical engagement that also overcome power imbalances between Western and African epistemologies.

Technological advancements, such as mass printing, have transformed how we store and interact with memories since their inception (Yasseri et al, 2022). There is no "I" without "We" and the notion of we and the collective saturates the IKS digital factory. Lecturers are trusted with the intention of teaching intrinsic value of nature; humans live inside of nature. Thus, there is no substitute for exploratory learning. Indigenous education is one education that solely learn by doing rather than learning through book learning. IKDTF insisted on the need to apply immersive and experiential learning. Our societies form, sustain, and function because of our intelligence (Yasseri et al, 2022). To evaluate the effectiveness of programs that teach indigenous values, the key criteria that should be considered is the level of student engagement and cultural connection. In the efficient planning of digital factory, the following questions must drive the lecturers:

- Which approach best enables students to create their own understanding of indigenous values in a culturally responsive way? Could this be achieved by engaging students in a collaborative project?
- How to engage students in collaborative projects with indigenous communities?
- How can lecturers apply principles of indigenous knowledge to promote environmental stewardship among students? Which land-based learning experiences are in place?

- When evaluating educational curricula that incorporate indigenous knowledge, which strategy is most effective for fostering respect and understanding among students?

These questions will be clearly answered when integrating indigenous voices and perspectives into lessons. Most judicial systems in Africa continue to destroy African histories, knowledge systems, and customs, hence this could be called the cultural crime of colonisation (Castellino, 2024). Traditional African knowledge used to be passed down by oral tradition, including folklore, music, stories, dances, proverbs, rituals, local language, myths, beliefs, and games. Thus, cultural values, agricultural expertise, local flora and fauna, medicinal and culinary activities, history of the earth, stars, and water systems, and artefacts are all important aspects to consider as digital factories. Raseroka (2008) viewed these as means of securing space for passing traditional African knowledge. These involved using oral tradition, where folklore, music, stories, dances, proverbs, rituals, local language, myths, beliefs, and games to secure relevant knowledge and information. These factories are dynamic, renewing itself and adapting to changing circumstances. It innovates internally and uses external knowledge to adapt to local conditions. The usefulness of digital factory varies across data wherein a large portion of data included the need for Supervisor responses to professional development varied across the data. Data indicated the need for volunteering in the digital factory to develop activities for the digital factory. Digital factories crafted new avenues of co-creation where the voice of Indigenous communities and knowledge holders in technological design are custodians of knowledge. Thus, digital factories needed extensive development opportunities in schools, and institutions of higher learning using the following characteristics:

- Clear and strategically grounded indigenous principles and how the concepts fit into the overall educational strategy.
- Valuable and measurable objectives should be our focus rather than outputs or activities.
- The scope is holistic, hence digital factory should be broad enough to have a significant influence. Teams must consequently have the authority and commitment to solve educational challenges beyond organisational silos.

FUTURE RESEARCH DIRECTIONS

The trajectory of IKS using digital transformation framework resolute within the scope of this study. Also, the architectural element brought into existence the place-based pedagogy which embodies the culture, society, lectures, students, abstract and visual places. The focus is digital preservation outside the outside geographical landscape (Masenya & Ngulube, 2018), but disseminating knowledge in context (Mhlongo & Ngulube, 2019). Future study will demonstrate empirical direction of this architectural path I intend to map how indigenous knowledge systems could evolve within the digital landscape. Urgency of this study is endorsed, since the fourth Industrial Revolution brought the new wave of epistemicides which threatens fragile cultures and knowledge (Traxler, 2019).This will involve exploring the balance between maintaining traditional wisdom and embracing digital innovations, and how place-based pedagogies, rooted in the subconscious mind's connection to environment and culture, could shape the future of education and digital transformation. IKS is used as catalysts in cracking relative digital skills (Magoro, 2024).

CONCLUSION

Integrating digital frameworks into education systems presents a transformative opportunity to sustain indigenous knowledge. To achieve this first governments must prioritize inclusive digital policies. Analysing traditional ecological practices develops critical evaluation skills and fosters problem-solving abilities. Second, educational institutions should adopt participatory methods to engage indigenous communities. At the end, research should focus on creating adaptive technologies that honour cultural nuances. In an interconnected world, digital framework education can ensure that indigenous knowledge not only survives but thrives as a dynamic and integral part of global heritage.

The call for indigenisation of knowledge in African education institutions is a call to acknowledge rich knowledge in African cultures. The factory embraced the community's political, economic, and ethical perspectives, bringing vitality to the neighbourhoods. Indigenous knowledge systems are rich repositories of cultural, environmental, and social wisdom that have been passed down through generations. However, globalization and modernity threaten their survival. This chapter explores the potential of digital framework education in sustaining indigenous knowledge by integrating technology into preservation, dissemination, and rejuvenation efforts. The chapter presented a conceptual model that highlights how digital tools and inclusive education frameworks can empower indigenous communities while fostering cross-generational and intercultural learning. The chapter explored how place-based pedagogies tap into the subconscious mind, using sensory, emotional, and experiential connections to the land and environment. In the conclusion, this tied to the way technology, while often abstract and detached, can still evoke the same deep, subconscious learning connections by creating virtual spaces that simulate or honour indigenous ways of knowing.

REFERENCES

Ajani, Y. A., Oladokun, B. D., Olarongbe, S. A., Amaechi, M. N., Rabiu, N., & Bashorun, M.

Badenoch, A., & Follmer, G. (2018). Transationalizing Radio Research: New Encounters with An old medium. In Follmer, G., & Badenoch, A. (Eds.), *Transnationalizing Radio Research New Approaches to an Old Medium* (pp. 11–32). DOI: 10.1515/9783839439135-001

Balogun, T. (2023). Digitization of Indigenous knowledge systems in Africa: The case of South Africa's National Recorded System (NRS). *Records Management Journal, 1*(33), 88–102. DOI: 10.1108/RMJ-10-2022-0029

Bhuda, M. T. (2021). Making a case for indigenous education systems in South Africa. *African Journal of Development Studies, 2021*(1, si2), 67–82. DOI: 10.31920/2634-3649/2021/SIv1a4

Buthelezi, S. C., Ocholla, D., & Dlamini, P. (2024). *Strategies for documenting and disseminating Indigenous Knowledge at a South African university.* South African. DOI: 10.4102/sajim.v26i1.1648

Cakata, Z. (2015). *In search of the absent voice: The status of indigenous languages in post-apartheid South Africa* (Doctoral dissertation, University of South Africa).

Cakata, Z. (2023). We want to speak to our children: Introducing Azibuye. In *Z. Cakata, N.Z. Radebe & M.B. Ramose in Azibuye emasisweni. Reclaiming our space and centring our knowledge.* Oasis. DOI: 10.4102/aosis.2023.BK364.01

Carvalho, L., Goodyear, P., & de Laat, M. (2016). Place, space, and networked learning. In *Place-based spaces for networked learning* (pp. 1–10). Routledge. DOI: 10.4324/9781315724485-1

Castellino, J. (2024). Colonial crime, environmental destruction and indigenous peoples: A roadmap to accountability and protection. In *Calibrating Colonial Crime* (pp. 97–119). Bristol University Press. DOI: 10.1332/policypress/9781529241822.003.0005

Chamunorwa, M. B., Winschiers-Theophilus, H., & Zaman, T. (2018). An Intermediary Database Node in the Namibian Communities Indigenous Knowledge Management System, in Digitisation of Culture: *Namibian and International Perspectives* (Singapore: Springer, 99. DOI: 10.1007/978-981-10-7697-8_7

Chand, A. V. M. (2018). Place Based Approach to plan for Resilient Cities: A local government perspective. *Procedia Engineering, 212*, 157–164. Doi.org/1.10.1016/j.proeng.2018.01.021. DOI: 10.1016/j.proeng.2018.01.021

Dlamini, P., & Ocholla, D. N. (2018). Information and Communication Technology Tools for Managing Indigenous Knowledge in KwaZulu-Natal Province, South Africa. *African Journal of Library Archives and Information Science, 28*(2), 137–153.

Du, J. T. (2017). Research on Indigenous People and the Role of Information and Communications. Technology in Development: *A Review of the Literature. Journal of the Australian Library and Information Association, 66*(4), 344–363. DOI: 10.1080/24750158.2017.1397857

Faldi, G., Fisher, A., & Moretto, L. (2021). Five points for conceptualising place-based approaches to African urban planning: an introduction. In Faldi, G., Fisher, A., & Moretto, L. (Eds.), *African Cities Through Local Eyes: Experiments in Place-Based Planning and Design* (pp. 1–27). Springer. DOI: 10.1007/978-3-030-84906-1_1

Gumbo, M. T. (2023). Digitisation of higher education and research: Raising inclusivity and equity issues for indigenous students. *South African Computer Journal = Suid-Afrikaanse Rekenaartydskrif, 35*(1), 149–163. DOI: 10.18489/sacj.v35i1.1107

Hayes, M. (2022). Place-Based Pedagogy and Teacher Education. In Peters, M. A. (Ed.), *Encyclopedia of Teacher Education*. Springer., DOI: 10.1007/978-981-16-8679-5_293

Keikelame, M. J., & Swartz, L. (2019). Decolonising research methodologies: Lessons from a qualitative research project, Cape Town, South Africa. *Global Health Action, 12*(1), 1561175. DOI: 10.1080/16549716.2018.1561175 PMID: 30661464

Kovach, M. (2018). Doing Indigenous Methodologies: A Letter to a Research Class. In N.K. Denzin and Y.S. Lincoln. *The SAGE Handbook of Qualitative Research.*

Maddox, R., & Ninomiya, M. M. E. (2025). Indigenous sovereignty in research and epistemic justice: Truth telling through research. *Global Public Health: An International Journal for Research, Policy and Practice, 20*(1), 2436436. DOI: 10.1080/17441692.2024.2436436 PMID: 39661944

Madsen, V. (2018). Transnational Encounters and Peregrinations of the Radio Documentary Imagination. In Follmer, G., & Badenoch, A. (Eds.), *Transnationalizing Radio Research, New Approaches to an Old Medium 83-100, Creative commons*. DOI: 10.1515/9783839439135-008

Magoro, K. (2024). *Indigenous knowledge systems situated and contextual learning: unlocking collective human digital capabilities and competencies for Global South communities in the digital era*. ICEGOV.

Makgopa, M. (2022). Implications of the fourth industrial revolution (4IR) on the development of indigenous languages of South Africa: Challenges and opportunities. *Handbook of Research on Protecting and Managing Global Indigenous Knowledge Systems,* 152-161.

Masenya, T. M. (2024). Revitalization and Digital Preservation of Indigenous Knowledge Systems for Sustainable Development of Indigenous Communities in South Africa. *The Serials Librarian, 84*(5-8), 86–102. DOI: 10.1080/0361526X.2023.2277962

Masenya, T.M. & Ngulube, P. (2021). Digital Preservation Systems and Technologies in South African Academic Libraries. *South African Journal of Information Management* 23(1), 1249. https://doi.org/ . v23i1.1249DOI: 10.4102/sajim

Mbilinyi, D., & Mwabungulu, E. (2020). The Fate of Indigenous Knowledge: The Role Played by Libraries in Tanzania. *Information Development, 36*(4), 489–02. DOI: 10.1177/0266666919871088

Mboti, N. (2015). May the Real Ubuntu Please Stand Up? *Journal of Media Ethics, 30*(2), 125–147. DOI: 10.1080/23736992.2015.1020380

McCarthy, G. (2007). Finding a Future for Digital Cultural Heritage Resources Using Contextual Information Frameworks. In Cameron, F., & Kenderdine, S. (Eds.), *Theorizing Digital Cultural Heritage: A Critical Discourse* (pp. 245–260). The MIT Press. DOI: 10.7551/mitpress/6751.003.0017

Mdhluli, D., Mokgoatsana, S., Kugara, S. L., & Vuma, L. (2021). Knowledge management: Preserving, managing, and sharing indigenous knowledge through a digital library HTS. *Teologiesstudies. Theological Studies*, *77*(2).

Mendoza, M. A. D., De La Hoz Franco, E., & Gómez Gómez, J. E. (2023). Technologies for the Preservation of Cultural Heritage—A Systematic Review of the Literature. *Sustainability (Basel)*, *15*(2), 1059. DOI: 10.3390/su15021059

Metz, T. (2020). African theories of meaning in life. A critical assessment. *South African Journal of Philosophy*, *39*(2), 113–126. DOI: 10.1080/02580136.2020.1770429

Mhlongo, M. (2021). A Framework for the Integration of Indigenous Knowledge into Libraries in South Africa. *IFLA Journal*, *47*(3), 75–85. DOI: 10.1177/03400352211018224

Mhlongo, M., & Ngulube, P. (2020). Resource Provision and Access to Indigenous Knowledge in Public Libraries in South Africa. *Information Development*, *36*(2), 271–287. DOI: 10.1177/0266666919841095

Mills, K. A., Davis-Warra, J., Sewell, M., & Anderson, M. (2016). Indigenous ways with literacies: Transgenerational, multimodal, placed, and collective. *Language and Education*, *30*(1), 1–21. DOI: 10.1080/09500782.2015.1069836

Msila, V. (2021). Student Teachers, Professional Development and Effectiveness: The Role of Resilience in Enhancing the Culture of Learning and Teaching. *Universal Journal of Educational Research*, *9*(10), 1794–1804. DOI: 10.13189/ujer.2021.091008

Ngulube, P. (2024). From Orality to Digital Assets: Managing Indigenous Knowledge in Africa in the Wake of the Open Science Movement. *The Serials Librarian*, *84*(5-8), 71–85. DOI: 10.1080/0361526X.2023.2296880

NIKSOPORTAL. (2024). *The National Indigenous Knowledge Management System*. https://nikso.dst.gov.za/

Ntuli, P. P. (2002) *Indigenous Knowledge Systems and the African Renaissance. Laying a foundation for the creation of counter-hegemonic discourses*. In C.A. Odora Hoppers, Indigenous Knowledge and the Integration of Knowledge Systems: Towards a Philosophy of Articulation. 285

Nyoni, J. (2025). Interrogation of decoloniality in higher education curriculum reform: Examining influences of the subconscious mind. In E.T. Woldegiorgis, D.Z. Atibuni, E. Tadesse, & Woldegiorgis, G. Woldegiorgis, *Higher Education Transformation in Africa* (pp. 86-106). Routledge

Odora Hoppers, C. A. (2002). Indigenous Knowledge Systems, Sustainable Livelihoods and the Intellectual Property System: A Peace Action Perspective. *Journal of Peacebuilding &Development*, *1*(1), 106–112. DOI: 10.1080/15423166.2002.373049356048

Ozdemir, D. (2022). A conceptual framework on the relationship of digital technology and art. [IJonSES]. *International Journal on Social and Education Sciences*, *4*(1), 121–134. DOI: 10.46328/ijonses.313

Plockery, F. D. (2015). Indigenous Knowledge Production, Digital Media and Academic Libraries in Ghana. *The Journal of Pan African Studies*, *8*(4), 32–44.

Raseroka, K. (2008). Information transformation Africa: Indigenous knowledge e Securing space in the knowledge society. *The International Information & Library Review*, *40*(4), 243–250. DOI: 10.1080/10572317.2008.10762789

Seroto, J. (2012). Indigenous education in the pre-colonial era. In J.J. Booyse, C.S. le Roux, J. Seroto, C.C. Wolhuter. *A history of schooling in South Africa. Method and Context.* Van Schaik.37-52

Sweller, N., Dascalu, M., Ponciano, L., Dascalu, M., & McNamara, D. (2019). Designing Virtual Reality for Indigenous Language Education. *Educational Technology Research and Development*, *67*(1), 171–195.

T. (2024). Revitalizing Indigenous Knowledge Systems via Digital Media Technologies for Sustainability of Indigenous Languages. *Preservation, Digital Technology & Culture*, 53(1), 35-44.

Thylstrup, N. B. (2018). Cultural memory in the Digital Age. In Follmer, G., & Badenoch, A. (Eds.), *Transnationalizing Radio Research New Approaches to an Old Medium* (pp. 183–198). DOI: 10.1515/9783839439135-017

Traxler, J. (2019) Only Connect: Indigenous digital learning, *Interaction Design and Architecture(s)*, 41, pp. 7-23. https://wlv.openrepository.com/handle/2436/622885

Tutu, D. (1999). *No future without forgiveness*. Random House. DOI: 10.1111/j.1540-5842.1999.tb00012.x

Winschiers-Theophilus, H. (2022). *Bridging Worlds*. Indigenous Knowledge in the Digital World., DOI: 10.1145/3529372.3538820

Yasseri, T., Gildersleve, P., & David, L. (2022). Collective Memory in the Digital Age. In O'Mara, S. (Ed.), *Collective Memory*. Elsevier. DOI: 10.1016/bs.pbr.2022.07.001

KEY TERMS AND DEFINITION

Indigenous Knowledge: It is knowledge recognised as valuable in all aspects of life where an alternative perspective to modern scientific approaches is offered. This knowledge is shaped by direct interaction with context and social practices.

Digital Transformation Framework: It can be both structured and unstructured approach responsible for reshaping, processing and preserving of culture, values and integrity while selecting what works best for an individual, society, project and new business ventures.

Cultural Preservation: Preservation refers to lifestyle that has been adopted or chosen as good for a particular context, generational achievements to be shared and improvised. Connections of the pasts is cascaded to future generations, promoting diversity and understanding in the present. In this regard sense of identity, pride and continuity is emphasised.

Technical Integration: Process of connecting various systems, application and technologies ensuring that different software, hardware, data sources and platforms communicate, share data and perform tasks in a co-ordination.

Chapter 25
Models and Frameworks for Digital Preservation of Indigenous Knowledge Systems

Sundari M. Shanmuga
https://orcid.org/0000-0001-5755-474X
BVRIT Hyderabad College of Engineering for Women, India

Sri Harshini Jupally
https://orcid.org/0009-0009-2693-3056
BVRIT Hyderabad College of Engineering for Women, India

Nainika Reddy Vempalli
https://orcid.org/0009-0007-7162-2949
BVRIT Hyderabad College of Engineering for Women, India

ABSTRACT

The digital preservation of Indigenous Knowledge Systems (IKS) is a critical endeavor that requires the integration of community-centric, collaborative, and decentralized models. Effective preservation frameworks are needed to address technical, ethical, and legal aspects while overcoming challenges related to technology, cultural sensitivity, sustainability, and capacity building. By leveraging digital technologies, we can ensure that IKS are preserved, respected, and transmitted to future generations, contributing to the richness and diversity of human knowledge. The ongoing collaboration between indigenous communities, researchers, and institutions is essential to achieving this goal, fostering a future where indigenous knowledge is valued and protected in the digital age.

INTRODUCTION

Indigenous Knowledge Systems (IKS) are the collection of traditions, wisdom, and practices developed by indigenous communities over centuries. Because they integrate cultural, ecological, social, and spiritual components into a cohesive whole, these systems are by nature holistic. In contrast to contemporary scientific methods, IKS are region-specific, context-specific, and the result of direct engagement with

DOI: 10.4018/979-8-3693-7964-6.ch025

the local environment. This knowledge is a great source of sustainable and adaptable tactics because it covers topics like land management, medicine, agriculture, spirituality, and social customs. Urbanization, climate change, and globalization are some of the current threats to IKS. Because of their exposure to contemporary lifestyles, the younger generation frequently disassociates themselves from customs. Additionally, there is a serious concern associated with the loss of indigenous languages, which serve as the main means of passing forward this information. Notwithstanding these obstacles, the significance of IKS is becoming more widely acknowledged in international discourse, especially in relation to climate resilience, sustainable development, and cultural heritage preservation. Efforts to conserve, record, and safeguard IKS for upcoming generations have resulted from this awareness.

Importance of Digital Preservation

A vital tactic for preserving Indigenous Knowledge Systems (IKS) in the face of modernity and the digital era is digital preservation. IKS may be recorded, preserved, and disseminated in ways that cut across national borders by utilizing digital technology, guaranteeing that present and future generations can continue to access this priceless legacy. The forces of urbanization and globalization, which have endangered traditional languages, cultural practices, and ways of life, are balanced by digital preservation.

Challenges in Preserving Indigenous Knowledge

There are many cultural and technological obstacles to the preservation of Indigenous7Knowledge Systems (IKS). IKS, which are frequently transmitted orally throughout generations or ingrained in customs, are in danger of disappearing as a result of urbanization, globalization, and the uprooting of indigenous populations. The survival of indigenous languages, which serve as the main means of passing down this information, is also under jeopardy due to these pressures, in addition to the continuation of traditional activities (Hewson & Holtman, 2022). The transmission of IKS to future generations is becoming more and more precarious as younger generations adopt more contemporary lives.

The absence of suitable and sustainable technology is one of the main obstacles to maintaining IKS. Many indigenous people live in isolated locations with little access to infrastructure and digital technologies. Using contemporary digital platforms to document, archive, and disseminate knowledge is challenging due to this technology gap. Furthermore, previous digital formats may become outdated due to the quick speed of technological advancement, endangering the long-term viability (Nkebem & Ofulue, 2022) of digitized IKS. Funding and technical know-how are two continuous resources that are frequently in short supply for maintaining and updating digital archives.

Enhanced Literature Survey

Indigenous Knowledge (IK) preservation has evolved from traditional oral transmission to digital documentation and AI-driven analysis. Various frameworks have been developed to ensure the longevity and accessibility of IK while maintaining cultural authenticity and community ownership. Table 1 compares the most widely used models for digital IK preservation.

Table 1. Comparative literature survey

Model/Framework	Approach	Key Features	Advantages	Challenges
Community-Centric Model	Led by Indigenous communities with a focus on self-governance and cultural sovereignty.	Digitization of oral histories by local elders	•Local ownership and control •Strengthens community participation	•Risk of knowledge silos •Requires long-term engagement
Decentralized Model	Uses blockchain and distributed networks for secure IK storage. | - Tamper-proof digital records	Tamper-proof digital records	•Smart contracts for access control •Open-source architecture	Resistant to censorship Potential resistance from traditional custodians
Collaborative Model	Partnerships between Indigenous communities, academic institutions, and government agencies.	AI and NLP for knowledge retrieval	Access to technical expertise Scalable and adaptable	Data ownership concerns Ethical issues in sharing IK

Defining Indigenous Knowledge

Through generations of direct engagement with their local environments, indigenous cultures have evolved and polished a vast body of wisdom, traditions, and beliefs that are collectively referred to as Indigenous Knowledge. Agriculture, medicine (Sundari, Penthala & Nayyar, 2024), environmental management, spirituality, social structure, and craftsmanship are just a few of the fields it covers. Deeply ingrained in a community's cultural and spiritual life, indigenous knowledge frequently acts as a manual for leading a sustainable and peaceful existence in balance with the environment. Indigenous knowledge is flexible, context-specific, and frequently passed down orally and through real-world experiences, in contrast to formal scientific information, which is usually recorded and categorized. Indigenous knowledge is holistic, which means it incorporates many facets of existence rather than separating topics like contemporary science. Indigenous agriculture (Tshabalala & Mthunzi, 2021) practices, for instance, frequently include ecological, spiritual, and social aspects, demonstrating a thorough awareness of the environment. Indigenous knowledge systems also place a strong emphasis on relationships, emphasizing balance and interconnection between humans, the natural world, and the cosmos which explained in Figure 1.

Figure 1. Indigenous practices

Traditional Methods of Knowledge Preservation

Indigenous Knowledge has traditionally been preserved through various non-digital methods, relying heavily on oral traditions, practical demonstration, and cultural practices. Oral transmission is the cornerstone of Indigenous Knowledge preservation, with elders and knowledge holders passing down wisdom to younger generations through stories, songs, proverbs, and rituals. Storytelling is an essential preservation technique that is used to teach and transmit morals, values, and useful information in addition to providing entertainment. Mythological stories are used by many indigenous societies to explain historical events, social institutions, and natural occurrences, creating a rich cultural legacy (Glover & Poitras, 2023). As venues for knowledge enactment, reinforcement, and transmission, rituals, ceremonies, and festivals are also essential to knowledge preservation. Knowledge is woven into the cultural fabric through these events, which frequently incorporate distinctive customs, symbols, and languages.

Challenges Associated with Traditional Methods of Indigenous Knowledge (IK) Preservation

Indigenous Knowledge (IK) has historically been preserved through non-digital means, such as oral traditions, practical demonstrations, and cultural practices. While these methods have sustained communities for generations, they also face several challenges in the modern era. The shift toward globalization, the decline of native speakers, and the lack of formal documentation threaten the survival of traditional knowledge systems. The key challenges associated with these preservation methods are outlined below:

Vulnerability to Loss Due to Oral Transmission

Oral traditions remain the primary means of passing down Indigenous Knowledge, with elders transmitting wisdom through storytelling, songs, proverbs, and rituals. However, oral transmission is highly susceptible to loss due to factors such as aging knowledge holders, declining interest among younger generations, and language extinction. Once elders pass away without transferring their knowledge, valuable cultural and historical insights can be permanently lost.

Lack of Formal Documentation and Standardization

Traditional Indigenous Knowledge is often preserved in informal, non-written forms, making it difficult to systematically archive or share across generations. Unlike written records, which can be stored and referenced over time, oral traditions and practical demonstrations do not provide a standardized or easily retrievable format. The absence of structured documentation limits accessibility, consistency, and scalability in preserving IK.

Influence of Globalization and Cultural Erosion

The increasing impact of globalization has contributed to the gradual erosion of Indigenous Knowledge systems. As younger generations become more engaged with mainstream education and digital technologies, they may prioritize modern knowledge over traditional wisdom. Additionally, urbanization and migration disrupt traditional community structures, leading to reduced opportunities for knowledge transmission through ceremonies, storytelling, and practical demonstrations.

Language Endangerment and Extinction

Many Indigenous Knowledge systems are closely tied to native languages, which serve as vehicles for transmitting cultural wisdom. However, a significant number of Indigenous languages are endangered, with some facing extinction. When a language disappears, the knowledge embedded within its linguistic expressions, metaphors, and terminologies also risks being lost, making preservation efforts even more challenging.

Limited Accessibility and Exclusivity

Traditional IK preservation methods are often community-specific, limiting their accessibility to outsiders, researchers, or future generations who may not be actively engaged in Indigenous practices. Some forms of knowledge are considered sacred and are restricted to specific groups within the community. While this exclusivity safeguards cultural integrity, it also makes broader preservation efforts difficult, particularly in digital or written forms.

Disruptions in Knowledge Transmission Due to Modern Lifestyles

Indigenous Knowledge transmission relies heavily on intergenerational learning through rituals, ceremonies, and storytelling. However, modern lifestyles, including increased reliance on formal education, migration to urban centres, and changing family structures, have disrupted these traditional learning pathways. Younger generations may have limited exposure to Indigenous teachings, reducing their ability to absorb and carry forward ancestral knowledge.

Difficulty in Adapting to Contemporary Knowledge Systems

Traditional methods of IK preservation often do not align with modern information management techniques. Indigenous Knowledge is typically holistic, context-dependent, and embedded within cultural and spiritual practices. In contrast, contemporary knowledge systems prioritize written records, scientific validation, and structured categorization. Bridging the gap between these knowledge paradigms remains a significant challenge.

Cultural Sensitivity and Ethical Considerations

Indigenous knowledge preservation necessitates a strong dedication to ethical and cultural sensitivity. Certain cultural norms are followed by indigenous cultures about who has the right to access knowledge, how it should be communicated, and what information is private or sacred. These procedures are essential to maintaining the integrity of Indigenous Knowledge since they are frequently based on spiritual beliefs, rituals, and past experiences. Respecting these limits and making sure that preservation initiatives don't unintentionally hurt or mislead the people concerned are key components of cultural sensitivity.

A key factor is data sovereignty, which highlights indigenous groups' right to manage and control their data. Guidelines for managing indigenous data are provided by ethical frameworks such as the CARE (communal Benefit, Authority to Control, Responsibility, and Ethics) principles, which emphasize the ethical treatment of information and its communal benefit. By ensuring that preservation initiatives are in line with the community's interests and cultural values, these frameworks aim to stop the commodification of Indigenous knowledge. Any preservation effort must continue to prioritize ethical issues in order to build relationships of trust and cooperation with indigenous groups.

Role of Language and Oral Traditions

Indigenous information Systems are centred on language and oral traditions (Ferreira & Pienaar, 2024), which are the main means of passing down information from one generation to the next. Indigenous languages have distinct worldviews, spiritual beliefs, and ecological understandings, making them more than merely means of communication. Specific terminology, concepts, and classifications pertaining to the local environment are unique to each language and frequently do not have direct translations in other languages. Because language loss frequently corresponds to the loss of cultural identity and traditional wisdom, language preservation is therefore a crucial component of protecting Indigenous knowledge.

DIGITAL PRESERVATION MODELS

Overview of Digital Preservation Concepts

The phrase "digital preservation" describes the methods and procedures used to protect digital content throughout time so that it can be accessed and used by future generations. This idea includes managing, storing, and safeguarding digital files in addition to digitization. The goal of digital preservation is to stop data loss via unintentional deletion, physical deterioration, and technological obsolescence. Digital preservation for Indigenous Knowledge Systems (IKS) entails establishing digital archives of customs, languages, ecological knowledge, and cultural assets. Communities may protect their legacy (Sundari, et al., 2024) from the dangers of globalization and cultural deterioration by storing and maintaining these materials in digital archives, databases, and repositories.

Key Technologies in Digital Preservation (Databases, AI, Blockchain, etc.)

A number of critical technologies are necessary for the digital preservation of Indigenous knowledge, and each one contributes differently to data security, accessibility, and authenticity. Databases are the fundamental mechanism for storing vast amounts of data and provide organized spaces for cataloguing Indigenous knowledge. Information of all kinds, from multimedia files to textual descriptions, can be arranged and managed using relational databases, document databases, and graph databases (Alessa & Kliskey, 2020). Digital repositories are frequently included with these databases, offering centralized platforms for maintaining, sharing, and accessing digital content.

Initiatives for digital preservation are also incorporating other technologies, such as Augmented Reality (AR) (Odeh & Akhigbe, 2023) for immersive cultural experiences and Geographic Information Systems (GIS) (Njagi & Mwangi, 2021) for mapping cultural landscapes. Figure 2 explains the new technologies. These tools not only document Indigenous Knowledge but also present it in ways that engage both indigenous communities and the broader public, fostering cultural awareness and understanding.

Figure 2. Technologies in digital preservation

Metadata Standards for Indigenous Knowledge

In order to properly define, arrange, and manage digital content, metadata is essential to the digital preservation of Indigenous knowledge. As the "data about data," it provides details about the origin, content, context, and rights of every digital record. For Indigenous Knowledge to be appropriately represented and continue to be available in the future, appropriate metadata standards are necessary. In order to properly classify and index Indigenous Knowledge, controlled vocabularies and ontologies are essential parts of metadata standards. Better searchability and interoperability between digital archives are made possible by these technologies (Barnhardt & Kawagley, 2022), which make it easier to describe indigenous terminology consistently and culturally appropriate. In order to preserve sensitive data and provide indigenous people control over who can access and utilize their knowledge, metadata is also important for controlling access rights and data sovereignty. To ensure that their viewpoints and knowledge systems are respected, developing metadata standards for Indigenous Knowledge necessitates close cooperation with indigenous populations. In order to acknowledge the diversity and complexity of Indigenous Knowledge Systems, this method entails developing adaptable frameworks that can be tailored to the unique requirements of various communities.

Long-Term Accessibility and Sustainability

When it comes to the digital preservation of Indigenous Knowledge Systems (IKS), sustainability and long-term accessibility are important objectives. Strategies are required to mitigate the danger of digital data obsolescence due to the rapid evolution of technology. To ensure data durability, effective preservation entails backward compatibility, adherence to open standards, and periodic migration to new formats. Although open-source software reduces reliance on proprietary platforms and offers flexibility, it needs regular technical assistance and training to stay competitive. Another important consideration is financial sustainability, since digital archives require steady funding for staff, infrastructure, and continuing upkeep. Collaborations with commercial businesses (Singh & Roy, 2023), governmental organizations, non-profits, and academic institutions can supply the resources and know-how required for long-term initiatives.

FRAMEWORKS FOR THE DIGITAL PRESERVATION OF INDIGENOUS KNOWLEDGE SYSTEMS

Community-Centred Models

The ethical digital preservation of Indigenous Knowledge Systems (IKS) is based on community-centred methods, which highlight the significance of indigenous participation, control, and decision-making at every stage of the preservation process. Since indigenous communities are the legitimate stewards of their cultural heritage, these approaches place a high priority on their active participation. Community-centred strategies explained in Figure 3, in contrast to traditional top-down methods (Alemu & Assefa, 2023), are cooperative and involve communities in all facets, from data administration and access control to planning and documentation.

Figure 3. Indigenous communities

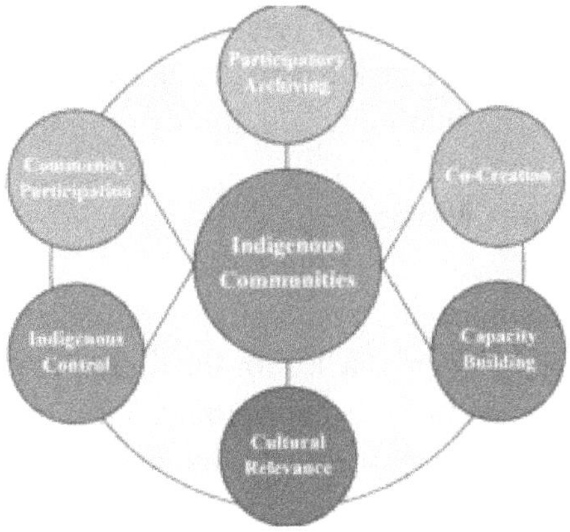

Participatory Archiving

A cooperative technique called "participatory archiving" enables indigenous groups to have an active role in the recording, conservation, and administration of their knowledge. This method is in contrast to conventional archival procedures, which sometimes entail outside scholars recording indigenous knowledge with little to no participation from the communities. Indigenous people are not only subjects in participatory archiving; they are also co-researchers, content providers, and decision-makers. By ensuring that digital archives represent indigenous viewpoints, beliefs, and cultural customs, this approach preserves the integrity and authenticity of knowledge that has been saved. By establishing trust between indigenous and non-indigenous partners (Park & Lee, 2024), participatory archiving promotes fair relationships and produces archives that are more sustainable and culturally relevant. This approach not only safeguards IKS but also strengthens cultural identity, resilience, and community pride, ensuring that the knowledge remains vibrant and relevant for future generations.

Co-Creation and Co-Ownership of Knowledge

The digital preservation of Indigenous Knowledge Systems (IKS) is guided by the concepts of co-creation and co-ownership, which place an emphasis on respect, equity, and cooperation. Co-creation is the process of documenting and digitizing traditional knowledge in collaboration with indigenous people and outside partners, including NGOs, academic scholars, and technological specialists. This collaboration guarantees that the preservation procedure appropriately reflects indigenous viewpoints and is in line with traditional values. Going one step further, co-ownership gives indigenous groups authority over the digital assets they have contributed to the creation of. The communal character of IKS, where information is frequently exchanged among members and passed down through generations, is acknowledged by this principle. It opposes Western ideas (Woods & Blanchard, 2020) of private property and individual ownership in favor of shared stewardship that honors customary custodianship responsibilities. Communities are empowered by co-ownership to control access, establish usage guidelines, and guarantee that their knowledge is shielded from exploitation and theft. By fostering fair collaborations and fostering trust, these guidelines make sure that efforts to preserve digital content are not exploitative but rather advantageous to both parties. Since indigenous people actively participate in the upkeep and upgrading of digital archives, co-creation and co-ownership promote sustainable preservation initiatives. These practices are crucial for creating ethical, inclusive, and culturally appropriate digital archives, ensuring that the preservation of IKS honors the autonomy and dignity of indigenous peoples.

Decentralized and Blockchain Models

Concerns regarding data quality, authenticity, and control are addressed by decentralized and blockchain models, which provide creative and safe approaches for the digital preservation of Indigenous Knowledge Systems (IKS). By transferring authority from centralized institutions to indigenous communities through decentralized technologies, these approaches enable them to autonomously manage their digital assets. A decentralized ledger technology called blockchain ensures that data is safe and unmodified by producing a transparent and impenetrable record of transactions. For IKS, where the legitimacy of saved knowledge is essential, this is especially beneficial. Additionally, benefit-sharing agreements can be facilitated by blockchain, giving indigenous people the ability to specify terms for the access and use of their knowledge, guaranteeing its ethical and equitable use.

In keeping with traditional governance and custodianship roles, decentralized platforms empower indigenous peoples to manage who can access and alter their data. These systems are perfect for long-term preservation because they are robust and independent of a single server or organization. A clear foundation for information management is provided by decentralized and blockchain models, which also present chances for open documentation of consent agreements, intellectual property rights, and cultural customs. But putting these technologies into practice calls for rigorous evaluation of ethical standards, technical capability, and cultural values. Decentralized and blockchain approaches, when applied carefully, can produce inclusive, safe, and long-lasting digital archives that respect the intricacy and authenticity of indigenous customs.

Benefits of Decentralization for Indigenous Knowledge

Decentralization improves community sovereignty, data integrity, and resilience, among other advantages for the preservation of Indigenous Knowledge Systems (IKS). Indigenous communities maintain authority over their digital assets in decentralized models, choosing how their knowledge is recorded, kept, and accessed. This lessens dependency on outside organizations that have traditionally controlled the conservation of indigenous cultural heritage, which frequently results in problems with cultural appropriation, exploitation, and distortion. Data sovereignty is given priority in decentralized systems, which enable communities to manage their information in accordance with customary norms and values. A crucial element of decentralization, blockchain technology produces unchangeable, tamper-proof records that guarantee the authenticity and transparency of stored data. Because communities can confirm how their information is used and shared, this helps foster confidence between indigenous and non-indigenous partners. Because digital archives are dispersed over numerous nodes, decentralization also increases their resilience by lowering the possibility of data loss via centralized control or technical malfunctions. Furthermore, fair benefit-sharing can be facilitated by decentralized models, giving communities the power to determine the conditions for any commercial or scholarly use of their knowledge.

Decentralized techniques are in line with ethical preservation ideals by giving indigenous peoples the majority of control, which promotes respectful and culturally appropriate framework for digital archives.

Use Cases in Indigenous Communities

In a variety of indigenous contexts, decentralized and blockchain technologies have demonstrated encouraging outcomes, offering safe and culturally relevant platforms for the digital preservation of Indigenous Knowledge Systems (IKS). Blockchain technology has been utilized in Australia to preserve and record traditional ecological knowledge, resulting in transparent, unchangeable records that are still under the authority of Aboriginal people. Some First Nations in Canada have created decentralized systems to oversee their digital archives, guaranteeing that only authorized personnel may access sensitive data. These projects have also looked into benefit-sharing arrangements using blockchain technology, which enables communities to impose restrictions on the commercial usage of indigenous knowledge, such sustainable resource management techniques or traditional medicines.

By lowering the dangers of exploitation and promoting trust, blockchain's distributed ledger system offers an open means of monitoring how knowledge is accessed, shared, and used. These use examples demonstrate how decentralized approaches can empower indigenous communities by granting them more authority over their data sovereignty and cultural heritage. They also show how cutting-edge digital technology can be used to conserve old knowledge systems in an ethical and efficient manner by fusing contemporary instruments with long-standing cultural customs. These examples demonstrate how inclusive and respectful preservation may be accomplished in the digital era and provide a model for future initiatives.

Intellectual Property and Rights Management

The ethical digital preservation of Indigenous Knowledge Systems (IKS) depends on the management of intellectual property and rights. Indigenous knowledge frequently functions under collective ownership, which reflects the shared character of cultural legacy that has been passed down through the ages. Western intellectual property regimes, which prioritize private property and business rights, are different from this. A culturally aware strategy that recognizes customary custodianship responsibilities and values the distinct viewpoints of indigenous groups is necessary for safeguarding IKS in a digital setting. Legal safeguards against unapproved use, exploitation, and misappropriation include cultural use agreements, traditional knowledge licensing, and community-driven norms. By outlining the conditions of access, use, and benefit-sharing, these agreements make sure that any commercial or scholarly exploitation complies with the standards and values of the community. Indigenous communities should be included in the development of community-driven protocols that take into account their governance systems and cultural norms. In addition to preventing cultural appropriation, good rights management promotes fair collaborations between indigenous and non-indigenous partners. By promoting data sovereignty, it guarantees that communities keep authority over the dissemination and application of their knowledge. By supporting respectful and reciprocal digital preservation initiatives, these frameworks go beyond simple legal obligations to uphold the integrity, authenticity, and dignity of Indigenous Knowledge Systems.

Protection of Traditional Knowledge

A cautious and culturally aware strategy that strikes a balance between the advantages of preservation and the necessity to maintain indigenous cultural assets is needed to protect traditional knowledge in digital environments. Strict guidelines for documentation, preservation, and sharing are necessary since many components of Indigenous Knowledge Systems (IKS) are regarded as sacred, private, or limited.

Legal safeguards, like traditional knowledge licenses, offer a flexible approach to access management, enabling communities to establish guidelines for who can use their knowledge and how. These agreements are necessary to stop traditional knowledge from being used, misrepresented, or commercialized without permission. Cultural appropriation should also be covered by ethical principles, which should set up safeguards against IKS being misrepresented or turned into a commodity for profit. Protection measures must be co-developed with indigenous communities, ensuring that they reflect traditional values and cultural expectations.

Data Sovereignty and Ownership Models

Important considerations in the digital preservation of Indigenous Knowledge Systems (IKS) are data sovereignty and ownership. The right of indigenous communities to manage own data, making sure that it is gathered, preserved, and disseminated in accordance with their cultural norms and values, is known as data sovereignty. This idea stems from the understanding that indigenous knowledge is a shared resource that is frequently owned by certain families, clans, or tribes and is administered in accordance with customary custodianship responsibilities. IKS ownership models need to take these collective viewpoints into account, unlike Western models that put individual liberties and business interests first.

Digital platforms under indigenous control can promote data sovereignty by giving communities the ability to control own archives, impose access limitations, and determine who is permitted to access, use, or contribute to the material. These platforms ought to be closely developed. These platforms should be developed in close consultation with the communities involved, ensuring that they align with traditional governance structures (Sundari et al., 2024). In addition to being morally and legally required, data sovereignty empowers indigenous peoples by granting them authority over the preservation and dissemination of their cultural legacy. By ensuring that digital preservation upholds Indigenous Knowledge Systems' autonomy, dignity, and integrity, it promotes a more inclusive and egalitarian online space.

Financial Models for Long-Term Digital Preservation

Sustaining digital preservation efforts requires robust financial backing. Various funding models have been proposed to ensure the longevity and accessibility of IK archives. Table 2 compares the key financial models for digital IK preservation.

Table 2. Comparison of financial models for digital IK preservation

Financial Model	Financial Model	Financial Model	Financial Model
Government Grants	Public funding for Indigenous heritage projects	• Cultural ministries allocate annual budgets for digitization • Reliable long-term funding • Supports large-scale digitization projects	• Limited flexibility in resource allocation • Political influence on funding priorities
NGO & Non-Profit Partnerships	Funding from global organizations (e.g., UNESCO, World Bank)	• Grants for Indigenous-led initiatives • Supports community-driven efforts • Provides expertise and resources	• Dependence on external organizations • Competitive grant application process
Public-Private Collaborations	Tech companies, universities, and Indigenous groups work together	• AI and cloud services for IK preservation • Access to cutting-edge technology • Scalable and cost-effective solutions	• Risk of corporate exploitation • Data ownership and profit-sharing concerns
Crowdfunding & Community Contributions	Platforms like GoFundMe and Kickstarter	• Small-scale community-based funding • Empowers Indigenous communities • Allows direct public participation	• Limited funding potential • Requires strong outreach and visibility

Alternative Technologies for Data Integrity and Sovereignty

- While blockchain provides immutability, other technologies can offer viable alternatives.
- Decentralized File Storage Systems (e.g., IPFS, Solid): Provide a way to store and distribute knowledge without dependence on centralized servers.
- Version Control Systems (e.g., Git-based repositories): Allow communities to track changes and revert access when needed without high computational costs.
- Cloud-Based Access with Local Backup Strategies: Ensures redundancy while giving communities control over how and when data is accessed.
- Example: The First Nations Information Governance Centre (FNIGC) in Canada is exploring multiple strategies, including cloud and local storage, to ensure data sovereignty.
- Collaborative Institutional Partnerships
- Libraries, museums, and universities work alongside Indigenous communities to co-develop digital archives.
- Example: The Digital Himalaya Project collaborates with Indigenous groups to digitize and share cultural resources ethically.

CASE STUDIES OF DIGITAL PRESERVATION IN INDIGENOUS COMMUNITIES

Successful Digital Preservation Projects

The potential of contemporary technology to preserve and disseminate Indigenous Knowledge Systems (IKS) has been shown by a number of successful digital preservation initiatives. These initiatives offer insightful illustrations of how to preserve, document, and display IKS in a way that upholds cultural norms and encourages knowledge accessibility. The Digital Song lines Project in Australia is one such project. This project creates immersive cultural landscapes using Geographic Information Systems (GIS), 3D visualization, and digital storytelling techniques. The initiative enables indigenous groups to share their cultural heritage while managing access to sensitive data by digitally mapping and displaying traditional stories and landscapes. It demonstrates how technology may be utilized to record Indigenous knowledge in its entirety, including its ecological and spiritual aspects, in addition to facts.

Collaborative Frameworks for Preservation

For Indigenous Knowledge Systems digital preservation initiatives to be effective, cooperation is essential. In order to develop thorough and culturally aware digital preservation techniques, collaborative frameworks (Dawson & Levy, 2022) bring together indigenous people, academic institutions, researchers, and technology vendors. These collaborations support fair and moral preservation initiatives by guaranteeing that indigenous voices continue to play a key role in the documentation and digitalization processes. The foundation of collaborative frameworks is frequently a community-centric approach, in which indigenous groups are included from the very beginning of project design through its execution and assessment. An example of a collaborative digital platform (Ramirez & Mari, 2021) that prioritizes the requirements of indigenous people is the Mukurtu Content Management System (CMS).

Challenges Encountered and Lessons Learned

Despite being crucial for preserving Indigenous knowledge, digital preservation efforts confront a number of obstacles that teach important lessons for similar programs in the future. To guarantee the viability and integrity of digital preservation initiatives, these difficulties which range from ethical conundrums and technical obstacles to sustainability issues need to be carefully considered. The technical gap in many indigenous communities is one frequent problem. Effective documentation and archiving of digital content may be hampered by a lack of technological know-how, dependable internet, and access to digital infrastructure. Investments in technology and capacity-building programs that enable communities to autonomously manage their digital assets are necessary to overcome this. One important way to close this gap has been to teach indigenous people digital skills and involve them in the technical aspects of preservation.

Technical Challenges in Digital Preservation of IKS

- Technological Obsolescence: Digital formats and storage mediums change rapidly, leading to the loss of critical information.

- Example: The BBC Domesday Project (1986) stored vast amounts of data on specialized laser-discs, which became obsolete within a decade, rendering the data inaccessible.
- Example: NASA's 1970s Lunar Orbiter tapes, which were stored on outdated magnetic tapes, required extensive efforts to recover due to the lack of compatible reading devices.
- Example: The MySpace data loss incident in 2019 resulted in the permanent deletion of over 50 million music files due to server migration failures, demonstrating the fragility of digital archives.
- Lack of Community Engagement: Without active participation from Indigenous communities, digital archives risk being incomplete, inaccurate, or disconnected from their intended cultural context.
- Example: The Rosetta Project, an effort to document global linguistic diversity, faced criticism for not involving Indigenous communities in decision-making, leading to concerns about misrepresentation.
- Example: The HRAF (Human Relations Area Files) faced scrutiny for archiving Indigenous knowledge without direct consultation with the communities involved, raising ethical concerns about data sovereignty and misuse.

TECHNOLOGICAL SOLUTIONS FOR INDIGENOUS KNOWLEDGE PRESERVATION

Digital Repositories and Databases

Indigenous Knowledge Systems (IKS) preservation relies heavily on digital repositories and databases. They provide an organized, easily accessible, and safe method for managing, preserving, and disseminating indigenous cultural material. An online database that stores digital items, such as text, music, video, and multimedia files, is called a digital repository. To ensure that their knowledge is available to future generations, these repositories give indigenous people a single location to gather, arrange, and preserve it. Digital repositories for Indigenous knowledge frequently place an emphasis on cultural sensitivity and adaptability, enabling communities to set access parameters and handle their material in accordance with their beliefs.

Open Access vs. Controlled Access

When it comes to the digital preservation of Indigenous knowledge, the argument between limited and free access is vital. When digital content is freely available to the public, anybody can read, distribute, or use it. This is known as open access. Controlled access, on the other hand, places limitations and calls for credentials or permits in order to access particular content. When deciding between open and regulated access, Indigenous Knowledge Systems must carefully weigh practical, cultural, and ethical considerations. By enabling indigenous groups to handle their digital assets in accordance with their cultural norms, controlled access provides a solution. It guarantees that those who possess knowledge have the power to decide who can access their data, when they can do so, and why. For knowledge that is regarded as sacred, restricted, or unique to a particular community member, controlled access is especially crucial. By preventing illicit usage, economic exploitation, and cultural misrepresentation, it protects Indigenous Knowledge's authenticity and integrity.

Indigenous Knowledge and Artificial Intelligence

With technologies that improve documentation, analysis, and accessibility, artificial intelligence (AI) is becoming more and more significant in the digital preservation of indigenous knowledge. Artificial intelligence (AI) tools like computer vision, machine learning, and natural language processing help expedite the digitalization of conventional information, making it simpler to handle and distribute across digital platforms. To guarantee that it is in line with the cultural norms and expectations of indigenous groups, the application of AI in this setting needs to be carefully considered. These oral records can be automatically translated and transcribed by AI-powered tools, opening them up to a larger audience. Indigenous languages can also be analyzed by natural language processing (NLP) (Uddagiri & Shanmuga Sundari, 2023) algorithms, which can find links and patterns in the data that might not be immediately obvious. By supporting language revitalization initiatives, endangered languages and dialects are preserved.

Use of Virtual and Augmented Reality in Knowledge Preservation

With their immersive and participatory approaches to cultural heritage, virtual reality (VR) and augmented reality (AR) (Cheng & Li, 2024) are becoming increasingly potent instruments for the presentation and preservation of indigenous knowledge. Through the creation of dynamic platforms for storytelling, education, and cultural renewal, VR and AR technologies present special chances to connect both indigenous people and the general public.

Through virtual reality (VR), users can experience scenarios that mimic historical landscapes, cultural settings, and spiritual sites. VR, for instance, can replicate customary dances, ceremonies, and rituals, giving viewers the impression that they are actually there. This captures the geographical and contextual components of Indigenous Knowledge that are frequently overlooked in traditional recordkeeping, in addition to maintaining its visual and aural components.

ETHICAL AND CULTURAL CONSIDERATIONS IN DIGITAL PRESERVATION

Consent and Community Involvement

Two fundamental tenets of the digital preservation of Indigenous Knowledge Systems (IKS) are consent and community involvement. To guarantee that indigenous communities have control over the documentation, storage, and dissemination of their knowledge, it is imperative to obtain their informed consent. Consent is an ethical commitment to upholding the autonomy and dignity of indigenous peoples, in addition to being a legal necessity. Building confidence and obtaining meaningful consent for preservation projects requires open and honest communication about their goals, scope, and any repercussions. Beyond consultation, community involvement necessitates active participation in every step of preservation, from data collecting and planning to maintenance and archiving.

By empowering indigenous people to direct the preservation process, this participatory approach makes sure that digital archives reflect their objectives and cultural values. Participation from community members, especially elders and knowledge holders, guarantees that customary contexts and interpretations are appropriately reflected in the record. Participation in the community also fosters the development of skills, allowing indigenous peoples to gain archival knowledge and digital literacy that they can use in future preservation initiatives.

Navigating Cultural Sensitivities and Knowledge Boundaries

Preserving Indigenous Knowledge Systems (IKS) requires navigating cultural sensitivities and knowledge boundaries. IKS usually incorporates knowledge that is holy, private, or limited to particular people within a community, in contrast to Western knowledge systems, which usually place a higher priority on open access and data exchange. In order to prevent sensitive or confidential information from reaching the wrong audiences, digital preservation initiatives must adhere to established cultural norms. Inappropriate management of sensitive information can undermine confidence between academics and indigenous populations, perpetuate past injustices, and cause cultural harm. In order to navigate these issues, it is essential to work with indigenous knowledge holders, who offer advice on what should be kept private and what can be given openly.

Policy Frameworks and Legal Considerations

In order to preserve Indigenous Knowledge Systems (IKS) digitally, policy frameworks and legal considerations are essential since they offer rules that guarantee the fair and moral treatment of traditional knowledge. Co-developing effective policies with indigenous populations should take into account their viewpoints on data sovereignty, ownership, and consent. In contrast to Western intellectual property rules, which place an emphasis on individual rights, traditional knowledge frequently functions under common ownership. Recognizing that IKS is a living legacy with profound spiritual, social, and environmental importance rather than a commodity, legal systems must be flexible enough to accommodate these distinct cultural settings. Cultural use agreements, traditional knowledge licensing, and intellectual property rights can all offer defence against unapproved usage, guaranteeing that digital IKS stays in the hands of its legitimate owners. Data sovereignty should also be covered by policies, with a focus on indigenous people' rights to control the collection, storage, and sharing of their digital assets. This entails establishing benefit-sharing plans, access mechanisms, and commercial use guidelines.

Maintaining Knowledge Integrity

The authenticity, context, and complexity of traditional knowledge must all be respected in order to preserve the integrity of Indigenous Knowledge Systems (IKS) during digital preservation. This means that digital preservation initiatives should respect the traditional frameworks, narratives, and categories in which indigenous knowledge was initially communicated rather than just translating it into Western scientific terminology. Maintaining accuracy and cultural relevance in the documentation process requires the participation of indigenous knowledge bearers. They can offer the background information required to guarantee that digital knowledge captures the community's distinct values, language, and worldview.

Instead than enforcing foreign standards, well-designed metadata, classification schemes, and digital platforms should integrate local terms and classifications. This keeps cultural subtleties from being lost and complicated ideas from being reduced to oversimplified interpretations. Since IKS is a living system that is always changing and growing, digital platforms must also be flexible enough to accept new data, updates, and changing interpretations. The cultural integrity of the information can also be preserved by adhering to customs during the documenting process, such as ceremonial observance or oral storytelling norms. In the end, maintaining knowledge integrity necessitates a dedication to community engagement, cultural sensitivity, and continuous communication between traditional and digital information systems.

Legal Frameworks for Digital IK Preservation

The legal landscape for IK preservation varies significantly across countries, creating both challenges and opportunities for Indigenous communities. Some nations explained in Table 3 have enacted strong legal protections, while others lack clear policies, leaving Indigenous data vulnerable to exploitation.

Table 3. Comparison of legal frameworks for digital IK preservation

Country/Region	Legal Protections for IK	Strengths	Legal Gaps & Challenges
United States	Native American Graves Protection and Repatriation Act (NAGPRA)	• Tribal Digital Sovereignty Policies I - Recognizes Indigenous data sovereignty • Provides legal recourse for misused cultural artefacts	• Limited enforcement in digital spaces • No federal laws specific to IK digitization
Canada	Indigenous Knowledge Protection Act (proposed)	First Nations Principles of OCAP (Ownership, Control, Access, and Possession) I - Strong focus on community governance	OCAP ensures Indigenous control over data I - No national legislation enforcing digital IK rights
Australia	Aboriginal and Torres Strait Islander Cultural Heritage Acts	Copyright protections for traditional knowledge Acknowledges Indigenous IP rights	State-level enforcement of cultural laws No uniform federal law for IK digitization
European Union	GDPR (General Data Protection Regulation)	UNESCO Indigenous Rights Initiatives I - Strong data protection framework Recognizes cultural heritage protection	GDPR does not fully address IK digital ownership

FUTURE DIRECTIONS IN THE DIGITAL PRESERVATION OF INDIGENOUS KNOWLEDGE

Emerging Technologies and Opportunities

Indigenous information Systems (IKS) can now be preserved thanks to emerging technology, which offer creative approaches to managing, disseminating, and documenting traditional information. Technological developments in areas such as digital archiving, blockchain, virtual and augmented reality (VR/AR), and artificial intelligence (AI) offer instruments that can improve IKS's accessibility, security, and participation. AI can help with automated transcribing, language revitalization, and large-scale dataset analysis, which aids in the organization and interpretation of complex data. In order to promote research and teaching while honoring cultural boundaries, machine learning algorithms can be trained to recognize patterns in cultural data. Users can connect with ancient customs, holy locations, and cultural narratives through immersive platforms made possible by VR and AR technologies.

Building Global Collaboration Networks

For Indigenous Knowledge Systems (IKS) to be effectively preserved digitally, international collaboration networks must be established. International relationships enable the sharing of best practices and creative solutions for preserving traditional knowledge by bringing together a variety of viewpoints, resources, and skills. These networks, which support a cooperative preservation effort, may consist of indigenous groups, educational institutions, cultural organizations, non-governmental organizations, governmental organizations, and corporate sector partners. Global networks improve the sustainability of digital preservation initiatives by offering chances for funding, technical assistance, resource sharing, and capacity building. Additionally, they promote intercultural communication, enabling the coexistence and mutual learning of indigenous and non-indigenous knowledge systems. In order to guarantee that indigenous viewpoints continue to play a prominent role in the preservation process, collaborative networks must to be founded on the values of equality, reciprocity, and mutual respect. This entails abiding by moral principles that give informed consent, data sovereignty, and benefit-sharing first priority. Indigenous communities from various regions can share their experiences, methods, and difficulties in conserving traditional knowledge by using international venues to enhance knowledge sharing. By emphasizing IKS's benefits to biodiversity protection, climate resilience, sustainable development, and cultural variety, these networks can raise awareness of the importance of IKS on a worldwide scale. Global networks can develop a more respectful, inclusive, and successful strategy for the digital preservation of Indigenous Knowledge Systems by encouraging cross-border cooperation.

Preserving Dynamic and Evolving Knowledge Systems

Indigenous Knowledge Systems (IKS) are dynamic and ever-changing, adjusting to shifting social situations, cultural settings, and environmental changes. Such living knowledge systems must be preserved through an adaptable and flexible strategy that permits ongoing additions and updates. Because IKS is dynamic, digital preservation platforms should be made to adapt to changes. This entails building digital tools, databases, and archives that can readily incorporate changes, updates, and various interpretations throughout time. To keep preserved information accurate and current, it is essential to involve indigenous groups in the upkeep and administration of digital archives. They can offer direction on how cultural narratives, ecological knowledge, and traditional behaviours adapt in response to novel issues like social transition or climate change.

Videos, audio recordings, and VR/AR experiences are examples of interactive and multimedia forms that effectively capture the dynamic characteristics of IKS and enable real-time user engagement with traditional information. The plurality of interpretations within IKS must also be respected by digital platforms, which acknowledge that many communities or people may have distinct viewpoints on the same idea. In order to promote a more inclusive view of IKS, this variety ought to be honored rather than homogenized. Digital preservation initiatives can provide more robust, current, and culturally sensitive archives that capture the diversity and adaptability of traditional knowledge by recognizing the living nature of Indigenous Knowledge Systems.

The Role of Education in Preservation Efforts

As a link between traditional and contemporary knowledge systems, education is essential to the digital preservation of Indigenous Knowledge Systems (IKS). Maintaining cultural continuity, particularly in the face of urbanization, globalization, and language loss, requires educating younger generations about the importance of IKS. By offering digital literacy training, educational programs can enable indigenous people to actively participate in the preservation process. In order to promote a comprehensive understanding of IKS, educational institutions such as schools, colleges, and community centres can act as centres for teaching both traditional and digital knowledge. By incorporating indigenous viewpoints into mainstream educational programs, prejudices are dispelled and a greater awareness for the diversity of human knowledge is fostered.

IKS can be made more widely available to a worldwide audience using digital technologies including online platforms, multimedia resources, and virtual learning environments, which can assist both official and informal education. Storytelling, mentoring, and educational initiatives run by Indigenous peoples help strengthen cultural identity and keep traditional knowledge alive and relevant. Additionally, education promotes capacity building by giving indigenous youth the tools they need to interact with emerging technology, manage digital archives, and produce multimedia material. Education can produce a new generation of knowledge keepers who can successfully navigate both worlds by fusing traditional wisdom with digital innovation, protecting the complexity and depth of Indigenous Knowledge Systems for future generations.

CONCLUSION

Indigenous Knowledge Systems (IKS) preservation calls for a multipronged strategy that incorporates community involvement, cultural sensitivity, and digital technologies. Successful digital preservation initiatives demonstrate the efficacy of decentralized, cooperative, and community-centered methods in preserving traditional knowledge while respecting indigenous authority. Care must be taken when addressing important issues including sustainability, ethical conundrums, cultural sensitivity, and technological obstacles. These difficulties show how important it is to establish trust, be open, and give indigenous populations' concerns first priority during the preservation process. Though they provide creative answers, emerging technologies like blockchain, AI, and VR/AR must be used carefully and in collaboration with indigenous groups to enhance rather than replace existing knowledge systems. Recognizing the dynamic nature of IKS, effective preservation strikes a compromise between conventional and digital techniques. Case studies show that sustainable preservation requires inclusive, respectful procedures that are informed by indigenous viewpoints. In addition to safeguarding priceless knowledge, digital preservation promotes biodiversity, sustainable development, and cultural diversity, all of which contribute to a more just and connected society. The knowledge gathered from these endeavours serves as a foundation for further projects, emphasizing the significance of moral, inclusive, and long-term preservation tactics. We can guarantee that IKS continues to be a thriving and esteemed component of our common global legacy by adopting these standards.

Recommendations for Best Practices

A number of best practices are necessary to sustain Indigenous Knowledge Systems (IKS) in an ethical and efficient manner. Prioritize community participation and consent first, making sure that indigenous viewpoints guide the preservation process all the way through. Utilize culturally relevant technologies that honor established knowledge systems, acknowledging that IKS is a dynamic system with strong cultural roots. Put in place strong policy frameworks that address ethical issues, data sovereignty, and intellectual property in order to stop deception and exploitation. In order to create fair and open partnerships, cooperation between indigenous and non-indigenous stakeholders is essential. Building capacity is essential, providing technical assistance, resources, and training to enable indigenous people to autonomously manage their digital archives. Long-term maintenance of access and security requires infrastructure and funding that are sustainable. Additionally, preservation tactics must to be flexible enough to take into account how IKS is changing and allow for updates as conventional knowledge expands. In order to foster intercultural understanding, education and awareness are essential, incorporating indigenous ideas into traditional classroom instruction and online resources. Digital preservation may provide inclusive, respectful, and culturally relevant archives by adhering to these best practices, guaranteeing that the diversity and integrity of IKS are respected and conserved for upcoming generations.

Final Thoughts on the Future of Indigenous Knowledge Preservation

The careful fusion of contemporary digital technologies with traditional knowledge is essential to the development of Indigenous Knowledge Systems (IKS). The cultural and ethical aspects of IKS must be given top priority in this integration to avoid making preservation exploitative or extractive. In order to respect indigenous communities' distinct knowledge systems, digital platforms must be customized to their particular demands and circumstances. As technology develops, IKS's dynamic character should be supported, enabling these living knowledge systems to develop organically. For indigenous knowledge to be valued, preserved, and disseminated in an ethical manner, cooperation, creativity, and education will be essential. By serving as a link between the traditional and modern worlds, digital preservation promotes a deeper comprehension of the many viewpoints held by people. It makes it possible for ecological knowledge, language, and cultural traditions to endure and be available to future generations. But in order to preserve authenticity and relevance, this process needs to be led by indigenous groups, whose knowledge, leadership, and cultural insights are essential. Giving indigenous people authority over their digital archives helps to preserve knowledge's accuracy and significance while preventing misuse or misunderstanding. Beyond simply preserving historical knowledge, the future of IKS preservation aims to promote a more inclusive and equitable society by appreciating and celebrating indigenous contributions now. We foster an environment where indigenous knowledge can thrive by adopting respectful digital preservation techniques, which promotes intercultural understanding and global sustainability. This well-rounded strategy guarantees that indigenous cultures' knowledge will not only be conserved but will also actively influence a varied and sustainable future for everybody.

REFERENCES

Alemu, B. W., & Assefa, K. D. (2023). Preservation of Ethiopian cultural heritage through digital repositories. *International Journal of Digital Curation, 18*(1), 92–108.

Alessa, L., & Kliskey, A. (2020). Indigenous knowledge systems and their integration with GIS technologies for land management. *Journal of Environmental Management, 260*, 110128.

Balogun, T., & Kalusopa, T. (2021). A framework for digital preservation of Indigenous knowledge system (IKS) in repositories in South Africa. *Records Management Journal, 31*(2), 176–196. DOI: 10.1108/RMJ-12-2020-0042

Barnhardt, R., & Kawagley, A. O. (2022). Traditional knowledge systems and education frameworks: Insights from Alaska Native communities. *Journal of Indigenous Education, 41*(3), 241–260.

Buell, M.-C., Ritchie, D., Ryan, K., & Metcalfe, C. D. (2020). Using Indigenous and Western knowledge systems for environmental risk assessment. *Ecological Applications, 30*(4), 1–10. DOI: 10.1002/eap.2146 PMID: 32338804

Cheng, Y., & Li, W. (2024). Addressing barriers to preserving Indigenous language databases in China. *Journal of Asian Cultural Studies, 19*(3), 149–165.

Dawson, P., & Levy, C. (2022). Archiving the Arctic: Challenges in preserving Inuit knowledge digitally. *The Polar Record, 58*(6), 385–401.

Ferreira, A., & Pienaar, M. (2024). Developing inclusive digital platforms for African oral traditions. *Journal of Knowledge Management, 28*(5), 289–310.

Glover, L., & Poitras, G. (2023). Indigenous data sovereignty and digital archiving frameworks: A Canadian perspective. *Archives and Manuscripts, 51*(4), 412–428.

Hewson, M. G., & Holtman, L. B. (2022). Incorporating Indigenous health knowledge into science curricula: Challenges and solutions. *African Journal of Research in Science. Technology and Mathematics Education, 26*(1), 78–98.

Kai Bol, G., & Van Niekerk, D. (2024). Application of African Indigenous Knowledge Systems for climate change management. *International Journal of Disaster Risk Reduction, 111*. Advance online publication. DOI: 10.1016/j.ijdrr.2024.103153

Masenya, T. M. (Ed.). (2023). *Digital Preservation and Documentation of Global Indigenous Knowledge Systems*. IGI Global., DOI: 10.4018/978-1-6684-7024-4

Njagi, P., & Mwangi, R. (2021). Digital storytelling as a tool for preserving Kenyan Indigenous knowledge. *International Journal of Communication, 15*(6), 1385–1400.

Nkebem, O., & Ofulue, S. (2022). Digitization challenges in preserving cultural knowledge: A study of Nigerian practices. *Journal of Library and Information Science, 56*(3), 312–330.

Park, S. M., & Lee, H. (2024). Culturally sensitive frameworks for archiving Indigenous knowledge in South Korea. *Asian Journal of Information Science, 32*(1), 57–72.

Ramirez, R., & Mari, F. (2021). Co-creating digital archives with Indigenous communities: The Mexican experience. *Cultural Heritage Journal*, *14*(2), 153–171.

Shanmuga Sundari, M., Penthala, H. R., Mogullapalli, A., & Ammangatambu, M. M. (2024). AI-Based Personalized Drug Treatment. In *Artificial Intelligence and Machine Learning in Drug Design and* (pp. 369–406). Development.

Singh, D., & Roy, B. (2023). Ethical concerns in digital preservation of Indigenous heritage in India. *Journal of Heritage Studies*, *30*(7), 445–460.

Sundari, M. S., Penthala, H. R., & Nayyar, A. (2024). Transforming Education through AI-Enhanced Content Creation and Personalized Learning Experiences. In *Impact of Artificial Intelligence on Society* (pp. 98–118). Chapman and Hall/CRC. DOI: 10.1201/9781032644509-7

Sundari, S., Penthala, H. R., Mogullapalli, A., Sukhavasi, V., & Nayyar, A. (2024). Women and Millets: Historical Perspective. In *The Role of Women in Cultivating Sustainable Societies Through Millets* (pp. 196-229). IGI Global.

Tshabalala, N., & Mthunzi, S. (2021). Evaluating long-term digital preservation policies for Indigenous knowledge in rural communities. *Information Development*, *38*(2), 117–135.

Uddagiri, C., & Shanmuga Sundari, M. (2023). Authorship Identification Through Stylometry Analysis Using Text Processing and Machine Learning Algorithms. In *Proceedings of Fourth International Conference on Computer and Communication Technologies* (pp. 573-581). Springer, Singapore. DOI: 10.1007/978-981-19-8563-8_55

Woods, M., & Blanchard, C. (2020). Sustainable digital platforms for Indigenous storytelling. *Journal of Digital Humanities*, *9*(4), 215–232.

KEY TERMS AND DEFINITIONS

Digital Preservation: The process of maintaining and protecting digital content over time to ensure future access.

Indigenous Knowledge Systems (IKS): Traditional knowledge and cultural practices passed down through generations within indigenous communities.

Intellectual Property Rights: Legal rights that protect the ownership and use of original creations, including indigenous knowledge.

Metadata Standards: Structured information that describes and provides context to digital content for better organization and retrieval.

Knowledge Misappropriation: Unauthorized use or exploitation of indigenous knowledge without consent or proper acknowledgment.

Compilation of References

Abah, J., Mashebe, P., & Denuga, D. D. (2015). Prospect of integrating African indigenous knowledge systems into the teaching of sciences in Africa. *American Journal of Educational Research*, *3*(6), 668–673.

Abbas, A. E., van Velzen, T., Ofe, H., van de Kaa, G., Zuiderwijk, A., & de Reuver, M. (2024). Beyond control over data: Conceptualizing data sovereignty from a social contract perspective. *Electronic Markets*, *34*(1), 20. DOI: 10.1007/s12525-024-00695-2

Absolon, K. (2010). Indigenous wholistic theory: A knowledge set for practice. *First Peoples Child & Family Review*, *5*(2), 74–87. DOI: 10.7202/1068933ar

Aça, M. (2023). The Role of Rituals and Ceremonies in the Identity Struggle of a Society that Has Lost Its Language: the Example of Lithuanian Tatars. Lietuvos istorijos studijos, (Spec/2023), 93-108.

Acosta Castellanos, P. M., & Queiruga-Dios, A. (2022). From environmental education to education for sustainable development in higher education: A systematic review. *International Journal of Sustainability in Higher Education*, *23*(3), 622–644. DOI: 10.1108/IJSHE-04-2021-0167

Adam, L. (2007, April). Information and communication technologies, knowledge management and Indigenous knowledge: Implications to livelihood of communities in Ethiopia. In *a Workshop on" The Role of ICT in Preserving and Disseminating Indigenous Knowledge", Addis Ababa.*

Adam, A. A., Othman, N., & Halim, A. A. (2022). Indigenous Knowledge Documentation: Perspectives of Dusun and Bajau communities in Kota Belud, Sabah, Malaysia. *International Journal of Academic Research in Business & Social Sciences*, *12*(10), 163–175. DOI: 10.6007/IJARBSS/v12-i10/15205

Adams, G. K. (2022). Does the argument that museums hold collections on behalf of the world still stand? Museums Association. https://www.museumsassociation.org/museums-journal/news/2020/01/17012020-does-universality-argument-still-stand/# Accessed 28/11/24.

Adebayo, C., Oluwaseye, J., & Adeyemo Cln, A. A. (2017). Documentation and dissemination of indigenous knowledge by library personnel in selected research institutes in Nigeria. *Library Philosophy and Practice (e-journal)*, 16-28.

Adebola, A. O. (2020). Protection of Craft Products Embodied in Cultural and Creative Industries in South Africa. In Sustainable Organizations-Models, Applications and New Perspectives. *IntechOpen.*

Adeola, O. (2023). Introduction: A Casebook of Indigenous Business Practices in Africa. I*n Casebook of Indigenous Business Practices in Africa* [Emerald Publishing Limited.]. *Apprenticeship, Craft, and Healthcare*, *1*, 1–14.

Adeoye-Olatunde, O. A., & Olenik, N. L. (2021). Research and scholarly methods: Semi-structured interviews. *Journal of the American College of Clinical Pharmacy: JAACP*, *4*(10), 1358–1367. DOI: 10.1002/jac5.1441

Adeyemo, A. A., & Adebayo, J. O. (2018). Documentation and Dissemination of Indigenous Knowledge by Library Personnel in Selected Research Institutes in Nigeria. Library Philosophy & Practice.

Adhikari, M. (2010). A total extinction confidently hoped for: The destruction of Cape San society under Dutch colonial rule, 1700–1795. *Journal of Genocide Research*, *12*(1-2), 19–44. DOI: 10.1080/14623528.2010.508274 PMID: 20941880

Adigun, G. O., Ajani, Y. A., & Enakrire, R. T. (2024). The intelligent libraries: Innovation for a sustainable knowledge system in the fifth (5th) Industrial Revolution. *Libri*, *74*(3), 211–223. DOI: 10.1515/libri-2023-0111

Afful-Arthur, P., Kwafoa, P. N. Y., Ampah-Johnston, M., & Mensah, V. (2022). Managing and accessing indigenous knowledge for national development: The role of academic libraries in Ghana. *Information Development*, *38*(4), 535–548. DOI: 10.1177/02666669211009916

African Charter on Human and Peoples' Rights. https://au.int/sites/default/files/treaties/36390-treaty-0011_-_african_charter_on_human_and_peoples_rights_e.pdf

Africa, S. S. (2018). Men, Women and Children: Findings of the Living Conditions. *Survey (London, England)*, *2014*(15).

Agan, W. (2013). Commercialization of traditional knowledge in South Africa: whether the existing intellectual property framework encourages commercialization. Cape Town: University of Cape Town. (Dissertation-PhD).

Agbehadji, I. E., Awuzie, B. O., & Ngowi, A. B. (2021). COVID-19 pandemic waves: 4IR technology utilisation in multi-sector economy. *Sustainability (Basel)*, *13*(18), 10168. DOI: 10.3390/su131810168

Agbeyangi, A., & Jere, N. (2024). IsiXhosa in the Digital Age: Navigating Language Preservation and Innovation (A Systematic Scoping Review). *IEEE Access*. https://gsarpublishers.com/journal-gjahss-home/

Agrawal, A. (1995). Dismantling the divide between indigenous and scientific knowledge. *Development and Change*, *26*(3), 413–439. DOI: 10.1111/j.1467-7660.1995.tb00560.x

Agrawal, A. (1995). Indigenous and scientific knowledge: Some critical comments. *Indigenous Knowledge and Development Monitor*, *3*(3), 3–6.

Agrawal, A. (2002). Indigenous knowledge and the politics of classification. *International Social Science Journal*, *54*(173), 287–297. DOI: 10.1111/1468-2451.00382

AIGI. (2018). Indigenous Data Sovereignty Communique. Australian Indigenous Governance Institute. https://aigi.org.au/news/indigenous-data-sovereignty-communique

Ajani, Y.A., Aladokun, B.D., Agboola, O.S. & Amaechi, M.N. (2024). Revitalizing Indigenous Knowledge Systems via Digital Media Technologies for Sustainability of Indigenous Languages. *Preservation, Digital technology & culture,* DOI:.DOI: 10.1515/pdtc-2023-0051

Ajani, Y. A., Baloyi, N. C., Chibike, J. J., Elisifa, Z., Lubambo, R. J., Mabika, M., & Xaba, B. B. B. M. (2024). *Digital Media and the Preservation of Indigenous Languages in Africa: Toward a Digitalized and Sustainable Society*. Rowman & Littlefield.

Ajani, Y. A., Oladokun, B. D., Olarongbe, S. A., Amaechi, M. N., Rabiu, N., & Bashorun, M.

Ajani, Y. A., Tella, A., & Dlamini, N. P. (2024). Indigenous language preservation and promotion through digital media technology in the Fourth Industrial Revolution. In *Digital Media and the Preservation of Indigenous Languages in Africa* (p. 49). Toward a Digitalized and Sustainable Society.

Akahome, J. E., & Wilson, E. F. (2024). Indigenous Entrepreneurship as a Pathway for Nollywood's Global Relevance and SDG 8 Intervention. In *Indigenous Entrepreneurship in Sub-Saharan Africa: Implications for Theory and Practice* (pp. 185–199). Springer Nature Switzerland. DOI: 10.1007/978-3-031-64924-0_8

Akena, F. A. (2012). Critical analysis of the production of Western knowledge and its implications for Indigenous knowledge and decolonization. *Journal of Black Studies*, *43*(6), 599–619. DOI: 10.1177/0021934712440448

Akinola, R., Pereira, L. M., Mabhaudhi, T., De Bruin, F. M., & Rusch, L. (2020). A review of indigenous food crops in Africa and the implications for more sustainable and healthy food systems. *Sustainability*, *12*(8), 3493. DOI: 10.3390/su12083493 PMID: 33520291

Akinwale, A.A. (2013). Digitisation of indigenous knowledge for natural resources management in Africa.

Akolgo-Azupogo, H., Rubens, A., & Bardy, R. (2021). Soft skills in developing economies: An African view on the hidden linkage between indigenous knowledge and business perspectives. *Journal of African Studies and Development*, *13*(3), 47–58. DOI: 10.5897/JASD2021.0625

Akpan, J., et al. (2004). Human rights and participation among southern African indigenous peoples. *IWGIA document,* (110), 184-201.

Akyol, G., & Avcı, A. B. (2023). Ai Applications in Cultural Heritage Preservation: Technological Advancements for the Conservation. https://hdl.handle.net/11147/13725

Alemu, B. W., & Assefa, K. D. (2023). Preservation of Ethiopian cultural heritage through digital repositories. *International Journal of Digital Curation*, *18*(1), 92–108.

Alessa, L., & Kliskey, A. (2020). Indigenous knowledge systems and their integration with GIS technologies for land management. *Journal of Environmental Management*, *260*, 110128.

Alharbi, S., Attiah, A., & Alghazzawi, D. (2022). Integrating blockchain with artificial intelligence to secure IoT networks: Future trends. *Sustainability (Basel)*, *14*(23), 16002. DOI: 10.3390/su142316002

Ali, A., Audi, M., & Roussel, Y. (2021). Natural resources depletion, renewable energy consumption and environmental degradation: A comparative analysis of developed and developing world. *International Journal of Energy Economics and Policy*, *11*(3), 251–260. DOI: 10.32479/ijeep.11008

Aliber, M., & Hall, R. (2012). Support for smallholder farmers in South Africa: Challenges of scale and strategy. *Development Southern Africa*, *29*(4), 548–562. DOI: 10.1080/0376835X.2012.715441

Allison-Cassin, S., Bateman, R., Bhardwaj, R. K., & Chisita, C. T. (2023). Introduction. *The Serials Librarian, 84*(5–8), 62–65. DOI: 10.1080/0361526X.2024.2313400

Al-Mansoori, F., & Hamdan, A. (2023). Integrating indigenous knowledge systems into environmental education for biodiversity conservation: A study of sociocultural perspectives and ecological outcomes. *AI. IoT and the Fourth Industrial Revolution Review, 13*(7), 61–74.

Altieri, M. A. (2009). Agroecology, small farms, and food sovereignty. *Monthly Review (New York, N.Y.), 61*(3), 102–113. DOI: 10.14452/MR-061-03-2009-07_8

Altieri, M. A., & Nicholls, C. I. (2017). *Agroecology: A transdisciplinary, participatory and action-oriented approach.* CRC Press.

Amaducci, S. (2022). Adapting the cultivation of industrial hemp (Cannabis sativa L.) to marginal lands: *A review.Global Change Biology. Bioenergy, 14*(9), 1004–1022. DOI: 10.1111/gcbb.12979

Amani, B., Ncube, C. B., & Rimmer, M. (2024). Introduction: 'the people's agenda': a history of intellectual property and sustainable development. In *The Elgar Companion to Intellectual Property and the Sustainable Development Goals* (pp. 1–36). Edward Elgar Publishing. DOI: 10.4337/9781803925233.00007

Ambler, C. (2020). The African Roots of Marijuana by Chris S. Duvall. *Bulletin of the History of Medicine, 94*(2), 303–304. DOI: 10.1353/bhm.2020.0045

Amechi, E. P. (2018). Bio-economy, patents, and the commercialisation of traditional knowledge on the medicinal uses of plants in South Africa. *SA Mercantile Law Journal, 30*(2), 251–279.

Amechi, E. P. (2018). Bio-economy, patents, and the commercialization of traditional knowledge on the medicinal uses of plants in South Africa. *SA Mercantile Law Journal, 30*(2), 251–279.

Amedie, J. (2015). *The impact of social media on society. Advanced writing: Pop Culture Intersections, 2*. Retrieved April, 19, 2021, from https://scholarcommons.scu.edu/engl_176/2/

Amelework, A. B., Bairu, M. W., Maema, O., Venter, S. L., & Laing, M. (2021). Adoption and promotion of resilient crops for climate risk mitigation and import substitution: A case analysis of cassava for South African agriculture. *Frontiers in Sustainable Food Systems, 5*, 617783. DOI: 10.3389/fsufs.2021.617783

Andanda, P. (2012). Striking a Balance between Intellectual Property Protection of Traditional Knowledge, Cultural Preservation and Access to Knowledge. *SECO/WTI Academic Cooperation Project Working Paper Series 2012/02.* Available at SSRN: http://dx.doi.org/DOI: 10.2139/ssrn.2718887

Anderson, J. E. (2015). Indigenous knowledge and intellectual property rights. In *International Encyclopaedia of the Social & Behavioural Sciences* (2nd ed., pp. 769–778). Elsevier Inc., DOI: 10.1016/B978-0-08-097086-8.64078-3

Anderson, J., & Christen, K. (2021). Traditional Knowledge labels: A tool for supporting indigenous data sovereignty. *The Journal of Documentation, 77*(2), 542–563.

Andrae, S. (2025). Data cooperatives to strengthen digital citizenship: Opportunities and risks. In Hawamdeh, M. Kh. M. (Ed.), *Digital citizenship and the future of AI engagement, ethics, and privacy* (pp. 533–568). IGI Global Publishing., DOI: 10.4018/979-8-3693-9015-3.ch018

Angioni, G. (2003). Indigenous Knowledge: Subordination and Localism. In G. Sanga & G. Ortalli (Ed), *Nature, Knowledge: Ethnoscience, cognition, and utility* (pp. 287- 296). New York: Oxford.

Apraku, A., Morton, J. F., & Gyampoh, B. A. (2021). Climate change and small-scale agriculture in Africa: Does indigenous knowledge matter? Insights from Kenya and South Africa. *Scientific African, 12*, e00821. DOI: 10.1016/j.sciaf.2021.e00821

Arcand, M. M., Bradford, L., Worme, D. F., Strickert, G. E., Bear, K., Johnston, A. B. D., Wuttunee, S. M., Gamble, A., & Shewfelt, D. (2020). Sowing a way towards revitalizing Indigenous agriculture: creating meaning from a forum discussion in Saskatchewan, Canada, *Canadian Science Publishing* 1840 Woodward Drive, Suite 1, Ottawa, ON K2C 0P7. 5: 619-641.

Ardhede, H. (2007). *Traditional Knowledge and the Patent System-Irreconcilable differences or a simple case of mistaken identity?* Asiwe, J.A.N. (2009). Needs assessment of cowpea production practices, constraints and utilization in South Africa. *African Journal of Biotechnology, 8*(20).

Argumedo, A. (2008). The Potato Park, Peru: Conserving agrobiodiversity in an Andean indigenous biocultural heritage area. *Protected landscapes and agrobiodiversity values, 1*, 45-58.

Arita, M. (2025). Data sovereignty and open sharing: Reconceiving benefit-sharing and governance of digital sequence information. *Data Science Journal, 24*(9), 1–5. DOI: 10.5334/dsj-2025-009

Arksey, H., & O'Malley, L. (2005). Scoping Studies: Towards a Methodological Framework. *International Journal of Social Research Methodology, 8*(1), 19–32. DOI: 10.1080/1364557032000119616

Arnold, S. E., Elisante, F., Mkenda, P. A., Tembo, Y. L. B., Ndakidemi, P. A., Gurr, G. M., Darbyshire, I. A., Belmain, S. R., & Stevenson, P. C. (2021). Beneficial insects are associated with botanically rich margins with trees on small farms. *Scientific Reports, 11*(1), 15190. DOI: 10.1038/s41598-021-94536-3 PMID: 34312457

Arsenault, R., Bourassa, C., Diver, S., McGregor, D., & Witham, A. (2019). Including indigenous knowledge systems in environmental assessments: Restructuring the process. *Global Environmental Politics, 19*(3), 120–132. DOI: 10.1162/glep_a_00519

Arslanbek, A., Malhotra, B., & Kaimal, G. (2022). Indigenous and traditional arts in art therapy: Value, meaning, and clinical implications. *The Arts in Psychotherapy, 77*, 101879. DOI: 10.1016/j.aip.2021.101879

Asante, M. K. (2016). *Afrocentricity and its critics.* Available at: https://science.jrank.org/pages/8216/Afrocentricity-Afrocentricity-Its-Critics.html. Accessed on 2 April 2016.

Asfaw, Z. (2001). The role of home gardens in the production and conservation of medicinal plants. Ethnobotany and Conservation of Medicinal Plants, 79–100.

Asheim, B. T., Smith, H. L., & Oughton, C. (2011). Regional innovation systems: Theory, empirics and policy. *Regional Studies, 45*(7), 875–891. DOI: 10.1080/00343404.2011.596701

Austin, B. J., Robinson, C. J., Mathews, D., Oades, D., Wiggin, A., Dobbs, R. J., Lincoln, G., & Garnett, S. T. (2019). An Indigenous-led approach for regional knowledge partnerships in the Kimberley region of Australia. *Human Ecology: an Interdisciplinary Journal, 47*(4), 577–588. DOI: 10.1007/s10745-019-00085-9

Awuzie, B., & Moghayedi, A. (2024). Harnessing 4IR/5IR technologies for improved procurement and delivery. In *Routledge Handbook of Construction Project Procurement and Delivery* (pp. 371–390). Routledge.

Ayanwale, A. B., Amusan, C. A., Adeyemo, V. A., & Oyedele, D. J. (2016). Analysis of household demand for underutilized indigenous vegetables. *International Journal of Vegetable Science*, *22*(6), 570–577. DOI: 10.1080/19315260.2015.1103350

Babalola, O. O., & Glick, B. R. (2012). Indigenous African agriculture and plant associated microbes: Current practice and future transgenic prospects. *Scientific Research and Essays*, *7*(28), 2431–2439.

Badenoch, A., & Follmer, G. (2018). Transationalizing Radio Research: New Encounters with An old medium. In Follmer, G., & Badenoch, A. (Eds.), *Transnationalizing Radio Research New Approaches to an Old Medium* (pp. 11–32). DOI: 10.1515/9783839439135-001

Baezner, M., & Robin, P. (2018). *Cyber sovereignty and data sovereignty*. CSS Cyber Défense Project.

Bagley, M. (2018). *Toward an effective indigenous knowledge protection regime case study of South Africa*. https://www.cigionline.org/search?search_api_fulltext=Bagley%20&sort_by=created

Balasubramanian, A. (2014). Classification of agricultural crops. Africa. *Agrekon*, *48*(4), 459–482.

Bali, A., & Kofinas, G. P. (2014). Voices of Caribou people: A participatory videography method to document and share local knowledge from the North-American human *Rangifer* systems. *Ecology and Society*, *19*(2), 1–13. DOI: 10.5751/ES-06327-190216

Balogun, T. (2023). Digitization of Indigenous knowledge systems in Africa: The case of South Africa's National Recorded System (NRS). *Records Management Journal*, *1*(33), 88–102. DOI: 10.1108/RMJ-10-2022-0029

Balogun, T., & Kalusopa, T. (2021). A framework for digital preservation of Indigenous knowledge system (IKS) in repositories in South Africa. *Records Management Journal*, *31*(2), 176–196. DOI: 10.1108/RMJ-12-2020-0042

Balogun, T., & Kalusopa, T. (2021). Developing a framework for web archiving of Indigenous Knowledge Systems (IKS) in selected repositories in South Africa. *In C. Mnubi-Mchombu.,* Nengomasha, C., & Uutoni, W. *(Ed.),Proceedings of the 24th Standing Conference of Eastern, Central and Southern African Library and Information Associations*, pp. 293- 300.

Banda, L. O. L., Banda, C. V., Banda, J. T., & Singini, T. (2024). Preserving cultural heritage: A community-centric approach to safeguarding the Khulubvi Traditional Temple Malawi. *Heliyon*, *10*(18), e37610. Advance online publication. DOI: 10.1016/j.heliyon.2024.e37610 PMID: 39347398

Banerjee, A. V., & Duflo, E. (2005). Growth theory through the lens of development economics. *Handbook of economic growth*, 1,473 -552.

Bannister, K. (2018). Biopiracy or fair biotrade? Ethical dilemmas in biodiversity conservation and commercial use of traditional knowledge. *Ethics in Science and Environmental Politics*, *18*(2), 123–140. DOI: 10.3354/esep00182

Banton, M. (2017). Displaced Archives in the National Archives of the United Kingdom. In Lowry, J. (Ed.), *Displaced Archives*. Routledge. DOI: 10.4324/9781315577609-4

Baporikar, N. (2022). Strategies to manage and preserve indigenous knowledge. In *Handbook of research on protecting and managing global indigenous knowledge systems* (pp. 207–222). IGI Global. DOI: 10.4018/978-1-7998-7492-8.ch012

Bareetseng, S. (2022). The worldwide herbal market: Trends and opportunities. *Journal of Biomedical Research & Environmental Sciences*, *2766*(5), 2276. DOI: 10.37871/jbres1482

Barman, P. (2024). The Ethical Frontier: Integrating IKS for Sustainable Economic Development. *MSW Management Journal*, *34*(1), 228–238.

Barnhardt, R., & Kawagley, A. O. (2022). Traditional knowledge systems and education frameworks: Insights from Alaska Native communities. *Journal of Indigenous Education*, *41*(3), 241–260.

Barnhardt, R., & Oscar Kawagley, A. (2005). Indigenous knowledge systems and Alaska Native waysof knowing. *Anthropology & Education Quarterly*, *36*(1), 8–23. DOI: 10.1525/aeq.2005.36.1.008

Baruah, T. D. (2012). Effectiveness of social media as a tool of communication and its potential for technology enabled connections: A micro-level study. *International Journal of Scientific and Research Publications*, *2*(5), 1–10.

Bastos, F. M. C. (2009). Art education in the spirit of Sankofa. *Art Education*, *62*(2), 1–5. DOI: 10.1080/00043125.2009.11519005

Basu, S., Roberts, J. A., Azam-Ali, S. N., & Mayes, S. (2007). Bambara groundnut. In *Pulses, Sugar and Tuber Crops* (pp. 159–173). Springer Berlin Heidelberg. DOI: 10.1007/978-3-540-34516-9_10

Batt, F. (2021). *The repatriation of African heritage: shutting the door on the imperialist narrative*. African Human Rights Yearbook.

Battiste, M. (2008). The decolonization of Aboriginal education: Dialogue, reflection, and action in Canada. Educational theories and practices from the majority world, 168-195.

Battiste, M. (2005). Indigenous knowledge: Foundations for first nations. *WINHEC: International Journal of Indigenous Education Scholarship*, (1), 1–17.

Battiste, M. (2013). *Decolonizing education: Nourishing the learning spirit*. Purich Publishing.

Battiste, M., & Henderson, J. H. (2000). *Protecting indigenous knowledge and heritage: A global challenge*. Purich. DOI: 10.59962/9781895830439

Bavikatte, K. S., Robinson, D., & Oliva, M. J. (2015). Biocultural community protocols: dialogues on the space within. *IK: Other Ways of Knowing*, pp.1-31.

Bavikatte, K., & Jonas, H. (2009). Bio-cultural community protocols: a community approach to ensuring the integrity of environmental law and policy. *Square Brackets*, (2).

Bawack, R., Roderick, S., Badhrus, A., Dennehy, D., & Corbett, J. (2025). Indigenous knowledge and information technology for sustainable development. *Information Technology for Development*, •••, 1–18. DOI: 10.1080/02681102.2025.2472495

Baxter, G., & Sommerville, I. (2011). Socio-technical systems: From design methods to systems engineering. *Interacting with Computers*, *23*(1), 4–17. DOI: 10.1016/j.intcom.2010.07.003

Baye, B., & Teshome, W. (2020). The role indigenous knowledge in agricultural farming practices: The case of Gonder and Gojam area, Amhara regional state, Ethiopia. *International Journal of Advanced Research in Biological Sciences*, *7*(12), 106–112.

Becerra-Fernandez, R. S. (2001). Organizational knowledge management: A contingency perspective. *Journal of Management Information Systems*, *18*(1), 23–55. DOI: 10.1080/07421222.2001.11045676

Bell, J., Kimberly, C., & Turin, M. (2013). After the Return: Digital Repatriation and the Circulation of Indigenous Knowledge Workshop Report. 1. 195–203. DOI: 10.3167/armw.2013.010112

Bender, K., Negi, N., & Fowler, D. N. (2010). Exploring the Relationship between Self-Awareness and Student Commitment and Understanding of Culturally Responsive Social Work Practice. *Journal of Ethnic & Cultural Diversity in Social Work*, *19*(1), 34–53. DOI: 10.1080/15313200903531990 PMID: 23255873

Berhane, G., Berhane, G., Ragasa, C., Abate, G. T., & Assefa, T. W. (2018). *The state of agricultural extension services in Ethiopia and their contribution to agricultural productivity*. Intl Food Policy Res Institution.

Berkes, F. (2009). Indigenous ways of knowing and the study of environmental change. *Journal of the Royal Society of New Zealand*, *39*(4), 151–156. DOI: 10.1080/03014220909510568

Berkes, F. (2012). *Sacred ecology*. Routledge. DOI: 10.4324/9780203123843

Berkes, F. (2021). Sacred ecology: Traditional ecological knowledge and resource management in the digital age. *Environmental Conservation*, *48*(1), 15–28.

Berry, E. M., Dernini, S., Burlingame, B., Meybeck, A., & Conforti, P. (2015). Food security and sustainability: Can one exist without the other? *Public Health Nutrition*, *18*(13), 2293–2302. DOI: 10.1017/S136898001500021X PMID: 25684016

Bese, D., Zwane, E., & Cheteni, P. (2021). The use of sustainable agricultural methods amongst smallholder farmers in the Eastern Cape province, South Africa. *African Journal of Science, Technology, Innovation and Development*, *13*(3), 261–271. DOI: 10.1080/20421338.2020.1724388

Béteille, A. (2013). What should we mean by "indigenous people"? In *Indigeneity in India* (pp. 19–31). Routledge.

Betts, R. F. (2012). Decolonization A brief history of the word. In Bogaerts, E., & Raben, R. (Eds.), *Beyond Empire and Nation: The Decolonization of African and Asian societies, 1930s-1970s*. Brills. DOI: 10.1163/9789004260443

Bhat, S. (2013). *Natural Resources Conservation Law*. Eastern Book Company.

Bhuda, M. T. (202). "Making a Case for Indigenous Education Systems in South Africa." *African Journal of Development Studies* (2), 67–82.

Bhuda, M. T., & Gumbo, M. T. (2024). Developments Towards Integrating African Indigenous Knowledge in South African Universities. Reprint. Pretoria. University of South Africa Press. DOI: 10.25159/UnisaRxiv/000082.v1

Bhuda, M. T., & Marumo, P. (2021). The role of art in sustaining the livelihoods of Amandebele women in South Africa. *In: Proceedings of the conference on the implications of Covid-19 on gender and behaviour in Africa*. Johannesburg: The Havilah Glo Marketing.

Bhuda, M.T. & Marumo, P. (2020). African traditional medicine and healing in South Africa: Challenges and prospects before and during COVID 19. *Ife Centre for Psychological Studies/Services, 18*, 16710

Bhuda, M.T. & Marumo, P. (2020). African traditional medicine and healing in South Africa: Challenges and prospects before and during COVID 19. *Ife Centre for Psychological Studies/Services,* 18, 16710.

Bhuda, M.T. (2021). The role of art in sustaining the livelihoods of Amandebele women in South Africa.

Bhuda, M. T. (2019). *The role of ethnomathematics in the cultural life of AmaNdebele women at Ekosini village in Mpumalanga Province. Mafikeng: North-West University*. Dessertation-M.IKS.

Bhuda, M. T. (2021). Making a case for indigenous education systems in South Africa. *African Journal of Development Studies, 2021*(1, si2), 67–82. DOI: 10.31920/2634-3649/2021/SIv1a4

Bhuda, M. T. (2023). The process towards cannabis commercialization by indigenous communities for medicinal purposes in South Africa. In *Rapid innovation and development in the global cannabis market* (pp. 231–245). IGI Global. DOI: 10.4018/978-1-6684-6346-8.ch013

Bhuda, M. T. (2024). Decolonising and indigenising African Indigenous research: Why Indigenous ways of knowing, being and living matter in research. In *Higher Education Transformation in Africa* (pp. 50–69). Routledge. DOI: 10.4324/9781003528319-4

Bhuda, M. T., & Kgwadi, B. (2023). The economic measures established to rebuild the South African cultural heritage tourism after the COVID-19 impact. *Gender & Behaviour, 21*(1), 21398–21412.

Bhuda, M. T., & Koitsiwe, M. (2021). Practicing rites of passage by Southern Ndzundza Ndebele during Covid-19: Is it safe for male initiation schools to operate during Covid-19? *Gender & Behaviour, 19*(1), 17280–17293.

Bhuda, M. T., & Koitsiwe, M. (2022). The Covid-19 Pandemic and The Challenges Encountered by The Southern Ndebele Crafts Entrepreneurs Due to Closure of The Tourism Sector. *Gender & Behaviour, 20*(1), 19188–19202.

Bhuda, M. T., & Marumo, P. (2020). African traditional medicine and healing in South Africa: Challenges and prospects before and during COVID 19. *Ife Centre for Psychological Studies/Services. PO Box 1548. Ile-Ife Osun State, Nigeria, 18*(4), 16710.

Bhuda, M., & Saurombe, T. (2022). Mechanisms of Ethnomathematical Knowledge Transmission amongst AmaNdebele women: Ways of Transmitting Indigenous Knowledge. *Journal of Indigenous Social Development, 11*(1), 97–119.

Bidwell, N. J. (2021). Decolonising in the Gaps: Community Networks and the Identity of African Innovation. In *Re-imagining Communication in Africa and the Caribbean* (pp. 97–115). Springer. DOI: 10.1007/978-3-030-54169-9_6

Bietti, E., Etxeberria, A., Mannan, M., & Wong, J. (2021). Data cooperatives in Europe: A legal and empirical investigation. The New School's Platform Cooperativism Consortium. https://cyber.harvard.edu/sites/default/files/2022-02/Data_Cooperatives_Europe-group2.pdf

Bijker, W. E., Hughes, T. P., & Pinch, T. (1987). *The Social Construction of Technological Systems: New Directions in the Sociology and History of Technology*. MIT Press.

Bijker, W. E., & Law, J. (Eds.). (1992). *Shaping technology/building society: Studies in sociotechnical change*. MIT Press.

Bird, S., Shobbrook, K., Flick, D., Carroll, S., Austin, A., & Lim, A. (2019). Technology for Indigenous Language Revitalization: The State of the Field. *Language, 95*(S1), e1–e29.

Bird, S., & Yibarbuk, D. (2024). Cntering the speech community. In *Proceedings of the 18th Conference of the European Chapter of the Association for Computational Linguistics* (Vol. 1, pp. 826–839). Long Papers., https://aclanthology.org/2024.eacl-long.50/

Biyela, N., Oyelude, A., & Haumba, E. (2016). Digital preservation of indigenous knowledge (IK) by cultural heritage institutions: A comparative study of Nigeria, South Africa and Uganda.

Black, B. E. (2020). *Creating a qualified cannabis workforce: How higher education can support cannabis career pathways*. Kansas State University.

Blacking, J. (1967). *Venda children's songs: A study in ethnomusicological analysis*. Witwatersrand University Press.

Blair, N. (2015). Researched to death: Indigenous peoples talkin'up our experiences of research. *International Journal of Qualitative Research, 8*(4), 463–478. DOI: 10.1525/irqr.2015.8.4.463

Bleibleh, S., & Awad, J. (2020). Preserving cultural heritage: Shifting paradigms in the face of war, occupation, and identity. *Journal of Cultural Heritage, 44*, 196–203. DOI: 10.1016/j.culher.2020.02.013

Blose, P. (2023). *The incorporation of education for sustainable development in Technology Education for an indigenous context* (Doctoral Thesis). https://uir.unisa.ac.za/handle/10500/30606

Blose, P., & Gumbo, M. T (2024). Developing a framework for integrating IKS indigenous knowledge systems in Technology Education with sustainable development principles. *African Journal of Research in Mathematics, Science and Technology Education*, 1-13. https://doi.org/DOI: 10.1080/18117295.2024.2356416

Bobo, Z. N. (2021). *Promoting traditional and indigenous foods in South Africa: a desktop review.* Doctoral dissertation, Stellenbosch: Stellenbosch University.

Boeckhout, M., Zielhuis, G. A., & Bredenoord, A. L. (2018). The FAIR guiding principles for data stewardship: Fair enough? *European Journal of Human Genetics : EJHG*, *26*(7), 931–936. DOI: 10.1038/s41431-018-0160-0 PMID: 29777206

Boelens, R., Perreault, T., & Vos, J. (2016). *Water justice*. Cambridge University Press.

Bolderston, A. (2008). Writing an effective literature review. *Journal of Medical Imaging and Radiation Sciences*, *39*(2), 86–92. DOI: 10.1016/j.jmir.2008.04.009 PMID: 31051808

Bonini, T., & Treré, E. (2024). *The everyday fight against platform power*. The MIT Press., DOI: 10.7551/mitpress/14329.001.0001

Borbale, M. K., Pohekar, A. U., Nistane, J. R., Nichit, N. S., Yesankar, T. I., & Kaware, A. G. (2021). Study of traditional method of weeding and fertilizing and their drawbacks. *International Research Journal of Engineering and Technology*, *8*(3), 236–242.

Borish, D., Cunsolo, A., Mauro, I., Dewey, C., & Harper, S. L. (2021). Moving Images, Moving Methods: Advancing documentary film for qualitative research. *International Journal of Qualitative Methods*, *20*, 1–14. DOI: 10.1177/16094069211013646

Borthakur, A. & Singh, P. (2021). Indigenous agricultural knowledge towards achieving sustainable agriculture. *Sustainable Agriculture Reviews 50: Emerging Contaminants in Agriculture*, 401-413.

Bostoen, K. (2018). *The bantu expansion*. Oxford University Press.

Botangen, K. A., Vodanovich, S., & Yu, J. (2017). Preservation of indigenous culture among indigenous migrants through social media: the Igorot peoples. *Proceedings of the 50th Hawaii International Conference on System Sciences*, 2303–2312. Retrieved May 27, 2022, from: https://doi.org/DOI: 10.24251/HICSS.2017.278

Botha, C. (2013). Patent protection of traditional knowledge relating to biological resources in South Africa. *Responsa Meridiana*.

Boussard, J., Choplin, G., Strickner, A., & Trouvé, A. (2010). *World Food Summit Plan of Action, World Food Summit,* 13-17 Novembre, Roma http://www. fao. org/wfs/index_en. Htm.

Boven, K., & Morohashi, J. (2002). *Best practices using indigenous knowledge*. Nuffic.

Bowen, G. A. (2009). Document analysis as a qualitative research method. *Qualitative Research Journal*, *9*(2), 27–40. DOI: 10.3316/QRJ0902027

Bowen, J., & Hinze, A. (2022). Participatory data design: Managing data sovereignty in IoT solutions. *Interacting with Computers*, *34*(2), 60–71. DOI: 10.1093/iwc/iwac031

Bowman, D. M., O'Brien, J. A., & Goldammer, J. G. (2020). Indigenous use of fire: The social and environmental benefits of traditional knowledge. *Proceedings of the National Academy of Sciences of the United States of America*, *117*(34), 20475–20484. DOI: 10.1073/pnas.2008734117

Bowman, D. N., & Bremner, L. (2024). Indigenous Data Sovereignty: Applying It By, With, For and Through Indigenous Evaluators and Evaluations. *The Canadian Journal of Program Evaluation*, *39*(2), 265–287. DOI: 10.3138/cjpe-2024-0039

Brandt, H. M., Freedman, D. A., Friedman, D. B., Choi, S. K., Seel, J. S., Guest, M. A., & Khang, L. (2016). Planting healthy roots: Using documentary film to evaluate and disseminate community-based participatory research. *Family & Community Health, 39*(4), 242–250. DOI: 10.1097/FCH.0000000000000120 PMID: 27536929

Bratianu, C. (2010). A critical analysis of Nonaka's model of knowledge dynamics. *Electronic Journal of Knowledge Management, 8*(2), 193–200.

Bratianu, C., & Orzea, I. (2010). Organizational knowledge creation. *Management. Marketing Challenges for Knowledge Society, 5*(3), 41–62.

Braun, V., & Clarke, V. (2006). Using thematic analysis in psychology. *Qualitative Research in Psychology, 3*(2), 77–101. DOI: 10.1191/1478088706qp063oa

Braun, V., & Clarke, V. (2012). *Thematic analysis*. American Psychological Association. DOI: 10.1037/13620-004

Breckenridge, K. (2014). The Politics of the Parallel Archive: Digital Imperialism and the Future of Record-Keeping in the Age of Digital Reproduction on JSTOR. *Journal of Southern African Studies, 40*(3), 499–519. https://www.jstor.org/stable/24566488. DOI: 10.1080/03057070.2014.913427

Briggs, J. (2005). The use of indigenous knowledge in development: Problems and challenges. *Progress in Development Studies, 5*(2), 99–114. DOI: 10.1191/1464993405ps105oa

Broadbent, M., & Weill, P. (2003). Improving business and information strategy alignment: Learning from the banking industry. *IBM Systems Journal, 32*(1), 162–179. DOI: 10.1147/sj.321.0162

Brofeldt, S., Argyriou, D., Turreira-García, N., Meilby, H., Danielsen, F., & Theilade, I. (2018). Community-based monitoring of tropical forest crimes and forest resources using information and communication technology – Experiences from Prey Lang, Cambodia. *Citizen Science: Theory and Practice, 3*(2), 1–14. DOI: 10.5334/cstp.129

Brondízio, E. S., Aumeeruddy-Thomas, Y., Bates, P., Carino, J., Fernández-Llamazares, Á., Ferrari, M. F., Galvin, K., Reyes-García, V., McElwee, P., Molnár, Z., Samakov, A., & Shrestha, U. B. (2021). Locally based, regionally manifested, and globally relevant: Indigenous and local knowledge, values, and practices for nature. *Annual Review of Environment and Resources, 46*(1), 481–509. DOI: 10.1146/annurev-environ-012220-012127

Bronson, K. (2022). *The immaculate conception of data: Agribusiness, activists, and their shared politics of the future*. McGill-Queen's University Press.

Brown, P. T., Wilson, D., West, K., Escott, K.-R., Basabas, K., Ritchie, B., ... (2023). Māori algorithmic sovereignty: Idea, principles, and use. *arXiv*. https://doi.org//arXiv.2311.15473DOI: 10.48550

Brown, J., Knol, D., Prevost-Derbecker, S., & Andrushko, K. (2007). Housing for Indigenous Youth in the Inner City of Winnipeg. *First Peoples Child & Family Review, 3*(2), 56–64. DOI: 10.7202/1069464ar

Brown, P., Smith, J., & Hunter, L. (2022). Blockchain and cultural burning: Managing Indigenous fire practices. *Journal of Australian Environmental Management, 29*(2), 134–147.

Brush, S. B. (2005). Protecting traditional agricultural knowledge. *Wash. UJL & Pol'y*, *17*, 59.

Buell, M.-C., Ritchie, D., Ryan, K., & Metcalfe, C. D. (2020). Using Indigenous and Western knowledge systems for environmental risk assessment. *Ecological Applications*, *30*(4), 1–10. DOI: 10.1002/eap.2146 PMID: 32338804

Bühler, M. M., Calzada, I., Cane, I., Jelinek, T., Kapoor, A., Mannan, M., Mehta, S., Mookerje, V., Nübel, K., Pentland, A., Scholz, T., Siddarth, D., Tait, J., Vaitla, B., & Zhu, J. (2023). Unlocking the power of digital commons: Data cooperatives as a pathway for data sovereign, innovative and equitable digital communities. *Digital*, *3*(3), 146–171. DOI: 10.3390/digital3030011

Bullas, J. (2014). An exploratory study of Indian University students' use of social networking websites: Implications for the workplace. *Business Communication 32 Quarterly*, 72(1), 105-110.

Burke, M., & Kraut, R. (2013). Using Facebook after losing a job: Differential benefits of strong and weak ties. *InProceedings of the 2013 conference on Computer supported cooperative work*. 1419–1430. DOI: 10.1145/2441776.2441936

Burrel, T. (1999) *Burrell's South African Patent and Design Law*. Durban: Butterworths

Bushe, B. (2019). The causes and impact of business failure among small to micro and medium enterprises in South Africa. Africa's public service delivery and performance review, 7(1), 1-26.

Buthelezi, N. N. (2014). Indigenous knowledge systems and agricultural rural development in South Africa: Past and present perspectives. *Indilinga*, *13*(2), 231–250.

Buthelezi, S. C., Ocholla, D., & Dlamini, P. (2024). *Strategies for documenting and disseminating Indigenous Knowledge at a South African university*. South African. DOI: 10.4102/sajim.v26i1.1648

Bvenura, C., & Afolayan, A. J. (2015). The role of wild vegetables in household food security in South Africa: A review. Food Research International, 76, pp.1001-1011.change adaptation decisions to enhance food security in Sub-Saharan Africa? *Agricultural Systems*, *127*, 161–177.

Cabe, M. (2023). 'How Rhodes must fall amplified calls to decolonise'. https://newint.org/features/2023/08/21/how-rhodes-must-fall-amplified-calls-decolonize. Accessed on 29 Nov. 2024.

Cakata, Z. (2015). *In search of the absent voice: The status of indigenous languages in post-apartheid South Africa* (Doctoral dissertation, University of South Africa).

Cakata, Z. (2023). We want to speak to our children: Introducing Azibuye. In *Z. Cakata, N.Z. Radebe & M.B. Ramose in Azibuye emasisweni. Reclaiming our space and centring our knowledge*. Oasis. DOI: 10.4102/aosis.2023.BK364.01

Cannon, S. E., Moore, J. W., Adams, M. S., Degai, T., Griggs, E., Griggs, J., Marsden, T., Reid, A. J., Sainsbury, N., Stirling, K. M., Barnes, A. A. Y. S., Benson, R., Burrows, D., Chamberlin, G. R., Charley, B., Dick, D., Duncan, A. T., Liddle, K. K. M., Paul, M., & Wilson, K. B. (2024). Taking care of knowledge, taking care of salmon: Towards Indigenous data sovereignty in an era of climate change and cumulative effects. *Facets*, *9*, 1–21. DOI: 10.1139/facets-2023-0135

Capgemini Research Institute. (2020). *Digital Mastery: How organizations have progressed in their digital transformations over the past two years.* Available at: https://www.capgemini.com/insights/research-library/digital-mastery/

Carlson, B., & Dreher, T. (2018). Introduction: Indigenous innovation in social media. *Media International Australia, Incorporating Culture & Policy, 169*(1), 16–20. DOI: 10.1177/1329878X18803798

Carpenter, J.. (2017). *Digital access for language and culture in first Nations communities.* Heiltsuk Arts UBC.

Carroll, S. R., Duarte, M., & Liboiron, M. (2024). Indigenous data sovereignty. *Data & Society. 4*, 207–223. https://datasociety.net/wp-content/uploads/2024/04/Keywords_Indigenous_Data_Sovereignty_Carroll_Duarte_Liboiron_04242024.pdf

Carroll, S. R., Garba, I., Figueroa-Rodríguez, O. L., Holbrook, J., Lovett, R., Materechera, S., Parsons, M., Raseroka, K., Rodriguez-Lonebear, D., Rowe, R., Sara, R., Walker, J. D., Anderson, J., & Hudson, M. (2020). The CARE principles for Indigenous data governance. *Data Science Journal, 19*(43), 1–12. DOI: 10.5334/dsj-2020-043

Carroll, S. R., Garba, I., Plevel, R., Small-Rodriguez, D., Hiratsuka, V. Y., Hudson, M., & Garrison, N. A. (2022). Using indigenous standards to implement the CARE principles: Setting expectations through tribal research codes. *Frontiers in Genetics, 13*, 823309. DOI: 10.3389/fgene.2022.823309 PMID: 35386282

Carroll, S. R., Herczog, E., Hudson, M., Russell, K., & Stall, S. (2021). Operationalizing the CARE and FAIR Principles for Indigenous data futures. *Scientific Data, 8*(1), 108. DOI: 10.1038/s41597-021-00892-0 PMID: 33863927

Carroll, S. R., Rodriguez-Lonebear, D., & Martinez, A. (2019). Indigenous data governance: Strategies from United States native nations. *Data Science Journal, 18*(1), 31. DOI: 10.5334/dsj-2019-031 PMID: 34764990

Carvalho, L., Goodyear, P., & de Laat, M. (2016). Place, space, and networked learning. In *Place-based spaces for networked learning* (pp. 1–10). Routledge. DOI: 10.4324/9781315724485-1

Castellino, J. (2024). Colonial crime, environmental destruction and indigenous peoples: A roadmap to accountability and protection. In *Calibrating Colonial Crime* (pp. 97–119). Bristol University Press. DOI: 10.1332/policypress/9781529241822.003.0005

Cavalli-Sforza, L. L. (1994). *The history and geography of human genes*. Princeton UP.

Cayla, J., & Eckhardt, G. M. (2008). Asian Brands and the Shaping of a Transnational Imagined Community. *The Journal of Consumer Research, 35*(2), 216–230. DOI: 10.1086/587629

Chambers, J., Wyatt, S., Blackstock, K., & Renard, Y. (2021). Indigenous Data Sovereignty: Foundations for Indigenous knowledge governance. *Journal of Global Information Management, 29*(3), 172–189.

Chamunorwa, M. B., Winschiers-Theophilus, H., & Zaman, T. (2018). An Intermediary Database Node in the Namibian Communities Indigenous Knowledge Management System, in Digitisation of Culture: *Namibian and International Perspectives* (Singapore: Springer, 99. DOI: 10.1007/978-981-10-7697-8_7

Chand, A. V. M. (2018). Place Based Approach to plan for Resilient Cities: A local government perspective. *Procedia Engineering*, *212*, 157–164. Doi.org/1.10.1016/j.proeng.2018.01.021. DOI: 10.1016/j.proeng.2018.01.021

Chappell, M. J., & LaValle, L. A. (2011). Food security and biodiversity: Can we have both? An agroecological analysis. *Agriculture and Human Values*, *28*(1), 3–26. DOI: 10.1007/s10460-009-9251-4

Chaputula, A. H., Abdullah, H., & Mwale, B. (2020). Proliferation of social media in academic libraries: Use of WhatsApp as a platform for providing library services. *Library Management*, *41*(8/9), 717–729. DOI: 10.1108/LM-04-2020-0075

Chaturvedi, A. (2023). *Importance of media in preserving and promoting cultural heritage*. Medium.

Chawane, M. (2016). The Development of Afrocentricity: A Historical Survey. *Yesterday and Today*, *16*(16), 78–99. DOI: 10.17159/2223-0386/2016/n16a5

Cheney, G., Noyes, M., Do, E., Vieta, M., Azkarraga, J., & Michel, C. (2023). *Cooperatives at work*. Emerald Publishing Limited., DOI: 10.1108/9781838678258

Cheng, Y., & Li, W. (2024). Addressing barriers to preserving Indigenous language databases in China. *Journal of Asian Cultural Studies*, *19*(3), 149–165.

Cheng, Z., Wang, H., Xiong, W., Zhu, D., & Cheng, L. (2021). Public–private partnership as a driver of sustainable development: Toward a conceptual framework of sustainability-oriented PPP. *Environment, Development and Sustainability*, *23*(1), 1043–1063. DOI: 10.1007/s10668-019-00576-1

Chennells, R. (2023). Implementing international frameworks for the protection of indigenous knowledge: Challenges and opportunities. *International Journal of Cultural Property*, *30*(1), 41–65.

Cherid, M. I. (2021). Ain't Got Enough Money to Pay Me Respect: Black fishing, Cultural Appropriation, and the Commodification of Blackness. *Cultural Studies, Critical Methodologies*, *21*(5), 359–364. DOI: 10.1177/15327086211029357

Cheteni, P., & Umejesi, I. (2023). Application of indigenous knowledge and sustainable agriculture by small-scale farmer households and entrepreneurs in South Africa. *International Journal of Development and Sustainability*, *12*(1), 1–18.

Chibarabada, T. P., Modi, A. T., & Mabhaudhi, T. (2019). Water use of selected grain legumes in response to varying irrigation regimes. *Water S.A.*, *45*(1), 110–120.

Chibememe, G., (2014). *Review of National Laws and Policies that support or undermine Indigenous Peoples and Local communities-Zimbabwe*. Available at: https://naturaljustice.org/wp-content/uploads/2015/09/Zimbabwe-Legal-Review.pdf (Accessed 23 December, 2024).

Chigwada, J., & Chiparausha, B. (2020). Efficacy of culture houses and centres in the acquisition, preservation, and dissemination of indigenous knowledge in Zimbabwe. In: Information Resources Management Association (ed.) *Indigenous Studies: Breakthroughs in Research and Practice*. Hershey, PA: IGI Global, 498–518. DOI: 10.4018/978-1-7998-0423-9.ch027

Chikaire, J. Osuagwu, C.O. Ihenacho, R.A. Oguegbuchulam, M.N. Ejiogu-Okereke, N. & Obi, K.U.(2012). Indigenous knowledge system: The need for reform and the way forward. *Global advanced research journal of agricultural science*, 1(8),201-209.

Chikezie, P. C., Ogidi, C. O., & Anosike, C. A. (2020). Artificial intelligence in traditional medicine research. *Journal of Ethnopharmacology*, *262*, 113254.

Chilisa, B. (2012). *Indigenous research methodologies*. University of Botswana: SAGE.

Chilisa, B. (2023). Indigenous research methodologies in the digital age. *Qualitative Inquiry*, *29*(2), 217–231.

Chinsembu, K. C. (2015). 'Bioprospecting for 'green diamonds': Medicinal plants used in the management of HIV/AIDS related conditions'. In Chinsembu, K.C., Cheikhyoussef, A., Mumbengegwi, D., Kandawa-Schulz, M., Kasanda, Choshi, D. and Kazembe, L (eds) *Indigenous Knowledge of Namibia*, Windhoek: UNAM Press, 9-40.

Chinsembu, K. C. (2016). Ethnobotanical study of medicinal plants used to treat sexually transmitted infections in Northern Namibia (Oshikoto region). *Journal of Ethnobiology and Ethnomedicine*, *12*(1), 1–12. DOI: 10.1186/s13002-016-0087-2

Chipangura, N., & Seabela, M. G. (2024). Community Collaborations and Social Biographies of Museum Collections from Colonial Contexts: Meanings of Zulu Beadwork. *Museum Worlds*, *12*(1), 16–30. DOI: 10.3167/armw.2024.120103

Chisita, C. T. (2011) Role of libraries in promoting the dis-semination and documentation of indigenous agricultural information: case study of Zimbabwe. *In: World Library and Information Congress: 77th IFLA General Conference and Assembly*. Retrieved May 10, 2023, from http://conference. ifla. org/past-wlic/2011/78-chisita-en

Chisita, C. T. (2024). The Library Amid an Epistemic Storm: Research Descriptions on Indigenous Knowledge Services. In Evaluating Indigenous African Tradition for Cultural Reconstruction and Mind Decolonization (pp. 26-44). IGI Global.

Chisita, C. T., Durodolu, O. O., & Rusero, A. M. (2025). Data capitalism in the milieu of the surveillance economy: What can libraries do? *IFLA Journal*, **•••**, 03400352241286170. DOI: 10.1177/03400352241286170

Chisita, C. T., Enakrire, R. T., Durodolu, O. O., Tsabedze, V. W., & Ngoaketsi, J. M. (Eds.). (2021). *Handbook of research on records and information management strategies for enhanced knowledge coordination*. IGI Global. DOI: 10.4018/978-1-7998-6618-3

Chiwara, D. (2019). Documentation: A Security Tool for the Identification and Repatriation of Illicitly Trafficked Objects from Museums with Particular Reference to the National Gallery of Zimbabwe. *Heritage*, *2*(1), 390–399. DOI: 10.3390/heritage2010027

Choudhury, A., Sengupta, D., Ramsay, M., & Schlebusch, C. (2021). Bantu-speaker migration and admixture in Southern Africa. *Human Molecular Genetics*, *30*(R1), R56–R63. DOI: 10.1093/hmg/ddaa274 PMID: 33367711

Christen, K. (2022). Opening archives: Respectful repatriation in the digital age. *The American Archivist, 85*(1), 60–83.

Christen, K. (2023). Indigenous knowledge and digital archives: Rethinking preservation paradigms in the age of intellectual property. *International Journal of Cultural Property, 30*(1), 1–25.

Christen, K., & Cooney, C. (2022). Mukurtu CMS: A platform for indigenous digital heritage management. *Digital Scholarship in the Humanities, 37*(3), 645–661.

Christie, M. (2022). Digital boundary objects: Working with indigenous knowledge across ontological divides. *Information Communication and Society, 25*(5), 631–646.

Chukwuokolo, J.C. (2009). Afrocentrism or Eurocentrism. *New Journal of African Studies,* 32.

Chukwuokolo, J. C. (2010). Afrocentrism or Eurocentrism: The Dilemma Of African Development. *OGI-RISI a New Journal of African Studies, 6*(1). Advance online publication. DOI: 10.4314/og.v6i1.52333

Cindi, L. (2021). *Incorporating African indigenous knowledge systems into the Basic Education curriculum: Experiences from two schools in the Gauteng and KwaZulu-Natal provinces, South Africa.* (Master's dissertation). University of Free State, Bloemfontein.

Cindi, L. (2021). Incorporating African Indigenous Knowledge Systems into the Basic Education curriculum: Experiences from two schools in the Gauteng and KwaZulu-Natal Provinces, South Africa. University of the Free state

Clark, D. (2004). Apprentice and applied theoretical knowledge. *Educational Philosophy and Theory, 36*(5), 509–521. DOI: 10.1111/j.1469-5812.2004.087_1.x

Clarke, V., Braun, V.& Hayfield, N. (2015). Thematic analysis. *Qualitative psychology: A practical guide to research methods,* 222(2015), 248.

Clarke, V., & Braun, V. (2017). Thematic analysis. *The Journal of Positive Psychology, 12*(3), 297–298. DOI: 10.1080/17439760.2016.1262613

Clark, K. B., & Fujimoto, T. (1991). *Product development performance: Strategy, organization, and management in the world auto industry.* Harvard Business Press.

Clayton, S., Czellar, S., Nartova-Bochaver, S., Skibins, J. C., Salazar, G., Tseng, Y. C., Irkhin, B., & Monge-Rodriguez, F. S. (2021). Cross-cultural validation of a revised environmental identity scale. *Sustainability (Basel), 13*(4), 1–12. DOI: 10.3390/su13042387

Cleveland, D. A., & Murray, S. C. (1997). The world's crop genetic resources and the rights of indigenous farmers. *Current Anthropology, 38*(4), 477–516. DOI: 10.1086/204641

Cloete, P. C. (2010). *Economic growth and development through agriculture: the case of the North West Province of South Africa* (Doctoral dissertation, University of the Free State).

Cloete, P. C., & Idsardi, E. F. (2013). Consumption of indigenous and traditional food crops: Perceptions and realities from South Africa. *Agroecology and Sustainable Food Systems, 37*(8), 902–914. DOI: 10.1080/21683565.2013.805179

Cobo, J. R. M. (1986). *Study of the problem of discrimination against indigenous populations.*

Cock, I. E., & Van Vuuren, S. F. (2020). The traditional use of southern African medicinal plants in the treatment of viral respiratory diseases: A review of the ethnobotany and scientific evaluations. *Journal of Ethnopharmacology*, *262*, 113194. DOI: 10.1016/j.jep.2020.113194 PMID: 32730880

Connolly, D., Nam, S., & Goodman, K. (2022). Solving old problems or making new ones? Blockchain technology for the protection of refugees and migrants. *Journal of Human Rights*, *22*(2), 109–134. DOI: 10.1080/14754835.2022.2100984

Constitution of the Republic of South Africa. (1996). Pretoria. Government printer.2013 reprint.

Convention on Biological Diversity. (1992). *Text of the Convention.* Retrieved from https://www.cbd.int/convention/text/

Conway, P. (2021). Preservation in the age of digital obsolescence. *Digital Library Perspectives*, *37*(1), 79–95.

Copeland, A. (2025). Engagement for Sustainable Communities: Leading, Collaborating and Supporting Roles for Librarians. *Public Library Quarterly*, •••, 1–6. DOI: 10.1080/01616846.2025.2464515

Corbett & Boddington. (2011). Copyright Law and the Digitisation of Cultural Heritage. SSRN Electronic Journal. DOI: 10.2139/ssrn.1806809

Cornell, S., & Kalt, J. P. (2007). *The state of the Native nations: Conditions under U.S. policies of self-determination.* Oxford University Press.

Correa, C. M. (2001). Traditional knowledge and intellectual property. *Geneva: The Quaker United Nations Office (QUNO), 17.*

Corti, E., & Storto, C. (2000). Knowledge creation in small manufacturing firms during product innovation: An empirical analysis of cause–effect relationships among its determinants. *Enterprise and Innovation Management Studies*, *1*(3), 245–263. DOI: 10.1080/14632440010023190

Couldry, N., & Mejias, U. A. (2019a). *The costs of connection: How data is colonizing human life and appropriating it for capitalism.* Stanford University Press.

Couldry, N., & Mejias, U. A. (2019b). Data colonialism: Rethinking big data's relation to the contemporary subject. *Television & New Media*, *20*(4), 336–349. DOI: 10.1177/1527476418796632

Couldry, N., & Mejias, U. A. (2023). The decolonial turn in data and technology research: What is at stake and where is it heading? *Information Communication and Society*, *26*(4), 786–802. DOI: 10.1080/1369118X.2021.1986102

Couture, S., & Toupin, S. (2019). What does the notion of "sovereignty" mean when referring to the digital? *New Media & Society*, *21*(10), 2305–2322. DOI: 10.1177/1461444819865984

Crawford, C. M., & Di Benedetto, A. (2008). *New products management* (9th ed.). McGraw-Hill Irwin.

Creswell, J. W., Hanson, W. E., Clark Plano, V. L., & Morales, A. (2007). Qualitative research designs: Selection and implementation. *The Counseling Psychologist*, *35*(2), 236–264. DOI: 10.1177/0011000006287390

Creswell, J. W., & Poth, C. N. (2016). *Qualitative inquiry and research design: Choosing among five approaches*. Sage publications.

Cruz, A. G. B., Seo, Y., & Scaraboto, D. (2024). Between Cultural Appreciation and Cultural Appropriation: Self-Authorizing the Consumption of Cultural Difference. *The Journal of Consumer Research*, *50*(5), 962–984. DOI: 10.1093/jcr/ucad022

Cyber and Data Protection Act [*chapter 12:07*] Available at: https://zimlii.org/akn/zw/act/2021/5/eng@2022-03-11 (Accessed 23 December, 2024).

D1, interviewed on 19 November, 2024

Da Silva, C., Pereira, F., & Amorim, J. P. (2024). The integration of indigenous knowledge in school: A systematic review. *Compare: A Journal of Comparative Education*, *54*(7), 1210–1228. DOI: 10.1080/03057925.2023.2184200

Daes, E. I. A. (1995). *Standard-setting activities: Evolution of standards concerning the rights of indigenous people. United Nations. Sub-Commission on Prevention of Discrimination and Protection of Minorities*. Working Group on Indigenous Populations.

Daes, E. I. A. (2008). An overview of the history of indigenous peoples: Self-determination and the United Nations. *Cambridge Review of International Affairs*, *21*(1), 7–26. DOI: 10.1080/09557570701828386

Dahlström, A. N. (2009). The Two-Way Appropriation of Indigenous Knowledge: Environmental Management Policies and the Laponia Process. *Journal of Northern Studies*, *2*, 39–57.

Dalal, A. K. (2011). Folk wisdom and traditional healing practices: Some lessons for modern psychotherapies. *Foundations of Indian psychology, 2*, 21-35.

Dansu, T. (2021). Impact of Indigenous Knowledge System and Practices (IKSP) on Secondary School Students' Comprehension and Entrepreneurial Skill in Chemistry Concepts. *Available atSSRN* 4348210.

Daraz, B. (2023). *Cultural capital and academic advising for international students* (Doctoral dissertation, Fordham University).

Das, D., & Semaan, B. (2022). Collaborative Identity Decolonization as Reclaiming Narrative Agency: Identity Work of Bengali Communities on Quora. In *CHI Conference on Human Factors in Computing Systems (CHI '22)*. New Orleans, LA, USA. ACM, New York, NY, USA. DOI: 10.1145/3491102.3517600

Dasgupta, R., Dhyani, S., Basu, M., Kadaverugu, R., Hashimoto, S., Kumar, P., Johnson, B. A., Takahashi, Y., Mitra, B. K., Avtar, R., & Mitra, P. (2021). Exploring indigenous and local knowledge and practices (ILKPs) in traditional jhum cultivation for localizing sustainable development goals (SDGs): A case study from Zunheboto district of Nagaland, India. *Environmental Management*, •••, 1–13. PMID: 34342685

David-Chavez, D. M., & Gavin, M. C. (2018). A global assessment of Indigenous community engagement in climate research. *Environmental Research Letters*, *13*(12), 123005. DOI: 10.1088/1748-9326/aaf300

David, J. O. (2024). Decolonizing climate change response: African indigenous knowledge and sustainable development. *Frontiers in Sociology*, *9*, 1456871. DOI: 10.3389/fsoc.2024.1456871 PMID: 39640009

Dawadi, S. (2020). Thematic analysis approach: A step-by-step guide for ELT research practitioners. *Journal of NELTA*, *25*(1-2), 62–71. DOI: 10.3126/nelta.v25i1-2.49731

Dawson, P., & Levy, C. (2022). Archiving the Arctic: Challenges in preserving Inuit knowledge digitally. *The Polar Record*, *58*(6), 385–401.

de Castro Leal, D., Duarte, A. M. B., Krüger, M., & Strohmayer, A. (2021). Into the Mine: Wicked Reflections on Decolonial Thinking and Technologies. *In C&T'21:Proceedings of the 10th International Conference on Communities & Technologies-Wicked Problems in the Age of Tech*. 269–280. DOI: 10.1145/3461564.3461578

de la Porte, B., & Higgs, R.De la Porte & Higgs. (2019). Challenges in digitisation of cultural heritage material in the Western Cape, South Africa. *South African Journal of Information Management*, *21*(1). Advance online publication. DOI: 10.4102/sajim.v21i1.1104

De Villiers, K. (2020). *Expanding the Product Range of Reaboka Foundation Through Action-based Research*. University of Johannesburg.

Deacon, H. J., & Deacon, J. (1999). *Human Beginnings in South Africa: Uncovering the Secrets of the Stone Age. In David Philip*. Cape Town, South Africa.

Deaton, B. J., & Scholz, A. (2022). Food security, food insecurity, and Canada's national food policy: Meaning, measures, and assessment. *Outlook on Agriculture*, *51*(3), 303–312. DOI: 10.1177/00307270221113601

Deci, E. L., Olafsen, A. H., & Ryan, R. M. (2017). Self-determination theory in work organizations: The state of a science. *Annual Review of Organizational Psychology and Organizational Behavior*, *4*(1), 19–43. DOI: 10.1146/annurev-orgpsych-032516-113108

Dei, G. J. S. (2000). Rethinking the role of Indigenous knowledges in the academy. *International Journal of Inclusive Education*, *4*(2), 111–132. DOI: 10.1080/136031100284849

Dei, G. J. S. (2014). Indigenizing the curriculum: The case of the African University. In Emeagwali, G., & Dei, G. J. S. (Eds.), *African Indigenous Knowledge and the Disciplines* (pp. 165–180). Sense Publishers.

Dei, G. S. (2012). Indigenous anti-colonial knowledge as "heritage knowledge" for promoting Black/African education in diasporic contexts. *Decolonization*, *1*(1), 102–119.

Denison, J. A., & Wotshela, L. (2012). An overview of indigenous, indigenised and contemporary water harvesting and conservation practices in South Africa. *Irrigation and Drainage*, *61*(S2), 7–23. DOI: 10.1002/ird.1689

Denzin, N. K. (2012). Triangulation 2.0. *Journal of Mixed Methods Research*, *6*(2), 80–88. DOI: 10.1177/1558689812437186

Department of Basic Education. (2011). *National Curriculum Statement: Curriculum and Assessment Policy Statement (CAPS) Technology Grades 7 – 9*. Government Printers.

Department of Science and Technology. (2013). *The Bio-economy Strategy*.

Department Science and Technology. (RSA). (2011). Indigenous Knowledge System Policy. https://www.dsti.gov.za/images/pdfs/IKS_Policy%20PDF.pdf

Dewey, W. J. (2006). Repatriation of a Great Zimbabwe Bird. *Presentation to the Society of Africanist Archaeologists (SAFA).*safa.rice.edu/WorkArea/DownloadAsset.aspx. (Accessed 28/11/24).

Dhlamini, N., & Kaya, H. O. (2016). Environmental Security, Indigenous Knowledge Systems and implications for food security in South Africa. *Journal of Human Ecology (Delhi, India)*, *53*(2), 135–140. DOI: 10.1080/09709274.2016.11906965

Dlamini, P. N. (2017). Use of information and communication technology tools to capture, store and disseminate indigenous knowledge: A literature review. In Ngulube, P. (Ed.), *Handbook of Research on Theoretical Perspectives on Indigenous Knowledge Systems in Developing Countries*. IGI Global Publication. DOI: 10.4018/978-1-5225-0833-5.ch010

Dlamini, P., & Ocholla, D. N. (2018). Information and communication technology tools for managing indigenous knowledge in KwaZulu-Natal Province, South Africa. *African Journal of Library Archives and Information Science*, *28*(2), 137–153.

Dlamini, P., & Ocholla, D. N. (2018). Information and Communication Technology Tools for Managing Indigenous Knowledge in KwaZulu-Natal Province, South Africa. *African Journal of Library Archives and Information Science*, *28*(2), 137–153.

Docrat, T. F., Abdul, N. S., & Marnewick, J. L. (2024). The phytotherapeutic potential of commercial South African medicinal plants: Current knowledge and future prospects. *Physical Sciences Reviews*, *9*(1), 271–289. DOI: 10.1515/psr-2022-0136

Donkor, F. K., & Mearns, K. (2022). Harnessing Indigenous knowledge systems for enhanced climate change adaptation and governance: Perspectives from Sub-Saharan Africa. In *Indigenous knowledge and climate governance: A Sub-Saharan African perspective* (pp. 181–191). Springer International Publishing. DOI: 10.1007/978-3-030-99411-2_14

Donohue, D. (2020). Culture, cognition, and college: How do cultural values and theories of intelligence predict students' intrinsic value for learning? *Journal of Culture and Values in Education*, *4*(1), 1–14. DOI: 10.46303/jcve.2020.3

Dorji, T., Rinchen, K., Morrison-Saunders, A., Blake, D., Banham, V., & Pelden, S. (2024). Understanding How Indigenous Knowledge Contributes to Climate Change Adaptation and Resilience: A Systematic Literature Review. *Environmental Management*, *74*(6), 1101–1123. DOI: 10.1007/s00267-024-02032-x PMID: 39215837

Dosono, B., & Semaan, B. (2020). Decolonizing Tactics as Collective Resilience: Identity Work of AAPI Communities on Reddit. *Proceedings of the ACM on Human-Computer Interaction 4,* CSCW1, 1–20. DOI: 10.1145/3392881

Dowling, D. O., & Miller, K. J. (2019). Immersive Audio Storytelling: Podcasting and Serial Documentary in the Digital Publishing Industry. *Journal of Radio & Audio Media*, *26*(1), 167–184. DOI: 10.1080/19376529.2018.1509218

Du Bois, M. (2013). Protection of intellectual property rights as human rights in international law. *South African Yearbook of International Law*, *38*(1), 92–122.

Du Toit, D. C. (2011). Department of Agriculture, Forestry and Fisheries. *Food security*. Pretoria.

Duarte, M. E., & Belarde-Lewis, M. (2021). Imagining: Creating Spaces for Indigenous Ontologies. *The Journal of Documentation*, *77*(5), 1053–1066.

Dube, I., Makola, Z. S., Ntoyanto, N., & Msiza, C. (2023). Investigating Small-scale Farmers and Barriers to Entry in Cannabis Cultivation in South Africa. *Politeia*, *42*(1), 277–290. DOI: 10.25159/2663-6689/15002

Du, J. T. (2017). Research on Indigenous People and the Role of Information and Communications. Technology in Development: *A Review of the Literature*. *Journal of the Australian Library and Information Association*, *66*(4), 344–363. DOI: 10.1080/24750158.2017.1397857

Dunbar, T. & Scrimgeour, M. (2017). LSIC: Procedural ethics through an Indigenous ethical lens. *Indigenous Children Growing Up Strong: A Longitudinal Study of Aboriginal and Torres Strait Islander Families,* 61-78.

Duncan, J. (2023). Data protection beyond data rights: Governing data production through collective intermediaries. *Internet Policy Review*, *12*(3). Advance online publication. DOI: 10.14763/2023.3.1722

Duncan, N. B. (1995). Capturing of information technology infrastructure: A study of resource characteristics and their measure. *Journal of Management Information Systems*, *12*(2), 37–57. DOI: 10.1080/07421222.1995.11518080

Dury, J., Schaller, N., Garcia, F., Reynaud, A., & Bergez, J. E. (2012). Models to support cropping plan and crop rotation decisions. A review. *Agronomy for Sustainable Development*, *32*(2), 567–580. DOI: 10.1007/s13593-011-0037-x

Dutfield, G. (2000). *Intellectual Property Rights, Trade, and Biodiversity*. Earthscan. DOI: 10.4324/9781849776233

Duval, E. (2022). Metadata for digital preservation: A cross-cultural perspective. *The Journal of Documentation*, *78*(2), 454–472.

Dzobo, N. K. (1976). The courage to be an African in education. *The Oguaa Educator*, *7*(1), 132–138.

Eason, A. E., Braddy, L. M., & Fryberg, S. A. (2018). Reclaiming Representations & Interrupting the Cycle of Bias Against Native Americans. *Daedalus*, *147*(2), 70–81. DOI: 10.1162/DAED_a_00491

Edgar, B. A., Eva, M. P., & Lee, K. S. (2022). Disappearance of African Indigenous Knowledge of Water Conservation and Management in Limpopo Province of South Africa: An IKS Perspective. *African Journal of Development Studies*, *12*(2).

Egan, D., Hjerpe, E. E., & Abrams, J. (Eds.). (2011). *Why people matter in ecological restoration*. Island Press. DOI: 10.5822/978-1-61091-039-2_1

Ehret, C. (1972). Bantu origins and history: Critique and interpretation. *Transafrican Journal of History*, *2*(1), 1–9.

Eitzen, D. (2018). The Duties of Documentary in a Post-Truth Society. In *Cognitive Theory and Documentary Film*. Palgrave Macmillan. DOI: 10.1007/978-3-319-90332-3_6

Ekene, A. K., Lee, K. S., Decide, M. T., & Thobejane, T. D. (2023). Placing Indigenous Knowledge at the Core of Production and Marketing of Medicinal Plant (Mpesu) in Zimbabwe and South Africa. *In Casebook of Indigenous Business Practices in Africa* [Emerald Publishing Limited.]. *Apprenticeship, Craft, and Healthcare*, *1*, 167–185.

Ekobi, G.A., Mboh, L. & Tanga, P. (2023). Contribution of Indigenous Knowledge to Agricultural Growth in South Africa: A Case of Disaneng Community in the Ratlou Local Municipality. *African Journal of Inter/Multidisciplinary Studies, 5*(1), 1-12.

Eliazer Nelson, A.R.L. Ravichandran, K. & Antony, U. (2019). The impact of the Green.

Elmore, R. W. (1996). Our relationship with the ecosystem and its impact on agriculture. *Journal of Production Agriculture*, *9*(1), 42–45. DOI: 10.2134/jpa1996.0042

Emmanuel, O. C., & Babalola, O. O. (2022). Amaranth production and consumption in South Africa: The challenges of sustainability for food and nutrition security. *International Journal of Agricultural Sustainability*, *20*(4), 449–460. DOI: 10.1080/14735903.2021.1940729

Engstrom, T., Lobo, E. H., Watego, K., Nelson, C., Wang, J., Wong, H., Kim, S. L., Oh, S. I., Lawley, M., Gorse, A.-D., Ward, J., & Sullivan, C. (2024). Indigenous data governance approaches applied in research using routinely collected health data: A scoping review. *npj. Digital Medicine*, *7*(1), 68. DOI: 10.1038/s41746-024-01070-3 PMID: 38491156

Ens, E. J., Finlayson, M., Preuss, K., Jackson, S., & Holcombe, S. (2015). Australian approaches for managing 'country' using Indigenous and non-Indigenous knowledge. *Ecological Management & Restoration*, *16*(2), 84–92. DOI: 10.1111/emr.12120

Escobar, A. (2011). *Encountering development: The making and unmaking of the Third World.* Princeton University Press.

Escobar, A. (2018). *Designs for the pluriverse: Radical interdependence, autonomy and the making of worlds*. Duke University Press. DOI: 10.1215/9780822371816

Eshun, G. (2011). *Ecotourism development in Ghana: A postcolonial study with focus on Boabeng-Fiema monkey sanctuary and Kakum National Park*. University of Leicester.

Etikan, I., Musa, S. A., & Alkassim, R. S. (2016). Comparison of convenience sampling and purposive sampling. *American Journal of Theoretical and Applied Statistics*, *5*(1), 1–4. DOI: 10.11648/j.ajtas.20160501.11

Evans, I. (2023). *Bureaucracy and race: Native administration in South Africa* (Vol. 53). Univ of California Press. DOI: 10.2307/jj.5973059

Eveleigh, A. (2021). Funding models for digital preservation initiatives: Sustainability challenges and opportunities. *Archival Science*, *21*(2), 143–165.

Eyo, N. A. (2024). Influence of documentary films on learning cultural awareness among university students in Uyo, Nigeria. *Quantum Journal of Social Sciences and Humanities*, *5*(2), 151–171. DOI: 10.55197/qjssh.v5i2.359

Eyong, C. T. (2007). Indigenous knowledge and sustainable development in Africa: Case study on Central Africa. *Indigenous knowledge systems and development: Relevance for Africa.Tribes and Tribals, 1*, 121–139.

Ezeanya-Esiobu, C. (2019). *Indigenous knowledge and education in Africa*. Springer Nature. DOI: 10.1007/978-981-13-6635-2

Faldi, G., Fisher, A., & Moretto, L. (2021). Five points for conceptualising place-based approaches to African urban planning: an introduction. In Faldi, G., Fisher, A., & Moretto, L. (Eds.), *African Cities Through Local Eyes: Experiments in Place-Based Planning and Design* (pp. 1–27). Springer. DOI: 10.1007/978-3-030-84906-1_1

FAO. (2009). The State of Food Insecurity in the World. Economic crises-impacts and lessons learned, FAO, Roma. *Agriregionieuropa (Ancona)*, •••, 30.

FasterCapital. (2024). *Cultural Exchange: Bridging Nations: BRICS and Cultural Exchange*. Nad Al Sheba.

Felch, J., & Frammoline, R. (2011). *Chasing Aphrodite: The Hunt for Looted Antiquities at the World's Richest Museum*. Mifflin Press.

Fernandez, M. E. (1994). Gender and indigenous knowledge. *Indigenous knowledge and development monitor* 2(3), 6-7.

Fernández-Llamazares, Á., Lepofsky, D., Lertzman, K., Armstrong, C. G., Brondizio, E. S., Gavin, M. C., Lyver, P. O. B., Nicholas, G. P., Pascua, P. A., Reo, N. J., & Reyes-García, V. (2021). Scientists' warning to humanity on threats to indigenous and local knowledge systems. *Journal of Ethnobiology, 41*(2), 144–169. DOI: 10.2993/0278-0771-41.2.144

Ferranti, P., Berry, E., & Jock, A. (2018). *Encyclopedia of food security and sustainability*. Elsevier.

Ferreira, A., & Pienaar, M. (2024). Developing inclusive digital platforms for African oral traditions. *Journal of Knowledge Management, 28*(5), 289–310.

Figueiredo, B., Larsen, H. P., & Bean, J. (2021). The Cosmopolitan Servicescape. *Journal of Retailing, 97*(2), 267–287. DOI: 10.1016/j.jretai.2020.09.001

Fink, A. (2024). Data cooperative. *Internet Policy Review, 13*(2). Advance online publication. DOI: 10.14763/2024.2.1752

Finnegan, R. (2022). Oral traditions and digital archives: Methodological challenges and opportunities. *Journal of Folklore Research, 59*(1), 1–24.

First Nations Information Governance Centre. (2014). Ownership, control, access, and possession (OCAP): The path to First Nations information governance. First Nations Information Governance Centre. https://achh.ca/wp-content/uploads/2018/07/OCAP_FNIGC.pdf

Flemmer, R., & Schilling-Vacaflor, A. (2015). Unfulfilled promises of the consultation approach: The limits to effective Indigenous participation in Bolivia's and Peru's extractive industries. *Third World Quarterly, 37*(1), 172–188. DOI: 10.1080/01436597.2015.1092867

Flick, U. (2017). *The Sage handbook of qualitative data collection*. Sage publications.

Flint, M. T., & William, A. (1989). *Intellectual Property the new law*. Butterworths.

Floridi, L. (2020). The fight for digital sovereignty: What it is, and why it matters, especially for the EU. *Philosophy & Technology*, *33*(3), 369–378. DOI: 10.1007/s13347-020-00423-6 PMID: 35194548

Folarin, J. (2005). The impact of knowledge sharing, organizational capacity and partnership quality on IS outsourcing success. *Information & Management*, *38*(5), 323–335.

Foley, J. M. (2012). *Oral tradition and the Internet: Pathways of the mind*. University of Illinois Press. DOI: 10.5406/illinois/9780252037184.001.0001

Food and Agricultural Organization. (2002). *The state of Food insecurity in the World*. FOA.

Food and Agriculture Organization (FAO). (2021). *Globally Important Agricultural Heritage Systems (GIAHS)*. Retrieved from https://www.fao.org/giahs/en/

Foot, K. A. (2014). Cultural-historical activity theory: Exploring a theory to inform practice and research. *Journal of Human Behavior in the Social Environment*, *24*(3), 329–347. DOI: 10.1080/10911359.2013.831011

Ford, J. D., Cameron, L., Rubis, J., Maillet, M., Nakashima, D., Willox, A. C., & Pearce, T. (2016). Including Indigenous knowledge and experience in IPCC assessment reports. *Nature Climate Change*, *6*(4), 349–353. DOI: 10.1038/nclimate2954

Ford, J. D., Maillet, M., Pouliot, V., Meredith, T., Cavanaugh, A., & Labbe, J. (2016). Adaptation and Indigenous peoples in the Arctic: An analysis of best practices in climate adaptation planning. *Environmental Research Letters*, *11*(9), 093004. DOI: 10.1088/1748-9326/11/9/093004

Ford, J. D., Smit, B., & Wandel, J. (2020). Climate change adaptation in Indigenous communities. *Environmental Research Letters*, *15*(8), 083001. DOI: 10.1088/1748-9326/ab9b90

Fox, J.. (2018). Shifting cultivation and climate change in Southeast Asia. *Nature Climate Change*, *8*(1), 1–3.

Francis, D., & Webster, E. (2019). Poverty and inequality in South Africa: Critical reflections. *Development Southern Africa*, *36*(6), 788–802. DOI: 10.1080/0376835X.2019.1666703

Francis, J., Nyamukondiwa, P., & Hagenmeier, C. (2022). How can rural-based universities harness the Fourth Industrial Revolution to strengthen community engagement beyond national borders? *Journal of Community Engagement and Scholarship*, *13*(2). Advance online publication. DOI: 10.54656/CFHX2494

Friend, C. (2004). *Social contract theory*. Internet Encyclopaedia of Philosophy.

Frison, C., Parks, L., & Tsioumani, E. (2022). Biocultural Community Protocols: Making Space for Indigenous and Local Cultures in Access and Benefit Sharing? In *Access and Benefit Sharing of Genetic Resources, Information and Traditional Knowledge* (pp. 177-190). Routledge.

Furness, M., & Trautner, B. (2020). Reconstituting social contracts in conflict-affected MENA countries: Whither Iraq and Libya? *World Development*, *135*, 105085. DOI: 10.1016/j.worlddev.2020.105085

Gadgil, M., & Berkes, F. (1991). Traditional resource management systems. *Resource Management and Optimization*, *8*(3), 127–141.

Gadgil, M., Berkes, F., & Folke, C. (1993). Indigenous knowledge for biodiversity conservation. *Ambio*, *22*(2-3), 151–156. PMID: 33566330

Galloway, D. (2004). *Sankofa, Sasa and Zamani: The Ongoing Influence of Spirituals on African American Poetry*. http://ctl.du

Galvin, K. A.. (2020). Climate change and pastoralism: Traditional coping mechanisms and sustainable adaptation in Maasai communities. *Global Environmental Change*, *64*, 102–110.

Gao, W., Rao, V. R., & Zhou, M. (2001). *Plant genetic resources conservation and use in China*. Bioversity International.

Garba, I., Sterling, R., Plevel, R., Carson, W., Cordova-Marks, F. M., Cummins, J., Curley, C., David-Chavez, D., Fernandez, A., Hiraldo, D., Hiratsuka, V., Hudson, M., Jäger, M. B., Jennings, L. L., Martinez, A., Yracheta, J., Garrison, N. A., & Carroll, S. R. (2023). Indigenous peoples and research: Self-determination in research governance. *Frontiers in Research Metrics and Analytics*, *8*, 1272318. Advance online publication. DOI: 10.3389/frma.2023.1272318 PMID: 38033627

Gebashe, F. C., Naidoo, D., Amoo, S. O., & Masondo, N. A. (2022). Cosmeceuticals: A newly expanding industry in South Africa. *Cosmetics*, *9*(4), 77–89. DOI: 10.3390/cosmetics9040077

Gegeo, D. W., & Watson-Gegeo, K. A. (2022). Participatory methodologies in indigenous knowledge documentation: Reflections from the Solomon Islands. *Qualitative Research*, *22*(4), 498–515.

Gepts, P. (2004). Who owns biodiversity, and how should the owners be compensated? *Plant Physiology*, *134*(4), 1295–1307. DOI: 10.1104/pp.103.038885 PMID: 15084724

German Ethics Council. (2017). *Big Data and Health: Data Sovereignty as the Shaping of Informational Freedom (Executive Summary & Recommendations)*. German Ethics Council.

Ghosh, P. K. (2006). Biodiversity conservation in India: A legal perspective. *Indian Journal of Environmental Law*, *7*(1), 5–18.

Giannossa, L. C., Muntoni, I. M., Laviano, R., & Mangone, A. (2020). Building a step-by-step result in archaeometry. Raw materials, provenance and production technology of Apulian Red Figure pottery. *Journal of Cultural Heritage*, *43*, 242–248. DOI: 10.1016/j.culher.2019.11.002

Gibson, K. (2019). Digital Preservation, Social Justice, and the Role of Information Professionals. *Library Trends*, *68*(2), 155–172.

Giliomee, H. B., & Mbenga, B. (2007). *New History of South Africa*. Tafelberg.

Glavič, P. (2020). Identifying key issues of education for sustainable development. *Sustainability (Basel)*, *12*(16), 1–18. DOI: 10.3390/su12166500

Glover, L., & Poitras, G. (2023). Indigenous data sovereignty and digital archiving frameworks: A Canadian perspective. *Archives and Manuscripts*, *51*(4), 412–428.

Gonçalves, C. D. B. Q., Schlindwein, M. M., & Martinelli, G. D. C. (2021). Agroforestry systems: A systematic review focusing on traditional indigenous practices, food and nutrition security, economic viability, and the role of women. *Sustainability (Basel)*, *13*(20), 11397. DOI: 10.3390/su132011397

González-Tennant, E. (2021). Mobile technologies and indigenous knowledge documentation: A comparative analysis. *Information Technology for Development*, *27*(1), 81–99.

Gover, K. (2015). Settler–state political theory, 'CANZUS'and the UN declaration on the rights of indigenous peoples. *European Journal of International Law*, *26*(2), 345–373. DOI: 10.1093/ejil/chv019

Government of India. (2023). *National policy on tribal development*. Ministry of Tribal Affairs.

Government of the Republic of Zimbabwe. (2023). As amended up to 20th June, 2023 https://veritaszim.net/sites/veritas_d/files/Constitution%20Consolidated%20%282023%29.pdf

Graeber, D., & Wengrow, D. (2021). *The dawn of everything: A new history of humanity*. Macmillan.

Graham, J., & Fortier, J. (2005). *Indigenous governance: Emerging trends and ongoing challenges*. Institute on Governance. https://iog.ca/docs/IndigenousGovernance.pdf

Green, B. N., Johnson, C. D. & Adams, A. (2001). Writing narrative literature reviews for peer-reviewed journals: secrets of the trade. *Journal of sports chiropractic and rehabilitation,* 15(1), 5-19.

Green, B. N., Johnson, C. D., & Adams, A. (2001). Writing narrative literature reviews for peer-reviewed journals: Secrets of the trade. *Journal of Chiropractic Medicine*, *5*(3), 101–117. DOI: 10.1016/S0899-3467(07)60142-6 PMID: 19674681

Greenberg, J. H. (1972). Linguistic Evidence Regarding Bantu Origins. *Journal of African History*, *13*(2), 189–216. DOI: 10.1017/S0021853700011427

Grenier, L. (1998). *Working with indigenous knowledge: a guide for researchers*. International Development Research Centre Books.

Greyling, B. (2010). *A model for community participation in African libraries to preserve indigenous knowledge*. Retrieved May 13, 2023, from https://www.inst.at/trans/17Nr/8-2/8-2_greyling.htm

Greyling, E., & Zulu, S. (2010). Content development in an indigenous digital library: A case study in community participation. *IFLA Journal*, *36*(1), 30–39. DOI: 10.1177/0340035209359570

Grobler, M. (2018). The ethical challenges of the Fourth Industrial Revolution. *South African Journal of Science*, *114*(1/2).

Gruzd, A., & Staves, K. (2011). Trends in scholarly use of online social media. Position paper presented at the workshop on changing dynamics of scientific collaboration, *the 44th Annual Hawaii International Conference on System Sciences (HICSS)*. Retrieved August 24, 2022, from: https://dalspace.library.dal.ca/handle/10222/14427

Guarda, T., Portela, F., & Diaz-Nafria, J. M. (Eds.). (2023). *Advanced Research in Technologies, Information, Innovation and Sustainability: Third International Conference, ARTIIS 2023, Madrid, Spain, October 18–20, 2023, Proceedings, Part I*. Springer Nature

Guba, E. G., & Lincoln, Y. S. (1994). Competing paradigms in qualitative research. *Handbook of qualitative research, 2*(163-194), 105.

Gumbo, M. T. (2023). Digitisation of higher education and research: Raising inclusivity and equity issues for indigenous students. *South African Computer Journal = Suid-Afrikaanse Rekenaartydskrif, 35*(1), 149–163. DOI: 10.18489/sacj.v35i1.1107

Gunderson, F., Lancefield, R. C., & Woods, B. (Eds.). (2019). *The Oxford handbook of musical repatriation.* Oxford University Press.

Guo, J. (2025). Digital cultural heritage challenges, solutions, and future directions: edited by Vinay Kukreja, Amitoj Singh, Deepinder Kaur, and Jagpuneet Kaur Bajwa, London and New York, CRC Press, 2025, 320 pp., $154.80 (hardback), ISBN 978-1-032-63054-0.

Gupta, A. D. (2012). Way to study indigenous knowledge and indigenous knowledge system. *Research Journal of Recent Sciences. ISSN, 2277*, 2502. Available online at: www.isca.in, www.isca.me

Gupta, A. (2000). *Rewarding traditional knowledge and contemporary grassroots creativity.* Indian Institute of Management.

Gupta, A. K. (2002). Value addition to local resources and institutions. *Electronic Journal of Sustainable Development, 1*(1), 43–55.

Gupta, R. K., Gautam, J. N., & Khare, V. P. (2014). Awareness and use of social media applications among library staff of power sector organizations. [ALIS]. *Annals of Library and Information Studies, 61*(4), 320–331.

Gupta, V. K. (2022). Traditional Knowledge Digital Library: A defensive mechanism against biopiracy. *Journal of Intellectual Property Rights, 27*(3), 245–260.

Gururani, K., Sood, S., Kumar, A., Joshi, D. C., Pandey, D., & Sharma, A. R. (2021). Mainstreaming Barahnaja cultivation for food and nutritional security in the Himalayan region. *Biodiversity and Conservation, 30*(3), 551–574. DOI: 10.1007/s10531-021-02123-9 PMID: 33526962

Guttenbrunner, M., & Rauber, A. (2022). Emulation strategies for digital preservation: Lessons from a decade of practice. *International Journal of Digital Curation, 17*(1), 1–20.

Gwandure, C., & Lukhele-Olorunju, P. (2023). Women's use of indigenous knowledge in Africa. *Social Sciences & Humanities Open, 8*(1), 100741. DOI: 10.1016/j.ssaho.2023.100741

Gwerevende, S., & Mthombeni, Z. M. (2023). Safeguarding Intangible Cultural Heritage: Exploring the Synergies in the Transmission of Indigenous Languages, Dance and Music Practices in Southern Africa. *International Journal of Heritage Studies, 29*(5), 398–412. DOI: 10.1080/13527258.2023.2193902

Haimson, O. L., Brubaker, J. R., Dombrowski, L., & Hayes, G. R. (2015). Disclosure, stress, and support during gender transition on Facebook. *InProceedings of the 18th ACM Conference on Computer Supported Cooperative Work & Social Computing.* 1176–1190. DOI: 10.1145/2675133.2675152

Hall, R. (2004). A political economy of land reform in South Africa. *Review of African Political Economy, 31*(100), 213–227. DOI: 10.1080/0305624042000262257

Hamadani, H., Rashid, S. M., Parrah, J. D., Khan, A. A., Dar, K. A., Ganie, A. A., Gazal, A., Dar, R. A., & Ali, A. (2021). Traditional farming practices and its consequences. *Microbiota and Biofertilizers. Ecofriendly Tools for Reclamation of Degraded Soil Environs*, *2*, 119–128.

Hamwaalwa, N. (n.d). Indigenous knowledge management: Roles of public library and archives in preserving and providing access to indigenous knowledge (IK) in Namibia: Case study of 'Preserve Namibia Indigenous Knowledge' project. https://www.scecsal.org/publications/papers2021/scecsal_papers2021_hamwaalwa.pdf

Hangshing, J. & Laloo, B. (2021). Role of libraries in preservation and accessibility of indigenous knowledge: A study of University Libraries in Northeast India. *Library Philosophy and Practice (e-journal)*. 5254.

Hanson, V. J. (2020). Cannabis policy reform: Jamaica's experience. In *Legalizing Cannabis* (pp. 375–389). Routledge. DOI: 10.4324/9780429427794-21

Hardy, K., & Maurushat, A. (2021). Blockchain and the protection of Indigenous Knowledge. Journal of Law. *Information, and Science*, *28*(1), 123–145.

Harmsworth, G., Awatere, S., & Robb, M. (2023). Indigenous ecological knowledge and digital ethnography: A case study from Aotearoa New Zealand. *Environmental Management*, *71*(2), 285–300. PMID: 36639605

Hart, M. (2007). Indigenous knowledge and research: The míkiwáhp as a symbol for reclaiming our knowledge and ways of knowing. *First Peoples Child & Family Review: A Journal on Innovation and Best Practices in Aboriginal Child Welfare Administration, Research. Policy & Practice*, *3*(1), 83–90.

Hasan, M., & Abdullah, H. M. (2015). Plant genetic resources and traditional knowledge: emerging needs for conservation. *Plant genetic resources and traditional knowledge for food security*, pp.105-120.

Haumba, E. N., & Kaddu, S. (2017). Documenting and disseminating agricultural indigenous knowledge for sustainable food security in Uganda. *University of Dar es Salaam.Library Journal*, *12*(1), 66–86.

Havas, K., & Salman, M. (2011). Food security: Its components and challenges. *International Journal of Food Safety, Nutrition and Public Health*, *4*(1), 4–11. DOI: 10.1504/IJFSNPH.2011.042571

Hayes, M. (2022). Place-Based Pedagogy and Teacher Education. In Peters, M. A. (Ed.), *Encyclopedia of Teacher Education*. Springer., DOI: 10.1007/978-981-16-8679-5_293

Heritage Chairs of Australia and New Zealand. (HCOANZ). (2020). Dhawura Ngilan: a vision for Aboriginal and Torres Strait Islander heritage in Australia and the best practice standards in cultural heritage management and legislation, *Australian Government Department of Agriculture, Water and the Environment,* Canberra.

Hewson, M. G., & Holtman, L. B. (2022). Incorporating Indigenous health knowledge into science curricula: Challenges and solutions. *African Journal of Research in Science. Technology and Mathematics Education*, *26*(1), 78–98.

Hickey, S. (2011). The politics of social protection: What do we get from a 'social contract' approach? *Canadian Journal of Development Studies*, *32*(4), 426–438. DOI: 10.1080/02255189.2011.647447

Hiddink, G. A., Termorshuizen, A. J., & van Bruggen, A. H. (2010). Mixed cropping and suppression ofs oilborne diseases. *Genetic Engineering, biofertilisation, soil quality and organic farming,* pp.119-146.

Higgins, S. (2022). Preservation planning for digital cultural heritage: A framework for sustainable approaches. *International Journal of Digital Curation, 17*(2), 56–72.

Hignett, S., & McDermott, H. (2013). Qualitative methodology for Ergonomics. In Wilson, J. R., & Sharples, S. (Eds.), *Evaluation of Human Work. A practical ergonomics methodology* (4th ed.). CRC Press.

Hildreth, P., Kimble, C., & Wright, P. (2000). Communities of practice in the distributed international environment. *Journal of Knowledge Management, 4*(1), 27–38. DOI: 10.1108/13673270010315920

Hill, R., Grant, C., George, M., Robinson, C. J., & Sparrow, A. (2012). A typology of Indigenous engagement in Australian environmental management: Implications for knowledge integration and social-ecological system sustainability. *Ecology and Society, 17*(1), 23. DOI: 10.5751/ES-04587-170123

Himmelmann, N., Thieberger, N., & Wittenburg, P. (2023). Multimodal documentation of endangered languages: Methodological advances and technological applications. *Language Documentation and Conservation, 17*(1), 125–147.

Hinton, L., Huss, L., Roche, G., & Tsosie, E. (2016). Language Revitalization in the Digital Age: Promises and Challenges. In *The Routledge Handbook of Language Revitalization, edited by L. Hinton, L. Huss, and G. Roche, 371-84*. Routledge.

Hirsch, D. D. (2013). The glass house effect: Big Data, the new oil, and the power of analogy. *Me. L. Rev., 66*, 373.

Hlangwani, E., Adebiyi, J. A., Doorsamy, W., & Adebo, O. A. (2020). Processing, characteristics and composition of umqombothi (a South African traditional beer). *Processes (Basel, Switzerland), 8*(11), 1451–1481. DOI: 10.3390/pr8111451

Hlangwani, E., Hal, P. H. V., Moganedi, K. L., & Dlamini, B. C. (2023). The future of African wild fruits–a drive towards responsible production and consumption of the marula fruit. *Frontiers in Sustainable Food Systems, 7*, 1294437. DOI: 10.3389/fsufs.2023.1294437

Hlengiwe, M. (2023). *From Beadwork to Africanfuturism: Exploring Maxhosa Africa's SS22 Collection* (Master's thesis, University of Pretoria (South Africa) Indigenous knowledge system (IKS) in repositories in South Africa. *Records Management Journal, 31*(2), 176-196

Hobbes, T. (1651). *Leviathan*. London: Penguin Books (reprint 1985).

Hohmann, J., & Weller, M. (2018). The UN Declaration on the Rights of Indigenous Peoples: a commentary, Oxford commentaries on international law. Oxford: Oxford University Press, pp. 63-86. DOI: 10.1093/law/9780199673223.001.0001

Holden, C. J. (2017). The spread of Bantu languages, farming, and pastoralism in sub-equatorial Africa. In Mapping our ancestors (pp. 247-268). Routledge. indigenous ways of knowing. *The Journal of Pan African Studies, 2*(4), 184–202.

Holste, J. S., & Fields, D. (2010). Trust and tacit knowledge sharing and use. *Journal of Knowledge Management*, *14*(1), 128–140. DOI: 10.1108/13673271011015615

Hoppers, O. (Ed.). (2002a). *Indigenous Knowledge and the Integration of Knowledge Systems: Towards a philosophy of articulation*. New Africa Books.

Hossain, K., & Ballardini, R. (2021). Protecting Indigenous Traditional Knowledge Through a Holistic Principle-Based Approach. *Nordic Journal of Human Rights*, *39*(1), 51–72. DOI: 10.1080/18918131.2021.1947449

Hudson, M., Carroll, S. R., Anderson, J., Blackwater, D., Cordova-Marks, F. M., Cummins, J., David-Chavez, D., Fernandez, A., Garba, I., Hiraldo, D., Jäger, M. B., Jennings, L. L., Martinez, A., Sterling, R., Walker, J. D., & Rowe, R. K. (2023). Indigenous Peoples' rights in data: A contribution toward Indigenous Research Sovereignty. *Frontiers in Research Metrics and Analytics*, *8*, 1173805. DOI: 10.3389/frma.2023.1173805 PMID: 37215248

Hughes, H. (2003, December). What is remembered and what forgotten: a decade of redefining culture and heritage for tourism in South Africa. In *Conference on Developing Cultural Tourism,* University of Nottingham, Nottingham, UK (pp. 470-479).

Hummel, P., Braun, M., Tretter, M., & Dabrock, P. (2021). Data sovereignty: A review. *Big Data & Society*, *8*(1), 2053951720982012. DOI: 10.1177/2053951720982012

Hunter, J. (2005). The Role of Information Technologies in Indigenous Knowledge Management. *Australian Academic and Research Libraries*, *36*(2), 109–124. DOI: 10.1080/00048623.2005.10721252

Hunter, J., Koopman, B., & Szeredi, P. (2023). Digital divide-conscious design: Developing technologies for indigenous knowledge documentation. *Information Technology for Development*, *29*(1), 70–88.

Huntington, H. P. (2005). 'We dance around in a ring and suppose.' Academic engagement with traditional knowledge. *Arctic Anthropology*, *42*(1), 29–32. DOI: 10.1353/arc.2011.0101

Huntington, H. P., Quakenbush, L., & Nelson, M. (2023). Documenting traditional ecological knowledge in the Arctic: Applications for climate change adaptation. *Arctic Science*, *9*(1), 143–162.

Hu, X., Ng, J., & Lee, J. H. (2019). VR creation experience in cultural heritage education: A preliminary exploration. *Proceedings of the Association for Information Science and Technology*, *56*(1), 422–426.

Huynh, C. V., Pham, T. G., Nguyen, T. Q., Nguyen, L. H. K., Tran, P. T., Le, Q. N. P., & Nguyen, M. T. H. (2020). Understanding indigenous farming systems in response to climate change: An investigation into soil erosion in the mountainous regions of Central Vietnam. *Applied Sciences (Basel, Switzerland)*, *10*(15), 5091. DOI: 10.3390/app10155091

Ibnouf, F. O. (2012). The Value of women's' indigenous knowledge in food processing and preservation for achieving household food security in rural Sudan. *Journal of Food Research*, *1*(1), 238–253. DOI: 10.5539/jfr.v1n1p238

IFLA. (2002). *IFLA Statement on indigenous traditional knowledge*. Retrieved June 10, 2024, from https://www.ifla.org/publications/ifla-statement-on-indigenous-traditional-knowledge

Iloka, N. G. (2016). Indigenous knowledge for disaster risk reduction: An African perspective. *Jàambá*, *8*(1), 1–7. DOI: 10.4102/jamba.v8i1.272 PMID: 29955294

Indigenous Knowledge Act no 6 of 2019, Government Gazette, Cape Town, 19 August 2019

Indigenous Knowledge Systems Policy. (2004), Department of Science and Technology, Pretoria International Labour Organization (ILO) Convention 169: https://www.ilo.org/dyn/normlex/en/f?p=NORMLEXPUB:12100:0:NO:P12100_ILO_CO DE:C169

Ingram, J. (2011). A food systems approach to researching food security and its interactions with global environmental change. *Food Security*, *3*(4), 417–431. DOI: 10.1007/s12571-011-0149-9

Intergovernmental Panel on Climate Change (IPCC). (2019). *Special report on climate change and land.* https://www.ipcc.ch/srccl/

International Open Data Charter. (2015). Available at: https://opendatacharter.org/wp-content/uploads/2023/12/opendatacharter-charter_F.pdf (Accessed 8 December 2024).

International Union for the Protection of New Varieties of Plants. (1991). *UPOV Convention.* Retrieved from https://www.upov.int/

Irani, L., Vertesi, J., Dourish, P., Philip, K., & Grinter, R. E. (2010). Postcolonial computing: a lens on design and development. *InProceedings of the SIGCHI conference on human factors in computing systems.* 1311–1320. DOI: 10.1145/1753326.1753522

Iredale, G. W. Y. N. E. T. H. (2021). What are the different types of blockchain technology? 101Blockchains. Retrieved from https://101blockchains.com/types-of-blockchain

Irion, K. (2012). Government cloud computing and national data sovereignty. *Policy and Internet*, *4*(3-4), 40–71. DOI: 10.1002/poi3.10

Isah, A., Bashorun, M. T., & Omopupa, K. T. (2012). Libraries and preservation of indigenous knowledge in developing countries: The Nigeria experience. In *Library and Information Science in Developing Countries* (pp. 96–106). Contemporary Issues. DOI: 10.4018/978-1-61350-335-5.ch008

Isern, N., & Fort, J. (2019). Assessing the importance of cultural diffusion in the Bantu spread into South Eastern Africa. *PLoS One*, *14*(5), e0215573. DOI: 10.1371/journal.pone.0215573 PMID: 31067220

Israel, B. A., Schulz, A. J., Parker, E. A., & Becker, A. B. (1998). Review of Community-Based Research: Assessing Partnership Approaches to Improve Public Health. *Annual Review of Public Health*, *19*(1), 173–202. DOI: 10.1146/annurev.publhealth.19.1.173 PMID: 9611617

Iwara, I. O., & Kilonzo, B. M. (2022). Towards a conceptual understanding of an effective rural-based entrepreneurial University in South Africa. *Social Sciences (Basel, Switzerland)*, *11*(9), 388. DOI: 10.3390/socsci11090388

Jacob, R., & Alexander, D. (1993). *A guide to intellectual property*. Sweet & Maxwell.

Jafari, Z. (2023). The Role of AI in Supporting Indigenous Languages.https://www.researchgate.net/publication/379040413_The_Role_of_AI_in_Supporting_Indigenous_Languages

Jain, P. (2014). Indigenous knowledge management in Botswana using ICT applications. In *Concepts and Advances in Information Knowledge Management* (167-191). DOI: 10.1533/9781780634357.2.167

Jain, S. K. (1991). Ethnobotany and research on medicinal plants in India. *Ethnobotany: Principles and Applications*, 185–203.

Jain, S. K. (1991). *Dictionary of Indian folk medicine and ethnobotany*. Deep Publications.

Jaleel, S., & Verghis, A. M. (2015). Knowledge creation in constructivist learning. *Universal Journal of Educational Research*, *3*(1), 8–12. DOI: 10.13189/ujer.2015.030102

Janke, T. (2021). Indigenous intellectual property rights in the digital age. *International Journal of Cultural Property*, *28*(1), 1–25.

Janke, T. (2023). Indigenous Cultural and Intellectual Property (ICIP) protocols in the digital environment. *International Journal of Cultural Property*, *30*(2), 157–178.

Janos, A. C. (2000). *East Central Europe in the Modern World*. University of California Press.

Jarke, M., Otto, B., & Ram, S. (2019). Data sovereignty and data space ecosystems. *Business & Information Systems Engineering*, *61*(5), 549–550. DOI: 10.1007/s12599-019-00614-2

Jena, A. K., Deuri, R., Sharma, P., & Singh, S. P. (2018). Underutilized vegetable crops and their importance. *Journal of Pharmacognosy and Phytochemistry*, *7*(5), 402–407.

Jessen, T. D., Ban, N. C., Claxton, N. X., & Darimont, C. T. (2022). Contributions of indigenous knowledge to ecological and evolutionary understanding. *Frontiers in Ecology and the Environment*, *20*(2), 93–101. DOI: 10.1002/fee.2435

John, H. C., Egbeyemi, T. A., & Oniyide, F. D. (2020). Social networking technologies in libraries and the role of librarians as agent for repositioning Nigerian education. *International Journal of Library and Information Studies*, *10*(2), 75–84.

Johnson, B. & Lundvall, B.A. (1994). The learning economy. *Journal of industry studies*, 1(2), 23- 42.

Johnson, T. G. (2019). Existing conceptual models of arts and culture: an inventory. *The Rural Policy Research Institute (RUPRI)–University of Iowa, Iowa City*. Dostupno na: http://www. rupri. org/wp-content/uploads/Conceptual-Mo- dels-of-Arts-and-Culture. pdf (pristupljeno: 20. srpnja 2021).

Jonas, H., Shrumm, H., & Bavikatte, K. (2010). Biocultural community protocols and conservation pluralism. *Policy Matters*, *17*, 102–112.

Jones, M. (2024). *'O Cannabis': A History of Cannabis Legalization in Canada* (Doctoral dissertation, University of Victoria).

Jones, P. L., Mahelona, K., Duncan, S., & Leoni, G. (2023). Kia tangata whenua: Artificial intelligence that grows from the land and people.

Jones, A. D., Ngure, F. M., Pelto, G., & Young, S. L. (2013). What are we assessing when we measure food security? A compendium and review of current metrics. *Advances in Nutrition*, *4*(5), 481–505. DOI: 10.3945/an.113.004119 PMID: 24038241

Jones, J., Seet, P. S., Acker, T., & Whittle, M. (2021). Barriers to grassroots innovation: The phenomenon of social-commercial-cultural trilemmas in remote indigenous art centres. *Technological Forecasting and Social Change*, *164*, 119583. DOI: 10.1016/j.techfore.2019.02.003

Jongwe, A. I., Moroz, P. W., Gordon, M., & Anderson, R. B. (2020). Strategic alliances in firm-centric and collective contexts: Implications for indigenous entrepreneurship. *Economies*, *8*(2), 31–57. DOI: 10.3390/economies8020031

Joseph, R. K. (2010). The Nagoya Protocol (Joseph, 2010) on Access and Benefit Sharing: International treaty on genetic resources comes into force. *Current Science*, *100*(10), 1441–1442.

Juma, K., & Agyei-Mensah, S. (2021). AI ethics and Indigenous Knowledge: Examining intellectual property challenges in Africa. *African Journal of Science, Technology, Innovation and Development*, *13*(3), 225–236.

Jumba, H., Bundi, D., & John, J. (2020). Use of ICTs in Knowledge Management for Enhanced Institutional Sustainability. *Regional Journal of Information and Knowledge Management*, *5*(1), 1–17. DOI: 10.70759/kpbgc061

Junior, A. A., Gana, D., Adukeye, E., & Ahuchaogu, C. (2023). Unlocking the Socioeconomic Advantages of Cannabis Sativa L. Legalization in African Countries: An In-Depth Review. *Global Journal of Agricultural Research*, *11*(1), 32–52. DOI: 10.37745/gjar.2013/vol11n13252

Kai Bol, G., & Van Niekerk, D. (2024). Application of African Indigenous Knowledge Systems for climate change management. *International Journal of Disaster Risk Reduction*, *111*. Advance online publication. DOI: 10.1016/j.ijdrr.2024.103153

Kale, P., & Singh, H. (2007). Building firm capabilities through learning: The role of the alliance learning process in alliance capability and firm-level alliance success. *Strategic Management Journal*, *28*(10), 981–1000. DOI: 10.1002/smj.616

Kalland, A. (2000). Indigenous knowledge: prospects and limitations. *Indigenous environmental knowledge and its transformations: critical anthropological perspectives*, 316.

Kamau, E. C., Fedder, B., & Winter, G. (2010). The Nagoya Protocol (Joseph, 2010) on Access to Genetic Resources and Benefit Sharing: What is new and what are the implications for provider and user countries and the scientific community? *Law, Environment & Development Journal*, *6*(3), 246–262.

Kamwi, J. (2017). Indigenous knowledge and cultural heritage: Challenges and opportunities for preservation and management. *Journal of Cultural Heritage*, *23*, 1–9.

Kandolo, K. M., & Ngibe, M. (2024). Exploring financial knowledge influences on the economic empowerment of women entrepreneurs in Kwazulu-Natal, South Africa. *International Journal of Research in Business & Social Science*, *13*(8), 46–61. DOI: 10.20525/ijrbs.v13i8.3683

Kanyandago, P. (2007). From Negative to Positive Universality: Anthropological and Ethical Implication for Africa. *In Universality; From Theory to Practice: An Intercultural and Interdisciplinary Debate about Facts, Possibilities, Lies and Myths, 25th Colloquium of the Swiss Academy of Humanities and Social Sciences*. Fribourg: Academic Press, 316.

Kaplan, I. (2022). The Case against the Universal Museum. https://www.artsy.net/article/artsy-editorial-the-case-against-the-universal-museum (Accessed 28/11/24).

Kaplan, A. M., & Haenlein, M. (2010). Users of the world, unite! the challenges and opportunities of social media. *Business Horizons*, *53*(1), 59–68. DOI: 10.1016/j.bushor.2009.09.003

Kapoor, A., Tait, J., & Vaitl, B. (2024). How to build a data cooperative: A practitioner's handbook. https://data2x.org/wp-content/uploads/2024/06/Data-Cooperative-Handbook.pdf

Kaya, H. O., & Koitsiwe, M. (2016). African Indigenous Knowledge Systems and Natural Disaster Management in North West Province, South Africa. *Journal of Human Ecology (Delhi, India)*, *53*(2), 101–105. DOI: 10.1080/09709274.2016.11906961

Kazmi, T., Ghafoor, A., Chaudhry, A. A., & Khan, S. E. (2014). Farmers beliefs about indigenous farming practices and sustainable agricultural development. *Pakistan Journal of Agricultural Research*, *27*(1).

Keikelame, M. J., & Swartz, L. (2019). Decolonising research methodologies: Lessons from a qualitative research project, Cape Town, South Africa. *Global Health Action*, *12*(1), 1561175. DOI: 10.1080/16549716.2018.1561175 PMID: 30661464

Keitumetse, S. (2006). UNESCO 2003 convention on intangible heritage: Practical implications for heritage management approaches in Africa. *South African Archaeological Bulletin*, *61*(184), 166–171. DOI: 10.2307/20474924

Kelsey Leonard (Shinnecock Nation), Stephanie Russo (Native Village of Kluti-Kaah), Andrew Martinez (O'odham, Diegueño, and Yoeme), Lauren McElroy (2023). Our Common Agenda Global Digital Compact March 2023: CARE Statement for Indigenous Data Sovereignty (Scottish Canadian)

Kepe, T. (2003). Cannabis sativa and rural livelihoods in South Africa: Politics of cultivation, trade and value in Pondoland. *Development Southern Africa*, *20*(5), 605–615. DOI: 10.1080/0376835032000149252

Kgomoeswana, K. B., & Makgahlela, L. (2024). Key challenges faced when preserving records in traditional councils during the 4th Industrial Revolution. In *Multidisciplinary Approach to Information Technology in Library and Information Science* (pp. 62–80). IGI Global.

Khan, S. (2021). Rural land reform and local economic development through agri-parks: A case study of Harry Gwala Municipality in KwaZulu-Natal. *The New Political Economy of Land Reform in South Africa*, 179-197.

Khoddami, A., Messina, V., Vadabalija Venkata, K., Farahnaky, A., Blanchard, C. L., & Roberts, T. H. (2023). Sorghum in foods: Functionality and potential in innovative products. *Critical Reviews in Food Science and Nutrition*, *63*(9), 1170–1186. DOI: 10.1080/10408398.2021.1960793 PMID: 34357823

Khor, M. (2002). *Intellectual Property, Biodiversity, and Sustainable Development*. Zed Books.

Khumalo, S. B., & Mapotse, T. A. (2023). Building modern technology innovation on indigenous knowledge systems in Technology Education. In Gumbo, M. T., & Williams, P. J. (Eds.), *Indigenous Technology Knowledge Systems: Contemporary issues in Technology Education*. Springer., DOI: 10.1007/978-981-99-1396-1_4

Kiarie, E. (2024). *Cultural Heritage Preservation in the Face of Globalization. International Journal of Humanity and Social Sciences*. Catholic University of Eastern Africa.

Kipuri, N. (1999). The human rights situation of indigenous peoples in Africa. *Indigenous Affairs*, (1), 18–25.

Kitchenham, B. (2004). *Procedures for Performing Systematic Reviews*. Keele University.

Kiwa, F. J., Chimhundu, H., Tsvere, M., & Nleya, S. (2023). Transmission of Traditional Culture through Digital Technology in Zimbabwe. *East African Journal of Education and Social Sciences*, *4*(3), 1–6. DOI: 10.46606/eajess2023v04i03.0271

Kock, K. T. (2020). The influence of culture on the utilization of traditional leafy vegetables among VaTsonga and Ba-Pedi women of Bushbuckridge, Mpumalanga Province (Doctoral dissertation, North-West University (South Africa)).

Kogut, B., & Zander, U. (1992). Knowledge of the firm, combinative capabilities, and the replication of technology. *Organization Science*, *3*(3), 383–397. DOI: 10.1287/orsc.3.3.383

Kom, Z., Nethengwe, N. S., Mpandeli, S., & Chikoore, H. (2023). Indigenous knowledge indicators employed by farmers for adaptation to climate change in rural South Africa. *Journal of Environmental Planning and Management*, *66*(13), 2778–2793. DOI: 10.1080/09640568.2022.2086854

Koomen, O., & Koopman, K. J. (2023). How the technological epoch and COVID-19 silenced indigenous knowledge in the university curriculum. In *Decolonizing the South African University: Towards Curriculum as Self Authentication* (pp. 57–75). Springer Nature Switzerland.

Kopnina, H. (2020). Education for the future? Critical evaluation of education for sustainable development goals. *The Journal of Environmental Education*, *51*(4), 280–291. DOI: 10.1080/00958964.2019.1710444

Koptie, S. W. (2010). Indigenous Self-Discovery: "Being Called to Witness". *First Peoples Child & Family Review*, *5*(1), 114–125. DOI: 10.7202/1069068ar

Korat, H. & R. Mathukia. (2022). *Effect of natural farming, organic farming and conventional farming on soil physical, chemical and biological properties.*

Kostauli, M. R. (2024). *Towards the co-management of natural resources in protected areas in South Africa: a study of the Silaka nature reserve, in Eastern Cape province* (Doctoral dissertation, University of the Witwatersrand, Johannesburg).

Kothari, A., & Anuradha, R. V. (1999). Biodiversity and Intellectual Property Rights: Can the Two Co-Exist? *J. Int'l Wildlife L. & Pol'y*, *2*, 252.

Kovach, M. (2018). Doing Indigenous Methodologies: A Letter to a Research Class. In N.K. Denzin and Y.S. Lincoln. *The SAGE Handbook of Qualitative Research.*

Kovach, M. (2021). Indigenous methodologies: Characteristics, conversations, and contexts in the digital age. *Qualitative Inquiry*, *27*(2), 165–180.

Kovach, M. (2021). *Indigenous methodologies: Characteristics, conversations, and contexts*. University of Toronto press.

Kristin, B., Kelly, S., Travis, H., Joseph, L., & Lori, C. (2017). Retail food environments, shopping experiences, First Nations and the provincial Norths. *Health Promotion and Chronic Disease Prevention in Canada : Research, Policy and Practice*, *37*(10), 333–341. DOI: 10.24095/hpcdp.37.10.03 PMID: 29043760

Krubu, D. E., & Osawaru, K. E. (2011). The impact of information and communication technology (ICT) in Nigerian university libraries. *Library philosophy and Practice, 2011*(583), 1-19.

Kruger, S. (2024, June). Transcultural leadership and sustainable development in the digital era: Navigating the 4IR in South Africa. In *International Conference on Society 5.0* (pp. 207–217). Cham: Springer Nature Switzerland.

Kruger, S., & Steyn, A. A. (2024). Leveraging technology adoption to navigate the 4IR towards a future-ready business: A systematic literature review. *Engineering Reports*, *6*(5), e12762. DOI: 10.1002/eng2.12762

Krupa KL & Grimm KT. (2021). Digital Repatriation as a Decolonizing Practice in the Archaeological Archive. ATD, VOL18(ISSUE1/2)

Kugedera, A. T., Sakadzo, N., Museva, T., Kokerai, L. K., Muridzi, G., & Musara, A. (2021). Role and commercialisation of indigenous knowledge in the control of Covid-19 pandemic. *The Dyke*, *15*(2), 43–57.

Kuhnlein, H., Eme, P., & Larrinoa, Y. D. (2019). Indigenous food systems: contributions to sustainable food systems and sustainable diets. In Sustainable diets: linking nutrition and food systems (pp. 64-78). Wallingford UK: CABI. (ASAP). DOI: 10.1079/9781786392848.0064

Kukutai, T., Russo Carroll, S., & Walter, M. (2020*). Indigenous Data Sovereignty. Indigenous World.* https://researchcommons.waikato.ac.nz/server/api/core/bitstreams/6796content (Accessed 10 March 2025)

Kukutai, Tahu; McIntosh, Tracey; Boulton, Amohia; Durie, Meihana; Foster, Meika; Hutchings, Jessica; et al. (2024). Te Pūtahitanga: A Tiriti-Led Science-Policy Approach for Aotearoa New Zealand. AgResearch. Report. **https://doi.org/**DOI: 10.57935/AGR.26001496.v1

Kukutai, T. (2023). Indigenous data sovereignty—A new take on an old theme. *Science*, *382*(6674), eadl4664. Advance online publication. DOI: 10.1126/science.adl4664 PMID: 38033063

Kukutai, T., & Cormack, D. (2018). Census 2018 and Implications for Māori. *New Zealand Population Review*, *44*, 131–151.

Kukutai, T., & Taylor, J. (2016). Indigenous data sovereignty. In Kukutai, T., & Taylor, J. (Eds.), *Indigenous data sovereignty: Towards on agenda* (pp. 1–22). ANU Press., DOI: 10.22459/CAEPR38.11.2016.01

Kukutai, T., & Taylor, J. (2016). *Indigenous data sovereignty: Toward an agenda.* ANU press. DOI: 10.22459/CAEPR38.11.2016

Kukutai, T., & Taylor, J. (2022). Indigenous data sovereignty: Implications for data governance in the digital age. *Big Data & Society*, *9*(1), 1–15.

Kumar, A. (2013). Ethno-botanical study of wild vegetables used by rural communities of Kanaujidistrict, Uttar Pradesh, India. *Emirates Journal of Food and Agriculture*, *25*(10), 760–766. DOI: 10.9755/ejfa.v25i10.16403

Kumar, G., & Chander, H. (2017). Documentation of indigenous agricultural implements, practices and other conservation techniques in subtropical climatic zone of Shivalik hills, North Western Himalaya. *J. Biol. Chem. Chron*, *3*(2), 15–23.

Kumari, M., Pandey, S., Giri, V. P., Chauhan, P., Mishra, N., Verma, P., Tripathi, A., Singh, S. P., Bajpai, R., & Mishra, A. (2022). Integrated approach for technology transfer awareness of traditional knowledge for upliftment of circular bioeconomy. In *Biomass, Biofuels, Biochemicals* (pp. 613–636). Elsevier. DOI: 10.1016/B978-0-323-89855-3.00005-4

Kumar, M., & Kerr, J. M. (2013). Democratic assertions: The making of India's Recognition of Forest Rights Act. *Development and Change*, *44*(3), 433–455.

Kumar, R., Singh, S., & Pradhan, R. (2019). Blockchain in agriculture: Preserving knowledge and innovation. *International Journal of Technology Management & Sustainable Development*, *18*(1), 78–95.

Kumar, S., Kumar, N., & Vivekadhish, S. (2016). Millennium development goals (MDGS) to sustainable development goals (SDGS): Addressing unfinished agenda and strengthening sustainable development and partnership. *Indian Journal of Community Medicine : Official Publication of Indian Association of Preventive & Social Medicine*, *41*(1), 1. DOI: 10.4103/0970-0218.170955 PMID: 26917865

Kwet, M. (2019). Digital Colonialism: US Empire and the New Imperialism in the Global South. *Race & Class*, *60*(4), 3–26. DOI: 10.1177/0306396818823172

Labadi, S. (2024). The World Heritage Convention at 50: Management, credibility and sustainable development. *Journal of Cultural Heritage Management and Sustainable Development*, *14*(4), 750–764. DOI: 10.1108/JCHMSD-05-2022-0077

Laird, S. A., & Wynberg, R. P. (2018). A fair share? Sharing the benefits of botanical knowledge. *Nature Plants*, *4*(9), 605–609.

Lalonde, D. (2021). Does Cultural Appropriation Cause Harm? *Politics, Groups & Identities*, *9*(2), 329–346. DOI: 10.1080/21565503.2019.1674160

Lam, D. P., Hinz, E., Lang, D. J., Tengö, M., von Wehrden, H., & Martín-López, B. (2020). Indigenous and local knowledge in sustainability transformations research: A literature review. *Ecology and Society*, *25*(1), 23–45. DOI: 10.5751/ES-11305-250103

Langer, J. (2023). From acting to understanding: The comparative development of meaning. In *The nature and ontogenesis of meaning* (pp. 191–213). Routledge. DOI: 10.4324/9781003428152-9

Lara, S. W., Tsiami, A., & Cross, P. (2023). Discovering and mapping colloquial terminologies describing underutilized and neglected food crops—A comprehensive review. *Foods*, *12*(12), 2428. DOI: 10.3390/foods12122428 PMID: 37372639

Larson, A. M., & Pulhin, J. M. (2012). Enhancing forest tenure reforms through more responsive governance. *Conservation & Society*, *10*(2), 103–113. DOI: 10.4103/0972-4923.97482

Larson, N., Laska, M. N., & Neumark-Sztainer, D. (2020). Food insecurity, diet quality, home food availability, and health risk behaviors among emerging adults: Findings from the EAT 2010–2018 study. *American Journal of Public Health, 110*(9), 1422–1428. DOI: 10.2105/AJPH.2020.305783 PMID: 32673120

Lave, J., & Wenger, E. (1991). *Situated Learning: Legitimate Peripheral Participation*. Cambridge University Press. DOI: 10.1017/CBO9780511815355

Le Roux, C. J. B. (2003). Tapping indigenous knowledge on the world-wide web. *Indilinga, 2*(1), 107–113.

Leal Filho, W., Barbir, J., Gwenzi, J., Ayal, D., Simpson, N. P., Adeleke, L., Tilahun, B., Chirisa, I., Gbedemah, S. F., Nzengya, D. M., Sharifi, A., Theodory, T., & Yaffa, S. (2022). The role of indigenous knowledge in climate change adaptation in Africa. *Environmental Science & Policy, 136*, 250–260. DOI: 10.1016/j.envsci.2022.06.004

Lee, H., & Choi, B. (2003). Knowledge management enablers, processes, and organizational performance: An integrative view and empirical examination. *Journal of Management Information Systems, 20*(1), 179–228. DOI: 10.1080/07421222.2003.11045756

Lefèvre, T. (2018). Big data in forensic science and medicine. *Journal of Forensic and Legal Medicine, 57*, 1–6. DOI: 10.1016/j.jflm.2017.08.001 PMID: 29801942

Leonard, W. Y. (2017). Producing language reclamation by decolonising 'language'. *In Leonard, W.Y & De Korne, H. (eds) Language Documentation and Description,* 14. London: EL Publishing. pp. 15-36.

Leonardi, P. M. (2012). Materiality, sociomateriality, and socio-technical systems: What do these terms mean? How are they related? Do we need them? *Information and Organization, 22*(3), 139–149.

Lerario, A. (2022). The role of built heritage for sustainable development goals: From statement to action. *Heritage, 5*(3), 2444–2463. DOI: 10.3390/heritage5030127

Leroy, J. L., Ruel, M., Frongillo, E. A., Harris, J., & Ballard, T. J. (2015). Measuring the food access dimension of food security: A critical review and mapping of indicators. *Food and Nutrition Bulletin, 36*(2), 167–195. DOI: 10.1177/0379572115587274 PMID: 26121701

Letshaba, R. K. (2024). *Developing a framework for the implementation of knowledge management within skills development providers in South Africa* (Doctoral dissertation, North-West University (South Africa).

Levis, C., Costa, F. R. C., Bongers, F., Peña-Claros, M., Clement, C. R., & Ter Steege, H. (2017). Persistent effects of pre-Columbian plant domestication on Amazonian forest composition. *Science, 355*(6328), 925–931. DOI: 10.1126/science.aal0157 PMID: 28254935

Lewis, J. E.. (2020). *Indigenous protocol and artificial intelligence position paper*. Indigenous Protocol and Artificial Intelligence Working Group.

Lewis, J. L., & Sheppard, S. R. (2006). Culture and communication: Can landscape visualization improve forest management consultation with indigenous communities? *Landscape and Urban Planning, 77*(3), 291–313. DOI: 10.1016/j.landurbplan.2005.04.004

Leyew, Z. (2024). Indigenous knowledge practices in danger of extinction due to cultural and environmental changes: The case of Gojjam, Amhara region. *Ethiopian Journal of Language. Culture and Communication*, *9*(2), 30–75.

Lightfoot, S. (2016). *Global Indigenous Politics: A Subtle Revolution*. Routledge. DOI: 10.4324/9781315670669

Lilomaiava-Doktor, S. I. (2020). Oral traditions, cultural significance of storytelling, and Samoan understandings of place or fanua. *Native American and Indigenous Studies*, *7*(1), 121–151.

Lim, J. B. Y. (2025). *Digital media interventions in Southeast Asia*. Palgrave Macmillan. DOI: 10.1007/978-3-031-78211-4

Litheko, A. (2021). Small tourism business growth through eco-tourism in Mahikeng, South Africa. *African Journal of Hospitality, Tourism and Leisure*, *10*(1), 256–269. DOI: 10.46222/ajhtl.19770720-100

Littletree, S., Belarde-Lewis, M., & Duarte, M. (2020). Centering relationality: A conceptual model to advance indigenous knowledge organization practices. *Knowledge Organization*, *47*(5), 410–426. DOI: 10.5771/0943-7444-2020-5-410

Liu, J. (2019). The impact of Bantu migration on agriculture in sub-Saharan Africa. *International Journal of New Developments in Engineering and Society*, *3*(2), 28–34.

Liu, Y., Weng, C., Li, H., & Sanga, G. (2019). Social media-based digital storytelling for preserving oral tradition: A case study in Tibetan Culture. *Digital Scholarship in the Humanities*, *34*(1), 165–179.

Li, X., Kim, J. S., & Lee, T. J. (2021). Collaboration for community-based cultural sustainability in island tourism development: A case in Korea. *Sustainability (Basel)*, *13*(13), 1–17. DOI: 10.3390/su13137306

Local Contexts. (2021). Grounding Indigenous rights. Local Contexts. https://localcontexts.org

Locke, J. (1689). *Two Treatises of Government*. New Haven: Yale University Press (reprint 2003).

Loh, J., Harmon, D., & Tok, S. K. (2022). Digital Indigenous Knowledge Exchange: A multi-stakeholder approach to climate change adaptation. *Climate and Development*, *14*(5), 485–500.

Lor, P. J. (2003). National libraries and the digital divide. *Mousaion*, *21*, 62–78.

Lovett, R., Lee, V., Kukutai, T., Cormack, D., Rainie, S., & Walker, J. (2019). Good Data Practices for Indigenous Data Sovereignty and Governance. In Daly, A., Devitt, S. K., & Mann, M. (Eds.), *Good Data* (pp. 26–36). Institute of Network Cultures.

Lubinga, S. N., Maramura, T. C., & Masiya, T. (2023). Adoption of Fourth Industrial Revolution: challenges in South African higher education institutions. http://hdl.handle.net/2263/94027

Lucarelli, R. (2023). From Virtual Reality to virtual restitution: How 3D-Egyptology can contribute to decolonizing the field and the question of digital copies vs the original. In Palladino, C., & Bodard, G. (Eds.), *Can't Touch This*. Ubiquity Press., DOI: 10.5334/bcv.i

Luedicke, M. K., Craig, J. T., & Markus, G. (2010). Consumer Identity Work as Moral Protagonism: How Myth and Ideology Animate a Brand-Mediated Moral Conflict. *The Journal of Consumer Research*, *36*(6), 1016–1032. DOI: 10.1086/644761

Lugo-Morin, D. R. (2020). Indigenous communities and their food systems: A contribution to the current debate. *Journal of Ethnic Foods*, *7*(1), 1–10. DOI: 10.1186/s42779-019-0043-1

Lum, N. R. (2013). Community perceptions of Cannabis sativa (dagga) in the Mzintlava location, Ingquza Hill Local Municipality, Eastern Cape province [master's dissertation]. *Walter Sisulu University, Umtata.*

Luthria, D. L., & Pastor-Corrales, M. A. (2006). Phenolic acids content of fifteen dry edible bean (Phaseolus vulgaris L.) varieties. *Journal of Food Composition and Analysis*, *19*(2-3), 205–211. DOI: 10.1016/j.jfca.2005.09.003

Lwoga, E. D., Ngulube, P., & Stilwell, C. (2017). Indigenous knowledge management practices in indigenous organizations in South Africa and Tanzania. In: Ngulube P (ed), *Handbook of research on social, cultural, and educational considerations of indigenous knowledge in developing countries*. IGI Global, 181–201. Retrieved October 14, 2021, from: https://doi.org/DOI: 10.4018/978-1-5225-0838-0.ch010

Lwoga, E. T., & Ngulube, P. (2008). Managing indigenous and exogenous knowledge through information and communication technologies for agricultural development and achievement of the UN Millennium Development Goals, In B. Njobvu S. Koopman (Eds.), *Libraries and information services towards the attainment of the UN Millennium Development Goals*. Berlin: Walter de Gruyter.

Lwoga, E. T., Ngulube, P., & Stilwell, C. (2010). Managing indigenous knowledge for sustainable agricultural development in developing countries: Knowledge management approaches in the social context. *The International Information & Library Review*, *42*(3), 174–185. DOI: 10.1080/10572317.2010.10762862

Lwoga, E. T., Ngulube, P., & Stilwell, C. (2022). Indigenous knowledge and technologies in African educational contexts: A digital library perspective. *Information Development*, *38*(1), 105–121.

M1, interviewed on 20 November, 2024

Mabhaudhi, T., Chibarabada, T. P., Chimonyo, V. G. P., Murugani, V. G., Pereira, L. M., Sobratee, N., Govender, L., Slotow, R., & Modi, A. T. (2018). Mainstreaming underutilized indigenous and traditional crops into food systems: A South African perspective. *Sustainability*, *11*(1), 172. DOI: 10.3390/su11010172 PMID: 37681213

Macaulay, A. C. (2007). Promoting participatory research by family physicians. *Annals of Family Medicine*, *5*(6), 557–560. DOI: 10.1370/afm.755 PMID: 18025494

MacNeil, M. (2008). An Epidemiologic Study of Aboriginal Adolescent Risk in Canada: The Meaning of Suicide. *Journal of Child and Adolescent Psychiatric Nursing*, *21*(1), 3–12. DOI: 10.1111/j.1744-6171.2008.00117.x PMID: 18269407

Maddox, R., & Ninomiya, M. M. E. (2025). Indigenous sovereignty in research and epistemic justice: Truth telling through research. *Global Public Health: An International Journal for Research, Policy and Practice*, *20*(1), 2436436. DOI: 10.1080/17441692.2024.2436436 PMID: 39661944

Maditsi, M. E. (2018). *The role of traditional leaders in postapartheid local governance in South Africa: a case study of the Bapong 2 community, North West Province, South Africa* (Doctoral dissertation, North-West University (South Africa)).

Maditsi, M. E., & Materechera, S. A. (2021). The Role of Indigenous Communities of Practice in Fostering Community Engagement and Partnerships for Sustainable Development in Africa. In *Sustainable Development in Africa* (pp. 3–18). Springer. DOI: 10.1007/978-3-030-74693-3_1

Madsen, V. (2018). Transnational Encounters and Peregrinations of the Radio Documentary Imagination. In Follmer, G., & Badenoch, A. (Eds.), *Transnationalizing Radio Research, New Approaches to an Old Medium 83-100, Creative commons*. DOI: 10.1515/9783839439135-008

Magni, G. (2017). Indigenous knowledge and implications for the sustainable development agenda. *European Journal of Education*, *52*(4), 437–447. DOI: 10.1111/ejed.12238

Magoro, B. K. (2021). *Implications of youth unemployment on socio-economic development in South Africa: A case of Mankweng Township in Limpopo Province* (Doctoral dissertation).

Magoro, K. (2024). *Indigenous knowledge systems situated and contextual learning: unlocking collective human digital capabilities and competencies for Global South communities in the digital era*. ICEGOV.

Mahachi, G. (2011). *Zimbabwe Report on the UNESCO Workshop on Prevention and Fight Against Illicit Traffic of Cultural Goods in Southern Africa: Current Situation and Way Forward*; UNESCO: Windhoek, Namibia, 14–15 September 2011.

Maina, C. K. (2012). Traditional knowledge management and preservation: Intersections with Library and Information Science. *The International Information & Library Review*, *44*(1), 13–27. DOI: 10.1080/10572317.2012.10762911

Mainardes, E. W., Alves, H., & Raposo, M. (2010). An exploratory research on the stakeholders of a university. *Journal of Management and strategy, 1*(1), 76.

Makananise, F. O., & Madima, S. E. (Eds.). (2025). *Decolonising digital media and indigenisation of participatory epistemologies*. Routledge.

Makazhe, E. C., & Maramura, T. C. (2023, September). The role of Fourth Industrial Revolution (4IR) technologies in achieving sustainable development goals. In *The Global Conference on Entrepreneurship and the Economy in an Era of Uncertainty* (pp. 985–998). Springer Nature Singapore.

Makgamatha, M. G., & Moikanyane, K. K. (2024). The role of community development projects in tackling socio-economic issues in South Africa. *Management, Business, Administration and Legal*, 256.

Makgopa, M. (2022). Implications of the fourth industrial revolution (4IR) on the development of indigenous languages of South Africa: Challenges and opportunities. *Handbook of Research on Protecting and Managing Global Indigenous Knowledge Systems,* 152-161.

Makuse, S. M., & Mbhenyane, X. G. (2011). Health benefits and omega-3-fatty acid content of selected indigenous foods in the Limpopo Province, South Africa. *Indilinga, 10*(2), 182–194.

Malabadi, R. B., Kolker, K. P., & Chalannavar, R. K. (2023). Medical Cannabis sativa (Marijuana or Drug type); The story of discovery of Δ9-Tetrahydrocannabinol (THC). *International Journal of Innovation Scientific Research and Review, 5*(3), 4134–4143.

Malaccorto, R. B. (2023). *ODC's contribution to the Global Digital Compact- Report Open Data and the Right to Privacy in a global digital strategy.* Available at: https://www.un.org/techenvoy/sites/www.un.org.techenvoy/files/GDC-submission_Open-Data-Charter.pdf

Malapane, O. L., Chanza, N., & Musakwa, W. (2024). Transmission of indigenous knowledge systems under changing landscapes within the Vhavenda community, South Africa. *Environmental Science & Policy, 161*, 103861. DOI: 10.1016/j.envsci.2024.103861

Malapane, O. L., Musakwa, W., & Chanza, N. (2024). Indigenous agricultural practices employed by the Vhavenda community in the Musina local municipality to promote sustainable environmental management. *Heliyon, 10*(13), e33713. DOI: 10.1016/j.heliyon.2024.e33713 PMID: 39040358

Malatji, T., & Makgamatha, M. G. (2024). An Assessment on The Effect of Immigrant Entrepreneurship on Local Economic Development in South Africa: A Literature Analysis. [Public Administration Journal]. *Jurnal Administrasi Publik, 14*(2), 92–103.

Malgwi, P. G., Otubelu, B. N., & Sadiq, H. (2023). Enhancing library services in the Fourth Industrial Revolution (4IR): The innovative role of emerging technologies.

Malobane, M. E., Nciizah, A. D., Mudau, F. N., & Wakindiki, I. I. (2020). Tillage, crop rotation and crop residue management effects on nutrient availability in a sweet sorghum-based cropping system in marginal soils of South Africa. *Agronomy (Basel), 10*(6), 776–886. DOI: 10.3390/agronomy10060776

Maluleka, R. J. (2020). The status of traditional healing in the Limpopo province of South Africa. *Hervormde Teologiese Studies, 76*(4), 36–43. DOI: 10.4102/hts.v76i4.6103

Mama, M. L. (2020). *Promoting agriculture amongst the black youth of South Africa: the role the sector has to play in creating employment, alleviating poverty, and achieving equitable economic growth* (Doctoral dissertation, Stellenbosch: Stellenbosch University).

Mangla, S. K., Kazançoğlu, Y., Yıldızbaşı, A., Öztürk, C., & Çalık, A. (2022). A conceptual framework for blockchain-based sustainable supply chain and evaluating implementation barriers: A case of the tea supply chain. *Business Strategy and the Environment, 31*(8), 3693–3716. DOI: 10.1002/bse.3027

Manyaga, M. B., Goldman, G. A., & Thomas, P. (2024). Sustaining SMEs through Indigenous Knowledge Systems: Exploring opportunities and challenges. *Southern African Journal of Entrepreneurship and Small Business Management, 16*(1), 1–12. DOI: 10.4102/sajesbm.v16i1.882

Mapara, J. (2009). Indigenous knowledge systems in Zimbabwe: Juxtaposingpost-colonialal theory. *The Journal of Pan African Studies, 3*(1), 139–155.

Mapara, J. (2022). Indigenous knowledge across the curriculum: An epistemic and cognitive shifting process. In Muyambo, T., Mapangisana-Hlatywayo, A., & Sithole, P. (Eds.), *Re-imagining indigenous knowledge and practices in 21st century Africa: Debunking myths and misconceptions for conviviality and sustainability* (pp. 41–68). Langaa Publishers.

Mapiye, O., Chikwanha, O. C., Makombe, G., Dzama, K., & Mapiye, C. (2020). Livelihood, food, an nutrition security in Southern Africa: What role do indigenous cattle genetic resources play? *Diversity*, *12*(2), 74. DOI: 10.3390/d12020074

Mariam, H. (2009). The Return of the Axum Obelisk: The cultural benefits of the return of the axum obelisk. *Museum International*, *61*(1), 48–51. DOI: 10.1111/j.1468-0033.2009.01684.x

Marinova, D., & Raven, M. (2006). Indigenous knowledge and intellectual property: A sustainability agenda. *Journal of Economic Surveys*, *20*(4), 587–605. DOI: 10.1111/j.1467-6419.2006.00260.x

Marley, I. R. (2012). Investigating the appropriateness of the theory of organisational knowledge creation as a management model for practice-led research. *Journal of Literacy Criticism. Comparative Linguistic and Literacy Studies*, *31*(1), 1–10.

Marnewick, J. L., Venter, I., Rautenbach, F., & Piek, H. (2019). Rooibos herbal tea: An optimal cup an its consumers. *Health SA Gesondheid*, *24*(1), 1–9. PMID: 31934411

Maroyi, A. (2012). Use of traditional veterinary medicine in Nhema communal area of the Midlands province, Zimbabwe. *African Journal of Traditional, Complementary, and Alternative Medicines : AJTCAM*, *9*(3), 315–322. DOI: 10.4314/ajtcam.v9i3.3 PMID: 23983361

Maroyi, A. (2017). Diversity of use and local knowledge of wild and cultivated plants in the Eastern Cape province, South Africa. *Journal of Ethnobiology and Ethnomedicine*, *13*(1), 1–16. DOI: 10.1186/s13002-017-0173-8 PMID: 28789692

Marra, M. (2004). The contribution of evaluation to socialization and externalization of tacit knowledge: The case of the World Bank. *Evaluation*, *10*(3), 263–283. DOI: 10.1177/1356389004048278

Marwala, T., & Hurwitz, E. (2017). *Artificial intelligence and economic theory: skynet in the market* (Vol. 1). Springer International Publishing.

Masango, C. A. (2010). Indigenous Traditional Knowledge Protection: Prospects in South Africa's Intellectual Property Framework? *South African Journal of Library and Information Science*, *76*(1). Advance online publication. DOI: 10.7553/76-1-88

Masango, M. J. S. (2002). Leadership in the African Context: Words on Leadership. *Verbum et Ecclesia*, *23*(3), 707–718. DOI: 10.4102/ve.v23i3.1234

Masekoameng, C. M. (2022). *Digital preservation and access to the South African documentary national heritage at the National Library of South Africa. Doctoral dissertation*, University of Limpopo. http://ulspace.ul.ac.za/handle/10386/3898

Masekoameng, M. (2007). *Indigenous knowledge systems in food gathering and production in selected rural communities in Sekhukhune* District of the Limpopo Province. Polokwane: University of Limpopo.

Masenya, T. M., & Chisita, T. C. (2025). Deterritorialization and Reterritorialization of Indigenous Knowledge: Reclamation of Misappropriated Indigenous Heritage Among Ethnocultural Communities, Southern Africa. *In: Ngulube, P. Harnessing Indigenous Epistemologies for Sustainable Progress*. (pp. 173-196). IGI Global Publishers.

Masenya, T.M. & Ngulube, P. (2021). Digital Preservation Systems and Technologies in South African Academic Libraries. *South African Journal of Information Management* 23(1), 1249. https://doi.org/ . v23i1.1249DOI: 10.4102/sajim

Masenya, T. M. (2021). Digital Preservation of Indigenous Knowledge in South African Rural Communities. In Tshifhumulo, R., & Makhanikhe, T. (Eds.), *J: Handbook of Research on Protecting and Managing Global Indigenous Knowledge*. IGI Global Scientific Publishing.

Masenya, T. M. (2022). Decolonization of Indigenous Knowledge Systems in South Africa: Impact of Policy and Protocols. *International Journal of Knowledge Management*, *18*(1), 1–22. DOI: 10.4018/IJKM.310005

Masenya, T. M. (2023). Valorization and Digital Preservation of Indigenous Knowledge Systems in South African Indigenous Communities: Best Practices in the Digital Transformation Era. In Masenya, T. M. (Ed.), *Digital Preservation and documentation of global indigenous knowledge systems*. IGI Global Publishers. DOI: 10.4018/978-1-6684-7024-4.ch002

Masenya, T. M. (2024). Revitalization and digital preservation of indigenous knowledge systems for sustainable development of indigenous communities in South Africa. *The Serials Librarian*, •••, 1–17. DOI: 10.1080/0361526X.2023.2277962

Masenya, T. M. (2024). Revitalization and Digital Preservation of Indigenous Knowledge Systems for Sustainable Development of Indigenous Communities in South Africa. *The Serials Librarian*, *84*, 5–8, 86–102.

Masenya, T. M. (Ed.). (2023). *Digital Preservation and Documentation of Global Indigenous Knowledge Systems*. IGI Global., DOI: 10.4018/978-1-6684-7024-4

Masolo, D. (2010). *Self and Community in a Changing World*. Indiana University Press. DOI: 10.2979/5989.0

Massawe, F., Mayes, S., & Cheng, A. (2016). Crop diversity: An unexploited treasure trove for food security. *Trends in Plant Science*, *21*(5), 365–368. DOI: 10.1016/j.tplants.2016.02.006 PMID: 27131298

Matenga, E. (2011). *The Soapstone Birds of Great Zimbabwe: Archaeological Heritage, Religion and Politics in Postcolonial Zimbabwe and the return of Cultural Property*. Uppsala University.

Materer, S. M., Valdivia, C., & Gilles, J. L. (2002). *Indigenous knowledge systems: Characteristics and importance to climatic uncertainty* (No. 1717-2016-140130).

Mathieson, S., Black, K., Allin, L., Hooper, H., Penlington, R., Mcinnes, L., & Anderson, E. (2023). New academics' experiences of induction to teaching: Using Cultural Historical Activity Theory (CHAT) to understand and improve induction experiences. *The International Journal for Academic Development*, •••, 1–15.

Matiure, P. (2023). Preserving the Tangible Material Culture of the Shona Traditional Music Legacy: An Applied Ethnomusicological Report. *Collections*, *19*(4), 553–570. Advance online publication. DOI: 10.1177/15501906231189216

Matubatuba, M. G. (2022). *The contribution of small, medium and micro enterprises to rural development in the Matatiele local municipality, South Africa* (Doctoral dissertation, North-West University (South Africa).

Mavengahama, S. (2013). The contribution of indigenous vegetables to food security and nutrition within selected sites in South Africa.

Mavhungu, T. J., Nesamvuni, A. E., Tshikolomo, K. A., Mpandeli, N. S., & van Niekerk, J. (2022). Productivity and profitability of sweet potato (ipomoea batatas l.), dry bean (Phaseolus Vulgaris) and maize (Zea mays l.) as selected field crops in irrigated smallholder agricultural enterprises (ISAEs) in Vhembe District, Limpopo Province, South Africa. *Technium Soc. Sci. J.*, *29*, 683–699. DOI: 10.47577/tssj.v29i1.5932

Mawere, J. (2023). *Exploring The Efficacy of Customary Tenure as A Tool for Rural Development and Food Security in Vhembe District of South Africa: An Indigenous Knowledge System Perspective* (Doctoral dissertation)

Mawere, J. (2024). Green energy solutions in Africa: Rediscovering indigenous knowledge for the Fourth Industrial Revolution. In *Contributions of Africa's Indigenous Knowledge to the Wave of Digital Technology: Decolonial Perspectives* (pp. 159–177). IGI Global.

Mawere, J., & Mukonza, R. M. (2024). Empowering marginalised communities: Leveraging indigenous knowledge for sustainable energy development in South Africa. *Sustainable Development*. Advance online publication. DOI: 10.1002/sd.3243

Mawere, J., & Walter Tshamano, N. (2023). *The Role of African Indigenous Knowledge Systems in Achieving Restorative Justice among the Vhavenda of South Africa*. IntechOpen., DOI: 10.5772/intechopen.1002677

May, C. (2006). *The Global Political Economy of Intellectual Property Rights*. Routledge.

Mazarire, G. (2004). Ethical considerations for Zimbabwean archives and the digital challenge. *Innovation.*, *24*(1). Advance online publication. DOI: 10.4314/innovation.v24i1.26445

Mazzocchi, F. (2022). Diving deeper into the concept of 'cultural heritage'and its relationship with epistemic diversity. *Social Epistemology*, *36*(3), 393–406. DOI: 10.1080/02691728.2021.2023682

Mbah, M., Ajaps, S., & Molthan-Hill, P. (2021). A systematic review of the deployment of indigenous knowledge systems towards climate change adaptation in developing world contexts: Implications for climate change education. *Sustainability (Basel)*, *13*(9), 4811. DOI: 10.3390/su13094811

Mbatha, B. (2013). Unpacking four sequential modes of knowledge conversion in managing indigenous knowledge. *Indilinga*, *12*(2), 171–187.

Mbenyane, X. G. (2017). Indigenous foods and their contribution to nutrient requirements. *The South African Journal of Clinical Nutrition : SAJCN : the Official Journal of the South African Society of Parenteral and Enteral Nutrition and the Association for Dietetics in South Africa*, *30*(4), 5–7.

Mbhele, T. P., & Eric, N. (2023). The role of supply chain technology in the growth of the cannabis industry. In *Rapid Innovation and Development in the Global Cannabis Market* (pp. 194–214). IGI Global. DOI: 10.4018/978-1-6684-6346-8.ch011

Mbilinyi, D., & Mwabungulu, E. (2020). The Fate of Indigenous Knowledge: The Role Played by Libraries in Tanzania. *Information Development, 36*(4), 489–02. DOI: 10.1177/0266666919871088

Mboti, N. (2015). May the Real Ubuntu Please Stand Up? *Journal of Media Ethics, 30*(2), 125–147. DOI: 10.1080/23736992.2015.1020380

McCallum, I. (2015). Use of social media by the library: Current practices and future opportunities. A White Paper from Taylor and Francis. *The Australian Library Journal, 64*(2), 161–162. DOI: 10.1080/00049670.2015.1040364

McCarthy, G. (2007). Finding a Future for Digital Cultural Heritage Resources Using Contextual Information Frameworks. In Cameron, F., & Kenderdine, S. (Eds.), *Theorizing Digital Cultural Heritage: A Critical Discourse* (pp. 245–260). The MIT Press. DOI: 10.7551/mitpress/6751.003.0017

McCarty, T. L. (2018). Community-Based Language Planning Perspectives from Indigenous Language Revitalization. In Hinton, L., Huss, L., & Roche, G. (Eds.), *The Routledge handbook of language revitalization* (pp. 22–35). Routledge. DOI: 10.4324/9781315561271-4

McConkey, J. (2004). Knowledge and Acknowledgement: 'Epistemic Injustice' as a Problem of Recognition. *Politics, 24*(3), 198–205. DOI: 10.1111/j.1467-9256.2004.00220.x

McDonough, S. (2013). Decolonizing Methodologies: Research and Indigenous Peoples by Linda Tuhiwai Smith. *Collaborative Anthropologies, 6*(1), 458–464. DOI: 10.1353/cla.2013.0001

McGuire-Adams, T. D. (2019). Paradigm shifting: Centering Indigenous research methodologies, an Anishaabe perspective. *Qualitative Research in Sport, Exercise and Health, 12*(10), 34–47.

McLean, M., Warner, B., Markham, R., Fischer, M., Walker, J., Klein, C., Hoeberechts, M., & Dunn, D. C. (2023). Connecting conservation & culture: The importance of indigenous knowledge in conservation decision-making and resource management of migratory marine species. *Marine Policy, 155*, 105582. DOI: 10.1016/j.marpol.2023.105582

McManis, C. R. (2007). *Biodiversity and the Law: Intellectual Property, Biotechnology and Traditional Knowledge. London; Sterling*. Earthscan.

Mdhluli, D., Mokgoatsana, S., Kugara, S. L., & Vuma, L. (2021). Knowledge management: Preserving, managing, and sharing indigenous knowledge through a digital library HTS. *Teologiesstudies. Theological Studies, 77*(2).

Mdhluli, T. D., Mokgoatšana, S., Kugara, S. L., & Vuma, L. (2021). Knowledge management: Preserving, managing and sharing indigenous knowledge through digital library. *Hervormde Teologiese Studies, 77*(2), a6795. DOI: 10.4102/hts.v77i2.6795

Mdletshe, S. T. C. (2023). *Viability of Government-Funded Broiler Production: Lessons from Northern KwaZulu-Natal, South Africa* (Doctoral dissertation, University of Fort Hare).

Mearns, M. A., Du Toit, A. S. A., & Mukuka, G. (2006). Conservation of Indigenous knowledge at cultural villages: An exploratory study. *Mousaion*, *24*(1), 23–50.

Meighan, P. J. (2021). Decolonizing the digital landscape: The role of technology in Indigenous language revitalization. *Alternative*, *17*(3), 397–405. DOI: 10.1177/11771801211037672

Mejias, U. A., & Couldry, N. (2024). *Data grab: The new colonialism of Big Tech and how to fight back.* University of Chicago Press. DOI: 10.7208/chicago/9780226832319.001.0001

Mendoza, M. A. D., De La Hoz Franco, E., & Gómez Gómez, J. E. (2023). Technologies for the Preservation of Cultural Heritage—A Systematic Review of the Literature. *Sustainability (Basel)*, *15*(2), 1059. DOI: 10.3390/su15021059

Merino, R. (2018). Re-politicizing participation or reframing environmental governance? Beyond Indigenous' prior consultation and citizen participation. *World Development*, *111*, 75–83. DOI: 10.1016/j.worlddev.2018.06.025

Merriam, S. B. (Ed.). (2019). *Qualitative research in practice: Examples for discussion and analysis.* John Wiley & Sons.

Mesec, B. (1998). *Uvod v kvalitativno raziskovanje v socialnem delu.* Ljubljana: Visoka šola za socialno delo.

Metz, T. (2020). African theories of meaning in life. A critical assessment. *South African Journal of Philosophy*, *39*(2), 113–126. DOI: 10.1080/02580136.2020.1770429

Meyer, A. (2012). Suid-Afrika se oerverlede. *In F. Pretorius F (ed.), Geskiedenis van Suid-Afrika: van voortye tot vandag,* pp. 17–36, Tafelberg, Cape Town.

Mgwenya, L. I., Agholor, I. A., Ludidi, N., Morepje, M. T., Sithole, M. Z., Msweli, N. S., & Thabane, V. N. (2025). Unpacking the Multifaceted Benefits of Indigenous Crops for Food Security: A Review of Nutritional, Economic and Environmental Impacts in Southern Africa. *WORLD (Oakland, Calif.)*, *6*(1), 16. DOI: 10.3390/world6010016

Mhandara, L. (2020). Community-based reconciliation in practice and lessons for the National Peace and Reconciliation Commission of Zimbabwe. *African Journal on Conflict Resolution*, *20*(1), 35–58.

Mhlanga, S., & Molokwane, T. (2020). AI for Indigenous language preservation: A case study in Southern Africa. *SA Journal of Linguistics*, *12*(2), 45–58.

Mhlongo, M., & Ngulube, P. (2018). Public libraries as conduits for indigenous knowledge in South Africa. *Library Philosophy and Practice*, (e-journal). Retrieved June 23, 2020, from: https://digitalcommons.unl.edu/libphilprac/1757/

Mhlongo, M. (2021). A Framework for the Integration of Indigenous Knowledge into Libraries in South Africa. *IFLA Journal*, *47*(3), 75–85. DOI: 10.1177/03400352211018224

Mhlongo, M., & Ngulube, P. (2020). Resource Provision and Access to Indigenous Knowledge in Public Libraries in South Africa. *Information Development*, *36*(2), 271–287. DOI: 10.1177/0266666919841095

Michaela, R. (2024). Restoring Taranaki Maunga: 'Religion' and Conservation in Aotearoa New Zealand. Open Access Te Herenga Waka-Victoria Unziversity of Wellington. Thesis. https://doi.org/DOI: 10.26686/wgtn.26860153

Micheli, M., Farell, E., Carballa-Smichowski, B., Posada-Sánchez, M., Signorelli, S., & Vespe, M. (2023). Mapping the landscape of data intermediaries: Emerging models for more inclusive data governance. [Report]. Publications Office of the European Union. https://data.europa.eu/doi/10.2760/8943

Mignolo, W. D. (2018). On pluriversality and multipolar world order: Decoloniality after decolonization; dewesternization after the Cold War. In Reiter, B. (Ed.), *Constructing the pluriverse: The geopolitics of knowledge* (pp. 90–116). Duke University Press. DOI: 10.1215/9781478002017-005

Mijatović, D., Van Oudenhoven, F., Eyzaguirre, P., & Hodgkin, T. (2013). The role of agricultural biodiversity in strengthening resilience to climate change: Towards an analytical framework. *International Journal of Agricultural Sustainability, 11*(2), 95–107. DOI: 10.1080/14735903.2012.691221

Milan, S., & Treré, E. (2019). Big data from the south(s): Beyond data universalism. *Television & New Media, 20*(4), 319–335. DOI: 10.1177/1527476419837739

Milius, D. (2009). Justifying intellectual property in traditional knowledge. *Intellectual Property Quarterly, 2*, 185–216.

Miller, K. (2021). Radical proposal: Data cooperatives could give us more power over our data. Stanford University Human-Centered Artificial Intelligence, Oct 20, 2021. https://hai.stanford.edu/news/radical-proposal-data-cooperatives-could-give-us-more-power-over-our-data

Mills, K. A., Davis-Warra, J., Sewell, M., & Anderson, M. (2016). Indigenous ways with literacies: Transgenerational, multimodal, placed, and collective. *Language and Education, 30*(1), 1–21. DOI: 10.1080/09500782.2015.1069836

Ministry of Environment. (2002). Forest and Climate Change. Biological Diversity Act, 2002. Retrieved from https://www.moef.gov.in/

Mishra, M. P., Joshi, B., Agarwal, P., & Varshney, I. (2024). Sustainable development goals–Role of indigenous knowledge. *Educational Administration: Theory and Practice, 30*(5), 4521–4527.

Mishra, M. P., Joshi, B., Agarwal, P., & Varshney, I. (2024). Sustainable Development Goals–Role Of Indigenous Knowledge. *Educational Administration: Theory and Practice, 30*(5), 4521–4527.

Misselhorn, A., Aggarwal, P., Ericksen, P., Gregory, P., Horn-Phathanothai, L., Ingram, J., & Wiebe, K. (2012). A vision for attaining food security. *Current Opinion in Environmental Sustainability, 4*(1), 7–17. DOI: 10.1016/j.cosust.2012.01.008

Mistry, J., & Berardi, A. (2016). Bridging Indigenous and scientific knowledge. *Science, 352*(6291), 1274–1275. DOI: 10.1126/science.aaf1160 PMID: 27284180

Mji, G., Kalenga, R., Ned, L., Alperstein, M., & Banda, D. (2017). Indigenous knowledge exclusion in education systems of Africans: Impact of beingness and becoming an African. In Ngulube, P. (Ed.), *Handbook of research on social, cultural, and educational considerations of indigenous knowledge in developing countries* (pp. 36–59)., DOI: 10.4018/978-1-5225-0838-0.ch003

Mkabela, Q. (2005). Using the Afrocentric Method in Researching Indigenous African Culture. *Qualitative Report*, *10*(1), 178–189.

Mkapa, B. (2004). Indigenous knowledge-a local pathway to Global Development: In World Bank, Indigenous knowledge. Local pathways to Global Development. Marking Five Years of the World Bank Indigenous Knowledge for Development Program, World Bank. http://www. worldbank. org/afr/ik/ikcomplete. pdf

Mkhize, P. (2024, September). Navigating the Fourth Industrial Revolution: Generational differences, COVID-19 impact, and knowledge sharing in the modern workplace. In *European Conference on Innovation and Entrepreneurship* (Vol. 19, No. 1, pp. 515–522). DOI: 10.34190/ecie.19.1.2783

Mlambo, R., & Matfunjwa, M. (2024). The use of technology to preserve indigenous languages of South Africa. *Literator-Journal of Literary Criticism, Comparative Linguistics and Literary Studies, 45*(1), 2007. https://hdl.handle.net/10520/ejc-literat_v45_n1_a2007

Mlambo, V. H., Thusi, X., Shoba, M., Mlambo, H., Mnguni, H., & Mbongwa, L. (2024). Harnessing digital technology for economic development in Africa. In *Contributions of Africa's Indigenous Knowledge to the Wave of Digital Technology: Decolonial Perspectives* (pp. 232–259). IGI Global. DOI: 10.4018/978-1-6684-7851-6.ch009

Mncayi, N. P. (2020). *An analysis of underemployment amongst young people in South Africa: The case of university graduates* (Doctoral dissertation, North-West University (South Africa).

Moahi, K. H. (2007). Globalization, knowledge economy and the implication for indigenous knowledge. *International Journal of Information Ethics*, *7*, 55–62. DOI: 10.29173/irie13

Modi, A. T. (2003). What do subsistence farmers know about indigenous crops and organic farming? Preliminary experience in KwaZulu-Natal. *Development Southern Africa*, *20*(5), 675–684. DOI: 10.1080/0376835032000149306

Mogajane, N. M. (2022). Digital technology impact in the access, preservation, and curation of indigenous artefacts at Mafikeng Museum (Doctoral dissertation, North-West University, South Africa).

Mohamedbhai, G. (2013). Indigenous knowledge must be harvested for development. *Traditional African Clinic*, *10*(262), 12–13.

Mokgomo, M. N., Chagwiza, C., & Tshilowa, P. F. (2022). The impact of government agricultural development support on agricultural income, production and food security of beneficiary small-scale farmers in South Africa. *Agriculture*, *12*(11), 1760. DOI: 10.3390/agriculture12111760

Mokuku, T., & Janse van Rensburg, E. (1997). Reflecting on socially transformative environmental literacy for Lesotho. *Southern African Journal of Environmental Education*, *17*, 31–39.

Molebatsi, J. K. (2023). *African identity in architecture: Guiding principles for the architectural design of traditional health practices in Gauteng, South Africa* (Doctoral dissertation, University of Pretoria (South Africa).

Molefe, T., & Mkhize, T. (2021). Blockchain technology as a solution for Indigenous Knowledge preservation in South Africa. *Journal of African Studies*, *47*(2), 101–116.

Mooney, P. (1996). The parts of life: Agricultural biodiversity, indigenous knowledge, and the role of the third system. *Development Dialogue*, *1*(96), 73–94.

Morgan, H. (2022). Conducting a qualitative document analysis. *The Qualitative Report*, *27*(1), 64–77. DOI: 10.46743/2160-3715/2022.5044

Morris, A. S., Silk, J. S., Steinberg, L., Myers, S. S., & Robinson, L. R. (2007). The Role of the Family Context in the Development of Emotion Regulation. *Social Development*, *16*(2), 361–388. DOI: 10.1111/j.1467-9507.2007.00389.x PMID: 19756175

Morris, M. H., Santos, S. C., & Neumeyer, X. (2020). Entrepreneurship as a solution to poverty in developed economies. *Business Horizons*, *63*(3), 377–390. DOI: 10.1016/j.bushor.2020.01.010

Moseley, C. (2022). Atlas of the world's languages in danger: Digital documentation priorities. *Language Documentation and Conservation*, *16*(1), 1–28.

Mosimege, M. (2004). Mathematical knowledge and its use in daily activities of workers at South African cultural villages. In *10th International Congress on Mathematical Education,* Copenhagen, Denmark.

Mosoti, Z., & Masheka, B. (2010). Knowledge management: The case for Kenya. *The Journal of Language. Technology and Entrepreneurship in Africa*, *2*(1), 107–133.

Mousavi, S. R., & Eskandari, H. (2011). A general overview on intercropping and its advantages in sustainable agriculture. *Journal of Applied Environmental and Biological Sciences*, *1*(11), 482–486.

Moyo, O.N. & Moyo, O.N. (2021). Performing Africanity Southern African Immigrants' Perspectives on Ubuntu. *Africanity and Ubuntu as Decolonizing Discourse*, 151-189.

Msila, V. (2021). Student Teachers, Professional Development and Effectiveness: The Role of Resilience in Enhancing the Culture of Learning and Teaching. *Universal Journal of Educational Research*, *9*(10), 1794–1804. DOI: 10.13189/ujer.2021.091008

Msuya, J. (2007). Challenges and opportunities in the protection and preservation of indigenous knowledge in Africa. *International Journal of Information Ethics*, *7*, 1–8.

Msuya, J. (2007). Challenges and Opportunities in the Protection and Preservation of Indigenous Knowledge in Africa. *International Journal of Information Ethics*, *7*, 346–353. DOI: 10.29173/irie38

Muchineripi, P.C. (2008). Feeding Five Thousand: The Case for Indigenous Crops, in Zimbabwe.

Mudau, F. N., Chimonyo, V. G. P., Modi, A. T., & Mabhaudhi, T. (2022). Neglected and underutilized crops: A systematic review of their potential as food and herbal medicinal crops in South Africa. *Frontiers in Pharmacology*, *12*, 809866. DOI: 10.3389/fphar.2021.809866 PMID: 35126143

Mugari, D. (2021). *Bioprospecting of indigenous species berchemia discolor and its socio-economic prospects in Vhembe District Municipality* (Doctoral dissertation).

Mugwisi, T., Jiyane, G. V., & Fombad, M. C. (2018). Public libraries as facilitators of information services: A case study of selected libraries in KwaZulu-Natal. *Information Development*, *34*(1), 31–43. DOI: 10.1177/0266666916672718

Muleba, J. L. I. N. (2020). *Understanding the use patterns of medicinal plants: investigating three ethnobotanical hypotheses in the Mpumalanga Province, South Africa*. University of Johannesburg.

Mumford, D. (2021). Data colonialism: Compelling and useful, but whither epistemes? *Information Communication and Society*, *25*(10), 1511–1516. DOI: 10.1080/1369118X.2021.1986103

Munro, K., & Bilbrough, P. (2018). An ecology of relationships: Tensions and negotiations in documentary filmmaking practice as research. *Media Practice and Education*, *19*(3), 256–269. DOI: 10.1080/25741136.2018.1511361

Murove, M. F. (2018). Indigenous Knowledge Systems Discourse and Inclusionality: An Afro-centric Quest for Recognition in a Globalised World. *Journal for the Study of Religion*, *31*(1), 159–176. DOI: 10.17159/2413-3027/2018/v31n1a9

Murray, J. S. (1970). Social Factors in the Development of Agriculture in the Bantu Areas with Special Reference to Land Tenure Systems. *Agrekon*, *9*(1), 29–34. DOI: 10.1080/03031853.1970.9524422

Mushi, G., Ndhlovu, T., & Makalima, K. (2019). Digital tools for cultural heritage preservation: Exploring AI's potential in Africa. *Journal of Cultural Heritage Studies*, *8*(4), 361–375.

Musiani, F. (2025). Reassessing "infrastructuring digital sovereignty": Digital self-determination as a set of infrastructure-embedded practices. *Frontiers in Communication*, *10*, 1562072. Advance online publication. DOI: 10.3389/fcomm.2025.1562072

Mutanga, C., Mlambo, M., & Sithole, P. (2018). Harnessing AI in traditional agriculture: A Zimbabwean perspective. *Agrárinformatika Folyóirat*, *9*(3), 14–22.

Mutengwa, C. S., Mnkeni, P., & Kondwakwenda, A. (2023). Climate-smart agriculture and food security in southern Africa: A review of the vulnerability of smallholder agriculture and food security to climate change. *Sustainability (Basel)*, *15*(4), 2882. DOI: 10.3390/su15042882

Muyambo, T. (2019). Indigenous Knowledge Systems of the Ndau people of Manicaland province in Zimbabwe: A case study of *Bota reshupa*. PhD Thesis, University of KwaZulu Natal, Pietermaritzburg, South Africa.

Muyambo, T. M., & Marashe, J. (2020). Indigenous knowledge systems and sustainable development: a case of Zunde raMambo (*Isiphala seNkosi*) as food security in Chipinge, Zimbabwe. *Indilinga: Journal of Indigenous Knowledge Systems*19(2), 232-244. https://journals.co.za/toc/linga/19/2

Mwangi, A. W. (2023). *Development and Quality Evaluation of a Functional Food From Cultured Milk Supplemented With Baobab (Adansonia Digitata L.) Fruit Pulp* (Doctoral dissertation, University of Nairobi).

Nakamoto, S. (2008). Bitcoin: A peer-to-peer electronic cash system. Retrieved from. https://bitcoin .org/bitcoin.pdf

Nakamura, N., Galla, C. K., & Snowberry, K. (2022). Collaborative documentation of Maya traditional ecological knowledge: Community engagement and capacity building. *Journal of Ethnobiology and Ethnomedicine*, *18*(1), 1–18. PMID: 34980177

Nakanyete, N.F., Matengu, K.K. & Diez, J.R. (2024). 6 Requirements for community-company partnerships in non-timber forest product trade: The case of San communities in northern Namibia. *Indigenous Knowledge for Sustainable Livelihoods: Evaluating Non-Timber Forest Product Value Chains for San Communities in Northern Namibia*, 84.

Nakashima, D., Krupnik, I., & Rubis, J. (2012). *Indigenous knowledge for climate change assessment and adaptation*. Cambridge University Press.

Nakata, M. (2002). Indigenous Knowledge and the Cultural Interface: Underlying Issues at the Intersection of Knowledge and Information Systems. *IFLA Journal*, *28*(5-6), 281–291. DOI: 10.1177/034003520202800513

Nakata, M. (2007). *Disciplining the savages: Savaging the disciplines*. Aboriginal Studies Press.

Nakata, M. (2021). Indigenous knowledge and the cultural interface: Underlying issues at the intersection of knowledge and information systems. *IFLA Journal*, *47*(3), 380–392.

Nakata, M., Byrne, A., Nakata, V., & Gardiner, G. (2021). Mapping the relationship between Indigenous knowledge, information systems, and technologies. *Information Communication and Society*, *24*(10), 1483–1501.

Nakaziba, S., & Ngulube, P. (2024). Harnessing digital power for relevance: Status of digital transformation in selected university libraries in Uganda. *Collection and Curation*, *43*(2), 33–44. DOI: 10.1108/CC-11-2023-0034

National Research Council. Global Affairs, Board on Science, Technology for International Development and Ad Hoc Panel of the Advisory Committee on Technology Innovation, (1989). *Lost crops of the Incas: little-known plants of the Andes with promise for worldwide cultivation*. National Academies Press.

National Youth Development Agency. (2020-2030). https://www.nyda.gov.za/Portals/0/downloads/National%20Youth%20Policy%202030.pdf. (Accessed 6/10/2024)

Naudé, W. (2018). Brilliant technologies and brave entrepreneurs. *Journal of International Affairs*, *72*(1), 143–158.

Ncube, C. B. (2024). Access to science, technology, and innovation: intellectual property, human rights, and sustainable development. In The Elgar Companion to Intellectual Property and the Sustainable DOI: 10.4337/9781803925233.00030

Ncube, B. (2018). Insights into indigenous knowledge strategies for coping and adapting to drought in agriculture: A Karoo scenario, South Africa. *Indilinga*, *17*(1), 92–108.

Ndlovu-Gatsheni, S. J. (2018). *Epistemic freedom in Africa: Deprovincialization and decolonization*. Routledge. DOI: 10.4324/9780429492204

Ndung'u, N., Paul, S., & Waema, T. (2019). Bridging the digital divide in Africa: AI and the rural challenge. *The African Journal of Information Systems*, *11*(2), 65–83.

Ndwandwe, S. (2013). *The Contribution of indigenous knowledge practices to household food production and food security: A case of UKhahlamba local municipality, South Africa*. University of KwaZulu-Natal.

Ndzendze, B., Singh, A., Timm, S., Meyers, C., Malatji, M. J., Malatji, M. J., ... & Manyana, Z. (2024). The 4IR and the humanities in South Africa: Perspectives on innovation, power and potentialities.

Nelson, R. (1997). How new is new growth theory? *Challenge*, *40*(5), 29–58. DOI: 10.1080/05775132.1997.11471992

Nesamvuni, C., Steyn, N. P., & Potgieter, M. J. (2001). Nutritional value of wild, leafy plants consumed by the Vhavenda. *South African Journal of Science*, *97*(1), 51–54.

Netting, R. M. (1993). *Smallholders, householders: farm families and the ecology of intensive, sustainable agriculture*. Stanford University Press. DOI: 10.1515/9781503622067

Neuman, W. (2014). *Social research methods: Qualitative and quantitative approaches* (7th ed.). Pearson Education.

Ngenge, G. A. (2003). *Gender equity and new information and communication technologies*. Retrieved May 07, 2022, from: www.sdnp.undp.org/sdncmr/fawecam/ngenge.htm

Ngoepe, M., & Bhebhe, S. (2023). *Indigenous Archives in Postcolonial Contexts: Recalling the Past in Africa*. Taylor & Francis. DOI: 10.4324/9781003277989

Ngoepe, M., & Netshakhuma, S. (2018). Archives in the trenches: Repatriation of African National Congress liberation archives in diaspora to South Africa. *Archival Science*, *18*(1), 51–71. DOI: 10.1007/s10502-018-9284-9

Ngomane, T.S. (2023). Fostering women entrepreneurship in the tourism industry: a look into obtacles and opportunities of women selling arts and crafts around the Kruger national park. *International Journal of Research in Business and Social Science (2147-4478)*, *12*(10), 175-181.

Ngulube, P. (2002). Managing and Preserving Indigenous Knowledge in the Knowledge Management Era: Challenges and opportunities for information professionals. *Information Development*, *18*(2), 95–101. DOI: 10.1177/026666602400842486

Ngulube, P. (2003). Using the SECI knowledge management model and other networks to communicate and manage tacit indigenous knowledge. *Innovation*, *27*(1), 21–30.

Ngulube, P. (2015). Towards a cartography of indigenous knowledge systems in library and information science training and education in Anglophone eastern and southern Africa. *Indilinga*, *14*(2), 145–168.

Ngulube, P. (2024). From Orality to Digital Assets: Managing Indigenous Knowledge in Africa in the Wake of the Open Science Movement. *The Serials Librarian*, *84*(5-8), 71–85. DOI: 10.1080/0361526X.2023.2296880

Nhambura, C. (2023). Indigenous People and the Protection of Intellectual Property Rights (IPRs) in Zimbabwe. In Barnabas, S. G. (Ed.), *Indigenous and Minority Populations - Perspectives From Scholars and Writers across the World*. Nigerian Turkish Nile University. DOI: 10.5772/intechopen.108167

NIKSOPORTAL. (2024). *The National Indigenous Knowledge Management System*. https://nikso.dst.gov.za/

Niyitunga, E. B. (Ed.). (2024). *Contributions of Africa's Indigenous Knowledge to the Wave of Digital Technology: Decolonial Perspectives*. IGI Global. DOI: 10.4018/978-1-6684-7851-6

Njagi, P., & Mwangi, R. (2021). Digital storytelling as a tool for preserving Kenyan Indigenous knowledge. *International Journal of Communication*, *15*(6), 1385–1400.

Njeme, C., Goduka, N. I., & George, G. (2014). Indigenous leafy vegetables (imifino, morogo, muhuro) in South Africa: A rich and unexplored source of nutrients and antioxidants. *African Journal of Biotechnology*, *13*(19), 1933–1942. DOI: 10.5897/AJB2013.13320

Nkebem, O., & Ofulue, S. (2022). Digitization challenges in preserving cultural knowledge: A study of Nigerian practices. *Journal of Library and Information Science*, *56*(3), 312–330.

Nkondo, G. (2012). Indigenous Knowledge Systems in the 21st century: Perspectives on possibilities and challenges in South Africa. *South African Journal of Education*, *32*(2), 185–200.

Nkosi, N. N., Mostert, T. H. C., Dzikiti, S., & Ntuli, N. R. (2020). Prioritization of indigenous fruit tree species with domestication and commercialization potential in KwaZulu-Natal, South Africa. *Genetic Resources and Crop Evolution*, *67*(6), 1567–1575. DOI: 10.1007/s10722-020-00932-5

Nkosi, T., Aboginije, A., Mashwama, N., & Thwala, W. (2020, March). Harnessing Fourth Industrial Revolution (4IR) for improving poor universities infrastructure in developing countries: A review. In *Proceedings of the International Conference on Industrial Engineering and Operations Management Dubai,* UAE (pp. 10–12).

Nonaka, I. (1994). A dynamic theory of organizational knowledge creation. *Organization Science*, *5*(1), 14–37. DOI: 10.1287/orsc.5.1.14

Nonaka, I., & Konno, N. (1998). The concept of "Ba": Building a foundation for knowledge creation. *California Management Review*, *3*(40), 40–54. DOI: 10.2307/41165942

Nonaka, I., & Takeuchi, H. (1995). *The knowledge-creating company: How Japanese companies create the dynamics of innovation*. Oxford University Press. DOI: 10.1093/oso/9780195092691.001.0001

Nonaka, I., Umemoto, K., & Sasaki, K. (1998). Three tales of knowledge-creating companies. In von Krogh, G., Roos, J., & Kleine, D. (Eds.), *Knowing in firms. Understanding, managing and measuring knowledge* (pp. 146–172). Sage. DOI: 10.4135/9781446280256.n7

Nonaka, I., & Von Krogh, G. (2009). Perspective-tacit knowledge and knowledge conversion: Controversy and advancement in organizational knowledge creation theory. *Organization Science*, *20*(3), 635–652. DOI: 10.1287/orsc.1080.0412

Norms, Q.C., 2012. *Community Biocultural Protocols.*

North-West University. (2020). NWU partners with department to promote Indigenous Knowledge Act. https://news.nwu.ac.za/create-news

Notsi, L. (2012). African indigenous farming methods used in the cultivation of African indigenous vegetables: A comparative study of Tsitas Nek (Lesotho) and Mabeskraal village (South Africa). *Conference on Strategies to Overcome Poverty and Inequality: Towards Carnegie III at University of Cape Town,* South Africa.

Noyoo, N. (2007). Indigenous knowledge systems and their relevance for sustainable development: A case of Southern Africa. *Tribes and Tribals*, *1*, 167–172.

Nthutang, S. T. (2015). *Issues affecting e-learning practices at Northwest University, Mafikeng Campus* (Doctoral dissertation).

Ntlangani, B. S. (2021). *An evaluation of community tourism organisations' role in small, medium and micro enterprises (SMME) tourism development in the eThekwini region* (Doctoral dissertation).

Ntlhabo, M. A. (2022). *Navigating between the sacred and the profane: Mohokare sacred sites, spiritual tourism and the challenges in the formal heritage sector* (Doctoral dissertation, University of the Free State).

Ntuli, P. P. (2002) *Indigenous Knowledge Systems and the African Renaissance. Laying a foundation for the creation of counter-hegemonic discourses.* In C.A. Odora Hoppers, Indigenous Knowledge and the Integration of Knowledge Systems: Towards a Philosophy of Articulation. 285

Nyeleka, S., Taruvinga, A., Zhou, L., & Mopipi, K. (2019). Income and Food Security Status Among rural Women Indigenous Knowledge Based Crafters and Non-crafters: Evidence from Rural South Africa. *International Journal of Food Engineering*, *5*(1).

Nyong, A., Adesina, F., & Osman Elasha, B. (2007). The value of indigenous knowledge in climate change mitigation and adaptation strategies in the African Sahel. *Mitigation and Adaptation Strategies for Global Change*, *12*(5), 787–797. DOI: 10.1007/s11027-007-9099-0

Nyoni, J. (2025). Interrogation of decoloniality in higher education curriculum reform: Examining influences of the subconscious mind. In E.T. Woldegiorgis, D.Z. Atibuni, E. Tadesse, & Woldegiorgis, G. Woldegiorgis, *Higher Education Transformation in Africa* (pp. 86-106). Routledge

Nyota, S., & Mapara, J. (2008). *Shona Traditional children's games and play songs as Nyota, S., Mapara, J. and Centre for Advanced Studies of African Society, (2007). Language as indigenous knowledge.* Centre for Advanced Studies of African Society.

Nzewi, O. E. S. (2010). *The use of performance composition on African music instruments for effective classroom music education in Africa.* University of Pretoria.

O'Donohue, J. (2007). *The Problem of Africa. From Negative to Positive Universality: Anthropological and Ethical Implication for Africa.* Academic Press.

Odora Hoppers, C. (2021). Research on Indigenous knowledge systems: The search for cognitive justice. *International Journal of Lifelong Education*, *40*(4), 310–327. DOI: 10.1080/02601370.2021.1966109

Odora Hoppers, C. A. (2002). Indigenous knowledge systems, sustainable livelihoods and the intellectual property system: A peace action perspective. *Journal of Peacebuilding & Development*, *1*(1), 106–112. DOI: 10.1080/15423166.2002.373049356048

Odora Hoppers, C. A. (Ed.). (2002). *Indigenous Knowledge and the integration of knowledge systems: Towards a philosophy of articulation*. New African Books.

Ogar, E., Pecl, G., & Mustonen, T. (2020). Science Must Embrace Traditional and Indigenous Knowledge to Solve Our Biodiversity Crisis. *One Earth*, *3*(2), 162–165. DOI: 10.1016/j.oneear.2020.07.006

Ogedengbe, D. E., James, O. O., Afolabi, J. O. A., Olatoye, F. O., & Eboigbe, E. O.Damilola Emmanuel Ogedengbe1Oladapo Olakunle JamesJennifer Osayawe Atu AfolabiFunmilola Olatundun OlatoyeEmmanuel Osamuyimen Eboigbe. (2023). Human resources in the era of the Fourth Industrial Revolution (4IR): Strategies and innovations in the global south. *Engineering Science & Technology Journal*, *4*(5), 308–322. DOI: 10.51594/estj.v4i5.617

Oguamanam, C. (2018). ABS: Big data, data sovereignty and digitization: A new Indigenous research landscape. Genetic Resources, Justice and Reconciliation: Canada and Global Access and Benefit Sharing, ed (Cambridge: Cambridge University Press, 2018). *Ottawa Faculty of Law Working Paper,* (2019-14).

Oguamanam, C. (2004). Protecting indigenous knowledge in international law: Solidarity beyond the nation-state. *Law Text Culture*, *8*(8), 191–230. DOI: 10.14453/ltc.858

Oguamanam, C. (2020). Indigenous Peoples, Data Sovereignty, and Self-Determination: Current Realities and Imperatives. *The African Journal of Information and Communication*, *16*(26). Advance online publication. DOI: 10.23962/10539/30360

Oguamanam, C. (2022). Transition to the Fourth Industrial Revolution: Africa's Science, Technology and Innovation Framework and Indigenous Knowledge Systems. *African Journal of Legal Studies*, *15*(1), 1–37. DOI: 10.1163/17087384-bja10058

Oguamanam, C., & Jain, V. (2017). Access and benefit sharing, Canadian and Aboriginal research ethics policy after the Nagoya Protocol: Digital DNA and transformations in biotechnology. *Forthcoming, JELP*, *3*(1), 79–112.

Ogunmodede, T. A., & Ebijuwa, A. S. (2013). Problems of conservation and preservation of library resources in African academic libraries: A review of literature. Greener Journal of Social Sciences, 3(1), 050-057.

Ogur, E. O. (2024, September). The Fourth Industrial Revolution: Opportunities and challenges for Kenya. In *Proceedings of the Sustainable Research and Innovation Conference* (pp. 76–87).

Ohaja, M., & Anyim, C. (2021). Rituals and embodied cultural practices at the beginning of life: African perspectives. *Religions*, *12*(11), 1024. DOI: 10.3390/rel12111024

Ojo, T. A., & Ndzendze, B. (2025). Women and the 4IR in Africa. In *African Women in the Fourth Industrial Revolution* (pp. 8–35). Routledge.

Okonkwo, E. (2017). Challenges and opportunities in cultural heritage preservation in Nigeria. *Journal of Cultural Heritage*, *24*(3), 27–34.

Okuonghae, O. (2018). Librarians' awareness of social media usage for informal scientific communication in university libraries in south-south. *Library Philosophy and Practice. (e-journal)*, 1–19. Retrieved May 14, 2022, from: https://www.thefreelibrary.com/Librarians%27+Use+Of+Social+Media+For+Informal+Scientific+Communication...-a0539387789

Oladokun, B. D., Abdulahi, F., & Adeyinka, T. (2024). Bridging the Digital Divide: Empowering Nigerian Universities through Technological Advancements in Academic Libraries Conference: https://www.researchgate.net/publication/382947230_Bridging_the_Digital_Divide_Empowering_Nigerian_Universities_through_Technological_Advancements_in_Academic_Libraries

Oliver, P. (2006). *Purposive Sampling*. Sage Publishers.

Oliver, W. H. (2014). Reconsidering the skin colour (race) of Jesus and his ancestors. *Journal of Early Christian History*, *4*(2), 95–115. DOI: 10.1080/2222582X.2014.11877306

Olum, S., Okello-Uma, I., Tumuhimbise, G. A., Taylor, D., & Ongeng, D. (2017). The relationship between cultural norms and food security in the Karamoja Sub-Region of Uganda. *Journal of Food and Nutrition Research*, *5*(6), 427–435. DOI: 10.12691/jfnr-5-6-10

Omodan, B. I., & Diko, N. (2021). A Conceptualisation of Ubuntugogy as a Decolonial Pedagogy in Africa. *Journal of Culture and Values in Education*, *4*(2), 95–104. DOI: 10.46303/jcve.2021.8

Omole, O. H., & Olusegun, A. (2019). Preserving African traditional medicine using AI. *African Journal of Health Sciences*, *26*(3), 112–120.

Omona, J. (2013). Sampling in qualitative research: Improving the quality of research outcomes in higher education. *Makerere Journal of Higher Education*, *4*(2), 169–185. DOI: 10.4314/majohe.v4i2.4

Ongena, G., Hossain, F., & Steijn, C. (2020). Mobile technology for cultural heritage. *Digital Humanities Quarterly*, *14*(2), 1–20.

Onyancha, O. B. (2022). Indigenous knowledge, traditional knowledge and local knowledge: What is the difference? An informetrics perspective. *Global Knowledge, Memory and Communication*, *73*(3), 237–257. DOI: 10.1108/GKMC-01-2022-0011

Oosthuizen, J. H., Ungerer, M., & Volschenk, J. (2023). A Fourth Industrial Revolution Integrated Intelligence Taxonomy for Top Management. *J. Contemp Management*, *20*(1), 404–443. DOI: 10.35683/jcm1018.202

Opedes, H., Nedala, S., Mücher, C. A., Baartman, J. E., & Mugagga, F. (2023). How can drones uncover land degradation hotspots and restoration hopespots? An integrated approach in the Mount Elgon region with community perceptions. *Land (Basel)*, *13*(1), 1. https://www.mdpi.com/2073-445X/13/1/1#. DOI: 10.3390/land13010001

Open Data Cube (ODC). (2020). *An Open Source Geospatial Data Management & Analysis Platform.*

Orr, A., Mwema, C., Gierend, A. & Nedumaran, S.(2016). Sorghum and millets in Eastern and Southern Africa: facts, trends and outlook

Osagie, O., & Oladokun, B. (2024). Usefulness of artificial intelligence to safeguard records in libraries: A new trend. *Southern African Journal of Security, 2*, 13-pages. https://doi.org/DOI: 10.25159/3005-4222/16803

Osman, A. (2010). *Indigenous knowledge in Africa: Challenges and opportunities. Inaugural Lecture*. University of the Free State.

Ostrom, E. (1990). *Governing the Commons: The Evolution of Institutions for Collective Action*. Cambridge University Press. DOI: 10.1017/CBO9780511807763

Otang-Mbeng, W. D. T., & Muloche, E. Kola & Ndhlovu, P.T. (2023). The Role of Indigenous Knowledge Systems in Sustainable Utilisation and Conservation of Medicinal Plants. *In Sustainable Uses and Prospects of Medicinal Plants,* 245–54. Boca Raton: CRC Press.

Ouzman, S. (2004). Silencing and sharing southern African indigenous and embedded knowledge. In *Indigenous Archaeologies* (pp. 195–212). Routledge.

Owiny, S. A., Mehta, K., & Maretzki, A. N. (2014). The use of social media technologies to create, preserve, and disseminate Indigenous knowledge and skills to communities in East Africa. *International Journal of Communication*, *8*(10), 234–247. Retrieved March 05, 2020, from https://ijoc.org/index.php/ijoc/article/view/1667

Owusu-Ansah, F. E., & Mji, G. (2013). African indigenous knowledge and research. *African Journal of Disability*, *2*(1), 1–5. DOI: 10.4102/ajod.v2i1.30 PMID: 28729984

Ozdemir, D. (2022). A conceptual framework on the relationship of digital technology and art. [IJonSES]. *International Journal on Social and Education Sciences*, *4*(1), 121–134. DOI: 10.46328/ijonses.313

P1, interviewed on 23 November, 2024 E1, interviewed on 24 November, 2024

Padeli, W., Mustafa, W. A., Pangil, F., Abd Kadir, K., & Dwita, V. (2025). Knowledge management and the Fourth Industrial Revolution (4IR): A recent systematic review. *Journal of Advanced Research in Applied Sciences and Engineering Technology*, *51*(2), 18–33. DOI: 10.37934/araset.51.2.1833

Panda, T., Mishra, N., Pradhan, B. K., & Mohanty, R. B. (2015). Diversity of leafy vegetables and its significance to rural households of Bhadrak District, Odisha, India. *Scientia Agrícola*, *11*(3), 114–124.

Pandor, N. (2015). Protection, Promotion, Development and Management of Indigenous Knowledge Systems Bill, 2014. (Bill Notice 243 of 2015). *Government gazette,* 38574: 20 March.

Pansoni, S., Tiribelli, S., Paolanti, M., Di Stefano, F., Frontoni, E., Malinverni, E. S., & Giovanola, B. (2023). Artificial intelligence and cultural heritage: Design and assessment of an ethical framework. *The International Archives of the Photogrammetry, Remote Sensing and Spatial Information Sciences*, *48*, 1149–1155. DOI: 10.5194/isprs-archives-XLVIII-M-2-2023-1149-2023

Parangu, R. N. A., & Salim, T. A. (2018). Indigenous knowledge preservation of oral Literature "Hahiwang" in West Lampung. *PEOPLE: International Journal of Social Sciences*, *4*(2), 1221–1232. DOI: 10.20319/pijss.2018.42.12211232

Park, S. M., & Lee, H. (2024). Culturally sensitive frameworks for archiving Indigenous knowledge in South Korea. *Asian Journal of Information Science*, *32*(1), 57–72.

Parra-Moyano, J. (2024). Data entrepreneurs of the world, unite! How business leaders should react to the emergence of data cooperatives. *The Capco Institute Journal of Financial Transformation, 59*, 20–25. https://www.capco.com/Capco-Institute/Journal-59-innovation-and-control

Paruk, Z. (2022). *Exploring indigenous beliefs of black Africans in the South African workplace* (Doctoral dissertation, University of Pretoria (South Africa).

Patel, S. K., Sharma, A., & Singh, G. S. (2020). Traditional agricultural practices in India: An approach for environmental sustainability and food security. *Energy, Ecology & Environment*, *5*(4), 253–271. DOI: 10.1007/s40974-020-00158-2

Paterson, C. (2009). *Prohibition & resistance: a socio-political exploration of the changing dynamics of the Southern African cannabis trade, c. 1850-the present* (Doctoral dissertation, Rhodes University).

Pautasso, M. (2013). Ten simple rules for writing a literature review. *PLoS Computational Biology*, *9*(7), e1003149. DOI: 10.1371/journal.pcbi.1003149 PMID: 23874189

Payne, L., & O'Neil, J. K. (2019). Cultural sustainability in higher education. In Leal Filho, W. (Ed.), *Encyclopedia of sustainability in higher education*. Springer., DOI: 10.1007/978-3-319-63951-2_109-1

Peltokorpi, V., Nonaka, I., & Kodama, M. (2007). NTT DoCoMo's launch of I-mode in the Japanese mobile phone market: A knowledge creation perspective. *Journal of Management Studies*, *44*(1), 50–72. DOI: 10.1111/j.1467-6486.2007.00664.x

Perullo, A. (2018). *Digital Repatriation: Copyright Policies*. Fair Use, and Ethics.

Pescaru, M. (2018). The importance of the socialization process for the integration of the child in the society. *Revista Universitară de Sociologie*, *14*(2), 18–26.

Petrarca, D. M., & Hughes, J. M. (2014). Mobilizing knowledge via documentary filmmaking—Is the academy ready? *McGIill Journal of Education* 49(3). https://mje.mcgill.ca/article/viewFile/9074/7009

Petzold, J., Andrews, N., Ford, J. D., Hedemann, C., & Postigo, J. C. (2020). Indigenous knowledge on climate change adaptation: A global evidence map of academic literature. *Environmental Research Letters*, *15*(11), 113007. DOI: 10.1088/1748-9326/abb330

Pham, L. (2018). A review of key paradigms: Positivism, interpretivism and critical inquiry. *School of Education, Med Program,* 1–7. Retrieved July 20, 2021, from: DOI: 10.13140/RG.2.2.13995.54569

Pham, L. T. M. (2018). *Qualitative approach to research a review of advantages and disadvantages of three paradigms: Positivism, interpretivism and critical inquiry*. University of Adelaide.

Philbeck, T., & Davis, N. (2018). The fourth industrial revolution. *Journal of International Affairs*, *72*(1), 17–22. https://www.jstor.org/stable/26588339

Philip, K., Irani, L., & Dourish, P. (2012). Postcolonial computing: A tactical survey. *Science, Technology & Human Values*, *37*(1), 1, 3–29. DOI: 10.1177/0162243910389594

Philipson, D. W. (1977). *The later prehistory of the Eastern and Southern Africa*. Heinemann Educational Books.

Pichop, G. N., Abukutsa-Onyango, M., Noorani, A., & Nono-Womdim, R. (2014). Importance of indigenous food crops in tropical Africa: Case study. In *XXIX International Horticultural Congress on Horticulture: Sustaining Lives, Livelihoods and Landscapes (IHC2014): 1128*, 315-322.

Pimentel, D., & Burgess, M. (2014). An environmental, energetic and economic comparison of organic and conventional farming systems. Integrated Pest Management, Springer: 141-166. DOI: 10.1007/978-94-007-7796-5_6

Pinhasi, R., & Stock, J. T. (Eds.). (2011). *Human bioarchaeology of the transition to agriculture*. John Wiley & Sons. DOI: 10.1002/9780470670170

Pinstrup-Andersen, P. (2009). Food security: Definition and measurement. *Food Security*, *1*(1), 5–7. DOI: 10.1007/s12571-008-0002-y

Pisolkar, Y. (2024). Cultural Heritage Management and Sustainable Development: Major Themes and Research Trajectories, *Journal of Electrical Science*. DOI: 10.52783/jes.3224

Pitawanakwat, B. (2018). Strategies and methods for Anishina abemowin revitalization. *Canadian Modern Language Review*, *74*(3), 460–482. DOI: 10.3138/cmlr.4058

Plockery, F. D. (2015). Indigenous Knowledge Production, Digital Media and Academic Libraries in Ghana. *The Journal of Pan African Studies*, *8*(4), 32–44.

Plockey, F. D. (2014). The role of Ghana public libraries in the digitization of indigenous knowledge: Issues and prospects. *The Journal of Pan African Studies*, *6*(10), 20–36.

Plockey, F. D. D. (2015). Indigenous knowledge production, digital media and academic libraries in Ghana. *The Journal of Pan African Studies*, *8*(4), 32–44.

Podolecka, A. (2018). Vusamazulu Credo Mutwa–Sangoma, African sage, fraud or a new age shaman? *Studies in African Languages and Cultures*, *52*, 131–153. DOI: 10.32690/SALC52.6

Ponge, A. (2011). Integrating indigenous knowledge for food security: perspectives from the millennium village project at Bar-Sauri in Nyanza Province in Kenya. *International Conference on Enhancing Food Security in the Eastern and Horn of Africa regions, held at the Imperial Royale Hotel in Kampala, Uganda*

Pongel, A. (2011). Integrating Indigenous Knowledge for Food Security: Perspectives from the Millennium Village Project at in Nyanza Province in Kenya. *International Conference on Enhancing Food Security in the Eastern and Horn of Africa regions*. Kampala, Uganda.

Posey, D. A. (1990). Intellectual property rights: And just compensation for indigenous knowledge. *Anthropology Today*, *6*(4), 13–16. DOI: 10.2307/3032735

Posey, D. A., & Dutfield, G. (1996). *Beyond intellectual property*. International Development Research Centre.

Posey, D. A., & Dutfield, G. (1996). *Beyond Intellectual Property: Toward Traditional Resource Rights for Indigenous Peoples and Local Communities*. IDRC.

Posey, D. A., & Dutfield, G. (1996). *Beyond intellectual property: Toward traditional resource rights for Indigenous peoples and local communities*. International Development Research Centre.

Postill, J., & Pink, S. (2022). Social media ethnography: The digital researcher in a messy web. *Information Communication and Society*, *25*(3), 328–345.

Poulton, C., Dorward, A., & Kydd, J. (1998). The revival of smallholder cash crops in Africa: Public and private roles in the provision of finance. Journal of International Development. *The Journal of the Development Studies Association*, *10*(1), 85–103.

Prabhakar, S. V. R., & Rani, S. M. (2017). Influence of social networking sites on library and information centers. [IJLIS]. *International Journal of Library and Information Science*, *6*(1), 83–87.

Pretty, J., & Vodouhe, S. (1997). Using rapid or participatory rural appraisal. In Swanson, B., Bentz, R. and Sofranko, A. (eds.) 1997. *Improving agricultural extension, 47-55*: A reference manual. Food and agriculture organisation of the United Nations, Rome. Retrieved May 20, 2022, from: https://www.fao.org/docrep/W5830E/w5830e08.htm

Priyadarshini, P., & Abhilash, P. C. (2019). Promoting tribal communities and indigenous knowledge as potential solutions for the sustainable development of India. *Environmental Development*, *32*, 100459. DOI: 10.1016/j.envdev.2019.100459

Priya, R. M., & Rabindra, K. M. (2010). Documenting indigenous traditional knowledge in Odish*a*. *Orissa Review*, •••, 99–103. Retrieved June 24, 2022, from https://www.academia.edu/76494585/Documenting_Indigenous_Traditional_Knowledge_In_Odisa

Protection, P.Development and Management of Indigenous Knowledge Act. (2019). https://www.gov.za/sites/default/files/gcis_document/201908/42647198act6of2019protectpromodevelopmanagementindigenousknowledgeact.pdf

Puoane, T. Matwa, P. Hughes, G. & Bradley, H.A. (2006). Socio-cultural factors influencing food consumption patterns in the black African population in an urban township in South Africa.

Puri, R. K. (2021). Systematic approaches to indigenous knowledge documentation: Methods and ethics. *Current Anthropology*, *62*(2), 168–187.

Qizi, L. R. M., & Kamalovich, B. K. (2025). Blockchain Technology Usage on Intellectual Property Rights. *International Journal for the Semiotics of Law*, *38*(2), 363–380. DOI: 10.1007/s11196-024-10224-1

Quan-Baffour, K. P. (2012). Sankofa: 'Gazing Back' to Indigenous Knowledge and Skills for Socio-economic Development of Ghana. *Studies of Tribes and Tribals*, *10*(1), 1–5. Advance online publication. DOI: 10.1080/0972639X.2012.11886637

Quijano, A. (2000). Coloniality of power and eurocentrism in Latin America. *International Sociology*, *15*(2), 215–232. DOI: 10.1177/0268580900015002005

Radjawali, I., & Pye, O. (2021). Drones for indigenous mapping: Enhancing indigenous peoples' participation in forest management. *Development and Change*, *52*(4), 868–897.

Raihan, A., & Bijoy, T. R. (2023). A review of the industrial use and global sustainability of Cannabis sativa. *Global Sustainability Research, 2*(4), 1–29. DOI: 10.56556/gssr.v2i4.597

Raimondo, D. (2009). *Red list of South African plants*. South African National Biodiversity Institute.

Rainie, S., (2019). Indigenous data sovereignty. Issues in open data: Indigenous data sovereignty. In Davies, T., Walker, S., Rubinstein, M., & Perini, F. (Eds.), *The state of open data: Histories and horizons* (pp. 300–319). African Minds and International Development Research Centre., http://stateofopendata.od4d.net

Rajah, S. S. (2019). Conceptualising community engagement through the lens of African indigenous education. *Perspectives in Education, 37*(1), 1–14.

Ramaano, A. I. (2023). *The prospects of using the tourism industry to advance community livelihoods in Musina Municipality, Limpopo Province* (Doctoral dissertation).

Ramakrishnan, P. S. (2007). Traditional forest knowledge and sustainable forestry: A north-east India perspective. *Forest Ecology and Management, 249*(1-2), 91–99. DOI: 10.1016/j.foreco.2007.04.001

Rambe, P., & Maime, R. (2023). The impact of 4IR technologies on venture creation and technology commercialisation: Insights and exemplars from an emerging economy context. In *Small Business and Entrepreneurial Development in Africa: A Route to Sustained Economic Development* (pp. 149–178). Springer Nature Switzerland. DOI: 10.1007/978-3-031-37675-7_7

Ramirez, R., & Mari, F. (2021). Co-creating digital archives with Indigenous communities: The Mexican experience. *Cultural Heritage Journal, 14*(2), 153–171.

Rampedi, I. T. (2010). Indigenous plants in the Limpopo Province: *Potential for their commercial beverage production*. Doctoral dissertation, University of South Africa.

Rana, V., & Azeez, G. K. (2025). Indigenous data sovereignty in Australian higher education: Paving the way for First Nations' self-determination. *Journal of Higher Education Policy and Management*, 1–9. Advance online publication. DOI: 10.1080/1360080X.2025.2469920

Rankoana, S. A. (2021). Indigenous plant foods of Dikgale community in South Africa. *Journal of Ethnic Foods, 8*(1), 1–8. DOI: 10.1186/s42779-021-00080-w

Rao, M. (2003). *Biodiversity and Sustainable Development: Policies and Practices*. National Book Trust.

Raseroka, K. (2008). Information transformation Africa: Indigenous knowledge e Securing space in the knowledge society. *The International Information & Library Review, 40*(4), 243–250. DOI: 10.1080/10572317.2008.10762789

Rathore, S., & Rai, S. (2024). Interplay of Intellectual Property Rights, Technological Innovation, and Sustainable Food Security: A Comprehensive Analysis. In *E-Financial Strategies for Advancing Sustainable Development: Fostering Financial Inclusion and Alleviating Poverty* (pp. 351–361). Springer Nature Switzerland. DOI: 10.1007/978-3-031-67523-2_21

Ravishankar, M. N., Pan, S. L., & Myers, M. D. (2013). Information technology offshoring in India: A postcolonial perspective. *European Journal of Information Systems*, *22*(4), 387–402. DOI: 10.1057/ejis.2012.32

Reeve, J. (2002). *Self-determination theory applied to educational settings.* Revolution on indigenous crops of India. *Journal of Ethnic Foods*, *6*(1), 1–10.

Rehman, M., Fahad, S., Du, G., Cheng, X., Yang, Y., Tang, K., Liu, L., Liu, F. H., & Deng, G. (2021). Evaluation of hemp (Cannabis sativa L.) as an industrial crop: A review. *Environmental Science and Pollution Research International*, *28*(38), 52832–52843. DOI: 10.1007/s11356-021-16264-5 PMID: 34476693

Reid, A. J., Eckert, L. E., Lane, J. F., Young, N., Hinch, S. G., Darimont, C. T., Cooke, S. J., Ban, N. C., & Marshall, A. (2014). Two-eyed seeing: An Indigenous framework to transform fisheries research and management. *Fish and Fisheries*, *22*(1), 243–261. DOI: 10.1111/faf.12516

Reid, R. S., Fernández-Giménez, M. E., & Galvin, K. A. (2014). Dynamics and resilience of pastoralist societies. *Annual Review of Environment and Resources*, *39*, 217–242. DOI: 10.1146/annurev-environ-020713-163329

Reijerkerk, D., & Nyitray, K. J. (2023). Moving indigenous knowledge boundaries: A Review of Library and Archival Collection Literature since the Enactment of the Native American Graves Protection and Repatriation Act (NAGPRA). *Collection Management*, *48*(1), 22–47. DOI: 10.1080/01462679.2022.2033144

Resta, P. (2022). Digital divide in indigenous communities: Challenges and opportunities. *Information Technology for Development*, *28*(1), 60–78.

Resta, P., Laferrière, T., & Lamon, M. (2022). Digital technologies in indigenous communities: Implications for educational equity. *Education and Information Technologies*, *27*(4), 5097–5117.

Reyes-Garciá, V., Tofighi-Niaki, A., Austin, B. J., Benyei, P., Danielsen, F., Fernández-Llamazares, Á., Sharma, A., Soleymani-Fard, R., & Tengö, M. (2022). Data sovereignty in community-based environmental monitoring: Toward equitable environmental data governance. *Bioscience*, *72*(8), 714–717. DOI: 10.1093/biosci/biac048 PMID: 35923191

Ricaurte, P. (2019). Data epistemologies, the coloniality of power, and resistance. *Television & New Media*, *20*(4), 350–365. DOI: 10.1177/1527476419831640

Richard, T.T. (2014). Indigenous farming methods in the Mambilla Plateau of Nigeria: the need for a sustainable maintenance culture policy. *Science Journal of Sociology and Anthropology*, 2014.

Richards, P. (2023). *Indigenous agricultural revolution: ecology and food production in West Africa.* Routledge. DOI: 10.4324/9781003383734

Rifat, M. R., Toriq, T., & Ahmed, S. I. (2020). Religion and Sustainability: Lessons of Sustainable Computing from Islamic Religious Communities. Proceedings of the ACM on Human-Computer Interaction 4, CSCW2 (2020), 1–32.

Rima, I. H. (2004). Increasing returns, new growth theory, and the classicals. *Journal of Post Keynesian Economics*, *27*(1), 171–184.

Rindos, D. (2013). *The origins of agriculture: an evolutionary perspective*. Academic Press.

Rivera, D., Allkin, R., Obón, C., Alcaraz, F., Verpoorte, R., & Heinrich, M. (2014). What is in a name? The need for accurate scientific nomenclature for plants. *Journal of Ethnopharmacology*, *152*(3), 393–402. DOI: 10.1016/j.jep.2013.12.022 PMID: 24374235

Robbins, P., Hunt, H. H., Pelegri, F., & Gilbert, J. (2023). Sovereign genes: Wildlife conservation, genetic preservation, and Indigenous data sovereignty. *Frontiers in Conservation Science*, *4*, 1099562. Advance online publication. DOI: 10.3389/fcosc.2023.1099562

Roberts, J. S., & Montoya, L. N. (2023). In consideration of Indigenous data sovereignty: Data mining as a colonial practice. *arXiv*. https://doi.org//arXiv.2309.10215DOI: 10.1007/978-3-031-47451-4_13

Robinson, D. F. (2014). *Biodiversity, access and benefit-sharing: Global case studies*. Routledge., DOI: 10.4324/9781315882819

Rogers, R. A. (2006). From Cultural Exchange to Transculturation: A Review and Reconceptualization of Cultural Appropriation. *Communication Theory*, *16*(4), 474–503. DOI: 10.1111/j.1468-2885.2006.00277.x

Rose, C. M. (2003). Property as the keystone right? Some reflections on the IP/biodiversity interface. *Arizona Law Review*, *45*, 445–470.

Rosenberg, M. (2021). Thoughts about food security, food loss and waste and what has to be done. *AIMS Agriculture and Food*, *6*(3), 797–798. DOI: 10.3934/agrfood.2021048

Rothenberg, J. (2023). Digital preservation through migration: A computational approach to format obsolescence. *International Journal of Digital Curation*, *18*(1), 1–22.

Rousseau, J.-J. (1762). *Du Contrat Social: Ou, Principes du Droit Politique*. Chez Marc Michel Rey.

Roy, M., & Basu, R. (2020). Indigenous environmental knowledge and climate adaptation in India. *Current Science*, *119*(3), 467–475.

Rozi, S., & Taufik, Z. (2020). Adaptation of religion and local wisdom in global environmental issues in Indonesia. *Religious: Jurnal Studi Agama-Agama Dan Lintas Budaya*, *4*(3), 191–203. DOI: 10.15575/rjsalb.v4i3.9593

Ruelle, M. (2015). *Human-plant ecology of an afromontane agricultural landscape: diversity, knowledge, and food sovereignty in Debark, Ethiopia*.

Ruijer, E. (2021). Designing and implementing data collaboratives: A governance perspective. *Government Information Quarterly*, *38*(4), 101612. DOI: 10.1016/j.giq.2021.101612

Rule, P., & Yazan, B. (2024). Knowledge and knowing in case study research: towards a relational epistemology. In *Handbook of Case Study Research in the Social Sciences* (pp. 31–50). Edward Elgar Publishing. DOI: 10.4337/9781803920320.00012

Russell, S. J., & Norvig, P. (2016). Artificial intelligence: a modern approach. *Pearson.* Ofosu-Asare, Y. (2024). Cognitive imperialism in artificial intelligence: counteracting bias with indigenous epistemologies. *AI & Society*, •••, 1–17. DOI: 10.1007/s00146-024-02065-0

Ryan, R. M., & Deci, E. L. (2024). Self-determination theory. In *Encyclopedia of quality of life and well- being research* (pp. 6229–6235). Springer International Publishing.

S1, interviewed on 21 November, 2024 J1, interviewed on 23 November, 2024

Saeed, J. (2024). Guardians of the Data: Government Use of AI and IoT in the Digital Age.

Saffer, C. (2024). Secure from exploitation. Cultural artwork protected by blockchain. The Australian Jewish News, 2024).

Salas-Pilco, S. Z. (2025). Information technology (IT) and indigenous knowledge (IK) in Latin America: Pro-indigenous, para-indigenous, and per-indigenous IT–IK integration for development. *Information Technology for Development*, •••, 1–24. DOI: 10.1080/02681102.2025.2462122

San Nicolas-Rocca, T & Parrish, J. (2013). Capturing and Conveying Chamorro Cultural Knowledge Using Social Media. *International Journal of Knowledge Management (IJKM), IGI Global,* 9(3), 1-18.

Santoro, A. (2022). Agricultural heritage systems and agrobiodiversity. *Biodiversity and Conservation, 31*(10), 2231–2241. DOI: 10.1007/s10531-022-02460-3

Santoro, G., Bresciani, S., & Papa, A. (2020). Collaborative modes with cultural and creative industries and innovation performance: The moderating role of heterogeneous sources of knowledge and absorptive capacity. *Technovation, 92*, 102040. DOI: 10.1016/j.technovation.2018.06.003

Santos, J., Silva, M. J., & Costa, P. (2021). Virtual reality as a tool for Indigenous knowledge preservation. *Digital Cultural Heritage*, *9*(2), 145–162.

Sarkhel, J. K. (2017). Strategies of Indigenous knowledge management in libraries. *Qualitative and Quantitative Methods in Libraries, 5*(2), 427-439.Stats, S. A. (2011). Statistics south Africa. *Formal census, 1*, 79-106.

Sarkhel, J. K. (2017). Strategies of indigenous knowledge management in libraries. *Qualitative and Quantitative Methods in Libraries*, *5*(2), 427–439.

Satika, R., Salahuddin, , Kasman, , & Suraya, . (2024). Implementation of Public Service Reform In Enhancing The Welfare Of Bima District Community: Challenges And Opportunities. *KOMUNITAS, 15*(1), 91–106. DOI: 10.20414/komunitas.v15i1.9989

Saurombe, A. (2022). Critical evaluation of Traditional Knowledge Digital Library (TKDL): Achievements and limitations. *Journal of Intellectual Property Rights*, *27*(4), 301–315.

Schaffrick, M., Perreault, M. L., Jones, A. M. P., & Illes, J. (2023). Understanding and Rebalancing: A Rapid Scoping Review of Cannabis Research Among Indigenous People. *Cannabis and Cannabinoid Research*, *8*(3), 426–433. DOI: 10.1089/can.2022.0251 PMID: 36454240

Scheyvens, R., Carr, A., Movono, A., Hughes, E., Higgins-Desbiolles, F., & Mika, J. P. (2021). Indigenous tourism and the sustainable development goals. *Annals of Tourism Research, 90*, 103260. DOI: 10.1016/j.annals.2021.103260

Schlebusch, C. M., & Jakobsson, M. (2018). Tales of human migration, admixture, and selection in Africa. *Annual Review of Genomics and Human Genetics*, *19*(1), 405–428. DOI: 10.1146/annurev-genom-083117-021759 PMID: 29727585

Schneider, A. (2003). On 'Appropriation'. A Critical Reappraisal of the Concept and Its Application in Global Art Practices. *Social Anthropology*, *11*(2), 215–229. DOI: 10.1017/S0964028203000156

Schneider, E. A. (1985). Ndebele mural art. *African Arts*, *18*(3), 60–67. DOI: 10.2307/3336357

Schoenbrun, D. L. (1993). We Are What We Eat: Ancient Agriculture Between the Great Lakes1. *Journal of African History*, *34*(1), 1–31. DOI: 10.1017/S0021853700032989

Schönfeldt, H. C., Hall, N. & Pretorius, B. (2017). Nutrition-sensitive agricultural development for food security in Africa: a case study of South Africa. *International Development*, 1.

Schroeder, D., Chennells, R., Louw, C., Snyders, L., & Hodges, T. (2020). The Rooibos benefit sharing agreement–Breaking new ground with respect, honesty, fairness, and care. *Cambridge Quarterly of Healthcare Ethics*, *29*(2), 285–301. DOI: 10.1017/S0963180119001075 PMID: 31685054

Schulze, A., & Hoegl, M. (2006). Knowledge creation in new product development projects. *Journal of Management*, *32*(2), 210–236. DOI: 10.1177/0149206305280102

Schwab, K. (2016). PDF The Fourth Industrial Revolution By.

Schwab, K. (2018). "The Fourth Industrial Revolution" Encyclopaedia Britannica Anniversary Edition: 250 Years of Excellence (1768–2018).

Schwab, K. (2017). *The fourth industrial revolution*. Crown Currency.

Schwartzman, S., Zimmerman, B., & Kormos, C. (2019). Indigenous lands, protected areas, and slowing climate change. *Conservation Biology*, *33*(4), 700–711. DOI: 10.1111/cobi.13329

Seko, J., Bain, E., & Maponya, P. (2021). Assessing the Impact of Indigenous Knowledge Systems on Sustainable Agriculture: A Case Study of the Selected Communities in the City of Tshwane Metropolitan, Gauteng Province, South Africa. Sustainable Bioeconomy: *Pathways to Sustainable Development Goals*, pp.183-208.

Semali, L. M., & Asino, T. I. (2013). Decolonizing cultural heritage of indigenous people's knowledge from images in global films. *Deconization: Indigeneity & Society*, *2*(2), 25–56.

Semenya, S. S., & Potgieter, M. J. (2014). Bapedi traditional healers in the Limpopo Province, South Africa: Their socio-cultural profile and traditional healing practice. *Journal of Ethnobiology and Ethnomedicine*, *10*(1), 1–12. DOI: 10.1186/1746-4269-10-4 PMID: 24410790

Sen, A., & Nagendra, H. (2019). The role of environmental placemaking in shaping contemporary environmentalism and understanding land change. *Journal of Land Use Science*, *14*(4-6), 410–424. DOI: 10.1080/1747423X.2020.1720841

Senanayake, S. G. J. N. (2006). Indigenous Knowledge as a Key to Sustainable Development. *Journal of Agricultural Sciences*, *2*(1), 87–94. DOI: 10.4038/jas.v2i1.8117

Seroto, J. (2012). Indigenous education in the pre-colonial era. In J.J. Booyse, C.S. le Roux, J. Seroto, C.C. Wolhuter. *A history of schooling in South Africa. Method and Context.* Van Schaik.37-52

Shahmohamadloo, R. S., Febria, C. M., Fraser, E. D., & Sibley, P. K. (2021). The sustainable agriculture imperative: A perspective on the need for an agrosystem approach to meet the United Nations Sustainable Development Goals by 2030. *Integrated Environmental Assessment and Management, 18*(5), 1199–1205. DOI: 10.1002/ieam.4558 PMID: 34821459

Shanmuga Sundari, M., Penthala, H. R., Mogullapalli, A., & Ammangatambu, M. M. (2024). AI-Based Personalized Drug Treatment. In *Artificial Intelligence and Machine Learning in Drug Design and* (pp. 369–406). Development.

Sharma, A. (2021). Decolonizing international relations: Confronting erasures through Indigenous knowledge systems. *International Studies, 58*(1), 25–40. DOI: 10.1177/0020881720981209

Sharma, A., & Saha, B. (2020). Blockchain-enabled transparency in Indigenous agricultural practices. *Indian Journal of Sustainable Development, 14*(3), 215–230.

Sharma, I. P., Kanta, C., Dwivedi, T., & Rani, R. (2020). *Indigenous agricultural practices: A supreme key to maintaining biodiversity.* Microbiological Advancements for Higher Altitude Agro-Ecosystems & Sustainability.

Shiva, V. (1997). *Biopiracy: The Plunder of Nature and Knowledge.* South End Press.

Shizha, E. (Ed.). (n.d.). *(2016). African Indigenous knowledge and the sciences: Journeys into the past and the present. Boston: Sense Publishers United Nations (UN).* Draft Declaration on the Rights of Indigenous Peoples. Prepared by the Aboriginal and TorresStrait Islander Commission. [Online], http://www.gcc.ca/Political_Issues/international/

Shrumm, H., & Jonas, H. (2012). *Biocultural Community Protocols: A Toolkit for Community Facilitators.* Natural Justice.

Shu, C., Page, A. L., Gao, S., & Jiang, X. (2012). Managerial ties and firm innovation: Is knowledge creation a missing link? *Journal of Product Innovation Management, 29*(1), 125–143. DOI: 10.1111/j.1540-5885.2011.00883.x

Shylon, F. (2014). Repatriation of Antiquities to sub-Saharan Africa: The Agony and the Ecstasy. *Antiquity, 19*, 121–132.

Sianturi, M., Lee, J.-S., & Cumming, T. M. (2025). The synergy between Indigenous ways of knowing, being, and doing and technology in culturally responsive home-school partnerships. *Information Technology for Development*, 1–24. Advance online publication. DOI: 10.1080/02681102.2025.2465435

Sibiya, A., van der Westhuizen, J. & Sibiya, B. (2023). Challenges Experienced by SMMEs and Interventions by the South African National and Provincial Government: A Literature Review. *African Journal of Inter/Multidisciplinary Studies,* 5(1), 1-11.

Sibiya, P. (2023). *Predicting Loan Defaults in Development Financing in South Africa.* (Master's dissertation). University of Cape Town, Cape Town.

Sibuye, R., Uys, M.-T., Cocchiaro, G., & Lorenzen, J. (2012). The Bushbuckridge BCP: Traditional Healers Organise for ABS in South Africa in Participatory Learning and Action, Biodiversity and Culture: Exploring Community Protocols, Rights and Consent.http://pubs.iied.org/pdfs/G03403.pdf

Siddiqui, S. A., Singh, P., Khan, S., Fernando, I., Baklanov, I. S., Ambartsumov, T. G., & Ibrahim, S. A. (2022). Cultural, social and psychological factors of the conservative consumer towards legal cannabis use—A review since 2013. *Sustainability (Basel)*, *14*(17), 10993. DOI: 10.3390/su141710993

Siems, M. (2019). The Law and Ethics of 'Cultural Appropriation. *International Journal of Law in Context*, *15*(4), 408–423. DOI: 10.1017/S1744552319000405

Sillitoe, P. (2006). Knowing the land: Soil and land resource evaluation and indigenous knowledge. *Soil Use and Management*, *14*(4), 188–251. DOI: 10.1111/j.1475-2743.1998.tb00148.x

Sillitoe, P., & Marzano, M. (2022). Ethnobotanical knowledge documentation in Papua New Guinea: Digital approaches and indigenous participation. *Journal of Ethnobiology and Ethnomedicine*, *18*(1), 1–20. PMID: 34980177

Silva, M. J., Santos, J., & Rocha, L. (2020). Virtual heritage: Documenting Indigenous cultural practices through immersive technology. *Heritage Science*, *8*(1), 45–62.

Simango, S. S. (2022). *Examining the impact of business incubation on the growth of tourism SMMEs: the case of Pilanesberg business Incubator Programme*. PBIP.

Simatele, M., & Kabange, M. (2022). Financial inclusion and Intersectionality: A case of business funding in the South African informal sector. *Journal of Risk and Financial Management*, *15*(9), 380. DOI: 10.3390/jrfm15090380

Simelane, S. P., Hansen, C., & Munghemezulu, C. (2021). The use of remote sensing and GIS for land use and land cover mapping in Eswatini: A review. *South African Journal of Geomatics*, *10*(2), 181–206. DOI: 10.4314/sajg.v10i2.13

Simeone, T. (2004). *Indigenous traditional knowledge and intellectual property rights*. Available: https://www.parl.gc.ca/information/library/PRBpubs/prb0338-e.htm. (Accessed 10 September 2009).

Simeone, T. (2004). Indigenous Traditional Knowledge and Intellectual Property Rights' Political and Social Affairs Division. Government of Canada, Ottawa: Library of Parliament Publications. (Accessed, 20 October 2014).

Simpson, L. R. (2004). Anticolonial strategies for the recovery and maintenance of indigenous knowledge. *American Indian Quarterly*, *28*(3), 3–4, 373–384. DOI: 10.1353/aiq.2004.0107

Singh, D., & Roy, B. (2023). Ethical concerns in digital preservation of Indigenous heritage in India. *Journal of Heritage Studies*, *30*(7), 445–460.

Singh, G., & Pandey, N. (2017). Role and impact of media on society: A sociological approach with respect to demonetisation. *IMPACT: International Journal of Research in Humanities. Arts and Literature*, *5*(10), 127–136.

Singh, R. B. (2000). Indigenous knowledge and sustainable development in the Indian Himalayan Region. *Mountain Research and Development, 20*(2), 156–160.

Sinha, R. K., Bhatia, S., & Vishwakarma, R. (2018). Indigenous knowledge systems and sustainable development: A case study of the Bishnoi community. *Indian Journal of Traditional Knowledge, 17*(2), 298–305.

Sithole, J. (2007). The Challenges Faced by African Libraries and Information Centres in Documenting and Preserving Indigenous Knowledge. *IFLA Journal, 33*(2), 117–123. DOI: 10.1177/0340035207080304

Sixaba, Z., & Rogerson, C. M. (2023). 'Transformation' and Tourism Small Firm Development in South Africa: Evidence from Eastern Cape Province. *Studia Periegetica, 41*(1), 83–108. DOI: 10.58683/sp.378

Slater, J. (2019). Sankofa-the need to turn back to move forward: Addressing reconstruction challenges that face Africa and South Africa today. *Studia Historiae Ecclesiasticae, 45*(1), 1–24. DOI: 10.25159/2412-4265/4167

Small Enterprise Development Agency. (2020). s://www.seda.org.za/Publications/Publications/SEDA%20Integrated%20report%20-%20DSBD%2024%20Feb%2020.pdf. (Accessed 15/11/2024)

Smith, A. B. (2019). Ancient and traditional agriculture, pastoralism, and agricultural societies in sub-Saharan Africa. In Oxford Research Encyclopedia of Environmental Science.

Smith, A., Ndwandwe, S., & Dlamini, N. (2020). Ethical considerations in the use of Blockchain for Indigenous Knowledge preservation. *South African Journal of Information Management, 22*(1), 1–10.

Smith, L. (1999). *Decolonizing methodologies: Research and indigenous peoples*. Zed Books.

Smith, L. T. (2012). *Decolonizing Methodologies: Research and Indigenous Peoples*. Zed Books.

Smith, L. T. (2021). *Decolonising methodologies: Research and Indigenous peoples* (3rd ed.). Zed Books. DOI: 10.5040/9781350225282

Smith, L. T. (2021). Decolonizing methodologies: Research and indigenous peoples in the digital age. *Qualitative Inquiry, 27*(1), 141–156.

Smyth, B. M., Martin, M. A., & Downing, M. (Eds.). (2025). *The Routledge handbook of human research ethics and integrity in Australia*. Routledge, Taylor & Francis Group.

Snipp, C. M. (2016). What does data sovereignty imply: what does it look like? *Indigenous Data Sovereignty: Toward an agenda*. Available at: https://www.scielo.org.za/scielo.php?pid=S2077-72132020000200003&script=sci_arttext

Snowball, J., Collins, A., & Nwauche, E. (2021). Ethics, values and legality in the restoration of cultural artefacts: The case of South Africa. *International Journal of Cultural Policy, 28*(5), 513–530. DOI: 10.1080/10286632.2021.1995377

Sobhy, H. (2021). The lived social contract in schools: From protection to the production of hegemony. *World Development, 137*(1), 1–15. DOI: 10.1016/j.worlddev.2020.104986

Socha, T., Chambers, L., Zahaf, M., Abraham, R., & Fiddler, T. (2011). Food availability, food store management, and food pricing in a northern community First Nation community. *International Journal of Humanities and Social Science*, *1*(11), 49–61.

Sohail, M. T., & Chen, S. (2022). A systematic PLS-SEM approach on assessment of indigenous knowledge in adapting to floods; A way forward to sustainable agriculture. *Frontiers in Plant Science*, *13*, 990785. DOI: 10.3389/fpls.2022.990785 PMID: 36092446

Solow, R. M. (1994). Perspectives on growth theory. *The Journal of Economic Perspectives*, *8*(1), 45–54. DOI: 10.1257/jep.8.1.45

Sousa, E. C., & Raizada, M. N. (2020). Contributions of African crops to American culture and beyond: The slave trade and other journeys of resilient peoples and crops. *Frontiers in Sustainable Food Systems*, *4*, 586340. DOI: 10.3389/fsufs.2020.586340

Soylu, D., & Şahin, A. (2024). The Role of AI in Supporting Indigenous Languages. *AI and Tech in Behavioral and Social Sciences*, *2*(4), 11–18. DOI: 10.61838/kman.aitech.2.4.2

Spiegel, A. D. (2020). Struggling with tradition in South Africa: the multivocality of images of the past. In *Social Construction of the Past* (pp. 185–202). Routledge. DOI: 10.4324/9780203754412-17

Spiekermann, M. (2019). Data marketplaces: Trends and monetisation of data goods. *Inter Economics*, *54*(4), 208–216. DOI: 10.1007/s10272-019-0826-z

State of Tourism Report. (2022/21). https://www.tourism.gov.za/AboutNDT/Publications/Annual%20State%20of%20Tourism%20Report%202020-21.pdf (Accessed 17/11/2024).

Statistics South Africa. (2017). *Poverty trends in South Africa, No. 03-10–06*. Statistics South Africa.

Stephenson, M., Hamid, M. F. S., Peter, A., Sauvant, K. P., Seric, A., & Tajoli, L. (2021). More and better investment now! How unlocking sustainable and digital investment flows can help achieve the SDGs. *Journal of International Business Policy*, *4*(1), 152–167. DOI: 10.1057/s42214-020-00094-2

Stikkers, K. W. (2008). An outline of methodological Afrocentrism, with particular application to the thought of W.E.B. DuBois. *The Journal of Speculative Philosophy*, *1*(22), 40–49. DOI: 10.2307/25670688

Stilwell, C. (2010). Understanding indigenous knowledge: Bridging the knowledge gap through a knowledge creation model for agricultural development. *South African Journal of Information Management*, *12*(1), 1–8.

Subramanian, S. M., & Pisupati, B. (2010). *Traditional Knowledge in Policy and Practice: Approaches to Development and Human Well-being*. United Nations University Press.

Sukhera, J. (2022). Narrative reviews: Flexible, rigorous, and practical. *Journal of Graduate Medical Education*, *14*(4), 414–417. DOI: 10.4300/JGME-D-22-00480.1 PMID: 35991099

Sundari, S., Penthala, H. R., Mogullapalli, A., Sukhavasi, V., & Nayyar, A. (2024). Women and Millets: Historical Perspective. In *The Role of Women in Cultivating Sustainable Societies Through Millets* (pp. 196-229). IGI Global.

Sundari, M. S., Penthala, H. R., & Nayyar, A. (2024). Transforming Education through AI-Enhanced Content Creation and Personalized Learning Experiences. In *Impact of Artificial Intelligence on Society* (pp. 98–118). Chapman and Hall/CRC. DOI: 10.1201/9781032644509-7

Suryadi, B., Larasati, A. S., Verniastika, R., & Firdaus, S. N. (2022). Harnessing Fourth Industrial Revolution (4IR) technologies for climate change mitigation in the Asia-Pacific region.

Swanson, T. (1995). The reliance of northern economies on southern biodiversity: Biodiversity as information. *Ecological Economics*, *17*(1), 1–8. DOI: 10.1016/0921-8009(95)00101-8

Sweller, N., Dascalu, M., Ponciano, L., Dascalu, M., & McNamara, D. (2019). Designing Virtual Reality for Indigenous Language Education. *Educational Technology Research and Development*, *67*(1), 171–195.

T. (2024). Revitalizing Indigenous Knowledge Systems via Digital Media Technologies for Sustainability of Indigenous Languages. *Preservation, Digital Technology & Culture*, 53(1), 35-44.

Taherdoost, H. (2016). Sampling methods in research methodology; How to choose a sampling technique for research. *SSRN*, *5*(2), 18–27. DOI: 10.2139/ssrn.3205035

Takaidza, S. (2023). *Indigenous South African Food: Nutrition and Health Benefits*. Vaal University of Technology.

Takazira, L., & Dzingirayi, P. (2024). The Use of African Indigenous Knowledge Systems in Climate Change adaptation in Zimbabwe. *East African Journal of Education and Social Sciences*, *5*(6), 90–97. DOI: 10.46606/eajess2024v05i06.0417

Tamuno, P. (2017). New human rights concept for old African problems: An analysis of the challenges of introducing and implementing indigenous rights in Africa. *Journal of African Law*, *61*(3), 305–332. DOI: 10.1017/S0021855317000249

Tangwa, G., & Tiebere, F. (2019). Digital preservation strategies for Indigenous Knowledge systems in Africa. *International Journal of Cultural Policy*, *25*(4), 451–469.

Tariq, M. U. (2025). Bridging Generations: Digital Transformation and Its Multifaceted Impact Across Societal Sectors. In Anshari, M., Almunawar, M., & Ordóñez de Pablos, P. (Eds.), *Impacts of Digital Technologies Across Generations* (pp. 1–22). IGI Global Scientific Publishing., DOI: 10.4018/979-8-3693-6366-9.ch001

Tariq, M. U. (2025). Empowering Learning Through Networked and Connected Education: Case Studies in Digital Engagement. In Gierhart, A. (Ed.), *Cases on Enhancing P-16 Student Engagement With Digital Technologies* (pp. 169–198). IGI Global Scientific Publishing., DOI: 10.4018/979-8-3693-5633-3.ch007

Tariq, M. U. (2025). Leveraging Libraries and Information Centers for Startup Innovation and Growth. In Gupta, V., & Gupta, C. (Eds.), *Navigating Strategic Partnerships for Sustainable Startup Growth* (pp. 1–28). IGI Global Scientific Publishing., DOI: 10.4018/979-8-3693-4066-0.ch001

Tariq, M. U. (2025). The Belmont Report: Guiding Ethical Principles in Human Research. In Throne, R. (Ed.), *IRB, Human Research Protections, and Data Ethics for Researchers* (pp. 245–268). IGI Global Scientific Publishing., DOI: 10.4018/979-8-3693-3848-3.ch010

Tauli-Corpuz, V. (2022). Indigenous networks for knowledge exchange: Global experiences and lessons learned. *International Indigenous Policy Journal, 13*(2), 1–25.

Taylor, E.C. (2008). Sharing of benefits or enclosure of the commons? investigating the compatibility of national and local access and benefit sharing mechanisms in Peru.

Taylor, J. L.. (2019). A relational approach to designing social technologies that foster use of the Kuku Yalanji language. *Proceedings of the 31st Australian conference on human computer-interaction* (pp. 161–72). DOI: 10.1145/3369457.3369471

Taylor, J. R., & Duodu, K. G. (2017). Sorghum and millets: Grain-quality characteristics and management of quality requirements. In *Cereal grains* (pp. 317–351). Woodhead Publishing. DOI: 10.1016/B978-0-08-100719-8.00013-9

Te Mana Rarunga. (2018). Principles of Māori data sovereignty. https://www.otago.ac.nz/__data/assets/pdf_file/0014/321044/tmr-maori-data-sovereignty-principles-october-2018-832194.pdf

Terchilova, E. (2022). Intellectual property rights and indigenous knowledge: Identifying incompatibilities and potential solutions. *Journal of Intellectual Property Law, 29*(1), 67–94.

The Constitution of the Republic of South Africa. (1996) Available at: https://www.saflii.org/za/legis/num_act/cotrosa1996423/. (Accessed 23 December 2024).

Thompson, K. L., Lantz, T., & Ban, N. (2020). A review of Indigenous knowledge and participation in environmental monitoring. *Ecology and Society, 25*(2), art10. DOI: 10.5751/ES-11503-250210

Thornton, P. K., Ericksen, P. J., Herrero, M., & Challinor, A. J. (2014). Climate variability and vulnerability to climate change: A review. *Global Change Biology, 20*(11), 3313–3328. DOI: 10.1111/gcb.12581 PMID: 24668802

Tihapi, W. R. (2004). *The use of intellectual property as a form of information warfare with specific reference to indigenous knowledge*: A critical analysis and proposed solution. (PhD Thesis), University of Pretoria, South Africa.

Tikai, P., & Kama, A. (2010). A study of indigenous knowledge and its role to sustainable agriculture in Samoa. *Open Journal of Social Sciences, 3*(1), 65–79.

Tirivangasi, H. M., & Tayengwa, D. (2017). Indigenous knowledge systems and food security in South Africa: Is land reform a prerequisite? *Journal of Human Ecology (Delhi, India), 57*(3), 118–124.

Tisdell, C. A. (Ed.). (2005). *Economics of environmental conservation* (Vol. 1). Edward Elgar Publishing. United Nations Environment Programme (UNEP). (2008) Indigenous Knowledge in Disaster Management in Africa. Nairobi, Kenya.

Tiwari, B. K. (2013). Indigenous knowledge and sustainable development. *Journal of North East India Studies, 3*(2), 1–8.

Tobin, B. (2004). *The Role of Customary Law in Access and Benefit-sharing and Traditional Knowledge Governance: Perspectives from Andean and Pacific Island Countries*. UNU-IAS.

Toki, V. (2024). *Indigenous rights, climate change and governance: Measuring success and data*. Edward Elgar., DOI: 10.4337/9781803924984

Torre, D., Chennamaneni, A., & Rodriguez, A. (2023). Privacy-preservation techniques for IoT devices: A systematic mapping study. *IEEE Access : Practical Innovations, Open Solutions, 11*, 16323–16345. DOI: 10.1109/ACCESS.2023.3245524

Traxler, J. (2019) Only Connect: Indigenous digital learning, *Interaction Design and Architecture(s),* 41, pp. 7-23. https://wlv.openrepository.com/handle/2436/622885

Trček, D. (2022). Cultural heritage preservation by using blockchain technologies. *Heritage Science, 10*(6), 6. Advance online publication. DOI: 10.1186/s40494-021-00643-9

Treleaven, P., Brown, R. G., & Yang, D. (2017). Blockchain technology in finance. *Computer, 50*(9), 14–17. DOI: 10.1109/MC.2017.3571047

Trigo, A., Marta-Costa, A., & Fragoso, R. (2021). Principles of sustainable agriculture: Defining standardized reference points. *Sustainability (Basel), 13*(8), 4086. DOI: 10.3390/su13084086

Trist, E., & Bamforth, K. (1951). Some social and psychological consequences of the longwall method of coal-getting. *Human Relations, 4*(1), 3–38. DOI: 10.1177/001872675100400101

Trucco, F., & Tranel, P. J. (2011). Amaranthus. In Wild Crop Relatives: Genomic and Breeding Resources: Vegetables (pp. 11-21). Berlin, Heidelberg: Springer Berlin Heidelberg.Palada, M.C. and Chang, L.C. (2003). Suggested cultural practices for jute mallow. *International Cooperator Guide,* 2(14), 1-4

Tsai, C. (2020). The Impact of the Fourth Industrial Revolution on Knowledge Management. *Journal of Knowledge Management, 24*(2), 291–307.

Tsai, M., & Lee, K. (2006). A study of knowledge internalization: From the perspective of learning cycle theory. *Journal of Knowledge Management, 10*(3), 57–71. DOI: 10.1108/13673270610670858

Tsetselelane, M. D., Sekgothe, M., Stewart, K. L., & Vuma, L. (2021). Knowledge management: Preserving, managing and sharing indigenous knowledge through digital library. *Theological Studies, 77*(2).

Tshabalala, N., & Mthunzi, S. (2021). Evaluating long-term digital preservation policies for Indigenous knowledge in rural communities. *Information Development, 38*(2), 117–135.

Tshidzumba, S. E. (2020). *Nutritional composition of traditional mixed-dishes eaten by elderly women in Vhembe District* (Doctoral dissertation).

Tshikovhi, N. (2021). A theory of indigenous entrepreneurship in the modern African state: The case of South Africa and Zimbabwe. *International Journal of African Renaissance Studies-Multi-. Inter-and Transdisciplinary, 16*(2), 21–60.

Tshishonga, N. (2019). The legacy of apartheid on democracy and citizenship in post-apartheid South Africa: An inclusionary and exclusionary binary? *African Journal of Development Studies, 9*(1), 167–191. DOI: 10.31920/2075-6534/2019/9n1a8

Turner, N. J., & Turner, K. L. (2007). *Traditional food systems, erosion and renewal in Northwestern.* North America.

Tutu, D. (1999). *No future without forgiveness*. Random House. DOI: 10.1111/j.1540-5842.1999.tb00012.x

Udayakumar, P., & Anandan, R. (2024). Get Started with IoT. In *Design and Deploy Microsoft Defender for IoT*. Appress., DOI: 10.1007/979-8-8688-0239-3_1

Uddagiri, C., & Shanmuga Sundari, M. (2023). Authorship Identification Through Stylometry Analysis Using Text Processing and Machine Learning Algorithms. In *Proceedings of Fourth International Conference on Computer and Communication Technologies* (pp. 573-581). Springer, Singapore. DOI: 10.1007/978-981-19-8563-8_55

UN, I. (1992). Convention on biological diversity. *Treaty Collection*.

UNDP. (2004). *Biodiversity and Development: A Global Assessment of National Biodiversity Strategies and Action Plans*. Retrieved from https://www.undp.org/

Unesco. (2009). The draft of the National Policy on Living Heritage in South Africa. https://www.maropeng.co.za/uploads/files/National_Policy_on_South_African_Living_Heritage__ICH.pdf

UNESCO. (2018). *Indigenous knowledge and cultural heritage*. UNESCO Publishing.

UNESCO. (n.d.). *Richtersveld Cultural and Botanical Landscape*. UNESCO World Heritage Centre. Retrieved from https://whc.unesco.org/en/list/1265

UNESCO. 2021. World Health Convention. World Heritage and Indigenous People.

United Nations (UN). (2007). *United Nations Declaration on the Rights of Indigenous Peoples (UNDRIP)*.https://www.un.org/development/desa/indigenouspeoples/declaration-on-the-rights-of-indigenous-peoples.html

United Nations General Assembly. (2007). United Nations Declaration on the Rights of Indigenous Peoples. United Nations., Retrieved October 2, 2007, from.

United Nations High Commissioner for Refugees (UNHCR). (2018). *Global Report*. United Nations General Assembly.

United Nations. (2007). *United Nations Declaration on the Rights of Indigenous Peoples: Resolution/ Adopted by the General Assembly*, 2 October 2007, A/RES/61/295.

United Nations. (2008). United Nations declaration on the rights of Indigenous peoples. chrome- https://www.un.org/development/desa/indigenouspeoples/wp-content/uploads/sites/19/2018/11/UNDRIP_E_web.pdf

United Nations. (2020). Sustainable Development Goals. https://sdgs.un.org/goals

United Nations. (2021). Scientific and Cultural Organisation (UNESCO). Cultural heritage and sustainable development. https://unesdoc.unesco.org/

Upreti, P. N. (2021). The Role of National and International Intellectual Property Law and Policy in Reconceptualising the Definition of Investment. *IIC; International Review of Industrial Property and Copyright Law*, *52*(2), 103–136. DOI: 10.1007/s40319-020-01009-7

Usoro, A., Sharratt, M. W., Tsui, E., & Shekhar, S. (2007). Trust as an antecedent to knowledge sharing in virtual communities of practice. *Knowledge Management Research and Practice*, *5*(3), 199–212. DOI: 10.1057/palgrave.kmrp.8500143

Uusiku, N. P., Oelofse, A., Duodu, K. G., Bester, M. J., & Faber, M. (2010). Nutritional value of leafy vegetables of sub-Saharan Africa and their potential contribution to human health: A review. *Journal of Food Composition and Analysis*, *23*(6), 499–509. DOI: 10.1016/j.jfca.2010.05.002

Uwaoma, P. U., Eboigbe, E. O., Kaggwa, S., Akinwolemiwa, D. I., & Eloghosa, S. O. (2023). Ecological economics in the age of 4IR: Spotlight on sustainability initiatives in the global south. *International Journal of Advanced Economics*, *5*(9), 271–284. DOI: 10.51594/ijae.v5i9.647

Vallerand, R.J. Pelletier, L.G. & Koestner, R. (2008). Reflections on self-determination theory.

van den Broek, T., & van Veenstra, A. F. (2018). Governance of big data collaborations: How to balance regulatory compliance and disruptive innovation. *Technological Forecasting and Social Change*, *129*, 330–338. DOI: 10.1016/j.techfore.2017.09.040

Van der Hoeven, M., Osei, J., Greeff, M., Kruger, A., Faber, M., & Smuts, C. M. (2013). Indigenous and traditional plants: South African parents' knowledge, perceptions and uses and their children's sensory acceptance. *Journal of Ethnobiology and Ethnomedicine*, *9*(1), 78. DOI: 10.1186/1746-4269-9-78 PMID: 24274394

Van der Merwe, A., & Chitiga, L. (2020). Technology, education, and the preservation of Indigenous Knowledge. *African Journal of Education and Technology*, *15*(3), 25–40.

Van der Merwe, J. D., Cloete, P. C., & Van der Hoeven, M. (2016). Promoting food security through indigenous and traditional food crops. *Agroecology and Sustainable Food Systems*, *40*(8), 830–847. DOI: 10.1080/21683565.2016.1159642

Van Wyk, B. E. (2011). The potential of South African plants in the development of new medicinal products. *South African Journal of Botany*, *77*(4), 812–829. DOI: 10.1016/j.sajb.2011.08.011

Vansina, J. (1992). Population movements and emergence of new socio-political forms in Africa. *General History of Africa*, *5*, 46–73.

Vasconcellos, D., Parker, P. D., Hilland, T., Cinelli, R., Owen, K. B., Kapsal, N., Lee, J., Antczak, D., Ntoumanis, N., Ryan, R. M., & Lonsdale, C. (2020). Self-determination theory applied to physical education: A systematic review and meta-analysis. *Journal of Educational Psychology*, *112*(7), 1444–1469. DOI: 10.1037/edu0000420

Venter, P., Malemela, K., Mbazima, V., Mampuru, L. J., Muller, C. J., & Riedel, S. (2021). An RP-LC- UV-TWIMS-HRMS and Chemometric Approach to Differentiate between Momordica balsamina Chemotypes from Three Different Geographical Locations in Limpopo Province of South Africa. *Molecules (Basel, Switzerland)*, *26*(7), 1896. DOI: 10.3390/molecules26071896 PMID: 33801575

Veresiu, E., & Giesler, M. (2018). Beyond Acculturation: Multiculturalism and the Institutional Shaping of an Ethnic Consumer Subject. *The Journal of Consumer Research*, *45*(3), 553–570. DOI: 10.1093/jcr/ucy019

Verhoef, G. (2023). *The public domain and the legal protection of Indigenous Knowledge Systems* (Doctoral dissertation, North-West University (South Africa).

Verhoef, G. (2023). *The public domain and the legal protection of Indigenous Knowledge Systems* (Doctoral dissertation, North-West University (South Africa).).

Vermeylen, S. (2007). Contextualizing 'fair' and 'equitable': The San, Hoodia, and benefit-sharing. *Local Environment*, *12*(4), 423–436. DOI: 10.1080/13549830701495252

Verran, H., Christie, M., & Anbins, B. (2021). Ontological transparency and indigenous knowledge representation in digital environments. *Information Communication and Society*, *24*(5), 627–643.

Vilakazi, M., & Moore, C. (2023). The Black Industrialist Programme's prospects to the advancement of developmental state in South Africa. *Africa's Public Service Delivery and Performance Review*, *11*(1), 602. DOI: 10.4102/apsdpr.v11i1.602

Viljoen, D. (2020*). Business framework for managing the sustainability of emerging farmers in South Africa* (Doctoral dissertation, North-West University (South Africa).

Virginia, P., & Pallavi, S. (2016). Role of women's indigenous knowledge in food processing and preservation for enhancing household food security in Allahabad District. *World Journal of Pharmacy and Pharmaceutical Sciences*, *5*(11), 729–743.

Virtanen, P. K. (2015). Indigenous social media practices in Southwestern Amazonia. *Alternative*, *11*(4), 350–362. DOI: 10.1177/117718011501100403

Vorster, H. J., Stevens, J. B., & Steyn, G. J. (2008). Production systems of traditional leafy vegetables: Challenges for research and extension. *South African Journal of Agricultural Extension*, *37*, 85–96.

Wadley, L., Hodgskiss, T., & Grant, M. (2009). Implications for complex cognition from the hafting of tools with compound adhesives in the middle stone age, South Africa. *Proceedings of the National Academy of Sciences of the United States of America*, *106*(24), 9590–9594. DOI: 10.1073/pnas.0900957106 PMID: 19433786

Wagner, A., & de Clippele, M.-S. (2023). Safeguarding Cultural Heritage in the Digital Era – A Critical Challenge. *International Journal for the Semiotics of Law - Revue internationale de Sémiotique juridique*, 36, 1915–1923. https://doi.org/SpringerLinkDOI: 10.1007/s11196-023-10040-z

Walter, M., & Carroll, S. R. (2021). Indigenous data sovereignty, governance and the link to Indigenous policy. In Walter, M., Kukutai, T., Carroll, S. R., & Rodriguez-Lonebear, D. (Eds.), *Indigenous data sovereignty and policy* (pp. 1–20). Routledge., DOI: 10.4324/9780429273957

Walter, M., Lovett, R., Maher, B., Williamson, B., Prehn, J., Bodkin-Andrews, G., & Lee, V. (2021). Indigenous data sovereignty in the era of big data and open data. *The Australian Journal of Social Issues*, *56*(2), 143–156. DOI: 10.1002/ajs4.141

Wang, J., & Tian, G. (2023). Sustainability of Forest Eco-Products: Comprehensive Analysis and Future Research Directions. *Forests*, *14*(10), 2008–2220. DOI: 10.3390/f14102008

Wang, S., & Noe, R. A. (2010). Knowledge sharing: A review and directions for future research. *Human Resource Management Review*, *20*(2), 115–131. DOI: 10.1016/j.hrmr.2009.10.001

Wanjiku, J. (2024). The Impact of Blockchain Technology on ART Authentication and Provenance Kiu Publication Extension. Kampala International University, Ugandahttps://www.researchgate.net/publication/382878671_The_Impact_of_Blockchain_Technology_on_ART_Authentication_and_Provenance

Warren, D. M. (1991). *Using Indigenous knowledge in agricultural development* (No. 127, p. 46pp).

Webber, H. Gaiser, T. & Ewert, F. (2014). What role can crop models play in supporting climate

Weibel, S. (2022). Dublin Core and indigenous knowledge: Towards culturally sensitive metadata. *International Journal of Metadata, Semantics and Ontologies*, *17*(1), 1–15.

Weinberger, K., & Pichop, G. N. (2009). Marketing of African Indigenous Vegetables along Urban and Peri-Urban Supply Chains in Sub-Saharan Africa1. In *African Indigenous Vegetables in Urban Agriculture* (pp. 225–244). Routledge.

Weinberger, M. F. (2015). Dominant Consumption Rituals and Intragroup Boundary Work: How Non-Celebrants Manage Conflicting Relational and Identity Goals. *The Journal of Consumer Research*, *42*(3), ucv020. DOI: 10.1093/jcr/ucv020

Welford, L., Venter, S., Dohse, C., & Chibaya, I. (2015). Harvesting from the Tree of Life: responsible commercialization of baobab in South Africa and Malawi. In *Ecological sustainability for non-timber Forest products* (pp. 90–102). Routledge.

Wendland, W. (2022). Protecting traditional knowledge: The WIPO Intergovernmental Committee on Intellectual Property and Genetic Resources, Traditional Knowledge and Folklore. *International Journal of Cultural Property*, *29*(2), 117–136.

Wenger, E., McDermott, R. A., & Snyder, W. (2002). *Cultivating communities of practice: A guide to managing knowledge*. Harvard Business Press.

Werner, K. (2023). Who is indigenous in Africa? The concept of indigeneity, its impacts, and progression. *Millennium*, *51*(2), 379–404. DOI: 10.1177/03058298221135666

Wesley, A., & Faminow, M. (2014). Background paper: research and development and extension services in agriculture and food security. *Asian Development Bank Economics Working Paper Series*, (425).

Whitaker, A., Bracegirdle, A., de Menil, S., Gitlitz, M. A., & Saltos, L. (2020). Art, antiquities, and blockchain: New approaches to the restitution of cultural heritage. *International Journal of Cultural Policy*, *27*(3), 312–329. DOI: 10.1080/10286632.2020.1765163

WHO. https://www.who.int/activities/implementation-of-the-WHO-traditional-medicine-strategy-2014-2023. Accessed 16 November 2024

Wildwood, C. (2003). *Mood Enhancing PLANTS*. Bloomsbury publishing plc.

Williams, D. H., & Shipley, G. P. (2023). Indigenous research methodologies: Challenges and opportunities for broader recognition and acceptance. *Open Journal of Social Sciences*, *11*(5), 467–500. DOI: 10.4236/jss.2023.115030

Willoughby, S. (2019). In Davies, T., Walker, S., Rubinstein, M., & Perini, F. (Eds.), *Open data and the environment. The State of Open Data: Histories and Horizons* (pp. 103–118).

Wilson, K., Bowles, N., & Bell, L. (2021). Indigenous-led digital heritage initiatives: Lessons from the Indigitization project. *Information & Culture*, *56*(3), 217–238.

Wilson, S. (2020). *Research is ceremony: Indigenous research methods*. Fernwood publishing.

Wingara. M.N, (2018). Indigenous Data Sovereignty Communique Indigenous Data Sovereignty Summit 20th June 2018, Canberra, ACT. Retrieved from Maiam Nayri Wingara.(n.d.) Principles.

Winschiers-Theophilus, H. (2022). *Bridging Worlds*. Indigenous Knowledge in the Digital World., DOI: 10.1145/3529372.3538820

WIPO. (2002). *Traditional Knowledge and Intellectual Property*. World Intellectual Property Organization. Retrieved from https://www.wipo.int/

Woods, M., & Blanchard, C. (2020). Sustainable digital platforms for Indigenous storytelling. *Journal of Digital Humanities*, *9*(4), 215–232.

World Bank. (2002). *Indigenous Knowledge: Local Pathways to Global Development*. Retrieved from https://www.worldbank.org/

World Economic Forum (WEF). (2024). Advancing data equity: An action-oriented framework. September 2024. https://www3.weforum.org/docs/WEF_Advancing_Data_Equity_2024.pdf

World Food Summit. (1996). Rome Declaration on World Food Security.

World Health Organisation. (WHO) (2023). Traditional medicine has a long history of contributing to conventional medicine and continues to hold promise. https://www.who.int/news-room/feature-stories/detail/traditional-medicine-has-a-long-history-of-contributing-to-conventional-medicine-and-continues-to-hold-promise

World Health Organization (WHO). (2013). *Traditional Medicine Strategy 2014-2023*. WHO.

World Health Organization. (2023). Traditional medicine has a long history of contributing to conventional medicine and continues to hold promise. *World Health Organization [Internet]. [cited 25 Feb 2024]. Available:*https://www. who.int/newsroom/feature-stories/detail/traditional-medicine-has-a-long-history-o*contributing-to-conventional-medicine-and-continues-to-hold-promise*.

Wynberg, R., & Chennells, R. (2009). Indigenous peoples, consent, and benefit sharing: Lessons from the San-Hoodia case. *Environmental Sciences*, *6*(2), 139–156.

Xia, Y., Zhang, M., Tsang, D. C., Geng, N., Lu, D., Zhu, L., Igalavithana, A. D., Dissanayake, P. D., Rinklebe, J., Yang, X., & Ok, Y. S. (2020). Recent advances in control technologies for non-point source pollution with nitrogen and phosphorous from agricultural runoff: Current practices and future prospects. *Applied Biological Chemistry*, *63*(1), 1–13. DOI: 10.1186/s13765-020-0493-6

Xu, M., David, J. M., & Kim, S. H. (2018). The fourth industrial revolution: Opportunities and challenges. *International journal of financial research, 9*(2), 90-95. URL: https://doi.org/DOI: 10.5430/ijfr.v9n2p90

Yamagata-Lynch, L. C. (2007). Confronting analytical dilemmas for understanding complex human interactions in design-based research from a cultural—Historical activity theory (CHAT) framework. *Journal of the Learning Sciences*, *16*(4), 451–484. DOI: 10.1080/10508400701524777

Yasseri, T., Gildersleve, P., & David, L. (2022). Collective Memory in the Digital Age. In O'Mara, S. (Ed.), *Collective Memory*. Elsevier. DOI: 10.1016/bs.pbr.2022.07.001

Yassine, J. (2018). IP Rights and Indigenous Rights: Between Commercialization and Humanization of Traditional Knowledge. *San Diego Int'l LJ*, *20*, 71.

Yin, R. K. (2017). *Case study research and applications: Design and methods*. Sage publications.

Young, J. O. (2011). Cultural appropriation and the arts. *Journal of Cultural Economics*, (35), 233–236.

Young, J. O., & Brunk, C. G. (2009). *The Ethics of Cultural Appropriation*. Wiley-Blackwell. DOI: 10.1002/9781444311099

Yumagulova, L., Parsons, M., Yellow Old Woman-Munro, D., Dicken, E., Lambert, S., Vergustina, N., Scott, J. C., Michell, P., & Black, W. (2023). Indigenous perspectives on climate mobility justice and displacement-mobility-immobility continuum. *Climate and Development*, •••, 1–18. DOI: 10.1080/17565529.2023.2227158

Yunnus, F. (2017). *Preservation of indigenous knowledge (IK) by public libraries in Westcliff, Chatsworth, Durban*. Retrieved July 20, 2020, from https://core.ac.uk/display/154920975

Zadik, M. (1986). Teaching horticulture with a human perspective. *NACTA Journal*, *30*(1), 59.

Zarkadakis, G. (2020). 'Data trusts' could be the key to better AI. Harvard Business Review. November 10, 2020. https://hbr.org/2020/11/data-trusts-could-be-the-key-to-better-ai

Zega, A., Ndraha, A. B., Laoli, D., Zebua, R. D., & Telaumbanua, B. V. (2025). Designing and managing deep sea biodiversity. In *Technological advancements for deep sea ecosystem conservation and exploration* (pp. 99–128). IGI Global Scientific Publishing.

Zhang, G., Chen, X., Law, R., & Zhang, M. (2020). Sustainability of heritage tourism: A structural perspective from cultural identity and consumption intention. *Sustainability (Basel)*, *12*(21), 1–17. DOI: 10.3390/su12219199

Zhang, Y., & Li, Y. (2017). Traditional ecological knowledge and sustainable development. *Sustainability*, *9*(2), 222.

Zhou, Z., & Chibaya, T. (2024). The opportunities and challenges of harnessing digital technologies. In *Tourism and Hospitality for Sustainable Development: Technological Innovations and Development Realities* (1), 111.

Zidny, R., Sjöström, J., & Eilks, I. (2023). Indigenous Knowledge and Science and Technology Education. In Akpan, B., Cavas, B., & Kennedy, T. (Eds.), *Contemporary Issues in Science and Technology Education. Contemporary Trends and Issues in Science Education* (Vol. 56). Springer., DOI: 10.1007/978-3-031-24259-5_12

Zidny, R., Sjöström, J., & Eilks, I. A. (2020). Multi-Perspective Reflection on How Indigenous Knowledge and Related Ideas Can Improve Science Education for Sustainability. *Science & Education*, *29*(1), 145–185. DOI: 10.1007/s11191-019-00100-x

Ziff, B. H., & Pratima, V. R. (1997). Introduction to Cultural Appropriation: A Framework for Analysis," *in Borrowed Power: Essays on Cultural Appropriation,* ed. Bruce H. Ziff and Pratima V. Rao, New Jersey: Rutgers University Press, 1–29.

Zimu-Biyela, A. N. (2016). *The management and preservation of indigenous knowledge in Dlangubo Village in KwaZulu-Natal, South Africa* (Doctoral dissertation, University of South Africa).

Zimu-Biyela, A. N. (2016). *The management and preservation of indigenous knowledge in Dlangubo village in KwaZulu-Natal, South Africa.* (Doctoral dissertation, University of South Africa), South Africa.

Zimu-Biyela, A. N. (2016). The management and preservation of indigenous knowledge in Dlangubo Village in KwaZulu-Natal, South Africa. Doctoral dissertation, University of South Africa.

Zoannos, N., Chourdaki, P., & Assimakopoulos, N.UNESCO. (2023). Can UNESCO Use Blockchain to Ensure the Intangible Cultural Heritage of Humanity? *Heritage*, *6*(3), 3232–3255. DOI: 10.3390/heritage6030171

Zoll, M. (2003). Intellectual property rights and bioprospecting: Policies for conserving biodiversity. *Environmental Policy and Governance*, *13*(5), 367–380.

Zondi, N. T. B., Ngidi, M. S. C., Ojo, T. O., & Hlatshwayo, S. I. (2022). Impact of market participation of indigenous crops on household food security of smallholder farmers of South Africa. *Sustainability*, *14*(22), 15194. DOI: 10.3390/su142215194 PMID: 39659967

Zwane, E. M., & Masipa, M. P. (2021). The importance of indigenous knowledge system for food security and conservation of natural resources. *Indilinga*, *20*(1), 124–140.

About the Contributors

Tlou Maggie Masenya holds PhD in Information Science from University of South Africa (UNISA) and Masters in Information Technology from University of Pretoria (UP). She has 11 years of working experience as a senior lecturer in the field of Information Science, Information Systems and Technology, at various universities in South Africa. She was also appointed as a Head of Department in Information Technology at Durban University of Technology in 2023-2024. She is currently an Associate Professor in the Department of Information Systems at Durban University of Technology, South Africa. She is a visionary leader and distinguished researcher with extensive and an impressive national and international research standing, and has published 25 book chapters, 25 articles in peer reviewed journals and also presented numerous papers nationally and internationally within the broad niche area of Digital preservation and Data Curation, Records and Knowledge Management, Technopreneurship, Digital Entrepreneurship, Cloud Computing, Innovative Pedagogies, Disruptive Technologies and Indigenous Knowledge Systems, which had a positive impact on her scholarly growth. She recently edited two books, and the first book is entitled "Innovative technologies for enhancing knowledge access in academic libraries" published by IGI-Global Scientific Publishing in July 2022. The second book is entitled "Digital preservation and Documentation of Global Indigenous Knowledge Systems" also published by IGI-Global Scientific Publishing in August 2023. She is currently editing two books: the first book entitled "'Revaluation and preservation of Indigenous Knowledge Systems in Modern Society"' and the second book entitled " Resurgence and revalorization of Indigenous Knowledge Systems in the Contemporary Society, all to be published by IGI Global Scientific Publishing in May 2025. She leads research projects and mentors emerging academics and researchers, and.serves on numerous academic/ university committees and boards.

Monicca Thulisile Bhuda holds PhD in Social Sciences with Indigenous Knowledge Systems (IKS) from the North-West University (NWU), with 6 years' experience of teaching within the IKS, culture and heritage fields. I am a coordinator of a programme that focuses on African identities, Heritage and History at the University of Mpumalanga.Furthermore, I am also a culture and heritage lecturer within the same University. Currently as an academic, Ihave 24 academic publications on IKS and presented 31 abstracts in diverse conferences. I am co supervisingstudents who are registered for master's degrees in IKS with NWU. I have also been declared a national expert ofIKS by the Minister of Sport, Arts and Culture in the republic of South Africa and I serve in a national panel ofexperts as the youngest member. In 2023, I have been awarded by the South African National Heritage Council (NHC) under the category of Voice of heritage. The North-West University in 2023 also awarded me with an alumni excellence award within the education category. I have been featured numerous times on TV, radio and newsletters as an IKS and culture expert.

Princess Blose is the Lecturer for Technology Education in the Department of Science and Technology Education at the University of South Africa. Her academic background includes BEd (Hons), MEd, PhD in Technology Education. She has developed a keen interest in researching and learning more about Indigenous Knowledge Systems and Education for Sustainable Development in the context of teaching the subject of Technology. As a young scholar, she is also interested in the incorporation of Indigenous Knowledge in Graphic Design Teachings. Her future endeavours include researching digitalization in Technology within the African context.

Kgakgamatso Suprice Boikanyo obtained her Bachelor of Art in Indigenous Knowledge Systems (BA Hons in IKS) and Her Masters in Indigenous Knowledge Systems (MIKS) both from NWU Mafikeng Campus. she is a Junior Lecturer at the North West University Mahikeng campus at the Indigenous Knowledge Systems Centre under Nurturing Emerging Scholars Program(NESP). Kgakgamatso lectures 1st year Modules based on Indigenous Knowledge Systems, Innovation and Sustainable Development. She was born and bred in Dinokana Village, Ikageleng Section, Zeerust town.

Samuel Chabata is a Senior Curator of Ethnography and the National Museums and Monuments of Zimbabwe. He is based at the Zimbabwe Museum of Human Sciences in Harare.

Margaret Chauke is a senior lecturer in the Department of Inclusive Education at the University of South Africa (Unisa). Prior to joining Unisa in 2019, she served as an Institutional Developmental Support Officer (IDSO) with the Gauteng Department of Education. Her expertise lies in inclusive education, disabilities, and intervention strategies to support learners facing barriers to learning and development. Her PhD focused on "The experiences of educators' in management of inclusive classrooms". Dr Chauke is an active member of both local and international organisations, including Education Association of South Africa (EASA) and the International Society for the Study of Behavioural Development (ISSBD). She has presented at numerous international and national conferences and contributed to scholarly articles and book chapters.

Pauline Chiripanhura is an Archaeologist and Lecturer in the GZU Archaeology and Heritage Unit. She holds a PhD in Archaeology from the University of Cape Town, South Africa. Her interests are in African Archaeology, Paleopathology and heritage management. Prior to lecturing, she served as a heritage manager for National Museums and Monuments of Zimbabwe for 8 years. During her tenure with NMMZ, she researched and developed a number of exhibitions among them a permanent exhibition in the Beit Gallery- Mutare Museum which chronicled the lives of the Eastern Shona people.

Collence Takaingenhamo Chisita (PhD) is a National Research Foundation of South Africa (NRF) C2-rated researcher, lecturer, author and editor of books and several journal articles in high impact journals and book chapters. He is based at the Durban University of Technology in the Department of Information Systems. He is also a Research Fellow at the University of South Africa's College of Human Sciences in the Department of Information Science. Collence Chisita has been working with different local and international organizations in the area of LIS. He has collaborated with other researchers from the different IFLA Special interest groups on issues that affect libraries in the modern era. Chisita has written journal articles, book chapters, edited journals and produced conference papers and participated and shared his knowledge on topical issues on libraries and related institutions and emerging technologies through multimedia platforms.

Petros N. Dlamini, (PhD) is a Senior Lecturer in the Department of Information Studies in the Faculty of Humanities and Social Sciences at the University of Zululand, South Africa. He holds a PhD and MLIS in Library and Information Science from University of Zululand in 2016 and 2004. He has published several papers in accredited journals and book chapters. He has supervised several masters and PhD students to completion in the domain.

Mishack T. Gumbo is an NRF C1-Rated Researcher and a Research Professor of Indigenous Technology Knowledge Systems Education in the Department of Science and Technology Education at the University of South Africa. He is Editor-in-Chief of Indilinga: African Journal of Indigenous Knowledge Systems. He is a well-published scholar. He leads research projects and mentors developing academics. He serves on numerous academic committees and boards. He achieved the following between 2019 and 2022: External Recognition Award; NRF C1 Rating; NRF C1 Rating Award; Recognized Among the Top 10 000 Influential Scientists in Africa in the AD Scientific Index 2021; Competitive Programme for Rated Researchers; and Appointed as Research Professor.

Tasnim Jahan is currently working as Assistant Professor at Ramaiah University of Applied Sciences and have previously worked a Teaching Assistant of Law at Symbiosis Law School, Hyderabad and as a Research Associate at National Law School, Bengaluru. She is also pursuing her PhD in Intellectual Property Rights from West Bengal National University of Juridical Sciences. Having profound interest in Academia and research, she has presented papers in various National and International Conferences and have few reputed publications under her name.

Sri Harshini Jupally is currently pursuing BTech in Computer Science and Engineering from the BVRIT Hyderabad College of Engineering for Women.

Xholiso Ennocent Khazamula received his professional degree in Indigenous Knowledge Systems from the University of Venda in 2021. In 2024, he was hired as a heritage officer under the Living Heritage Sub directorate at the National Department of Sport, Arts and Culture. He is currently perusing a Masters in Heritage Studies at the University of the Witwatersrand, Johannesburg. He writes and presents passionately on issues of Indigenous Knowledge management and promotion.

Motheo Koitsiwe is a proud South African, the first African indigenous scholar to have completed successfully the IKS PhD thesis in African indigenous astronomy (bolepa - dinaledi) of Batswana in Botswana and South Africa. Dr Koitsiwe is the Director of Indigenous Knowledge Systems Centre (IKS Center), Faculty of Natural and Agricultural Sciences (FNAS), North - West University (Mahikeng Campus)

Destriman Laoli was born in Onozikho, Gunungsitoli City, precisely on Nias Island, North Sumatra Province on December 10, 1987. The author is a permanent lecturer at the Aquatic Resources Study Program, Faculty of Science and Technology, University of Nias. The author completed his undergraduate education in the Aquaculture Study Program at the University of Riau. Riau University and continued his Masters Education in the Marine Science Study Program Marine Science Study Program Concentration in

Aquaculture at Riau University. The author lectures on Fundamentals of aquatic microbiology, Disease of Aquatic Organisms, Aquatic Productivity. Aquatic Organisms, Aquatic Productivity.

Mothusiotsile Edwin Maditsi is a lecturer and Community Engagement Coordinator at the North-West University's (NWU) Indigenous Knowledge Systems (IKS) Centre. He holds a BA (Hons) in Indigenous Knowledge Systems and a Master of Indigenous Knowledge Systems degree (MIKS) both from NWU. He has participated in and facilitated some community engagement (CE) activities for and on behalf of the IKS Centre at National, provincial, and local level. Some of the CE activities include the IK Act no 6 of 2019 Public awareness workshops, the Biocultural Community Protocols public awareness workshops and IK Regulations public consultations across all nine provinces of the country. Otsile lectures BIKS modules from 1st to 4th year level and he also supervises honors and master's research projects. He is a member of South African Higher Education Community Engagement Forum (SAHECEF) and he has presented at both international and national conferences. He works with the Department of Science and Innovation (DSI) on projects such as the IK-Led smart village, IK Regulations, IKSDC project and a host of other projects at national level. In his PhD he is focusing on decolonizing university community engagement (UCE) through indigenous communities of practice (iCoPs) using the NWU as a case study. Since assuming his position as a lecturer, he has shown dedication and commitment in decolonizing the teaching, learning, research, and innovation activities within the higher education sector. He is focused on achieving scholarship of engagement which is at the peak of higher education debates. He has published several journal articles and book chapters with various publishers and some which are high index journals. For his hard work on various national projects and programmes with communities, Otsile was appointed as a technical expert by the Gauteng Commission on Traditional Leadership Disputes and Claims which he served for a period of 1 year. He was appointed as a technical expert to the commission. He has recently been appointed by the IKS Consortium as a deputy chairperson of the committee on IKS institutionalization in South African universities.

Nthabiseng Maepa is an experienced Project Coordinator and Manager with a demonstrated history of working in the higher education industry. Skilled in coordination, Management, Electronic Data Capture (EDC), Microsoft Excel, Microsoft Word, Data Entry, Microsoft Office and Research Laboratory Management. Strong research professional with a Bachelor's degree focused on Nutrition and Physiology from North-West University/Noordwes-Universiteit.

Lindiwe Carol Mthethwa is a Senior Lecturer at Educational Foundations, Faculty of Education, University of Zululand, South Africa. She is a leader of the digital championship in the Faculty of Education and has written articles based on infusing technology into teaching and learning. Her passion is on handling case studies in philosophy, sociology, educational leadership and management, decolonization strategies, indigenous knowledge and climate change education together with innovative strategies for teaching and learning.

Hangwani Tshisevhe HamiltonMuedi holds a PhD Degree in Plant Pathology from the University of Free State. He practiced in the field of resistance breeding with special focus on bacterial and fungal diseases of dry beans for over seven years. He briefly worked as a Senior Plant Breeder at the University of South Africa (UNISA) before joining the North West Department of Agriculture and Rural Development as a Scientific Manager, a position he still holds currently. His dedication to improving the growth and development of the agricultural sector has earned him an appointment as an Extraordinary Senior Researcher at the Indigenous Knowledge Systems Centre of the North West University to assist with supervision of postgraduate students. Dr Muedi supervises postgraduate students at other institutions of higher learning such as UNISA and Tshwane University of Technology (TUT). He has contributed a total of six journal publications, publications in other platforms including numerous peer reviewed conference presentations.

Ayler Beniah Ndraha is a government scholar and practitioner who has made valuable contributions to the field of Human Resource Management. He is known as someone who combines deep academic knowledge with rich practical experience, thus providing deep insights and practical solutions in human resource planning. human resource planning. With a strong educational background, Dr. Ayler Beniah Ndraha earned a Bachelor of Applied Science in Government (S.STP) and a Master of Science (M.Si) degree as well as a Doctor of Government Science (DIP), all of which were taken at the Institute of Domestic Government (IPDN) Jatinangor-Bandung. His experience working in government agencies such as Bappeda, BKPSDM, Kominfo, and Kesbangpol makes his expertise in government science provide a solid foundation for understanding organizational dynamics and governance in collaboration with empirical practices. His contribution to the Human Resource Management book can be seen through the chapter titled "Human Resource Planning." In this section, the author brings an in-depth understanding of how to manage important aspects of HR planning, including workforce needs analysis, forecasting, and human resource development strategies. His emphasis on integration between theory and practice provides valuable guidance for readers in facing the complex challenges of human resource management. The author is currently active as a civil servant lecturer with a functional position of Lektor at Nias University. Although still relatively young, the author has also contributed to various seminars and scientific discussions both on a regional, national and even international scale sharing his valuable views on the dynamics of government and human resource management. His expertise and dedication in developing government science as well as his deep knowledge in human resource management make him an important figure in inspiring and guiding future generations in this field.

Kamogelo Noge is a masters student at the North-West University, Mahikeng Campus. He holds a Bachelor of Indigenous Knowledge Systems Degree from the NWU. He participated in activities that involves community based research and community engagement project which seek to valurize IKS within South Africa and beyond. He has also actively participated in the recognition of prior learning (RPL) pilot project which is apearheaded by the NWU and the Department of Science, Technology and Innovation. He also facilitated in student assistance program where he was metoring undergraduate students from 1st to 3rd year level. Currently he is working on completing his Masters degree which is focused on Cannabis as a potential socioeconomic resource that can benefit communities. He has published with accredited journals and has presented research papers nationally and internationally.

Modjadji Mandy Rasehlomi is a Bachelor of Indigenous Knowledge Systems (BIKS) Cum Laude graduate from North-West University. She is currently a Master of Indigenous Knowledge Systems (MIKS) students focusing on indigenous foods and crops as a source of nutrition and food security. Mandy has been engaged by the Department of Science and Innovation (DSI) in conjunction with the NWU to participate in the National project focusing on a sui generis law that seeks to protect, promote, develop, and manage indigenous knowledge, known to the public as the IK Act no 6 of 2019. The work was but not limited to the IK Act and Biocultural Community Protocol public awareness workshops, and public consultations on the IK regulations across South African Provinces. Mandy is also a member of the Golden Key International Honor Society. In her endevours she seeks to advance a decolonial agenda within education and also provide a nutritional solution to various communities using clinic gardens, a project she envisages for her PhD.

Vanessa Risimati is a dedicated postgraduate student at North-West University (NWU), Mahikeng campus, where she is furthering her academic journey in the field of Indigenous Knowledge Systems. She holds a Bachelor's degree in Indigenous Knowledge Systems from NWU, which provided her with a comprehensive understanding of the rich and diverse traditions, practices, and philosophies inherent in indigenous cultures. Currently she is pursuing a Master's degree in Indigenous Knowledge Systems at NWU, where she continues to deepen her expertise and engage with contemporary scholarly approaches to the preservation, revitalization, and application of indigenous knowledge. Her academic work reflects a commitment to the advancement of indigenous studies, contributing to the broader discourse surrounding the importance of indigenous knowledge in modern society.

Shashikant Saurav is an Assistant Professor of Law (Senior Scale) at the School of Law, Ramaiah University of Applied Sciences. He specializes in criminal laws, with research interests in transgender rights, women's rights, gender neutrality, and consumer protection. He holds a BA LLB (Hons.) from Aligarh Muslim University and an LLM from the Central University of South Bihar, where he authored a dissertation on the efficacy of rape laws in India. He qualified UGC-NET (Law) in December 2019 and is currently pursuing a PhD at Manipal Academy of Higher Education, Bengaluru. Before joining Ramaiah University, he served as an Assistant Professor at Symbiosis Law School, Hyderabad for three and a half years and at the Indian Institute of Legal Studies for one and a half years. As a passionate academician, he has presented research papers at national and international conferences and has several reputed publications to his name. He believes in challenging students to test the limits of their intellectual abilities. In his free time, he enjoys reading fiction and writing monologues.

Thato Hazel Seemise is a Bachelor of Indigenous Knowledge Systems (BIKS) Cum Laude graduate. She is currently a Master of Indigenous Knowledge Systems (MIKS) at North-West University (NWU), Mafikeng Campus focusing on African indigenous agricultural practices that can be utilized by farmers to drive a sustainable agriculture culture leading to food security. Thato is a Golden Key International Honor Society member at the above institution, and she has served as a student assistant under the Faculty of Natural and agricultural Science (FNAS), specifically placed at the IKS Centre since June 2020. She is now appointed as a Student Instructor. Thato is also engaged in a collaborative activity between the Department of Science and Innovation (DSI) and NWU as a facilitator and a report writer for the IK Act public awareness and IK Regulations public consultations. She has been engaged in this work from November 2020 to the current year (2024). Her goal is to strive for excellence and always dedicate her time to work and achieve more within the space of IKS in the country.

Muhammad Usman Tariq has more than 16+ year's experience in industry and academia. He has authored more than 200+ research articles, 110+ case studies, 115+ book chapters and several books other than 4 patents. He is founder and CEO of The Case HQ, a unique repository for courses, narrative and video case studies. He has been working as a consultant and trainer for industries representing six sigma, quality, health and safety, environmental systems, project management, and information security standards. His work has encompassed sectors in aviation, manufacturing, food, hospitality, education, finance, research, software and transportation. He has diverse and significant experience working with accreditation agencies of ABET, ACBSP, AACSB, WASC, CAA, EFQM and NCEAC. Additionally, Dr. Tariq has operational experience in incubators, research labs, government research projects, private sector startups, program creation and management at various industrial and academic levels. He is Certified Higher Education Teacher from Harvard University, USA, Certified Online Educator from HMBSU, Certified Six Sigma Master Black Belt and has been awarded PFHEA, SFSEDA, SMIEEE, and CMBE.

Arisman Telaumbanua is a dedicated academic and researcher in the field of engineering education. He holds a Master's degree in Engineering Education from Padang State University and now plays an important role as a lecturer at Nias University. In addition to his responsibilities as a lecturer, Arisman is also the Secretary of the Institute for Research and Community Service. His contributions in research cover various important aspects of education and technology. Arisman also sees great opportunities for the younger generation to utilize digital technology as a tool for innovation and the creation of career opportunities in the digital era. He is committed to making a real impact in the development of education and human resources, particularly in the Nias region and beyond. Through teaching, research and community service, Arisman hopes to create a generation that is ready to compete and adapt in an ever-changing world.

Dermawan Zebua, S.T., M.T., IPP was born in Oladano Village, Idanogawo District, Nias Regency. The author completed elementary school in 2010 in Nias Regency, junior high school in 2013, and high school in 2016 in Mentawai Islands Regency. The author completed his Bachelor of Civil Engineering in 2020, Master of Civil Engineering program in 2022 in Surabaya City, and Professional Engineer program in 2024 in Jakarta. The author received his Engineer Registration Certificate as a Professional Engineer Pratama in 2024. The author has worked as a Structural Engineer at HanmiGlobal Co., Ltd. from South Korea. Currently, the author works as a Permanent Lecturer at Nias University and has worked as a Permanent Lecturer at Seruyan Polytechnic and an Extraordinary Lecturer at Bakrie University Jakarta. The author has extensive interest and knowledge in the field of Civil Engineering, especially in the structure and management of construction projects. In addition, the author is also active in research and development activities and is involved in several research projects in the field of Civil Engineering. The author hopes to continue contributing to the development of science and technology in Indonesia through his career as an academic.

Okniel Zebua was born in Gunungsitoli on October 05, 1985. The author is a permanent lecturer at the Aquatic Resources Study Program, Faculty of Science and Technology, Nias University. He teaches English and Education Management courses. Completed his undergraduate education in the English Education study program at the Faculty of Language and Arts Education (FPBS) at the Gunungsitoli Institute of Teacher Training and Education (IKIP) in 2008 and continued his Masters in the Department of Educational Management Administration at Universitas Kristen Indonesia (UKI) in 2014 and graduated in 2016. The author actively writes articles in several scientific journals.

Asokhiwa Zega is a scholar and practitioner in the field of Aquatic Resources who has made valuable contributions to the management of deep-sea biodiversity. He is recognized as someone who combines deep academic knowledge with rich practical experience, thus being able to provide profound insights and innovative solutions in the conservation of deep-sea resources. With a strong educational background, Asokhiwa Zega is currently a student of the Aquatic Resources Study Program at Nias University's Faculty of Science and Technology. His experience in writing for Sinta-indexed journals shows his dedication and expertise in quality scientific research. His contribution to the book chapter titled “Designing and Managing Deep Sea Biodiversity” is seen through an in-depth and detailed chapter on how to design and manage deep sea biodiversity. In this chapter, the author brings an in-depth understanding of various scientific techniques and methods, such as the use of underwater cameras and pressure sensors, to monitor and sustainably manage deep-sea ecosystems. His emphasis on integrating theory and practice provides valuable guidance for readers facing the complex challenges of deep-sea conservation. Although still a student, Asokhiwa Zega has contributed to numerous regional and national scientific seminars and discussions, sharing his valuable insights on aquatic resource dynamics and management. His expertise and dedication to developing aquatic resource science and his in-depth knowledge of deep-sea biodiversity management make him an important figure in inspiring and mentoring future generations in this field.

Index

A

B

C

D

E

F

G

H

I

K

L

M

N

P

R

S

www.ingramcontent.com/pod-product-compliance
Lightning Source LLC
LaVergne TN
LVHW080617300926
849162LV00005B/1192